AF812666

Hoai An Le Thi · Hoai Minh Le ·
Quang Thuan Nguyen
Editors

Advances in Data Science and Optimization of Complex Systems

Proceedings of the International Conference on Applied Mathematics and Computer Science – ICAMCS 2024, Volume 1

Springer

Editors
Hoai An Le Thi
University of Lorraine
Metz, France

Hoai Minh Le
University of Lorraine
Metz, France

Quang Thuan Nguyen
International School
Vietnam National University
Hanoi, Vietnam

ISSN 2367-3370 ISSN 2367-3389 (electronic)
Lecture Notes in Networks and Systems
ISBN 978-3-031-90605-3 ISBN 978-3-031-90606-0 (eBook)
https://doi.org/10.1007/978-3-031-90606-0

© The Editor(s) (if applicable) and The Author(s), under exclusive license
to Springer Nature Switzerland AG 2025

This work is subject to copyright. All rights are solely and exclusively licensed by the Publisher, whether the whole or part of the material is concerned, specifically the rights of translation, reprinting, reuse of illustrations, recitation, broadcasting, reproduction on microfilms or in any other physical way, and transmission or information storage and retrieval, electronic adaptation, computer software, or by similar or dissimilar methodology now known or hereafter developed.
The use of general descriptive names, registered names, trademarks, service marks, etc. in this publication does not imply, even in the absence of a specific statement, that such names are exempt from the relevant protective laws and regulations and therefore free for general use.
The publisher, the authors and the editors are safe to assume that the advice and information in this book are believed to be true and accurate at the date of publication. Neither the publisher nor the authors or the editors give a warranty, expressed or implied, with respect to the material contained herein or for any errors or omissions that may have been made. The publisher remains neutral with regard to jurisdictional claims in published maps and institutional affiliations.

This Springer imprint is published by the registered company Springer Nature Switzerland AG
The registered company address is: Gewerbestrasse 11, 6330 Cham, Switzerland

If disposing of this product, please recycle the paper.

Lecture Notes in Networks and Systems 1311

Series Editor

Janusz Kacprzyk, *Systems Research Institute, Polish Academy of Sciences, Warsaw, Poland*

Advisory Editors

Fernando Gomide, *Department of Computer Engineering and Automation—DCA, School of Electrical and Computer Engineering—FEEC, University of Campinas—UNICAMP, São Paulo, Brazil*
Okyay Kaynak, *Department of Electrical and Electronic Engineering, Bogazici University, Istanbul, Türkiye*
Derong Liu, *Department of Electrical and Computer Engineering, University of Illinois at Chicago, Chicago, USA*
Institute of Automation, Chinese Academy of Sciences, Beijing, China
Witold Pedrycz, *Department of Electrical and Computer Engineering, University of Alberta, Alberta, Canada*
Systems Research Institute, Polish Academy of Sciences, Warsaw, Poland
Marios M. Polycarpou, *Department of Electrical and Computer Engineering, KIOS Research Center for Intelligent Systems and Networks, University of Cyprus, Nicosia, Cyprus*
Imre J. Rudas, *Óbuda University, Budapest, Hungary*
Jun Wang, *Department of Computer Science, City University of Hong Kong, Kowloon, Hong Kong*

The series "Lecture Notes in Networks and Systems" publishes the latest developments in Networks and Systems—quickly, informally and with high quality. Original research reported in proceedings and post-proceedings represents the core of LNNS.

Volumes published in LNNS embrace all aspects and subfields of, as well as new challenges in, Networks and Systems.

The series contains proceedings and edited volumes in systems and networks, spanning the areas of Cyber-Physical Systems, Autonomous Systems, Sensor Networks, Control Systems, Energy Systems, Automotive Systems, Biological Systems, Vehicular Networking and Connected Vehicles, Aerospace Systems, Automation, Manufacturing, Smart Grids, Nonlinear Systems, Power Systems, Robotics, Social Systems, Economic Systems and other. Of particular value to both the contributors and the readership are the short publication timeframe and the world-wide distribution and exposure which enable both a wide and rapid dissemination of research output.

The series covers the theory, applications, and perspectives on the state of the art and future developments relevant to systems and networks, decision making, control, complex processes and related areas, as embedded in the fields of interdisciplinary and applied sciences, engineering, computer science, physics, economics, social, and life sciences, as well as the paradigms and methodologies behind them.

Indexed by SCOPUS, EI Compendex, INSPEC, WTI Frankfurt eG, zbMATH, SCImago.

All books published in the series are submitted for consideration in Web of Science.

For proposals from Asia please contact Aninda Bose (aninda.bose@springer.com).

Preface

The International Conference on Applied Mathematics and Computer Science (ICAMCS 2024) was held on December 20–21, 2024, in Hanoi-Vietnam, in honor of Professors Pham Dinh Tao and Le Thi Hoai An for the 40th birthday of DC (Difference of Convex functions) programming and DCA (DC Algorithm), a crucial field of nonconvex optimization. This field was created in 1985 by Pham Dinh Tao in its preliminary state, and then the intensive research of Pham Dinh Tao and Le Thi Hoai An has led to decisive developments since 1993 and has now become classic and increasingly popular worldwide. For 40 years from their birthday, these theoretical and algorithmic tools have been greatly enriched, thanks to a lot of their applications, by researchers and practitioners in the world, to model and solve nonconvex programs from many fields of applied sciences. The ever-growing number of works using DC programming and DCA proves their power and their key role in nonconvex programming/global optimization and many areas of applications.

ICAMCS 2024 is organized by International School–Vietnam National University Hanoi; Hanoi University of Science and Technology, Vietnam; Laboratoire de Conception, Optimisation et Modélisation des Systèmes, Université de Lorraine, France; in collaboration with Institute of Optimization CTOPTIMAL, Vietnam. The conference aims to share and consolidate the latest developments across interconnected topics in Computer Science and Applied Mathematics, while exploring cutting-edge advancements with applications in diverse areas.

The Program Committee has received a total of 158 submissions. Each paper was peer-reviewed by at least two members of the International Program Committee and the International Reviewer board. After the review process, 81 papers were selected for oral presentation and publication in this book. The selected papers cover several topics in Applied Mathematics and Computer Science, and they are divided into nine parts: Mathematical programming and optimization; DC Programming & DCA; Operations research and decision making; Economics & Finance; Engineering Systems; Autonomous systems; Information theory and Information security; and Machine Learning - Algorithms and Applications.

The scientific program of the conference includes three plenary lectures by world-class speakers and oral presentations of 81 selected papers. Two special sessions are dedicated to DC programming and DCA, featuring three invited talks and five contributed talks. The conference opens with the ceremony honoring Professors Pham Dinh Tao and Le Thi Hoai An, followed by the plenary lecture on forty years of developments of DC programming and DCA.

We would like to express our sincere gratitude to everyone who contributed to the success of the conference and this proceedings book. Our first heartfelt thanks go to the members of the International Program Committee and reviewers for their hard work under tight deadlines, ensuring the highest quality of the selected papers. We are also deeply grateful to the plenary speakers for their insightful presentations, as well as to the authors for their valuable contributions and to all participants who enriched the conference's success. Special thanks are extended to the Organizing Committee members for their tireless dedication in making this conference a resounding success.

In particular, the first editor would like to express his gratitude to the organizing institutions, especially the International School–Vietnam National University Hanoi, for its initiative in organizing this conference in honor of Professor Pham Dinh Tao and herself.

Finally, we cordially thank Springer, especially Prof. Janusz Kacprzyk and Dr. Thomas Ditzinger, for their supports in publishing this book.

October 2024

Hoai An Le Thi
Hoai Minh Le
Quang Thuan Nguyen

Organization

ICAMCS 2024 is co-organized by International School–Vietnam National University Hanoi; Hanoi University of Science and Technology, Vietnam; Laboratoire de Conception, Optimisation et Modélisation des Systèmes, Université de Lorraine, France; and in collaboration with Institute of Optimization CTOPTIMAL, Vietnam.

Steering Commitee

Trung Thanh Le	International School–Vietnam National University Hanoi

International Scientific Committe

Yaroslav D. Sergeyev (Chair)	University of Calabria, Italy

The International Scientific Committe consists of about 70 members of 15 countries all the world over.

Organizing Chairs

Hoai Minh Le	University of Lorraine, France
Quang Thuan Nguyen	International School–Vietnam National University Hanoi

Contents

DC Programming and DCA

Operations Research and Decision Making

Economics and Finance

Engineering Systems

Mathematical Programming and Optimization

On Solving Difficult Combinatorial Optimization Problems Efficiently - A New Framework

Tiru S. Arthanari(✉)

The University of Auckland, Auckland, New Zealand
t.arthanari@auckland.ac.nz, tiru111@gmail.com

Abstract. Generally, it is conjectured that problems in the same class as the Symmetric Traveling Salesman Problem (STSP) are difficult problems. Pedigrees are combinatorial objects defined and studied by the author, and they are in $1-1$ correspondence with the symmetric Traveling salesman tours. Recently the author published the book, entitled Pedigree Polytopes ([1]), bringing together known results on pedigree polytopes, in addition to showing that membership in pedigree polytopes can be checked by implementing a strongly polynomial framework. This article explains the framework and outlines how the author has a Python implementation of the framework for checking the following: Given $X \in Q^{\tau_n}$, is $X \in conv(P_n)$, where $\tau_n = \binom{n}{3} - 1$, is the cardinality of the set, $\{(i,j,k)|1 \leq i < j < k, k = 4, \ldots, n\}$, and $conv(P_n)$ is the convex hull of pedigrees for a given $n > 3$. This implementation uses open-source Python libraries like NetworkX ([7]) and a linear programming package, like Gurobi ([5]). The framework given in the book shows that the framework can be implemented in strongly polynomial time using Tardos's approach ([11]). The framework given by Tardos is theoretical, as cautioned by Tardos, "probably should be considered a purely theoretical contribution." ([11, p. 251]. Future implementations will deploy algorithms for solving a combinatorial LP problem as a subroutine. This talk on this new beginning is fitting with the occasion of celebrating DC Programming, which opened new ways to solve difficult integer and combinatorial optimization problems forty years ago.

Keywords: Hamiltonian Cycles · Pedigree polytope · Multistage Insertion · membership problem · STSP polytope · Computational Complexity · Python implementation · Gurobi

1 Introduction

Given a polyhedral convex set C and a point X, deciding whether the point is a member of the convex set is called the membership problem and if X is not in the convex set, identifying a violated defining inequality is called the separation problem. A linear optimization problem over a polyhedral convex set gives

© The Author(s), under exclusive license to Springer Nature Switzerland AG 2025
H. A. Le Thi et al. (Eds.): ICAMCS 2024, LNNS 1311, pp. 3–14, 2025.
https://doi.org/10.1007/978-3-031-90606-0_1

rise to a linear programming problem. However, to use the finite precision polynomial algorithm due to [8] to solve that linear programming problem, we are required to solve the separation problem efficiently. Khachiyan's work generated considerable enthusiasm to study polytopes corresponding to combinatorial optimization problems. A major work in this area is the book, *Geometric algorithms and combinatorial optimization* [4]. This brings out the connections between the efficiencies of solving linear optimization, separation, and membership problems.

In Geometric Algorithms [4] the construction of Yudin and Nemirovsky [12] is used to establish conditions for the existence of a polynomial separation algorithm for a bounded convex body. The approach in [4] uses the ellipsoid algorithm twice. Given a polytope P, a, in the interior of P and a $X \notin P$, Maurras has given under certain conditions [10], a simple construction to identify a violated facet of the polytope, P, whose supporting hyperplane separates X from P. This uses a polynomial number of calls to an oracle checking membership in P. Thus, if we can have a polynomial algorithm for solving the membership problem, we can have a polynomial algorithm for the linear optimization over the polytope under study. This was the motivation for the author to study the membership problem for the pedigree polytopes.

However, if the number of linear constraints in the formulation is exponential in the size of the data used to describe the problem, we do not get a polynomial-time algorithm using the ellipsoidal approach for the optimization problem under consideration.

It is in this context, that a fresh look at the alternative combinatorial structure underlying the travelingsalesman problem becomes relevant. We introduce one such combinatorial object called, pedigree, related to tours, in Subsect. 1.1. The study of the polyhedral structure of the pedigree polytope throws new light on the complexity of solving the symmetric traveling salesman problem.

1.1 Importance of Studying Membership in Pedigree Polytope

In this subsection, we present an alternative polyhedral representation of the $STSP$, using the definition of pedigrees. Some required notations and definitions are introduced here. (Other preliminaries and notations are given in Sect. 2).

Let n be an integer, $n \geq 3$. Let V_n be a set of *vertices*. Assuming, without loss of generality, that the vertices are numbered in some fixed order, we write $V_n = \{1, \ldots, n\}$. Let $E_n = \{(i,j)|i,j \in V_n, i < j\}$ be the set of *edges*. The cardinality of E_n is denoted by $p_n = n(n-1)/2$. Let $K_n = (V_n, E_n)$ denote the complete graph of n vertices. We denote the elements of E_n by e where $e = (i,j)$. We also use the notation ij for (i,j). Unlike the usual practice, an edge is assumed to be written with $i < j$.

A subset H of E_n is called a *Hamiltonian cycle* in K_n if it is the edge set of a simple cycle in K_n, of length n. We also call such a Hamiltonian cycle a $n-tour$ in K_n. At times we represent H by the vector $(1 i_2 \ldots i_n 1)$ where $(i_2 \ldots i_n)$ is a permutation of $(2 \ldots n)$, corresponding to H. Let Q_n denote the standard $STSP$ polytope, given by

$$Q_n = conv(\{X_H : X_H \text{ is the characteristic vector of } H \in \mathcal{H}_n\})$$

where $\mathcal{H}_n$ denotes the set of all *Hamiltonian cycles* (or $n - tours$) in K_n.

In polyhedral combinatoric approaches, generally, Q_n is studied while solving $STSP$ (see [9]).

Given $H \in \mathcal{H}_{k-1}$, the operation *insertion* is defined as follows: Let $e = (i, j) \in H$. Inserting k in e is equivalent to replacing e in H by $\{(i, k), (j, k)\}$ obtaining a $k - tour$. When we denote H as a subset of E_{k-1}, then inserting k in e gives us a $H' \in \mathcal{H}_k$ such that,

$$H' = (H \cup \{(i, k), (j, k)\}) \setminus \{e\}.$$

We write $H \overrightarrow{\ e, k\ } H'$.

Similarly, the inverse operation shrinking can be defined.

Definition 1. *[Pedigree]* $W = (e_4, \ldots, e_n) \in E_3 \times \ldots \times E_{n-1}$ *is called a* pedigree *if and only if there exists a* $H \in \mathcal{H}_n$ *such that* H *is obtained from the* $3 - tour$ *by the sequence of insertions, viz.,*

$$3 - tour \overrightarrow{\ e_4, 4\ } H^4 \ \ldots \ H^{n-1} \overrightarrow{\ e_n, n\ } H.$$

For example, $W = ((2, 3), (1, 2), (3, 4))$ is a pedigree as it yields, by insertion operations, the 6-tour $H = \{(1, 3), (3, 6), (4, 6), (2, 4), (2, 5), (1, 5)\}$. But $W = ((2, 3), (1, 2), (1, 4))$ is not a pedigree, as the decision to insert 6 in $(1, 4)$ is not possible as the 5-tour that results from the insertion decisions for $4, 5$ is $\{(1, 3), (3, 4), (2, 4), (2, 5), (1, 5)\}$. The pedigree W is referred to as the pedigree of H.

The pedigree of H can be obtained by shrinking H sequentially to the $3-tour$ and noting the edge created at each stage. We then write the edges obtained in the reverse order of their occurrence, to obtain the corresponding pedigree. Motivation for the definition of pedigree and its connection to MI- formulation [3] are given in [2]. We denote by $P_{MI}(n)$, the set of solutions to the integer relaxation of the MI- formulation. (This LP formulation, is called MI- relaxation.)

Let the elements of E_n be labelled as follows: $(i, j) \in E_n$, has the label, $l_{ij} = p_{j-1} + i$. This means, edges $(1, 2), (1, 3), (2, 3) \in E_3$ are labeled, $1, 2$, and 3 respectively. Once the elements in E_{n-1} are labelled then the elements of $E_n \setminus E_{n-1}$ are labelled in increasing order of the first coordinate, namely i.

Let the set of all pedigrees, corresponding to $H \in \mathcal{H}_n$ be denoted by $\mathcal{P}_n$. For any $4 \leq k \leq n$, given an edge $e \in E_{k-1}$, with edge label l, we can associate a $0 - 1$ vector, $\mathbf{x}(e) \in B^{p_{k-1}}$, such that, $\mathbf{x}(e)$ has a 1 in the lth coordinate, and zeros else where. That is, $\mathbf{x}(e)$ is the indicator of e.

The number of coordinates in X, is $\sum_{k=4}^{n} p_{k-1}$, and is denoted by τ_n, notice that, $\tau_n = \binom{n}{3} - 1$. Similarly, we can associate a $X = (\mathbf{x}_4, \ldots, \mathbf{x}_n) \in B^{\tau_n}$, the characteristic vector of the pedigree W, where $(W)_k = e_k$, the $(k-3)^{rd}$ component of W, $4 \leq k \leq n$ and $\mathbf{x}_k$ is the indicator of e_k. We use the notation, X/k to denote the first $k - 3$ components of X. Thus, $X/k = (\mathbf{x}_4, \ldots, \mathbf{x}_k)$.

Let $P_n = \{X \in B^{\tau_n} : X \text{ is the characteristic vector of the pedigree } W \in \mathcal{P}_n\}$.

Thus there is a one-to-one correspondence between $H \in \mathcal{H}_n$ and $X \in P_n$. Consider the convex hull of P_n. We call this the *pedigree polytope*, denoted by $conv(P_n)$.

The recent book by the author, *Pedigree Polytopes*, apart from bringing together published results on pedigree polytopes, has the major contribution, proving the following theorem:

Theorem 1 (Main Theorem) *Given* $n, X \in P_{MI}(n)$, *and* $X/n-1 \in conv(P_{n-1})$. *Then* $X \in conv(P_n)$ *if and only if there exists a solution for the multi-commodity flow problem*[1], *with* $z^* = \sum_{s \in \mathcal{S}_k} v^s = z_{max}$.

The complexity of checking this necessary and sufficient condition is also estimated in the book leading to the result:

Theorem 2 (Efficient Checking of Membership). *Given* $n, X \in Q^{\tau_n}$, *there exists a strongly polynomial time algorithm for checking whether* $X \in conv(P_n)$.

This article briefly introduces concepts required to grasp the proofs of the two theorems mentioned above for a clear understanding of the Framework and its validity and complexity correctly. In addition, provides a `Python` implementation of the Framework. It is obvious that this result has far-reaching consequences in combinatorial optimization, and so in computational complexity theory, as it implies $STSP$ is efficiently solvable, as shown in Chapter 7 of [1] or it establishes $NP = P$.

Next, we outline the structure of this paper.

Section 2 gives additional preliminaries and notation used. Section 3 describes briefly the construction of the Layered Network (N_k, R_k, μ). Section 4 defines and studies the solutions to a multicommodity flow Problem to check the membership in the pedigree polytope. This section deals with a necessary and sufficient condition for membership in the pedigree polytope. Section 5 gives the details of how some of the steps in the Framework are implemented using Python packages and libraries. Finally, Sect. 6 provides concluding remarks and discussions on future research paths.

2 Preliminaries and Notations

Let R denote the set of reals. Similarly Q, Z, N denote the rationals, integers, and natural numbers respectively, and B stands for the binary set of $\{0, 1\}$. Let R_+ denote the set of nonnegative reals. Similarly, the subscript $_+$ is understood with rationals. Let R^d denote the set of d -tuples of reals. Similarly the superscript ^{d} is understood with rationals, etc. Let $R^{m \times n}$ denote the set of $m \times n$ real matrices. For a subset $F \subset E_n$ we write the *characteristic* vector of F by $x_F \in R^{p_n}$ where $x_F(e) = 1$ if $e \in F$, and 0 otherwise.

For a subset $S \subset V_n$ we write

$E(S) = \{ij | ij \in E, i, j \in S\}$.. Given $u \in R^{p_n}$, $F \subset E_n$, we define, $u(F) = \sum_{e \in F} u(e)$. For any $S \subset V_n$, let $\delta(S)$ denote the set of edges in E_n with one end in S and the other in $S^c = V_n \setminus S$. For $S = \{i\}$, we write $\delta(\{i\}) = \delta(i)$.

---

[1] Problem $MCF(n)$, see page 131 of [1].

2.1 MI Formulation

Let x_{ijk} correspond to the insertion decision of k in $e = (i, j)$. That is, $x_{ijk} = 1$ if we decide to insert k in (i, j), otherwise it is zero. We use $x_k(e)$ equivalently. Let $C_{ijk} = c_{ik} + c_{jk} - c_{ij}$. Here C_{ijk} gives the incremental cost of inserting k between (i, j), where c_{ij} is the cost of visiting city j from city i.

Problem 1 (MI- Relaxation).

$$minimise \sum_{k=4}^{n} \sum_{1 \le i < j \le k-1} C_{ijk} x_{ijk}$$

subject to

$$\sum_{1 \le i < j \le k-1} x_{ijk} = 1, \; 4 \le k \le n \tag{1}$$

$$\sum_{k=4}^{n} x_{ijk} \le 1, \; 1 \le i < j \le 3 \tag{2}$$

$$-\sum_{r=1}^{i-1} x_{rij} - \sum_{s=i+1}^{j-1} x_{isj} + \sum_{k=j+1}^{n} x_{ijk} \le 0, \; 4 \le j \le n-1; 1 \le i < j \tag{3}$$

$$-\sum_{r=1}^{i-1} x_{rin} - \sum_{s=i+1}^{n-1} x_{isn} \le 0, \; i = 1, \ldots, n-1. \tag{4}$$

$$x_{ijk} \ge 0, \; 4 \le k \le n; \quad 1 \le i < j \le k-1 \tag{5}$$

Constraints 4 are redundant. Let the slack variables corresponding to all inequality constraints be given by $\mathbf{u} = (u_{ij})$. This relaxation is used in the proofs of the important theorems mentioned in the Subsect. 1.1.

2.2 Rigid, Dummy Arcs in a Capacited Transportation Problem

Consider a balanced transportation problem, in which, some arcs called the *forbidden* arcs are not available for transportation. We call the problem of finding whether a feasible flow exists in such an incomplete bipartite network, a Forbidden Arcs Transportation (FAT) problem. This could be viewed as a capacited transportation problem, as well. When any arc (α, β) has the capacity restriction, it is specified by $f_{\alpha\beta} \le c_{\alpha\beta}$, as an additional restriction. This problem has many appearances in this paper. In addition, identifying arc flows that remain the same in all feasible solutions to a FAT problem are called *frozen flows*, and are important to identify. We use a Frozen Flow Finding (FFF) algorithm, like that given by Guesfield in [6]. Such arcs an arc is called a *Rigid arc.* If the rigid flow is zero the arc is called a *dummy arc.*

3 Construction of the Layered Network

In this section, we define the layered network N_k and a set of pedigrees R_k with fixed positive weight μ_P for $P \in R_k$, for a given $X \in P_{MI}(n)$ and for $3 < k \leq n$. Given $X/k \in conv(P_k)$, this network is used in showing whether or not $X/k+1 \in conv(P_{k+1})$.

We denote the node set of N_k by $\mathcal{V}(N_k)$ and the arc set by $\mathcal{A}(N_k)$. Let $v = [k : e]$ denote a node in the $(k-3)^{rd}$ layer corresponding to an edge $e \in E_{k-1}$. Let $x(v) = x_k(e)$ for $v = [k : e]$.

3.1 Construction of the Network for K = 4

If $X/5 \notin conv(P_5)$ we can conclude that $X \notin conv(P_n)$.

Let

$$V_{[r]} = \{v | v = [r+3 : e], e \in E_{r+2}, x(v) > 0\}.$$

The node name $[r+3 : e]$ alludes to the insertion decision corresponding to the stage r; that is, the edge e is used for the insertion of $r+3$. First, we define the nodes in the network N_4.

$$\mathcal{V}(N_4) = V_{[1]} \bigcup V_{[2]}.$$

And

$$\mathcal{A}(N_4) = \{(u,v) | u = [4 : e_\alpha] \in V_{[1]}, v = [5 : e_\beta] \in V_{[2]}, e_\alpha \in G(e_\beta)\}.$$

Let $x(w)$ be the capacity on a node $w \in V_{[r]}, r = 1, 2$. Capacity on an arc $(u, v) \in \mathcal{A}(N_4)$ is $x(u)$.

Given this network, we consider a flow feasibility problem of finding a nonnegative flow defined on the arcs that saturate all the node capacities and violates no arc capacity, that is, solve the F_4 problem. Thus, the construction of the layered network proceeds from solving a bipartite flow feasibility problem for stage 1, involving layers 1 and 2. If F_4 is infeasible we conclude $X \notin conv(P_n)$, and the construction of the network stops. When F_4 is feasible we use the FFF algorithm (or any such) and identify the set of rigid arcs, $\mathcal{R}$. If there are any dummy arcs in $\mathcal{R}$ we delete them from the set of arcs and update. For every rigid arc with a positive frozen flow, we replace the capacity of the arc with the frozen flow. And include the corresponding arc (pedigree, $P \in conv(P_5)$ in the set R_4 with μ_P, the frozen flow as the weight for P. We reduce the node capacities of the head and tail nodes of the rigid arc P by μ_P, for each $P \in R_4$, and delete nodes with zero capacity, along with the arcs incident with them. Consider the sub-network obtained as N_4. Let $\bar{x}(v)$ be the updated capacities of node v in the network. Now we say (N_4, R_4, μ) is *well-defined.* Observe that either the node set in N_4 is $\emptyset$ or $R_4 = \emptyset$, but not both.

Example 1. Consider $X \in P_{MI}(5)$ given by

$$\mathbf{x}_4 = (0, 3/4, 1/4);$$
$$\mathbf{x}_5 = (1/2, 0, 0, 0, 0, 1/2);$$

It can be verified that F_4 is feasible and f given by $f_{([4:1,3],[5:1,2])} = 1/4, f_{([4:2,3],[5:1,2])} = 1/4$, and $f_{([4:1,3],[5:3,4])} = 1/2$ does it. And so, we conclude $X \in conv(P_5)$. But no arc is rigid in F_4, is obvious. Therefore, (N_4 = the network of $F_4, R_4 = \emptyset, \mu = 0$). However, X' given by

$$\mathbf{x'}_4 = (0, 3/4, 1/4);$$
$$\mathbf{x'}_5 = (0, 1/4, 0, 0, 1/4, 1/2);$$

is such that, $X' \in P_{MI}(5)$ but F_4 is not feasible. So $X' \notin conv(P_5)$.

Example 2. Consider $X = (0, \frac{1}{3}, \frac{2}{3}, 0, \frac{1}{6}, 0, \frac{1}{6}, \frac{1}{3}, \frac{1}{3})$. We have F_4 feasible but all arcs are rigid and so ($N_4 = \emptyset, R_4 = \{((1,3)(1,4)), ((1,3)(3,4)), ((2,3)(1,3)), ((2,3)(3,4)), ((2,3)(2,4))\}, \mu = (\frac{1}{6}, \frac{1}{6}, \frac{1}{6}, \frac{1}{6}, \frac{1}{3})$)

3.2 Overview of the Membership Checking in Pedigree Polytope

If (N_{k-1}, R_{k-1}, μ) is well defined and $k < n$ we proceed further. Next, consider a bipartite flow feasibility problem for stage $k-3$, called F_k. If F_k is infeasible we conclude $X \notin conv(P_n)$, and the construction of the network stops. Otherwise, we proceed to construct (N_k, R_k, μ). Next, we seek a feasible solution to a multicommodity flow problem. If we have a feasible solution we have evidence that $X/k \in conv(P_k)$, and we say the network is well-defined. If (N_k, R_k, μ) is not well-defined, we conclude $X \notin conv(P_n)$, and stop. Otherwise proceed until we reach $k = n-1$, constructing (N_k, R_k, μ), and concluding $X \in conv(P_n)$ or not.

This construction process is explained in Fig. 1 and the following discussion elaborates on these steps.

4 Construction of the Layered Network for k > 4

Given (N_{k-1}, R_{k-1}, μ) is well-defined, we proceed to construct (N_k, R_k, μ) recursively. Firstly we define the node-set,

$$\mathcal{V}(N_k) = \mathcal{V}(N_{k-1}) \bigcup R_{k-1} \bigcup V_{[k-2]}. \tag{6}$$

In $\mathcal{V}(N_k)$ any pedigree, P in R_{k-1} is *shurnk* to become a new node with capacity μ_P. And we have the new sinks corresponding to $V_{[k-2]}$. The new origin node P introduced corresponding to a rigid pedigree ensures that the weight on the extensions of P is exactly equal to μ_P. We shall have an example to illustrate the importance of these additional origins. Now consider the links between layers $k-3$ and $k-2$. Any link, L can give rise to an arc in the network N_k depending on the solution to a max flow problem defined on a subnetwork (called restricted network) derived from N_{k-1} and the link L. The motivation for the restricted network stems from the fact, any link $L = ([k : (s,t)], [k+1 : (i,j)])$ to extend a pedigree $P \in P_k$, we require (1) a generator of (s,t) as well as that of (i,j) must be present in P; (2) (s,t) and (i,j) should not be present in $P/k-1$ and (3) P must end in (s,t). So we need not consider paths in N_{k-1} that

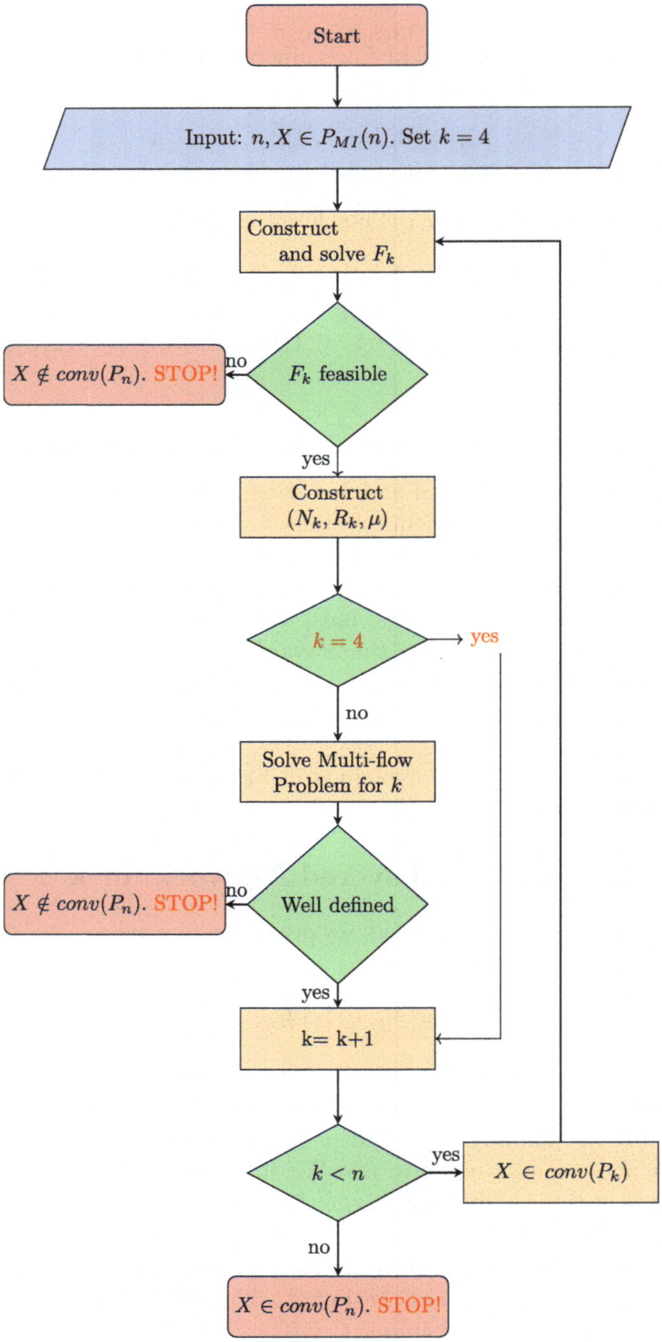

Fig. 1. Flowchart showing the Membership Checking in $conv(P_n)$

- are not ending in $[k : (s,t)]$.
- are having $[q : (s,t)]$ or $[r : (i,j)]$ for some $q, r < k$.
- are not having a generator node of (s,t) or a generator node of (i,j).

The restricted network $N_{k-1}(L)$ defined for a link L ensures that it does not have such paths in it, by deleting a set of nodes from N_{k-1}.

To determine the capacity $C(L)$ along a link L, we consider the problem of finding the maximal flow in $N_{k-1}(L)$ satisfying non-negativity of flow, flow conservation at nodes, and capacity restrictions on the available nodes and arcs. We find $C(L)$ for each link L.

Besides, if the path that brings this flow $C(L)$ for any L is unique, we save the corresponding path, named $P_{unique}(L)$, against L.

Now we are in a position to define the FAT problem, called F_k.

Definition 2 **(F_k).** *Consider a forbidden arc transportation problem with*

$$\begin{array}{lrl}
O -- & Origins] & : u = [k : e_\alpha] \in V_{[k-3]} \text{ and } P \in R_{k-1} \\
a -- & Availability] & : \bar{x}(u) \text{ for } u \text{ and } \mu_P \text{ for } P \\
D -- & Sinks] & : v = [k+1 : e_\beta] \in V_{[k-2]} \\
b -- & Demand] & : x(v) \text{ for } v \\
\mathcal{A} -- & Arcs] & : \{(u,v) \text{ such that, either } L = (e_\alpha, e_\beta) \text{ is a link and } C(L) > 0 \\
 &] & : \text{ or } u = P \text{ and } v = [k+1 : e_\beta] \text{ and } (P, e_\beta) \text{ is a pedigree, that is, } (P, e_\beta) \in P_{k+1}\} \\
C -- & Capacity] & : C_{u,v} = C(L) \text{ or } \mu_P.
\end{array}$$

If F_k is feasible we apply the *Frozen* Flow *Finding* algorithm and identify the set $\mathcal{R}$ and the dummy subset of arcs in that. Update $\mathcal{A}$ by deleting the dummy arcs. Update the capacity of any rigid arc, $C(L)$, to the frozen flow. Recall that the frozen flow is a positive constant, in every feasible solution to F_k.

If $L = (u,v)$ is rigid and the path is unique for L and $(P_{unique}(L), [k+1 : e_\beta])$ does not correspond to a pedigree then we can conclude that $X/k+1 \notin conv(P_{k+1})$.

4.1 Completing the Construction of the Layered Network

Here we are in a position to complete the construction of the network (N_k, R_k, μ). Construction of R_k is driven by the rigid arcs in F_k. If the rigid flow is from a unique path then that contributes a rigid pedigree to R_k. Similarly, if the rigid flow is from an origin in F_k that is a shrunken node, $P \in R_{k-1}$ then that contributes a rigid pedigree to R_k. After updating node and arc capacities and deleting nodes and arcs with zero capacity, thus, we have constructed recursively the network, (N_k, R_k, μ). Our next task is to check that (N_k, R_k, μ) is well-defined. To declare the network (N_k, R_k, μ) to be well-defined, we need to have evidence that $X/k+1 \in conv(P_{k+1})$. This was easy for $k = 4$. Even though pedigree paths are bringing the flow along each arc in F_k, there could be conflicts arising out of the simultaneous capacity restrictions on these flows in the network N_{k-1}. This calls for ensuring that these restrictions are not violated. The multicommodity flow problem defined in [1] does precisely this check. The proofs of theorems stated in Sect. 1.1 relate to this multicommodity flow problem and validate a polynomial framework to check membership in pedigree polytopes. Chapters 5 & 6 of the recent book, *Pedigree Polytopes*, contain the proofs and complexity checks [1].

5 Python Implementation of the Framework

This section outlines how the framework (1) given in Subsect. 3.2 is implemented using Python packages and libraries on graph theoretical algorithms, such as NetworkX ([7]), and any LP solver, like Gurobi ([5]).

Python Package: The package implementing the Framework is called, `checking4membership`. From this package, we run `memcheck` to check the membership of a given X in the pedigree polytope for a given n. For want of space Python implementation of various steps in the framework is not detailed here. However, how to prepare the input for the package and how to install the package is briefed. Firstly, input the constant EPSILON, which we need regarding the precision with which we do the computations. For instance, EPSILON = 0.000000000000001, is currently used for rounding off a given x to 0 or 1.

5.1 Input Preparation

Next, we discuss how the data is input to the package created by the author, `checking4membership`, that implements the framework. We are given, n, and $X = (x_{ijk})$ a vector from $R_+^{\tau_n}$.

We assume X is given by specifying only the positive coordinates of X, and exploits the sparsity in X. We specify a positive $x_{i,j,k}$ by three rows, the first row gives the edge (i,j), the second row has k and the third row has the value, $x_{i,j,k}$. For instance, given $n = 5$, and $X = (1/3, 0, 2/3, 0, 0, 1/5, 0, 1/2, 3/5)$ is represented by the matrix:

$$\begin{pmatrix} (1,2) & (2,3) & (2,3) & (2,4) & (3,4) \\ 4 & 4 & 5 & 5 & 5 \\ 1/3 & 2/3 & 1/5 & 1/2 & 3/5 \end{pmatrix}$$

Notice that for each k the values add to 1. The edges are ordered according to edge labels, that is, $(i,j) = p_{j-1} + i$ where $p_j = j(j-1)/2$. And here $\tau_5 = p_3 + p_4 = 3 + 6 = 9$.

A problem instance is given by a text file, say, `myproblem.txt`.

`get_problem()` is the first module executed to have the input values read in from a text file,(ex.: 'problem6.txt'). The main module can select a text file from your computer as directed by you using a GUI. These details are explained in the ReadMe.md file when you install `checking4membership`.

Prepare a text file (ex.: `'prob_in_6.txt'`) using your favourite editor and store it in a folder, to be selected as `problemfile`. This file contains line [1] `prob_name:str()` ex.: `'Sixth_problem'`., line [2] $n : int$, ex.: 7, line [3] row:str() having i, j for edge (i, j), each edge separated by space, ex.: 1,2 1,3, ... 4,5 6,7, line [4] col:str() having k for node k separated by space, ex.: 4 4 5 5 5 ... 7 7 7, line [5] data:str() having x_{ijk} separated by space having x_{ijk} values, ex.: 1 .5 .5 ..... .5 .25, line [6] Could contain `True` or `False` or empty, depending on whether X is checked to be in $P_{MI}(n)$ or not, or this check has not been done yet, respectively.

5.2 Note on Installing the Package `checking4membership`

Once the input file, say `problem_one.txt`, giving your instance of X is prepared as explained above, store it in a folder/directory and name it, say `myproblems`. You may store your other files giving different instances in this folder as well. In addition, create a folder for output files for the given instance, say with name `output_one`. Once you have `Python` running on your computer, install `checking4membership` from `Testpypi` repository. And follow instructions as they appear.

6 Concluding Remarks

In this paper, we consider the membership problem of the pedigree polytope. A necessary and sufficient condition for membership in the pedigree polytope is shown as the existence of feasible multicommodity flow with maximum possible value in a recursively constructed layered network. An algorithmic framework for checking membership is given and the computational complexity is shown to be strongly polynomial in n. This estimate uses only loose upper bounds on the computational burden involved in the different steps. The implication of the paper's main result is far-reaching as this implies we can have a polynomial algorithm to solve the pedigree optimization problem, or equivalently we can solve the symmetric traveling salesman problem with such an algorithm. These results appear in the recent book *Pedigree Polytopes* [1]. The main contribution of this paper is creating a Python package for implementing the efficient checking of membership in pedigree polytopes. This package is called `checking4membership` and the `memcheck` module from this package does the job. This package is in its initial stages of development, and later versions will be made available through the Python packages repository TestPYPI.

Acknowledgments. The author thanks the organizers of ICAMCS2024 for inviting me to this conference celebrating 40 years of DC programming.

Disclosure of Interests. The author has no competing interests to declare that are relevant to the content of this article.

References

1. Arthanari, T.S.: Pedigree Polytopes: New Insights on Computational Complexity of Combinatorial Optimisation Problems. Springer Nature, Singapore (2023)

2. Arthanari, T.S.: On pedigree polytopes and Hamiltonian cycles. Discret. Math. **306**(14), 1474–1492 (2006)
3. Arthanari, T.S., Usha, M.: An alternate formulation of the symmetric traveling salesman problem and its properties. Discret. Appl. Math. **98**(3), 173–190 (2000)
4. Grötschel, M., Lovász, L., Schrijver, A.: Geometric algorithms and combinatorial optimization, vol. 2. Springer Science & Business Media (2012)
5. Gurobi Optimization, LLC: Gurobi Optimizer Reference Manual (2024), https://www.gurobi.com
6. Gusfield, D.: A graph theoretic approach to statistical data security. SIAM J. Comput. **17**(3), 552–571 (1988)
7. Hagberg, A., Swart, P.J., Schult, D.A.: Exploring network structure, dynamics, and function using networkx. Tech. rep, Los Alamos National Laboratory (LANL), Los Alamos, NM (United States) (2008)
8. Khachiyan, L.G.: A polynomial algorithm in linear programming. In: Doklady Akademii Nauk. vol. 244, pp. 1093–1096. Russian Academy of Sciences (1979)
9. Lawler, E.L.: The traveling salesman problem: a guided tour of combinatorial optimization. Wiley-Interscience Series in Discrete Mathematics (1985)
10. Maurras, J.F.: From membership to separation, a simple construction. Combinatorica **22**, 531–536 (2002)
11. Tardos, É.: A strongly polynomial algorithm to solve combinatorial linear programs. Oper. Res. **34**(2), 250–256 (1986)
12. Yudin, D.B., Nemirovskii, A.S.: Informational complexity and efficient methods for the solution of convex extremal problems. Matekon **13**(2), 22–45 (1976)

Variants of Inertial Algorithms for Nonmonotone Equilibrium Problems

Van Dinh Bui(✉), Hy Duc Manh, Tran Thi Huyen Thanh, and Thanh Tuyen Bien

Le Quy Don Technical University, 236 Hoang Quoc Viet Street, Hanoi, Vietnam
{vandinhb,ducmanhh,huyenthanhtt,thanhtuyenb}@lqdtu.edu.vn

Abstract. This paper proposes two variants of inertial algorithms designed to solve equilibrium problems in real Hilbert spaces, without assuming any generalized monotonicity or the Lipschitz-type condition of the bifunctions defining the problems. Our strategy replaces the shrinking projection step commonly used in existing algorithms with a projection onto a set defined by the feasible region and an appropriate half-space. Under the same assumptions for solving such problems, we show that the sequences generated by the proposed algorithms converge weakly and strongly to a solution of the equilibrium problem, respectively. Furthermore, some numerical examples are included to illustrate the effectiveness of the proposed algorithms.

Keywords: Nonmonotone equilibrium problem · inertial algorithm · Armijo linesearch · Strong convergence · Weak convergence

1 Introduction

In a real Hilbert space $\mathbb{H}$ with the inner product $\langle \cdot, \cdot \rangle$ and the induced norm $\|\cdot\|$, given a nonempty closed convex subset C of $\mathbb{H}$ and a bifunction defined on C, that is, $f : C \times C \to \mathbb{R}$ satisfying $f(x, x) = 0$ for all $x \in C$, the equilibrium problem associated with C and f (denoted as $\mathrm{EP}(C, f)$) can be stated as follows:

$$\text{Find } u \in C \text{ such that } f(u, y) \geq 0, \ \forall y \in C.$$

The equilibrium problem, in this form, can be found in the papers of Muu and Oettli [12] and Blum and Oettli [3]. However, it had appeared earlier in the works Ky Fan [8], and is known as the Ky Fan inequality. The equilibrium problem associated with $\mathrm{EP}(C, f)$, sometimes called the dual equilibrium problem or Minty equilibrium problem (denoted as $\mathrm{MEP}(C, f)$) [4,11], is to find $u \in C$ such that

$$f(y, u) \leq 0, \ \forall y \in C.$$

The solution sets of $\mathrm{EP}(C, f)$ and $\mathrm{MEP}(C, f)$ are denoted by S_E and S_M, respectively. It is pointed out in [3,12] that equilibrium problems include many important problems, such as optimization problems, fixed point problems, saddle point

© The Author(s), under exclusive license to Springer Nature Switzerland AG 2025
H. A. Le Thi et al. (Eds.): ICAMCS 2024, LNNS 1311, pp. 15–26, 2025.
https://doi.org/10.1007/978-3-031-90606-0_2

problems, complementarity problems, variational inequality problems, and Nash equilibrium problems in noncooperative games. The existence of a solution to EP(C, f) when C is a convex and compact set, and $f(x, y)$ is quasiconvex in y, can be found in the work of Ky Fan [8]. In the case where C is a noncompact set and the bifunction f possesses or does not possess generalized monotonicity, the existence of a solution to EP(C, f) can be found in [2,4] and the references quoted therein.

One of the most important directions for studying equilibrium problems is to build efficient algorithms to find a solution of EP(C, f). By exploiting the structure of the constraint set C and the function f, many methods have been proposed for solving equilibrium problems in the literature, such as gap function methods [9], proximal point methods [10], and projection methods [7,16,17]. Among these, projection methods play an important role because they allow us to solve equilibrium problems using convex optimization. These methods often require convexity in the second variable and, in particular, the monotonicity or generalized monotonicity of the bifunction f. More specifically, if f is strongly monotone (that is, there exists $\tau > 0$ such that $f(x, y) + f(y, x) \le -\tau\|x - y\|^2$ for all $x, y \in C$) and satisfies the Lipschitz-type condition on C (that is, there exist $L_1, L_2 > 0$ such that $f(x, y) + f(y, z) \ge f(z, z) - L_1\|x - y\|^2 - L_2\|y - z\|^2$ for all $x, y, z \in C$), then the single projection method converges linearly to the unique solution of EP(C, f) [9,13]. However, this method may not converge when f is only monotone on C. To overcome this drawback, the double projection method, or extragradient method, has been proposed to find a solution of EP(C, f) [17]. It is noted that if f is pseudomonotone on C (that is, for all $x, y \in C$, $f(x, y) \ge 0$ implies $f(y, x) \le 0$) and satisfies the Lipschitz-type condition on C, then the double projection method converges weakly to a solution of EP(C, f), provided that S_E is a nonempty set. When f is pseudomonotone on C and does not satisfy the Lipschitz-type condition, linesearch procedures are incorporated into the double projection methods to develop algorithms for EP(C, f) [7,17]. When f is nonmonotone and non-Lipschitz on C, modified double projections with linesearches, known as shrinking projections, are proposed to solve EP(C, f) [6].

On the other hand, an inertial-type algorithm was first proposed by Polyak [15] as an acceleration process for solving smooth convex minimization problems. An inertial-type algorithm can be considered a two-step iterative method in which the next iterate is computed using the previous two iterates. It is well known that incorporating an inertial term in an algorithm may speed up the rate of convergence. Consequently, the inertial effect has drawn attention from researchers and plays an important role in various methods for solving optimization problems, variational inequality problems, and equilibrium problems when the bifunction f is generalized monotone [1,19].

Very recently, Deng et al. [5] proposed combining the shrinking projection method with an inertial term to develop an algorithm for finding a solution to nonmonotone and non-Lipschitz-type equilibrium problems. However, to apply this algorithm, at each iteration, we have to solve an optimization problem in which the constraint set is the intersection of the constraint set from the previous iteration and a half-space. This may lead to large computational costs, especially when dealing with high-dimensional spaces and complicated constraint sets.

Based on these facts and recent works [18], in this paper, we propose two inertial algorithms for finding a solution of EP(C, f), where the bifunction is nonmonotone and non-Lipschitz type on C. Notably, our algorithms do not rely on the utilization of shrinking projection methods commonly employed in the existing literature. The rest of the paper is organized as follows. In Sect. 2, we recall some preliminary results. The proposed algorithms and their convergence are presented in Section 3. Section 4 is devoted to presenting some numerical examples to illustrate the efficiency of the proposed algorithms.

2 Preliminaries

Let Ω be an open convex set containing C in which $C \subset \mathbb{H}$ is a nonempty, closed and convex set. The bifunction $f : \Omega \times \Omega \to \mathbb{R}$ satisfies:

(A1) f is weakly continuous on $\Omega \times \Omega$, that is for every $x, y \in C$ and sequences $\{x^k\}, \{y^k\}$ in C converging weakly to x and y, respectively, then $f(x^k, y^k)$ converges to $f(x, y)$ as $k \to \infty$.

(A2) $f(x, .)$ is convex, lower semicontinuous and subdifferentiable on C for all $x \in C$;

Lemma 1 (Proposition 4.3 [20]). *Let $f : \Omega \times \Omega \to \mathbb{R}$ be a function satisfying assumptions (A1) and (A2). Let $\bar{x}, \bar{y} \in \Omega$ and $\{x^k\}, \{y^k\}$ be two sequences in Ω converging weakly to $\bar{x}, \bar{y}$, respectively. Then, for any $\epsilon > 0$, there exists $\eta > 0$ and $k_\epsilon \in \mathbb{N}$ such that $\partial_2 f(x^k, y^k) \subset \partial_2 f(\bar{x}, \bar{y}) + \frac{\epsilon}{\eta}B$, for every $k \geq k_\epsilon$, where B denotes the closed unit ball in $\mathbb{H}$.*

Lemma 2 (Lemma 2.6 [6]). *Suppose the bifunction f satisfies the assumptions (A1) and (A2). If $\{x^k\} \subset C$ is bounded, $\rho > 0$ and $\{y^k\}$ is a sequence such that $y^k = \arg\min\{f(x^k, y) + \frac{\rho}{2}\|y - x^k\|^2 : y \in C\}$ then $\{y^k\}$ is bounded.*

Lemma 3 ([21]). *Let C be a nonempty closed convex subset of $\mathbb{H}$. Let $\{x^k\}$ be a sequence in $\mathbb{H}$ and $u \in \mathbb{H}$. If any weak limit point of $\{x^k\}$ belongs to C and $\|x^k - u\| \leq \|u - P_C(u)\|$ for all k, then $x^k \to P_C(u)$.*

Lemma 4 (Theorem 2.1 [1]). *Let $\{\phi_k\}, \{\delta_k\}$ and $\{\alpha_k\}$ be nonnegative sequences such that $\phi_{k+1} \leq \phi_k + \alpha_k(\phi_k - \phi_{k-1}) + \delta_k, \forall k \in \mathbb{N}$, $\Sigma_{k=1}^{\infty}\delta_k < \infty$, and there exists a real number α satisfying $0 \leq \alpha_k \leq \alpha < 1, \forall k \in \mathbb{N}$. Then the following properties hold:*

(i) $\Sigma_{k=1}^{\infty}[\phi_k - \phi_{k-1}]_+ < \infty$ where $[t]_+ = \max\{t, 0\}$;
(ii) There exists $\phi^* \geq 0$ such that $\lim_{k\to\infty} \phi_k = \phi^*$.

Lemma 5 (Opial's Lemma [14]). *Let $A \subset \mathbb{H}$ be a nonempty set and $\{x^k\}$ be a sequence in $\mathbb{H}$. Suppose that*

(i) for each point $x^* \in A$, there exists $\lim_{k\to\infty} \|x^k - x^*\|$;
(ii) every weak limit point of $\{x^k\}$ belongs to A as $k \to \infty$.

Then, it follows that $\{x^k\}$ converges weakly to a point in A as $k \to \infty$.

3 Main Results

Now, we propose inertial algorithms for solving EP(C, f), where the bifunction f is nonmonotone and non-Lipschitz-type on C, and we do not use the embedding methods found in the literature.

Algorithm 1.
Initialization. Pick $x^{-1}, x^0 = x^g \in C$, choose sequences $\{\rho_k\} \subset [\alpha, \beta]$ with $0 < \alpha \leq \beta$, $\{\eta_i\} \subset (0,1)$ strictly decreasing to zero, and $\{\epsilon_k\} \subset [0, +\infty)$ such that $\sum_{k=0}^{+\infty} \epsilon_k < +\infty$. Choose parameters $\mu \in (0,1)$, $\lambda \in [0,1)$.
Iteration k ($k = 0, 1, 2, ...$). Having x^{k-1}, x^k do the following steps:
Step 1. Compute $u^k = P_C(x^k + \lambda_k(x^k - x^{k-1}))$, where $\lambda_k \in [0, \hat{\lambda}_k]$ with

$$\hat{\lambda}_k = \begin{cases} \min\{\lambda, \frac{\epsilon_k}{\|x^k - x^{k-1}\|}\} & \text{if } x^k \neq x^{k-1}, \\ \lambda & \text{otherwise.} \end{cases}$$

Step 2. Compute $y^k = \arg\min\left\{\rho_k f(u^k, y) + \frac{1}{2}\|y - u^k\|^2 : y \in C\right\}$.
If $y^k = u^k$ then stop, $u^k \in S_E$. Otherwise, do Step 3.
Step 3. Find θ_k be the largest number in the sequence $\{\eta_i\}$ such that

$$\begin{cases} z^k = (1 - \theta_k)u^k + \theta_k y^k, \\ f(z^k, u^k) - f(z^k, y^k) \geq \frac{\mu}{\rho_k}\|y^k - u^k\|^2. \end{cases} \tag{1}$$

Step 4. Compute $v^k = P_{D_k}(u^k)$, where
$D_k = C \cap H_{t_k}$, $t_k \in \text{argmax}\{d(u^k, H_j) : 0 \leq j \leq k\}$.
$H_k = \{x \in \mathbb{H} : \langle \xi^k, x - u^k \rangle + f(z^k, u^k) \leq 0\}$, $\xi^k \in \partial_2 f(z^k, u^k)$.
Step 5. Compute $x^{k+1} = P_{A_k}(x^g)$, where $A_k = C_k \cap Q_k \cap C$ with
$C_k = \{x \in \mathbb{H} : \|x - v^k\| \leq \|x - u^k\|\}$, $Q_k = \{x \in \mathbb{H} : \langle x - x^k, x^g - x^k \rangle \leq 0\}$.
Go to Step 1 with k is replaced by $k + 1$.

Firstly, we need the following lemma to prove the convergence of Algorithm 1.

Lemma 6. *At iteration k, if $y^k \neq u^k$, then we have:*

(i) The linesearch is well-defined;
(ii) $f(z^k, u^k) > 0$.

The proof of this lemma is quite similar to the proof of Lemma 4.5 in [17] so we obmit it. The following theorem gives us the strong convergence of Algorithm 1.

Theorem 1. *Suppose that f satisfies two assumptions (A1) - (A2), and $S_M \neq \emptyset$. Then $\{x^k\}$, $\{u^k\}$ and $\{v^k\}$ generated by Algorithm 1 converges strongly to a solution of EP(C, f).*

Proof. We divide this proof into three statements as the following.

Statement 1. Six sequences $\{x^k\}, \{u^k\}, \{y^k\}, \{z^k\}, \{\xi^k\}, \{v^k\}$ are bounded.

Since $x^{k+1} \in A_k \subset Q_k$, we have

$$\|x^{k+1} - x^k\|^2 \le \|x^{k+1} - x^g\|^2 - \|x^g - x^k\|^2, \tag{2}$$

Take $p \in S_M$, we have $f(y,p) \le 0,\ \forall y \in C$. Especially,

$$f(z^k, p) \le 0,\ \forall k. \tag{3}$$

Since $\xi^k \in \partial_2 f(z^k, u^k)$, we have

$$f(z^k, p) \ge f(z^k, u^k) + \langle \xi^k, p - u^k \rangle. \tag{4}$$

From (3) and (4), we get $f(z^k, u^k) + \langle \xi^k, p - u^k \rangle \le 0,\ \forall k$. This means that $p \in (\cap_{j=0}^{\infty} H_j) \cap C$. Now, to prove that $(\cap_{j=0}^{\infty} H_j) \cap C \subset \cap_{j=0}^{\infty} A_j$, we will show that $(\cap_{j=0}^{\infty} H_j) \cap C \subset C_k, \forall k$ and $(\cap_{j=0}^{\infty} H_j) \cap C \subset Q_k, \forall k$.

Firstly, take $p \in (\cap_{j=0}^{\infty} H_j) \cap C$. Then, $p \in C \cap H_{t_k} = D_k$ and $P_{D_k}(p) = p$. We have

$$\|v^k - p\| = \|P_{D_k}(u^k) - P_{D_k}(p)\| \le \|u^k - p\|. \tag{5}$$

Hence, $p \in C_k,\ \forall k$.

Secondly, we show that $(\cap_{j=0}^{\infty} H_j) \cap C \subset Q_k, \forall k$ by induction method. If $k = 0$, we have $(\cap_{j=0}^{\infty} H_j) \cap C \subset Q_0 = \mathbb{H}$. Suppose that $(\cap_{j=0}^{\infty} H_j) \cap C \subset Q_k$. Take $p \in (\cap_{j=0}^{\infty} H_j) \cap C$, we have to show that $p \in Q_{k+1}$. Indeed, since $x^{k+1} = P_{A_k}(x^g)$, we obtain

$$\langle x - x^{k+1}, x^g - x^{k+1} \rangle \le 0,\ \forall x \in A_k = C_k \cap Q_k \cap C. \tag{6}$$

We have $(\cap_{j=0}^{\infty} H_j) \cap C \subset Q_k$ (induction assumption) and $p \in (\cap_{j=0}^{\infty} H_j) \cap C$, so $p \in Q_k$. On the other hand, $(\cap_{j=0}^{\infty} H_j) \cap C \subset C_k$. Therefore, $p \in C_k \cap Q_k \cap C = A_k$. Replacing $x = p$ into (6), we have $\langle p - x^{k+1}, x^g - x^{k+1} \rangle \le 0$. Therefore, $p \in Q_{k+1}$. In conclusion, we have $S_M \subset (\cap_{j=0}^{\infty} H_j) \cap C \subset \cap_{j=0}^{\infty} A_j$.

Since $Q_k = \{x \in \mathbb{H} : \langle x - x^k, x^g - x^k \rangle \le 0\}$, we have $x^k = P_{Q_k}(x^g)$. Furthermore, $p \in S_M \subset \cap_{j=0}^{\infty} A_j = \cap_{j=0}^{\infty}(C_j \cap Q_j \cap C)$, we have $p \in Q_k$. Therefore, we obtain

$$\|x^k - x^g\| \le \|p - x^g\|. \tag{7}$$

Hence, there exists $\lim_{k\to\infty} \|x^k - x^g\|$ and $\{x^k\}$ is bounded.

From the definition of $\hat{\lambda}_k$ and $\lambda_k \in [0, \hat{\lambda}_k]$, we have

$$\lambda_k \|x^k - x^{k-1}\| \le \hat{\lambda}_k \|x^k - x^{k-1}\| \le \epsilon_k. \tag{8}$$

From the definition of u^k and the nonexpansive property of projection, we get

$$0 \le \|u^k - x^k\| \le \|x^k + \lambda_k(x^k - x^{k-1}) - x^k\| = \|\lambda_k(x^k - x^{k-1})\| \le \epsilon_k,$$

in which the last inequality is derived from (8). Because $\Sigma_{k=0}^{+\infty} \epsilon_k < +\infty$, so we have $\epsilon_k \to 0$ as $k \to \infty$. Hence, we get $\lim_{k\to\infty} \|u^k - x^k\| = 0$. Combining this fact, the boundedness of the sequence $\{x^k\}$ and $\|u^k\| \le \|u^k - x^k\| + \|x^k\|$, we deduce that $\{u^k\}$ is also bounded.

By (5), we also have the boundedness of $\{v^k\}$. By Lemma 2, $\{y^k\}$ and $\{z^k\}$ are bounded. By Lemma 1, $\{\xi^k\}$ is also bounded.

Statement 2. $\{x^k\}, \{u^k\}$ and $\{v^k\}$ converges strongly to $x^* = P_{\cap_{j=0}^{\infty} A_j}(x^g)$. Replacing $p = x^*$ into (7), we have

$$\|x^k - x^g\| \le \|P_{\cap_{j=0}^{\infty} A_j}(x^g) - x^g\|. \tag{9}$$

Now, we show that any weak accumulation of $\{x^k\}$ belongs to $\cap_{j=0}^{\infty} A_j$. Indeed, let $\bar{x}$ be a weak accumulation of $\{x^k\}$. Then, there exists $\{x^{k_l}\} \subset \{x^k\}$ and $x^{k_l} \rightharpoonup \bar{x}$ as $l \to \infty$.

On the other hand, from (2) and the fact that there exists $\lim_{k\to\infty} \|x^k - x^g\|$, we deduce $\lim\limits_{k\to\infty} \|x^{k+1} - x^k\| = 0$. Combining with the fact in Statement Statement 2 that $\lim_{k\to\infty} \|u^k - x^k\| = 0$, we get

$$\lim_{k\to+\infty} \|x^{k+1} - u^k\| = 0. \tag{10}$$

Because $x^{k_l} \rightharpoonup \bar{x}$ as $l \to \infty$, we also get $u^{k_l} \rightharpoonup \bar{x}$ as $l \to \infty$.

Now, we have to prove that $\bar{x} \in \cap_{j=0}^{\infty} A_j$. From definition of t_k, we have

$$0 \le d(u^k, H_i) \le d(u^k, H_{t_k}), \quad \text{for each } \; i = 0, 1, ..., k. \tag{11}$$

Because $x^{k+1} \in A_k \subset C_k$, we obtain $\|x^{k+1} - v^k\| \le \|x^{k+1} - u^k\|$. So,

$$\lim_{k\to\infty} \|x^{k+1} - v^k\| = 0. \tag{12}$$

On the other hand, we have $0 \le \|u^k - v^k\| \le \|u^k - x^{k+1}\| + \|x^{k+1} - v^k\|$. Combining with (10) and (12), we obtain $\lim\limits_{k\to\infty} d(u^k, D_k) = \lim\limits_{k\to\infty} \|u^k - v^k\| = 0$. By definition of t_k, $D_k = C \cap H_{t_k}$ and $\|\xi^k\| \le M$ we have $0 \le \frac{|f(z^k,u^k)|}{M} \le \frac{|f(z^k,u^k)|}{\|\xi^k\|} = d(u^k, H_k) \le d(u^k, H_{t_k}) \le d(u^k, D_k)$. Therefore, $\lim\limits_{k\to\infty} d(u^k, H_k) = \lim\limits_{k\to\infty} d(u^k, H_{t_k}) = \lim\limits_{k\to\infty} f(z^k, u^k) = 0$.

Combining with (11), we have $\lim\limits_{k\to\infty} d(u^k, H_i) = 0$ for any fixed $i \ge 0$. So, we obtain $\lim\limits_{l\to\infty} d(u^{k_l}, H_i) = 0$. Set $t^{k_l} := P_{H_i}(u^{k_l})$. It is clear that $\lim\limits_{l\to\infty} \|u^{k_l} - t^{k_l}\| = 0$. We have for all $y \in \mathbb{H}$

$$\langle t^{k_l} - \bar{x}, y\rangle = \langle t^{k_l} - u^{k_l}, y\rangle + \langle u^{k_l} - \bar{x}, y\rangle \le \|t^{k_l} - u^{k_l}\|.\|y\| + \langle u^{k_l} - \bar{x}, y\rangle.$$

Combining with $\lim\limits_{l\to\infty} \|u^{k_l} - t^{k_l}\| = 0$ and $u^{k_l} \rightharpoonup \bar{x}$ as $l \to \infty$, we also have $t^{k_l} \rightharpoonup \bar{x}$ as $l \to \infty$. Furthermore, $\{t^{k_l}\} \subset H_i$ and H_i is weakly closed (because H_i is closed convex in $\mathbb{H}$). Hence, $\bar{x} \in H_i, \forall i$. Similiary, $\bar{x} \in C$. Therefore, $\bar{x} \in (\cap_{j=0}^{\infty} H_j) \cap C \subset \cap_{j=0}^{\infty} A_j$. Combining this fact and (9), by Lemma 3, we obtain $x^k \to P_{\cap_{j=0}^{\infty} A_j}(x^g)$. Since $\lim\limits_{k\to\infty} \|u^k - x^k\| = 0$, we get that $u^k \to P_{\cap_{j=0}^{\infty} A_j}(x^g)$. Because $\lim\limits_{k\to\infty} \|u^k - v^k\| = 0$, we also get that $v^k \to P_{\cap_{j=0}^{\infty} A_j}(x^g)$.

Statement 3. x^* is a solution of EP(C, f).
By definition of y^k and applying the optimality condition, we obtain $0 \in \rho_k \partial_2 f(u^k, y^k) + (y^k - u^k) + N_C(y^k)$. Therefore, there exists $w \in \partial_2 f(u^k, y^k)$ such that $-\rho_k w - (y^k - u^k) \in N_C(y^k)$. This means that $\langle -\rho_k w - (y^k - u^k), y - y^k \rangle \leq 0$. By definition of subdifferentiable $w \in \partial_2 f(u^k, y^k)$, we get $f(u^k, y) - f(u^k, y^k) \geq \langle w, y - y^k \rangle$, $\forall y \in C$.

From two above inequalities, we obtain

$$\rho_k [f(u^k, y) - f(u^k, y^k)] \geq \langle y^k - u^k, y^k - y \rangle, \ \forall y \in C. \tag{13}$$

We have

$$f(z^k, z^k) = f(z^k, (1 - \theta_k) u^k + \theta_k y^k) \leq (1 - \theta_k) f(z^k, u^k) + \theta_k f(z^k, y^k), \tag{14}$$

From (1) and (14) we get $f(z^k, u^k) \geq \theta_k [f(z^k, u^k) - f(z^k, y^k)] \geq \frac{\theta_k \mu}{\rho_k} \|y^k - u^k\|^2 \geq \frac{\theta_k \mu}{\beta} \|y^k - u^k\|^2$. Combining with $\lim_{k \to \infty} f(z^k, u^k) = 0$, we have

$$\lim_{k \to \infty} \theta_k \|y^k - u^k\|^2 = 0. \tag{15}$$

Now we consider two distinct cases.

Case 1. $\limsup_{k \to \infty} \theta_k > 0$. Then there exists $\bar{\theta} > 0$ and a subsequence $\{\theta_{k_l}\} \subset \{\theta_k\}$ such that $\theta_{k_l} > \bar{\theta}$ for all l. From (15), we get $\lim_{l \to \infty} \|u^{k_l} - y^{k_l}\| = 0$. Remember that $u^{k_l} \to x^*$, it follows that $y^{k_l} \to x^*$. From (13), replacing k by k_l, we get $\rho_{k_l}[f(u^{k_l}, y) - f(u^{k_l}, y^{k_l})] \geq -\|y^{k_l} - u^{k_l}\|.\|y^{k_l} - y\|$. Letting $l \to \infty$, using (A1), $0 < \alpha \leq \rho_{k_l} \leq \beta$, $u^{k_l} \to x^*$, $y^{k_l} \to x^*$, we get $f(x^*, y) \geq 0$, $\forall y \in C$. Therefore, x^* is a solution of EP(C, f).

Case 2. $\lim_{k \to \infty} \theta_k = 0$. From the boundedness of $\{y^k\}$, it deduces that there exists $\{y^{k_l}\} \subset \{y^k\}$ such that $y^{k_l} \rightharpoonup \bar{y}$ as $l \to \infty$. Replacing y by u^{k_l} in (13), we get $f(u^{k_l}, y^{k_l}) \leq -\frac{1}{\rho_{k_l}} \|y^{k_l} - u^{k_l}\|^2$. Suppose that $\theta_{k_l} = \eta_{i_{k_l}}$ satisfies the linesearch (1). Then, for $i_{k_l} - 1$, we have $f((1 - \eta_{i_{k_l}-1}) u^{k_l} + \eta_{i_{k_l}-1} y^{k_l}, u^{k_l}) - f((1 - \eta_{i_{k_l}-1}) u^{k_l} + \eta_{i_{k_l}-1} y^{k_l}, y^{k_l}) < \frac{\mu}{\rho_{k_l}} \|y^{k_l} - u^{k_l}\|^2$. Combining two above inequalities, we get

$$\begin{aligned} f(u^{k_l}, y^{k_l}) &\leq -\frac{1}{\rho_{k_l}} \|y^{k_l} - u^{k_l}\|^2 \\ &< \frac{1}{\mu} [f((1 - \eta_{i_{k_l}-1}) u^{k_l} + \eta_{i_{k_l}-1} y^{k_l}, y^{k_l}) - f((1 - \eta_{i_{k_l}-1}) u^{k_l} + \eta_{i_{k_l}-1} y^{k_l}, u^{k_l})]. \end{aligned} \tag{16}$$

According to the algorithm, we have $\eta_{i_{k_l}-1} \to 0$ (because $\eta_{i_{k_l}} = \theta_{k_l} \to 0$) and $u^{k_l} \rightharpoonup x^*$, $y^{k_l} \rightharpoonup \bar{y}$ as $l \to \infty$. It implies that $z^{k_l, m_{k_l}-1} \rightharpoonup x^*$ as $l \to \infty$. Besides that $\{\frac{1}{\rho_{k_l}} \|y^{k_l} - u^{k_l}\|^2\}$ is bounded, without loss of generality we may assume that $\lim_{l \to \infty} \frac{1}{\rho_{k_l}} \|y^{k_l} - u^{k_l}\|^2$ exists. Therefore, letting $l \to \infty$ in (16) and using (A1), we get $f(x^*, \bar{y}) \leq -\lim_{l \to \infty} \frac{1}{\rho_{k_l}} \|y^{k_l} - u^{k_l}\|^2 \leq \frac{1}{\mu} f(x^*, \bar{y})$.

Since $\mu \in (0,1)$, therefore, we have $f(x^*, \bar{y}) = 0$ and $\lim_{l\to\infty} \frac{1}{\rho_{k_l}} \|y^{k_l} - u^{k_l}\|^2 = 0$. Because $0 < \alpha \le \rho_{k_l} \le \beta$, so we obtain $\lim_{l\to\infty} \|y^{k_l} - u^{k_l}\|^2 = 0$.

By the Case 1, it is immediate that x^* is a solution of EP(C, f). □

In some cases, we want to design a weakly convergent algorithm to solve our problem. This often happens, especially when we work in a finite-dimensional Hilbert space. The following algorithm shows that if we remove the hybrid cutting technique in Algorithm 1, we can obtain a weakly convergent algorithm.

Algorithm 2.
Initialization. Pick $x^{-1}, x^0 = x^g \in C$, choose parameters $\{\rho_k\} \subset [\alpha, \beta]$ with $0 < \alpha \le \beta$, $\mu \in (0,1)$, $\lambda \in [0,1)$. Take sequences $\{\eta_i\} \subset (0,1)$ strictly decreasing to zero and $\{\epsilon_k\} \subset [0, +\infty)$ such that $\sum_{k=0}^{+\infty} \epsilon_k < +\infty$.
Iteration k $(k = 0, 1, 2, \ldots)$. Having x^{k-1}, x^k do the following steps:

Step 1. Compute $u^k = P_C(x^k + \lambda_k(x^k - x^{k-1}))$, where $\lambda_k \in [0, \hat{\lambda}_k]$ with

$$\hat{\lambda}_k = \begin{cases} \min\{\lambda, \frac{\epsilon_k}{\|x^k - x^{k-1}\|}\} & \text{if } x^k \neq x^{k-1}, \\ \lambda & \text{otherwise.} \end{cases}$$

Step 2. Compute $y^k = \arg\min \left\{ \rho_k f(u^k, y) + \frac{1}{2}\|y - u^k\|^2 : y \in C \right\}$.
If $y^k = u^k$, stop and u^k is a solution of EP(C, f). Otherwise, do Step 3.

Step 3. Find θ_k be the largest number in the sequence $\{\eta_i\}$ such that

$$\begin{cases} z^k = (1 - \theta_k)u^k + \theta_k y^k, \\ f(z^k, u^k) - f(z^k, y^k) \ge \frac{\mu}{\rho_k}\|y^k - u^k\|^2. \end{cases}$$

Step 4. Compute $x^{k+1} = P_{D_k}(u^k)$, where

$$D_k = C \cap H_{t_k},\ t_k \in \text{argmax}\{d(u^k, H_j) : 0 \le j \le k\}.$$
$$H_k = \{x \in \mathbb{H} : \langle \xi^k, x - u^k \rangle + f(z^k, u^k) \le 0\},\ \xi^k \in \partial_2 f(z^k, u^k).$$

Go to Step 1 with k is replaced by $k + 1$.

The following theorem give us the convergence of Algorithm 2.

Theorem 2. *Suppose that f satisfies two assumptions (A1) - (A2), and $S_M \neq \emptyset$. Then $\{x^k\}$, $\{u^k\}$ generated by Algorithm 2 converges weakly to a solution of EP(C, f).*

Proof. We divide this proof into four statements as the following.

Statement 1. Five sequences $\{x^k\}, \{u^k\}, \{y^k\}, \{z^k\}$ and $\{\xi_k\}$ are bounded. Indeed, by proving similarly to Statement 1 of Theorem 1, we also have $S_M \subset (\cap_{j=0}^{\infty} H_j) \cap C$. It implies that $(\cap_{j=0}^{\infty} H_j) \cap C$ is nonempty. Take $p \in (\cap_{j=0}^{\infty} H_j) \cap C$. By the definition of $\{u^k\}$ and the nonexpansive property of projection, we have

$$\|u^k - p\|^2 \le \|x^k - p\|^2 + \lambda_k^2 \|x^k - x^{k-1}\|^2 + 2\lambda_k \langle x^k - p, x^k - x^{k-1} \rangle. \quad (17)$$

On the other hand, we have

$$2\langle x^k - p, x^k - x^{k-1}\rangle \quad = \|x^k - p\|^2 + \|x^k - x^{k-1}\|^2 - \|x^{k-1} - p\|^2. \quad (18)$$

Replacing (18) into (17), we obtain

$$\|u^k - p\|^2 \quad \leq \|x^k - p\|^2 + \lambda_k(\|x^k - p\|^2 - \|x^{k-1} - p\|^2) + 2\lambda_k\|x^k - x^{k-1}\|^2, \quad (19)$$

in which the second inequality is derived from $\lambda_k^2 \leq \lambda_k$.

Since $x^{k+1} = P_{D_k}(u^k)$ and $p \in (\cap_{j=0}^{\infty} H_j) \cap C \subset C \cap H_{t_k} = D_k$, so we have

$$\|x^{k+1} - p\|^2 \leq \|u^k - p\|^2. \quad (20)$$

From (19) and (20), we get

$$\|x^{k+1} - p\|^2 - \|x^k - p\|^2 \leq \lambda_k(\|x^k - p\|^2 - \|x^{k-1} - p\|^2) + 2\lambda_k\|x^k - x^{k-1}\|^2.$$

Set $\phi_k = \|x^k - p\|^2$ and $\delta_k = 2\lambda_k\|x^k - x^{k-1}\|^2$. Noting the fact that $\Sigma_{k=0}^{\infty}\lambda_k\|x^k - x^{k-1}\|^2 \leq \Sigma_{k=0}^{\infty}\hat{\lambda}_k\|x^k - x^{k-1}\|^2 \leq \Sigma_{k=0}^{\infty}\epsilon_k < +\infty$ and $0 \leq \lambda_k \leq \hat{\lambda}_k \leq \lambda < 1$, by Lemma 4, we deduce that $\lim_{k\to\infty}\|x^k - p\|$ exists for every $p \in (\cap_{j=0}^{\infty} H_j) \cap C$. Hence, the sequence $\{x^k\}$ is bounded. By proving similarly to Statement 1 of Theorem 1, we also have the boundedness of the sequence $\{u^k\}$ and $\lim_{k\to\infty}\|u^k - x^k\| = 0$.

By Lemma 2, $\{y^k\}$ and $\{z^k\}$ are bounded. By Lemma 1, $\{\xi^k\}$ is also bounded.

Statement 2. We have to prove that $\lim_{k\to\infty} d(u^k, H_k) = \lim_{k\to\infty} d(u^k, H_{t_k}) = \lim_{k\to\infty} f(z^k, u^k) = \lim_{k\to\infty}\|x^{k+1} - u^k\| = 0$. Indeed, we have for all $p \in (\cap_{j=0}^{\infty} H_j) \cap C \subset D_k$

$$\|x^{k+1} - p\|^2 = \|u^k - p\|^2 - \|x^{k+1} - u^k\|^2 + 2\langle x^{k+1} - u^k, x^{k+1} - p\rangle.$$

It follows that

$$\|x^{k+1} - u^k\|^2 = \|u^k - p\|^2 - \|x^{k+1} - p\|^2 + 2\langle x^{k+1} - u^k, x^{k+1} - p\rangle \quad (21)$$

From (19), (21), and the fact that $x^{k+1} = P_{D_k}(u^k)$ and $p \in D_k$. we have

$$\|x^{k+1} - u^k\|^2 \leq \|x^k - p\|^2 - \|x^{k+1} - p\|^2 + \lambda_k(\|x^k - p\|^2 - \|x^{k-1} - p\|^2) \quad (22)$$
$$+ 2\lambda_k\|x^k - x^{k-1}\|^2.$$

Because there exists $\lim_{k\to\infty}\|x^k - p\|$ and $0 \leq \lambda_k \leq \lambda < 1$, we have $\lim_{k\to\infty}\left(\|x^k - p\|^2 - \|x^{k+1} - p\|^2 + \lambda_k(\|x^k - p\|^2 - \|x^{k-1} - p\|^2)\right) = 0$. From the definition of $\hat{\lambda}_k$ and $\lambda_k \in [0, \hat{\lambda}_k]$, we have $\lambda_k\|x^k - x^{k-1}\|^2 \leq \hat{\lambda}_k\|x^k - x^{k-1}\|^2 \leq \epsilon_k\|x^k - x^{k-1}\| \leq \epsilon_k(\|x^k\| + \|x^{k-1}\|)$. Because $\lim\limits_{k\to\infty}\epsilon_k = 0$ and $\{x^k\}$ is bounded, we have $\lim\limits_{k\to\infty}\lambda_k\|x^k - x^{k-1}\|^2 = 0$. Thus, from (22), we conclude that $\lim_{k\to\infty} d(u^k, D_k) = \lim_{k\to\infty}\|x^{k+1} - u^k\| = 0$.

By definition of t_k, $D_k = C \cap H_{t_k}$ and $\|\xi^k\| \le M$ we have $0 \le \frac{|f(z^k,u^k)|}{M} \le \frac{|f(z^k,u^k)|}{\|\xi^k\|} = d(u^k, H_k) \le d(u^k, H_{t_k}) \le d(u^k, D_k)$. Therefore, $\lim_{k\to\infty} d(u^k, H_k) = \lim_{k\to\infty} d(u^k, H_{t_k}) = \lim_{k\to\infty} f(z^k, u^k) = 0$.

Statement 3. $\{x^k\}$ and $\{u^k\}$ converges weakly to $x^* \in (\cap_{j=0}^{\infty} H_j) \cap C$.
We will show that any weak accumulation of $\{x^k\}$ belongs to $(\cap_{j=0}^{\infty} H_j) \cap C$. This is proven in exactly the same way as in Statement 2 of Theorem 1. By Statement 1, there exists $\lim_{k\to\infty} \|x^k - p\|$ for each $p \in (\cap_{j=0}^{\infty} H_j) \cap C$. By Lemma 5, it holds that $\{x^k\}$ converges weakly to $x^* \in (\cap_{j=0}^{\infty} H_j) \cap C$. Because $\lim_{k\to\infty} \|x^k - u^k\| = 0$, we also have $u^k \rightharpoonup x^*$.

Statement 4. x^* is a solution of EP(C, f). The proof is completely similar to Statement 3 in Theorem 1, we have $x^* \in S$. □

4 Numerical Examples

In this section, we present some numerical examples to illustrate the convergence of the proposed algorithms and compare them with the Algorithm of Deng et al. [5]. The numerical experiments are implemented in MATLAB R2015 and performed on a Desktop DELL Intel (R), Core (TM) i7-9700 CPU, 3.00 Ghz, Ram 16.0 GB. To solve nonlinear optimization problems we use the fmincon function, and for quadratic programming problems, we use the quadprog function.

Example 1. In this example, we consider the nonmonotone equilibrium probblem arising in Nash-Cournot oligopolistic electricity market equilibrium model which was investigated in some research papers, for instance, [5–7]. In this model, the bifunction $f : \mathbb{R}^6 \times \mathbb{R}^6 \to \mathbb{R}^6$ is given by $f(x, y) = [(A + B)x + By + a]^T (y - x) + c(y) - c(x),\ \forall x, y \in \mathbb{R}^6$, where the feasible set $C = \{x \in \mathbb{R}^6 | a \le x \le b\}$, A is a nonpositive semidefinite matrix, B is a symmetric positive semidefinite matrix, $a = (0, 0, 0, 0, 0, 0)^T$, $b = (80, 80, 50, 55, 30, 40)^T$, and $c(x)$ is a nonsmooth convex function. See [6] for more details about the values of A, B and $c(x)$. In this case, the bifunction is nonmonotone and nonsmooth. We use Algorithm 2 and Algorithm WIEA in [5] to compute, the parameters is given as follows. We choose $\epsilon_k = \dfrac{30}{k^{1.1}}$ in both algorithms. In Algorithm WIEA, we take $\alpha = 0.01, \gamma = 1.5, \rho = 0.5$. We choose $\rho_k = 2, \mu = 0.4, \lambda = 0.01$ in Algorithm 2. The initial points are $x^0 = x^g = (60, 20, 40, 15, 10, 10), x^1 = 0.9x^0$ in both algorithms. To terminate the algorithms we use the stopping criteria $Err = \dfrac{\|x^{k+1} - x^k\|}{\max\{1, \|x^k\|\}} < 10^{-4}$. The computation results are reported in Table 1 with some parameters θ^k in Algorithm WIEA and η_k in Algorithm 2.

In the next example, we consider the convergence behavior of Algorithm 1 with Algorithm SIEA given in [5].

Table 1. Computed results for Example 1.

θ^k	Alg. WIEA		η_k	Alg. 2	
	CPU(s)	Iter.		CPU(s)	Iter.
0.99^k	6.27	85	0.99^k	4.83	87
0.9^k	5.52	86	0.9^k	4.52	100
0.5^k	6.06	95	0.5^k	4.52	98
0.3^k	6.59	105	$\frac{1}{k}$	4.13	91
0.1^k	19.31	265	$\frac{1}{\sqrt{k}}$	3.39	78

Example 2. Let $C = [-1,1] \times [-1,1]$. The bifunction $f : C \times C \to \mathbb{R}$ is given by $f(x,y) = |x_1 + x_2| \left[(y_1^2 - x_1^2) + (y_2^2 - x_2^2)\right], \forall x = (x_1,x_2)^T, y = (y_1,y_2)^T \in C$. It can be seen that f is not pseudomonotone on C and $S_E = \left\{(t,-t)^T | t \in [-1,1]\right\}$ and $S_M = \left\{(0,0)^T\right\}$. We use Algorithm 1 and SIEA in [5] to compute. The parameters and starting point are given as follows. $\epsilon_k = \frac{1}{k^{1.2}}$ in both algorithms. In Algorithm WIEA, $\alpha = 0.1, \gamma = 0.1, \rho = 0.5$ and $\rho_k = 2, \mu = 0.4, \lambda = 0.4$ in Algorithm 1. $x^0 = x^g = (0,0.5)^T, x^1 = (0,0.3)^T$ in Algorithms SIEA and $x^0 = x^g = (0,0.3)^T, x^{-1} = (0,0.5)^T$ in Algorithm 2. The stopping criteria is $Err = \frac{\|x^{k+1} - x^k\|}{\max\{1, \|x^k\|\}} < 10^{-4}$. The results are shown in Table 2 with some parameters θ^k in Algorithm SIEA and η_k in Algorithm 1.

Table 2. Experiment for Example 2.

θ^k	Alg. SIEA		η_k	Alg. 2	
	CPU(s)	Iter.		CPU(s)	Iter.
0.9^k	3.09	69	$\frac{1}{k^2}$	1.13	16
0.7^k	8.09	177	$\frac{1}{k}$	1.93	32
0.5^k	9.41	213	$\frac{1}{\sqrt{k}}$	0.98	13

5 Conclusion

We have introduced two new inertial algorithms to find a solution of the non-monotone and non-Lipschitz equilibrium problems in Hilbert spaces. Especially, we haven't used the shrinking projections found in methods used for solving such problems in the literature. The strong and weak convergence of the proposed algorithms has been obtained, and some numerical examples are provided to illustrate the efficiency of the algorithms.

References

1. Alvarez, F., Attouch, H.: An inertial proximal method for maximal monotone operators via discretization of a nonlinear oscillator with damping. Set-Valued Anal. **9**, 3–11 (2001)
2. Bigi, G., Castellani, M., Pappalardo. M., Passacantando, M.: Nonlinear programming techniques for equilibria. Springer (2019)
3. Blum, E., Oettli, W.: From optimization and variational inequalities to equilibrium problems. Math Student. **63**, 127–149 (1994)
4. Castellani, M., Giuli, M.: Refinements of existence results for relaxed quasimonotone equilibrium problems. J. Glob. Optim. **57**, 1213–1227 (2013)
5. Deng, L., Hu, R., Fang, Y.: Fast inertial extragradient algorithms for solving non-Lipschitzian equilibrium problems without monotonicity condition in real Hilbert spaces, J. Comput. Appl. Math. **423** (2023)
6. Dinh, B.V., Kim, D.S.: Projection algorithms for solving nonmonotone equilibrium problems in Hilbert space. J. Comput. Appl. Math. **302**, 106–117 (2016)
7. Dinh, B.V., Muu, L.D.: A projection algorithm for solving pseudo-monotone equilibrium problems and its application to a class of bilevel equilibria. Optim. **54**, 559–575 (2015)
8. Fan, K.: A minimax inequality and applications. In: Shisha, O. (ed.) Inequality III, pp. 103–113. Academic Press, New York (NY) (1972)
9. Mastroeni, G.: Gap function for equilibrium problems. J. Glob. Optim. **27**(4), 411–426 (2003)
10. Moudafi, A.: Proximal point algorithm extended to equilibrium problem. J. Nat. Geometry **15**, 91–100 (1999)
11. Muu, L.D.: Stability properties of a class of variational Inequalities. Math. Operationsforsch. Stat. Ser. Optim. **15**, 347–351 (1984)
12. Muu, L.D., Oettli, W.: Convergence of an adaptive penalty scheme for finding constrained equilibria. Nonlinear Anal. TMA **12**, 1159–1166 (1992)
13. Muu, L.D., Quoc, T.D.: Regularization algorithms for solving monotone Ky Fan inequalities with application to a Nash-Cournot equilibrium model. J. Optim. Theory Appl. **142**, 185–204 (2009)
14. Opial, Z.: Weak convergence of the sequence of successive approximations for nonexpansive mappings. Bull. Amer. Math. Soc. **73**, 591–597 (1967)
15. Poliak, P.T.: Some methods of speeding up the convergence of iterative methods. Zh. Vychisl. Mat. Mat. Fiz. **4**(5), 791–803 (1964)
16. Quoc, T.D., Anh, P.N., Muu, L.D.: Dual extragradient algorithms extended to equilibrium problems. J. Glob. Optim. **52**, 139–159 (2012)
17. Quoc, T.D., Muu, L.D., Nguyen, V.H.: Extragradient algorithms extended to equilibrium problems. Optim. **57**, 749–776 (2008)
18. Tuyen, B.T., Manh, H.D., Dinh, B.V.: Inertial algorithms for solving nonmonotone variational inequality problems. Taiwanese J. Math. **28**(2), 397–421 (2024)
19. Vinh, N.T., Muu, L.D.: Inertial extragradient algorithms for solving equilibrium problems. Acta Math. Vietnam **44**, 639–663 (2019)
20. Vuong, P.T., Strodiot, J.J., Nguyen, V.H.: Extragradient methods and linesearch algorithms for solving Ky Fan inequalities and fixed point problems. J. Optim. Theory Appl. **155**, 605–627 (2013)
21. Yanes, C.M., Xu, H.K.: Strong convergence of the CQ method for fixed point iteration processes. Nonlinear Anal. TMA. **64**, 2400–2411 (2006)

Recovering Linear Causal Models with Latent Variables via Cholesky Factorization of Covariance Matrix

Yunfeng Cai[1(✉)], Xu Li[2], Mingming Sun[1], and Ping Li[3]

[1] Beijing Institute of Mathematical Sciences and Applications, Beijing 101408, China
yunfengcai09@gmail.com
[2] Cognitive Computing Lab, Baidu Research, Beijing 100193, China
[3] VecML Inc., Bellevue, WA 98004, USA

Abstract. Discovering the causal relationship via recovering the directed acyclic graph (DAG) structure from the observed data is a well-known challenging combinatorial problem. When there are latent variables, the problem becomes even more difficult. In this paper, we first propose a DAG structure recovering algorithm, which is based on the Cholesky factorization of the covariance matrix of the observed data. The algorithm is fast and easy to implement and has theoretical grantees for exact recovery. On synthetic and real-world datasets, the algorithm is significantly faster than previous methods and achieves the state-of-the-art performance. Furthermore, under the equal error variances assumption, we incorporate an optimization procedure into the Cholesky factorization based algorithm to handle the DAG recovering problem with latent variables. Numerical simulations show that the modified "Cholesky + optimization" algorithm is able to recover the ground truth graph in most cases and outperforms existing algorithms.

Keywords: Causal discovery · Linear model · Cholesky factorization

1 Introduction

Learning the causal relationships is a fundamental problem and has many applications in biology, machine learning, medicine, and economics. However, in many cases, not all causal variables relevant to the observed features have been measured. For example, in healthcare domains, there may be numerous unobserved factors such as gene expression; In financial markets, stock returns may be causally related but may also be confounded or mediated by a complicated network of unmeasured economic and political factors; Self-reported family history and diets may also leave out some important information.

Most causal discovery approaches focus on the situation without latent variables. For example, search-based algorithms [2,6,8,15] generally adopt a score (e.g., BIC score [22], Cholesky score [31], remove-fill score [28], Clustering Information Criterion [19]) to measure the fitness of graphs over data and then search

© The Author(s), under exclusive license to Springer Nature Switzerland AG 2025
H. A. Le Thi et al. (Eds.): ICAMCS 2024, LNNS 1311, pp. 27–39, 2025.
https://doi.org/10.1007/978-3-031-90606-0_3

over the legal DAG (directed acyclic graph) space to find the structure that achieves the highest score. However, exhaustive search over the legal DAG space is infeasible when p is large. Those algorithms go in quest of a trade-off between performance and time complexity.

Topology order search algorithms [5,9–11,20] decompose the DAG learning problem into two phases: (i) Topology order learning via conditional variance of the observed data; (ii) Graph estimation depends on the learned topology order. Those algorithms reduce the computation complexity into polynomial time and are guaranteed to recover the DAG structure under proper assumptions. Since Zheng *et al.* [33] proposed an approach that converts the traditional combinatorial optimization problem into a continuous program, many methods have been proposed, e.g., [17,28,32]. To overcome the constraints of conventional methodologies, researchers have introduced algorithms that leverage reinforcement learning [29] or flow-based techniques [23], aiming to mitigate the stringent assumptions inherent in traditional approaches.

Causal discovery with latent variables is less studied. The FCI [27], RFCI [7], and ICD [24] methods can distinguish the observed variables and the latent variables that are confounders of the two observed ones and can recover up to a partial ancestral graph (PAG), which is an equivalence class of the true causal graph, even in the situation that the causal graph is able to exactly recovered. Some methods focus on special topology cases for non-Gaussian models, e.g., [4, 14,26]. Other works [25,30] relax the special cases to more general non-Gaussian partially observable DAGs.

Contributions. This paper studies the linear causal models via Cholesky factorization. For the case when all variables are observed, we propose a Cholesky factorization based method (namely, CDCF) to discover the causal structure. The time complexity of CDCF is $O(p^3)$ (p is the number of nodes), which is the fastest method so far. Theoretically, we show that CDCF is able to recover the DAG exactly, under standard assumptions. Sample complexity can be also obtained. In addition, we propose a novel algorithm (namely, CDCF$^+$) to deal with the linear causal models with latent variables. Numerical simulations show that CDCF and CDCF$^+$ achieve the state-of-the-art performance.

2 Recover Linear Causal Models via Cholesky

2.1 Preliminaries and Assumptions

Assume that the observed data is entailed by a DAG $\mathcal{G} = (V, E)$, where V, E are the sets of nodes and edges, respectively. Each node v_i corresponds to a random variable X_i. The data matrix is given by $\boldsymbol{X} = [\mathbf{x}_1, ..., \mathbf{x}_p] \in \mathbb{R}^{n \times p}$, where $p = |V|$, $\mathbf{x}_i$ is consisted of n i.i.d observations of the random variable X_i, for $i = 1, \ldots, p$. The joint distribution of X is $P(X) = \prod_{i=1}^{p} P(X_i|Pa_{\mathcal{G}}(X_i))$, where $Pa_{\mathcal{G}}(X_i) \equiv \{X_j|(v_i, v_j) \in E\}$ is the set of parents of node X_i.

Given $\boldsymbol{X}$, we seek to recover the latent DAG topology structure for the joint probability distribution [13,21]. Generally, X is modeled via a structural equation model (SEM) with the form

$$X_i = f_i(Pa_{\mathcal{G}}(X_i)) + N_i, \quad i = 1, ..., p,$$

where f_i is an arbitrary function representing the relation between X_i and its parents, N_i is the jointly independent noise variable.

In this paper, we focus on the linear SEM defined by

$$X_i = [X_1, \ldots, X_p]\boldsymbol{w}_i + N_i, \quad i = 1, ..., p, \tag{1}$$

where $\mathbf{w}_i \in \mathbb{R}^p$ is a weighted column vector. Let $\boldsymbol{W} = [\boldsymbol{w}_1, \ldots, \boldsymbol{w}_p] \in \mathbb{R}^{p \times p}$ be the weighted adjacency matrix, $\boldsymbol{N} = [\boldsymbol{n}_1, \ldots, \boldsymbol{n}_p] \in \mathbb{R}^{n \times p}$ be the noise matrix, where $\boldsymbol{n}_i$ is n i.i.d observations of the noise variable N_i. The linear SEM model (1) can be rewritten as

$$\boldsymbol{X} = \boldsymbol{X}\boldsymbol{W} + \boldsymbol{N}. \tag{2}$$

As proposed in [16,18], a graph is DAG if and only if the corresponding weighted adjacent matrix $\boldsymbol{W}$ can be decomposed into

$$\boldsymbol{W} = \boldsymbol{P}\boldsymbol{T}\boldsymbol{P}^{\mathrm{T}}, \tag{3}$$

where $\boldsymbol{P}$ is a permutation matrix, $\boldsymbol{T}$ is a strict upper triangular matrix, i.e., $\boldsymbol{T}_{ij} = 0$ for all $i \geq j$.

Layer Decomposition. The *layer decomposition* [9] of a DAG can be obtained as follows: let $\mathcal{L}_0 \equiv \emptyset$, $\mathcal{A}_j \equiv \cup_{m=0}^{j}\mathcal{L}_j$; for $j > 0$, $\mathcal{L}_j$ is the set of all source nodes in the subgraph formed by removing the nodes in $\mathcal{A}_{j-1}$. For example, the graph in Fig. 1 has three layers with $\mathcal{L}_1 = \{node1, node4\}$, $\mathcal{L}_2 = \{node2, node5\}$, $\mathcal{L}_3 = \{node3, node6, node7\}$. Furthermore, it holds (3) with

$$\boldsymbol{W} = \begin{bmatrix} 0&1&0&0&0&0&0\\ 0&0&1&0&0&1&0\\ 0&0&0&0&0&0&0\\ 0&1&0&0&1&0&0\\ 0&0&0&0&0&1&1\\ 0&0&0&0&0&0&0\\ 0&0&0&0&0&0&0 \end{bmatrix}, \quad \boldsymbol{P} = \begin{bmatrix} 1&0&0&0&0&0&0\\ 0&0&1&0&0&0&0\\ 0&0&0&0&1&0&0\\ 0&1&0&0&0&0&0\\ 0&0&0&1&0&0&0\\ 0&0&0&0&0&1&0\\ 0&0&0&0&0&0&1 \end{bmatrix}, \quad \boldsymbol{T} = \begin{bmatrix} 0&0&1&0&0&0&0\\ 0&0&1&1&0&0&0\\ 0&0&0&0&1&1&0\\ 0&0&0&0&0&1&1\\ 0&0&0&0&0&0&0\\ 0&0&0&0&0&0&0\\ 0&0&0&0&0&0&0 \end{bmatrix}.$$

Note that $\boldsymbol{T}$ is a strict block upper triangular matrix. For a general r-layer DAG, there exists a permutation matrix $\boldsymbol{P}$ such that (3) holds with $\boldsymbol{T}$ having a strict block upper triangular form:

$$\boldsymbol{T} = [\boldsymbol{T}_{ij}] = \begin{bmatrix} 0 & \boldsymbol{T}_{12} & \boldsymbol{T}_{13} & \ldots & \boldsymbol{T}_{1r} \\ 0 & 0 & \boldsymbol{T}_{23} & \ldots & \boldsymbol{T}_{2r} \\ 0 & 0 & & \ddots & \vdots \\ \vdots & & \ddots & & \boldsymbol{T}_{r-1,r} \\ 0 & \ldots & \ldots & 0 & 0 \end{bmatrix}, \tag{4}$$

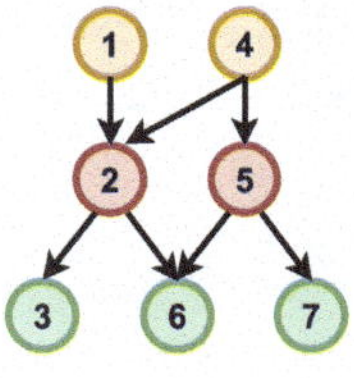

Fig. 1. A three-layer DAG

where $\boldsymbol{T}_{ij} \in \mathbb{R}^{p_i \times p_j} = 0$ for $1 \leq i \leq j \leq r$, p_i equals the number of nodes in $\mathcal{L}_i$ for $1 \leq i \leq r$, each column of $\boldsymbol{T}_{i,i+1}$ is nonzero (otherwise, the corresponding node belongs to the ith layer). The decomposition (3) with $\boldsymbol{T}$ having the strict block upper triangular form (4) is *unique* up to all within layer permutations (the Markov equivalence class).

Throughout the rest of this paper, we will make the following assumptions:

A1. The noise variables are centered, uncorrelated, and have finite variances, i.e.,

$$\mathbb{E}(N_i) = 0, \ \boldsymbol{\Sigma}_{nn} = [\mathbb{E}(N_i N_j)] = \mathrm{diag}(\sigma_1^2, \ldots, \sigma_p^2) < \infty.$$

A2. Let the adjacency matrix $\boldsymbol{W}$ be decomposed as in (3) with $\boldsymbol{T}$ of form (4). Assume $\Delta \equiv \min_{\substack{j<k \\ i_s \in \mathcal{I}_j, i_t \in \mathcal{I}_k}} \sigma_{i_t}^2 - \sigma_{i_s}^2 + \|[\boldsymbol{U}]_{i_s:i_t-1,i_t}\|^2 > 0$, where

$$\underline{\boldsymbol{i}} = [i_1, \ldots, i_p] = [1, \ldots, p]\boldsymbol{P}, \quad \mathcal{I}_j = \{\sum_{i=1}^{j-1} p_i + 1, \ldots, \sum_{i=1}^{j} p_i\}, \tag{5a}$$

$$\boldsymbol{U} = \mathrm{diag}(\sigma_{i_1}, \ldots, \sigma_{i_p})(\boldsymbol{I} - \boldsymbol{T})^{-1}. \tag{5b}$$

Assumption **A2** is essentially the same as the identifiability condition in [20, Thm. 2] and [9, Thm. B2]. In the rest of this section, we assume that there are no latent variables.

2.2 Algorithm

It follows from (2), (3) and (5a) that

$$[X_{i_1}, \ldots, X_{i_p}] = [N_{i_1}, \ldots, N_{i_p}](\boldsymbol{I} - \boldsymbol{T})^{-1}. \tag{6}$$

Using (6) and **A1**, we have

$$[\boldsymbol{\Sigma}_{xx}]_{\underline{\boldsymbol{i}},\underline{\boldsymbol{i}}} = (\boldsymbol{I} - \boldsymbol{T})^{-\mathrm{T}} [\boldsymbol{\Sigma}_{nn}]_{\underline{\boldsymbol{i}},\underline{\boldsymbol{i}}} (\boldsymbol{I} - \boldsymbol{T})^{-1} = \boldsymbol{U}^{\mathrm{T}} \boldsymbol{U}, \tag{7}$$

where $\boldsymbol{U}$ is defined in (5b). Note that $\boldsymbol{U}$ is upper triangular, hence (7) is simply the Cholesky factorization of the permuted covariance matrix $[\boldsymbol{\Sigma}_{xx}]_{\underline{\boldsymbol{i}},\underline{\boldsymbol{i}}}$. The task becomes to find the permutation $\underline{\boldsymbol{i}} = [i_1, \ldots, i_p]$ and an upper triangular matrix $\widetilde{\boldsymbol{U}}$ such that $\widetilde{\boldsymbol{U}}^{\mathrm{T}} \widetilde{\boldsymbol{U}}$ is a good approximation of $[\boldsymbol{\Sigma}_{xx}]_{\underline{\boldsymbol{i}},\underline{\boldsymbol{i}}}$.

Let $\widetilde{\boldsymbol{\Sigma}}_{xx}$ be an estimator for the population covariance matrix $\boldsymbol{\Sigma}_{xx}$. Assume that $\underline{\boldsymbol{i}}_{k-1} \equiv [i_1, \ldots, i_{k-1}]$ and $\widetilde{\boldsymbol{U}}_{k-1} \equiv \widetilde{\boldsymbol{U}}_{1:k-1,1:k-1}$ are settled, and we have

$$[\widetilde{\boldsymbol{\Sigma}}_{xx}]_{\underline{\boldsymbol{i}}_{k-1},\underline{\boldsymbol{i}}_{k-1}} = \widetilde{\boldsymbol{U}}_{k-1}^{\mathrm{T}} \widetilde{\boldsymbol{U}}_{k-1}. \tag{8}$$

Next, we show how to find i_k and $\widetilde{\boldsymbol{U}}_k$. For the time being, let us assume i_k is known, we show how to compute the last column of $\widetilde{\boldsymbol{U}}_k$. Let $\widetilde{\boldsymbol{U}}_k = \begin{bmatrix} \widetilde{\boldsymbol{U}}_{k-1} & \boldsymbol{\beta}_k \\ 0 & \alpha_k \end{bmatrix}$ with $\alpha_k \in \mathbb{R}$, $\boldsymbol{\beta}_k \in \mathbb{R}^{k-1}$. Then using (8), we get

$$[\widetilde{\boldsymbol{\Sigma}}_{xx}]_{\underline{\boldsymbol{i}}_k,\underline{\boldsymbol{i}}_k} = \widetilde{\boldsymbol{U}}_k^{\mathrm{T}} \widetilde{\boldsymbol{U}}_k = \begin{bmatrix} \widetilde{\boldsymbol{U}}_{k-1}^{\mathrm{T}} \widetilde{\boldsymbol{U}}_{k-1} & \widetilde{\boldsymbol{U}}_{k-1}^{\mathrm{T}} \boldsymbol{\beta}_k \\ \boldsymbol{\beta}_k^{\mathrm{T}} \widetilde{\boldsymbol{U}}_{k-1} & \alpha_k^2 + \|\boldsymbol{\beta}_k\|^2 \end{bmatrix}.$$

Algorithm 1. Causal Discovery via Cholesky Factorization (CDCF)

1: **input**: The covariance matrix $\widetilde{\boldsymbol{\Sigma}}_{xx} \in \mathbb{R}^{p \times p}$.
2: **output**: A permutation $\widetilde{\boldsymbol{P}}$ and an upper triangular matrix $\widetilde{\boldsymbol{U}}_p^{-1}$.
3: Set $\underline{\boldsymbol{i}} = [1, 2, \ldots, p]$;
4: $\ell = \operatorname{argmin}_{1 \le j \le p} [\widetilde{\boldsymbol{\Sigma}}_{xx}]_{i_j, i_j}$, exchange i_1 and i_ℓ in $\underline{\boldsymbol{i}}$;
5: Set $\alpha_1 = \sqrt{[\widetilde{\boldsymbol{\Sigma}}_{xx}]_{i_1, i_1}}$, $\widetilde{\boldsymbol{U}}_1 = \alpha_1$, $\widetilde{\boldsymbol{U}}_1^{-1} = \frac{1}{\alpha_1}$;
6: **for** $k = 2, 3, \ldots, p$ **do**
7:     **for** $j = k, k+1, \ldots, p$ **do**
8:         Compute α_j, $\boldsymbol{\beta}_j$ via (9);
9:     **end for**
10:     $\ell = \operatorname{argmin}_{k \le j \le p} \alpha_j^2$, exchange i_k and i_ℓ in $\underline{\boldsymbol{i}}$;
11:     $\widetilde{\boldsymbol{U}}_k = \begin{bmatrix} \widetilde{\boldsymbol{U}}_{k-1} & \boldsymbol{\beta}_\ell \\ 0 & \alpha_\ell \end{bmatrix}$, $\widetilde{\boldsymbol{U}}_k^{-1} = \begin{bmatrix} \widetilde{\boldsymbol{U}}_{k-1}^{-1} & -\frac{1}{\alpha_\ell} \widetilde{\boldsymbol{U}}_{k-1}^{-1} \boldsymbol{\beta}_\ell \\ 0 & \frac{1}{\alpha_\ell} \end{bmatrix}$;
12: **end for**
13: Set $\widetilde{\boldsymbol{P}} = \boldsymbol{I}(:, \underline{\boldsymbol{i}})$, $\boldsymbol{I}$ is the order p identity matrix.

It follows immediately that

$$\boldsymbol{\beta}_k = \widetilde{\boldsymbol{U}}_{k-1}^{-\mathrm{T}} [\widetilde{\boldsymbol{\Sigma}}_{xx}]_{\underline{\boldsymbol{i}}_{k-1}, i_k}, \quad \alpha_k = \sqrt{[\widetilde{\boldsymbol{\Sigma}}_{xx}]_{i_k, i_k} - \|\boldsymbol{\beta}_k\|^2}. \tag{9}$$

By (9), once i_k is settled, we can obtain the last column of $\widetilde{\boldsymbol{U}}_k$. The task remains to select i_k from $\{1, \ldots, p\} \setminus \{i_1, \ldots, i_{k-1}\}$. In this paper, we compute α_j and $\boldsymbol{\beta}_j$ by (9) for all possible j ($i_j \in \{1, \ldots, p\} \setminus \{i_1, \ldots, i_{k-1}\}$). Then determine i_k by

$$i_k = \operatorname*{argmin}_{k \le j \le p} \alpha_j^2. \tag{10}$$

The intuition behind (10) is that α_j^2 is the empirical conditional variance of X_j, thus, $\boldsymbol{X}_{i_k}$ is chosen as the one with the smallest conditional variance. One may also consider some alternatives, say choose i_k via the sparsity of $\widetilde{\boldsymbol{U}}_{k-1}^{-1} \boldsymbol{\beta}_k$, or take both the sparsity and the conditional variance into account.

The overall algorithm is summarized in Algorithm 1. Recall (5b), we can obtain the adjacent matrix $\widetilde{\boldsymbol{W}}$ by truncating and permuting $\widetilde{\boldsymbol{U}}_p^{-1}$ $\operatorname{diag}(\alpha_1, \ldots, \alpha_p)$.

Time Complexity. On line 8, we only need to compute the last entry of $\boldsymbol{\beta}_j$ at the cost of $\mathcal{O}(p)$ at worst since the other entries are available from the previous steps. Therefore, the overall time complexity of Algorithm 1 is $\mathcal{O}(p^3)$. Additionally, the inner loop (lines 7 to 9) can be made in parallel, which makes the algorithm friendly to run on GPU and suitable for large scale calculations.

Remark 1. Algorithm 1 differs from other methods (e.g., [5,9–11,20]) in that, it combines the topology order search stage and the graph estimation stage together, rather than two-stage: first topology order search then graph estimation, which leads to the reduction of the time complexity.

2.3 Exact Recovery

Theorem 1 (Exact recovery of the ordering). *For the linear SEM model* (1), *assume* (3), **A1** *and* **A2**. *Let* $\widetilde{\boldsymbol{\Sigma}}_{xx}$ *be an estimator for* $\boldsymbol{\Sigma}_{xx} = [\mathbb{E}(X_j X_k)]$ *and* $\max_i |[\boldsymbol{\Sigma}_{xx}]_{ii} - [\widetilde{\boldsymbol{\Sigma}}_{xx}]_{ii}| \leq \epsilon_d$ *for some* $\epsilon_d > 0$. *If* $\epsilon_d < \frac{\Delta}{4}$, *Algorithm 1 recovers the ordering exactly.*

Sample Complexity. According to Theorem 1, we can obtain the sample complexity for the exact recovery of the ordering by $\max_i \|[\boldsymbol{\Sigma}_{xx}]_{ii} - [\widetilde{\boldsymbol{\Sigma}}_{xx}]_{ii}\| \leq \Delta/4$. Specifically, X_i is a linear combination of N_i's and $\mathbb{E}(X_i) = 0$, $\mathrm{Var}(X_i) = [\boldsymbol{\Sigma}_{xx}]_{ii}$. Then we can obtain the sample complexity $\mathcal{O}(\log(p/\epsilon)\mathcal{M}^2)$, where $\mathcal{M} = \max_i [\boldsymbol{\Sigma}_{xx}]_{ii}$. In Table 1, we compare the sample complexity with others'.

Table 1. Sample complexity for the ordering. The last column represents the sample number that recovers the DAG w.p. $\geq 1-\epsilon$, p is the nodes number, r represents the level of the graph, q is the maximum in-degree, d is the maximum total degree, m represents the m'th moment bounded noise, $g(x) = x/\log x$.

Algorithm	Data	Function	Noise Type	$\boldsymbol{X}$	Sample Complexity
NPVAR [9]	–	(Non)-linear Lip-continuous	–	–	$\mathcal{O}((rp/\epsilon)^{1+p/2})$
EV [5]	$n > p$	Linear	Sub-Gaussian	$\lambda_{\min} > 0$	$\mathcal{O}(p^2 \log(p/\epsilon)\mathcal{M}^2)$
	$n < p$	Linear	Sub-Gaussian	$\lambda_{\min} > 0$	$\mathcal{O}(q^2 \log(p/\epsilon)\mathcal{M}^2)$
LISTEN [11]	–	Linear	Sub-Gaussian	–	$\mathcal{O}(d^4 \log(\frac{p}{\sqrt{\epsilon}})\mathcal{M}^2)$
	–	Linear	Bounded moment	–	$\mathcal{O}(d^4 (\frac{p^2}{\epsilon})^{1/m}\mathcal{M}^2)$
US [20]	$n > p$	Linear	Gaussian	$\lambda_{\min} > 0$ $\lambda_{\max} < \infty$	$g^{-1}(\mathcal{O}(\log(p/\epsilon)\mathcal{M}^2))$
CDCF (ours)	–	Linear	–	–	$\mathcal{O}(\log(p/\epsilon)\mathcal{M}^2)$

3 Learning Linear Causal Models with Latent Variables

3.1 The Effects of Latent Variables

When there are latent variables, the covariance matrix of the observed variables is essentially a principal submatrix of the overall covariance matrix. To get a clue on how the Cholesky factor changes, we consider the case when there is one latent variable.

Let $\boldsymbol{\Sigma}_+ = \begin{bmatrix} \boldsymbol{\Sigma}_{11} & \boldsymbol{x} & \boldsymbol{\Sigma}_{12} \\ \boldsymbol{x}^{\mathrm{T}} & d^2 & \boldsymbol{y}^{\mathrm{T}} \\ \boldsymbol{\Sigma}_{12}^{\mathrm{T}} & \boldsymbol{y} & \boldsymbol{\Sigma}_{22} \end{bmatrix}$ be a symmetric positive definite matrix, where $\boldsymbol{x} \in \mathbb{R}^{p_1}$, $\boldsymbol{y} \in \mathbb{R}^{p_2}$, $\boldsymbol{\Sigma}_{11} \in \mathbb{R}^{p_1 \times p_1}$, $\boldsymbol{\Sigma}_{22} \in \mathbb{R}^{p_2 \times p_2}$. Let the Cholesky factorization of $\boldsymbol{\Sigma}_+$ be $\boldsymbol{\Sigma}_+ = \boldsymbol{U}_+^{\mathrm{T}} \boldsymbol{U}_+$, where $\boldsymbol{U}_+ = \begin{bmatrix} \boldsymbol{U}_{11} & \widehat{\boldsymbol{x}} & \boldsymbol{U}_{12} \\ 0 & \widehat{d} & \widehat{\boldsymbol{y}}^{\mathrm{T}} \\ 0 & 0 & \widehat{\boldsymbol{U}}_{22} \end{bmatrix}$, $\boldsymbol{U}_{11} \in \mathbb{R}^{p_1 \times p_1}$ and $\widehat{\boldsymbol{U}}_{22} \in \mathbb{R}^{p_2 \times p_2}$ be both upper triangular. By simple calculations, we have

$$\widehat{\boldsymbol{x}} = \boldsymbol{U}_{11}^{-\mathrm{T}} \boldsymbol{x}, \quad \widehat{d} = \sqrt{d^2 - \|\widehat{\boldsymbol{x}}\|^2}, \quad \widehat{\boldsymbol{y}} = (\boldsymbol{y} - \boldsymbol{U}_{12}^{\mathrm{T}} \widehat{\boldsymbol{x}})/\widehat{d}. \tag{11}$$

Denote $\boldsymbol{v} = \boldsymbol{U}_{11}^{-1}\widehat{\boldsymbol{x}}$, $\boldsymbol{w} = \widehat{\boldsymbol{U}}_{22}^{-\mathrm{T}}\widehat{\boldsymbol{y}}$. It also holds that

$$\boldsymbol{U}_{+}^{-1} = \begin{bmatrix} \boldsymbol{U}_{11}^{-1} & -\frac{1}{d}\boldsymbol{v} & \frac{1}{d}\boldsymbol{v}\boldsymbol{w}^{\mathrm{T}} - \boldsymbol{U}_{11}^{-1}\boldsymbol{U}_{12}\widehat{\boldsymbol{U}}_{22}^{-1} \\ 0 & \frac{1}{d} & -\frac{1}{d}\boldsymbol{w}^{\mathrm{T}} \\ 0 & 0 & \widehat{\boldsymbol{U}}_{22}^{-1} \end{bmatrix}. \tag{12}$$

Now we remove the $p_1 + 1$st row and the corresponding column of $\boldsymbol{\Sigma}_{+}$, we get $\boldsymbol{\Sigma} = \begin{bmatrix} \boldsymbol{\Sigma}_{11} & \boldsymbol{\Sigma}_{12} \\ \boldsymbol{\Sigma}_{12}^{\mathrm{T}} & \boldsymbol{\Sigma}_{22} \end{bmatrix}$. For the Cholesky factor of $\boldsymbol{\Sigma}$, we have the following results.

Proposition 1. *Let the Cholesky factorization of $\boldsymbol{\Sigma}$ be $\boldsymbol{\Sigma} = \boldsymbol{U}^{\mathrm{T}}\boldsymbol{U}$. It holds that*

(a) $\boldsymbol{U} = \begin{bmatrix} \boldsymbol{U}_{11} & \boldsymbol{U}_{12} \\ 0 & \boldsymbol{U}_{22} \end{bmatrix}$, *where $\boldsymbol{U}_{22}$ satisfies $\boldsymbol{U}_{22}^{\mathrm{T}}\boldsymbol{U}_{22} = \widehat{\boldsymbol{U}}_{22}^{\mathrm{T}}\widehat{\boldsymbol{U}}_{22} + \widehat{\boldsymbol{y}}\widehat{\boldsymbol{y}}^{\mathrm{T}}$.*

(b) $\boldsymbol{U}^{-1} = \begin{bmatrix} \boldsymbol{U}_{11}^{-1} & -\boldsymbol{U}_{11}^{-1}\boldsymbol{U}_{12}\boldsymbol{U}_{22}^{-1} \\ 0 & \boldsymbol{U}_{22}^{-1} \end{bmatrix}$ *where $\boldsymbol{U}_{22}^{-1} = \widehat{\boldsymbol{U}}_{22}^{-1}\boldsymbol{L}^{-1}$, $\boldsymbol{L}$ is the Cholesky factor of $\boldsymbol{I} + \boldsymbol{w}\boldsymbol{w}^{\mathrm{T}}$, i.e., $\boldsymbol{L}^{\mathrm{T}}\boldsymbol{L} = \boldsymbol{I} + \boldsymbol{w}\boldsymbol{w}^{\mathrm{T}}$.*

(c) *The kth diagonal entry of $\boldsymbol{U}_{22}$ can be given by $[\boldsymbol{U}_{22}]_{kk} = \sqrt{\frac{1+\|\boldsymbol{w}_{1:k}\|^2}{1+\|\boldsymbol{w}_{1:k-1}\|^2}}[\widehat{\boldsymbol{U}}_{22}]_{kk}$.*

In Proposition 1, (a) tells that the $(1,1)$, $(1,2)$ blocks of $\boldsymbol{U}$ are the same as the $(1,1)$, $(1,3)$ blocks of $\boldsymbol{U}_{+}$, respectively, $\boldsymbol{U}_{22}$ is a rank-1 update of Cholesky factorization. (b) tells how the blocks of $\boldsymbol{U}^{-1}$ relate with the blocks of $\boldsymbol{U}_{+}^{-1}$. The $(1,1)$ block of $\boldsymbol{U}^{-1}$ is the same as the $(1,1)$ block of $\boldsymbol{U}_{+}^{-1}$. Recall that $\boldsymbol{v}$ and $\boldsymbol{w}$ determine the in-going and out-going edges of the $p_1 + 1$st node, respectively, we expect $\boldsymbol{v}$ and $\boldsymbol{w}$ to be sparse. For all $1 \le i < j < p_2$, $\boldsymbol{G}_{ij} \ne 0$ if and only if $\boldsymbol{w}_i\boldsymbol{w}_j \ne 0$. Hence $\boldsymbol{G}$ (also its $\boldsymbol{G}^{-1}$) are sparse. We thus claim that the $(1,2)$ block of $\boldsymbol{U}^{-1}$ can be obtained from the $(1,3)$ block of $\boldsymbol{U}_{+}^{-1}$ by a sparse rank-1 update followed by a sparse matrix multiplication; the $(2,2)$ block of $\boldsymbol{U}^{-1}$ can be obtained from the $(3,3)$ block of $\boldsymbol{U}_{+}^{-1}$ by a sparse matrix multiplication. (c) tells that whenever $\boldsymbol{w}_k \ne 0$, $[\boldsymbol{U}_{22}]_{kk} > [\widehat{\boldsymbol{U}}_{22}]_{kk}$ since $\frac{1+\|\boldsymbol{w}_{1:k}\|^2}{1+\|\boldsymbol{w}_{1:k-1}\|^2} > 1$. This indicates the variance of the child node becomes larger when its parent nodes are missing.

According to Proposition 1 and the discussion above, we may make the following claims: **C1** The $(1,2)$ and $(2,2)$ blocks of $\boldsymbol{U}^{-1}$ become denser compared with the $(1,3)$ and $(3,3)$ blocks of $\boldsymbol{U}_{+}^{-1}$; **C2** The variance for the child node of latent nodes becomes larger.

In order to recover DAG with latent variables, we have to detect the latent variables and identify their in-going and out-going edges. In this paper, we propose to detect latent variables via **C2** and identify in-going and out-going edges via **C1**. Note that **C2** alone is insufficient to detect latent variables since we only know the variance becomes larger. Thus, we make an equi-variance assumption to simplify the problem:

A3. The noise variables are centered, uncorrelated and have equal variance σ^2.

Remark 2. The essence of detecting latent variables is to determine where a latent node is missing and how it is connected to other nodes. To do so, one need to carefully design some criterion: whenever the criterion is violated, there are latent nodes. **A3** is perhaps the simplest criterion to accomplish the task. One may also consider exploring various alternatives, such as employing graph structures to refine the criterion. This approach might focus on the sparsity pattern of the graph or examine the in and out degree within the graph.

3.2 Algorithm

To proceed, we assume an estimation $\widehat{\sigma}$ for σ in **A3** is available. This assumption is not restrictive since when a root node is observed, we can estimate $\widehat{\sigma}$ via the sample variance of the root node.

At beginning, we observe $q < p$ nodes. Input $\widetilde{\boldsymbol{\Sigma}}_{xx} \in \mathbb{R}^{q\times q}$ into Algorithm 1, then it works as there were no latent variables until it encounters a child of latent variables. So, we may detect a latent variable when we find a diagonal entry of $\widetilde{\boldsymbol{U}}_q$ larger than $\widehat{\sigma}$, then we may determine its connection with other nodes via a sparse encouraging optimization problem. The above procedure can be used repeatedly to handle the case when there are multiple latent variables.

Next, we present the flowchart of our second algorithm – CDCF$^+$:

S1 Perform Algorithm 1 with input $\widetilde{\boldsymbol{\Sigma}}_{xx} \in \mathbb{R}^{q\times q}$, output $\widetilde{\boldsymbol{P}}$ and $\widetilde{\boldsymbol{U}}_q^{-1}$.

S2 Check the diagonal entries of $\widetilde{\boldsymbol{U}}_q^{-1}$: if all entries are approximately $\frac{1}{\sigma}$, then there is no (identifiable) latent variable; otherwise, find the first entry that is not approximately $\frac{1}{\sigma}$ and let it be the jth.

S3 Insert a variable between $j-1$st and jth variables and determine its connections with others.

S4 Update $\widetilde{\boldsymbol{\Sigma}}_{xx}$ by appending one row and column to it, then go to step **S1**.

In what follows we give details for **S3** and **S4**. Define

$$\boldsymbol{U}(\boldsymbol{S}) \equiv \widehat{\sigma}(\boldsymbol{I}-\boldsymbol{S})^{-1}, \qquad \boldsymbol{C}(\boldsymbol{S}) \equiv \boldsymbol{U}(\boldsymbol{S})^{\mathrm{T}}\boldsymbol{U}(\boldsymbol{S}), \tag{13}$$

where $\boldsymbol{S}$ is strictly upper triangular. Let $\underline{\boldsymbol{i}} = [1,\ldots,q]\widetilde{\boldsymbol{P}}$ and we parameterize the Cholesky factor of $[\widetilde{\boldsymbol{\Sigma}}_{xx}]_{\underline{\boldsymbol{i}},\underline{\boldsymbol{i}}}$ by $\boldsymbol{U}(\boldsymbol{S})$. Note that all diagonal entries of $\boldsymbol{U}(\boldsymbol{S})$ are $\widehat{\sigma}$. The reason for having such a requirement is that after inserting a variable, we expect that the variances of all variables are approximately σ^2. Note that $\boldsymbol{S}$ should be sparse, and the resulting covariance matrix should be close to the observed covariance matrix. We thus determine $\boldsymbol{S}$ by

$$\min_{\boldsymbol{S}} \|[\boldsymbol{C}(\boldsymbol{S})]_{\underline{\boldsymbol{j}}^c,\underline{\boldsymbol{j}}^c} - [\widetilde{\boldsymbol{\Sigma}}_{xx}]_{\underline{\boldsymbol{i}}^c,\underline{\boldsymbol{i}}^c}\|_F^2 + \mu\|\boldsymbol{S}\|_1, \tag{14}$$

where $\underline{\boldsymbol{i}}^c$ and $\underline{\boldsymbol{j}}^c$ are two index sets that pick all observed indices of $\widetilde{\boldsymbol{\Sigma}}_{xx}$ and $\boldsymbol{C}(\boldsymbol{S})$, respectively, $\mu \geq 0$ is a parameter.

Let $\boldsymbol{S}$ be the solution to (14). We compute $\boldsymbol{C}(\boldsymbol{S})$ by (13), then update $\widetilde{\boldsymbol{\Sigma}}_{xx}$ as follows:

$$\widetilde{\boldsymbol{\Sigma}}_{xx} = \begin{bmatrix} \widetilde{\boldsymbol{\Sigma}}_{xx} & \boldsymbol{z} \\ \boldsymbol{z}^{\mathrm{T}} & d^2 \end{bmatrix}, \tag{15}$$

where $\boldsymbol{z} = [\boldsymbol{x}^{\mathrm{T}}, \boldsymbol{y}^{\mathrm{T}}]^{\mathrm{T}}$, $\boldsymbol{x} = [\boldsymbol{C}(\boldsymbol{S})]_{1:j-1,j}$, $\boldsymbol{y} = [\boldsymbol{C}(\boldsymbol{S})]_{j+1:q,j}$, $d^2 = [\boldsymbol{C}(\boldsymbol{S})]_{j,j}$.

4 Numerical Experiments

In the experiments, the augmented covariance matrix $\widetilde{\boldsymbol{\Sigma}}_{xx} = \frac{1}{n}\boldsymbol{X}^{\mathrm{T}}\boldsymbol{X} + \lambda\boldsymbol{I}$ is used to estimate $\boldsymbol{\Sigma}_{xx}$, where λ is a parameter.

4.1 Experiments for CDCF

We evaluate CDCF on simulated graphs from two well-known ensembles of random graph types: Erdös-Rényi (ER) [12] and Scale-free (SF) [3]. The average edge number per node is denoted after the graph type, e.g., ER2 represents two edges per node on average. After the graph structure is settled, we assign uniformly random edge weights. We generate the observation data $\boldsymbol{X}$ from the linear Gaussian SEM.

Our baselines include: NOTEARS [33], DAG-GNN [32], CORL [29], NPVAR [9], and EQVAR [5]. Other methods perform no better than the baseline EQVAR.

Table 2 presents the structural Hamming distance (SHD) of baseline methods and our method on 3000 samples ($n = 3000$). Nodes number p is noted in the first column. Graph type and edge level are noted in the second column. We only report the SHD of different algorithms due to page limitations. And we find that other metrics such as true positive rate (TPR), false discovery rate (FDR), false positive rate (FPR), and F1 score have similar comparative performance with SHD. We also test bottom-up EQVAR, which is equivalent to LISTEN. The result is worse than top-down EQVAR (EV-TD) in this synthetic experiment, so we do not include the result in the table. For $p = 1000$ graphs, we only report the result of EV-TD and CDCF since other algorithms spend too much time.

Figure 2 plots the SHD results tested on 100 nodes graph recovering from different sample sizes. In the low dimension setting ($p < n$), we choose EV-TD, LISTEN (LTN), and LISTEN with Cholesky estimator (LTN-CH) as baselines. In the high dimension setting ($p > n$), we choose high dimension top-down (EV-HTD) and LISTEN (LTN) as baselines. We can see that CDCF achieves significantly better performance compared with previous baselines. In most cases, CDCF can reconstruct the DAG structure exactly according to the observing data, while the other algorithms fail to recover the ground truth graphs.

Table 3 shows the running time, tested on a 2.3 GHz single Intel Core i5 CPU. As illustrated in the table, CDCF is dozens or hundreds of times faster than EV-TD and LISTEN and tens of thousands of times faster than NOTEARS.

Table 2. Results of 50, 100, 1000 nodes on 3000 linear Gaussian SEM samples.

Nodes	Graph	NOTEARS	DAG-GNN	CORL-2	NPVAR	EV-TD	CDCF
50	ER2	$38.6_{10.8}$	$30.6_{8.3}$	$17.9_{10.6}$	$0.4_{0.5}$	$\mathbf{0.0_{0.0}}$	$\mathbf{0.0_{0.0}}$
	ER5	$67.8_{7.5}$	$93.2_{109.4}$	$64.8_{13.1}$	$0.6_{0.8}$	$0.1_{0.3}$	$\mathbf{0.0_{0.0}}$
	SF2	$3.5_{1.6}$	$79.3_{93.2}$	$\mathbf{0.0_{0.0}}$	$1.1_{1.0}$	$\mathbf{0.0_{0.0}}$	$\mathbf{0.0_{0.0}}$
	SF5	$20.1_{14.3}$	$89.2_{99.2}$	$20.8_{10.1}$	$1.0_{0.9}$	$\mathbf{0.0_{0.0}}$	$\mathbf{0.0_{0.0}}$
100	ER2	$72.6_{23.5}$	$66.2_{19.2}$	$18.6_{5.7}$	$2.1_{1.2}$	$\mathbf{0.0_{0.0}}$	$\mathbf{0.0_{0.0}}$
	ER5	$170.3_{34.2}$	$236.4_{36.8}$	$164.8_{17.1}$	$2.3_{1.2}$	$0.2_{0.4}$	$\mathbf{0.1_{0.3}}$
	SF2	$2.3_{1.3}$	$156.8_{21.2}$	$\mathbf{0.0_{0.0}}$	$3.0_{1.41}$	$\mathbf{0.0_{0.0}}$	$\mathbf{0.0_{0.0}}$
	SF5	$90.2_{34.5}$	$165.2_{22.0}$	$10.8_{6.1}$	$2.7_{0.9}$	$0.1_{0.3}$	$\mathbf{0.0_{0.0}}$
1000	ER2	–	–	–	–	$0.4_{0.5}$	$\mathbf{0.1_{0.3}}$
	ER5	–	–	–	–	$21.8_{3.8}$	$\mathbf{8.9_{4.2}}$
	SF2	–	–	–	–	$\mathbf{0.0_{0.0}}$	$\mathbf{0.0_{0.0}}$
	SF5	–	–	–	–	$0.3_{0.5}$	$\mathbf{0.0_{0.0}}$

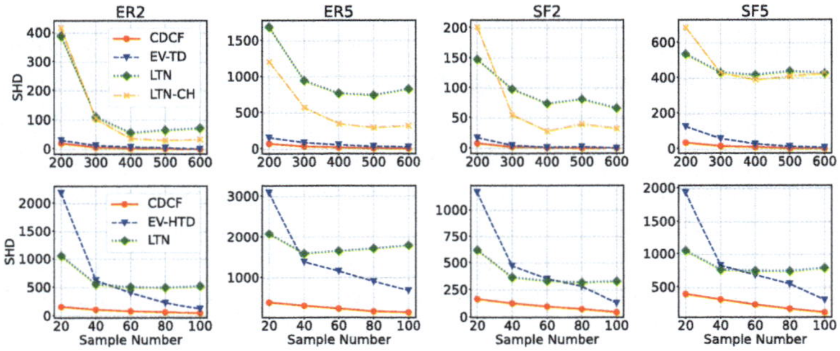

Fig. 2. SHD tested on 100 nodes graph recovering from different sample numbers.

Table 3. Running time (seconds) on 30 and 100 nodes over 3000 samples.

	30				100			
	ER2	ER5	SF2	SF5	ER2	ER5	SF2	SF5
CDCF	**0.004**	**0.005**	**0.004**	**0.005**	**0.017**	**0.016**	**0.016**	**0.017**
EV-TD	0.19	0.16	0.12	0.12	14.42	12.88	15.04	14.78
LISTEN	0.26	0.13	0.13	0.14	13.97	13.41	13.42	15.43
EV-HTD	8.27	7.48	6.72	12.50	260.74	302.36	241.59	387.92
DAG-GNN	49.15	49.02	38.44	41.03	137.25	238.71	158.13	187.21
NPVAR	84.24	82.57	108.37	109.13	9867.96	9084.78	10667.88	10173.89
NOTEARS	78.19	597.16	51.57	306.31	3237.8	1803.30	880.19	4159.82
CORL1	17573.08	18799.21	16422.11	16588.30	–	–	–	–

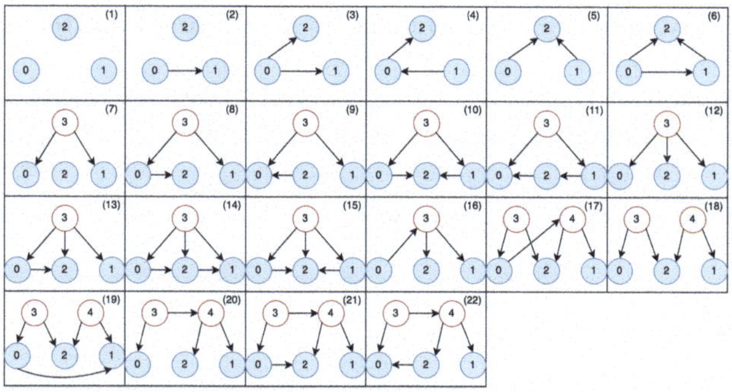

Fig. 3. Illustration of 22 identifiable graph types. The solid nodes represent the observed variables, and the hollow nodes represent the latent variables.

Table 4. Average SHD on partially observable models.

samp.	meth.	Graph Type																					
		1	2	3	4	5	6	7	8	9	10	11	12	13	14	15	16	17	18	19	20	21	22
5e3	B-S	**0.0**	1.6	1.0	2.2	**0.0**	3.8	3.0	4.4	3.0	4.6	4.5	3.5	4.6	3.8	6.6	4.6	**5.6**	6.0	6.1	5.4	5.1	5.7
	OUR	**0.0**	**0.0**	**0.0**	**0.0**	**0.0**	**0.0**	**0.0**	**0.0**	**0.0**	**0.0**	**0.0**	**0.0**	**0.0**	**0.0**	**0.0**	**0.0**	10.0	**0.0**	**4.8**	**0.0**	**1.3**	**1.8**
1e4	B-S	0.1	1.2	1.5	1.2	**0.0**	4.4	3.0	4.0	3.0	4.8	4.7	5.0	5.0	4.8	6.2	4.3	**6.6**	6.0	6.0	5.5	5.3	4.2
	OUR	**0.0**	**0.0**	**0.0**	**0.0**	**0.0**	**0.0**	**0.0**	**0.0**	**0.0**	**0.0**	**0.0**	**0.0**	**0.0**	**0.0**	**0.0**	**0.0**	10.0	**0.0**	**5.0**	**0.0**	**1.2**	**1.9**

4.2 Experiments for CDCF$^+$

We compare CDCF$^+$ with BIC guided search (B-S) algorithm mentioned in [1]. Their results show 22 different identifiable (up to re-indexing of the latent variables) typologies with three observable nodes and 0–2 unobservable nodes, which are shown in Fig. 3. Following their settings, we test CDCF$^+$ on the 22 graphs. We randomly generate 10 datasets for each graph and provide the average results in Table 4. For each random dataset, the result is recorded as the best SHD score among the latent variable permutations.

We compare CDCF$^+$ with B-S [1] since B-S is the most recent baseline, and it is applicable to equal variance and Gaussian noise cases without specific topology conditions. Table 4 shows that our algorithm achieves significantly better performance and does not depend on the specified latent structures.

5 Conclusion

We proposed two algorithms to handle the DAG recovery problems with full/partial observations, respectively. The first algorithm CDCF has theoretical guarantees for exact recovery and is better than the existing methods in both time and sample complexities. Experimental results on synthetic datasets and real-world data demonstrate the efficiency and effectiveness of CDCF. The second algorithm CDCF$^+$ is able to reveal latent variables and return a sparse DAG. Numerical results indicate that CDCF$^+$ recovers most of the topology structures exactly, and the performance is significantly better than previous algorithms.

The full version of this paper can be found at https://arxiv.org/abs/2311.00674, where you can find theoretical proofs and additional experimental results.

Disclosure of Interests. Most of this work was done at CCL, Baidu Research.

References

1. Adams, J., Hansen, N., Zhang, K.: Identification of partially observed linear causal models: graphical conditions for the non-gaussian and heterogeneous cases. In: NeurIPS, pp. 22822–22833. virtual (2021)
2. Aragam, B., Zhou, Q.: Concave penalized estimation of sparse gaussian bayesian networks. J. Mach. Learn. Res. **16**, 2273–2328 (2015)
3. Barabási, A.L., Albert, R.: Emergence of scaling in random networks. Science **286**(5439), 509–512 (1999)
4. Cai, R., Xie, F., Glymour, C., Hao, Z., Zhang, K.: Triad constraints for learning causal structure of latent variables. In: NeurIPS, pp. 12863–12872. Vancouver, Canada (2019)
5. Chen, W., Drton, M., Wang, Y.S.: On causal discovery with an equal-variance assumption. Biometrika **106**(4), 973–980 (2019)
6. Chickering, D.M.: Optimal structure identification with greedy search. J. Mach. Learn. Res. **3**, 507–554 (2002)
7. Colombo, D., Maathuis, M.H., Kalisch, M., Richardson, T.S.: Learning high-dimensional directed acyclic graphs with latent and selection variables. Ann. Stat. 294–321 (2012)
8. Friedman, N., Koller, D.: Being bayesian about network structure. A bayesian approach to structure discovery in bayesian networks. Mach. Learn. **50**(1-2), 95–125 (2003)
9. Gao, M., Ding, Y., Aragam, B.: A polynomial-time algorithm for learning nonparametric causal graphs. In: NeurIPS. virtual (2020)
10. Ghoshal, A., Honorio, J.: Learning identifiable Gaussian Bayesian networks in polynomial time and sample complexity. In: NIPS, pp. 6457–6466. Long Beach, CA (2017)
11. Ghoshal, A., Honorio, J.: Learning linear structural equation models in polynomial time and sample complexity. In: AISTATS, pp. 1466–1475. Playa Blanca, Lanzarote, Canary Islands, Spain (2018)
12. Gilbert, E.N.: Random graphs. Ann. Math. Stat. **30**(4), 1141–1144 (1959)
13. Hoyer, P.O., Janzing, D., Mooij, J.M., Peters, J., Schölkopf, B.: Nonlinear causal discovery with additive noise models. In: NIPS, pp. 689–696. Vancouver, Canada (2008)
14. Hoyer, P.O., Shimizu, S., Kerminen, A.J., Palviainen, M.: Estimation of causal effects using linear non-gaussian causal models with hidden variables. Int. J. Approx. Reason. **49**(2), 362–378 (2008)
15. Lv, Y., Miao, J., Liang, J., Chen, L., Qian, Y.: BIC-based node order learning for improving bayesian network structure learning. Front. Comput. Sci. **15**(6), 156337 (2021)

16. McKay, B.D., Oggier, F.E., Royle, G.F., Sloane, N., Wanless, I.M., Wilf, H.S., et al.: Acyclic digraphs and eigenvalues of (0, 1)-matrices. J. Integer Seq. **7**(2), 1–5 (2004)
17. Ng, I., Zhu, S., Fang, Z., Li, H., Chen, Z., Wang, J.: Masked gradient-based causal structure learning. In: SDM, pp. 424–432. Alexandria, VA (2022)
18. Nicholson, V.A.: Matrices with permanent equal to one. Linear Algebra Appl. **12**(2), 185–188 (1975)
19. Niu, X., Li, X., Li, P.: Learning cluster causal diagrams: an information-theoretic approach. In: IJCAI, pp. 4871–4877. Vienna, Austria (2022)
20. Park, G.: Identifiability of additive noise models using conditional variances. J. Mach. Learn. Res. **21**, 75:1–75:34 (2020)
21. Peters, J., Janzing, D., Schölkopf, B.: Elements of Causal Inference: Foundations and Learning Algorithms. The MIT Press, Cambridge (2017)
22. Peters, J., Mooij, J.M., Janzing, D., Schölkopf, B.: Causal discovery with continuous additive noise models. J. Mach. Learn. Res. **15**(1), 2009–2053 (2014)
23. Ren, S., Li, D., Li, P.: Causal effect prediction with flow-based inference. In: ICDM, pp. 1167–1172. Orlando, FL (2022)
24. Rohekar, R.Y., Nisimov, S., Gurwicz, Y., Novik, G.: Iterative causal discovery in the possible presence of latent confounders and selection bias. In: NeurIPS, pp. 2454–2465. virtual (2021)
25. Salehkaleybar, S., Ghassami, A., Kiyavash, N., Zhang, K.: Learning linear non-gaussian causal models in the presence of latent variables. J. Mach. Learn. Res. **21**, 39:1–39:24 (2020)
26. Shimizu, S., Hoyer, P.O., Hyvärinen, A.: Estimation of linear non-gaussian acyclic models for latent factors. Neurocomputing **72**(7–9), 2024–2027 (2009)
27. Spirtes, P., Meek, C., Richardson, T.: An algorithm for causal inference in the presence of latent variables and selection bias. Comput. Causation Discovery **21**, 1–252 (1999)
28. Squires, C., Amaniampong, J., Uhler, C.: Efficient permutation discovery in causal dags. arXiv preprint arXiv:2011.03610 (2020)
29. Wang, X., et al.: Ordering-based causal discovery with reinforcement learning. In: IJCAI, pp. 3566–3573. Virtual Event/Montreal, Canada (2021)
30. Xie, F., Cai, R., Huang, B., Glymour, C., Hao, Z., Zhang, K.: Generalized independent noise condition for estimating latent variable causal graphs. In: NeurIPS. virtual (2020)
31. Ye, Q., Amini, A.A., Zhou, Q.: Optimizing regularized cholesky score for order-based learning of bayesian networks. IEEE Trans. Pattern Anal. Mach. Intell. **43**(10), 3555–3572 (2021)
32. Yu, Y., Chen, J., Gao, T., Yu, M.: DAG-GNN: DAG structure learning with graph neural networks. In: ICML, pp. 7154–7163. Long Beach, CA (2019)
33. Zheng, X., Aragam, B., Ravikumar, P., Xing, E.P.: DAGs with NO TEARS: continuous optimization for structure learning. In: NeurIPS, pp. 9492–9503. Montréal, Canada (2018)

Computing Sparse Generalized Inverses for Least-Squares

Marcia Fampa[1] and Jon Lee[2](✉)

[1] Federal University of Rio de Janeiro, Rio de Janeiro, Brazil
fampa@cos.ufrj.br
[2] University of Michigan, Ann Arbor, USA
jonxlee@umich.edu

Abstract. The Moore-Penrose (M-P) pseudoinverse is a key object in matrix theory and applications. The M-P pseudoinverse is characterized by the four well-known M-P properties. But not all of these properties are needed for the use of it in applications like least-squares fitting. When a matrix is not full rank, there can be much sparser generalized inverses than the M-P pseudoinverse that solve the least-squares problem for arbitrary response vectors. Besides sparsity and even structured sparsity, we are interested in low-rank and low-norm solutions. We approach the problem of generating such generalized inverses using a variety of optimization methods, including linear optimization, second-order-cone optimization, first-order methods, and local-search based approximation algorithms.

Keywords: Moore-Penrose pseudoinverse · generalized inverse · least squares · sparse · convex optimization · ADMM · local search

1 Introduction

The Moore-Penrose (M-P) pseudoinverse has a prominent place in matrix theory and applications. It is well-known that the M-P pseudoinverse is beautifully characterized by the four M-P properties. But not all of these properties are needed for the use of it in applications like least-squares fitting. In particular, when a matrix is not full rank, as is common in modern applications, there can be much sparser (and even block structured, for explainabilty) generalized inverses than the M-P pseudoinverse that solve the least-squares problem for arbitrary response vectors. Besides sparsity and structured sparsity, we are interested in low-rank and low-norm solutions, for further explainability and numerical stability, respectively. So, we attack the problem of generating such generalized inverses using optimization methods. Our techniques include: linear optimization, second-order-cone optimization, local-search based approximation methods, the alternating direction method of multipliers (ADMM), and accelerations of these ideas via structural results on generalized inverses.

© The Author(s), under exclusive license to Springer Nature Switzerland AG 2025
H. A. Le Thi et al. (Eds.): ICAMCS 2024, LNNS 1311, pp. 40–51, 2025.
https://doi.org/10.1007/978-3-031-90606-0_4

This work is a survey of results that are scattered across our papers [6,10, 14,15,18]; also see [7,8,17]. Here, we mainly concentrate on generalized inverses that are relevant to the least-squares problem.

Organization. In Sect. 2, we give background and some basic results on generalized inverses, emphasizing the Moore-Penrose properties. In Sect. 3 (respectively, Sect. 4), we describe formulations and algorithms for inducing sparsity (resp., row sparsity) for some relevant types of generalized inverses. In Sect. 5, we give an approximation algorithm that enforces maximum row-sparsity, while seeking to minimize a genuine norm. In Sect. 6, we make some concluding remarks.

2 Background on Generalized Inverses

2.1 The M-P Pseudoinverse and the M-P Properties

Probably the most important "generalized inverse" is the well-known Moore-Penrose (M-P) pseudoinverse. It has a convenient definition via the compact singular-value decomposition.

Definition 1 (see [13]). For a square invertible diagonal matrix Σ, we let the Moore-Penrose (M-P) pseudoinverse be $\Sigma^\dagger := \Sigma^{-1}$. For any rank-$r > 0$ matrix $A_{m\times n} = U_{m\times r}\Sigma_{r\times r}V_{n\times r}^\top$ given by its compact singular-value decomposition, the M-P pseudoinverse $A^\dagger$ is defined as $A^\dagger := V\Sigma^{-1}U^\top$. Finally, $0_{m\times n}^\dagger := 0_{n\times m}$.

The following elementary results (all easily verified) are quite useful.

Proposition 2
(i) $(A^\dagger)^\dagger = A$.
(ii) $(A^\top)^\dagger = (A^\dagger)^\top$.
(iii) If A has full column rank, then $A^\dagger = (A^\top A)^{-1}A^\top$.
(iv) If A has full row rank, then $A^\dagger = A^\top(AA^\top)^{-1}$.

Our main point of departure is the following.

Theorem 3 (see[13]). *For $A \in \mathbb{R}^{m\times n}$, $A^\dagger$ is the unique $H \in \mathbb{R}^{n\times m}$ satisfying:*

$$AHA = A \quad \text{(generalized inverse)} \tag{P1}$$
$$HAH = H \quad \text{(reflexive)} \tag{P2}$$
$$(AH)^\top = AH \quad \text{(ah-symmetric)} \tag{P3}$$
$$(HA)^\top = HA \quad \text{(ha-symmetric)} \tag{P4}$$

The terms "generalized inverse" and "reflexive" are completely standard (see [1]). We use "ah-symmetric" (respective, "ha-symmetric") as convenient mnemonics to indicate that AH (respectively, HA) is symmetric.

In fact, $A^\dagger$ has a nice optimality property among generalized inverses:

Theorem 4 (see [1], Exer. 18, p. 112). *$A^\dagger$ is the unique solution of* $\min\{\, \|H\|_F \ :\ \text{P1} \,\}$.

The following is trivial, but we state it here for later convenience.

Proposition 5

(i) If P1 holds, then $A^\top H^\top A^\top = A^\top$ (that is, P1 holds when A is replaced by $A^\top$ and H is replaced by $H^\top$).
(ii) If P2 holds, then $H^\top A^\top H^\top = H^\top$ (that is, P2 holds when A is replaced by $A^\top$ and H is replaced by $H^\top$).
(iii) If P3 holds, then $H^\top A^\top = (H^\top A^\top)^\top$ (that is, P4 holds when A is replaced by $A^\top$ and H is replaced by $H^\top$).
(iv) If P4 holds, then $A^\top H^\top = (A^\top H^\top)^\top$ (that is, P3 holds when A is replaced by $A^\top$ and H is replaced by $H^\top$).

It is easy to check that $H := A^\dagger$ satisfies the four M-P properties. That $A^\dagger$ is the *unique* solution follows from the next result, which we prepare for by setting some notation. From the full singular-value decomposition $A =: U\Sigma V^\top$ with

$$\Sigma =: \begin{bmatrix} \underset{r\times r}{D} & \underset{r\times(n-r)}{0} \\ \underset{(m-r)\times r}{0} & \underset{(m-r)\times(n-r)}{0} \end{bmatrix},$$

with D diagonal, let $H \in \mathbb{R}^{n\times m}$ and $\Gamma := V^\top H U$, where

$$\Gamma =: \begin{bmatrix} \underset{r\times r}{X} & \underset{r\times(m-r)}{Y} \\ \underset{(n-r)\times r}{Z} & \underset{(n-r)\times(m-r)}{W} \end{bmatrix}.$$

Then $H = V\Gamma U^\top$.

Theorem 6 ([1], p. 208, Ex.14 (no proof); also see [15]).

(i) P1 is equivalent to $X = D^{-1}$.
(ii) If P1 is satisfied, then P2 is equivalent to $ZDY = W$.
(iii) If P1 is satisfied, then P3 is equivalent to $Y = 0$.
(iv) If P1 is satisfied, then P4 is equivalent to $Z = 0$.

A corollary of Theorem 6 is the celebrated Theorem 3.

Remark 7. We note that all of the M-P properties except P2 are linear in the variable H. The nonlinearity of P2 is reflected in the nonlinearity of $ZDY = W$.

In fact, P2 has an important meaning for generalized inverses:

Theorem 8 (see e.g. [16]). *P1* $\Rightarrow \operatorname{rank}(H) \geq \operatorname{rank}(A)$, *with equality if and only if P2 holds.*

2.2 Application of the M-P Pseudoinverse to Least-Squares And Basic Results

Least-Squares Estimation (LSE). $A^\dagger$ gives a least-squares solution to an over-determined system of equations, $A\theta = b$. Specifically, $\hat{\theta} := A^\dagger b$ solves $\min_\theta \{\|A\theta - b\|_2\}$. Typically, dense or even sparse A leads to a dense $A^\dagger$. So we may seek instead matrices H that behave similarly to $A^\dagger$ but are sparse; a key motivating situation is when we have many b and a single huge A, and we are concerned about the work to compute the matrix-vector products Hb. In fact, for this application, any ah-symmetric generalized inverse can be used instead of $A^\dagger$; see Corollary 12 below.

The following result is useful and is easy to check.

Lemma 9. $A^\top A A^\dagger = A^\top$.

From this lemma, we can obtain the following result.

Theorem 10 (see e.g. [16])

(i) P1+P3 $\Rightarrow AH = AA^\dagger$, the orthogonal projector onto the range of A.
(ii) P1+P4 $\Rightarrow HA = A^\dagger A$, the orthogonal projector onto the range of $A^\top$.

As we have observed in Remark 7, P2 is the only M-P property that is nonlinear in H. But now we can easily see that because of Theorem 10, P2 becomes linear under P1 and either P3 or P4:

Corollary 11
(i) P1+P3 $\Rightarrow$ P2 is equivalent to $HAA^\dagger = H$.
(ii) P1+P4 $\Rightarrow$ P2 is equivalent to $A^\dagger AH = H$.

Using Lemma 9 and Theorem 10.*i*, we can establish the following key result.

Corollary 12 (see e.g. [16]). *If H satisfies P1+P3, then $\hat{\theta} := Hb$ solves the least-squares problem $\min_\theta \{\|A\theta - b\|_2\}$ for every b.*

3 Inducing Sparsity

3.1 Inducing Sparsity for Left and Right Inverses

$H_{n\times m}$ is a *left inverse* (respectively, *right inverse*) of $A_{m\times n}$ if $HA = I_n$ (respectively, $AH = I_m$). If H is a left and right inverse, then it is the ordinary matrix inverse. Expanding on Proposition 2.*iii-iv*, we have

Proposition 13
(iii) If A has full column rank, then $A^\dagger = (A^\top A)^{-1}A^\top$ and is a left inverse of A. If A does not have full column rank, then it has no left inverse.
(iv) If A has full row rank, then $A^\dagger = A^\top(AA^\top)^{-1}$ and is a right inverse of A. If A does not have full row rank, then it has no right inverse.

It is easy to check the following.

Proposition 14
(i) If H is a left inverse, then H satisfies P1+P2+P4.
(ii) If H is a right inverse, then H satisfies P1+P2+P3.

The converses of the results in Proposition 14 do not hold:

Example 15. The M-P pseudoinverse of

$$A = \begin{pmatrix} 1 & 0 \\ 0 & 0 \end{pmatrix},$$

is itself, but it is neither a left nor right inverse of itself.

[3–5] introduced the idea of inducing sparsity for left and right inverses. When A has full column (respectively, row) rank, A may have more left (respectively, right) inverses than just $A^\dagger$, and this can be a basis for seeking sparse left (respectively, right) inverses.

Left inverses have a nice property for linear models. The expected mean-squared error based on a solution determined by a left inverse H is proportional to the Frobenius norm of H. Formally, we have the following result which is easy to check.

Theorem 16 (see [5]). *For the linear model $b = A\theta + \epsilon$, with $\mathbb{E}[\epsilon] = 0$, $\mathrm{Cov}[\epsilon] = \sigma^2 I$, we have $\mathbb{E}\left[\|\theta - Hb\|_2^2\right] = \sigma^2 \|H\|_F^2$ for every left inverse H.*

Remark 17. Recalling Theorem 4, that $A^\dagger = \mathbf{argmin}\,\{\, \|H\|_F \,:\, \mathrm{P1} \,\}$, and that $A^\dagger$ is a left inverse (when A is full column rank), we conclude that $\|A^\dagger\|_F$ is a lower bound on the mean-squared error for left inverses, and we can evaluate the mean-squared error of any left inverse by calculating its Frobenius norm (and comparing it to $\|A^\dagger\|_F$).

For an "overdetermined" case as would be typical for least-squares applications, where A is assumed to have full column rank, [5] defines a (hopefully) *sparse left pseudoinverse inverse* via the linear-optimization problem

$$\begin{aligned} &\min\{\, \|H\|_1 \,:\, HA = I_n \,\} \\ &= \min\left\{ \sum_{(i,j)\in n\times m} t_{ij} \,:\, t_{ij} \ge h_{ij},\ t_{ij} \ge -h_{ij},\ \forall ij \in n\times m;\ HA = I_n \right\}. \end{aligned}$$

For an "underdetermined" case where A is assumed to have full row rank, [5] defines a (hopefully) *sparse right pseudoinverse inverse* via the linear-optimization problem

$$\begin{aligned} &\min\{\, \|H\|_1 \,:\, AH = I_m \,\} \\ &= \min\left\{ \sum_{(i,j)\in n\times m} t_{ij} \,:\, t_{ij} \ge h_{ij},\ t_{ij} \ge -h_{ij},\ \forall ij \in n\times m;\ AH = I_m \right\}. \end{aligned}$$

3.2 Inducing Sparsity for Generalized Inverses Based on the M-P Properties

We are interested in calculating useful sparse generalized inverses for matrices A having neither full row nor full column rank. In such a situation, the methods of Sect. 3.1 do not apply. Instead, we consider minimizing $\|H\|_1$ over subsets of the M-P properties. Specifically, we define the following optimization problems.

$$\min\{\, \|H\|_1 \ :\ \text{P1}\,\}, \qquad (P_1^1)$$

$$\min\{\, \|H\|_1 \ :\ \text{P1+P3}\,\}, \qquad (P_{13}^1)$$

$$\min\{\, \|H\|_1 \ :\ \text{P1+P2+P3}\,\}, \qquad (P_{123}^1)$$

These problems can all be re-formulated as linear-optimization problems, using the well-known technique for handling minimization of a sum of absolute values. In particular, recall Corollary 11, that P2 becomes linear under P1+P3; hence P_{123}^1 can be re-formulated as a linear optimization problem. Owing to Proposition 5, we do not discuss further the analogous P_{14}^1 (respectively, P_{124}^1) because we can equivalently work with P_{13}^1 (respectively, P_{123}^1) with respect to $A^\top$. This covers most of the subsets of the M-P properties that contain P1. We of course do not take all of the properties together, because then the only feasible solution is $A^\dagger$. The one subset that we do not directly or indirectly address is P1+P2, and this is because then the feasible region is nonconvex (see [9–11] for some ideas on how to attack it).

Already, we can see that extreme solution if these linear-optimization problems guarantee us some sparsity:

Proposition 18 ([18]). *Suppose that $A \in \mathbb{R}^{n\times n}$ has rank r.*

(i) Extreme solutions of the linear-optimization problem P_1^1 have at most r^2 nonzeros.
(ii) Extreme solutions of the linear-optimization problem P_{13}^1 have at most rm nonzeros.
(iii) Extreme solutions of the linear-optimization problem P_{123}^1 have at most $rm + (m-r)(n-r)$ nonzeros.

3.3 Comparing Approaches for Least-Squares

We want to find an H so that $\hat{\theta} := Hb$ "solves" $A\theta \approx b$, for all b.

$H := A^\dagger$ does that in the data-centric LSE sense. And in fact any ah-symmetric generalized inverse does that (see Corollary 12). There is no better $\hat{\theta}$ (even among those that do not arise as Hb for some H), measuring via least-squares fit, for all response data b. So we have room to work here for sparser H than $A^\dagger$, without losing anything on LSE.

Additionally, $H := A^\dagger$ does that in a model-centric MSE sense (see Remark 17). We have a linear model $b = A\theta + \epsilon$, and we want to minimize $\mathbb{E}\left[\|\theta - Hb\|_2^2\right]$, which is an average squared error. Among all $\hat{\theta} := Hb$ where H is a *left inverse*, there is no other H that is at least this good, measuring by MSE. So we only have room to work here for sparser H than $A^\dagger$, if we are willing concede something on MSE. We also necessarily concede on LSE because P3 will not be satisfied for any left inverse except $A^\dagger$ (due to Proposition 14.i and Theorem 3). We could perhaps consider relaxing the left-inverse constraint to P1, but then we lose the direct connection between MSE and $\|H\|_F$ (see Theorem 16).

3.4 ADMM for Inducing Sparsity

Using Theorem 6, we have the following result.

Theorem 19 ([15]). *$\mathcal{P}_{123}^1$ can be efficiently reformulated as*

$$z(\mathcal{P}_{123}^1) := \min_{Z\in\mathbb{R}^{(n-r)\times r}} \left\|V_1 D^{-1} U_1^\top + V_2 Z U_1^\top\right\|_1 . \qquad (\mathcal{P}_{123}^1)$$

We can solve $\mathcal{P}_{123}^1$ as a smaller linear-optimization problem than what we would get from $\mathcal{P}_{123}^1$. But even more efficiently, we have an ADMM algorithm for $\mathcal{P}_{123}^1$ with closed-form subproblem updates, first described in [14]. Introducing $E \in \mathbb{R}^{n\times m}$, we rewrite $\mathcal{P}_{123}^1$ as

$$\min_{Z,E}\left\{\|E\|_1 \ : \ E = V_1 D^{-1} U_1^\top + V_2 Z U_1^\top\right\}.$$

The associated augmented Lagrangian function is

$$\mathcal{L}_\rho(Z,E,\Lambda) := \|E\|_1 + \frac{\rho}{2}\left\|V_1 D^{-1} U_1^\top + V_2 Z U_1^\top - E + \Lambda\right\|_F^2 ,$$

where $\rho > 0$ is the penalty parameter and $\Lambda \in \mathbb{R}^{n\times m}$ is the scaled Lagrangian multiplier, that is $\Lambda := \Theta/\rho$, where Θ is the Lagrangian multiplier. This leads to the ADMM iterations:

$$\begin{aligned} Z^{k+1} &:= \operatorname{argmin}_Z \ \mathcal{L}_\rho(Z, E^k, \Lambda^k),\\ E^{k+1} &:= \operatorname{argmin}_E \ \mathcal{L}_\rho(Z^{k+1}, E, \Lambda^k),\\ \Lambda^{k+1} &:= \Lambda^k + V_1 D^{-1} U_1^\top + V_2 Z^{k+1} U_1^\top - E^{k+1}. \end{aligned}$$

Proposition 20. $Z^{k+1} := \operatorname{argmin}_{\mathrm{Z}} \left\{\|\mathrm{J} - \mathrm{V}_2 \mathrm{Z} \mathrm{U}_1^\top\|_{\mathrm{F}}^2\right\} = \mathrm{V}_2^\top \mathrm{J} \mathrm{U}_1$, *where* $J := E^k - V_1 D^{-1} U_1^\top - \Lambda^k$.

Based on the well-known closed-form solution for the so-called "orthonormal case" of the "lasso problem"; see [12], for example], we have the following result.

Proposition 21 (**see** [2], Sec. 4.4.3). *The solution of*

$$E^{k+1} := \operatorname{argmin}_{\mathrm{E}} \left\{ \|\mathrm{E}\|_1 + \tfrac{\rho}{2}\|\mathrm{E} - \mathrm{Y}\|_{\mathrm{F}}^2 \right\},$$

where $Y := V_1 D^{-1} U_1^\top + V_2 Z^{k+1} U_1^\top + \Lambda^k$, *is given by*

$$E_{ij}^{k+1} := \mathcal{S}_{1/\rho}(Y_{ij}),\ for\ i = 1,\ldots,n \ and\ j = 1,\ldots,m,$$

where the "soft thresholding operator" $\mathcal{S}$ is defined as

$$\mathcal{S}_\kappa(a) := \begin{cases} a - \kappa, & a > \kappa;\\ 0, & |a| \le \kappa;\\ a + \kappa, & a < -\kappa. \end{cases}$$

4 Inducing Row-Sparsity

Even nicer than a sparse H is a "row-sparse" H, that is an H with many rows that are all zero. Such an H leads to a sparse $\hat{\theta} := Hb$, and is quite fast to compute, without even relying on sparse data structures. We can measure the number of nonzero rows by $\|H\|_{2,0} := \sum_{i=1}^{n} \|H_{i\cdot}\|_0 = \| (\|H_{1\cdot}\|_2, \|H_{2\cdot}\|_2, \ldots \|H_{n\cdot}\|_2) \|_0$. Of course any norms could replace the inner 2-norms, but the 2-norm leads to the nicely-behaved surrogate

$$\|H\|_{2,1} := \sum_{i=1}^{n} \|H_{i\cdot}\|_2 = \| (\|H_{1\cdot}\|_2, \|H_{2\cdot}\|_2, \ldots \|H_{n\cdot}\|_2) \|_1 .$$

4.1 Inducing Row-Sparsity for Generalized Inverses Based on the M-P Properties

To induce row sparsity for a generalized inverse H, we consider

$$\min\{ \|H\|_{2,1} \ : \ \text{P1} \}. \qquad (P_1^{2,1})$$

We will see and exploit that optimal solutions H of $P_1^{2,1}$ always satisfy P1+P3. It will be especially revealing to consider a more general optimization problem

$$\min\{ f_{\text{row}}(\|H_{1\cdot}\|_2, \|H_{2\cdot}\|_2, \ldots, \|H_{n\cdot}\|_2) \ : \ \text{P1} \}, \qquad (P_1^{f_{\text{row}}})$$

where $f_{\text{row}} : \mathbb{R}_+^n \to \mathbb{R}$ is increasing in each argument.

From Theorem 6, we can establish the following result.

Theorem 22 ([15]). *Suppose that H is an optimal solution to $P_1^{f_{\text{row}}}$, then H satisfies P2+P3.*

As a consequence of Theorems 6 and 22, we have the following result.

Theorem 23 ([15]). *$\mathcal{P}_1^{2,1}$ can be efficiently reformulated as the second-order-cone optimization problem*

$$z(\mathcal{P}_1^{2,1}) := \min_{Z \in \mathbb{R}^{(n-r)\times r}} \left\|V_1 D^{-1} + V_2 Z\right\|_{2,1} . \qquad (\mathcal{P}_1^{2,1})$$

4.2 ADMM for Inducing Row-Sparsity of Generalized Inverses

Next, we give an ADMM algorithm for $\mathcal{P}_1^{2,1}$, with closed-form subproblem updates, first described in [14]. Introducing $E \in \mathbb{R}^{n\times r}$, we rewrite $\mathcal{P}_1^{2,1}$ as

$$\min_{Z,E} \left\{ \|E\|_{2,1} \ : \ E = V_1 D^{-1} + V_2 Z \right\}.$$

The associated augmented Lagrangian function is

$$\mathcal{L}_\rho(Z, E, \Lambda) := \|E\|_{2,1} + \frac{\rho}{2} \left\|V_1 D^{-1} + V_2 Z - E + \Lambda\right\|_F^2 ,$$

where $\rho > 0$ is the penalty parameter and $\Lambda \in \mathbb{R}^{n\times r}$ is the scaled Lagrangian multiplier. This leads to the ADMM iterations:

$$\begin{aligned}
Z^{k+1} &:= \operatorname{argmin}_Z \mathcal{L}_\rho(Z, E^k, \Lambda^k),\\
E^{k+1} &:= \operatorname{argmin}_E \mathcal{L}_\rho(Z^{k+1}, E, \Lambda^k),\\
\Lambda^{k+1} &:= \Lambda^k + V_1 D^{-1} + V_2 Z^{k+1} - E^{k+1}.
\end{aligned}$$

Proposition 24. $Z^{k+1} := \operatorname{argmin}_{\mathrm{Z}} \left\{ \|\mathrm{J} - \mathrm{V_2 Z}\|_{\mathrm{F}}^2 \right\} = \mathrm{V}_2^\top \mathrm{J}$, *where* $J := E^k - V_1 D^{-1} - \Lambda^k$.

Based on the well-known closed-form solution for the so-called "orthonormal case" of the "group lasso problem", we have the following result.

Proposition 25 (see [19], Prop. 1; [2], Sec. 6.4.2)**.** *The solution of*

$$E^{k+1} := \operatorname{argmin}_{\mathrm{E}} \left\{ \|\mathrm{E}\|_{2,1} + \tfrac{\rho}{2} \|\mathrm{E} - \mathrm{Y}\|_{\mathrm{F}}^2 \right\},$$

where $Y := V_1 D^{-1} + V_2 Z^{k+1} + \Lambda^k$, *is given by*

$$E_{i\cdot}^{k+1} := \begin{cases} \frac{\|Y_{i\cdot}\|_2 - 1/\rho}{\|Y_{i\cdot}\|_2} Y_{i\cdot}, & \text{if } 1/\rho < \|Y_{i\cdot}\|_2\,;\\ 0, & \text{otherwise.} \end{cases}$$

5 Block Solutions and Local Search

Extending an idea from [18], we have the following.

Theorem 26 (Generalized column-block construction). *Let* $A \in \mathbb{R}^{m\times n}$. *For* T, *an ordered subset of elements from* $\{1,\ldots,n\}$, *let* $\hat{A} := A[:,T]$ *be the* $m \times |T|$ *submatrix of* A *formed by columns* T. *Let* $\hat{H}$ *be any ah-symmetric generalized inverse of* $\hat{A}$, *and let* H *be the* $n \times m$ *matrix* $\hat{H}$ *having all rows equal to zero, except rows* T, *which are given by* $\hat{H}$.

(i) Then H *is ah-symmetric.*
(ii) Moreover, if $\operatorname{rank}(\hat{A}) = \operatorname{rank}(A)$, *then* H *is a generalized inverse.*
(iii) Moreover, if $\hat{H}$ *is reflexive, then* H *is reflexive.*

Corollary 27 (Column-block construction; see[18]**).** *For* $A \in \mathbb{R}^{m\times n}$, *let* $r := \operatorname{rank}(A)$. *For* T, *an ordered subset of* r *elements from* $\{1,\ldots,n\}$, *let* $\hat{A} := A[:,T]$ *be the* $m \times r$ *submatrix of* A *formed by columns* T. *If* $\operatorname{rank}(\hat{A}) = r$, *let*

$$\hat{H} := \hat{A}^\dagger = (\hat{A}^\top \hat{A})^{-1} \hat{A}^\top.$$

The $n \times m$ *matrix* H *with all rows equal to zero, except rows* T, *which are given by* $\hat{H}$, *is an ah-symmetric reflexive generalized inverse of* A *having at most* rm *nonzeros.*

Theorem 28 **([6,14,18]).** *Let A be an $m \times n$, rank-r matrix, and let S be any ordered subset of r elements from $\{1,\ldots,m\}$ indexing linearly-independent rows of A. Choose $\epsilon \geq 0$, and let $\tilde{A} := A[S,T]$ be a $(1+\epsilon)$-local maximizer for the absolute determinant on the set of $r \times r$ nonsingular submatrices of $A[S,:]$, which can be calculated efficiently. Then the H built via column-block construction over $\hat{A} := A[:,T]$, is an ah-symmetric reflexive generalized inverse (having at most rm nonzeros), satisfying*

(i) $\|H\|_1 \leq r(1+\epsilon)\|H^1_{\mathrm{opt}}\|_1$, *where H^1_{opt} is a 1-norm minimizing generalized inverse of A.*
(ii) $\|H\|_{2,1} \leq r(1+\epsilon)\|H^{2,1}_{\mathrm{opt}}\|_{2,1}$, *where $H^{2,1}_{\mathrm{opt}}$ is a 2,1-norm minimizing generalized inverse of A.*

For rational A, this local search runs in polynomial time in the size of the bit encoding of A and with ϵ-dependence being a factor of $1+\frac{1}{\epsilon}$ (see [6]).

Remark 29. We note that Theorem 28.*ii* (proof in [14]) does not follow directly from 28.i (proof in [18]). For $H \in \mathbb{R}^{n\times m}$, we have $\|H\|_{2,1} \leq \|H\|_1 \leq \sqrt{m}\|H\|_{2,1}$. So, for our constructed H, we have $\|H\|_{2,1} \leq \|H\|_1 \leq r(1+\epsilon)\|H^1_{\mathrm{opt}}\|_1 \leq r(1+\epsilon)\|H^{2,1}_{\mathrm{opt}}\|_1 \leq r\sqrt{m}(1+\epsilon)\|H^{2,1}_{\mathrm{opt}}\|_{2,1}$. And we should note that m can be quite large in least-squares applications.

Remark 30. In the least-squares context, $\hat{\theta} := Hb$ exactly optimizes the fit $A\theta \approx b$, for any ah-symmetric generalized inverse H. Now, keep in mind that in the least-squares setting, we are assuming that $n \gg m$. And further, we assume that $m > r$. If H is dense, then per right-hand side b, we need nm (scalar) multiplications to form the least-squares solution $\hat{\theta} := Hb$. But if H is reflexive (i.e., rank r), then we can write $H = L_{n\times r}R_{r\times m}$, and calculate $\hat{\theta} := Hb = L(Rb)$ with $rm + nr$ multiplications. But this gives us no sparsity guarantee on $\hat{\theta}$, and it would cost us a huge linear-optimization problem to exactly minimize the 1-norm in this context. But with our column-block construction, we can do better: in this case $\hat{\theta} := Hb = \hat{H}_{r\times m}b$, so we need only rm multiplications. Additionally, we get a sparse $\hat{\theta}$, and with our local search, we can approximately and efficiently calculate such an H having low 1-norm and low 2,1-norm.

6 Concluding Remarks

We mainly concentrated on ah-symmetric reflexive generalized inverses, with full details and proof presented in the source papers; in particular, [14,15,18]. We also have two other ADMM algorithms, aimed at aggressively limiting the 2,0-norm; see [14]. Related results for other types of generalized inverses can be found in [6,18]. A comprehensive treatment will be in our forthcoming monograph. Computational results for the algorithms that we presented can be found in [7,8,10,14,15,18].

Acknowledgments. M. Fampa was supported in part by CNPq grant 307167/2022-4. J. Lee was supported in part by AFOSR grant FA9550-22-1-0172. We are particularly grateful to Gabriel Ponte and Luze Xu, our co-authors on much of the work that we surveyed in this manuscript.

Disclosure of Interests. The authors have no competing interests to declare that are relevant to the content of this article.

References

1. Ben-Israel, A., Greville, T.: Generalized Inverses: Theory and Applications, 2 edn. Springer, Berlin, Heidelberg (2003). https://doi.org/10.1007/b97366
2. Boyd, S., Parikh, N., Chu, E., Peleato, B., Eckstein, J.: Distributed optimization and statistical learning via the alternating direction method of multipliers. Found. Trends Mach. Learn. **3**(1), 1–122 (2011). https://doi.org/10.1561/2200000016
3. Dokmanić, I., Gribonval, R.: Beyond Moore-Penrose Part I: generalized inverses that minimize matrix norms (2017). http://arxiv.org/abs/1706.08349
4. Dokmanić, I., Gribonval, R.: Beyond Moore-Penrose Part II: the sparse pseudoinverse (2017). https://hal.inria.fr/hal-01547283/file/pseudo-part2.pdf
5. Dokmanić, I., Kolundžija, M., Vetterli, M.: Beyond Moore-Penrose: sparse pseudoinverse. In: ICASSP 2013 (38th International Conference on Acoustics, Speech, and Signal Processing), pp. 6526–6530. IEEE (2013). https://doi.org/10.1109/ICASSP.2013.6638923
6. Fampa, M., Lee, J.: On sparse reflexive generalized inverses. Oper. Res. Lett. **46**(6), 605–610 (2018). https://doi.org/10.1016/j.orl.2018.09.005
7. Fampa, M., Lee, J., Ponte, G.: Trading off 1-norm and sparsity against rank for linear models using mathematical optimization: 1-norm minimizing partially reflexive ah-symmetric generalized inverses. Open J. Math. Optim. **2**, 1–14 (2021). https://doi.org/10.5802/ojmo.6
8. Fampa, M., Lee, J., Ponte, G., Xu, L.: Experimental analysis of local searches for sparse reflexive generalized inverses. J. Glob. Optim. **81**, 1057–1093 (2021). https://doi.org/10.1007/s10898-021-01087-y
9. Fuentes, V.K.: On Computing Sparse Generalized Inverses and Sparse-Inverse/Low-Rank Decompositions. Ph.D. thesis, University of Michigan (2019). https://hdl.handle.net/2027.42/149803
10. Fuentes, V.K., Fampa, M., Lee, J.: Sparse pseudoinverses via LP and SDP relaxations of Moore-Penrose. In: CLAIO 2016, pp. 343–350 (2016). https://marciafampa.com/pdf/CLAIO_2016_Proceedings_FuentesFampaLee.pdf
11. Fuentes, V.K., Fampa, M., Lee, J.: Diving for sparse partially-reflexive generalized inverses. In: Le Thi, H.A., Le, H.M., Pham Dinh, T. (eds.) Optimization of Complex Systems: Theory, Models, Algorithms and Applications, LNCS, pp. 89–98. Springer, Berlin, Heidelberg (2020). https://doi.org/10.1007/978-3-030-21803-4_9
12. Gauraha, N.: Introduction to the LASSO. Resonance **23**(4), 439–464 (2018). https://doi.org/10.1007/s12045-018-0635-x
13. Penrose, R.: A generalized inverse for matrices. Math. Proc. Camb. Philos. Soc. **51**, 406–413 (1955). https://doi.org/10.1017/S0305004100030401
14. Ponte, G., Fampa, M., Lee, J., Xu, L.: Good and fast row-sparse ah-symmetric reflexive generalized inverses (2024).https://arxiv.org/abs/2401.17540

15. Ponte, G., Fampa, M., Lee, J., Xu, L.: On computing sparse generalized inverses. Oper. Res. Lett. **52**, 107058 (2024). https://doi.org/10.1016/j.orl.2023.107058
16. Rohde, C.: Contributions to the Theory, Computation and Application of Generalized Inverses. Ph.D. thesis, North Carolina State University, Raleigh (May 1964). https://www4.stat.ncsu.edu/~boos/library/mimeo.archive/ISMS_1964_392.pdf
17. Xu, L., Fampa, M., Lee, J.: 1-norm minimization and minimum-rank structured sparsity for symmetric and ah-symmetric generalized inverses: rank one and two (2020). to appear In: Fields Institute Communications volume on Data Science and Optimization, A. Deza, S. Gupta, S. Pokutta, eds. https://arxiv.org/abs/2010.11406
18. Xu, L., Fampa, M., Lee, J., Ponte, G.: Approximate 1-norm minimization and minimum-rank structured sparsity for various generalized inverses via local search. SIAM J. Optim. **31**(3), 1722–1747 (2021). https://doi.org/10.1137/19M1281514
19. Yuan, M., Lin, Y.: Model selection and estimation in regression with grouped variables. J. Roy. Stat. Soc. Series B: Stat. Methodol. **68**(1), 49–67 (2006). https://doi.org/10.1111/j.1467-9868.2005.00532.x

Dynamic Regret for Online Proximal Newton's Method

Chang He(✉), Zhaoye Pan, and Bo Jiang

Research Institute for Interdisciplinary Sciences, School of Information Management and Engineering, Shanghai University of Finance and Economics, Shanghai, China
ischanghe@gmail.com

Abstract. In this paper, we consider online optimization problems with nonconvex composite objective functions. These problems are particularly relevant in applications such as subspace tracking and sparse recovery, which are crucial in the field of machine learning. By leveraging techniques such as online Newton's update and scaled proximal mapping, we propose an inexact online proximal Newton's method. Furthermore, we establish a dynamic regret bound that scales linearly with the cumulative variation between time optima. This bound retains the favorable theoretical properties of online Newton's method for smooth optimization. Finally, we conduct numerical experiments to empirically validate and demonstrate the effectiveness of the proposed method.

Keywords: Online Optimization · Time-varying Optimization · Newton's Method

1 Introduction

Online optimization [1,2] has emerged as a prominent research area in recent decades, demonstrating its effectiveness in modeling and solving sequential decision-making problems. To provide a general overview, the framework of online optimization can be described as follows:

$$\min_{x_t \in \mathbb{R}^n} \ F_t(x_t), \ t = 0, 1, \ldots, T, \tag{1.1}$$

where T is the time horizon. At each time step t, the learner selects an action x_t. After the learner plays x_t, the adversary reveals the loss function F_t, and the learner incurs a loss $F_t(x_t)$. The goal of the learner is to minimize the total incurred cost over the finite time horizon. Mathematically, a widely used performance measure is the *static regret* [1,2], defined by:

$$\text{S-Regret}(T) = \sum_{t=1}^{T} F_t(x_t) - \min_{x \in \mathbb{R}^n} \sum_{t=1}^{T} F_t(x), \tag{1.2}$$

which quantifies the difference between the learner's cumulative loss and that of the best action. This paradigm has received significant attention due to its

© The Author(s), under exclusive license to Springer Nature Switzerland AG 2025
H. A. Le Thi et al. (Eds.): ICAMCS 2024, LNNS 1311, pp. 52–63, 2025.
https://doi.org/10.1007/978-3-031-90606-0_5

practical effectiveness in various fields such as online advertising [3], network resource allocation [4] and power systems [5]. In this paper, we focus on the following online nonconvex *composite* optimization

$$F_t(x_t) := f_t(x_t) + r_t(x_t), \tag{1.3}$$

where the loss function $f_t(\cdot)$ is smooth but may be nonconvex, and $r_t(\cdot)$ is a convex but not necessarily differentiable penalty function or regularizer. The presence of non-differentiable regularizers, such as the ℓ_1 norm, improves the sparsity and robustness of the model, yielding practical benefits in applications like online robust PCA [6] and sparse recovery [7].

The static regret (1.2) compares against a fixed benchmark, implicitly assuming the existence of a reasonably good decision across all iterations. However, this may not hold in non-stationary environments where the underlying distribution of online functions changes [8,9]. To cope with non-stationary environments, we focus on minimizing the *dynamic regret* introduced by [2]

$$\text{D-Regret}(T) = \sum_{t=1}^{T} F_t(x_t) - F_t(x_t^*), \tag{1.4}$$

where $x_t^* = \arg\min_{x_t \in \mathbb{R}^n} F_t(x_t), \ \forall t = 0, 1, \ldots, T$ represents the optimal solution in each time. In contrast to the fixed benchmark in static regret, dynamic regret compares the learner's actions to the optimal action at each time t and is more suitable in non-stationary environments. It is known that in the worst case, a sub-linear dynamic regret is not attainable unless imposing certain regularities of the comparator sequence or the function sequence [9,10]. Therefore, one generally uses the following natural problem-dependent quantities to upper bound dynamic regret.

- *Path-length* [2]: the cumulative variation between time optima

$$\mathcal{P}_T := \sum_{t=0}^{T-1} \|x_{t+1}^* - x_t^*\|; \tag{1.5}$$

- *Squared path-length* [11]: the cumulative squared variation between time optima

$$\mathcal{S}_T := \sum_{t=0}^{T-1} \|x_{t+1}^* - x_t^*\|^2; \tag{1.6}$$

- *Function variation* [9]: the variation over consecutive function values

$$\mathcal{V}_T := \sum_{t=0}^{T-1} \sup_x |F_{t-1}(x) - F_t(x)|. \tag{1.7}$$

1.1 Related Work

There is a rich body of work in the literature on online optimization that focuses on the study of dynamic regret. For general objective functions without a composite structure, dynamic regret was first introduced by Zinkevich [2] in the context of online gradient descent, establishing a bound of $\mathcal{O}(\sqrt{T}(1+\mathcal{P}_T))$ for convex objective functions. For strongly convex functions, Mokhtari et al. [12] improved the bound of online gradient descent to $\mathcal{O}(\mathcal{P}_T)$. Later, Yang et al. [13] demonstrated that the $\mathcal{O}(\mathcal{P}_T)$ rate is also attainable for convex and smooth objective functions. Besides, Besbes et al. [9] showed that online gradient descent with a restart strategy can achieve a dynamic regret of $\mathcal{O}(T^{2/3}\mathcal{V}_T^{1/3})$ when the function variation $\mathcal{V}_T$ is known in advance. Mulvaney-Kemp et al. [14] further explored dynamic regret under the Polyak-Łojasiewicz (PL) condition for nonconvex objective functions.

The above works make only a single query to the gradient of the objective functions at each time step. When more information is accessible, recent studies have shown that dynamic regret bounds can be further improved. On one side, when the learner has access to the gradients of the objective functions multiple times, Zhang et al. [11] proposed the online multiple gradient descent method, proving a dynamic regret bound of $\mathcal{O}(\min\{\mathcal{P}_T, \mathcal{S}_T\})$ for strongly convex and smooth functions. Subsequently, Zhao and Zhang [8] presented an enhanced analysis, showing that the online multiple gradient descent method achieves a rate of $\mathcal{O}(\min\{\mathcal{P}_T, \mathcal{S}_T, \mathcal{V}_T\})$. On the other side, when the learner can access both gradients and Hessians of the objective functions, Lesage-Landry et al. [15] developed an online Newton's method, establishing an $\mathcal{O}(1+\mathcal{P}_T)$ bound. Lupien and Lesage-Landry [16] subsequently extended this approach to online optimization problems with time-varying linear equality constraints.

In the context of online composite optimization (1.3), the relevant studies are comparatively underexplored. Dixit et al. [17] proposed an inexact online proximal gradient descent method for the case where the smooth component $f_t(\cdot)$ is strongly convex, establishing a dynamic regret bound of $\mathcal{O}(1+\mathcal{P}_T)$. A more complex scenario was analyzed by Ajalloeian et al. [18]. Bastianello and DallAnese [19] investigated online proximal gradient descent in a distributed setting over networks. For the case of convex functions $f_t(\cdot)$, Zhao et al. [20] refined the analysis of proximal online gradient descent and provided a general lower bound on dynamic regret. However, all of these analyses focus solely on online convex composite optimization, raising the natural question:

Can we develop an efficient online optimization method for the nonconvex composite objective functions with a dynamic regret bound guarantee?

1.2 Our Contributions

In this work, we answer this question firmly by proposing an inexact **O**nline **P**roximal **N**ewton's method. The main contributions are summarized as follows:

1. For online optimization, we are *the first* to study the case of nonconvex composite objective functions. Specifically, we do not require the smooth part to be convex or globally Lipschitz continuous. Instead, we only need the Hessian matrix $\nabla^2 f_t(x_t^*)$ at the optimal solution to be nondegenerate at each time step, along with local Lipschitz continuity in the neighborhood of the optimum.
2. At each time step, we develop an inexact online proximal Newton's update by extending the approaches of [15] and [21]. Unlike the absolute error used in [17], we introduce a relative error for solving the scaled proximal subproblem inexactly. By carefully choosing the relative error, we demonstrate that the dynamic regret bound is linear in the cumulative variation, achieving the same bound as the online Newton's method for online smooth optimization.
3. Finally, we empirically evaluate the performance of the inexact online proximal Newton (OPN) method in online logistic regression with ℓ_1 penalty and in target tracking with ℓ_1 penalty. Using both synthetic and real datasets, our numerical results show that the inexact OPN method outperforms online proximal gradient descent.

Notation. Given a positive definite matrix $H \in \mathbb{R}^{n \times n}$, we denote the weighted norm as $\|x\|_H := \sqrt{x^\top H x}$, $\forall x \in \mathbb{R}^n$. The standard Euclidean norm and ℓ_1 norm are represented by $\|\cdot\|$ and $\|\cdot\|_1$, respectively. For the sake of simplicity, throughout the remainder of this paper, we use the notations $g_t(x_t) := \nabla f_t(x_t)$ and $H_t(x_t) := \nabla^2 f_t(x_t)$ to refer to the gradient and Hessian of the smooth function $f_t(\cdot)$ in the time t. We use $\mathcal{N}(\mu, \sigma^2)$ to denote the normal distribution with mean μ, and variance σ^2.

2 Preliminaries

In this section, we propose the inexact online proximal Newton's update and present several technical lemmas crucial for establishing the dynamic regret bound.

2.1 Scaled Proximal Mapping

To begin, we introduce the scaled proximal mapping [21,22]. This concept serves as the foundation for formulating our inexact online proximal Newton's update.

Definition 1. *Let r be a convex function and H be a positive definite matrix, then the scaled proximal mapping of r at x is*

$$\operatorname{prox}_r^H(x) := \arg\min_{z \in \mathbb{R}^n} \; r(z) + \frac{1}{2}\|z - x\|_H^2. \tag{2.1}$$

The scaled proximal mapping shares several key properties with the standard proximal mappings, including the inexpensiveness in the weighted norm.

Lemma 1 (Lemma 2 in [22]). *Given a positive definite matrix $H \in \mathbb{R}^{n\times n}$, then the scaled proximal mapping is single-valued and satisfies the following property:*

$$\| \operatorname{prox}_r^H(x) - \operatorname{prox}_r^H(y)\|_H \le \|x-y\|_H, \; \forall x, y \in \mathbb{R}^n.$$

2.2 Inexact Online Proximal Newton's Update

As the scaled proximal mapping in (2.1) involves the weighted norm, the solution is typically not available in closed form. Therefore, in practical applications, computing the scaled proximal mapping (2.1) becomes unrealistic. To overcome this issue, we introduce a relative error ϵ_t and define the inexact online proximal Newton's update below

$$x_{t+1} \approx_{\epsilon_t} \operatorname{prox}_{r_t}^{H_t(x_t)}(x_t - H_t^{-1}(x_t)g_t(x_t)) \tag{2.2}$$

for $t = 0, 1, \ldots, T-1$, where $\approx_{\epsilon_t}$ means

$$\|x_{t+1} - \operatorname{prox}_{r_t}^{H_t(x_t)}(x_t - H_t^{-1}(x_t)g_t(x_t))\| \le \epsilon_t.$$

Note that the right-hand side of (2.2) equivalents as

$$\arg\min_{z\in\mathbb{R}^n} \; r(z) + g_t(x_t)^\top(z-x_t) + \frac{1}{2}(z-x_t)^\top H_t(x_t)(z-x_t),$$

and hence we do not need to compute the inverse of $H_t(x_t)$. By applying the implementations of proximal Newton-type methods such as glmnet [23] and QUIC [24], the subproblem (2.2) can be solved efficiently.

To analyze the property of inexact online proximal Newton's update (2.2), we make the following assumptions as [15].

Assumption 2.1. *For the smooth part $f_t(\cdot)$, we assume that*

- *the Hessian matrix at the optimal solution is nondegenerate, i.e. there exists a constant $\mu > 0$ such that $H_t(x_t^*) \succeq \mu I$ for all $t = 0, 1, \ldots, T-1$;*
- *the Hessian matrix is locally Lipschitz continuous around optimum, i.e. $\exists \beta, L > 0$ such that for $\|x - x_t^*\| \le \beta$, we have $\|H_t(x) - H_t(x_t^*)\| \le L\|x - x_t^*\|$ for all $t = 0, 1, \ldots, T-1$.*

Based on the above assumption, we are able to provide a lower bound for $H_t(x)$ whenever x is in a small neighborhood of x_t^*.

Lemma 2. *Suppose the Assumption 2.1 holds. If the point x satisfies $\|x - x_t^*\| \le \gamma := \min\{\beta, \frac{\mu}{3L}\}$, then $H_t(x) \succeq mI$ for all $t = 0, 1, \ldots, T-1$, where $m = \frac{2}{3}\mu$.*

The proof follows a similar approach to Lemma 2 in [15], and we omit it here. By the above Lemma, $H_t(x_t)$ is positive definite once x_t is sufficiently close to x_t^*, and then the scaled proximal mapping subproblem (2.2) is well-defined. Therefore, we characterize the property of the inexact online proximal Newton's update, which extends the convergence analysis of proximal Newton's method [21] to the realm of online optimization.

Lemma 3. *Suppose the Assumption 2.1 holds. If the point x_t satisfies $\|x_t - x_t^*\| \leq \gamma$, then the inexact online proximal Newton's update x_{t+1} satisfies*

$$\|x_{t+1} - x_t^*\| \leq \frac{L}{2m}\|x_t - x_t^*\|^2 + \epsilon_t. \tag{2.3}$$

Proof. By the inexact online proximal Newton's update mechanism (2.2), it follows

$$\|x_{t+1} - x_t^*\| \leq \| \operatorname{prox}_{r_t}^{H_t(x_t)}(x_t - H_t^{-1}(x_t)g_t(x_t)) - x_t^*\| + \epsilon_t.$$

Since $H_t(x_t) \succeq mI$ and $x_t^* = \arg\min_{x_t \in \mathbb{R}^n} F_t(x_t)$, combing with the Lemma 1, we have

$$\begin{aligned}
&\| \operatorname{prox}_{r_t}^{H_t(x_t)}(x_t - H_t^{-1}(x_t)g_t(x_t)) - x_t^*\| \\
\leq\ & \frac{1}{\sqrt{m}}\| \operatorname{prox}_{r_t}^{H_t(x_t)}(x_t - H_t^{-1}(x_t)g_t(x_t)) - x_t^*\|_{H_t(x_t)} \\
=\ & \frac{1}{\sqrt{m}}\| \operatorname{prox}_{r_t}^{H_t(x_t)}(x_t - H_t^{-1}(x_t)g_t(x_t)) - \operatorname{prox}_{r_t}^{H_t(x_t)}(x_t^* - H_t^{-1}(x_t)g_t(x_t^*))\|_{H_t(x_t)} \\
\leq\ & \frac{1}{\sqrt{m}}\|x_t - x_t^* + H_t^{-1}(x_t)(g_t(x_t^*) - g_t(x_t))\|_{H_t(x_t)} \\
\leq\ & \frac{1}{m}\|H_t(x_t)(x_t - x_t^*) - g_t(x_t) + g_t(x_t^*)\| \\
\leq\ & \frac{L}{2m}\|x_t - x_t^*\|^2,
\end{aligned}$$

which completes the proof.

3 Inexact Online Proximal Newton's Method

In this section, we introduce the inexact online proximal Newton's method in Algorithm 1. To establish an upper bound of the dynamic regret, we first assume that the variation between consecutive time steps will not be excessive since the loss function $F_t(\cdot)$ in each time generally exhibits similarity.

Algorithm 1 Inexact OPN Method

1: **Input:** Initial point $x_0 \in \mathbb{R}^n$ satisfying $\|x_0 - x_0^*\| \leq \gamma$
2: **for** $t = 0, 1, \ldots, T-1$ **do**
3: **Action** x_t
4: **Observe** $g_t(x_t)$, $H_t(x_t)$ and ϵ_t
5: **Update**

$$x_{t+1} \approx_{\epsilon_t} \operatorname{prox}_{r_t}^{H_t(x_t)}(x_t - H_t^{-1}(x_t)g_t(x_t))$$

6: **end for**

Assumption 3.1. *The maximum variation between two time steps is uniformly upper bounded, i.e.*

$$\max_{t=0,\ldots,T-1} \|x_{t+1}^* - x_t^*\| \leq \frac{4}{5}\gamma - \frac{L}{2m}\gamma^2. \tag{3.1}$$

Recall $0 \leq \gamma \leq \frac{\mu}{3L}$ and $m = \frac{2}{3}\mu$, which implies the right-hand side is nonnegative, thereby ensuring the inequality (3.1) holds. The following Lemma demonstrates Line 5 in Algorithm 1 is well-defined for all time steps, enabling the sequential updates of the inexact online proximal Newton's method.

Lemma 4. *Suppose the Assumption 2.1 and Assumption 3.1 hold. Let the inexactness parameters satisfy $\epsilon_t \leq \frac{1}{5}\|x_t - x_t^*\|, \forall t$. For each time $t = 0, 1, \ldots, T-1$, we have*

$$H_t(x_t) \succeq mI.$$

Proof. By Lemma 2, it is sufficient to validate $\|x_t - x_t^*\| \leq \gamma$ for all t. Now we apply induction on t, for the time $t+1$, we have

$$\|x_{t+1} - x_{t+1}^*\| \leq \|x_{t+1} - x_t^*\| + \|x_t^* - x_{t+1}^*\|. \tag{3.2}$$

Substituting the inequality (2.3) into (3.2), it follows

$$\begin{aligned}\|x_{t+1} - x_{t+1}^*\| \leq& \frac{L}{2m}\|x_t - x_t^*\|^2 + \epsilon_t + \|x_t^* - x_{t+1}^*\| \\ \leq& \frac{L}{2m}\gamma^2 + \frac{1}{5}\gamma + \max_{t=0,\ldots,T-1} \|x_{t+1}^* - x_t^*\| \\ \leq& \gamma,\end{aligned}$$

which completes the proof.

Remark 1. The initialization (Line 1 in Algorithm 1) plays a crucial role in our analysis. Combining with the quadratic convergence (Lemma 3) of the online proximal Newton's method ensures that all iterates remain close to the optimal solution at the current time step. Consequently, despite the nonconvex nature of the objective function, the online proximal Newton's method enjoys favorable local convergence properties throughout all iterations.

With the guarantee of Lemma 4, the results of Lemma 3 can be applied and then we could provide a dynamic regret bound for Algorithm 1.

Theorem 1. *Suppose the Assumption 2.1 and Assumption 3.1 hold, and provided that $\exists l > 0$ such that for x satisfying $\|x - x_t^*\| \leq \gamma$, we have $\|F_t(x) - F_t(x_t^*)\| \leq l\|x - x_t^*\|$ for all t. In each time t, by setting the relative error satisfy $\epsilon_t \leq \frac{1}{5}\|x_t - x_t^*\|$, the dynamic regret of the inexact online proximal Newton's method is upper bounded as*

$$\textit{D-Regret}(T) \leq \mathcal{O}\left(\mathcal{P}_T + 1\right).$$

Proof. From Lemma 3, summing the following inequality

$$\|x_{t+1} - x_{t+1}^*\| \leq \|x_{t+1} - x_t^*\| + \|x_t^* - x_{t+1}^*\|$$

from $t = 0$ to $T - 1$ implies

$$\begin{aligned}
\sum_{t=0}^{T-1} \|x_{t+1} - x_{t+1}^*\| \leq & \sum_{t=0}^{T-1} \|x_{t+1} - x_t^*\| + \sum_{t=0}^{T-1} \|x_t^* - x_{t+1}^*\| \\
\leq & \sum_{t=0}^{T-1} \left(\frac{L}{2m} \|x_t - x_t^*\|^2 + \epsilon_t \right) + \mathcal{P}_T \\
= & \sum_{t=1}^{T} \left(\frac{L}{2m} \|x_t - x_t^*\|^2 + \epsilon_t \right) + \frac{L}{2m} \left(\|x_0 - x_0^*\|^2 - \|x_T - x_T^*\|^2 \right) \\
& + \epsilon_0 - \epsilon_T + \mathcal{P}_T.
\end{aligned}$$

where the second inequality comes from Lemma 3. Note that

$$\begin{aligned}
\|x_{t+1} - x_{t+1}^*\| \leq & \|x_{t+1} - x_t^*\| + \|x_{t+1}^* - x_t^*\| \\
\leq & \frac{L}{2m} \|x_t - x_t^*\|^2 + \epsilon_t + \frac{4}{5}\gamma - \frac{L}{2m}\gamma^2,
\end{aligned}$$

and since the initial point satisfies $\|x_0 - x_0^*\| \leq \gamma$, we can conclude by applying induction on t

$$\|x_{t+1} - x_{t+1}^*\| \leq \frac{L}{2m}\gamma^2 + \frac{1}{5}\gamma + \frac{4}{5}\gamma - \frac{L}{2m}\gamma^2 \leq \gamma.$$

Consequently, it follows

$$\begin{aligned}
\sum_{t=1}^{T} \left(\frac{L}{2m} \|x_t - x_t^*\|^2 + \epsilon_t \right) \leq & \sum_{t=1}^{T} \left(\frac{L}{2m} \|x_t - x_t^*\| + \frac{1}{5} \right) \|x_t - x_t^*\| \\
\leq & \sum_{t=1}^{T} \left(\frac{L}{2m}\gamma + \frac{1}{5} \right) \|x_t - x_t^*\| \\
\leq & \frac{1}{2} \sum_{t=1}^{T} \|x_t - x_t^*\|,
\end{aligned}$$

where the last ineuqality holds due to that $\gamma = \min\left\{\beta, \frac{\mu}{3L}\right\}$. Therefore, we have

$$\sum_{t=0}^{T-1} \|x_{t+1} - x_{t+1}^*\| \leq \mathcal{P}_T + \frac{1}{2}\sum_{t=1}^{T} \|x_t - x_t^*\| + \underbrace{\frac{L}{2m} \left(\|x_0 - x_0^*\|^2 - \|x_T - x_T^*\|^2 \right) + \epsilon_0 - \epsilon_T}_{:=C}.$$

Rearranging terms leads

$$\frac{1}{2}\sum_{t=1}^{T} \|x_t - x_t^*\| \leq \mathcal{P}_T + C,$$

and recall the definition of the dynamic regret (1.4), we complete the proof by

$$\text{Regret}(T) \leq l \sum_{t=1}^{T} \|x_t - x_t^*\| \leq O\left(\mathcal{P}_T + 1\right).$$

4 Numerical Results

In this section, we evaluate the efficiency of the **I**nexact **O**nline **P**roximal **N**ewton's method (IOPN) through two numerical examples: online logistic regression with ℓ_1 penalty and target tracking with ℓ_1 penalty. We compare its performance with the **O**nline **P**roximal **G**radient **D**escent (OPGD), considering both constant stepsize [17] and decaying stepsize [20]. The experiments are done on a laptop with an Intel(R) Core(TM) i7 CPU running at 1.80 GHz, 16.0 GB of RAM, and Windows 10 operating system. Under the online setting, problem data is received sequentially over time. In each time, new data samples are observed and added to the training dataset.

4.1 Online Logistic Regression with ℓ_1 Penalty

We start by testing our method on online regularized logistic regression [25], where the smooth part is globally strongly convex to satisfy the nondegeneracy condition in Assumption (2.1). The loss function in each time t is defined as

$$\min_{x_t \in \mathbb{R}^n} \log\left(1 + \exp\left(-y_t A_t^\top x_t\right)\right) + \frac{\lambda_1}{2}\|x_t\|^2 + \lambda_2 \|x_t\|_1,$$

where A_t, y_t are the training set constructed at the current time and λ_1, λ_2 are the hyperparameters.

For the synthetic dataset, we follow the experimental setup from [20]. We randomly generate labels $y_t \in \{1, -1\}$ and create the corresponding A_t by sampling from a time-varying distribution. Specifically, when $y_t = 1$, we sample from $\mathcal{N}((1 + 0.01\sin(t)), 0.01)$; otherwise, we sample from $\mathcal{N}((-1 + 0.01\sin(t)), 0.01)$. Hyperparameters are set as $\lambda_1 = 0.02$ and $\lambda_2 = 0.01$. For the real data, we use the dataset *ijcnn* consisting of 35000 samples and 22 features. To simulate the online environment, we start with 500 samples in the initial time and introduce 20 more samples in each subsequent time. Hyperparameters are $\lambda_1 = 0.1$ and $\lambda_2 = 0.001$. In both experiments, we select the online proximal gradient descent stepsize as follows: a constant stepsize of 0.01 and a decaying stepsize of $\frac{0.02}{\sqrt{t}}$. The dynamic regret bounds for each method are shown in Fig. 1 and Fig. 2.

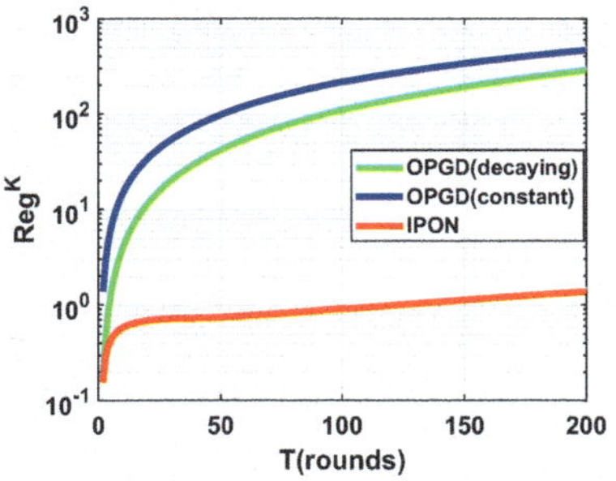

Fig. 1. Synthetic Data

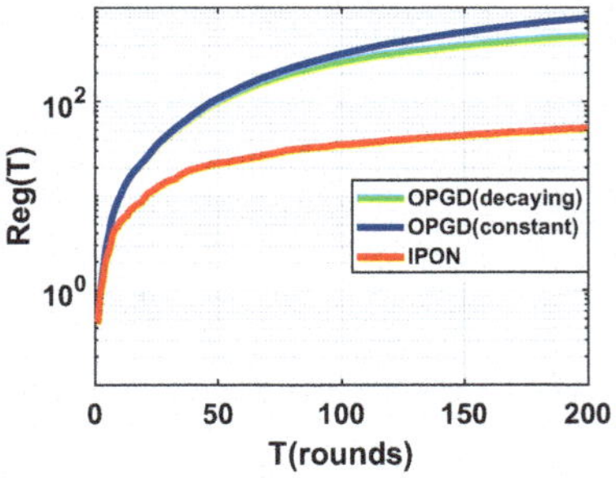

Fig. 2. Real Data

4.2 Target Tracking with ℓ_1 Penalty

To demonstrate the performance of our method for online nonconvex nonsmooth optimization, we focus on localizing the moving target problem [26,27] with ℓ_1 penalty. At time t, each sensor positioned at a_i provides a range measurement d_i based on the target's location x_t:

$$d_i = \|x_t - a_i\| + \omega_i,\ i = 1, \cdots, m,$$

where ω_i is the Gaussian noise. Therefore, our goal is to minimize the nonconvex least-squares problem below

$$\min_{x_t \in \mathbb{R}^n} \sum_{i=1}^{m} (\|x_t - a_i\| - d_i)^2 + \lambda \|x_t\|_1.$$

Our experimental setup follows a similar manner of [15], with the penalty parameter set as $\lambda = 10^{-4}$. For OPGD, we use a constant step size of 0.05 and a decaying step size of $\frac{0.1}{\sqrt{t}}$. It is worth noting that there is no theoretical guarantee on the OPGD for such an objective function, we test it as it could handle the nonsmooth regularizer. The numerical results in terms of dynamic regret and tracking trajectory comparisons are depicted in Fig. 3 and Fig. 4, respectively.

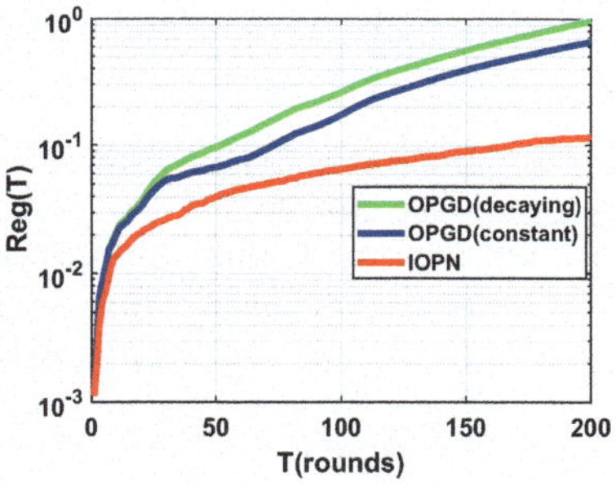

Fig. 3. Dynamic Regret

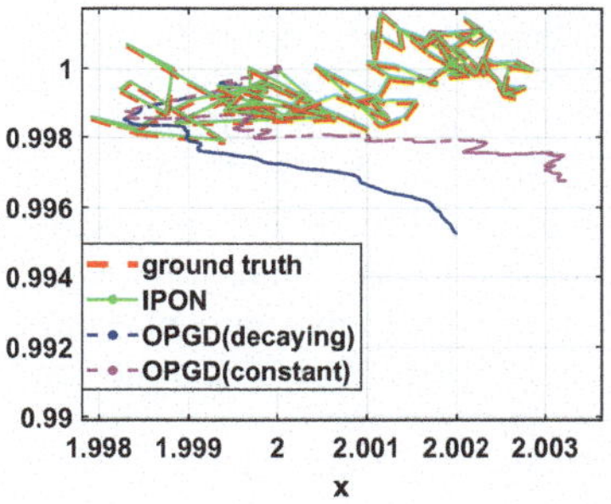

Fig. 4. Tracking Trajectory

5 Conclusion

In this work, we introduced the inexact online proximal Newton's method for online nonconvex nonsmooth optimization. Theoretically, we proved that our method achieves the same dynamic regret bound as the online Newton's method for smooth online optimization. Numerically, we validated its effectiveness using online logistic regression with an ℓ_1 penalty and target tracking with ℓ_1 penalty to validate the effectiveness of our proposed method. In the future, it would be interesting to explore the online quasi-Newton updates and globalization of online Newton-type methods.

References

1. Hazan, E., et al.: Introduction to online convex optimization. Found. Trends® Optim. **2**(3-4), 157–325 (2016)
2. Zinkevich, M.: Online convex programming and generalized infinitesimal gradient ascent. In: Proceedings of the 20th International Conference on Machine Learning (icml-03), pp. 928–936, 2003
3. McMahan, H.B., et al.: Ad click prediction: a view from the trenches. In: Proceedings of the 19th ACM SIGKDD International Conference on Knowledge Discovery and Data Mining, pp. 1222–1230, 2013
4. Yu, H., Neely, M.J.: Learning-aided optimization for energy-harvesting devices with outdated state information. IEEE/ACM Trans. Netw. **27**(4), 1501–1514 (2019)
5. Lesage-Landry, A., Wang, H., Shames, I., Mancarella, P., Taylor, J.A.: Online convex optimization of multi-energy building-to-grid ancillary services. IEEE Trans. Control Syst. Technol. **28**(6), 2416–2431 (2019)
6. Feng, J., Xu, H., Yan, S.: Online robust pca via stochastic optimization. Adv. Neural Inf. Process. Syst. **26** (2013)
7. Angelosante, D., Bazerque, J.A., Giannakis, G.B.: Online adaptive estimation of sparse signals: where rls meets the ℓ_1-norm. IEEE Trans. Signal Process. **58**(7), 3436–3447 (2010)
8. Zhao, P., Zhang, L.: Improved analysis for dynamic regret of strongly convex and smooth functions. In: Learning for Dynamics and Control, pp. 48–59. PMLR, 2021
9. Besbes, O., Gur, Y., Zeevi, A.: Non-stationary stochastic optimization. Oper. Res. **63**(5), 1227–1244 (2015)
10. Jadbabaie, A., Rakhlin, A., Shahrampour, S., Sridharan, K.: Online optimization: competing with dynamic comparators. In: Artificial Intelligence and Statistics, pp. 398–406. PMLR, 2015
11. Zhang, L., Yang, T., Yi, J., Jin, R., Zhou, Z.H.: Improved dynamic regret for non-degenerate functions. Adv. Neural Inf. Process. Syst. **30** (2017)
12. Mokhtari, A., Shahrampour, S., Jadbabaie, A., Ribeiro, A.: Online optimization in dynamic environments: improved regret rates for strongly convex problems. In: 2016 IEEE 55th Conference on Decision and Control (CDC), pp. 7195–7201. IEEE, 2016
13. Yang, T., Zhang, L., Jin, R., Yi, J.: Tracking slowly moving clairvoyant: optimal dynamic regret of online learning with true and noisy gradient. In: International Conference on Machine Learning, pp. 449–457. PMLR, 2016
14. Mulvaney-Kemp, J., Park, S., Jin, M., Lavaei, J.: Dynamic regret bounds for online nonconvex optimization, 2021
15. Lesage-Landry, A., Taylor, J.A., Shames, I.: Second-order online nonconvex optimization. IEEE Trans. Autom. Control **66**(10), 4866–4872 (2020)
16. Lupien, J.-L., Lesage-Landry, A.: An online newton's method for time-varying linear equality constraints. IEEE Control Syst. Lett. **7**, 1423–1428 (2023)
17. Dixit, R., Bedi, A.S., Tripathi, R., Rajawat, K.: Online learning with inexact proximal online gradient descent algorithms. IEEE Trans. Signal Process. **67**(5), 1338–1352 (2019)
18. Ajalloeian, A., Simonetto, A., Dall'Anese, E.: Inexact online proximal-gradient method for time-varying convex optimization. In: 2020 American Control Conference (ACC), pp. 2850–2857. IEEE, 2020
19. Bastianello, N., Dall'Anese, E.: Distributed and inexact proximal gradient method for online convex optimization. In: 2021 European Control Conference (ECC), pp. 2432–2437. IEEE, 2021

20. Zhao, Y., Qiu, S., Li, K., Luo, L., Yin, J., Liu, J.: Proximal online gradient is optimum for dynamic regret: a general lower bound. IEEE Trans. Neural Netw. Learn. Syst. **33**(12), 7755–7764 (2021)
21. Lee, J.D., Sun, Y., Saunders, M.A.: Proximal newton-type methods for minimizing composite functions. SIAM J. Optim. **24**(3), 1420–1443 (2014)
22. Tran-Dinh, Q., Kyrillidis, A., Cevher, V.: Composite self-concordant minimization. J. Mach. Learn. Res. **16**(1), 371–416 (2015)
23. Friedman, J., Hastie, T., Höfling, H., Tibshirani, R.: Pathwise coordinate optimization, 2007
24. Hsieh, C.J., Dhillon, I., Ravikumar, P., Sustik, M.: Sparse inverse covariance matrix estimation using quadratic approximation. Adv. Neural Inf. Process. Syst. **24** (2011)
25. Foster, D.J., Kale, S., Luo, H., Mohri, M., Sridharan, K.: Logistic regression: the importance of being improper. In: Conference On Learning Theory, pp. 167–208. PMLR, 2018
26. Derenick, J., Spletzer, J., Hsieh, A.: An optimal approach to collaborative target tracking with performance guarantees. J. Intell. Rob. Syst. **56**, 47–67 (2009)
27. Cao, X., Cheng, P., Chen, J., Sun, Y.: An online optimization approach for control and communication codesign in networked cyber-physical systems. IEEE Trans. Ind. Inf. **9**(1), 439–450 (2012)

A Computational Study on a Parabolic-Hyperbolic Chemotaxis System for Modeling Angiogenesis by a Stochastic Interacting Particle Field Method

Jongwon D. Kim(✉) and Jack Xin

University of California, Irvine, Irvine, CA 92697, USA
{jongwodk,jack.xin}@uci.edu

Abstract. Angiogenesis is a biological process that plays an important role in tumor growth and metastasis. Many mathematical models have been used to study cancer cell invasion through numerical computation of PDEs (partial differential equations). In this proceeding, we study a Parabolic-Hyperbolic Keller-Segel (PHKS) system in one and two space dimensions. The model arises in the angiogenesis literature. To compute solutions to the PHKS system, we develop a stochastic interacting particle-field (SIPF) method where the PDE solutions are approximated as empirical measures of particles coupled with a smoother field (concentration of chemo-attractant) variable approximated by the classical spline interpolation method. We describe an algorithm for updating the stochastic particle positions and chemical concentration (field). We present analytical form and numerical simulations of self-similar solutions, and discuss challenges to be addressed with machine learning approaches. The numerical experiments show diffusive spreading behavior of the system from a Gaussian shaped initial data and zero flux boundary conditions. Finally, we provide preliminary results of a neural interpolator based on a deep convolutional network and discuss its potential in SIPF for computing higher dimensional solutions to the system.

Keywords: Parabolic-hyperbolic chemotaxis · stochastic particle field method · pattern generation · self-similar solutions · neural interpolation

1 Introduction

It is widely known that cancer is currently the second leading cause of death globally [1] and has raised a variety of clinical concerns in the modern age. Although the disease has been documented since long ago since the ancient Egyptians, attention to treatments did not start until 1775. However, it wasn't until the 1970s that mathematical models were developed to study different phases of solid tumor growth [2]. The incorporation of empirical data by biologists gave rise to study cancer through growth laws such as Gompertzian, logistic, and

© The Author(s), under exclusive license to Springer Nature Switzerland AG 2025
H. A. Le Thi et al. (Eds.): ICAMCS 2024, LNNS 1311, pp. 64–75, 2025.
https://doi.org/10.1007/978-3-031-90606-0_6

exponential growth [3]. Naturally as the digital age arose, so did emphasis on computational research for cancer.

Mathematical models are powerful tools to study a wide range of physical and biological phenomena. The field of cancer modeling includes various approaches from mechanistic models that explore the detailed biochemical mechanisms of diseases to the data-driven models that facilitate clinical decision-making [4]. Angiogenesis is the biological process of the formation of new blood vessels and provides a way for tumors to metastasize. This phenomenon has been widely studied by both clinical and computational scientists [17]. Keller and Segel (KS) first introduced the chemotaxis system of partial differential equations (PDEs) to model the movement of bacteria to a food source (chemoattractant) [5]. Since then, many PDE models have been used to study tumor growth models and their associated biological processes.

Many numerical methods have been proposed for KS chemotaxis systems. While traditional numerical methods such as finite difference (FDM) or finite volume apply to fully parabolic systems, newer models have emerged in recent times to address higher dimensional problems. Models employing stochastic differential equations (SDE) have been formulated to capture the stochastic behaviors of cancer cell migration and invasion, specifically addressing the variability in diffusion processes [6,18]. See also [7–9] for a particle method and a reinforced random walk analysis in the context of fully parabolic chemotaxis. In [10,11], a random particle blob method is shown to converge for the parabolic-elliptic KS (PEKS) system. In [12], a deep-learning study of chemotaxis and aggregation in 3D laminar and chaotic flows is initiated based on an interacting particle method with a kernel regularization technique for PEKS systems in a fluid environment.

In this paper, in the spirit of [14,15] for fully parabolic chemotaxis and related haptotaxis systems, we introduce a stochastic interacting particle-field (SIPF) algorithm for a parabolic-hyperoblic KS (PHKS) system motivated by angiogenesis [13]. Our method takes into account the coupled stochastic particle and field dynamics, where the field is the chemoattractant concentration. In our SIPF algorithm, we approximate the density of active particles by a sum of delta functions centered at the particle positions. The non-smoothness of particle representation goes into the field due to hyperbolicity. An interpolation is necessary from particle representation to compute the gradient of the field that drives the particle evolution. Our method is mesh free, easy to implement, and able to capture the diffusive behavior from the system as shown by a FDM comparison and self-similar solutions.

The rest of the paper is organized as follows: In Sect. 2, we review the PHKS system of equations analyzed in [13]. In Sect. 3, we describe the main SIPF algorithm that utilizes the theoretical SDE formulation of particle density as well as the need of a numerical interpolator to evolve the gradient of concentration field and particle positions. In Sect. 4, we show numerical results of the algorithm with a Gaussian shaped initial condition for density. Lastly, in Sect. 5 we discuss initial ideas of incorporating neural interpolation for future research on PHKS dynamics in higher dimensions.

2 Parabolic-Hyperbolic Keller Segel (PHKS) System

The PHKS system is given below:

$$\rho_t = \nabla \cdot (\mu \nabla \rho - \chi \rho \nabla c) \tag{1}$$

$$c_t = -c\rho \tag{2}$$

where ρ is the density of the bacteria and c is the concentration of the chemoattractant. The bacteria diffuse with mobility μ and drift in the direction of ∇c with velocity $\chi \nabla c$, where χ is called chemo-sensitivity. Theoretical analysis of well-posedness of this system is in [13] with the existence of a family of self-similar solutions for the 2D case. We will show a self-similar solution for comparing with numerical solutions in 1D later in Sect. 4.

3 SIPF Algorithm for PHKS

In this section we present the algorithm to solve the PHKS system with a stochastic interacting particle-field method. We consider a finite spatial domain $\Omega = [0, L]^d$ for $d = 1, 2$ and assume Neumann boundary conditions for ρ and c. As a discrete time algorithm, we discretize a time domain $[0, T]$ partitioned by $\{t_n\}_{0:n_T}$ where $t_0 = 0$ and $t_{n_T} = T$. We approximate the density ρ by particles given by

$$\rho_t \approx \frac{M_0}{P} \sum_{j=1}^{P} \delta(x - X_t^p) \tag{3}$$

where $P \gg 1$ is the number of particles and M_0 is the conserved mass of the system. At $t_0 = 0$, we sample P particles from the initial condition ρ_0. To present the algorithm, we rewrite the particle approximation by $\rho_n = \frac{M_0}{P} \sum_{j=1}^{P} \delta(x - X_n^p)$. At a given time step, our algorithm consists of two sub-steps: updating c and ρ.

Updating Chemical Concentration c: Let $\delta t = t_{n+1} - t_n > 0$ be the time step. We update c by the explicit Euler scheme:

$$c_{n+1} = c_n - \delta t \, c_n \, \rho_n. \tag{4}$$

Updating Density ρ: update the particle positions $\{X_n^p\}_{p=1:P}$ using an Euler-Maruyama scheme of the SDE:

$$X_{n+1}^p = X_n^p + \chi \, \nabla_x c(X_n^p, t_n) \, \delta t + \sqrt{2\mu\delta t} \, N_n^p \tag{5}$$

where N_n^p's are i.i.d. standard normal distributions with respect to Brownian paths in the SDE formulation.

Computing the spatial gradient $\nabla_x c(x, t_n)$ proves to be difficult by finite difference as the gradient may not be available at certain locations since ρ_{n-1} is coarsely defined through a histogram of particle positions. To circumvent this, we apply a spline interpolation of c_n to obtain a differentiable function defined everywhere in the spatial domain, then take gradient on the spline function as an approximation. We formulate the algorithm for any dimension below:

Algorithm 1: SIPF

input : $\Omega, T, \chi, \mu, P, M_0, \delta t, \rho_0, c_0$.
output: ρ_T, c_T.
Initialize $X_0^1, ..., X_0^P$ on Ω based on initial data ρ_0.
for $n = 0, ..., n_T$ **do**
  Bin particles $X_n^1, ..., X_n^P$ and define ρ_n according to (3)
  $\rho_n \leftarrow$ histogram(ρ_n)
  if $n = 0$ **then**
    $c_n \leftarrow c_0$
  else
    $c_{n+1} \leftarrow c_n - \delta t\, c_n\, \rho_n$, over each bin
    $c(x, t_n) \leftarrow$ Interpolator(c_n)
    Compute $\nabla_x c(x, t_n)$
    Update X_n to X_{n+1} by (5) with $\nabla_x c(x, t_n)$.
  end
end

4 Numerical Experiments

In this section, we present numerical findings of the particle method in the 1D case. For simplicity, we take $\chi = \mu = 1$, the conserved mass $M_0 = 1$, and a Gaussian initial condition for c. The initial condition for density ρ_0 is the normalized Gaussian function with mean 5 centered at $x = 50$ with 100 spatial bins. The computational domain is $\Omega = [0, 100]$. We then compute the density ρ via Algorithm 1 until time $T = 20.0$ with $P = 50000$. A numerical interpolator is necessary to solve for the spatial gradient of c. We use SciPy's PChipInterpolator in the 1D case and RectBivariateSpline in the 2D case [20]. To the best of our knowledge, these packages are the only SciPy interpolators that provide a function handle with the option to compute partial derivatives. Other interpolators do not have this attribute, making it difficult to compute the gradient accurately. Having such an object is crucial to maintain smoothness of the field variable. A comparison of the FDM and SIPF plots can be found in Fig. 1.

4.1 Self-similar Solutions

We present computation of self-similar solutions to the 1D system with similar spreading bahavior as the solutions of the initial boundary value problem from Gaussian shaped initial data (in Fig. 1). In 1D, the system (1)–(2) becomes:

$$\rho_t = (\rho_x - \rho c_x)_x \tag{6}$$

$$c_t = -c\rho \tag{7}$$

To derive self-similar solutions we assume that the self-similar variable is of the form $z = \frac{x}{t^\gamma}$. We seek self-similar solutions of the form

$$\rho(t,x) = t^{-\alpha}\bar{\rho}(z),\ c(t,x) = t^{-\beta}\bar{c}(z).$$

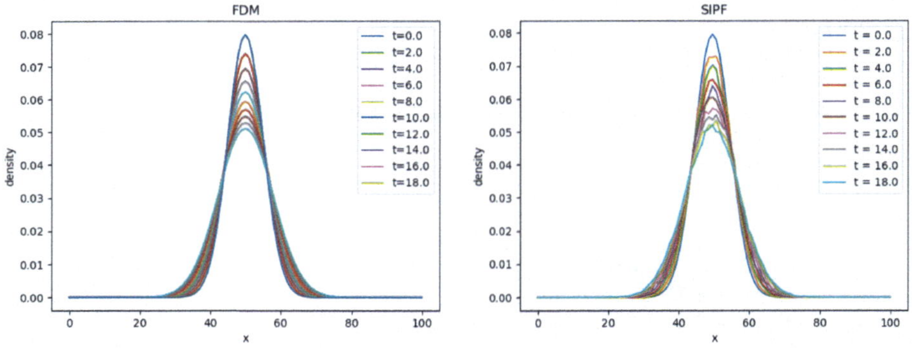

Fig. 1. FDM vs. SIPF at $T = 20.0, \delta t = 0.01, P = 50000$.

Taking the time derivative of ρ for the LHS (left hand side), we have

$$\rho_t = \frac{\partial}{\partial t}(t^{-\alpha}\bar{\rho}(z)) = -\alpha t^{-(\alpha+1)}\bar{\rho}(z) + t^{-\alpha}\bar{\rho}'(z)\frac{\partial z}{\partial t} \tag{8}$$

$$= -\alpha t^{-(\alpha+1)}\bar{\rho}(z) - \gamma z t^{-(\alpha+1)}\bar{\rho}'(z) \tag{9}$$

Likewise, $c_t = -\beta t^{-(\beta+1)}\bar{c}(z) + \gamma z t^{-(\beta+1)}\bar{c}'(z)$. Then computing the expression on the RHS (right hand side), we have

$$(\rho_x - \rho c_x)_x = t^{-(\alpha+2\gamma)}\bar{n}''(z) - t^{-(\alpha+\beta+2\gamma)}\bar{\rho}'(z)\bar{c}'(z) - t^{-(\alpha+\beta+2\gamma)}\bar{c}''(z)\bar{\rho}(z) \tag{10}$$

From the c_t equation we have

$$c_t = -\beta t^{-(\beta+1)}\bar{c}(z) - \gamma z t^{-(\beta+1)}\bar{c}'(z) \tag{11}$$

$$= -t^{-(\alpha+\beta)}\bar{\rho}(z)\bar{c}(z) \tag{12}$$

Balancing powers of t in this equation yield that $\alpha = 1$. Plugging this into the ρ_t equation yield that $\beta = 0, \gamma = 1/2$ so the self-similar solution in 1D is of the form $\rho(t,x) = \frac{1}{t}\bar{\rho}(\frac{x}{\sqrt{t}})$ and $c(t,x) = \bar{c}(\frac{x}{\sqrt{t}})$. Then rewriting the original system in terms of the self-similar variable z, we obtain the system of ODEs:

$$-\bar{\rho}(z) - \frac{z}{2}\bar{\rho}'(z) = \bar{\rho}''(z) - \bar{\rho}'(z)\bar{c}'(z) - \bar{c}''(z)\bar{\rho}(z) \tag{13}$$

$$\frac{z}{2}\bar{c}'(z) = \bar{\rho}(z)\bar{c}(z) \tag{14}$$

which simplifies to

$$\bar{\rho}''(z) - (\bar{c}'(z) - \frac{z}{2})\bar{\rho}'(z) - (\bar{c}''(z) - 1)\bar{\rho}(z) = 0 \tag{15}$$

$$\frac{z}{2}\bar{c}'(z) = \bar{\rho}(z)\bar{c}(z) \tag{16}$$

From Eq. (16), we can solve for $\bar{c}(z)$ to be

$$\bar{c}(z) = e^{\int 2\frac{\bar{\rho}(z)}{z}dz} \tag{17}$$

The self-similar solutions can be computed by numerically solving an ODE system derived from self-similar variable, see Fig. 2 showing diffusive behavior similar to Fig. 1.

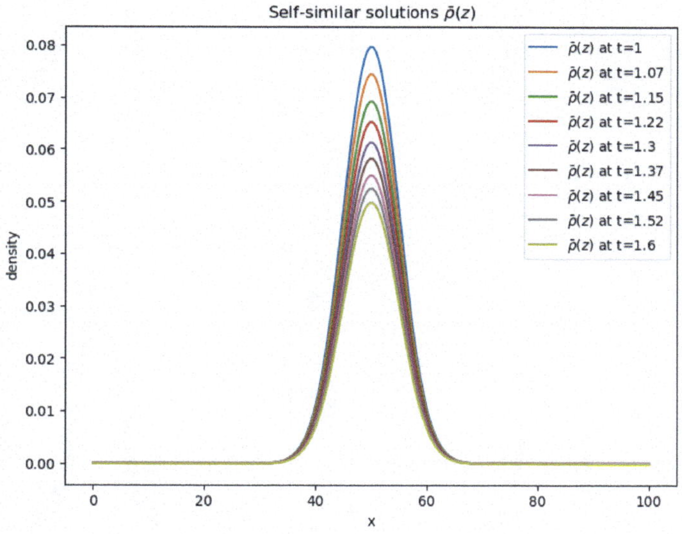

Fig. 2. Self-similar solutions for the 1D PHKS system at certain time-steps.

4.2 Discussion

In this subsection, we discuss the above results as well as some challenges that arise with the SIPF algorithm when the dimension increases. In the 2D case, the same initial data for density centered at (50, 50) and 100^2 spatial bins. A comparison of time steps between FDM and SIPF solutions in the 2D case can be found in Fig. 4. First, the SIPF result presents many fluctuations as a result of the stochastic nature of the method. We can resolve these fluctuations by increasing the number of particles. A plot showcasing the smoothing of the solutions with increased number of particles can be found in Fig. 3. This also true in the 2D case as shown in Fig. 6.

To demonstrate the efficacy of SIPF in 2D, we calculate the mean square error (MSE) when comparing to the FDM solution. Plots for different number of particles are shown in Fig. 5. While the error fluctuates across time, increasing the number of particles decreases the magnitude of the loss. However, because the algorithm recursively interpolates the field variable (concentration), this becomes

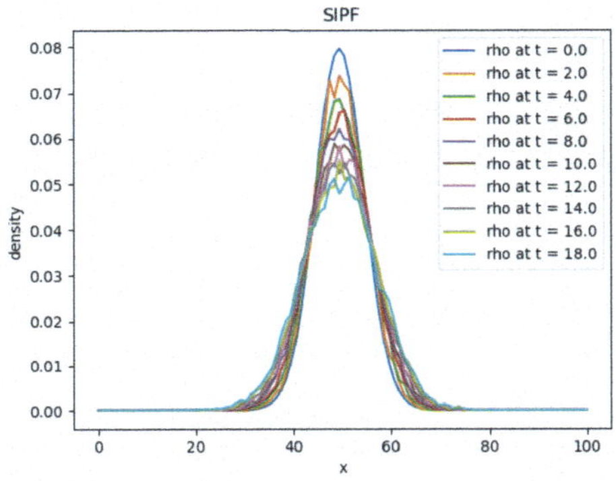

(a) P = 20000

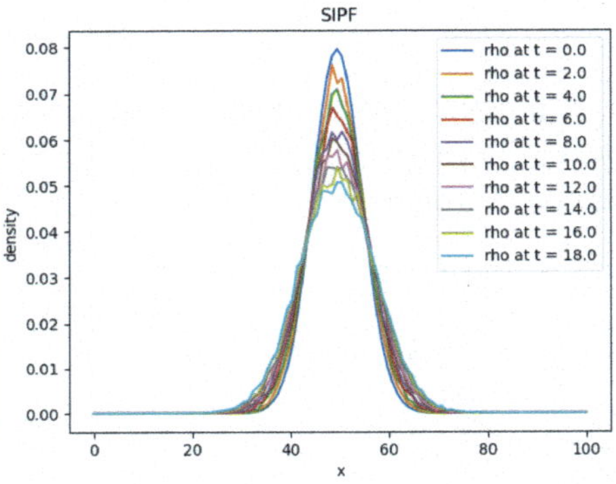

(b) P = 30000

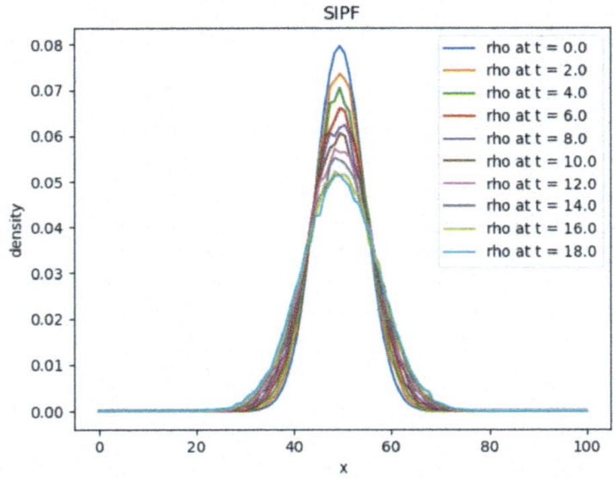

(c) P = 40000

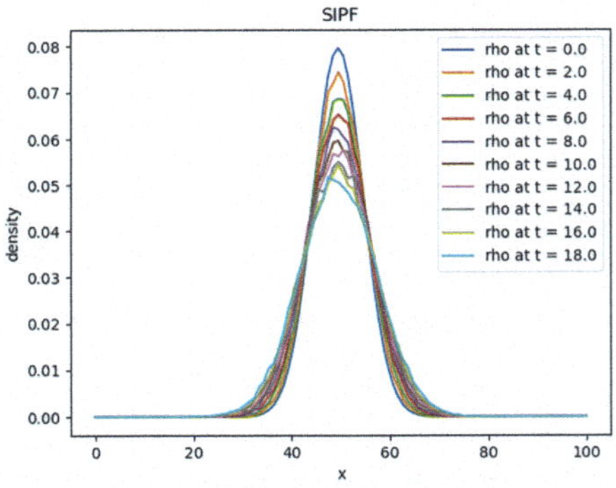

(d) P = 50000

Fig. 3. SIPF with different number of particles at $T = 20.0, \delta t = 0.01$.

computationally costly with classical interpolation methods. The FDM simulation takes around 60.62 s to complete while for small N, SIPF performs faster, taking around 24.70 s. The time it takes to complete the same simulation compounds as N grows large. For example, for $N = 10000$, the algorithm takes around 214.35 s to complete. While increasing the particles may smooth the numerical solution, the interpolation step slows down the algorithm significantly.

5 Neural Interpolation

In this section, we aim to address the interpolation issue. To do so, we consider neural interpolators. These interpolators arose in the context of image generation such as in [16] and [17]. The former explores a spatial interpolator using neural networks for geospatial data where data maybe sparse whereas the latter uses a combination of a convolutional neural network (CNN) and spatial transformer network (STN) to interpolate images. However, when using FunkNN [17],

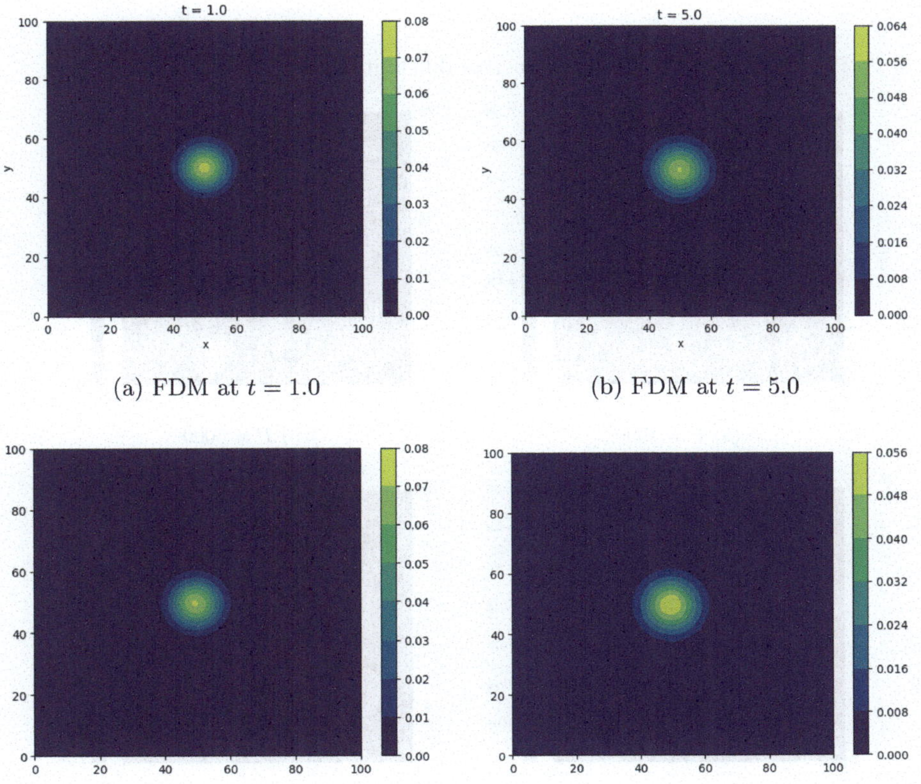

(a) FDM at $t = 1.0$ (b) FDM at $t = 5.0$

(c) SIPF at $t = 1.0$ (d) SIPF at $t = 5.0$

Fig. 4. Comparison of time steps of FDM and SIPF for the 2D PHKS system, $T = 5.0, \delta t = 0.1, P = 20000$.

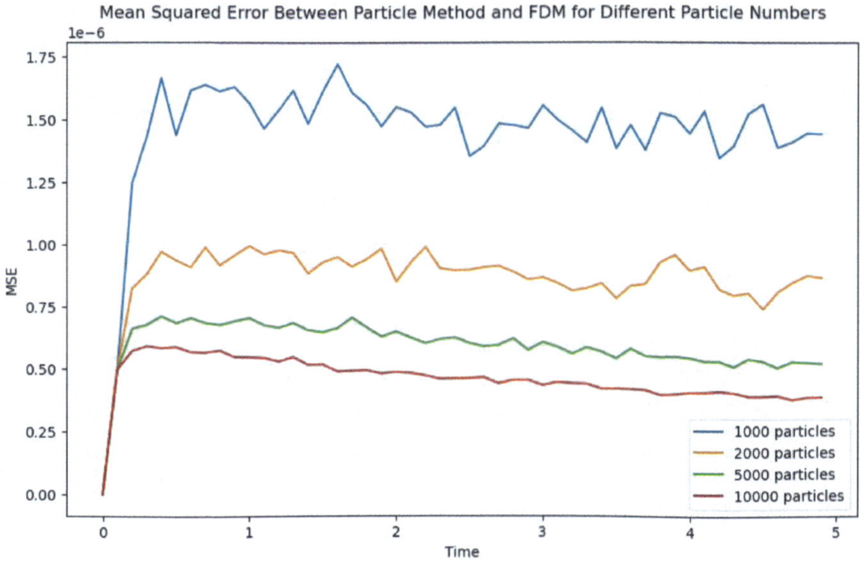

Fig. 5. MSE over time across different number of particles.

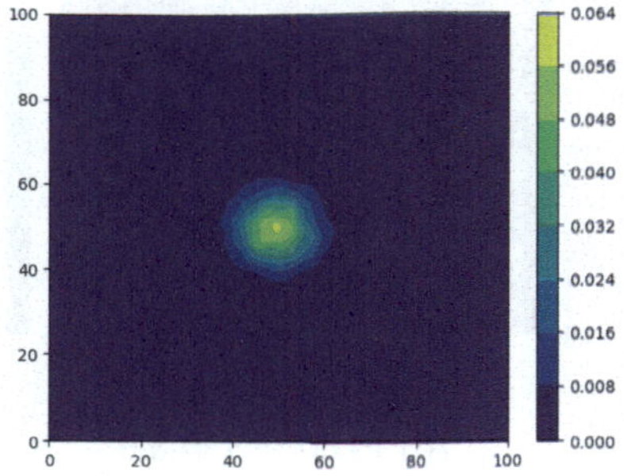

(a) $P = 1000$

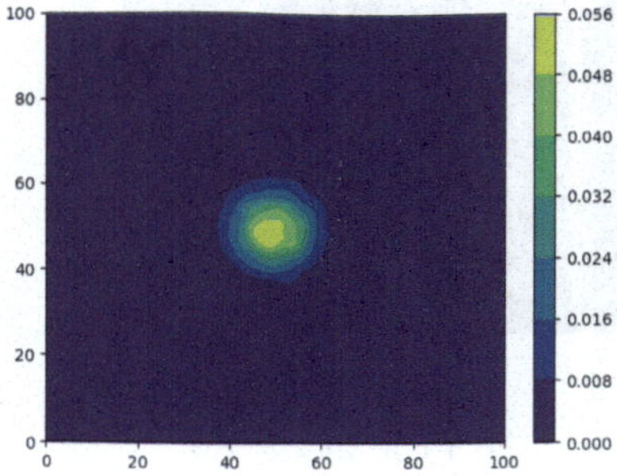

(b) $P = 5000$

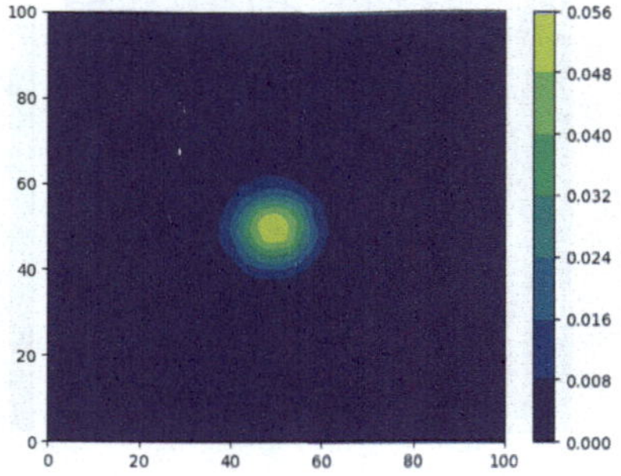

(c) $P = 10000$

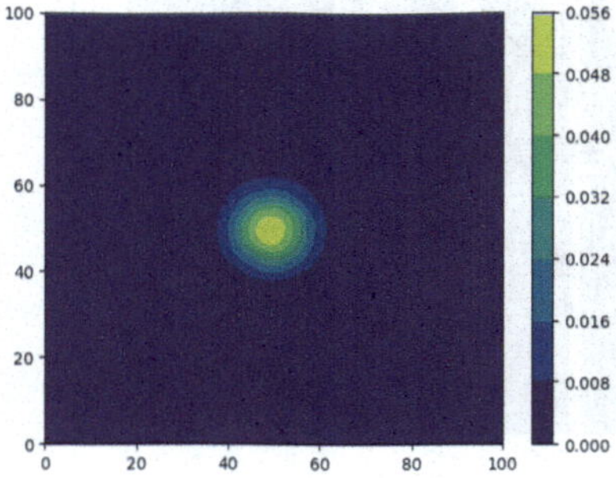

(d) $P = 20000$

Fig. 6. Comparison of last time steps of SIPF for the 2D PHKS system with different number of particles, $T = 5.0$, $\delta t = 0.1$.

a generative model is necessary to provide the network with training data. Such a generative model creates target data for the FunkNN module to interpolate. Successfully implementing an interpolator like FunkNN [17] remains to be explored when looking to simulate the PHKS system in higher dimensions.

As a proof of concept, we use high resolution FDM solution as target in lieu of a generative model as required by FunkNN. We interpolate the concentration field by a CNN architecture consisting of seven convolution layers. The mean squares error (MSE) between the neural interpolation and FDM solution is minimized at each epoch with Adam optimizer updating the CNN weights for 100 epochs. Figure 7 compares FDM, SIPF with classical and neural interpolations. The training performance (decay of MSE) of the CNN is in Fig. 8. In future work, we plan to train this CNN on synthetic data or self-similar solutions and apply it to SIPF and PHKS system.

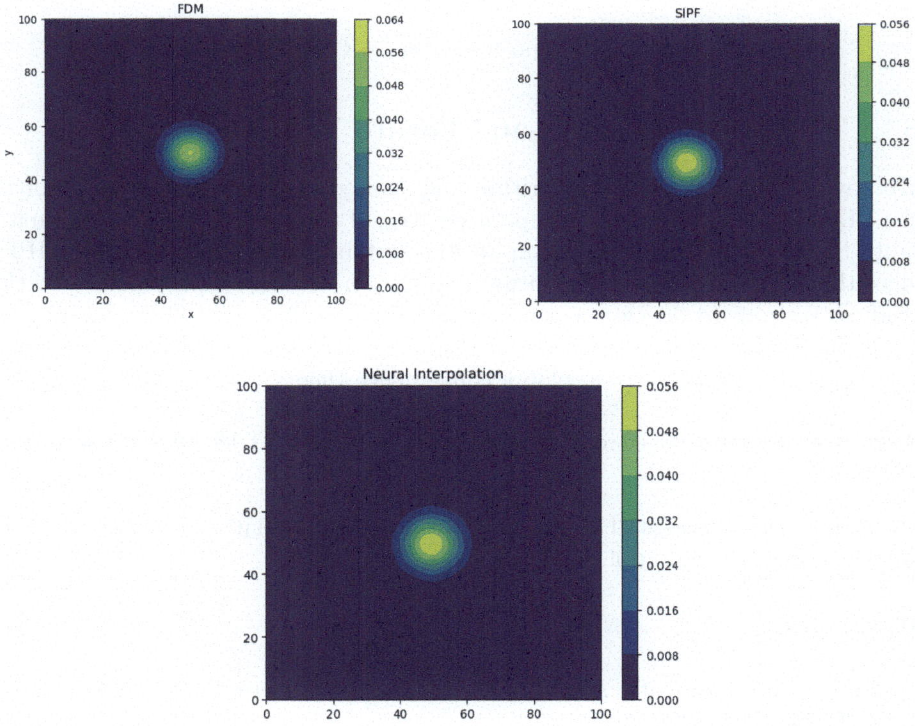

Fig. 7. Solution at last time step by FDM, SIPF w. classical/neural interpolations.

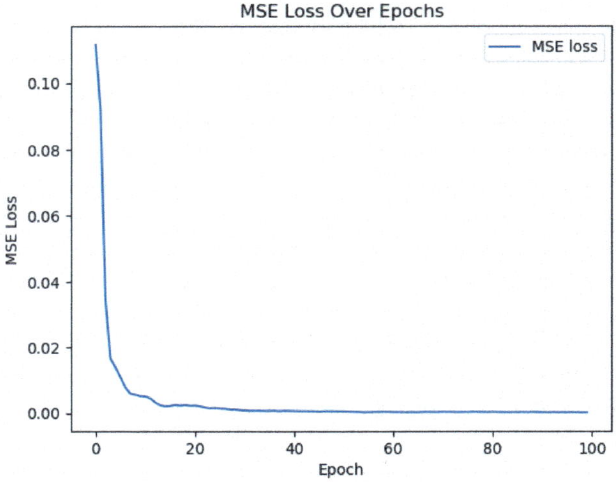

Fig. 8. Loss vs. epoch numbers in CNN training.

6 Concluding Remarks and Future Work

We have introduced a SIPF algorithm and showed preliminary results to solve the PHKS system. Given the comparative ease of implementation and its capability to capture the diffusive nature of the system, we conclude that the SIPF algorithm has the potential for future applications in angiogenic tumor growth studies. We aim to develop a neural interpolator in 2D and 3D by training a CNN on synthetic image data and self-similar solutions as both are low cost. Finally, a future goal is to develop a convergence theory for the SIPF method.

Acknowledgments. This research was partially supported by NSF grant DMS-2309520.

Disclosure of Interests. The authors have no competing interests to declare that are relevant to the content of this article.

References

1. Siegel, R.L., Giaquinto, A.N., Jemal, A.: Cancer statistics, 2024. CA Cancer J. Clin. **74**(1), 12–49 (2024)
2. Chaplain, M.: Avascular growth, angiogenesis and vascular growth in solid tumours: the mathematical modelling of the stages of tumour development. Math. Comput. Model. **23**(6), 47–87 (1996)
3. Retsky, M., Swartzendruber, D., Wardwell, R.H., Bame, P.D.: Is Gompertzian or exponential kinetics a valid description of individual human cancer growth? Med. Hypotheses **33**(2), 95–106 (1990)
4. Bekisz, S., Geris, L.: Cancer modeling: from mechanistic to data-driven approaches, and from fundamental insights to clinical applications. J. Comput. Sci. **46**, 101198 (2020)

5. Keller, E., Segel, L.: Initiation of slime mold aggregation viewed as an instability. J. Theor. Biol. **26**(3), 399–415 (1970)
6. Katsaounis, D., Chaplain, M., Sfakianakis, N.: Stochastic differential eqn modelling of cancer cell migration and tissue invasion. J. Math. Bio. **87**(1), 8 (2023)
7. Stevens, A.: The derivation of chemotaxis equations as limit dynamics of moderately interacting stochastic many-particle systems. SIAM J. Appl. Math. **61**(1), 183–212 (2000)
8. Stevens, A.: A stochastic cellular automaton modeling gliding and aggregation of myxobacteria. SIAM J. Appl. Math. **61**(1), 172–182 (2000)
9. Stevens, A., Othmer, H.: Aggregation, blowup, and collapse: the ABC's of taxis in reinforced random walks. SIAM J. Appl. Math. **57**(4), 1044–1081 (1997)
10. Liu, J.-G., Yang, R.: Propagation of chaos for the Keller-Segel equation with a logarithmic cut-off. Methods Appl. Anal. **26**(4), 319–348 (2019)
11. Liu, J.-G., Yang, R.: A random particle blob method for the Keller-Segel equation and convergence analysis. Math. Comput. **86**(304), 725–745 (2017)
12. Wang, Z., Xin, J., Zhang, Z.: A DeepParticle method for learning and generating aggregation patterns in multi-dimensional Keller-Segel chemotaxis systems. Nonlinear Phenomena Phys. D 134082 (2024)
13. Corrias, L., Perthame, B., Zaag, H.: A chemotaxis model motivated by angiogenesis. C. R. Acad. Sci. Paris Ser. **I**(336), 141–146 (2003)
14. Wang, Z., Xin, J., Zhang, Z.: A Novel Stochastic Interacting Particle-Field Algorithm for 3D Parabolic-Parabolic Keller-Segel Chemotaxis System, arXiv:2309.13554 (2023)
15. Hu, B., Wang, Z., Xin, J., Zhang, Z.: A Stochastic Interacting Particle-Field Algorithm for a Haptotaxis Advection-Diffusion System Modeling Cancer Cell Invasion, arXiv:2407.05626v1 (2024)
16. Zhan, J., et al.: A generalized spatial autoregressive neural network method for three-dimensional spatial interpolation. Geosci. Model Dev. **16**, 2777–2794 (2023)
17. Khorashadizadeh, A., Chaman, A., Debarnot, V., Dokmanió, I.: FunkNN: Neural Interpolation for Functional Generation, ICLR 2023
18. Hormuth, D., et al.: Biologically-based mathematical modeling of tumor vasculature and angiogenesis via time-resolved imaging data. Cancers (Basel) **13**(12), 3008 (2021). 16. https://doi.org/10.3390/cancers13123008
19. Capasso, V., Wieczorek, R.: A hybrid stochastic model of retinal angiogenesis. Math. Meth. Appl. Sci. **43**, 10578–10592 (2020). https://doi.org/10.1002/mma.6725
20. Virtanen, P., et al.: SciPy 1.0: fundamental algorithms for scientific computing in python. Nat. Methods **17**, 261–272 (2020)

Rethinking Linear Programming: A New Finite-Termination LP-Newton Approach

Yuki Matsuno[1(✉)] and Jianming Shi[2]

[1] Graduate School of Management, Tokyo University of Science, 1-11-2 Fujimi, Chiyoda-ku, 102-0071 Tokyo, Japan
8623702@ed.tus.ac.jp, yuhki908@gmail.com

[2] Department of Business Economics, Tokyo University of Science, 1-11-2 Fujimi, Chiyoda-ku, 102-0071 Tokyo, Japan
shi@rs.tus.ac.jp, jianmingshi.dr@gmail.com

Abstract. Fujishige et al. (2009) [2] proposed an LP-Newton type algorithm specifically designed to solve linear programming problems (LPs) by repeatedly applying Wolfe's algorithm. This approach involves iterative projections onto a particular case of a convex hull, represented as a zonotope, leveraging Wolfe's algorithm to systematically resolve the LP.

The computational complexity of the LP-Newton type algorithm remains unknown, while Wolfe's algorithm is known to have exponential complexityqueryPlease check and confirm if the authors Given and Family names have been correctly identified.. In this paper, we introduce an alternative algorithm that solves the same class of LPs without relying on Wolfe's algorithm. Our algorithm shares a key characteristic with the LP-Newton method, ensuring that it terminates after a finite number of iterations.

Keywords: LP-Newton method · Linear programming problems · minimum norm point

1 Introduction

Linear programming (LP) problems are fundamental in mathematical optimization and are widely applicable in various fields such as operations research, economics, engineering, and logistics. These problems involve optimizing a linear objective function subject to a set of linear equality and inequality constraints. Despite their simple formulation, solving LP problems efficiently is critical due to their widespread applications in practical scenarios.

The quest for a strongly polynomial-time algorithm for LP remains one of the major open problems in mathematical programming and theoretical computer science. While the simplex method and interior-point methods are widely used in practice, the former does not guarantee polynomial-time performance in the worst case, and the latter only guarantees weak polynomial-time complexity. This paper hopefully encourages researchers to explore more alternative algorithms.

© The Author(s), under exclusive license to Springer Nature Switzerland AG 2025
H. A. Le Thi et al. (Eds.): ICAMCS 2024, LNNS 1311, pp. 76–82, 2025.
https://doi.org/10.1007/978-3-031-90606-0_7

Fujishige et al. [2] proposed an LP-Newton type algorithm as a significant effort to address this challenge, inspired researchers in this field, providing an important contribution to the ongoing research. Their approach involves the use of Wolfe's algorithm [7] to iteratively project onto a zonotope, which is a special type of convex polytope. The method is innovative in its use of geometric properties of zonotopes to find the optimal solution to LP problems. However, the computational complexity of the approach remains unresolved, and Wolfe's algorithm itself is known to have exponential complexity [4]. There are other related studies [1,3,6] that also find the minimum norm point. It may be possible to use them as alternatives in this kind of research.

In this paper, we propose an alternative algorithm for solving LPs with box constraints on the variables. Unlike the LP-Newton method, our approach does not rely on Wolfe's algorithm, thereby potentially offering a more efficient solution strategy. Our algorithm retains the key feature of finite termination, ensuring that it converges to the optimal solution after a finite number of iterations. We also discuss the theoretical underpinnings of our approach and compare it with existing methods to highlight its advantages.

The main contributions of this research are as follows:

- We propose an alternative algorithm for solving linear programming problems with box constraints, which does not rely on Wolfe's algorithm.
- Our algorithm retains the desirable property of finite termination, ensuring that it reaches the optimal solution after a finite number of iterations
- Numerical experiments are supposed to be easily conducted to validate the performance of the proposed algorithm, highlighting its practical applicability in solving large-scale LP problems in some cases. We report the results of the preliminary experiments on the conference due to space limitations.

The remainder of this paper is organized as follows: In Sect. 2, we review Wolfe's algorithm and the LP-Newton method. In Sect. 3, we propose a new algorithm and provide a proof of its validity. Finally, in Sect. 4, we conclude the paper and discuss potential directions for future work.

2 LP-Newton Method and Wolfe's Algorithm

In this research we consider the following problem which was formed in Fujishige et al. [2]

$$\begin{aligned} \text{maximize} \quad & \boldsymbol{c}^T \boldsymbol{x} \\ \text{subject to} \quad & A\boldsymbol{x} = \boldsymbol{b} \\ & \boldsymbol{l} \leq \boldsymbol{x} \leq \boldsymbol{u} \end{aligned} \tag{2.1}$$

where $A \in \mathbb{R}^{m \times n}, \boldsymbol{b} \in \mathbb{R}^m, \boldsymbol{c} \in \mathbb{R}^n, \boldsymbol{l} \in \mathbb{R}^n, \boldsymbol{u} \in \mathbb{R}^n$ and $\boldsymbol{x} \in \mathbb{R}^n$. Let

$$\bar{A} = \begin{pmatrix} A \\ \boldsymbol{c}^\top \end{pmatrix} \in \mathbb{R}^{(m+1) \times n} \tag{2.2}$$

The box constraint is technically required. In practice, $\boldsymbol{l}$ can be $\mathbf{0}$, and $\boldsymbol{u}$ can be sufficiently large. An algorithm for solving the problem without the box constraint can be found in [5].

A zonotope $\bar{Z}$ is defined as follows

$$\bar{Z} = \left\{\bar{\boldsymbol{z}} \in \mathbb{R}^{m+1} \mid \bar{\boldsymbol{z}} = \bar{A}\boldsymbol{x}, \boldsymbol{l} \leq \boldsymbol{x} \leq \boldsymbol{u}\right\} \tag{2.3}$$

Therefore, the problem (2.1) turns to the follows

$$\text{maximize } \gamma \tag{2.4}$$

$$\text{subject to } \begin{pmatrix} \boldsymbol{b} \\ \gamma \end{pmatrix} \in \bar{Z} \tag{2.5}$$

where $\gamma \in \mathbb{R}$. Let

$$L = \left\{ \begin{pmatrix} \boldsymbol{b} \\ \gamma \end{pmatrix} \middle| \; \forall \gamma \in \mathbb{R} \right\},$$

which is a straight line parallel to the $(m+1)$th axis in $\mathbb{R}^{m+1}$.

Based on above formulation, solving linear programming problem (2.1) is equivalent to finding a point on the intersection of $\bar{Z}$ and L where the value γ^* of $(m+1)$th element of the point is maximal.

The LP-Newton method generates a sequence of points $\{\bar{\boldsymbol{b}}_k\}$ in L. Taking

$$\hat{x}_j = \begin{cases} u_j, & \text{if } c_j > 0 \\ l_j, & \text{otherwise} \end{cases} \tag{2.6}$$

and let $\hat{\boldsymbol{x}} = (\hat{x}_1, \ldots, \hat{x}_n)$, then $\gamma_0 = \boldsymbol{c}^\top \hat{\boldsymbol{x}}$ is an upper bound of the value of $(m+1)$th element of a point on the zonotope $\bar{Z}$.

The proposed algorithm begins with the initial point $\bar{\boldsymbol{b}}_0 = (\boldsymbol{b}; \gamma_0)$, which is equivalent to $(\boldsymbol{b}^\top, \gamma_0)^\top$ but formatted in a MATLAB-like command style. Suppose that $\bar{\boldsymbol{b}}_k$ is given at the kth iteration. LP-Newton method finds $\bar{\boldsymbol{z}}_{k+1} = \arg\min\{\|\boldsymbol{y} - \bar{\boldsymbol{b}}_k\| \mid \boldsymbol{y} \in \bar{Z}\}$ by using Wolfe's algorithm [7]. Note that when the Wolfe's algorithm terminates, we obtain a set J_k for several $\hat{\boldsymbol{x}}$ such that

$$\bar{\boldsymbol{z}}_{k+1} = \sum_{\hat{\boldsymbol{x}}_i \in J_k} \lambda_i \left(\bar{A}\hat{\boldsymbol{x}}_i\right), \tag{2.7}$$

$$\sum_{\text{same } i \text{ as } \hat{\boldsymbol{x}}_i \in J_k} \lambda_i = 1, \lambda_i > 0. \tag{2.8}$$

Then $\bar{\boldsymbol{b}}_{k+1}$ is updated by finding the intersection of following hyperplane and L

$$\bar{\boldsymbol{b}}_{k+1} = H_k(\bar{\boldsymbol{z}}_{k+1}) \cap L, \tag{2.9}$$

where $H_k(\bar{\boldsymbol{z}}_{k+1}) = \left\{\boldsymbol{x} \mid \bar{\boldsymbol{z}}_{k+1}^\top \boldsymbol{x} = \bar{\boldsymbol{z}}_{k+1}^\top \bar{\boldsymbol{z}}_{k+1}\right\}$.

2.1 LP-Newton Algorithm

The LP-Newton Algorithm can be expressed as follows

Algorithm 1 LP-Newton Algorithm

Require: $A \in \mathbb{R}^{m\times n}, \boldsymbol{b} \in \mathbb{R}^m, \boldsymbol{c} \in \mathbb{R}^n, \boldsymbol{l} \in \mathbb{R}^n, \boldsymbol{u} \in \mathbb{R}^n$.
Ensure: Output an optimal solution $\boldsymbol{x}^*$ to problem (2.1) , or declare `infeasible`.
Step 1: Find $\hat{\boldsymbol{x}}_0$ in (2.6). Set J_0 using (2.6). $\gamma_0 = \boldsymbol{c}^\top \hat{\boldsymbol{x}}_0, k = 0$.
Step 2: Let $\bar{\boldsymbol{b}}_k = (\boldsymbol{b}, \gamma_k)^\top$. Find $\bar{\boldsymbol{z}}_{k+1} = \arg\min\{\|\boldsymbol{y} - \bar{\boldsymbol{b}}_k\| \mid \boldsymbol{y} \in \bar{Z}\}$. Find $\bar{\boldsymbol{b}}_{k+1}$ by (2.9).
Write $\bar{\boldsymbol{z}}_{k+1} = (\boldsymbol{z}_{k+1}, \xi_{k+1})^\top$. Update J_k using (2.7).
Step 3: **If** $\bar{\boldsymbol{z}}_{k+1} = \bar{\boldsymbol{b}}_k$ **then** Output $\boldsymbol{x}^* = \sum_{\hat{\boldsymbol{x}}_i \in J_k} \lambda_i \hat{\boldsymbol{x}}_i$ and Stop and declare
`"An optimal solution is` $\boldsymbol{x}^*$`"`;
elseif $\xi_{k+1} > \gamma_k$, then Stop and declare `"The problem is infeasible"`.
Step 4: Let $\gamma_{k+1} := \xi_{k+1} - \dfrac{\|\boldsymbol{b} - \boldsymbol{z}_{k+1}\|^2}{\gamma_k - \xi_{k+1}}$, $k := k + 1$. Goto Step 2.

Fujishige et al. [2] shows that the LP-Newton algorithm terminates after a finite number of iterations.

2.2 Wolfe's Algorithm

Wolfe's algorithm plays a pivot rule in finding $\bar{z}_{k+1}$ in Step 2 in LP-Newton algorithm. Our algorithm shares several similarities with Wolfe's algorithm. To be self-contained we describe Wolfe's Algorithm below.

Algorithm 2 Wolfe's Algorithm

Require: A finite vertices set $P = \{\boldsymbol{p}_i \mid i \in I\} \subset \mathbb{R}^n$.
Ensure: $\boldsymbol{z}^* = \arg\min\{\|\boldsymbol{z}\| \mid \boldsymbol{z} \in \text{conv}(P)\}$.
Step 1: Pick a point $\boldsymbol{p} \in P$ and set $S := \{\boldsymbol{p}\}$, $\boldsymbol{z} := \boldsymbol{p}$.
Step 2: Find the point $\hat{\boldsymbol{p}} \in \arg\min\{\boldsymbol{z}^\top \boldsymbol{p}_i \mid \boldsymbol{p}_i \in P\}$.
Step 3: If $\boldsymbol{z}^\top \hat{\boldsymbol{p}} \geq \boldsymbol{z}^\top \boldsymbol{z}$. Then, Set $\boldsymbol{z}^* := \boldsymbol{z}$. Stop. Otherwise, Let $S := S \cup \{\hat{\boldsymbol{p}}\}$.
Step 4: Find $\boldsymbol{y} = \arg\min\{\|\boldsymbol{y}\| \mid \boldsymbol{y} \in \text{affine.hull}(S)\}$.
Step 5: If $\boldsymbol{y} \in \text{relat.int}(\text{conv.hull}(S))$ then, $\boldsymbol{z} := \boldsymbol{y}$, then Goto Step 2.
Step 6: Otherwise, find $\boldsymbol{w} = \arg\min\{\|\boldsymbol{w}' - \boldsymbol{y}\| \mid \boldsymbol{w}' \in [\boldsymbol{y}, \boldsymbol{z}] \cap \text{conv.hull}(S)\}$.
Step 7: Set $S' \subset S$ with $\boldsymbol{w} \in \text{relat.int}(\text{conv.hull}(S'))$. Set $S := S', \boldsymbol{z} := \boldsymbol{w}$.
Goto Step 4.

Wolfe's algorithm above terminates after finitely many iterations. To see this, we notice that $\boldsymbol{y}$ is updated at Step 4. Its norm is strictly decreasing Therefore, the same S never appears twice. In Step 7, the size of S is decreased by at least one, and the iterations are finite.

When Wolfe's algorithm is applied to LP-Newton method, the set P at Step 2 is not necessary to be given at beginning. To see this, we consider finding $\hat{\boldsymbol{p}}$

with $\bar{Z} = P$. At Step 3 $\min\{\boldsymbol{z}^\top \hat{\boldsymbol{p}}_i \mid \boldsymbol{p}_i \in \bar{Z}\} = \boldsymbol{z}^\top \bar{A}\bar{\boldsymbol{x}}$, where

$$\bar{x}_j = \begin{cases} u_j, \text{ if } (\boldsymbol{z}^\top \bar{A})_j < 0 \\ l_j, \text{ otherwise} \end{cases} \tag{2.10}$$

where $(\cdot)_j$ is the jth element of a vector. Note that the vector $\hat{\boldsymbol{p}}$ in Step 2 is obtained based on the sign of $\boldsymbol{z}^\top \bar{A}$ only when a zonotope $\bar{Z}$ is applied in LP-Newton method.

3 An Alternative New Algorithm

In this paper we assume that a feasible solution of (2.1) is given. This assumption implies that $L \cap \bar{Z} \neq \emptyset$, that is, there exists at least one point $\bar{\boldsymbol{b}}_0 \in L \cap \bar{Z}$. Similar with (2.7), we suppose that there exists an initial set S_0 such that

$$\bar{\boldsymbol{b}}_0 = \sum_{\hat{\boldsymbol{z}}_i \in S_0} w_i \hat{\boldsymbol{z}}_i, \tag{3.1}$$

$$\sum_{\text{same } i \text{ as } \hat{\boldsymbol{z}}_i \in S_0} w_i = 1, w_i > 0 \tag{3.2}$$

Our algorithm begins with this $\bar{\boldsymbol{b}}_0$ and generates a sequence $\{\bar{\boldsymbol{b}}_k\}$ on $L \cap \bar{Z}$ that converges to the optimal value, eventually lead us to obtain an optimal solution.

Our algorithm is formally proposed as follows.

Algorithm 3 A Wolfe's-Like Algorithm for Linear Programming

Require: $A \in \mathbb{R}^{m \times n}, \boldsymbol{b} \in \mathbb{R}^m, \boldsymbol{c} \in \mathbb{R}^n, \boldsymbol{l} \in \mathbb{R}^n, \boldsymbol{u} \in \mathbb{R}^n$.
Ensure: Output: An optimal solution $\boldsymbol{x}^*$ and maximal value γ^*.
Step 1: Set $\bar{\boldsymbol{b}}_0 \in L \cap \bar{Z}$ using (3.1). Let $(\boldsymbol{b}; \gamma_0)$ be the origin in $\mathbb{R}^{n+1}$. Set $k = 0$.
Step 2: Find $\hat{\boldsymbol{v}}_k = \arg\min\{\|\boldsymbol{z}\| \mid \boldsymbol{z} \in \text{affine.hull}(S_k)\}$.
Step 3: Find $\hat{\boldsymbol{z}}_k \in \arg\min\{\hat{\boldsymbol{v}}_k^\top \bar{\boldsymbol{z}}_j \mid \bar{\boldsymbol{z}}_j \in \bar{Z}\}$.
Step 4: `If` $\hat{\boldsymbol{z}}_k^\top \hat{\boldsymbol{v}}_k \geq \hat{\boldsymbol{v}}_k^\top \hat{\boldsymbol{v}}_k$, **then** let $\boldsymbol{x}^* = \sum_{\hat{\boldsymbol{x}}_i \in S_k} w_i \hat{\boldsymbol{x}}_i, \gamma^* = (\hat{\boldsymbol{b}}_k)_{m+1}$. Stop.
Step 5: `Else` let $\boldsymbol{y} = \arg\min\{\|\boldsymbol{z}\| \mid \boldsymbol{z} \in (\text{affine.hull}(S_k \cup \{\hat{\boldsymbol{z}}_k\})) \cap L\}$.
`If` $\boldsymbol{y} = \bar{\boldsymbol{b}}_k$. `Goto` Step 6.
`Else` $\boldsymbol{y} \neq \bar{\boldsymbol{b}}_k$ and $S_k := S_k \cup \{\hat{\boldsymbol{z}}_k\}$ `Goto` Step 7.
Step 6: A scheme to address the degeneracy $(\boldsymbol{y} = \bar{\boldsymbol{b}}_k)$.
Then set $k = k + 1$ and `Goto` Step 2.
Step 7: Let $\boldsymbol{w} = \arg\min\{\|\boldsymbol{w}' - \boldsymbol{y}\| \mid \boldsymbol{w}' \in (\text{conve.hull}(S_k)) \cap [\boldsymbol{y}, \bar{\boldsymbol{b}}_k]\}$.
Step 8: Set S'_k such that $\boldsymbol{w} \in \text{relat.int}(\text{conve.hull}(S'_k))$, $\bar{\boldsymbol{b}}_k = \boldsymbol{w}$ $S_k := S'_k$.
$k = k + 1$. `Goto` Step 2

Under the assumption that the scheme in Step 6 works well, we can prove the following result.

Theorem 1 (Monotonicity). The magnitude of the elements in the sequence $\{\bar{\boldsymbol{b}}_k\}$ generated by the proposed method is strictly monotonically decreasing. i,e., $\left(\bar{\boldsymbol{b}}_{k+1}\right)_{m+1} > \left(\bar{\boldsymbol{b}}_k\right)_{m+1}$ for k.

Moreover, next theorem declare that the Algorithm 3 finds an optimal solution to problem (2.1).

Theorem 2 (Optimality). Let $\boldsymbol{v}^*$ be the minimum norm point in the affine hull of vertex of the zonotope with set S_k. Suppose that the hyperplane $H(\boldsymbol{v}^*) = \left\{ \boldsymbol{x} \,\middle|\, \boldsymbol{v}^{*\top}\boldsymbol{x} = \boldsymbol{v}^{*\top}\boldsymbol{v}^* \right\}$ and γ^* is the intersection with $(m+1)$ axis. If $\bar{\boldsymbol{z}}_j^\top \boldsymbol{v}_k \geq \boldsymbol{v}_k^\top \boldsymbol{v}$ $({}^\forall \bar{\boldsymbol{z}}_j \in \bar{Z})$, then γ^* is the maximum intersection with the zonotope and the $(m+1)$ axis.

Proof. By contradiction assume that γ^* is not the maximum intersection point between the zonotope and the $(m+1)$ axis. Then there exists a hyperplane passing through $H(\boldsymbol{v}^*)$ and a half-space $H^-(\boldsymbol{v}^*)$. In this case, there exists a point $\bar{\boldsymbol{z}} \in \bar{Z}$ of the zonotope that satisfies $\bar{\boldsymbol{z}} \in H^-(\boldsymbol{v}^*)$. This contradicts the assumption. Therefore, γ^* is the maximum intersection point between the zonotope and the $(m+1)$ axis. □

Theorem 3 (Finiteness). If $\bar{\boldsymbol{b}}_k$ in the proposed Algorithm is monotonically increasing, then algorithm terminates in finitely many iterations.

Proof. Let $\{\bar{\boldsymbol{b}}_k\}$ be a sequence of points generated by the proposed method, where each $\bar{\boldsymbol{b}}_k$ is represented by a convex combination of affine independent end points $\bar{\boldsymbol{z}}_i \in \bar{Z}$ and set $S_k^{(i)}$ $(i = 1, 2 \ldots)$. From the monotonicity of the norm of $\bar{\boldsymbol{b}}_{k+1}$ and $\bar{\boldsymbol{b}}_k$ the k-th set $S_k^{(i)}$ $(i = 1, 2, \ldots)$ is not in the $(k+1)$-th set $S_{k+1}^{(j)}$ $(j = 1, 2, \ldots)$. Therefore, the proposed algorithm converges in a finite number of iterations. □

4 Conclusion

In this paper, we presented a novel algorithm for solving linear programming problems with box constraints, building upon the foundational work of Fujishige et al. [2]. Unlike their approach, which depends on Wolfe's algorithm, our method operates independently, offering a distinct alternative. We have proven the validity of our algorithm under the assumption that scheme in Step 6 works well. We think that our future work will provide the details in Step 6 and involve conducting numerical experiments to further assess its performance. Additionally, we aim to explore specific cases of linear programming where our algorithm can achieve faster a solution.

Acknowledgments. This study was partially supported by JSPS KAKENHI Grant Number 23K26326(23H01632 (2023)).

Disclosure of Interests. The authors of this article have no competing interests relevant to the content of their article.

References

1. Fujishige, S., Liu, X., Zhang, X.: An algorithm for solving the minimum-norm point problem over the intersection of a polytope and an affine set. J. Optim. Theory Appl. **105**, 113–147 (2000)
2. Fujishige, S., Hayashi, T., Yamashita, K., Zimmermann, U.: Zonotopes and the LP-Newton method. Optim. Eng. **10**, 193–205 (2009)
3. Fujishige, S., Sato, H., Zhan, P.: An algorithm for finding the minimum-norm point in the intersection of a convex polyhedron and a hyperplane. Jpn. J. Ind. Appl. Math. **11**, 245–264 (1994)
4. Loera, J., Haddock, J., Rademacher, A.: the minimum euclidean-norm point in a convex polytope: wolfe's combinatorial algorithm is exponential. SIAM J. Comput. **49**, 138–169 (2020)
5. Kitahara, T., Mizuno, S., Shi, J.: The LP-Newton method for standard form linear programming problems. Oper. Res. Lett. **41**, 426–429 (2013)
6. Sekitani, S., Yamamoto, Y.: A recursive algorithm for finding the minimum norm point in a polytope and a pair of' closest points in two polytopes. Math. Program. **6**, 233–249 (1993)
7. Wolfe, P.: Finding the nearest point in a polytope. Math. Program. **11**, 128–149 (2023)

The Effect of Marginal Probability in CMA-ES-PDM for Mixed-Integer Black-Box Optimization

Duc Manh Nguyen[1(✉)] and Ba Hung Tran[2]

[1] Department of Mathematics and Informatics, Hanoi National University of Education, Hanoi, Vietnam
nguyendm@hnue.edu.vn
[2] Nguyen Hue High School for the Gifted, Hanoi, Vietnam

Abstract. The CMA-ES-PDM is a variant of the CMA-ES designed for solving mixed-integer black-box optimization problems. In this method, elite points generated during the evolution process are selected to build sequential probability distribution models for the integer variables. These models are then partially used to generate the integer parts of candidate solutions in some next iterations. The performance of this method depends on a marginal probability, denoted by α, which must be carefully selected for each objective function. A higher α value typically results in a higher success rate but slower convergence. Conversely, a lower α value may lead to stagnation and a lower success rate. In this work, we propose a method to choose the value of α randomly in order to balance the exploration and exploitation of the algorithm. Additionally, the probability model is smoothed when certain intermediate values have zero probability, enhancing the sampling process for integer components. Numerical experiments on the MI-BBO benchmark problems demonstrate that these modifications improve both success rate and convergence speed on certain test functions.

Keywords: Evolutionary Computation · CMA-ES · CMA-ES-PDM · Mixed-Integer Black-Box Optimization

1 Introduction

Mixed-integer optimization problems, which involve both continuous and integer decision variables, are common in many real-world applications. These problems are classified as NP-hard, making them challenging to solve, even when both the objective function and the constraints are linear. In mixed-integer black-box optimization (MI-BBO) problems, the objective function and/or some constraints are presented in a black-box form, meaning they lack derivatives and are not explicitly defined in a functional form. The MI-BBO problems also have been received attention recently [5] since they appear in some applications such as, material design [11], topology optimization [4], placement optimization for CO2

© The Author(s), under exclusive license to Springer Nature Switzerland AG 2025
H. A. Le Thi et al. (Eds.): ICAMCS 2024, LNNS 1311, pp. 83–95, 2025.
https://doi.org/10.1007/978-3-031-90606-0_8

capture and storage [15], software tuning [13], optimizing the circuitry configuration of heat exchangers [19], optimal design of integrated circuits [3], and hyper-parameter optimization of machine learning [9]. Some algorithms have been designed for MI-BBO, for instances, the extended Mesh Adaptive Direct Search (MADS) methods [1], Liuzzi et al. [14] used line searches and proved global convergence of their MIDFO algorithm: the extended evolution strategies [12], surrogate model-based methods [2], CMA-ES with integer handling (CMA-ES-IM) [8], CMA-ES with lower-bounding marginal probability (CMA-ES with margin) [5], and the Distance-weighted eXponential Natural Evolution Strategy taking account of Implicit Constraint and Integer (DX-NES-ICI) [10].

In the CMA-ES-IM [8], Hansen proposed injecting mutations into the integer parts of some candidate solutions and modifyng the step-size adaption strategy in the CMA-ES for solving MI-BBO. However, the proposed approach's performance is limited on certain functions. In contrast, the CMA-ES-PDM [17] introduced an alternative method using the probability distribution models to generate the integer components of candidate solutions during the search process. In the method, after each period of consecutive iterations, some elite points among the generated ones are collected to build/update successively the probability distribution model of integer variables. The model is then partially used to generate the values for integer variables in the CMA-ES in subsequent iterations. In short, the main idea in CMA-ES-PDM is to replace the injection of mutation in CMA-ES-IM [8] by a more efficient perturbation process. The CMA-ES-PDM has demonstrated competitive results on several test problems when compared to the state-of-the-art algorithms, such as CMA-ES-IM [8], CMA-ES with margin [5], and DX-NES-ICI [10].

However, we observe that the marginal probability parameter of the models, denoted as α, needs to be adjusted for each specific problem to achieve better performance. For example, on the EllipsoidInt, the algorithm converges more quickly with $\alpha := \frac{1}{\lambda n}$ than with $\alpha := \frac{1}{n}$. Nevertheless, for the SphereLeadingOnes, the success rate decreases significantly with $\alpha := \frac{1}{\lambda n}$ as the problem dimension increases, while the success rate is 100% with $\alpha := \frac{1}{n}$ for all dimensions. In this work, we propose some modifications to CMA-ES-PDM to improve its performance. First, the value of α is selected randomly from a pre-defined set in order to balance the exploration and exploitation of the algorithm. Two strategies will be investigated for this selection. Second, the model is smoothed when certain intermediate values have zero probability to enhance the sampling process for integer components.

In the next section we review the perturbation techniques used in the CMA-ES for MI-BBO. The details of the modifications for CMA-ES-PDM will be presented in Sect. 3. The proposed method's performance is numerically investigated in Sect. 4. Some conclusions and future perspectives are discussed in Sect. 5.

2 Perturbation Methods in CMA-ES for MI-BBO

2.1 The CMA-ES-IM

The covariance matrix adaptation evolution strategy (CMA-ES) [6,7] is well known to be a powerful method in continuous black-box optimization. However, it is difficult to apply directly the CMA-ES to the MI-BBO by discretizing some elements of the continuous candidate solutions. Indeed, because of the plateau in the discretization, this simple method may not change the evaluation value by small variations in elements corresponding to the integer variables. Therefore, the stagnation occurs when the sample standard deviation of the dimension associated with an integer variable becomes much smaller than the granularity of the discretization [8]. In particular, the step-size tends to decrease in such situations, which enhances trapping the Gaussian distribution on the plateau of the integer variables. To address this problem, the CMA-ES-IM [8] proposed the injecting mutations into the integer components of candidate solutions and modifying the step-size adaption strategy in the CMA-ES. However, the effectiveness of this modification is relatively limited.

2.2 The CMA-ES-PDM: CMA-ES with Probability Distribution of Integer Variables

The CMA-ES-PDM [17] introduces a different method for perturbing the integer components of candidate solutions. This method employs probability distribution models for the integer variables, which are constructed sequentially during the search process. The following notations are used:

- $\texttt{stack}_x$ stores the points collected to build the probability distribution model for integer variables; $\texttt{stack}_f$ stores the objective function values corresponding to the points in $\texttt{stack}_x$.
- $\rho \in (0, 1)$ represents the proportion of best points selected from the set $\texttt{stack}_x$.
- $\texttt{select}_x$ represents the $\lfloor \rho \times |\texttt{stack}_x| \rfloor$ best points among those points stored in $\texttt{stack}_x$ based on the objective function values to construct the model. Their corresponding objective function values are stored in $\texttt{select}_f$.
- $K_j := \{v_1, v_2, ..., v_{|K_j|}\}$ is the ordered set of all possible discrete values for an integer variable $j \in n_{\text{int}}$.
- $1 - \alpha$ is the upper bound of the probability for each value of each integer variable. Here, $\alpha \in (0, 1)$ is referred to as the margin probability and is typically quite small.
- β is the probability of using the model to generate the integer parts for a candidate solution.
- T^{period} represents the number of consecutive iterations during which the model is applied to generate the integer parts of candidate solutions before updating the model.

The integer variables are assumed to be bounded. During the search process, the algorithm collects some generated points (stored in $\texttt{stack}_x$) and their corresponding objective function values (stored in $\texttt{stack}_f$). Then, it selects approximately $\lfloor \rho \times |\texttt{stack}_x| \rfloor$ of the best points (denoted by $\texttt{select}_x$), from $\texttt{stack}_x$ based on their objective function values, to construct the model. The probability distribution model for each integer variable $j \in n_{\text{int}}$ is computed as follows:

$$\texttt{Pr}\left[[\boldsymbol{x}]_j = v\right] := \frac{|\{\boldsymbol{x} \in \texttt{select}_x : [\boldsymbol{x}]_j = v\}|}{\sum\limits_{u \in K_j} |\{\boldsymbol{x} \in \texttt{select}_x : [\boldsymbol{x}]_j = u\}|}, \forall v \in K_j, \tag{1}$$

where $\texttt{Pr}\left[[\boldsymbol{x}]_j = v\right]$ denotes the probability that the j-th integer variable will take the value v. The notation $[\cdot]_j$ represents the j-th element of a vector.

This approach shares similarities with the Cross-Entropy method for combinatorial optimization [18,20]. However, instead of sampling from a predefined family of distributions, samples are collected during the algorithm's evolutions. Additionally, after each model-building phase, the algorithm resets $\texttt{stack}_x := \texttt{select}_x$ and $\texttt{stack}_f := \texttt{select}_f$, ensuring that the best patterns from the previous step are retained rather than clearing $\texttt{stack}_x$. To prevent premature convergence, an upper bound of probability, say $1 - \alpha$, is set for each value $v \in K_j$ in the distribution of all integer variables. Specifically, suppose that $K_j = \{v_1, v_2, ..., v_{|K_j|}\}$, and if $\texttt{Pr}[[\boldsymbol{x}]_j = v_h]$ exceeds $1-\alpha$, for some $v_h \in K_j$, then $\texttt{Pr}[[\boldsymbol{x}]_j = v_h]$ is reset to $1 - \alpha$, and the remaining probability α is geometrically distributed around the peak v_h.

Whenever the probability distribution model is available, the roulette wheel method is employed to generate the integer parts for certain candidate solutions $\boldsymbol{x}_i$ with a probability β over the next T^{period} consecutive iterations.

3 Some Modifications for CMA-ES-PDM

Algorithms 1 and 2 present the pseudo-code of CMA-ES-PDM with random margin. This is a modified version of CMA-ES-PDM for MI-BBO. Here, we partition the variable set into $n := n_{\text{co}} \cup n_{\text{int}}$, where n_{co} denotes the indices of continuous variables and n_{int} denotes the indices of integer variables. The red lines in the algorithms indicate modifications from the original algorithm. In this method, the marginal probability is chosen randomly. As discussed, the marginal probability α should be justified for each objective function. A large α (e.g., $\alpha = \frac{1}{n}$) results in slow convergence, while a small α (e.g., $\alpha = \frac{1}{\lambda n}$) may cause stagnation. We propose two strategies for selecting α randomly from the set of marginal probabilities $\texttt{P}_{\min}$, as detailed in Algorithms 1, Line 20, as follows

- *Strategy 1*: we set $\texttt{P}_{\min} = \{\frac{1}{\lambda n}, \frac{1}{n}\}$, and select the value for α randomly using a parameter τ: if $\texttt{rand}(0, 1) < \tau$ then $\alpha := \frac{1}{\lambda n}$; otherwise, $\alpha := \frac{1}{n}$. The parameter τ will be chosen later to balance the exploration and exploitation of the algorithm.
- *Strategy 2*: we set $\texttt{P}_{\min} = \{\frac{1}{\lambda n}, \frac{2}{\lambda n}, \ldots, \frac{\lambda}{\lambda n}\}$ and select the value for α randomly uniformly from this set.

Algorithm 1 CMA-ES-PDM with random margin

Input: $\boldsymbol{m}^{(0)} \in \mathbb{R}^n, \sigma^{(0)} \in \mathbb{R}_+$

Initialize: $\mathbf{C}^{(0)} = \boldsymbol{I}, \boldsymbol{p}_c^{(0)} = 0, \boldsymbol{p}_\sigma^{(0)} = 0, \lambda = \lfloor 4 + 3\ln(n) \rfloor$

Set up: $\mu, w_i, i = 1, ..., \mu, \mu_w, c_c, c_\sigma, c_1, c_\mu, d_\sigma$, $\texttt{stack}_x = \emptyset, \texttt{stack}_f = \emptyset$,

While not terminate

  for $i = 1, ..., \lambda$

    $\boldsymbol{x}_i = \boldsymbol{m}^{(k)} + \sigma^{(k)} \boldsymbol{y}_i, \boldsymbol{y}_i \sim \mathcal{N}(\mathbf{0}, \mathbf{C}^{(k)})$,

    //Using the probability distribution model with β

    if ($\texttt{model}$ exists) & ($\texttt{rand}(0, 1) < \beta$)

      Using the $\texttt{model}$ to generate the integer parts for the candidate solution $\boldsymbol{x}_i$.

    end

    $\hat{\boldsymbol{x}}_i = \texttt{ENCODING}(\boldsymbol{x}_i)$ //Encoding process

  end

  Evaluate the objective function values $f(\hat{\boldsymbol{x}}_1), ..., f(\hat{\boldsymbol{x}}_\lambda)$. Denote the f-sorted candidates as: $\{\boldsymbol{x}_{i:\lambda} : i = 1, ..., \lambda\}$ and $\{\boldsymbol{y}_{i:\lambda} : i = 1, ..., \lambda\}$ with respect to $f(\hat{\boldsymbol{x}}_{1:\lambda}) \leq f(\hat{\boldsymbol{x}}_{2:\lambda}) \leq ... \leq f(\hat{\boldsymbol{x}}_{\lambda:\lambda})$.

  $\boldsymbol{m}^{(k+1)} = \sum_{i=1}^{\mu} w_i \boldsymbol{x}_{i:\lambda} = \boldsymbol{m}^{(k)} + \sigma^{(k)} \boldsymbol{y}_w$, where $\boldsymbol{y}_w = \sum_{i=1}^{\mu} w_i \boldsymbol{y}_{i:\lambda}$

  $\boldsymbol{p}_\sigma^{(k+1)} = (1 - c_\sigma)\boldsymbol{p}_\sigma^{(k)} + \sqrt{(1 - (1 - c_\sigma)^2)}\sqrt{\mu_w}\left(\mathbf{C}^{(k)}\right)^{-\frac{1}{2}} \boldsymbol{y}_w$

  $\boldsymbol{p}_c^{(k+1)} = (1 - c_c)\boldsymbol{p}_c^{(k)} + h_\sigma \sqrt{(1 - (1 - c_c)^2)}\sqrt{\mu_w} \boldsymbol{y}_w$

  $\mathbf{C}^{(k+1)} = \left(1 - c_1 - c_\mu \sum_{i=1}^{\lambda} w_i + (1 - h_\sigma) c_1 c_c (2 - c_c)\right) \mathbf{C}^{(k)} + c_1 \boldsymbol{p}_c^{(k+1)} \boldsymbol{p}_c^{(k+1)^T} + c_\mu \sum_{i=1}^{\lambda} w_i^{\circ} \boldsymbol{y}_{i:\lambda} \boldsymbol{y}_{i:\lambda}^T$

  $\sigma^{(k+1)} = \sigma^{(k)} \times \exp\left(\frac{c_\sigma}{d_\sigma}\left(\frac{||\boldsymbol{I}_\sigma^{(k+1)} \boldsymbol{p}_\sigma^{(k+1)}||}{\mathbb{E}[||N(\mathbf{0}, \boldsymbol{I}_\sigma^{(k+1)})||]} - 1\right)\right)$

  Set $\texttt{stack}_x = \texttt{stack}_x \cup \{\hat{\boldsymbol{x}}_i : i = 1, ..., \lambda\}$, $\texttt{stack}_f = \texttt{stack}_f \cup \{f(\hat{\boldsymbol{x}}_1), ..., f(\hat{\boldsymbol{x}}_\lambda)\}$.

  //Build probability distribution model for integer variables

  if (($|\texttt{stack}_x| \geq \max(20n, 1000)$) & ($\texttt{model}$ is **NULL**)) **or** ($\texttt{model}$ exists & after each T^{period} consecutive iterations)

    • Selecting α randomly from a set of marginal probabilities $\mathrm{P}_{\min}$.

    • $\texttt{model}$ = **build_model**($\texttt{select}_x$, α)

    • Set $\texttt{stack}_x = \texttt{select}_x, \texttt{stack}_f = \texttt{select}_f$

  end

The CMA-ES-PDM using the strategy 1 and strategy 2 is denoted by *CMA-ES-PDM with RM1* and *CMA-ES-PDM with RM2* respectively.

Note that at certain iterations of the algorithm, the model for some integer variable $j \in n_{\text{int}}$ may resemble the distribution shown in Fig. 1a. That is, the probability for some intermediate values might be zero. In such cases, if the model is always used (i.e., $\beta = 1$), it could generate the value $[\boldsymbol{x}]_j = -2$ with a probability of 0.15, $[\boldsymbol{x}]_j = -1$ with a probability of 0.75, and $[\boldsymbol{x}]_j = 1$ with a probability of 0.10, but never the value $[\boldsymbol{x}]_j = 0$, which should appear in the optimal solution. To address this, the CMA-ES-PDM uses the parameter $\beta = 0.9$ to smooth the discrete distribution of the probability model by occasionally sampling the integer part according to a Gaussian distribution.

In this work, we propose an additional technique to directly smooth the model when the probability of intermediate values is missed, as illustrated in Fig. 1. Suppose this issue occurs with an integer variable $j \in n_{\text{int}}$ (see Fig. 1a).

Algorithm 2 Build Probability Distribution Model of Integer Variables

function model = **build_model**($\texttt{select}_x$, α)
    for $j \in n_{\text{int}}$
        • Using (1) to compute the probability distribution of the model: $\texttt{Pr}\left[[\boldsymbol{x}]_j = v\right], \forall v \in K_j$.
        • Smoothing out the model if there is intermediate values having zero probability in the model.
        • If there exits $v \in K_j$ satisfying $\texttt{Pr}[[\boldsymbol{x}]_j = v] > 1 - \alpha$ then redistribute the remaining probability α geometrically around the peak v.
    end
end

We fit the frequency data for the elements of K_j to a normal distribution to obtain a probability density function (pdf). We then compute the pdf values for each element in K_j and normalize these values to create a probability distribution for all elements in K_j (see Fig. 1b).

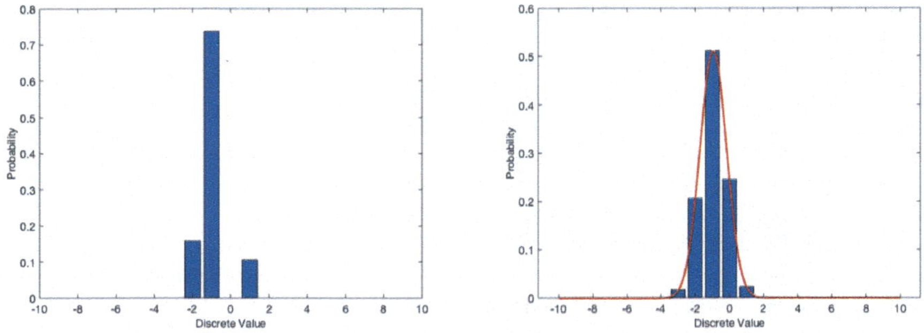

(a) The probability for $[\boldsymbol{x}]_j = 0$ is missed.

(b) The corrected distribution.

Fig. 1. The probability distribution for an integer variable $[\boldsymbol{x}]_j$ at certain iterations of the algorithm before and after correction: Using the distribution model in the left figure, we never generate the value $[\boldsymbol{x}]_j = 0$, which should appear in the optimal solution. The distribution is then corrected in the right figure.

4 Numerical Experiment

The proposed algorithm is written in matlab based on the source code of the CMA-ES-IM[1]. It is run on a Mac OS with 1.60 GHz Core(TM) i5-5250U CPU, 8 GB of RAM. The test functions which are defined in Table 1 will be used in this study include SphereOneMax, EllipsoidOneMax, SphereLeadingOnes, EllipsoidLeadingOnes, SphereInt, EllipsoidInt, ReversedEllipsoidInt, n_{int}-Tablet functions [5,10]. Additionally, we use the Rosenbrock function, and the signomial

[1] http://www.cmap.polytechnique.fr/~nikolaus.hansen.

P-sig function (see the library MINLPLib[2]). Note that in the first eight functions, except for the SphereLeadingOnes and EllipsoidLeadingOnes functions, the others are convex if the discrete variables are relaxed to continuous variables. The Rosenbrock function is non-convex while the P-sig function is convex non-separable. In this work, we consider the problem dimensions $n = 20, 40, 60, 80$ (except $n = 20, 30, 40$ for the P-sig function) and the indices for continuous and integer variables are $n_{\text{co}} := \{1, ..., \frac{n}{2}\}$ and $n_{\text{int}} := \{\frac{n}{2} + 1, ..., n\}$ respectively.

Table 1. The test functions.

- $f_{\text{SphereOneMax}}(\boldsymbol{x}) = \sum_{j=1}^{|n_{\text{co}}|} x_j^2 + |n_{\text{int}}| - \sum_{j=|n_{\text{co}}|+1}^{n} x_j$
- $f_{\text{EllipsoidOneMax}}(\boldsymbol{x}) = \sum_{j=1}^{|n_{\text{co}}|} \left(1000^{\frac{j-1}{|n_{\text{co}}|-1}} x_j\right)^2 + |n_{\text{int}}| - \sum_{j=|n_{\text{co}}|+1}^{n} x_j$
- $f_{\text{SphereLeadingOnes}}(\boldsymbol{x}) = \sum_{j=1}^{|n_{\text{co}}|} x_j^2 + |n_{\text{int}}| - \sum_{j=|n_{\text{co}}|+1}^{n} \prod_{i=|n_{\text{co}}|+1}^{j} x_i$
- $f_{\text{EllipsoidLeadingOnes}}(\boldsymbol{x}) = \sum_{j=1}^{|n_{\text{co}}|} \left(1000^{\frac{j-1}{|n_{\text{co}}|-1}} x_j\right)^2 + |n_{\text{int}}| - \sum_{j=|n_{\text{co}}|+1}^{n} \prod_{i=|n_{\text{co}}|+1}^{j} x_i$
- $f_{\text{SphereInt}}(\boldsymbol{x}) = \sum_{j=1}^{n} x_j^2$
- $f_{\text{EllipsoidInt}}(\boldsymbol{x}) = \sum_{j=1}^{n} \left(1000^{\frac{j-1}{n-1}} x_j\right)^2$
- $f_{\text{ReversedEllipsoidInt}}(\boldsymbol{x}) = \sum_{j=1}^{|n_{\text{co}}|} \left(1000^{\frac{|n_{\text{int}}|+j-1}{n-1}} x_j\right)^2 + \sum_{j=1}^{|n_{\text{int}}|} \left(1000^{\frac{j-1}{n-1}} x_{|n_{\text{co}}|+j}\right)^2$
- $f_{n_{\text{int}}\text{-Tablet}}(\boldsymbol{x}) = \sum_{j=1}^{|n_{\text{co}}|} (100x_j)^2 + \sum_{j=1}^{|n_{\text{int}}|} x_{|n_{\text{co}}|+j}^2$
- $f_{\text{Rosenbrock}}(\boldsymbol{x}) = \sum_{j=1}^{n-1} [100(x_{j+1} - x_j^2)^2 + (1 - x_j)^2]$
- $f_{\text{P-sig}}(\boldsymbol{x}) = a \prod_{j=1}^{n} x_j^{p_j} + \sum_{j=1}^{n} x_j$

The integer variables are binary in the first four functions. The integer variables in the next five functions can take values in $[-10, 10]$, while in the last function, a search range $[1, 10]$ is defined for all variables. Note that the optimal solution for the first four functions is $(\boldsymbol{x}_{\text{co}}, \boldsymbol{x}_{\text{int}}) = (\boldsymbol{0}, \boldsymbol{1})$. The optimal solution of the next five functions is $\boldsymbol{x} = \boldsymbol{0}$, while it is $\boldsymbol{x} = \boldsymbol{1}$ for the Rosenbrock function. The first nine functions have an optimal value of 0. The best objective value and the parameters a and $p_j, j = 1, 2, \ldots, n$ for the last function are presented in the library MINLPLib.

The variants of CMA-ES run with the default value of the population size, i.e., $\lambda = \lfloor 4 + 3\ln(n) \rfloor$. We also use the parameters of DX-NES-ICI as in [10], except for the population size λ which is set $\lambda = 12, 14, 16, 18$ for the dimension $n = 20, 40, 60, 80$ respectively. Additionally, the typical technique of handling box constraints in the CMA-ES algorithm is applied to the integer variables for all variants of CMA-ES. The penalty $||\boldsymbol{x}_i^{\text{feasible}} - \boldsymbol{x}_i||_2^2/n$ also is added to the evaluation value, where $\boldsymbol{x}_i^{\text{feasible}}$ is the nearest-neighbor feasible solution to the candidate $\boldsymbol{x}_i$ for the DX-NES-ICI algorithm on the P-sig function with the box constraint $[1, 10]$. The proposed algorithm is stopped when the minimum eigenvalue of $\sigma^2 \boldsymbol{C}$ is less than 10^{-30}, or the condition number of the covariance matrix exceeds 10^{14}. In this work, similar to CMA-ES-PDM, we set $\rho = 0.1, \beta = 0.9$, and $T^{\text{period}} = 20$.

The initial mean vector $\boldsymbol{m}^{(0)}$ is chosen randomly uniformly in $[1, 3]^n$, for the first eight functions. Additionally, we set $[\boldsymbol{m}^{(0)}]_j = \frac{1}{2}$ for all $j \in n_{\text{int}}$ for the first

[2] https://www.minlplib.org.

four functions. The initial mean vector $\boldsymbol{m}^{(0)} = (3, ..., 3)$ is set for the Rosenbrock while $\boldsymbol{m}^{(0)} = (1, ..., 1)$ is set for P-sig respectively. The initial step-size is $\sigma^{(0)} = 1$. In each setting, we conducted 100 run trials for each dimension. During a run, if the objective function value is smaller than the target $f_{\text{target}} = 10^{-10}$ (except $f_{\text{target}} = f_{\text{best}} + 10^{-10}$ for the last function), this run is considered to be successful. The success rate is the ratio of the number of successful trials to the total number of trials.

4.1 The Effect of τ in CMA-ES-PDM with RM1

We run CMA-ES-PDM with RM1 with the parameter $\tau \in \{0.1, 0.2, ..., 0.9\}$ on the first eight functions. Tables 2 and 3 present the success rates of the algorithm, and the median value of the number of function evaluations over successful runs on each test function for dimensions $n = 20, 40, 60$, and 80. We observe that the successful rates are 100% for the parameter $\tau = 0.1, 0.2, ..., 0.5$, and almost 100% for $\tau = 0.6$. The variation of τ seems to have strong effect on the success rate for the SphereLeadingOnes and n_{int}-Tablet functions. The large values of τ decreases the success rate on the functions especially in large dimensions. Additionally, the large value of τ decreases the number of function evaluations for almost all functions, except for the n_{int}-Tablet function.

From the tables, the value $\tau = 0.7$ seems to be a good choice for CMA-ES-PDM with RM1. Therefore, we will set $\tau = 0.7$ for CMA-ES-PDM with RM1 in the next comparison.

4.2 Comparative Results

Table 4 presents the performance of proposed variants of CMA-ES-PDM with some state-of-the-art algorithms for MI-BBO problems. Some numerical results in this table for CMA-ES with Margin, DX-NES-ICI, and CMA-ES-PDM are taken directly from [17].

On the First Eight Functions. Firstly, we observe that CMA-ES-PDM with RM1 ($\tau = 0.7$) performs better than CMA-ES-PDM with RM2 in almost all test cases. However, from the results in Table 3, CMA-ES-PDM with RM2's performance is the same as the case of $\tau = 0.5$ of CMA-ES-PDM with RM1. The CMA-ES-PDM with RM1 ($\tau = 0.7$) has a success rate of over 90%, while CMA-ES-PDM with RM2 has a success rate of 100% on all test functions. The proposed variants perform better than CMA-ES with Margin in almost all cases. Additionally, CMA-ES-PDM with RM1 ($\tau = 0.7$) runs much faster than CMA-ES-PDM with $\alpha = \frac{1}{n}$ in all test cases. This variant also performs much better than CMA-ES-PDM with $\alpha = \frac{1}{n\lambda}$ on the SphereLeadingOnes, ReversedEllipsoidInt, and n_{int}-Tablet functions. However, its performance is poorer on the EllipsoidOneMax and EllipsoidLeadingOne functions in high dimensions. This is because, for these two functions, the small value of α, say $\frac{1}{n\lambda}$, promotes faster convergence to the optimal solution than the larger values of α. The numerical results show the pertinence of the proposed modifications for CMA-ES-PDM. Nevertheless, the

Table 2. The effect of the parameter τ on the performance of CMA-ES-PDM with RM1: SR denotes the success rate, Fvals denotes the median value of the number of function evaluations over successful runs, and the values inside the parentheses represent the interquartile range (IQR).

Function	n	CMA-ES-PDM with RM1							
		$\tau = 0.1$		$\tau = 0.2$		$\tau = 0.3$		$\tau = 0.4$	
		Fvals	SR	Fvals	SR	Fvals	SR	Fvals	SR
SphereOneMax	20	3638 (384)	**1.00**	3554 (354)	**1.00**	3368 (342)	**1.00**	3326 (330)	**1.00**
	40	6805 (518)	**1.00**	6527 (398)	**1.00**	6287 (398)	**1.00**	6047 (480)	**1.00**
	60	9874 (728)	**1.00**	9466 (632)	**1.00**	9114 (488)	**1.00**	8802 (496)	**1.00**
	80	12591 (774)	**1.00**	12140 (672)	**1.00**	11741 (748)	**1.00**	11341 (604)	**1.00**
SphereLeadingOnes	20	3758 (474)	**1.00**	3596 (360)	**1.00**	3482 (276)	**1.00**	3380 (294)	**1.00**
	40	7427 (593)	**1.00**	7105 (563)	**1.00**	6902 (465)	**1.00**	6617 (563)	**1.00**
	60	10730 (664)	**1.00**	10434 (592)	**1.00**	10026 (616)	**1.00**	9602 (592)	**1.00**
	80	13900 (680)	**1.00**	13568 (782)	**1.00**	13024 (774)	**1.00**	12616 (706)	**1.00**
EllipsoidOneMax	20	13196 (1002)	**1.00**	12914 (804)	**1.00**	12404 (954)	**1.00**	12074 (672)	**1.00**
	40	43532 (1965)	**1.00**	42182 (1358)	**1.00**	40907 (1440)	**1.00**	39880 (1433)	**1.00**
	60	96722 (2720)	**1.00**	93298 (2480)	**1.00**	89938 (3312)	**1.00**	87298 (2256)	**1.00**
	80	174584 (4386)	**1.00**	168498 (3324)	**1.00**	161859 (3426)	**1.00**	156904 (3247)	**1.00**
EllipsoidLeadingOnes	20	14198 (924)	**1.00**	13628 (1188)	**1.00**	13094 (804)	**1.00**	12692 (642)	**1.00**
	40	47305 (2273)	**1.00**	45070 (1613)	**1.00**	43067 (1838)	**1.00**	41732 (1365)	**1.00**
	60	104338 (3144)	**1.00**	98826 (2752)	**1.00**	95178 (2616)	**1.00**	91386 (2296)	**1.00**
	80	186560 (4616)	**1.00**	178800 (4964)	**1.00**	170546 (4293)	**1.00**	163619 (3502)	**1.00**
SphereInt	20	3050 (258)	**1.00**	2978 (192)	**1.00**	2870 (216)	**1.00**	2822 (198)	**1.00**
	40	6107 (368)	**1.00**	6017 (315)	**1.00**	5800 (383)	**1.00**	5687 (383)	**1.00**
	60	8778 (520)	**1.00**	8594 (392)	**1.00**	8458 (408)	**1.00**	8242 (392)	**1.00**
	80	11350 (510)	**1.00**	11035 (476)	**1.00**	10831 (468)	**1.00**	10610 (493)	**1.00**
EllipsoidInt	20	8492 (630)	**1.00**	8048 (714)	**1.00**	7736 (612)	**1.00**	7364 (660)	**1.00**
	40	21797 (1125)	**1.00**	20762 (968)	**1.00**	19727 (1178)	**1.00**	18910 (855)	**1.00**
	60	38978 (1496)	**1.00**	37234 (1840)	**1.00**	35682 (1480)	**1.00**	33882 (1312)	**1.00**
	80	59545 (2151)	**1.00**	56629 (2406)	**1.00**	54113 (2457)	**1.00**	51878 (2389)	**1.00**
ReversedEllipsoidInt	20	7460 (690)	**1.00**	7232 (810)	**1.00**	7292 (726)	**1.00**	7076 (756)	**1.00**
	40	18887 (1523)	**1.00**	18595 (1290)	**1.00**	18167 (1373)	**1.00**	18092 (1560)	**1.00**
	60	34498 (2192)	**1.00**	34170 (2888)	**1.00**	32962 (2048)	**1.00**	32594 (1632)	**1.00**
	80	54581 (8849)	**1.00**	53323 (6741)	**1.00**	51861 (4658)	**1.00**	50832 (5670)	**1.00**
n_{int}-Tablet	20	5648 (774)	**1.00**	5696 (648)	**1.00**	5426 (594)	**1.00**	5522 (726)	**1.00**
	40	10840 (998)	**1.00**	10922 (818)	**1.00**	10750 (908)	**1.00**	10735 (878)	**1.00**
	60	15714 (1320)	**1.00**	15658 (1336)	**1.00**	15618 (1352)	**1.00**	15786 (1456)	**1.00**
	80	20844 (1624)	**1.00**	20972 (1386)	**1.00**	21057 (1581)	**1.00**	20674 (2006)	**1.00**

proposed variants are only better than DX-NES-ICI on the EllipsoidOneMax, EllipsoidLeadingOnes, EllipsoidInt functions in 60D and 80D.

On the Two Last Functions. It is clear that all algorithms have a low success rate on the Rosenbrock function, and a higher one on the P-sig function. It seems that the default population size is not compatible with the functions, perhaps a larger one would do better. We can see that the proposed variants perform better

Table 3. The effect of the parameter τ on the performance of CMA-ES-PDM with RM1: SR denotes the success rate, Fvals denotes the median value of the number of function evaluations over successful runs, and the values inside the parentheses represent the interquartile range (IQR).

Function	n	CMA-ES-PDM with RM1									
		$\tau = 0.5$		$\tau = 0.6$		$\tau = 0.7$		$\tau = 0.8$		$\tau = 0.9$	
		Fvals	SR	Fvals	SR	Fvals	SR	Fvals	SR	Fvals	SR
SphereOneMax	20	3182 (258)	**1.00**	3098 (294)	**1.00**	2972 (186)	**1.00**	2912 (192)	**1.00**	2822 (180)	**1.00**
	40	5942 (450)	**1.00**	5717 (398)	**1.00**	5500 (330)	**1.00**	5372 (330)	**1.00**	5222 (323)	**1.00**
	60	8418 (440)	**1.00**	8170 (408)	**1.00**	7978 (544)	**1.00**	7730 (336)	**1.00**	7490 (288)	**1.00**
	80	10882 (510)	**1.00**	10576 (536)	**1.00**	10313 (459)	**1.00**	9990 (289)	**1.00**	9641 (340)	**1.00**
SphereLeadingOnes	20	3242 (408)	**1.00**	3158 (246)	**1.00**	3074 (234)	**1.00**	2954 (276)	**1.00**	2900 (246)	**1.00**
	40	6422 (450)	**1.00**	6272 (578)	**1.00**	6002 (428)	**1.00**	5912 (458)	**1.00**	5732 (428)	**1.00**
	60	9330 (648)	**1.00**	9170 (688)	**1.00**	8866 (500)	0.99	8682 (672)	0.98	8506 (624)	0.82
	80	12183 (638)	**1.00**	12004 (765)	0.98	11834 (625)	0.91	11469 (1071)	0.78	11392 (1564)	0.46
EllipsoidOneMax	20	11780 (636)	**1.00**	11606 (558)	**1.00**	11258 (504)	**1.00**	11108 (486)	**1.00**	10916 (426)	**1.00**
	40	39017 (1530)	**1.00**	37930 (1073)	**1.00**	36902 (878)	**1.00**	36107 (960)	**1.00**	35612 (975)	**1.00**
	60	84562 (1864)	**1.00**	82218 (2512)	**1.00**	80458 (1672)	**1.00**	78466 (1376)	**1.00**	76610 (1392)	**1.00**
	80	152280 (3264)	**1.00**	147664 (2686)	**1.00**	143440 (2712)	**1.00**	140346 (1488)	**1.00**	136895 (1505)	**1.00**
EllipsoidLeadingOnes	20	12194 (600)	**1.00**	12026 (588)	**1.00**	11510 (540)	**1.00**	11282 (456)	**1.00**	11030 (516)	**1.00**
	40	40517 (1583)	**1.00**	39070 (1125)	**1.00**	37667 (1028)	**1.00**	36767 (1050)	**1.00**	36025 (1013)	**1.00**
	60	87882 (2392)	**1.00**	85002 (1664)	**1.00**	82018 (1544)	**1.00**	79666 (1424)	**1.00**	77442 (1184)	**1.00**
	80	157448 (3264)	**1.00**	151600 (2729)	**1.00**	146576 (2559)	**1.00**	142241 (2108)	**1.00**	138110 (1819)	**1.00**
SphereInt	20	2738 (198)	**1.00**	2702 (192)	**1.00**	2660 (168)	**1.00**	2594 (156)	**1.00**	2552 (192)	**1.00**
	40	5582 (345)	**1.00**	5470 (375)	**1.00**	5305 (308)	**1.00**	5222 (300)	**1.00**	5132 (255)	**1.00**
	60	7986 (416)	**1.00**	7874 (368)	**1.00**	7666 (336)	**1.00**	7506 (472)	**1.00**	7378 (288)	**1.00**
	80	10389 (417)	**1.00**	10100 (493)	**1.00**	10032 (451)	**1.00**	9811 (442)	**1.00**	9582 (340)	**1.00**
EllipsoidInt	20	7160 (648)	**1.00**	6932 (618)	**1.00**	6590 (468)	**1.00**	6344 (528)	**1.00**	6098 (396)	**1.00**
	40	18332 (930)	**1.00**	17492 (728)	**1.00**	16772 (833)	**1.00**	16262 (773)	**1.00**	15662 (848)	**1.00**
	60	32714 (1616)	**1.00**	31338 (1280)	**1.00**	30042 (1400)	**1.00**	29026 (1672)	**1.00**	27954 (1344)	**1.00**
	80	49897 (2278)	**1.00**	47874 (1666)	**1.00**	46183 (1785)	**1.00**	44619 (1292)	**1.00**	43046 (1573)	**1.00**
ReversedEllipsoidInt	20	7088 (744)	**1.00**	7040 (702)	**1.00**	6962 (804)	**1.00**	6962 (984)	**1.00**	7082 (918)	0.92
	40	17747 (1140)	**1.00**	17582 (1283)	**1.00**	17410 (1200)	**1.00**	17372 (1380)	**1.00**	17762 (1920)	0.96
	60	31698 (1808)	**1.00**	31490 (1720)	**1.00**	31506 (2264)	**1.00**	31258 (2632)	**1.00**	30978 (2088)	0.97
	80	50339 (5474)	**1.00**	48894 (3256)	**1.00**	48444 (4854)	**1.00**	48920 (4420)	**1.00**	49005 (3154)	**1.00**
n_{int}-Tablet	20	5432 (720)	**1.00**	5438 (648)	**1.00**	5264 (618)	**1.00**	5558 (702)	0.99	5414 (651)	0.95
	40	10652 (1320)	**1.00**	10690 (1103)	**1.00**	10832 (1065)	**1.00**	11132 (1088)	0.95	11462 (1620)	0.78
	60	15386 (1440)	**1.00**	15714 (1560)	**1.00**	16018 (1280)	0.98	16090 (1728)	0.98	17538 (2660)	0.71
	80	20836 (1649)	**1.00**	20853 (1811)	**1.00**	21150 (1734)	0.99	22442 (2355)	0.92	23870 (3349)	0.52

than CMA-ES with Margin on the Rosenbrock function, and provides a higher success rate than CMA-ES with Margin does on the P-sig function. Although, they have a smaller success rate than the standard CMA-ES-PDM with either $\alpha = \frac{1}{n}$ or $\alpha = \frac{1}{n\lambda}$, they generally runs faster than this algorithm. The proposed algorithms are similar to DX-NES-ICI on the Rosenbrock function, and only provide better success rate than DX-NES-ICI in large dimensions on the P-sig function.

Table 4. Comparative results: SR denotes the success rate, Fvals denotes the median value of the number of function evaluations over successful runs, and the values inside the parentheses represent the interquartile range (IQR). The bold fonts represent the best median values of the number of function evaluations and the best success rates among the methods.

Function	n	CMA-ES with Margin		DX-NES-ICI		CMA-ES-PDM with $\alpha = \frac{1}{n\lambda}$		CMA-ES-PDM with $\alpha = \frac{1}{n}$		CMA-ES-PDM with RM1 ($\tau = 0.7$)		CMA-ES-PDM with RM2	
		Fvals	SR	Fvals	SR	Fvals	SR	Fvals	SR	Fvals	SR	Fvals	SR
SphereOneMax	20	3876 (327)	**1.00**	**2136 (159)**	**1.00**	2756 (150)	**1.00**	3818 (306)	**1.00**	2972 (186)	**1.00**	3110 (288)	**1.00**
	40	8152 (630)	**1.00**	**4158 (182)**	**1.00**	5110 (360)	**1.00**	7120 (473)	**1.00**	5500 (330)	**1.00**	5777 (375)	**1.00**
	60	12256 (644)	**1.00**	**6312 (260)**	**1.00**	7434 (392)	**1.00**	10314 (656)	**1.00**	7978 (544)	**1.00**	8314 (384)	**1.00**
	80	17433 (935)	**1.00**	**8523 (378)**	**1.00**	9692 (459)	**1.00**	13305 (757)	**1.00**	10313 (459)	**1.00**	10712 (468)	**1.00**
SphereLeadingOnes	20	3978 (420)	**1.00**	**2220 (180)**	**1.00**	2858 (378)	**1.00**	3956 (474)	**1.00**	3074 (234)	**1.00**	3206 (240)	**1.00**
	40	8475 (611)	**1.00**	**4375 (273)**	**1.00**	6272 (743)	0.88	8177 (758)	**1.00**	6002 (428)	**1.00**	6317 (540)	**1.00**
	60	13416 (996)	**1.00**	**6728 (988)**	**1.00**	9266 (1292)	0.53	12218 (920)	**1.00**	8866 (500)	0.99	9258 (552)	**1.00**
	80	18861 (740)	**1.00**	**11259 (4163)**	**1.00**	12268 (3145)	0.14	15999 (1369)	**1.00**	11834 (625)	0.91	12149 (859)	**1.00**
EllipsoidOneMax	20	10938 (1209)	0.82	**10506 (1374)**	**1.00**	11276 (738)	**1.00**	14072 (1050)	**1.00**	11258 (504)	**1.00**	11732 (594)	**1.00**
	40	36195 (2460)	0.83	**32676 (6328)**	**1.00**	36115 (885)	**1.00**	45850 (1748)	**1.00**	36902 (878)	**1.00**	38695 (1230)	**1.00**
	60	77952 (5280)	0.79	89352(31036)	**1.00**	**77194 (1576)**	**1.00**	100602 (2968)	**1.00**	80458 (1672)	**1.00**	85050 (1944)	**1.00**
	80	136280 (8551)	0.82	307692 (172012)	0.84	**136444 (1649)**	**1.00**	182310 (4361)	**1.00**	143440 (2712)	**1.00**	152909 (2746)	**1.00**
EllipsoidLeadingOnes	20	11442 (1053)	**1.00**	**10386 (1269)**	**1.00**	11300 (582)	**1.00**	14978 (1074)	**1.00**	11510 (540)	**1.00**	12260 (690)	**1.00**
	40	41640 (2318)	**1.00**	**34062 (8729)**	**1.00**	36317 (986)	**1.00**	49142 (2963)	**1.00**	37667 (1028)	**1.00**	40330 (1418)	**1.00**
	60	91368 (4248)	**1.00**	131024 (75228)	**1.00**	**77418 (1048)**	**1.00**	108482 (3864)	**1.00**	82018 (1544)	**1.00**	88242 (2056)	**1.00**
	80	164619 (7229)	**1.00**	401634 (219726)	0.51	**137243 (1692)**	**1.00**	193794 (4310)	**1.00**	146576 (2559)	**1.00**	158306 (2669)	**1.00**
SphereInt	20	3780 (267)	**1.00**	**2124 (135)**	**1.00**	2522 (126)	**1.00**	4442 (384)	**1.00**	2660 (168)	**1.00**	2726 (192)	**1.00**
	40	7740 (473)	**1.00**	**4046 (189)**	**1.00**	5147 (225)	**1.00**	8627 (533)	**1.00**	5305 (308)	**1.00**	5522 (300)	**1.00**
	60	11464 (628)	**1.00**	**6120 (224)**	**1.00**	7506 (360)	**1.00**	12442 (696)	**1.00**	7666 (336)	**1.00**	7986 (328)	**1.00**
	80	14977 (672)	**1.00**	**8199 (311)**	**1.00**	9760 (370)	**1.00**	15923 (748)	**1.00**	10032 (451)	**1.00**	10304 (417)	**1.00**
EllipsoidInt	20	8256 (1044)	**1.00**	6348 (867)	**1.00**	**6338 (546)**	**1.00**	12884 (1362)	**1.00**	6590 (468)	**1.00**	7094 (504)	**1.00**
	40	23062 (2231)	**1.00**	17122 (2261)	**1.00**	16787 (1073)	0.99	32192 (2115)	**1.00**	**16772 (833)**	**1.00**	18422 (1013)	**1.00**
	60	43248 (3480)	**1.00**	33016 (5340)	**1.00**	**29186 (1504)**	0.98	56858 (2456)	**1.00**	30042 (1400)	**1.00**	32482 (1680)	**1.00**
	80	67498 (4603)	**1.00**	54684 (7610)	**1.00**	**44610 (1717)**	0.99	85385 (2873)	**1.00**	46183 (1785)	**1.00**	50110 (1938)	**1.00**
ReversedEllipsoidInt	20	14094 (3507)	0.38	**5538 (681)**	**1.00**	9158 (1581)	0.75	9608 (1326)	**1.00**	6962 (804)	**1.00**	7130 (654)	**1.00**
	40	37365 (4875)	0.21	**13279 (1449)**	**1.00**	21947 (2378)	0.63	23455 (2565)	**1.00**	17410 (1200)	**1.00**	17515 (968)	**1.00**
	60	69480 (12424)	0.18	**24704 (2432)**	**1.00**	37682 (3844)	0.69	40778 (4536)	**1.00**	31506 (2264)	**1.00**	31682 (1888)	**1.00**
	80	108392 (25415)	0.25	**39618 (3789)**	**1.00**	57666 (4930)	0.80	62834 (4896)	**1.00**	48444 (4854)	**1.00**	49736 (6103)	**1.00**
n_{int}-Tablet	20	10128 (2082)	0.71	**3876 (351)**	**1.00**	7598 (2136)	0.71	7334 (1122)	**1.00**	5264 (618)	**1.00**	5330 (666)	**1.00**
	40	23130 (2070)	0.23	**7623 (760)**	**1.00**	14905 (3780)	0.22	14605 (1545)	**1.00**	10832 (1065)	**1.00**	10690 (1005)	**1.00**
	60	35456 (3624)	0.31	**11664 (972)**	**1.00**	24274 (2840)	0.06	20530 (1672)	**1.00**	16018 (1280)	0.98	15506 (1416)	**1.00**
	80	49997 (6052)	0.22	**15959 (968)**	**1.00**	35906 (3451)	0.03	27194 (1887)	**1.00**	21150 (1734)	0.99	20470 (1437)	**1.00**
Rosenbrock	20	12826 (4263)	0.16	**6210 (2004)**	0.23	11114 (3186)	**0.31**	10430 (2643)	0.25	8654 (2712)	0.26	8354 (1740)	0.25
	40	56370 (17745)	0.03	**21748 (16331)**	0.30	38207 (5104)	0.33	37540 (5025)	0.36	27572 (6375)	**0.38**	27737 (6248)	0.29
	60	79728 (8592)	0.02	54747 (44932)	0.30	76898 (13408)	0.34	73290 (6672)	0.32	**54594 (10320)**	0.38	56466 (8576)	**0.46**
	80	221425 (0)	0.01	**65250 (46359)**	0.32	126805 (11110)	**0.45**	127961 (16601)	0.43	87824 (12971)	0.34	94505 (10387)	0.35
P-sig	20	3186 (366)	0.48	**2094 (825)**	**0.66**	5684 (1032)	**0.66**	5426 (1044)	**0.66**	4034 (705)	0.55	4154 (837)	0.57
	30	7168 (455)	0.28	**4928 (19453)**	0.38	24922 (2401)	0.48	25384 (4046)	**0.50**	15374 (1656)	0.35	17404 (2149)	0.45
	40	9922 (870)	0.20	**7042 (987)**	0.31	45002 (7757)	0.65	45602 (4830)	**0.70**	27857 (3113)	0.51	31382 (3071)	0.65

5 Conclusion

In this paper, we propose two modifications for the CMA-ES-PDM algorithm. Firstly, the marginal probability α is selected randomly from a pre-defined set. Secondly, when the probability model has certain intermediate values with zero probability, it is smoothed to enhance the sampling process. The numerical results show that the proposed methods are more robust, and generally run faster than the original CMA-ES-PDM in almost functions, except for the EllipsoidOneMax and EllipsoidLeadingOnes functions in large dimensions. They also perform better than CMA-ES with Margin in almost all test cases. However, they are still worse than DX-NES-ICI on some test functions.

In general, the variants of CMA-ES-PDM still perform worse on the non-convex and non-seperable functions. In future works, we aim to explore alternative techniques to enhance their performance, with a particular focus on leveraging larger population size (e.g., using the adaptive strategy in [16]) to address difficult problems such as the Rosenbrock and P-sig functions.

Acknowledgments. This work was partially supported by HNUE under Grant Number SPHN24-21 for the first author. The second author would like to thank the Vingroup Innovation Foundation (VinIF) for financial support.

References

1. Audet, C., Le Digabel, S., Tribes, C.: The mesh adaptive direct search algorithm for granular and discrete variables. SIAM J. Optim. **29**(2), 1164–1189 (2019)
2. Bliek, L., Guijt, A., Verwer, S., de Weerdt, M.: Black-box mixed-variable optimisation using a surrogate model that satisfies integer constraints. In: Proceedings of the Genetic and Evolutionary Computation Conference Companion, pp. 1851–1859. GECCO '21, Association for Computing Machinery, New York, NY, USA (2021)
3. Ciccazzo, A., Latorre, V., Liuzzi, G., Lucidi, S., Rinaldi, F.: Derivative-free robust optimization for circuit design. J. Optim. Theory Appl. **164**, 842–861 (2015)
4. Fujii, G., Takahashi, M., Akimoto, Y.: CMA-ES-based structural topology optimization using a level set boundary expression-application to optical and carpet cloaks. Comput. Methods Appl. Mech. Eng. **332**, 624–643 (2018)
5. Hamano, R., Saito, S., Nomura, M., Shirakawa, S.: CMA-ES with margin: lower-bounding marginal probability for mixed-integer black-box optimization. In: Proceedings of the Genetic and Evolutionary Computation Conference, pp. 639–647. GECCO '22, Association for Computing Machinery, New York, NY, USA (2022)
6. Hansen, N.: The CMA evolution strategy: a tutorial. CoRR **abs/1604.00772** (2016). http://arxiv.org/abs/1604.00772
7. Hansen, N., Ostermeier, A.: Completely derandomized self-adaptation in evolution strategies. Evol. Comput. **9**(2), 159–195 (2001)
8. Hansen, N.: A CMA-ES for mixed-integer nonlinear optimization. Technical report RR-7751, INRIA (2011). http://hal.inria.fr/inria-00362633/en/
9. Hazan, E., Klivans, A., Yuan, Y.: Hyperparameter optimization: a spectral approach. CoRR abs/1706.00764 (2017). http://arxiv.org/abs/1706.00764
10. Ikeda, K., Ono, I.: Natural evolution strategy for mixed-integer black-box optimization. In: Proceedings of the Genetic and Evolutionary Computation Conference, pp. 831–838. GECCO '23, Association for Computing Machinery, New York, NY, USA (2023)
11. Iyer, A., et al.: Data centric nanocomposites design via mixed-variable bayesian optimization. Mol. Syst. Des. Eng. **5**, 1376–1390 (2020)
12. Li, R., et al.: Mixed integer evolution strategies for parameter optimization. Evol. Comput. **21**, 29–64 (2013)
13. Liu, J., Ploskas, N., Sahinidis, N.V.: Tuning baron using derivative-free optimization algorithms. J. Glob. Optim. **74**, 611–637 (2019)
14. Liuzzi, G., Lucidi, S., Rinaldi, F.: Derivative-free methods for mixed-integer constrained optimization problems. J. Optim. Theory Appl. **164**, 933–965 (2015)
15. Miyagi, A., Akimoto, Y., Yamamoto, H.: Well placement optimization for carbon dioxide capture and storage via cma-es with mixed integer support. In: Proceedings of the Genetic and Evolutionary Computation Conference Companion, pp. 1696–1703. GECCO '18, Association for Computing Machinery, New York, NY, USA (2018)

16. Nguyen, D.M., Hansen, N.: Benchmarking cmaes-apop on the bbob noiseless testbed. In: Proceedings of the Genetic and Evolutionary Computation Conference Companion, pp. 1756–1763. GECCO '17, ACM, New York, NY, USA (2017)
17. Nguyen, D.M.: A combination of cma-es with probability distributions of integer variables for mixed-integer black-box optimization. In: Proceedings of the Genetic and Evolutionary Computation Conference Companion, pp. 415–418. GECCO '24 Companion, Association for Computing Machinery, New York, NY, USA (2024)
18. Nguyen, D.M., Le Thi, H.A., Pham Dinh, T.: Solving the multidimensional assignment problem by a cross-entropy method. J. Comb. Optim. **27**(4), 808–823 (2014)
19. Ploskas, N., Laughman, C., Raghunathan, A.U., Sahinidis, N.V.: Optimization of circuitry arrangements for heat exchangers using derivative-free optimization. Chem. Eng. Res. Des. **131**, 16–28 (2018). energy Systems Engineering
20. Rubinstein, R.Y., Kroese, D.P.: The Cross Entropy Method: A Unified Approach To Combinatorial Optimization, Monte-Carlo Simulation (Information Science and Statistics). Springer-Verlag, Berlin, Heidelberg (2004)

Solving Harmonic Balance Polynomial Systems of Duffing Oscillators Using Groebner Bases

Minh-Tuan Nguyen-Thai(✉) and Phu Minh Quang Nghiem

Hanoi University of Science and Technology, Dai Co Viet Str. 1, Hanoi, Vietnam
tuan.nguyenthaiminh@hust.edu.vn

Abstract. In this study, the harmonic balance method is applied to harmonically forced Duffing oscillator with a cubic nonlinearity of restoring force. The linear stiffness can be positive, zero or negative. The simplest forms of the governing equations are derived to ease latter steps. For an oscillator with nonlinearities of polynomial type only, the harmonic balance equations to be solved are of polynomial nature. To deal with a harmonic balance polynomial system, its Gröbner basis is computed to get an equivalent polynomial system whose unknowns can be found one by one. The adopted approach is used to find approximating symmetric periodic solution with up to harmonic order 3. It is shown that if the number of symbolic parameters of is low enough, the symbolic Gröbner basis can be computed. Some periodic solutions of period-1 and period-3 are reported.

Keywords: Duffing Oscillator · Groebner Basis · Harmonic Balance · Polynomial System

1 Introduction

For over a century since Duffing's book "Erzwungene schwingungen bei veränderlicher eigenfrequenz und ihre technische bedeutung" was published in 1918 [1], nonlinear oscillators with polynomial nonlinearities have been a great interest of researchers and engineers. It seems that this topic will continue to be studied for a long time because of not only the applications in engineering but also the richness in behaviors of such oscillators [2–4]. Some of the most common approaches to find periodic solutions of nonlinear oscillators are the shooting method [5], the cell-mapping method [6] and the harmonic balance method [7–9].

Due to the complexity of the problems and the numerousness of the solutions, a combination of methods or other improvements should be used to enhance the effectiveness of the calculation strategy. For example, the harmonic balance method can be combined with perturbation techniques [10]. One of the most important improvements that researchers are trying to make is how to effectively solve the algebraic system when applying the harmonic balance method, which is also called harmonic balance equations [7, 8]. In the cases of polynomial nonlinearities, the harmonic balance equations are a polynomial system. It motivates the idea of using the harmonic balance method together with Gröbner bases which is proposed by Grolet and Thouverez [11]. If a Gröbner basis

© The Author(s), under exclusive license to Springer Nature Switzerland AG 2025
H. A. Le Thi et al. (Eds.): ICAMCS 2024, LNNS 1311, pp. 96–109, 2025.
https://doi.org/10.1007/978-3-031-90606-0_9

is successfully found, all the roots of the polynomial system can be determined globally. However, to find a Gröbner basis requires much effort, and its success is not guaranteed. The study of Grolet and Thouverez has only given some results in some special cases including a free undamped system and systems with cyclic symmetry. Thus, there is still much room to continue in this direction.

In this study, forced Duffing systems with a cubic nonlinearity of restoring force are considered. There are three cases of linear stiffness: positive, zero and negative. The harmonic balance method is used to get harmonic balance polynomial systems. Then the Gröbner bases are found in some cases to solve the polynomial systems.

The main parts of the paper are structured as follows. Section 2 introduces the considered oscillators and their simplest forms. Section 3 briefly introduces Gröbner basis and its application in solving polynomial system. Section 4 introduces the concept of harmonic balance polynomial systems. Section 5 presents the harmonic balance polynomial systems of considered problems and some results related to their Gröbner bases.

2 Cubic Duffing Equations with Positive, Zero and Negative Linear Stiffness and Their Simplest Forms

Consider a forced Duffing oscillator

$$m\ddot{x} + d\dot{x} + k_1 x + k_3 x^3 = F\cos(\Omega t) \tag{1}$$

where six parameters m, d, k_1, k_3, F and Ω are constants. Three parameters m, F and Ω are positive. The number of parameters can be easily reduced by one by diving the whole equation by m. However, it can be reduced further by scaling both time and space variables.

2.1 Positive Linear Stiffness

In the case $k_1 > 0$, we use dimensionless time variable τ and dimensionless state variable s defined as follows.

$$\tau = \sqrt{\frac{k_1}{m}}t,\ s = \frac{k_1}{F}x \tag{2}$$

From Eqs. (2), we have

$$t = \sqrt{\frac{m}{k_1}}\tau,\ x = \frac{F}{k_1}s,\ \frac{1}{\mathrm{d}t} = \sqrt{\frac{k_1}{m}}\frac{1}{\mathrm{d}\tau} \tag{3}$$

Substituting Eqs. (3) into Eq. (1) yields

$$Fs'' + \frac{Fd}{\sqrt{mk_1}}s' + Fs + k_3\left(\frac{F}{k_1}\right)^3 s^3 = F\cos\left(\Omega\sqrt{\frac{m}{k_1}}\tau\right) \tag{4}$$

The dimensionless form of the equation is

$$s'' + \delta s' + s + \kappa s^3 = \cos(\eta\tau) \tag{5}$$

where the dimensionless parameters are defined as

$$\delta = \frac{d}{\sqrt{mk_1}}, \kappa = k_3 \frac{F^2}{k_1^3}, \eta = \Omega \sqrt{\frac{m}{k_1}} \tag{6}$$

Equation (5) is the simplest form of Eq. (1) when $k_1 > 0$. The number of parameters is reduced from six to three.

2.2 Positive Linear Stiffness

In the case $k_1 = 0$, we use dimensionless time variable τ and dimensionless state variable s defined as follows.

$$\tau = \Omega t, s = \frac{m\Omega^2}{F} x \tag{7}$$

From Eqs. (7), we have

$$t = \frac{1}{\Omega}\tau, x = \frac{F}{m\Omega^2} s, \frac{1}{\mathrm{d}t} = \Omega \frac{1}{\mathrm{d}\tau} \tag{8}$$

Substituting Eqs. (8) into Eq. (1) yields

$$Fs'' + \frac{Fd}{m\Omega} s' + k_3 \left(\frac{F}{m\Omega^2} \right)^3 s^3 = F\cos\tau \tag{9}$$

The dimensionless form of the equation is

$$s'' + \delta s' + \kappa s^3 = \cos\tau \tag{10}$$

where the dimensionless parameters are defined as

$$\delta = \frac{d}{m\Omega}, \kappa = k_3 \frac{F^2}{m^3\Omega^6} \tag{11}$$

Equation (11) is the simplest form of Eq. (1) when $k_1 = 0$. The number of parameters is reduced to two.

2.3 Negative Linear Stiffness

In the case $k_1 < 0$, we use dimensionless time variable τ and dimensionless state variable s defined as follows.

$$\tau = \sqrt{-\frac{k_1}{m}} t, s = -\frac{k_1}{F} x \tag{12}$$

From Eqs. (12), we have

$$t = \sqrt{-\frac{m}{k_1}} \tau, x = -\frac{F}{k_1} s, \frac{1}{\mathrm{d}t} = \sqrt{-\frac{k_1}{m}} \frac{1}{\mathrm{d}\tau} \tag{13}$$

Substituting Eqs. (3) into Eq. (1) yields

$$Fs'' + \frac{Fd}{\sqrt{-mk_1}}s' - Fs + k_3\left(-\frac{F}{k_1}\right)^3 s^3 = F\cos\left(\Omega\sqrt{-\frac{m}{k_1}}\tau\right) \tag{14}$$

The dimensionless form of the equation is

$$s'' + \delta s' - s + \kappa s^3 = \cos(\eta\tau) \tag{15}$$

where the dimensionless parameters are defined as

$$\delta = \frac{d}{\sqrt{-mk_1}}, \kappa = -k_3\frac{F^2}{k_1^3}, \eta = \Omega\sqrt{-\frac{m}{k_1}} \tag{16}$$

Equation (15) is the simplest form of Eq. (1) when $k_1 < 0$. The number of parameters is reduced from six to three.

3 Gröbner Basis and Polynomial System

Gröbner bases, named after Wolfgang Gröbner, are a powerful tool in computational algebraic geometry for solving systems of polynomial equations [12–14]. They provide a method for transforming a set of polynomials into a simpler equivalent set, from which the solutions can be more easily derived. In the context of nonlinear oscillators with polynomial nonlinearities, Gröbner bases offer a systematic approach to solving the polynomial systems that arise when applying the harmonic balance method.

Alphabetical order, or lexicographic order, is one of the several monomial orders that can be used when computing a Gröbner basis. When polynomials are arranged according to a lexicographic order, the system becomes easier to solve because this order prioritizes the leading terms in a way that simplifies back-substitution methods. With a lexicographic order, variables are eliminated step by step, similar to the Gaussian elimination process in linear algebra. In fact, Gröbner basis computation is considered as a generalization of Gaussian elimination [15]. This results in a structured reduction of the system that may enable a solving process that one variable at a time can be found. Such an arrangement reduces the complexity of solving the system, making lexicographic Gröbner bases especially advantageous in practical computations.

To compute a Gröbner basis for a given polynomial system, Buchberger's algorithm is commonly employed. This algorithm iteratively refines the initial set of polynomials by computing S-polynomials, which are used to eliminate leading terms and thus reduce the polynomials into a form where they generate the same ideal as the original set but with desirable computational properties. The algorithm continues until no further reductions are possible, resulting in a Gröbner basis. Despite its theoretical elegance, Buchberger's algorithm can be computationally expensive, especially for systems with many variables or high-degree polynomials. Even with constant improvements in the past decades [16–19], the required time to compute a Gröbner basis, especially a symbolic one, is not always reasonable. Therefore, its application requires careful consideration, particularly in complex systems like those encountered in nonlinear dynamics. The computation of Gröbner bases should be done using a computer [20].

4 Harmonic Balance Polynomial System

Approximate period-n solution of Eq. (5) or (10) or (15) is found in the following form

$$s = a_0 + \sum_{k=1}^{N} \left(a_k \cos\left(\frac{k\eta\tau}{n}\right) + b_k \sin\left(\frac{k\eta\tau}{n}\right)\right) \tag{17}$$

where N is the number of harmonic orders and coefficients a_0, a_k and b_k are unknowns. Denote

$$\boldsymbol{a} = [a_0, a_1, \ldots, a_N, b_1, b_2, \ldots, b_N]^{\mathrm{T}} \tag{18}$$

Substituting (17) into Eq. (5) or (10) or (15) and performing equivalent transformations yield

$$C_0(\boldsymbol{a}) + \sum_{k=1}^{M} \left(C_k(\boldsymbol{a}) \cos\left(\frac{k\eta\tau}{n}\right) + S_k(\boldsymbol{a}) \sin\left(\frac{k\eta\tau}{n}\right)\right) = 0 \tag{19}$$

where M is an integer, $M \geq N$, and functions $C_0(\boldsymbol{a})$, $C_k(\boldsymbol{a})$ and $S_k(\boldsymbol{a})$ are polynomials.

Equating two sides of Eq. (19) up to harmonic order N, one obtain a system of $2k + 1$ polynomials to be solved.

$$\boldsymbol{f}_N(\boldsymbol{a}) = [C_0(\boldsymbol{a}), C_1(\boldsymbol{a}), \ldots, C_N(\boldsymbol{a}), S_1(\boldsymbol{a}), S_2(\boldsymbol{a}), \ldots, S_N(\boldsymbol{a})]^{\mathrm{T}} = 0 \tag{20}$$

Equation (20) is normally called harmonic balance system of equations. Here, because of its polynomial nature, it is called harmonic balance polynomial system.

Similarly, to find symmetric period-$(2n-1)$ solution, the form of the approximate solution is

$$s = \sum_{k=1}^{N} \left(a_k \cos\left(\frac{(2k-1)\eta\tau}{2n-1} \right) + b_k \sin\left(\frac{(2k-1)\eta\tau}{2n-1} \right) \right) \tag{21}$$

Denote

$$\boldsymbol{a}_{\mathrm{sym}} = \left[a_1, a_3 \ldots, a_{2N-1}, b_1, b_3, \ldots, b_{2N-1} \right]^{\mathrm{T}} \tag{22}$$

The harmonic balance polynomial system to be solve is

$$\left[C_1(\boldsymbol{a}_{\mathrm{sym}}), C_3(\boldsymbol{a}_{\mathrm{sym}}), \ldots, C_{2N-1}(\boldsymbol{a}_{\mathrm{sym}}), S_1(\boldsymbol{a}), S_3(\boldsymbol{a}), \ldots, S_{2N-1}(\boldsymbol{a}_{\mathrm{sym}}) \right]^{\mathrm{T}} = 0 \tag{23}$$

5 Application to Cubic Duffing Equation

5.1 Application to Cubic Duffing Equation with Positive Linear Stiffness

Approximating symmetric period-1 solution with up to harmonic order 3, one gets the harmonic balance polynomial system as follows.

$$\left(1-\eta^2\right)a_1 + \delta\eta b_1 + 3\kappa\left(\frac{a_1^3}{4} + \frac{a_1^2 a_3}{4} + \frac{a_1 b_1^2}{4} + \frac{a_1 b_1 b_3}{2} + \frac{a_1 a_3^2}{2} + \frac{a_1 b_3^2}{2} - \frac{b_1^2 a_3}{4} \right) = 1 \tag{24}$$

$$\left(1-\eta^2\right)b_1 - \delta\eta a_1 + 3\kappa\left(\frac{a_1^2 b_1}{4} + \frac{a_1^2 b_3}{4} - \frac{a_1 b_1 a_3}{2} + \frac{b_1^3}{4} - \frac{b_1^2 b_3}{4} + \frac{b_1 a_3^2}{2} + \frac{b_1 b_3^2}{2} \right) = 0 \tag{25}$$

$$\left(1-9\eta^2\right)a_3 + 3\delta\eta b_3 + \kappa\left(\frac{a_1^3}{4} + \frac{3a_1^2 a_3}{2} - \frac{3a_1 b_1^2}{4} + \frac{3b_1^2 a_3}{2} + \frac{3a_3^3}{4} + \frac{3a_3 b_3^2}{4} \right) = 0 \tag{26}$$

$$\left(1-9\eta^2\right)b_3 - 3\delta\eta a_3 + \kappa\left(\frac{3a_1^2 b_1}{4} + \frac{3a_1^2 b_3}{2} - \frac{b_1^3}{4} + \frac{3b_1^2 b_3}{2} + \frac{3a_3^2 b_3}{4} + \frac{3b_3^3}{4} \right) = 0 \tag{27}$$

To find symbolic Gröbner basis, we consider a simpler case of $\eta = 1$. In that case, a polynomial of degree 9 in a_1 is found.

$$P_9(a_1) = c_0 + c_1 a_1 + c_2 a_1^2 + c_3 a_1^3 + c_4 a_1^4 + c_5 a_1^5 + c_6 a_1^6 + c_7 a_1^7 + c_8 a_1^8 + a_1^9 \tag{28}$$

where

$$c_0 = -\frac{32}{7203\kappa^7}(21618690048\delta^{12} + 264880521216\delta^{10} - 16929054720\delta^8\kappa + 9332431257760\delta^8 + 797247736\delta^6\kappa^2 + 34809151488\delta^6\kappa + 1015893983232\delta^6 - 98375680\delta^4\kappa^2 + 14136901632\delta^4\kappa + 181216\delta^2\kappa^3 + 115605504\delta^2\kappa^2 + 49\kappa^4)$$

$$c_1 = \frac{128}{21609\kappa^8}(216186900048\delta^{16} + 2648805212216\delta^{14} - 21733208064\delta^{12}\kappa + 933243125760\delta^{12} - 12357343296\delta^{10}\kappa^2 + 101096816640\delta^{10}\kappa + 1015893983232\delta^{10} - 12521602048\delta^8\kappa^2 - 17609785344\delta^8\kappa + 2264075424\delta^6\kappa^3 + 593014947840\delta^6\kappa^2 - 1622817\delta^4\kappa^4 - 304871424\delta^4\kappa^3 + 5083008\delta^2\kappa^4 + 1176\kappa^5)$$

$$c_2 = \frac{128}{21609\kappa^7}(570177146880\delta^{14} + 4445269524480\delta^{12} - 147703866880\delta^{10}\kappa + 13587217121280\delta^{10} - 32133184080\delta^8\kappa^2 + 882128322560\delta^8\kappa + 15238409748480\delta^8 - 835693824\delta^6\kappa^2 + 450153676800\delta^6\kappa + 490827120\delta^4\kappa^3 + 27630796800\delta^4\kappa^2 + 15435\delta^2\kappa^4 + 25267200\delta^2\kappa^3 - 9408\kappa^4)$$

$$c_3 = \frac{8}{21609\kappa^6}(8303629352960\delta^{12} + 32802587607040\delta^{10} - 1934848296960\delta^8\kappa + 70302386094080\delta^8 - 3995964000\delta^6\kappa^2 + 3246279229440\delta^6\kappa + 91430458490880\delta^6 - 702053376\delta^4\kappa^2 + 1272321146880\delta^4\kappa - 20187552\delta^2\kappa^3 + 10404495360\delta^2\kappa^2 - 2205\kappa^4)$$

$$c_4 = \frac{32}{7203\kappa^5}(260476952320\delta^{10} + 2115358720\delta^8 - 21600385280\delta^6\kappa - 587053137920\delta^6 - 7214445\delta^4\kappa^2 + 2436464640\delta^4\kappa - 76348160\delta^2\kappa^2 + 2646\kappa^3)$$

$$c_5 = \frac{64}{7203\kappa^4}(21676442400\delta^8 + 77350016080\delta^6 - 208004440\delta^4\kappa - 11291688960\delta^4 + 53949\delta^2\kappa^2 - 112788480\delta^2\kappa + 4704\kappa^2)$$

$$c_6 = \frac{2}{1029\kappa^3}(6647805440\delta^6 + 1060741120\delta^4 + 2798208\delta^2\kappa - 441\kappa^2)$$

$$c_7 = \frac{16}{49\kappa^2}(72553\delta^4 + 24768\delta^2 - 35\kappa)$$

$$c_8 = -\frac{2472\delta^2}{7\kappa}$$

For $\delta = 0.25$, $\kappa = 3$, polynomial (28) has one real root. The approximate solution is characterized by

$$a_1 = 0.724423, a_3 = 0.0486391, b_1 = 0.142862, b_3 = 0.0234695 \quad (29)$$

Figure 1a illustrates the time history of the approximate solution. The phase planes in Fig. 1b show that the approximate solution is close to the solution obtained using numerical integration.

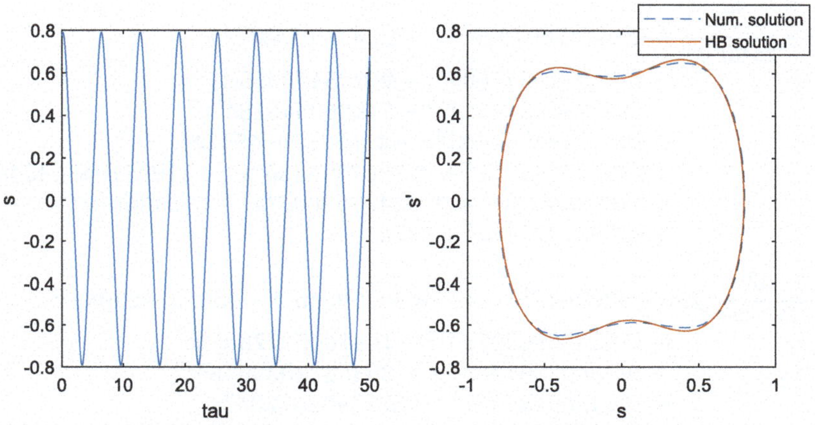

a) Time history b) Phase plane

Fig. 1. Period-1 motion in the case of positive linear stiffness and $\delta = 0.25$, $\kappa = 3$, $\eta = 1$.

5.2 Application to Cubic Duffing Equation with Zero Linear Stiffness

Approximating symmetric period-1 solution with up to harmonic order 3, one gets the harmonic balance polynomial system as follows.

$$3\kappa\left(\frac{a_1^3}{4} + \frac{a_1^2 a_3}{4} + \frac{a_1 b_1^2}{4} + \frac{a_1 b_1 b_3}{2} + \frac{a_1 a_3^2}{2} + \frac{a_1 b_3^2}{2} - \frac{b_1^2 a_3}{4}\right) - a_1 + \delta b_1 - 1 = 0 \quad (30)$$

$$3\kappa\left(\frac{a_1^2 b_1}{4} + \frac{a_1^2 b_3}{4} - \frac{a_1 b_1 a_3}{2} + \frac{b_1^3}{4} - \frac{b_1^2 b_3}{4} + \frac{b_1 a_3^2}{2} + \frac{b_1 b_3^2}{2}\right) - \delta a_1 - b_1 = 0 \quad (31)$$

$$\kappa\left(\frac{a_1^3}{4}+\frac{3a_1^2a_3}{2}-\frac{3a_1b_1^2}{4}+\frac{3b_1^2a_3}{2}+\frac{3a_3^3}{4}+\frac{3a_3b_3^2}{4}\right)-9a_3+3\delta b_3=0 \quad (32)$$

$$\kappa\left(\frac{3a_1^2b_1}{4}+\frac{3a_1^2b_3}{2}-\frac{b_1^3}{4}+\frac{3b_1^2b_3}{2}+\frac{3a_3^2b_3}{4}+\frac{3b_3^3}{4}\right)-3\delta a_3-9b_3= \quad (33)$$

Using Gröbner basis, a polynomial of degree 9 in a_1 is found with the following coefficients.

$$\begin{aligned}c_0=-\frac{32}{21609\kappa^8}&(-864747601920\delta^{14}+648560701440\delta^{12}\kappa\\&-118693195776\delta^{12}+9629517035520\delta^{10}\kappa\\&-416073845145\delta^{10}-51880854528\delta^8\kappa^2\\&+30412494950400\delta^8\kappa-434525766451\delta^8\\&+239174208\delta^6\kappa^3+82700779648\delta^6\kappa^2+319951843430\delta^6\kappa\\&+65558688\delta^4\kappa^3+73924099200\delta^4\kappa^2+653436\delta^2\kappa^4\\&+495521712\delta^2\kappa^3+147\kappa^5)\end{aligned}$$

$$\begin{aligned}c_1=\frac{32}{7203\kappa^8}&(288249200640\delta^{16}+424468905984\delta^{14}-130833623040\delta^{12}\kappa\\&+1782556803072\delta^{12}-16476457728\delta^{10}\kappa^2\\&-203660338176\delta^{10}\kappa+2835332038656\delta^{10}\\&-40445804800\delta^8\kappa^2-962234526720\delta^8\kappa\\&+1448419221504\delta^8+285349270\delta^6\kappa^3+114308610560\delta^6\kappa^2\\&-1066506144768\delta^6\kappa-216375\delta^4\kappa^4+171578928\delta^4\kappa^3\\&-24641366400\delta^4\kappa^2+815446\delta^2\kappa^4-165173904\delta^2\kappa^3\\&+1617\kappa^5)\end{aligned}$$

$$\begin{aligned}c_2=\frac{32}{7203\kappa^7}&(760236195840\delta^{14}+717141000192\delta^{12}-21741694720\delta^{10}\kappa\\&+2202965041152\delta^{10}-4284424544\delta^8\kappa^2+959174594560\delta^8\kappa\\&+2204335865856\delta^8+181603606\delta^6\kappa^2\\&+1235638208000\delta^6\kappa+69012132\delta^4\kappa^3+5101874688\delta^4\kappa^2\\&+20580\delta^2\kappa^4+538400520\delta^2\kappa^3-12495\kappa^4)\end{aligned}$$

$$\begin{aligned}c_3=\frac{8}{21609\kappa^6}&(8303629352960\delta^{12}+4278969729024\delta^{10}-2284942164480\delta^8\kappa\\&+10078697914368\delta^8-39959640000\delta^6\kappa^2\\&+2551952570880\delta^6\kappa+10317651304448\delta^6\\&-649046880\delta^4\kappa^2+2786915251200\delta^4\kappa-22070160\delta^2\kappa^3\\&+203217840\delta^2\kappa^2-2205\kappa^4-169932\kappa^3)\end{aligned}$$

$$\begin{aligned}c_4=-\frac{8}{7203\kappa^5}&(-1041907809280\delta^{10}-1170847508480\delta^8+99476028480\delta^6\kappa\\&+1106371834880\delta^6+28857780\delta^4\kappa^2+1307303712\delta^4\kappa\\&+465382680\delta^2\kappa^2-10437\kappa^3)\end{aligned}$$

$$c_5 = \frac{64}{2401\kappa^4}(722548080\delta^8 + 335825984\delta^6 - 9011548\delta^4\kappa - 487238512\delta^4 + 17983\delta^2\kappa^2 - 5667970\delta^2\kappa + 2450\kappa^2)$$

$$c_6 = \frac{2}{1029\kappa^3}(664780544\delta^6 + 252330240\delta^4 + 3392424\delta^2\kappa - 441\kappa^2)$$

$$c_7 = \frac{16}{49\kappa^2}(72553\delta^4 + 32225\delta^2 - 42\kappa)$$

$$c_8 = -\frac{2472\delta^2}{7\kappa}$$

For $\delta = 0.25, \kappa = 3$, the polynomial in a_1 has one real root. The approximate solution is characterized by

$$a_1 = 0.871615, a_3 = 0.0827474, b_1 = 0.225996, b_3 = 0.0592305 \quad (34)$$

Figure 2a illustrates the time history of the approximate solution. The phase planes in Fig. 2b show that the approximate solution is close to the solution obtained using numerical integration.

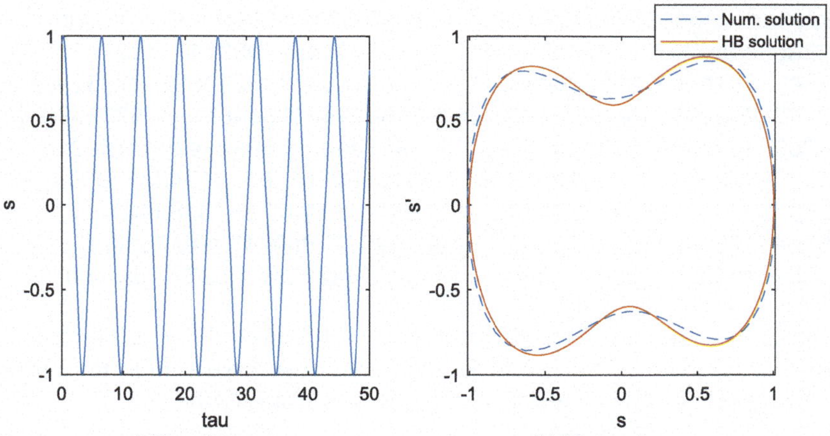

a) Time history b) Phase plane

Fig. 2. Period-1 motion in the case of zero linear stiffness and $\delta = 0.25$, $\kappa = 3$.

5.3 Application to Cubic Duffing Equation with Negative Linear Stiffness

Approximating symmetric period-3 solution with up to harmonic order 3, one gets the harmonic balance polynomial system as follows.

$$-\left(\eta^2+1\right)a_1+\delta\eta b_1+3\kappa\left(\frac{a_1^3}{4}+\frac{a_1^2a_3}{4}+\frac{a_1b_1^2}{4}+\frac{a_1b_1b_3}{2}+\frac{a_1a_3^2}{2}+\frac{a_1b_3^2}{2}-\frac{b_1^2a_3}{4}\right)=0 \tag{35}$$

$$-\left(\eta^2+1\right)b_1-\delta\eta a_1+3\kappa\left(\frac{a_1^2b_1}{4}+\frac{a_1^2b_3}{4}-\frac{a_1b_1a_3}{2}+\frac{b_1^3}{4}-\frac{b_1^2b_3}{4}+\frac{b_1a_3^2}{2}+\frac{b_1b_3^2}{2}\right)=0 \tag{36}$$

$$-\left(9\eta^2+1\right)a_3+3\delta\eta b_3+\kappa\left(\frac{a_1^3}{4}+\frac{3a_1^2a_3}{2}-\frac{3a_1b_1^2}{4}+\frac{3b_1^2a_3}{2}+\frac{3a_3^3}{4}+\frac{3a_3b_3^2}{4}\right)=1 \tag{37}$$

$$-\left(9\eta^2+1\right)b_3-3\delta\eta a_3+\kappa\left(\frac{3a_1^2b_1}{4}+\frac{3a_1^2b_3}{2}-\frac{b_1^3}{4}+\frac{3b_1^2b_3}{2}+\frac{3a_3^2b_3}{4}+\frac{3b_3^3}{4}\right)=0 \tag{38}$$

To find a Gröbner basis, we consider a specific case of $\eta = 3, \delta = 0.25, \kappa = 3$. In that case, a polynomial of degree 19 in a_1 is found with the following coefficients.

$$\begin{aligned}
&3372324133029357457244 16a_1^{19} - 26978593064234859657953 28a_1^{17} - \\
&19538068389773261458636 8a_1^{16} + 92535312456398244717527 04a_1^{15} + \\
&97154217910602835270041 6a_1^{14} - 16667160574862338816278528a_1^{13} - \\
&2227322272074050085322752a_1^{12} + 16890148103646130336235520a_1^{11} + \\
&2747036309793075137101824a_1^{10} - 9863060657686194280218624a_1^{9} - \\
&1748239970042094834106368a_1^{8} + 3415408361285796342372096a_1^{7} + \\
&537194548967098342763520a_1^{6} - 758896161462786480880896a_1^{5} - \\
&89526405780607489162560a_1^{4} + 100949537796024799634400a_1^{3} + \\
&1296879396652768398120 0a_1^{2} - 1453961982953720632375a_1 = 0
\end{aligned}$$

Among the real roots of the polynomial, $a_1 = 0$ leads to $b_1 = 0$, which means it is actually a period-1 solution. Another real root is associated with a stable period-3 motion, as checked using numerical integration. It is characterized by

$$a_1 = -0.229167, a_3 = -0.898175, b_1 = -0.0870788, b_3 = 0.0823799 \tag{39}$$

Figure 3 illustrates the approximate solution. It shows that the approximate solution is close to the solution obtained using numerical integration.

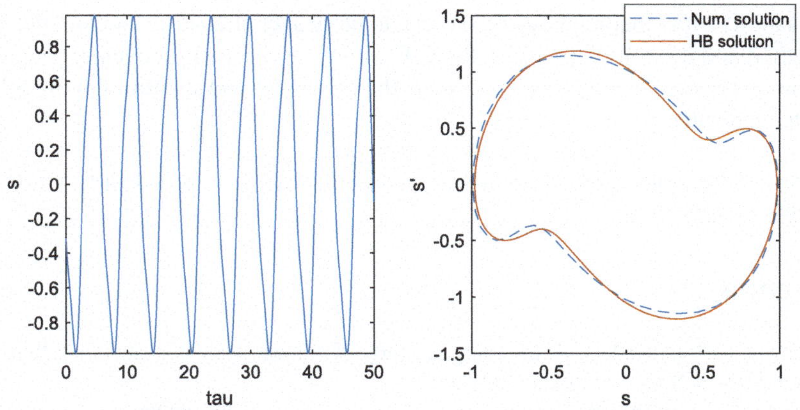

a) Time history b) Phase plane

Fig. 3. Period-3 motion in the case of negative linear stiffness and $\delta = 0.25$, $\kappa = 3$, $\eta = 3$.

6 Conclusions and Outlook

Harmonically forced Duffing oscillators with a cubic nonlinearity of restoring force are considered and simplified to minimize the number of parameters without loss of generality. In the cases of positive and negative linear stiffness, both simplest forms have three symbolic parameters. For a system with zero linear stiffness, the number of symbolic parameters is two.

The harmonic balance equations of the considered oscillators are of polynomial nature so that they are called harmonic balance polynomial systems. This paper focuses on finding approximate symmetric periodic solution with up to harmonic order 3. The associated polynomial systems have four equations, two for harmonic order 1 and two for harmonic order 3.

When the dimensionless angular frequency of the excitation is fixed as 1, the symbolic Gröbner basis for the case of positive linear stiffness and period-1 motion is found. The basis includes a univariate polynomial of degree 9. A numerical example shows that there is a stable period-1 motion agree with the result obtained using the proposed approach. For the oscillator of zero linear stiffness, the symbolic Gröbner basis for period-1 motion is found. The basis includes a univariate polynomial of degree 9. A numerical example shows that there is a stable period-1 motion agree with the result obtained using the proposed approach. Considering period-3 motions, for the case of negative linear stiffness, the basis includes a univariate polynomial of degree 19. A numerical example shows that there is a stable period-3 motion agree with the result obtained using the proposed approach.

The obtained results show that the combination of the harmonic balance method and Gröbner bases has a great potential in finding periodic motions of oscillators with polynomial nonlinearities.

In future studies, more complicated problems should be considered. Some examples are finding asymmetric motions and motions of higher period and dealing with multi-DOF oscillators. Improvements in calculation approach should also be considered, e.g., combinations with other methods such as the numerical continuation method, the shooting method and the cell-mapping method. The purposes of the improvement are to find the solutions more efficiently and to validate the existences and stabilities of the solution more conveniently.

Disclosure of Interests. The authors have no competing interests to declare that are relevant to the content of this article.

References

1. Kovacic, I., Brennan, M.J.: The Duffing Equation: Nonlinear Oscillators and Their Behaviour. John Wiley & Sons, Cambridge (2011)
2. Brennan, M.J., Kovacic, I., Carrella, A., Waters, T.P.: On the jump-up and jump-down frequencies of the duffing oscillator. J. Sound Vib. **318**(4–5), 1250–1261 (2008)
3. Kudryashov, N.A.: The generalized duffing oscillator. Commun. Nonlinear Sci. Numer. Simul. **93**, 105526 (2021)
4. Nguyen-Thai, M.T., Wulff, P., Gräbner, N., von Wagner, U.: On the dynamics of a 2-DOF nonlinear vibratory system with bistable characteristic and circulatory forces. Int. J. Non-Linear Mech. **150**, 104342 (2023)
5. Osborne, M.R.: On shooting methods for boundary value problems. J. Math. Anal. Appl. **27**(2), 417–433 (1969)
6. Hsu, C.S.: Cell-to-cell mapping: a method of global analysis for nonlinear systems (Vol. 64). Springer Science & Business Media (2013)
7. Krack, M., Gross, J.: Harmonic balance for nonlinear vibration problems. Springer International Publishing, Cham (2019)
8. Yan, Z., Dai, H., Wang, Q., Atluri, S.N.: Harmonic BalanceMethods: a review and recent developments. CMES – Comput. Model. Eng. Sci. **137**(2) (2023)
9. von Wagner, U., Lentz, L.: On the detection of artifacts in Harmonic Balance solutions of nonlinear oscillators. Appl. Math. Model. **65**, 408–414 (2019)
10. Lewis, A.P.: Approximations to limit cycles for a nonlinear multi-degree-of-freedom system with a cubic nonlinearity through combining the harmonic balance method with perturbation techniques. Int. J. Non-Linear Mech. **126**, 103590 (2020)
11. Grolet, A., Thouverez, F.: Computing multiple periodic solutions of nonlinear vibration problems using the harmonic balance method and Gröbner bases. Mech. Syst. Signal Process. **52**, 529–547 (2015)
12. Buchberger, B.: Bruno Buchberger's PhD thesis 1965: an algorithm for finding the basis elements of the residue class ring of a zero dimensional polynomial ideal. J. Symb. Comput. **41**(3–4), 475–511 (2006)
13. Adams, W.W., Loustaunau, P.: An introduction to Gröbner bases (Vol. 3). American Mathematical Society (2022)
14. Buchberger, B., Winkler, F. (Eds.): Gröbner Bases and Applications (Vol. 251). Cambridge University Press, Cambridge (1998)
15. Lazard, D.: Gröbner bases, Gaussian elimination and resolution of systems of algebraic equations. In: van Hulzen, J.A. (eds.) Computer Algebra. EUROCAL 1983. LNCS, vol. 162. Springer, Berlin, Heidelberg (1983)
16. Möller, H.M.: On the construction of Gröbner bases using syzygies. J. Symb. Comput. **6**(2–3), 345–359 (1988)
17. Montes, A.: A new algorithm for discussing Gröbner bases with parameters. J. Symb. Comput. **33**(2), 183–208 (2002)

18. Faugere, J.C.: A new efficient algorithm for computing Gröbner bases (F4). J. Pure Appl. Algebra **139**(1–3), 61–88 (1999)
19. Gao, S., Guan, Y., Volny IV, F.: A new incremental algorithm for computing Gröbner bases. In: Proceedings of the 2010 International Symposium on Symbolic and Algebraic Computation, pp. 13–19 (2010)
20. Faugère, J.C.: FGb: a library for computing Gröbner bases. In: International Congress on Mathematical Software, pp. 84–87. Springer, Berlin, Heidelberg (2010)

A Gradient Descent Approach for Multi-objective Picture-Fuzzy Stochastic Programming Problems

Truong Tuan Khang[1,2], Ta Anh Son[1], and Ngoc Thang Tran[1,2](✉)

[1] Faculty of Mathematics and Informatics, Hanoi University of Science and Technology, No. 1 Dai Co Viet, Hai Ba Trung, Hanoi, Vietnam
khang.tt232016M@sis.hust.edu.vn, {son.taanh,thang.tranngoc}@hust.edu.vn
[2] Institute for Digital Technology and Economy (BK Fintech), Hanoi University of Science and Technology, Hanoi, Vietnam

Abstract. This paper explores multi-objective stochastic fuzzy linear programming problems using picture-fuzzy theory to model parameter uncertainty and fuzziness. Picture-fuzzy theory provides a flexible way to handle uncertain data by representing acceptance, rejection, and hesitation degrees. Our study introduces a method to transform the initial stochastic fuzzy problem into a quasiconvex programming problem. Our study enhances computational efficiency and ensures algorithm convergence by applying a gradient descent approach rather than conventional heuristic methods. Our study provides theoretical proof using quasiconvex optimization to validate the proposed method, establishing a foundation for its convergence and effectiveness. To illustrate the method's feasibility and efficacy, the paper presents computational examples demonstrating its correctness and potential applications, particularly in economics and finance where uncertainty and fuzziness in market data are significant. The research opens new pathways for solving complex programming problems in uncertain environments.

Keywords: Multiobjective stochastic linear programming · Picture Fuzzy decision · Probability maximization · Gradient Descent · Variance covariance matrices

1 Introduction

This paper addresses the complex challenge of multi-objective stochastic linear programming, where the problem becomes more complicated due to the randomness not only of vector b in the inequality $A \leq b$, but also of matrix A, whereas previous studies have mentioned this but focused only on solving with the stochastic component b (see [4,5]). Such randomness in both components

---

T. T. Khang and T. A. Son—Contributing authors.

© The Author(s), under exclusive license to Springer Nature Switzerland AG 2025
H. A. Le Thi et al. (Eds.): ICAMCS 2024, LNNS 1311, pp. 110–122, 2025.
https://doi.org/10.1007/978-3-031-90606-0_10

significantly increases the complexity of the constraint space. Despite these complexities, we have successfully demonstrated that the overall feasible set remains convex, an essential property for applying advanced optimization techniques.

To effectively tackle this increased complexity, we propose a new approach using picture fuzzy theory (see [6]). This theory extends traditional fuzzy logic by incorporating degrees of acceptance, rejection, and hesitation, thus providing a richer framework for representing uncertainty. The improvements include the use of monotonic membership functions in picture fuzzy sets, ensuring that as the uncertainty in the parameters decreases, the membership values adjust in a predictable and consistent manner. This feature is crucial for maintaining the integrity of decision-making processes under uncertainty, especially in multi-objective situations. This is the key difference between our research and previous studies that have employed non-monotonic functions such as Gaussian (see [18]), trapezoidal (see [16,17]), etc.

Our paper further proposes a methodological transformation of the original stochastic fuzzy problem into a quasiconvex programming problem. To solve this type of problem, previous studies have used traditional heuristic methods (see [13,14]). Nonetheless, we have shown that this class of issues can be resolved using a gradient descent approach (see [12,19]), which is a robust and widely applied method in optimization problems. To ensure that the proposed gradient method can solve the proposed problem, we provide rigorous theoretical proof using quasiconvex optimization techniques. This proof establishes a solid foundation for the convergence and effectiveness of the algorithm, ensuring that the derived solutions are both theoretically robust and operationally viable. To demonstrate the practicality and accuracy of our approach, the paper includes computational examples that illustrate the validity of the methodology.

This problem is particularly relevant to the fields of economics and finance, where portfolio optimization and other financial instruments often grapple with significant uncertainty and ambiguity in market data. The objective function in this problem is the Sharpe Ratio (SR) (see [15]), which is a widely recognized measure for evaluating risk-adjusted returns, quantifying the ratio of expected returns to the portfolio's standard deviation. While SR is extensively used in financial applications and carries substantial economic relevance, its non-convex nature poses challenges when directly used in optimization problems. Nevertheless, this paper demonstrates that objective functions like SR can still be efficiently and globally solved using existing convex programming techniques (see also [12]). The global convergence of the solution is guaranteed.

Regarding the structure of this paper, we have organized it into key sections to facilitate a thorough understanding of the concepts and methodologies employed. In Section 2, we delve into the foundational theories of fuzzy logic and quasiconvex optimization, setting the stage for their application in complex stochastic problems. Section 3 is dedicated to presenting the general framework of the multi-objective stochastic linear optimization problem, detailing how we incorporate picture-fuzzy sets to model the inherent uncertainty. This section also includes the fuzzification process of both the objective functions and constraints to adequately capture the nuances of randomness in A and b. To demonstrate the practical application and effectiveness of our proposed method, we conclude Section 3 with computational example that highlight how our approach can be effectively applied.

2 Preliminaries: Concepts and Definitions

2.1 Picture Fuzzy

Given a universal set $\mathbb{U}$, a generalization of fuzzy sets that introduces more granularity and flexibility in representing uncertainty is called a picture fuzzy set of $\mathbb{U}$, denoted by $\tilde{S}$. This set is described by three distinct mappings: the positive membership function $\mu_{\tilde{S}} : \mathbb{U} \to [0,1]$, the neutral membership function $\eta_{\tilde{S}} : \mathbb{U} \to [0,1]$, and the negative membership function $\nu_{\tilde{S}} : \mathbb{U} \to [0,1]$. The definition of the picture fuzzy set $\tilde{S}$ is formalized as follows:

Definition 1 (see [6]). *Given the condition* $0 \leq \mu_{\tilde{S}}(\theta) + \eta_{\tilde{S}}(\theta) + \nu_{\tilde{S}}(\theta) \leq 1$ *for all* $\theta \in \mathbb{U}$*, the picture fuzzy set* $\tilde{S}$ *is given by:*

$$\tilde{S} = \{(\theta, \mu_{\tilde{S}}(\theta), \eta_{\tilde{S}}(\theta), \nu_{\tilde{S}}(\theta)) \mid \theta \in \mathbb{U}\} \tag{1}$$

We can now express the general optimization problem with the following formulation:

$$\begin{aligned} \min \quad & F(\boldsymbol{\theta}) \\ \text{s.t} \quad & z_t(\boldsymbol{\theta}) \leq 0, \quad t = 1, \ldots, k, \\ & \boldsymbol{\theta} \in \mathcal{X}, \end{aligned} \tag{2}$$

where $\mathcal{X}$ denotes a compact convex set ($\mathcal{X} \neq \emptyset$).

This study employs a fuzzy optimization model that integrates fuzzy components, treating the variables and parameters as exact values. This is termed flexible optimization (see [1]). Two categories of fuzzy components exist. The first concerns the objective function (indicated by ($\widetilde{\min}$ or $\widetilde{\max}$)), while the second applies to a fuzzy relation (such as ($\preceq, \simeq, \succeq$)). A fuzzy optimization problem comprising fuzzy elements can be expressed as follows:

$$\begin{aligned} \widetilde{\min} \quad & F(\boldsymbol{\theta}) \\ \text{s.t} \quad & z_t(\boldsymbol{\theta}) \preceq 0, \quad t = 1, \ldots, k, \\ & \boldsymbol{\theta} \in \mathcal{X} \end{aligned} \tag{3}$$

In this formulation, the symbols "$\widetilde{\min}$" and "$\preceq$" refer to the fuzzy versions of "minimize" and "less than or equal to," respectively, implying that the objective function should be minimized as much as feasible and that the constraints should be as well accepted as possible. Fuzzy optimization approaches with fuzzy objectives are widely utilized in decision-making for real-world applications.

2.2 Quasiconvexity

Definition 2 (Pseudoconvex function (see [7])). *Let $\mathcal{X}$ denote a non-empty convex set, and let $f : \mathbb{R}^n \to \mathbb{R}$ represent a differentiable function defined on $\mathcal{X}$. We characterize f as a pseudoconvex function on $\mathcal{X}$ If for each $\theta^1, \theta^2 \in \mathcal{X}$, the subsequent condition is satisfied:*

$$f(\theta^2) < f(\theta^1) \Rightarrow \langle \nabla f(\theta^1), \theta^2 - \theta^1 \rangle < 0. \tag{4}$$

If f is pseudoconvex, then the negative of f, i.e., $-f$, is referred to as a pseudoconcave function.

Definition 3 (Semistrictly quasiconvex function (see [2]). *Let $\mathcal{X} \subset \mathbb{R}^n$ be a convex set, and f be defined on $\mathcal{X}$. If for any $\theta^1, \theta^2 \in \mathcal{X}$ and $0 < \lambda < 1$, the condition*

$$f(\theta^1) > f(\theta^2) \tag{5}$$

implies that

$$f(\lambda\theta^1 + (1 - \lambda)\theta^2) < f(\theta^1), \tag{6}$$

then f is called semistrictly quasiconvex.

Definition 4 (Semistrictly quasiconvex programming problem). *The semistrictly quasiconvex programming problem is articulated as follows:*

$$\min f(\theta) \quad s.t. \quad \theta \in \mathcal{X}, \tag{SQP}$$

where $\mathcal{X} \subset \mathbb{R}^n$ is a convex set, and $f(\theta)$ is semistrictly quasiconvex on $\mathcal{X}$.

Proposition 1. *In the semistrictly quasiconvex programming problem (SQP), any local minimum is also a global minimum (see [1]). In light of this, convex programming methods are a powerful tool for solving this challenge.*

3 Methodology and Problem Formulation

3.1 Multiobjective Stochastic Linear Programming Problems

We focus on multiobjective stochastic linear programming in this part. The issue at hand can be stated as:

$$\max \overline{\boldsymbol{C}}\boldsymbol{\theta} = (\overline{\boldsymbol{c}}_1\boldsymbol{\theta}, \ldots, \overline{\boldsymbol{c}}_k\boldsymbol{\theta}) \quad \text{s.t.} \quad \overline{\boldsymbol{A}}\boldsymbol{\theta} \leq \overline{\boldsymbol{b}}, \quad \boldsymbol{\theta} \geq 0, \tag{MSLP}$$

where $\boldsymbol{\theta} = (\theta_1, \theta_2, \ldots, \theta_n)^T$ is an n-dimensional decision variable vector, $\overline{\boldsymbol{c}}_i = (\bar{c}_{i1}, \ldots, \bar{c}_{in})$ is an n-dimensional random variable vector, $\bar{c}_{i\ell}, \ell = 1, \ldots, n, i = 1, \ldots, k$ are Gaussian random variables, represented as $\mathcal{N}(m_{\bar{c}_{i\ell}}, \sigma_{i\ell\ell})$. The variance-covariance matrix $\boldsymbol{V}_{\bar{c}_i}, i = 1, \ldots, k$, describing the relationship between the Gaussian random variables, is defined as:

$$\boldsymbol{V}_{\bar{c}_i} = [\sigma_{izt}]_{n \times n}, i = 1, \ldots, k. \tag{7}$$

The predicted value of the row vector $\overline{\boldsymbol{c}}_i$ with respect to the random variable is represented by:

$$\boldsymbol{m}_{\overline{\boldsymbol{c}}_i} = \left(m_{\overline{c}_{i1}}, \cdots, m_{\overline{c}_{in}}\right), \tag{8}$$

and with the covariance matrix $\boldsymbol{V}_{\overline{\boldsymbol{c}}_i}$, the random variable vector $\overline{\boldsymbol{c}}$ can be expressed as:

$$\overline{\boldsymbol{c}}_i \sim \mathcal{N}\left(\boldsymbol{m}_{\overline{\boldsymbol{c}}_i}, \boldsymbol{V}_{\overline{\boldsymbol{c}}_i}\right). \tag{9}$$

The objective function $\overline{\boldsymbol{c}}_i\boldsymbol{\theta}$ also follows a Gaussian distribution, based on the features of Gaussian random variables in general:

$$\overline{\boldsymbol{c}}_i\boldsymbol{\theta} \sim \mathcal{N}\left(\boldsymbol{m}_{\overline{\boldsymbol{c}}_i}\boldsymbol{\theta}, \boldsymbol{\theta}^T\boldsymbol{V}_{\overline{\boldsymbol{c}}_i}\boldsymbol{\theta}\right), \tag{10}$$

Conventional mathematical programming techniques are inapplicable to the multiobjective stochastic linear programming (**MSLP**) issue because it incorporates random variables into the objective functions and restrictions. We deal with the limitations of (**MSLP**) as conditions that are confined by chance (see [5]) in order to solve this problem. A certain probability β_j or above, referred to as the constraint probability level, is required for the j-th constraint of (**MSLP**) to hold. The definition of the set $\mathcal{X}(\boldsymbol{\beta})$ that meets this requirement is:

$$\mathcal{X}(\boldsymbol{\beta}) = \left\{\boldsymbol{\theta} \in \mathbb{R}^n, \boldsymbol{\theta} \geq 0 \mid P\left(\overline{\boldsymbol{a}}_j\boldsymbol{\theta} \leq \overline{b}_j\right) \geq \beta_j, j = 1, \cdots, m\right\} \tag{11}$$

where $\overline{\boldsymbol{a}}_j = (\bar{a}_{j1}, \ldots, \bar{a}_{jn})$ is the j-th row vector of random variables from matrix $\overline{\boldsymbol{A}}$, and $\boldsymbol{\beta} = (\beta_1, \ldots, \beta_m)$.

To determine the deterministic equivalent constraints for the chance - constrained formulation, we first assume that only $\bar{b}_i$ is a random variable on the right-hand side, and that $\bar{a}_{ij} = a_{ij}$ is constant. Thus, we use the notation $\bar{a}_{ij} = a_{ij}$. Let $F_i(\tau)$ represent the cumulative distribution function of $\bar{b}_i$. Since

$$P\left(\sum_{j=1}^{n} a_{ij}\theta_j \leq \bar{b}_i\right) = 1 - F_i\left(\sum_{j=1}^{n} a_{ij}\theta_j\right), i = 1, \ldots, m \tag{12}$$

the chance constraints (11) can be rewritten as:

$$F_i\left(\sum_{j=1}^{n} a_{ij}\theta_j\right) \leq 1 - \beta_i, i = 1, \ldots, m \tag{13}$$

Let $K_{1-\beta_i}$ denote the maximum value of τ such that $\tau = F_i^{-1}(1 - \beta_i)$. Therefore, the inequality in (13) can be expressed as:

$$\sum_{j=1}^{n} a_{ij}\theta_j \leq K_{1-\beta_i}, i = 1, \ldots, m \tag{14}$$

We may describe the probability as follows if we assume that $\bar{b}_i$ is a normally distributed random variable with a mean of $m_{\bar{b}_i}$ and variance $\sigma^2_{\bar{b}_i}$:

$$P\left(\sum_{j=1}^{n} a_{ij}\theta_j \leq \bar{b}_i\right) = P\left(\frac{\bar{b}_i - m_{\bar{b}_i}}{\sigma_{\bar{b}_i}} \geq \frac{\sum_{j=1}^{n} a_{ij}\theta_j - m_{\bar{b}_i}}{\sigma_{\bar{b}_i}}\right) = 1 - \Phi\left(\frac{\sum_{j=1}^{n} a_{ij}\theta_j - m_{\bar{b}_i}}{\sigma_{\bar{b}_i}}\right), \tag{15}$$

where Φ represents the cumulative distribution function of the standard normal distribution $\mathcal{N}(0,1)$. Thus, the chance constraint (15) can be reformulated as:

$$\sum_{j=1}^{n} a_{ij}\theta_j \leq m_{\bar{b}_i} + \sigma_{\bar{b}_i}\Phi^{-1}(1-\beta_i), i = 1, \dots, m \tag{16}$$

where Φ^{-1} denotes the inverse cumulative distribution function of the standard normal distribution.

Now, in the scenario where both $\bar{b}_i$ and $\bar{a}_{ij}$ are normally distributed, we define $m_{\bar{b}_i}$ and $\sigma^2_{\bar{b}_i}$ as the mean and variance of $\bar{b}_i$, respectively, and $m_{\bar{a}_{ij}}$ and $V_{\bar{a}_i}$ as the mean and variance-covariance matrix of $\bar{a}_{ij}$. Assume that $\bar{b}_i$ and $\bar{a}_{ij}$ are independent; the random variable

$$\frac{\bar{b}_i - \bar{\boldsymbol{a}}_i\boldsymbol{\theta} - (m_{\bar{b}_i} - \boldsymbol{m}_{\bar{\boldsymbol{a}}_i}\boldsymbol{\theta})}{\sqrt{\sigma^2_{\bar{b}_i} + \boldsymbol{\theta}^T V_{\bar{\boldsymbol{a}}_i}\boldsymbol{\theta}}}, i = 1, \dots, m \tag{17}$$

adheres to a conventional Gaussian distribution Standard normal distribution with a mean of 0 and a variance of 1. Therefore, the probability $P(\bar{\boldsymbol{a}}_i\ \boldsymbol{\theta} \leq \bar{b}_i)$ can be reformulated as:

$$P\left(\frac{\bar{b}_i - \sum_{j=1}^{n}\bar{a}_{ij}\theta_j - \left(m_{\bar{b}_i} - \sum_{j=1}^{n} m_{\bar{a}_{ij}}\theta_j\right)}{\sqrt{\sigma^2_{\bar{b}_i} + \boldsymbol{\theta}^T V_{\bar{\boldsymbol{a}}_i}\boldsymbol{\theta}}} \geq \frac{-\left(m_{\bar{b}_i} - \sum_{j=1}^{n} m_{\bar{a}_{ij}}\theta_j\right)}{\sqrt{\sigma^2_{\bar{b}_i} + \boldsymbol{\theta}^T V_{\bar{\boldsymbol{a}}_i}\boldsymbol{\theta}}}\right) = 1 - \Phi\left(\frac{\sum_{j=1}^{n} m_{\bar{a}_{ij}}\theta_j - m_{\bar{b}_i}}{\sqrt{\sigma^2_{\bar{b}_i} + \boldsymbol{\theta}^T V_{\bar{\boldsymbol{a}}_i}\boldsymbol{\theta}}}\right), \tag{18}$$

which leads to the following transformed inequality for the chance constraint:

$$\sum_{j=1}^{n} m_{\bar{a}_{ij}}\theta_j - \Phi^{-1}(1-\beta_i)\sqrt{\sigma^2_{\bar{b}_i} + \boldsymbol{\theta}^T V_{\bar{\boldsymbol{a}}_i}\boldsymbol{\theta}} \leq m_{\bar{b}_i}, i = 1, \dots, m. \tag{19}$$

Proposition 2. *With constraint probability level* $\boldsymbol{\beta} = (\beta_1, \dots, \beta_m)$ *and* $0 \leq \beta_i \leq 0.5$, $\sum_{j=1}^{n} m_{\bar{a}_{ij}}\theta_j - \Phi^{-1}(1-\beta_i)\sqrt{\sigma^2_{\bar{b}_i} + \boldsymbol{\theta}^T V_{\bar{\boldsymbol{a}}_i}\boldsymbol{\theta}}$ *is convex and the constraint set* $\mathcal{X}(\boldsymbol{\beta})$ *is a convex set.*

According to the previously stated chance-constrained conditions (11), utilizing a probability maximization strategy for the objective function in (**MSLP**) allows us to reformulate the maximizing of the objective functions $\overline{\boldsymbol{C}}\boldsymbol{\theta}$, into maximizing the probability that each objective function $\overline{\boldsymbol{c}}_i\boldsymbol{\theta}$ meets or exceeds a certain acceptable threshold $\hat{f}_i$, which is referred to as the permissible objective level. This probability function is expressed as:

$$\kappa_i\left(\boldsymbol{\theta}, \hat{f}_i\right) \stackrel{\text{def}}{=} P\left(\bar{\boldsymbol{c}}_i\boldsymbol{\theta} \geq \hat{f}_i\right). \tag{20}$$

Consequently, the problem (**MSLP**) can be restructured as:

$$\max_{\boldsymbol{\theta}\in\mathcal{X}(\boldsymbol{\beta})} (\kappa_1(\boldsymbol{\theta}, \hat{f}_1), \ldots, \kappa_k(\boldsymbol{\theta}, \hat{f}_k)) \qquad \textbf{(MOP-P}(\hat{\boldsymbol{f}}, \beta))$$

From (10), the objective function $\kappa_i\left(\boldsymbol{\theta}, \hat{f}\right)$ can be rewritten as:

$$\begin{aligned}\kappa_i\left(\boldsymbol{\theta}, \hat{f}_i\right) &= P\left(\frac{\bar{\boldsymbol{c}}_i\boldsymbol{\theta} - E\left[\overline{\boldsymbol{c}}_i\right]\boldsymbol{\theta}}{\sqrt{\boldsymbol{\theta}^T \boldsymbol{V}_{\bar{\boldsymbol{c}}_i}\boldsymbol{\theta}}} \geq \frac{\hat{f}_i - E\left[\overline{\boldsymbol{c}}_i\right]\boldsymbol{\theta}}{\sqrt{\boldsymbol{\theta}^T \boldsymbol{V}_{\bar{\boldsymbol{c}}_i}\boldsymbol{\theta}}}\right) \\ &= 1 - \Phi\left(\frac{\hat{f}_i - E\left[\overline{\boldsymbol{c}}_i\right]\boldsymbol{\theta}}{\sqrt{\boldsymbol{\theta}^T \boldsymbol{V}_{\bar{\boldsymbol{c}}_i}\boldsymbol{\theta}}}\right)\end{aligned} \tag{21}$$

Given that Φ is a monotonically increasing function, the problem (**MOP-P**$(\hat{\boldsymbol{f}}, \beta)$) is equivalent to:

$$\min_{\boldsymbol{\theta}\in\mathcal{X}(\boldsymbol{\beta})} (\mathcal{S}_1(\boldsymbol{\theta}), \ldots, \mathcal{S}_k(\boldsymbol{\theta})) \qquad \textbf{(MOP-P1}(\hat{\boldsymbol{f}}, \boldsymbol{\beta}))$$

where $\mathcal{S}_i(\boldsymbol{\theta}) = -\frac{E[\overline{\boldsymbol{c}}_i]\boldsymbol{\theta} - \hat{f}_i}{\sqrt{\boldsymbol{\theta}^T \boldsymbol{V}_{\bar{\boldsymbol{c}}_i}\boldsymbol{\theta}}}, i = 1, \ldots, k$. Here, we observe that the objective function is no longer purely convex but has a more intricate structure, described as quasiconvex.

Proposition 3. *The objective functions $\mathcal{S}_i(x), i = 1, ..., k$ are pseudoconvex.*

Proof. Let φ_1 and φ_2 be two functions defined on a set X. If φ_1 is positive and concave, and φ_2 is positive and convex on X, with both φ_1 and φ_2 being differentiable on X, then the fractional function φ_1/φ_2 is pseudoconcave on X (refer to [2]). Observe that $E\left[\overline{\boldsymbol{c}}_i\right] - \hat{f}_i$ constitutes a positive linear expression, given that $\hat{f}_i$ is a constant, whereas $\sqrt{\boldsymbol{\theta}^T \boldsymbol{V}_{\bar{\boldsymbol{c}}_i}\boldsymbol{\theta}}$ is convex. Consequently, the function $\frac{E\left[\overline{\boldsymbol{c}}_i\right]\boldsymbol{\theta} - \hat{f}_i}{\sqrt{\boldsymbol{\theta}^T \boldsymbol{V}_{\bar{\boldsymbol{c}}_i}\boldsymbol{\theta}}}$ is pseudoconcave, and therefore $\mathcal{S}_i(\boldsymbol{\theta}) = -\frac{E\left[\overline{\boldsymbol{c}}_i\right]\boldsymbol{\theta} - \hat{f}_i}{\sqrt{\boldsymbol{\theta}^T \boldsymbol{V}_{\bar{\boldsymbol{c}}_i}\boldsymbol{\theta}}}$ is pseudoconvex. □

3.2 Problem (MSLP) with Fuzzy Decision Making

To address the issue (**MOP-P1**($\hat{\boldsymbol{f}}, \boldsymbol{\beta}$)), where (**MSLP**) is expressed using a probability maximization model, we present a picture fuzzy decision-making method in this section. The decision-maker needs to know their allowable goal level $\hat{\boldsymbol{f}} = (\hat{f}_1, \ldots, \hat{f}_k)$ in advance in order to solve (**MOP-P1**($\hat{\boldsymbol{f}}, \boldsymbol{\beta}$)). Both a smaller value of the function $\mathcal{S}_i(\boldsymbol{\theta})$ and a bigger value of the allowed goal level $\hat{f}_i$ are usually preferred by the decision-maker. The objective function $\mathcal{S}_i(\boldsymbol{\theta})$ rises as the allowed objective level $\hat{f}_i$ rises since these preferences are incompatible with one another.

We suggest the following multiobjective programming problem, which is seen as an obvious continuation of (**MOP-P1**($\hat{\boldsymbol{f}}, \boldsymbol{\beta}$)), from this point of view:

$$\min_{\boldsymbol{\theta} \in X(\boldsymbol{\beta}), \hat{f}_i \in \mathbb{R}, i=1,\ldots,k} (\mathcal{S}_1(\boldsymbol{\theta}), \ldots, \mathcal{S}_k(\boldsymbol{\theta}), -\hat{f}_1, \ldots, -\hat{f}_k) \qquad \textbf{(MOP-P2}(\hat{\boldsymbol{f}}, \boldsymbol{\beta}))$$

It is essential to recognize that in (**MOP-P1**($\hat{\boldsymbol{f}}, \boldsymbol{\beta}$)), the allowable objective levels $\hat{f}_i, i = 1, \cdots, k$ are not static values, but are regarded as decision variables to be optimized.

Considering the ambiguity in the decision-maker's evaluation, it is reasonable to deduce that the decision-maker has a picture fuzzy goal for each objective function and a picture fuzzy relation for each constraint in (**MOP-P1**($\hat{\boldsymbol{f}}, \boldsymbol{\beta}$)). Given these ambiguous objectives and relationships, the issue can be reformulated as the subsequent fuzzy multiobjective programming problem:

$$\begin{aligned} &\widetilde{\min}\ (\mathcal{S}_1(\boldsymbol{\theta}), \ldots, \mathcal{S}_k(\boldsymbol{\theta}), \hat{f}_1, \ldots, \hat{f}_k) \\ &\text{s.t. } g_i(\boldsymbol{\theta}) \preceq 0, i = 1, \ldots, m \qquad \textbf{(FMOP1)} \\ &\qquad \boldsymbol{\theta} \geq 0. \end{aligned}$$

where $g_i(\boldsymbol{\theta}) = \sum_{j=1}^{n} m_{\bar{a}_{ij}} \theta_j - \Phi^{-1}(1 - \beta_i) \sqrt{\sigma_{\bar{b}_i}^2 + \boldsymbol{\theta}^T V_{\bar{a}_i} \boldsymbol{\theta}} - m_{\bar{b}_i}$. Fuzzy decisions are quantified through the use of membership functions. Let us define membership functions for a objective function $\mathcal{S}_i(\boldsymbol{\theta})$ as $(\mu_{\mathcal{S}_i}(\mathcal{S}_i(\boldsymbol{\theta})), \eta_{\mathcal{S}_i}(\mathcal{S}_i(\boldsymbol{\theta})), \nu_{\mathcal{S}_i}(\mathcal{S}_i(\boldsymbol{\theta})))$, membership functions of a permissible objective level $\hat{f}_i$ as $(\mu_{\hat{f}_i}(\hat{f}_i), \eta_{\hat{f}_i}(\hat{f}_i), \nu_{\hat{f}_i}(\hat{f}_i))$, and membership functions of a constraints $g_j(\boldsymbol{\theta})$ as $(\mu_{g_j}(g_j(\boldsymbol{\theta})), \eta_{g_j}(g_j(\boldsymbol{\theta})), \nu_{g_j}(g_j(\boldsymbol{\theta})))$, respectively.

Proposition 4. *Consider a monotonically decreasing function $m(\cdot)$. The positive membership function μ with respect to a function or parameter z is expressed as:*

$$\mu(z) = \begin{cases} 0 & \textit{if } z \geq z^0, \\ m(z) & \textit{if } z^0 \geq z \geq z^1, \\ 1 & \textit{if } z \leq z^1 \end{cases}$$

where z^0 is the maximum value for which $\mu(z) = 0$ and z^1 is the minimum value for which $\mu(z) = 1$. On the other hand, for monotonically increasing functions $n(\cdot)$ and $k(\cdot)$, the neutral and negative membership functions η, ν with respect to *z are given as:*

$$\eta(z) = \begin{cases} a & \text{if } z \geq z^0, \\ n(z) & \text{if } z^0 \geq z \geq z^1, \\ 0 & \text{if } z \leq z^1 \end{cases}$$

$$\nu(z) = \begin{cases} b & \text{if } z \geq z^0, \\ k(z) & \text{if } z^0 \geq z \geq z^1, \\ 0 & \text{if } z \leq z^1 \end{cases}$$

Here, z^0 represents the minimum value of z when $\eta(z) = a, \nu(z) = b, (a+b \leq 1)$, while z^1 is the maximum value where z if $\eta(z) = 0, \nu(z) = 0$.

The issue (**FMOP1**) can thereafter be reformulated as the following multi-objective programming problem:

$$\begin{aligned} &\max \left(\mu_{S_i}\left(S_i(\boldsymbol{\theta})\right), \mu_{\hat{f}_i}(\hat{f}_i), \mu_{g_j}\left(g_j(\boldsymbol{\theta})\right)\right), i = 1, \ldots, k, j = 1, \ldots, m \\ &\min \left(\eta_{S_i}\left(S_i(\boldsymbol{\theta})\right), \eta_{\hat{f}_i}(\hat{f}_i), \eta_{g_j}\left(g_j(\boldsymbol{\theta})\right)\right), i = 1, \ldots, k, j = 1, \ldots, m \\ &\min \left(\nu_{S_i}\left(S_i(\boldsymbol{\theta})\right), \nu_{\hat{f}_i}(\hat{f}_i), \nu_{g_j}\left(g_j(\boldsymbol{\theta})\right)\right), i = 1, \ldots, k, j = 1, \ldots, m \\ &\text{s.t. } \boldsymbol{\theta} \geq 0. \end{aligned} \qquad \textbf{(FMOP2)}$$

Now, by defining $\dot{\mu} = 1 - \mu$, the problem (**FMOP2**) ecomes equivalent to:

$$\begin{aligned} \min \ &\left(\dot{\mu}_{S_i}\left(S_i(\boldsymbol{\theta})\right), \dot{\mu}_{\hat{f}_i}(\hat{f}_i), \dot{\mu}_{g_j}\left(g_j(\boldsymbol{\theta})\right), \right. \\ &\eta_{S_i}\left(S_i(\boldsymbol{\theta})\right), \eta_{\hat{f}_i}(\hat{f}_i), \eta_{g_j}\left(g_j(\boldsymbol{\theta})\right), \\ &\left.\nu_{S_i}\left(S_i(\boldsymbol{\theta})\right), \nu_{\hat{f}_i}(\hat{f}_i), \nu_{g_j}\left(g_j(\boldsymbol{\theta})\right)\right) \\ &i = 1, \ldots, k, j = 1, \ldots, m \\ \text{s.t. } &\boldsymbol{\theta} \geq 0. \end{aligned} \qquad \textbf{(FMOP3)}$$

This multiobjective semistrictly quasiconvex programming issue has been examined in several studies (see [8,9]). To circumvent the computational complexity of directly addressing this, we propose reformulating problem (**FMOP3**) as follows, inspired by [10]:

$$\begin{aligned} \min \max &\left(\dot{\mu}_{S_i}\left(S_i(\boldsymbol{\theta})\right), \dot{\mu}_{\hat{f}_i}(\hat{f}_i), \dot{\mu}_{g_j}\left(g_j(\boldsymbol{\theta})\right), \right. \\ &\eta_{S_i}\left(S_i(\boldsymbol{\theta})\right), \eta_{\hat{f}_i}(\hat{f}_i), \eta_{g_j}\left(g_j(\boldsymbol{\theta})\right), \\ &\left.\nu_{S_i}\left(S_i(\boldsymbol{\theta})\right), \nu_{\hat{f}_i}(\hat{f}_i), \nu_{g_j}\left(g_j(\boldsymbol{\theta})\right)\right) \\ &i = 1, \ldots, k, j = 1, \ldots, m \\ \text{s.t. } &\boldsymbol{\theta} \geq 0. \end{aligned} \qquad \textbf{(MP)}$$

This single-objective problem has garnered significant attention in the literature (see to [11]). The semistrictly quasiconvex characteristics of the objectives introduce complexity to the problem, rendering it tough and inadequately explored in prior research. This paper delineates the characteristics of the issue are as follows.

Proposition 5. *(MP) is categorized as a semistrictly quasiconvex programming issue.*

Thang et al. [12] have demonstrated the convergence of the universal solution in the realm of semistrictly quasiconvex programming and devised an algorithm utilizing the gradient direction approach. The efficacy of this approach has been confirmed by computational studies. This study employs the gradient direction method to address (**MP**), following the methodology of [12] and Proposition 5.

3.3 Examples

Example 1 (see [4]). To illustrate the efficacy of our suggested fuzzy decision-making methodology, we analyze the subsequent three-objective stochastic linear programming problem:

[MOSLP]

$$\max(\overline{\mathbf{c}}_1\boldsymbol{\theta}, \overline{\mathbf{c}}_2\boldsymbol{\theta}, \overline{\mathbf{c}}_3\boldsymbol{\theta}) \quad \text{s.t.} \boldsymbol{a}_j\boldsymbol{\theta} \leq \bar{b}_j, j = 1,2 \quad \boldsymbol{\theta} \geq 0 \tag{22}$$

where $\boldsymbol{\theta} = (\theta_1, \theta_2, \theta_3, \theta_4)^T$ is a 4-dimensional decision column vector. The coefficient row vectors $\boldsymbol{a}_j, j = 1,2$ are defined as $\boldsymbol{a}_1 = (7,3,4,6), \boldsymbol{a}_2 = (-5,-6,-7,-9)$. The Gaussian random variables $(\bar{b}_1, \bar{b}_2)$ with $b_1 \sim \mathrm{N}(27, 6^2)$, $b_2 \sim \mathrm{N}(-15, 7^2), \overline{\mathbf{c}}_i, i = 1,2,3$ are 4-dimensional Gaussian random variables, with mean vectors $\boldsymbol{E}[\overline{\mathbf{c}}_1] = (2,3,2,4), \boldsymbol{E}[\overline{\mathbf{c}}_2] = (10,-7,1,-2)$, $\boldsymbol{E}[\overline{\mathbf{c}}_3] = (-8,-5,-7,-14)$. The variance-covariance matrices $V_i, i = 1,2,3$, are as follows:

$$V_1 = \begin{pmatrix} 25 & -1 & 0.8 & -2 \\ -1 & 4 & -1.2 & 1.2 \\ 0.8 & -1.2 & 4 & 2 \\ -2 & 1.2 & 2 & 9 \end{pmatrix}, \quad V_2 = \begin{pmatrix} 16 & 1.4 & -1.2 & 1.4 \\ 1.4 & 1 & 1.5 & -0.8 \\ -1.2 & 1.5 & 25 & -0.6 \\ 1.4 & -0.8 & -0.6 & 4 \end{pmatrix}$$

$$V_3 = \begin{pmatrix} 4 & -1.9 & 1.5 & 1.8 \\ -1.9 & 25 & 0.8 & -0.4 \\ 1.5 & 0.8 & 9 & 2.5 \\ 1.8 & -0.4 & 2.5 & 36 \end{pmatrix}$$

For this issue, let's pretend the decision maker has $\boldsymbol{\beta} = (\beta_1, \beta_2) = (0.7, 0.7)$ as the levels of the constraint probability.

To validate the efficiency of our proposed method, we compare it with previous methods (see [4]), including fuzzy decision making and the probability maximization model.

	Proposed Methods	Yano's Method	Probability max.
λ^*	0.546523	0.536605	0.173128
f_1^*	30.1215	28.0819	25
f_2^*	-30.3785	-29.3565	-35
f_3^*	-31.6524	-32.7991	-35
$\mu_{f_1}(f_1^*)$	0.543432	0.536605	0.596050
$\mu_{f_2}(f_2^*)$	0.542875	0.536605	0.607070
$\mu_{f_3}(f_3^*)$	0.537543	0.536605	0.570710

The results from the proposed method demonstrate several advantages compared to the two previous methods (Yano's method and the probability maximization method). First, the λ^* value in the proposed method is 0.546523, higher than Yano's method (0.536605) and the probability maximization method (0.173128), indicating a better optimal performance. These results were achieved by employing a fuzzy relation method to extend the acceptable set, which led to a better optimization outcome. This approach allowed for more flexibility and precision in handling uncertainties, ultimately improving the overall performance.

4 Conclusion

This study introduces a multi-objective stochastic linear optimization model designed to maximize probability and threshold values, thus facilitating decision-making through picture fuzzy relations. We have established that the deterministic problem constitutes a semistrictly quasiconvex programming issue and have introduced a method utilizing strictly monotonic membership functions to convert it into a flexible variant, which is also demonstrated to be semistrictly quasiconvex. The picture fuzzy multi-objective stochastic optimization issue can be effectively resolved utilizing gradient descent techniques, which demand considerably less resources than the previously utilized genetic algorithms. This method possesses significant practical benefits and the prevalent presence of semistrictly quasiconvex functions in real-world applications, indicating its extensive applicability in diverse mathematical models incorporating semistrictly quasiconvex functions.

Acknowledgements. This work was funded by Rikkeisoft corporation and supported by Institute for Digital Technology and Economy (BK Fintech), Hanoi University of Science and Technology in the project BKFintech-2024.06.

References

1. Lodwick, W.A., Salles-Neto, L.L.: Flexible and generalized uncertainty optimization. Studies in Computational Intelligence, vol. 696. Springer, Cham (2021). https://doi.org/10.1007/978-3-030-61180-4_5
2. Avriel, M., Diewert, W.E., Schaible, S., Zang, I.: Generalized Concavity. Society for Industrial and Applied Mathematics, Philadelphia (2010). https://doi.org/10.1137/1.9780898719437
3. Penot, J.-P.: Glimpses upon quasiconvex analysis. ESAIM: Proc. **20**, 170–194 (2007). https://doi.org/10.1051/proc:072016
4. Yano, H.: Fuzzy decision making for multiobjective stochastic programming problems. Fuzzy Sets Syst. **296**, 97–111 (2016). https://doi.org/10.1016/j.fss.2015.08.018
5. Sakawa, M., Yano, H., Nishizaki, I.: Stochastic Linear Programming. In: Linear and Multiobjective Programming with Fuzzy Stochastic Extensions. International Series in Operations Research & Management Science, vol. 203. Springer, Boston, MA (2013). https://doi.org/10.1007/978-1-4614-9399-0_5
6. Cuong, B.: Picture fuzzy sets. J. Comput. Sci. Cybern. **30**(4), 10.15625/1813-9663/30/4/5032 (2015)
7. Mangasarian, O.L.: Pseudo-convex functions. In: Stochastic Optimization Models in Finance, pp. 23–32. Elsevier (1975)
8. Thang, T.N., Kim, N.: Outcome space algorithm for generalized multiplicative problems and optimization over the efficient set. JIMO **12**(4), 1417–1433 (2016)
9. Vuong, N.D., Thang, T.N.: Optimizing over Pareto set of semistrictly quasiconcave vector maximization and application to stochastic portfolio selection. J. Ind. Manag. Optim. **19**, 1999–2019 (2023)
10. Collette, Y., Siarry, P.: Multiobjective Optimization: Principles and Case Studies. Springer Science & Business Media (2004)
11. Tuoi, T.T.T., Khang, T.T., Anh, N.T.N., Thang, T.N.: Fuzzy portfolio selection with flexible optimization via quasiconvex programming. In: Lecture Notes in Networks and Systems, vol. 471. Springer (2022)
12. Thang, T.N., Hai, T.N.: Self-adaptive algorithms for quasiconvex programming and applications to machine learning. Comput. Appl. Math. **43**, 249 (2024). https://doi.org/10.1007/s40314-024-02764-w
13. Hoang, D.M., Thang, T.N., Tu, N.D., Hoang, N.V.: Stochastic linear programming approach for portfolio optimization problem. In: 2021 IEEE International Conference on Machine Learning and Applied Network Technologies (ICMLANT), pp. 1–4. IEEE, Soyapango, El Salvador (2021). https://doi.org/10.1109/ICMLANT53170.2021.9690552
14. Hoang, D.M., Xuan, T.V., Thang, T.N.: A multi-criteria fuzzy random crop planning problem using evolutionary optimization. In: Solanki, V.K., Quang, N.H. (eds.) RICE, pp. 49–52 (2021). ISBN: 978-83-962423-3-4
15. Sharpe, W.F.: The Sharpe ratio. J. Portf. Manag. **21**(1), 49–58 (1994). https://doi.org/10.3905/jpm.1994.409501
16. Akram, M., Ullah, I., Allahviranloo, T.: A new method to solve linear programming problems in the environment of picture fuzzy sets. Iran. J. Fuzzy Syst. **19**(6), 29–49 (2022). https://doi.org/10.22111/ijfs.2022.7208
17. Akram, M., Habib, A., Alcantud, J.: An optimization study based on Dijkstra algorithm for a network with trapezoidal picture fuzzy numbers. Neural Comput. Appl. **33**, 1329–1342 (2021). https://doi.org/10.1007/s00521-020-05034-y

18. Son, L.H., Van Viet, P., Van Hai, P.: Picture inference system: a new fuzzy inference system on picture fuzzy set. Appl. Intell. **46**, 652–669 (2017). https://doi.org/10.1007/s10489-016-0856-1
19. Thang, T.N., Hoang, D.M., Dung, N.V.: A neurodynamic approach for a class of pseudoconvex semivectorial bilevel optimization problems. Optim. Methods Softw. 1–28 (2024). https://doi.org/10.1080/10556788.2024.2380688

Optimizing the Path Planning by the Improved A* Algorithm

Tien Dung Vu, Vu Toan Thang, Quang Hung Tran, Han Quang Vu Ngo, and Doan Thanh Xuan(✉)

Hanoi University of Science and Technology, Hanoi, Vietnam
xuan.doanthanh@hust.edu.vn

Abstract. The A* algorithm is a widely used pathfinding and graph traversal algorithm known for its efficiency in finding the shortest path between two points. It combines the strengths of Dijkstra's Algorithm and Greedy Best-First Search by using both the actual distance traveled from the start point and an estimated distance to the goal, which makes it optimal and complete. A* is commonly applied in robotics, video games, and navigation systems due to its ability to find the most efficient route. Its performance depends heavily on the heuristic used, and while it is generally fast, it can be computationally expensive for large or complex maps. The drawback of the A* algorithm is that it examines each traversed point, resulting in low speed and high data storage requirements. This paper introduces an improved A* algorithm, which combines three techniques: expansion distance, reducing redundant points, and smoothing the path planning.

Keywords: Improved A* Algorithm · Path Planning · Trajectory Optimization

1 Introduction

Over the past decades, determining a collision-free path or trajectory for a moving entity, such as a robot, within a given environment (commonly referred to as path planning) has been the focus of extensive research [1–5]. Traditional path planning algorithms include the Genetic Algorithm (GA) [6, 7], Ant Colony Optimization (ACO) [8–10], Rapidly-exploring Random Tree (RRT) [11, 12], and the A-Star (A*) algorithm [13–15]. Among these methods, the A* algorithm is widely used for its graph-based approach, which iteratively converges toward an optimal path.

The path planning speed and path certainty are the main indicators of the A* algorithm's performance. The application of this technique still has several issues, such as redundant points and jagged routes, even after it has been thoroughly examined. The projected route's certainty is diminished by these circumstances. As a result, considerations of path planning speed and path smoothness are also used to study the algorithm's effectiveness and the mobile robot's speed. Reducing unnecessary points and smoothing the path should be included in the suggested algorithm in order to improve the resilience of the traditional A* algorithm. The purpose of this research study is to propose three improvements to the standard A* algorithm: smoothing, expansion distance,

© The Author(s), under exclusive license to Springer Nature Switzerland AG 2025
H. A. Le Thi et al. (Eds.): ICAMCS 2024, LNNS 1311, pp. 123–132, 2025.
https://doi.org/10.1007/978-3-031-90606-0_11

and removal of redundant points. The combination of the aforementioned three techniques to find a path for the robot is being presented for the first time in this study. Additionally, simulation experiments were carried out, which show that in terms of path planning performance, the modified A* algorithm performs better than the standard A* algorithm and its modifications (A* with redundant point removal, expansion distance, and smoothing).

2 Improved A Star Algorithm

2.1 Basic Theory of Traditional A* Algorithm

One of the most popular path planning methods is the A* algorithm, which was initially introduced and fully described in [13]. It is intended to be a heuristic search algorithm that explores all possible paths to find the least expensive one between the start and goal nodes. This algorithm is better than other blind search algorithms because it uses problem-specific heuristic information to direct its search [30]. It uses an Open list, a Closed list, and a heuristic function to estimate the distance on a 2D plane between any given node and the destination node.

A* is considered a best-first algorithm since it evaluates each configuration space cell based on the value of:

$$\mathrm{f(n)} = \mathrm{g(n)} + \mathrm{h(n)} \tag{1}$$

where: $h(n)$ indicates the predicted cost, as determined by the heuristic function, from node n to the destination node; $g(n)$ represents the cost of the path from the start node to the current node, n. The evaluation function for node n is $Z(n)$.

Once the method has accessed each nearby node of the currently evaluated node using $f(n)$, the node with the smallest $f(n)$ is selected as the next node in the path. One advantage of the A* method is that it makes it possible to add, modify, or adopt new distance measurements as standard distances.

2.2 Proposed Method

- **Expansion distance**

The traditional A* algorithm's path might go close to obstructions. Following this course increases the risk of a collaborative robot running into obstacles. As a result, when designing a path, it is essential to maintain a safe distance from obstructions. The idea of "expansion distance," or giving extra space around obstacles as a safety measure, was used in this study [31]. Robots use rasterized maps to plan their paths, and by employing a grid as its basic unit, the expansion distance goes beyond obstacles. The closest distance the path can approach obstacles is determined by the robot's speed, the size of the robot model, and the number of grids.

During path planning, both obstacles and nodes inside the expansion distance will not be visited. By serving as a "collision buffer" between the robot and obstacles, this expansion distance dramatically lowers the possibility of collisions occurring while the robot is moving. Therefore, expansion distance improves the resilience and efficiency of the algorithm. Since the method does not visit the expanded nodes, it essentially shrinks the map scale. As a result, the method becomes more efficient by reducing the total number of nodes it must traverse.

- **Redundant point elimination**

The improved approach first gets better by designing a local path ahead of time from the current node to the target node and then investigating the current node's surroundings. This local path is directly navigated if it is judged safe and free of collisions. The procedure also uses post-processing techniques to optimize the path that is produced, namely by straightening the local path, which minimizes the quantity and duration of local paths that are needed.

An improved version of the A* algorithm is used in this approach during the query phase of probabilistic motion planning [32, 33]. Preprocessing and querying are the two phases of probabilistic motion planning. In preprocessing, the robot creates randomly collision-free sample points in its workspace that are used as nodes in later phases.

Local path planning then creates secure, collision-free routes between these sample locations. By mapping these approaches into the robot's configuration space and taking into account shared limitations like energy optimization, acceleration, and velocity, these paths are verified for viability and collision avoidance. Hence, the safe local routes and collision-free sample points make up the elements of the probabilistic roadmap.

During the querying step, probabilistic motion planning finds a path that looks for and achieves a safe trajectory for the robot's movement from the starting point (Starting point) to the goal node (Ending point) using the upgraded A* algorithm. In order to build the probability map, this study sets configuration space constraints utilizing factors such as angular range, velocity range, acceleration, velocity deformation, and optimal inverse kinematics solutions.

The waypoints that the conventional A* algorithm determined to be the robot's path are calculated in the improved A* algorithm. The method determines if pathpoints pass past barriers by analyzing local connections to the next waypoints. A connection bypasses the intermediate waypoint if it avoids crossing over obstacles. While still ensuring a legitimate path from the starting point to the destination as stated in the original issue statement, this solution drastically reduces the number of checkpoints the robot must navigate.

The technique for removing redundant points is carried out in the following steps:

Step 1: Initialize i and j to n - 1 and i and j, respectively, to reflect the indices of the waypoints that are now being assessed along the robot's journey that are kept in the CLOSED list. From the starting point (p1) to the destination (pn), optimize the path.

Step 2: Define every local segment that connects point pi to point pj, where j is any value between $i + 1$ and $n - 1$.

Step 3: Check to see if the local portions run into problems. Go to Step 4 if there are collisions on every local path. If not, proceed to Step 5.

Step 4: Note the local segment that connects point pi to pi + 1. Next, increase i by 1 (i = i + 1).

Step 5: Set i = j, substitute j − 1 for i + 1, and store the local segment containing node pj with the highest j value.

Step 6: Determine whether i equals n − 1, which would suggest that point i is near the final point. In that case, move on to Step 7. Otherwise, go back to Step 2.

Step 7: Join the locations in the previously saved locally optimized path to form a sequential link of the globally optimized path.

- **Smoothing optimization**

Bezier curves can be used to smooth paths. Pierre Bezier, a French engineer, initially suggested the concept of a Bezier curve in 1962. It is a space curve with favorable geometric properties. It does not necessarily go through all of its control points, or defining data points, which is why it is frequently employed in computer graphics and computer-aided design. The Bezier curve that results from joining these points to form a convex polygon is also convex, setting it apart from other curves such as cubic splines or polynomials. This curve is smoother since it has fewer turning points.

Bezier curves' essential characteristics include:

1. Symmetry: Every curve coefficient has a reciprocal counterpart that is symmetrical.
2. Convex Hull: The curve's control points define the convex hull, which the curve always lies inside.
3. Endpoint Consistency: The Bezier curve's beginning and ending points are exactly coincident with the first and last control points.
4. Recursion: The Bezier curve's coefficients follow a recursive formula.

$$B_{i,n}(t) = (1 - t)B_{i,n-1}(t) + tB_{i-1,n-1}(t), \{i = 0, 1, \ldots, n\} \tag{2}$$

A Bezier curve's curvature changes smoothly from the beginning to the end because its higher-order derivatives are continuous. A parametric curve can be defined as a Bezier curve of degree n using degree n Bernstein basis polynomials:

$$P(t) = \sum_{i=1}^{n} p_i B_{i,n}(t), t \in [0, 1] \tag{3}$$

The normalized parameter in this case is denoted by t, the coordinate vector of the ith control point is represented by Pi(xi,yi)T, and the Bernstein basis polynomials, Bi,n, are the basis functions used in the Bezier curve formulation.

$$B_{i,n}(t) = C_n^i t^i = \frac{n!}{i!(n-i)!} t^i (1-t)^{n-i}, i = 0, 1, \ldots, n \tag{4}$$

The derivatives of the Bezier curve are affected by the control points. Equation 4 yields the Bezier curve's first derivative, which is given by Eq. 6. Moreover, it is feasible to compute higher-order derivatives of the Bezier curve.

$$\dot{P}(t) = \frac{dP(t)}{dt} = n \sum_{i=0}^{n-1} B_{i,n-1}(t)(P_{i+1} - P_i) \tag{5}$$

In two dimensions, the Bezier curve's curvature with respect to t is shown as follows:

$$k(t) = \frac{1}{R(t)} = \frac{\dot{P}_x(t)\ddot{P}_y(t) - \dot{P}_y(t)\ddot{P}_x(t)}{(\dot{P}_x^2(t) + \dot{P}_y^2(t))^{1.5}} \tag{6}$$

Bezier curves are concatenated to provide smooth routes in the context of robot path planning.

3 Results and Discussion

The experiment was programmed in Unity software on 50 × 50, 100 × 100, and 200 × 200 scale maps. Three techniques were performed sequentially: expansion distance, reducing redundant points, and smoothing the path planning. Obstacles were arranged randomly, with a size of 5 × 5. The expansion distance area is shown in blue and has an additional size of 1 grid unit. The experimental results are presented in Figs. 1, 2 and 3 and Tables 1, 2 and 3.

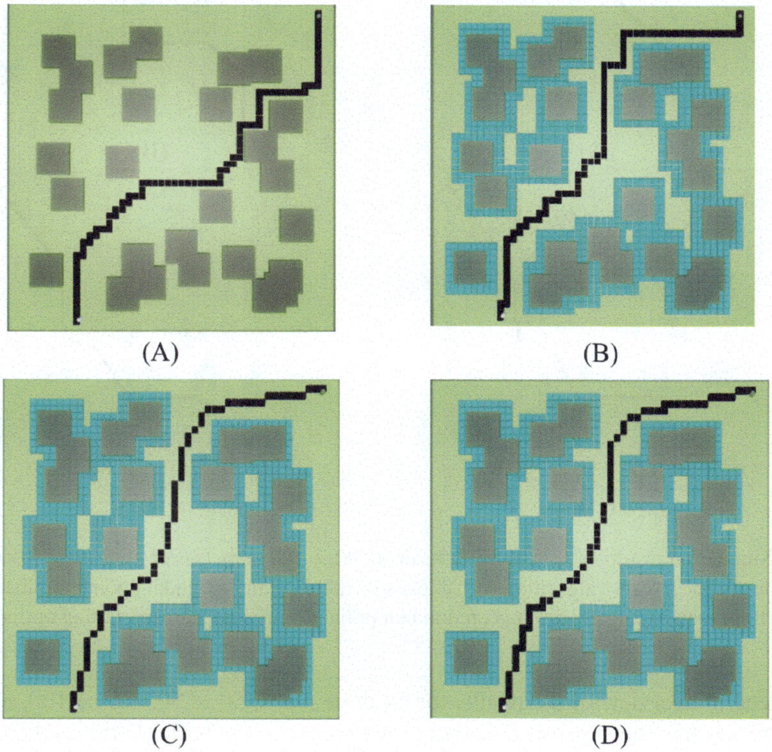

Fig. 1. Simulation results of four algorithms on a 50 × 50 map. (a) A* algorithm. (b) A* with expansion distance. (c) A* algorithm with expansion distance and redundant point elimination. (d) A* algorithm with expansion distance, redundant point elimination, and smoothing optimization.

Table 1. Simulation results of four algorithms on a 100 × 100 map

Indicators	A star algorithm	Plus Expansion Distance	Plus Redundant point elimination	Plus Smoothing optimization
Number of nodes	85	85	70	64
Number of right-angle turns	32	30	0	0
Max turning angle	90	90	45	-

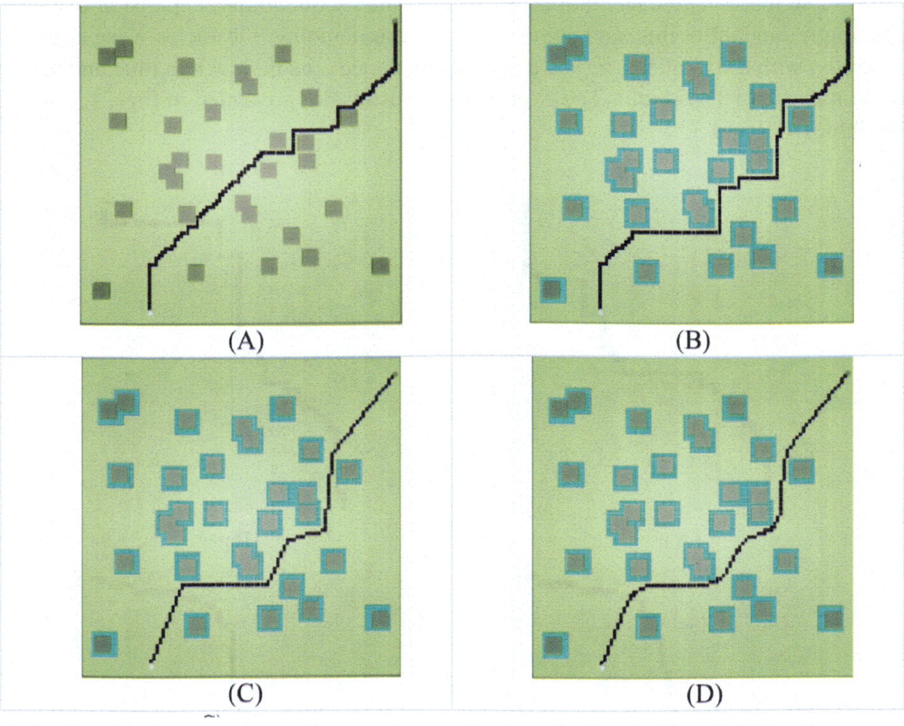

Fig. 2. Simulation results of four algorithms on a 100 × 100 map. (a) A* algorithm. (b) A* with expansion distance. (c) A* algorithm with expansion distance and redundant point elimination. (d) A* algorithm with expansion distance, redundant point elimination, and smoothing optimization.

A general overview of the results for all three maps, 50 × 50, 100 × 100, and 200 × 200, shows that the expansion distance only increases the buffer between the moving object and the obstacles to reduce the likelihood of collision, without significantly reducing the number of nodes traversed compared to the traditional A* algorithm. When the Redundant Point Elimination technique was applied, it significantly reduced the number of nodes traversed by 17.6%, 23.3%, and 29.5% for the 50 × 50, 100 × 100, and 200 × 200 maps, respectively, and eliminated all right angles, with the largest remaining

Table 2. Simulation results of four algorithms on a 100 × 100 map

Indicators	A star algorithm	Plus Expansion Distance	Plus Redundant point elimination	Plus Smoothing optimization
Number of nodes	169	167	128	125
Number of right-angle turns	86	78	0	0
Max turning angle	90	90	45	-

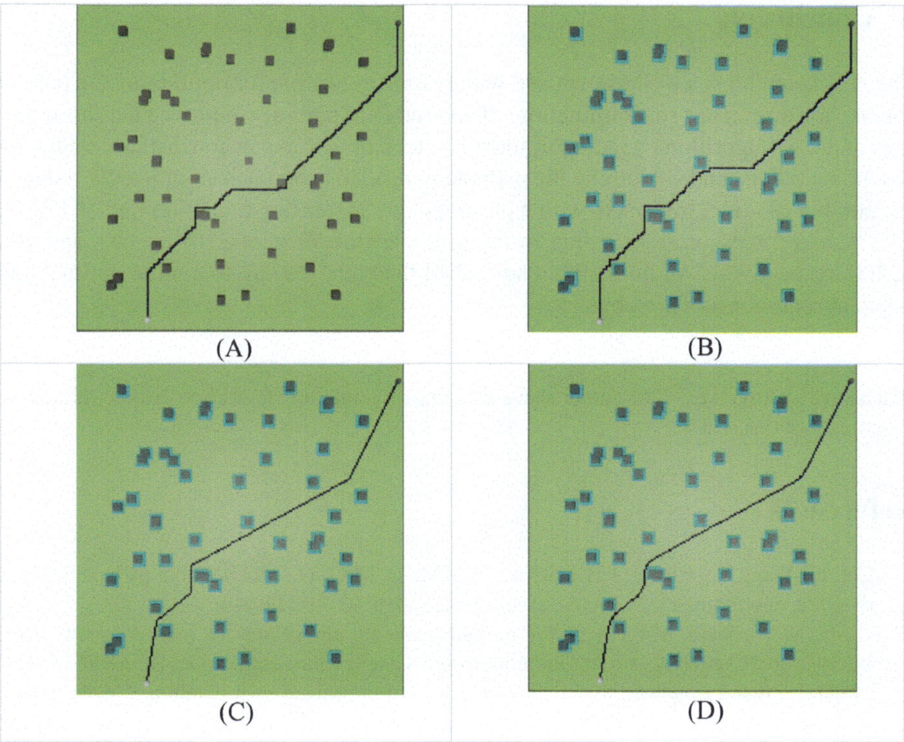

Fig. 3. Simulation results of four algorithms on a 200 × 200 map. (a) A* algorithm. (b) A* with expansion distance. (c) A* algorithm with expansion distance and redundant point elimination (d) A* algorithm with expansion distance, redundant point elimination and smoothing optimization.

angle being 45°. The Smoothing Optimization technique did not significantly change the number of nodes but helped smooth the path for the moving object.

Table 3. Simulation results of four algorithms on a 200 × 200 map

Indicators	A star algorithm	Plus Expansion Distance	Plus Redundant point elimination	Plus Smoothing optimization
Number of nodes	335	332	234	228
Number of right-angle turns	174	170	0	0
Max turning angle	90	90	45	-

4 Conclusion

The A* algorithm is a well-known and widely used algorithm for optimizing the path of robots; however, it has some limitations. This study has partially improved the shortcomings of the A* algorithm by incorporating three techniques: expansion distance, reducing redundant points, and smoothing the path planning. This approach significantly reduced the number of nodes traversed, with a progressively increasing reduction rate of 17.6%, 23.3%, and 29.5%, corresponding to the grid sizes of 50 × 50, 100 × 100, and 200 × 200, respectively. Additionally, it smoothed the path, leading to a more optimal and efficient route for moving objects.

Disclosure of Interests. The authors have no competing interests to declare that are relevant to the content of this article.

References

1. Cai, K., Wang, C., Cheng, J., De Silva, C.W., Meng, M.Q.-H.: Mobile robot path planning in dynamic environments: A survey. Instrumentation. **6**, 92–102 (2019)
2. Costa, M.M., Silva, M.F.: A survey on path planning algorithms for mobile robots. IEEE International Conference on Autonomous Robot Systems and Competitions (ICARSC) (2019)
3. Mandava, R.K., Bondada, S., Vundavilli, P.R.: An optimized path planning for the mobile robot using potential field method and PSO algorithm. Soft Computing for Problem Solving: SocProS 2017, Volume **2**. Springer (2019)
4. Wahab, M.N. Ab, Nefti-Meziani, S., Atyabi, A.: A comparative review on mobile robot path planning: Classical or meta-heuristic methods? Annual Reviews in Control **50**, 233–252 (2020)
5. Wang, B., Liu, Z., Li, Q., Prorok, A., Letters, A.: Mobile robot path planning in dynamic environments through globally guided reinforcement learning. IEEE Robotics. **5**(4), 6932–6939 (2020)
6. Elhoseny, M., Tharwat, A., Hassanien, A.E.: Bezier curve based path planning in a dynamic field using modified genetic algorithm. J. Computational Sci. **25**, 339–350 (2018)
7. Nazarahari, M., Khanmirza, E., Doostie, S.: Multi-objective multi-robot path planning in continuous environment using an enhanced genetic algorithm. Expert Syst. Appl. **115**, 106–120 (2019)

8. Kumar, P.M., Manogaran, G., Sundarasekar, R., Chilamkurti, N., Varatharajan, R.: Ant colony optimization algorithm with internet of vehicles for intelligent traffic control system. Comput. Netw. **144**, 154–162 (2018)
9. Ning, J., Zhang, Q., Zhang, C., Zhang, B.: A best-path-updating information-guided ant colony optimization algorithm. Inf. Sci. **433**, 142–162 (2018)
10. Maheshwari, P., Sharma, A.K., Verma, K.: Energy efficient cluster based routing protocol for WSN using butterfly optimization algorithm and ant colony optimization. Ad Hoc Netw. **110**, 102317 (2021)
11. Hauer, F., Tsiotras, P.: Deformable rapidly-exploring random trees. Robotics: Science and Systems (2017)
12. Hirakawa, T., Yamashita, T., Fujiyoshi, H.: Scene context-aware rapidly-exploring random trees for global path planning. 2019 IEEE International Conference on Pervasive Computing and Communications Workshops (PerCom Workshops) (2019)
13. Hart, P.E., Nilsson, N.J., Raphael, B.: A formal basis for the heuristic determination of minimum cost paths. IEEE Trans. Sys. Sci. **4**(2), 100–107 (1968)
14. Liu, C., Mao, Q., Chu, X., Xie, S.: An improved A-star algorithm considering water current, traffic separation and berthing for vessel path planning. Appl. Sci. **9**(6), 1057 (2019)
15. Sedighi, S., Nguyen, D.-V., Kuhnert, K.-D.: Guided hybrid A-star path planning algorithm for valet parking applications. 5th International Conference on Control, Automation and Robotics (ICCAR). IEEE (2019)
16. Jiang, D., Li, G., Tan, C., Huang, L., Sun, Y., Kong, J.: Semantic segmentation for multiscale target based on object recognition using the improved Faster-RCNN model. Futur. Gener. Comput. Syst. **123**, 94–104 (2021)
17. Hao, Z., Wang, Z., Bai, D., Tao, B., Tong, X., Chen, B.: Intelligent detection of steel defects based on improved split attention networks. Frontiers in Bioengineer. **9**, 810876 (2022)
18. Fei, T., Weiran, L., Meng, Z.: Digital twin five-dimensional model and ten domain applications. Comput. Integr. Manuf. Syst. **25**(1), 1–18 (2019)
19. Lu, Y., Liu, C., Kevin, I., Wang, K., Huang, H., Xu, X.: Digital Twin-driven smart manufacturing: Connotation, reference model, applications and research issues. Robotics Comp.-Integr. Manufact. **61**, 101837 (2020)
20. Jiang, D., Li, G., Sun, Y., Hu, J., Yun, J., Liu, Y.: Manipulator grabbing position detection with information fusion of color image and depth image using deep learning. J. Ambient Intel. Humanized Comp. **12**, 10809–10822 (2021)
21. Jiang, D., et al.: Gesture recognition based on binocular vision. Clust. Comput. **22**, 13261–13271 (2019)
22. Liao, S., et al.: Occlusion gesture recognition based on improved SSD. Concurrency Computation: Practice Experience **33**(6), e6063 (2021)
23. Li, B., et al.: Gesture recognition based on modified adaptive orthogonal matching pursuit algorithm. Clust. Comput. **22**, 503–512 (2019)
24. Li, C., Li, G., Jiang, G., Chen, D., Liu, H.: Surface EMG data aggregation processing for intelligent prosthetic action recognition. Neural Computing Applications **32**, 16795–16806 (2020)
25. https://www.universal-robots.com/. Last accessed 08 August 2024
26. Xiao, F., et al.: An effective and unified method to derive the inverse kinematics formulas of general six-DOF manipulator with simple geometry. Mechanism Machine Theory **159**, 104265 (2021)
27. Liu, Y., et al.: Grasping posture of humanoid manipulator based on target shape analysis and force closure. Alex. Eng. J. **61**(5), 3959–3969 (2022)
28. Yun, J., et al.: Self-adjusting force/bit blending control based on quantitative factor-scale factor fuzzy-PID bit control. Alex. Eng. J. **61**(6), 4389–4397 (2022)

29. Parak, R.: Unity3D Industrial Robotics - Universal Robots UR3. https://github.com/rparak/Unity3D_Robotics_UR. Last accessed 08 August 2024
30. Fu, B., et al.: An improved A* algorithm for the industrial robot path planning with high success rate and short length. Robotics Autonomous Systems. **106**, 26–37 (2018)
31. Wang, H., Lou, S., Jing, J., Wang, Y., Liu, W., Liu, T.: The EBS-A* algorithm: an improved A* algorithm for path planning. PLoS ONE **17**(2), e0263841 (2022)
32. Sánchez, G., Latombe, J.-C.: On delaying collision checking in prm planning: application to multi-robot coordination. The Int. J. Rob. Res. **21**(1), 5–26 (2002)
33. Clark, C.M.: Probabilistic road map sampling strategies for multi-robot motion planning. Robot. Auton. Syst. **53**(3–4), 244–264 (2005)

DC Programming and DCA

On Solving Discrete Stochastic Difference-of-Convex-Functions Optimization Problems

Hoai An Le Thi[1,2] and Thi My Le Le[1(✉)]

[1] Université de Lorraine, LCOMS, 57000 Metz, France
{hoai-an.le-thi,thi-my-le.le}@univ-lorraine.fr
[2] Institut Universitaire de France (IUF), Paris, France

Abstract. This paper investigates binary stochastic difference-of-convex-functions (DC) programs, which are optimization problems where the objective function is the expectation of a stochastic DC function based on a probability distribution with binary variables. Two approaches are proposed for this type of problem. The first approach involves building the sample average approximation (SAA) problem for the original problem and using DCA (DC Algorithm) and/or stochastic DCA for solving the SAA subproblem. The second approach develops a novel stochastic DCA on a continuous domain via the stochastic approximations (SA) technique.

Keywords: Discrete stochastic optimization · Binary stochastic DC program · DC programming · DCA · Stochastic DCA

1 Introduction

Nonconvex stochastic optimization with discrete variables involves minimizing the expected value of a nonconvex loss functions under some deterministic (or stochastic constraints) and discrete variables (e.g., binary or integer variables). This type of optimization problems is highly relevant in many areas like machine learning, combinatorial optimization, operations research, and decision-making under uncertainty. Common examples include scheduling, resource allocation, portfolio selection in finance, and machine learning models that involve discrete parameters (e.g., in neural architecture search or hyperparameter tuning). As nearly almost all nonconvex functions are DC (Difference of Convex functions) functions and DC programming constitutes the backbone of nonconvex programming, in this work we focus on a major class of nonconvex stochastic optimization with discrete variables, the so called Binary stochastic DC program. Its mathematical formulation has the form

$$\min\{F(x) : x \in \mathcal{X} := C \cap \{0, 1\}^n\}. \tag{1}$$

© The Author(s), under exclusive license to Springer Nature Switzerland AG 2025
H. A. Le Thi et al. (Eds.): ICAMCS 2024, LNNS 1311, pp. 135–146, 2025.
https://doi.org/10.1007/978-3-031-90606-0_12

Here $C \subseteq \mathbb{R}^n$ is a nonempty polyhedral convex set; $F : \mathbb{R}^n \to \mathbb{R}$ is the expectation of a loss function with respect to the probability distribution $\mathbb{P}$ of the complete probability space $(\Omega, \Sigma_\Omega, \mathbb{P})$, meaning

$$F(x) := \mathbb{E}_{s \simeq \mathbb{P}}[f(x, s)] = \int_\Omega f(x, s) d\mathbb{P},$$

where $f(\cdot, s)$ is a DC function, say $f(x, s) = g(x, s) - h(x, s)$ with g and h being lower semicontinuous convex functions. In general, the probability distribution $\mathbb{P}$ is unknown.

The problem (1) involves three primary challenges. One challenge is the stochasticity of the objective; the expected loss function is frequently either unknown, given the unknown probability distribution $\mathbb{P}$, or prohibitively expensive to compute exactly. The second challenge arises from the nonconvexity of the objective function. The third challenge come from the discrete nature of variables. Passing from continuity to discreteness the complexity of problems horribly increases and it requires a fundamental change in the methodology.

In the *continuous* framework, convex stochastic optimization has been thoroughly studied in the literature since the seminal work [15], while nonconvex stochastic optimization problems are much less worker. The major area of nonconvex stochastic optimization is DC stochastic optimization, which has been recently emerged in the literature, thanks to its important role in machine learning, operation research, statistic, and their applications in various domains such as logistic, finance, robotic, etc. The core of methodologies of the approaches developed in DC stochastic optimization is DCA (DC Algorithm), a powerful tool of nonconvex programming. DC programming and DCA, which constitute the backbone of nonconvex programming and global optimization, were introduced by Pham Dinh Tao in 1985. Since 1993 these tools have been significantly developed through the joint works of Le Thi Hoai An and Pham Dinh Tao, solidifying their place as fundamental and increasingly popular approaches in nonconvex programming (see [7–9, 13] and references therein). DCA is now widely regarded as a powerful approach for solving large-scale, nonconvex, and nonsmooth optimization problems. It has been successfully applied in many fields such as transport logistics, finance, data mining and machine learning, computational biology, robotics and computer vision, combinatorial optimization, cryptology, etc. (see e.g., [8,9] and references therein). Additionally, it is essential to observe that, as indicated in [8], appropriate DC decompositions and suitably equivalent DC reformulations enable DCA to recover all (or most) standard methods in convex (or nonconvex) programming. In the stochastic framework, Stochastic DCA (SDCA) was initially studied for some special classes of DC problems, such as large-sum and/or (partially) smooth, and was later extended to a very general class of stochastic nonsmooth DC programs. Le Thi et al. [3] first developed stochastic DCA for large-sum problems of nonconvex but L-smooth functions with $\ell_{2,0}$ regularization, and later extended to a more general class of large-sum nonsmooth DC programs [4]. Liu et al. [11] proposed a stochastic algorithm based on DCA to solve a special class of two-stage stochastic programs with linearly bi-parameterized quadratic recourse. Nitanda and Suzuki [12] proposed a stochastic proximal DCA for differentiable DC programs and provided for the first time non-asymptotic convergence rate to find ϵ-accurate solutions. Xu et al. [16] further improved this work by designing stochastic proximal DCA schemes for a larger class of DC programs where the smoothness condition can be partially relaxed. Later, [2] investigates four stochastic DCA for the most general class of stochastic DC optimization problems where several assumptions required in the previous works are relaxed. For a comprehensive review on nonconvex stochastic optimization, and more particularly DC stochastic optimization, the readers are referred to [2]. Very recently, stochastic DCA are investigated in [5] for a larger class of DC stochastic optimization - DC constrained DC stochastic programs.

In the *discrete* nonconvex stochastic framework, research was much less attractive than the continuous case. Especially, to our knowledge, this work is the first in the literature dealing with discrete Stochastic DC optimization.

The classical approaches for discrete nonconvex stochastic optimization consist of updating some common approaches in continuous framework to discrete context. Solving these problems requires a combination of stochastic approximation, heuristics, and specialized methods for handling discrete variables. To tackle stochasticity, methods the most commonly used include Sample Average Approximation, Stochastic Approximation, Sample Path Optimization, and Markov Decision Processes.

Sample Average Approximation (SAA): This is one of the most widely used methods for solving stochastic optimization problems. The basic idea is to replace the expectation in the objective function with an empirical average over a sample of the random variables. This results to a large sum problem of the form

$$\min\{F_k(x) = \frac{1}{k}\sum_{i=1}^{k} f(x, s_i) : x \in \mathcal{X} := C \cap \{0,1\}^n\}, \tag{2}$$

where $s_1, ..., s_k$ are i.i.d. random variables with probability distribution $\mathbb{P}$. Here the problem becomes a deterministic optimization problem but with *discrete* variables. SAA has been proven to be asymptotically consistent for discrete problems (see e.g. [1]).

Stochastic Approximation (SA). Stochastic Approximation (SA) methods, like the Robbins-Monro algorithm, are typically iterative procedures where a solution is updated based on noisy observations of gradients or objective function values. In the discrete setting, however, SA cannot use gradient information in the usual sense. Instead, finite difference methods or other heuristic adaptations are often applied to estimate the direction in which to move.

Sample Path Optimization (SPO). SPO generates sample paths (scenarios) of the random variable s and solves the optimization problem for each sample path. The solutions are then averaged to find an optimal solution that minimizes the expected objective function. Hence the stochastic problem is converted into a deterministic one for each sample path. This approach can be computationally expensive if a large number of sample paths are needed.

Markov Decision Processes (MDPs) can be used for the problem involving sequential decision-making under uncertainty to minimize the expected cost or maximize the expected reward. This technique is suited for problems with a temporal structure where decisions are made over time. Computational complexity increases with the number of stages and states.

To deal with nonconvexity and discrete variable, conventional Branch and Bound, Cutting Plan, meta heuristic methods for combinatorial optimization can be investigated. However, large-scale problems make exact methods computationally infeasible. It is well known that DCA is an efficient and scalable approach in nonconvex programming and combinatorial optimization.

The above mentioned methods provide robust approaches for minimizing the expected value of a stochastic function when the decision variables are discrete. The choice of method depends on factors like problem size, the nature of the randomness, and the computational resources available. The power of DCA and stochastic DCA in continuous framework suggested us to exploit them for discrete DC stochastic programs. In the first approach, the most commonly used SAA will be employed for stochastic approximation and deterministic DCA and/or stochastic DCA will be investigated for the SAA subproblem. In the second approach, the considered problem is formulated as a DC constrained stochastic DC program for which a novel stochastic DCA using SA technique will be developed.

This paper is a short version of our work on solving Stochastic DC optimization problems with binary variables. We first present the main idea of our different approaches, and then describe the corresponding algorithms and their convergence properties. An in-depth study will be presented in the complete works in which these algorithms will be explored for several practical problems in logistic.

2 Preliminary

2.1 The SAA Method for Discrete Stochastic Optimization Problem

Let us briefly introduce the SAA method for discrete stochastic optimization problems (see e.g. [1]). A discrete stochastic optimization problem takes the form

$$\min_{x \in \mathcal{S}} F(x) = \mathbb{E}_{s \simeq \mathbb{P}}[f(x,s)]. \tag{3}$$

Here s is a random vector having probability distribution $\mathbb{P}$ of the complete probability space $(\Omega, \Sigma_\Omega, \mathbb{P})$, $\mathcal{S}$ is a *finite* set (e.g., $\mathcal{S}$ can be a finite subset of $\mathbb{R}^n$ with integer coordinates), $f(x,s)$ is a real valued function of two (vector) variables x and s, and $\mathbb{E}_{s \simeq \mathbb{P}}[f(x,s)] = \int_\Omega f(x,s) d\mathbb{P}$.

Let $s_1, ..., s_k$ be independent, identically distributed (i.i.d.) realizations obtained from the probability distribution $\mathbb{P}$. Consider the sample average function

$$F_k(x) = \frac{1}{k} \sum_{i=1}^{k} f(x, s_i)$$

and the associated problem

$$\min_{x \in \mathcal{S}} F_k(x). \tag{4}$$

Problems (3) and (4) are referred to as the "true" (or expected value) and sample average approximation (SAA) problems, respectively.

Given that the feasible set $\mathcal{S}$ is finite, Problems (3) and (4) have nonempty sets of optimal solutions denoted $\mathcal{S}^*$ and $\widehat{\mathcal{S}}_k$, respectively. Let ϑ^* and $\widehat{\vartheta}_k$ denote the optimal values,

$$\vartheta^* := \min_{x \in \mathcal{S}} F(x) \quad \text{and} \quad \widehat{\vartheta}_k := \min_{x \in \mathcal{S}} F_k(x),$$

of the respective problems. For $\varepsilon \geq 0$, $\overline{x}$ is called an ε-optimal solution of (3) if $\overline{x} \in \mathcal{S}$ and $F(\overline{x}) \leq \vartheta^* + \varepsilon$. The sets of all ε-optimal solutions of (3) and (4) are denoted $\mathcal{S}^\varepsilon$ and $\widehat{\mathcal{S}}_k^\varepsilon$, respectively. It is clear that $\mathcal{S}^\varepsilon$ coincides with $\mathcal{S}^*$, and $\widehat{\mathcal{S}}_k^\varepsilon$ coincides with $\widehat{\mathcal{S}}_k$ when $\varepsilon = 0$.

The following proposition establishes convergence with probability one (w.p.1) of the above statistical estimators.

Proposition 1. *The following two properties hold: (i) $\widehat{\vartheta}_k \to \vartheta^*$ w.p.1 as $k \to \infty$, and (ii) for any $\varepsilon \geq 0$ the event $\{\widehat{\mathcal{S}}_k^\varepsilon \subset \mathcal{S}^\varepsilon\}$ happens w.p.1 for k large enough.*

The proof of the above proposition can be found in [1].

2.2 DCA for Deterministic DC Programs

Let $\Gamma_0(\mathbb{R}^n)$ denote the set (convex cone) of all lower semicontinuous proper convex functions on $\mathbb{R}^n$. A complete study of DC programming and DCA can be found in [7–9,13,14].

The standard DC program takes the form

$$\alpha := \inf\{f(x) := g(x) - h(x) : x \in \mathbb{R}^n\}, \tag{P_{dc}}$$

where $g, h \in \Gamma_0(\mathbb{R}^n)$. Such a function f is called DC, $g - h$ is DC decomposition, while g and h are DC components of f. Note that, a DC program with closed convex constraint $x \in C$ (C is a closed convex subset of $\mathbb{R}^n$) can be equivalently written as a standard DC program in such a way that $f = (g + \chi_C) - h$, where χ_C stands for the indicator function of C.

A point x^* is called a critical point of $g - h$, or a generalized Karush-Kuhn-Tucker (KKT) point of (P_{dc}) if $\partial g(x^*) \cap \partial h(x^*) \neq \emptyset$, or equivalently $0 \in \partial g(x^*) - \partial h(x^*)$, while it is called a strongly critical point of $g - h$ if $\emptyset \neq \partial h(x^*) \subset \partial g(x^*)$.

DCA is based on local optimality conditions and duality in DC programming, which introduces the nice and elegant concept of approximating a DC program by a sequence of convex ones: at each iteration l, DCA approximates the second DC component h by its affine minorization $h_l(x) = h(x^l) + \langle x - x^l, y \rangle$, with $y \in \partial h(x^l)$, and then solves the resulting convex program

$$\alpha_l := \inf\{g(x) - h_l(x) : x \in \mathbb{R}^n\}, \tag{P_l}$$

to get x^{l+1}. The standard DCA is formally described as follows.

Standard DCA Scheme

Initialization : Let $x^0 \in \text{dom}\partial h$ and set $l = 0$.

repeat

1. Compute the subgradient $y^l \in \partial h(x^l)$.
2. Solve the following convex program

$$x^{l+1} \in \arg\min\{g(x) - h_l(x) : x \in \mathbb{R}^n\}.$$

3. Set $l = l + 1$.

until Stopping criterion.

Convergences properties of the standard DCA can be found in the seminal works on DC programming and DCA [7,13,14].

3 Solving Binary Stochastic DC Programs

Let $(\Omega, \Sigma_\Omega, \mathbb{P})$ be a probability space. Let us consider the following problem:

$$\min\{F(x) := \mathbb{E}_s[f(x,s) = g(x,s) - h(x,s)] : x \in \mathcal{X} := C \cap \{0,1\}^n\} \tag{5}$$

where $C \subseteq \mathbb{R}^n$ is a nonempty polyhedral convex set; $g(\cdot, s), h(\cdot, s), s \in \Omega$ are continuous convex functions defined on $\mathbb{R}^n$. Throughout the paper, we assume that the expectations of $g(x, \cdot)$ and $h(x, \cdot)$ are finite for all $x \in \mathbb{R}^n$ and denoted by

$$G(x) := \mathbb{E}_s[g(x,s)], \quad H(x) := \mathbb{E}_s[h(x,s)].$$

Then, G, H are continuous convex functions on the whole space $\mathbb{R}^n$.

3.1 The First Approach: SAA Using DCA/stochastic DCA

Let $s_1, ..., s_k$ be independent, identically distributed (i.i.d.) realizations obtained from the probability distribution $\mathbb{P}$, we get the sample average approximation problem of (5):

$$\min \ \widehat{F}(x) = \frac{1}{k}\sum_{i=1}^{k} f(x, s_i), \quad \text{subject to } x \in \mathcal{X} = C \cap \{0,1\}^n. \tag{6}$$

As $f(\cdot, s)$ are DC functions, the problem (6) is a deterministic DC program with binary variables, which is known to be NP-hard. However, we can reformulate this problem as a standard DC program using the exact penalty techniques in DC programming [10]. Consequently, DCA, an efficient algorithm for DC programming, can handle the reformulated program.

Let $p(x)$ be the penalty function define by $p(x) = \sum_{i=1}^{n} x_i(1-x_i)$.

It is easy to see that p is finite concave on $\mathbb{R}^n$ and $p(x) \geq 0$ for all $x \in \overline{\mathcal{X}} := C \cap [0,1]^n$. Moreover, we have $\{x \in \mathbb{R}^n : x \in C,\ x \in \{0,1\}^n\} = \{x \in \overline{\mathcal{X}} : p(x) \leq 0\}$.

Hence, Problem (6) can be rewritten as

$$\min_{x \in \overline{\mathcal{X}}} \widehat{F}(x) = \frac{1}{k}\sum_{i=1}^{k} f(x, s_i), \quad \text{subject to} \quad p(x) \leq 0. \tag{7}$$

According to [10], there exists $\overline{\tau}$ such that Problem (7) and the following penalized problem are equivalent for all $\tau \geq \overline{\tau}$:

$$\min_{x \in \overline{\mathcal{X}}} \overline{F}(x) = \widehat{F}(x) + \tau p(x). \tag{8}$$

It is important to note that the number $\overline{\tau}$ is generally difficult to calculate in practice. Consequently, a common approach is to employ a relatively large τ at the beginning and gradually increase it in order to promote the solutions to be binary. Once the solution is binary, we stop increasing τ.

As $\widehat{F}$ is DC and p is concave, $\overline{F}$ is also DC, with the following DC decomposition:

$$\overline{F}(x) = \frac{1}{k}\sum_{i=1}^{k} f(x, s_i) + \tau p(x) = \frac{1}{k}\sum_{i=1}^{k} g(x, s_i) - \left[\frac{1}{k}\sum_{i=1}^{k} h(x, s_i) - \tau p(x)\right]. \tag{9}$$

By using the DC decomposition (9) of $\overline{F}$, we get the following DC formulation of (8):

$$\min\ \{\overline{G}(x) - \overline{H}(x)\ :\ x \in \mathbb{R}^n\}, \tag{10}$$

where $\overline{G}(x)$ and $\overline{H}(x)$ are convex functions defined by $\overline{G}(x) = \frac{1}{k}\sum_{i=1}^{k} g(x, s_i) + \chi_{\overline{\mathcal{X}}}(x)$, and $\overline{H}(x) = \frac{1}{k}\sum_{i=1}^{k} \overline{h}_i(x)$ with $\overline{h}_i(x) = h(x, s_i) - \tau p(x)$ for all $i = 1, ..., k$.

According to the generic DCA scheme described above, applying DCA to (10) amounts to computing two sequences $\{x^l\}$ and $\{y^l\}$ in the way that $y^l \in \partial\overline{H}(x^l)$ and x^{l+1} is an optimal solution of the following convex problem

$$\min\left\{\frac{1}{k}\sum_{i=1}^{k} g(x, s_i) - \langle y^l, x\rangle : x \in \overline{\mathcal{X}}\right\}.$$

Finally our SAA method using DCA for the SAA subproblem is described in Algorithm 1.

Algorithm 1. SAA method using DCA for solving the problem (5)

Initialization : Generate a sequence $s_1, ..., s_k$ i.i.d. samples; choose an initial point x^0, a penalty parameter $\tau_0 > 0$, a positive number $\theta > 0$, a sufficiently small positive number ε; set $l := 0$.
repeat (Solving the SAA subproblem (6) by DCA)
1. Compute $y^l = \frac{1}{k}\sum_{i=1}^k u_i^l - \tau_l \nabla p(x^l)$, where $u_i^l \in \partial h(x^l, s_i)$ for all $i = 1, ..., k$.
2. Compute a solution x^{l+1} of the convex program

$$\min\left\{\frac{1}{k}\sum_{i=1}^{k} g(x, s_i) - \langle y^l, x\rangle : x \in \overline{\mathcal{X}}\right\}. \tag{11}$$

3. **if** $p(x^{l+1}) > \varepsilon$ **then**
   $\tau_{l+1} = \tau_l + \theta$
   end if
4. $l = l + 1$.

until stopping condition.

The convergence properties of the sequence $\{x^l\}$ generated by Algorithm 1 are stated as follows.

Theorem 1. *Let the sequence $\{x^l\}$ be generated by Algorithm 1. The following holds:*

(i) There exists a $\tau^* \geq 0$ such that if $\tau_l > \tau^*$ and x^l is binary, then x^r remains binary for all $r \geq l$.
(ii) Every limit point $\overline{x}$ of the sequence $\{x^l\}$ is a critical point of the problem (10).

Proof. The property (i) is proved similarly as Proposition 1 in [6]. The property (ii) is a direct consequence of the convergence properties of DCA.

Remark 1. It is worthy to notice that to ensure that $\widehat{F}$ is a sufficiently accurate approximation of F, k must be sufficiently large. In that case, Problem (10) becomes a large-sum DC problem, which may be expensive to solve by a deterministic algorithm. Stochastic algorithms could be efficient for dealing with large scale setting.

Thus, we are motivated to substitute DCA in Algorithm 1 with stochastic DCA (SDCA) [4], as presented in Algorithm 2 below.

The fundamental idea of stochastic DCA involves updating, at each iteration, the subgradients of only a few randomly selected $\overline{h}_i$ while maintaining the subgradients of the other $\overline{h}_i$. Thus, the primary benefit of stochastic DCA compared to standard DCA is the computational reduction in the step of calculating a subgradient $\overline{H}$. The convergence of SDCA to a critical point is assured with probability one.

Algorithm 2. SAA method using SDCA for solving the problem (5)

Initialization: Generate a sequence $s_1, ..., s_k$ i.i.d. samples; choose an initial point x^0, $t_0 = \{1, ..., k\}$, a penalty parameter $\tau_0 > 0$, a positive number $\theta > 0$, a sufficiently small positive number ε; set $l := 0$.
repeat (Solving the SAA subproblem (6) by stochastic DCA)
1. Compute $u_i^l \in \partial h(x^l, s_i)$ if $i \in t_l$ and keep $u_i^l = u_i^{l-1}$ if $i \notin t_l$.
2. Compute a solution x^{l+1} of the convex program

$$\min\left\{\frac{1}{k}\sum_{i=1}^{k} g(x, s_i) - \left\langle \frac{1}{k}\sum_{i=1}^{k} u_i^l - \tau_l \nabla p(x^l), x\right\rangle : x \in \overline{\mathcal{X}}\right\}. \tag{12}$$

3. **if** $p(x^{l+1}) > \varepsilon$ **then**
   $\tau_{l+1} = \tau_l + \theta$
   end if
4. $l = l + 1$ and randomly choose a small subset $t_l \subset \{1, ..., k\}$.

until stopping condition.

The convergence properties of the sequence $\{x^l\}$ generated by Algorithm 2 are stated as follows [4].

Theorem 2. *Let the sequence $\{x^l\}$ be generated by Algorithm 2, with probability one, every limit point $\overline{x}$ of the sequence $\{x^l\}$ is a critical point of the problem* (10).

3.2 The Second Approach: SA Using Stochastic DCA for Stochastic DC Program with DC Constraints

We can express the binary variables $x \in \{0,1\}^n$ as a DC constraint as follows:

$$\{x \in \mathbb{R}^n : x \in \{0,1\}^n\} = \{x \in \mathbb{R}^n : p(x) = \sum_{i=1}^{n} x_i(1 - x_i) \leq 0, x \in [0,1]^n\}.$$

Thus, Problem (5) can be reformulated as:

$$\begin{aligned} \min \quad & F(x) = \mathbb{E}_s[g(x,s)] - \mathbb{E}_s[h(x,s)] \\ \text{subject to} \quad & p(x) = \sum_{i=1}^{n} x_i(1 - x_i) \leq 0, x \in \overline{\mathcal{X}} := C \cap [0,1]^n. \end{aligned} \tag{13}$$

We will make the following assumption in what follows:

Assumption 1. *For $x \in \mathbb{R}^n$, the functions $g(x,\cdot), h(x,\cdot)$ are bounded on Ω and the functions $g(\cdot,s), h(\cdot,s)$ are equicontinuous at each $x \in \mathbb{R}^n$, for $s \in \Omega$.*

In [5], a stochastic DC algorithm is proposed for DC constrained DC stochastic programs. This method can be employed to solve the problem (13).

Let us consider the penalty problem of (13):

$$\min_{x\in\overline{\mathcal{X}}} \widetilde{F}(x) = F(x) + \tau p(x) = \mathbb{E}_s[g(x,s)] - (\mathbb{E}_s[h(x,s)] - \tau p(x)), \tag{14}$$

where $\tau > 0$ is a penalty parameter.

According to [5], we propose the following stochastic DCA to solve the problem (13) with updating the penalty parameter. In this algorithm, at each lth iteration, n_l new samples are collected and combined with past samples to calculate all subgradients of $h(\cdot, s_{j,i})$ for $j = 1, ..., l,\ i = 1, ..., n_j$, at the current solution x^l, thus forming a stochastic approximation of a subgradient of H. Additionally, the function G is incrementally approximated at each iteration by integrating both the new samples and previous samples.

Algorithm 3. Stochastic DC algorithm for solving (13)

Initialization : an initial point x^1, a sequence of sample sizes $\{n_l\}$, a initial penalty parameter $\tau_1 > 0$, $\delta > 0$, a sufficiently small positive number ε; set $l = 1$.
repeat

0. Draw n_l samples $s_{l,1}, ..., s_{l,n_l} \overset{\text{iid}}{\simeq} \mathbb{P}$.
1. Compute $u^l_{j,i} \in \partial h(x^l, s_{j,i}), j = 1, ..., l,\ i = 1, ..., n_j$.
2. Compute a solution x^{l+1} of the convex program

$$\min\left\{\frac{1}{\sum_{j=1}^{l} n_j}\sum_{j=1}^{l}\sum_{i=1}^{n_j} g(x, s_{j,i}) - \left\langle \frac{1}{\sum_{j=1}^{l} n_j}\sum_{j=1}^{l}\sum_{i=1}^{n_j} u^l_{j,i} - \tau_l \nabla p(x^l), x\right\rangle : x \in \overline{\mathcal{X}}\right\}. \tag{15}$$

3. Penalty parameter update.
Compute $r_l := p(x^{l+1})$ and set

$$\tau_{l+1} = \begin{cases} \tau_l & \text{if either} \quad \tau_l \geq \|x^{l+1} - x^l\|^{-1} \text{ or } r_l \leq \varepsilon \\ \tau_l + \delta & \text{if} \quad \tau_l < \|x^{l+1} - x^l\|^{-1} \text{ and } r_l > \varepsilon \end{cases}$$

4. Set $l = l + 1$.
until stopping condition.

The convergence properties of Algorithm 3 are stated as follows [5].

Theorem 3. *Let a sequence of sample sizes* $\{n_l\}$ *be such that* $\sum_{l=1}^{\infty} \frac{n_l(1+\sqrt{\ln A_l})}{\sqrt{A_l}A_{l+1}} < +\infty$ $\left(\textit{with } A_l := \sum_{j=1}^{l} n_j\right)$. *Let* $\{x^l\}$ *be a sequence generated by Algorithm 3. Suppose that either* G *or* H *is differentiable on* $\mathbb{R}^n$. *Under Assumption 1, with probability* 1, *we have:*

a. Either every limit point x^* *of the sequence* $\{x^l\}$ *satisfies*

$$0 \in \{\nabla p(x^*)\} + \mathcal{N}(\overline{\mathcal{X}}, x^*)$$

if the penalty parameter sequence $\{\tau_l\}$ *is unbounded.*

b. Or $\sum_{l=1}^{\infty} \|x^{l+1} - x^l\|^2 < +\infty$ *and every limit point* x^* *of the sequence* $\{x^l\}$ *is a KKT point of the problem* (13) *if the penalty parameter sequence* $\{\tau_l\}$ *is bounded.*

4 Concluding Remarks

We have proposed two approaches for solving binary stochastic DC programs. The first involves the SAA technique while the second is based on SA method for dealing stochasticity.

In the SAA based approach, the stochastic objective function is replaced with an average over the sampled data, which results to a large sum DC problem with binary variables. The DCA and/or Stochastic DCA (SDCA) are then performed on this approximated problem, using exact penalty technique for tackling discrete variables. Each iteration of Algorithm 1 and/or Algorithm 2 corresponds to one iteration of DCA or SDCA. The two algorithms differ primarily in the first step of each iteration, specifically in how they compute the subgradient of the second DC component of the DC objective function - either using an exact subgradient (DCA) or an approximate (stochastic) subgradient (SDCA).

The SA-based approach directly addresses the original DC-constrained stochastic DC problem by penalizing the DC constraint and employing stochastic DCA to iteratively approach the solution. It estimates the subgradient of the second DC component and approximates the first DC component of the objective function based on random samples in each iteration.

When comparing Algorithm 2 and Algorithm 3 (two versions of stochastic DCA), we observe that the primaly different is the way to deal with stochasticity (SAA versus SA). However, they share the same optimization process: use the penalty technique in DC programming to reformulate the DC optimization problems with binary variables in the form of a DC program with continuous variables, and then apply stochastic DCA to solve the resulting problem.

More precisely, steps 1 and 2 in each iteration of these algorithms follow the same general principle of stochastic DCA. In step 1, an approximation of a subgradient for the second DC component of the objective function is computed, while in step 2, the convex subproblems (12) and (15), which involve the first DC component, take a similar form. However, the way to approximate the subgradient of the second DC component in step 1 and to approximate the first DC component in step 2 differ depending on whether the SAA or SA techniques are used for sampling data.

It is worth noting that the computational complexity in each iteration differs between the algorithms. While the cost per iteration in Algorithm 3 increases progressively with the number of iterations, in Algorithm 2, it remains constant. Consequently, Algorithm 3 typically requires more iterations to converge than Algorithm 2.

In the full paper, we will provide a more in-depth analysis of the convergence properties of the proposed algorithms and investigate their application to real-world problems, such as logistics. We will also present numerical experiments and empirical comparative results for these approaches.

References

1. Kleywegt, A.J., Shapiro, A., Homem-de Mello, T.: The sample average approximation method for stochastic discrete optimization. SIAM J. Optim. **12**(2), 479–502 (2002)
2. Le Thi, H.A., Huynh, V.N., Pham Dinh, T., Luu, H.P.H.: Stochastic difference-of-convex-functions algorithms for nonconvex programming. SIAM J. Optim. **32**(3), 2263–2293 (2022)
3. Le Thi, H.A., Le, H.M., Phan, D.N., Tran, B.: Stochastic dca for the large-sum of non-convex functions problem and its application to group variable selection in classification. In: International Conference on Machine Learning, pp. 3394–3403. PMLR (2017)
4. Le Thi, H.A., Le, H.M., Phan, D.N., Tran, B.: Stochastic dca for minimizing a large sum of dc functions with application to multi-class logistic regression. Neural Netw. **132**, 220–231 (2020)
5. Le Thi, H.A., Le, T.M.L., Huynh, V.N.: Stochastic difference-of-convex-functions (dc) algorithms for dc constrained stochastic programming. (submitted) (2024)
6. Le Thi, H.A., Pham Dinh, T.: A continuous approach for globally solving linearly constrained quadratic zero-one programming problems. Optimization **50**(1–2), 93–120 (2001)
7. Le Thi, H.A., Pham Dinh, T.: The dc (difference of convex functions) programming and dca revisited with dc models of real world nonconvex optimization problems. Ann. Oper. Res. **133**, 23–46 (2005)
8. Le Thi, H.A., Pham Dinh, T.: Dc programming and dca: thirty years of developments. Math. Program. **169**(1), 5–68 (2018)
9. Le Thi, H.A., Pham Dinh, T.: Open issues and recent advances in dc programming and dca. J. Global Optim. 1–58 (2023)
10. Le Thi, H.A., Pham Dinh, T., Huynh, V.N.: Exact penalty and error bounds in dc programming. J. Global Optim. **52**(3), 509–535 (2012)
11. Liu, J., Cui, Y., Pang, J.S., Sen, S.: Two-stage stochastic programming with linearly bi-parameterized quadratic recourse. SIAM J. Optim. **30**(3), 2530–2558 (2020)
12. Nitanda, A., Suzuki, T.: Stochastic difference of convex algorithm and its application to training deep boltzmann machines. In: Artificial Intelligence and Statistics, pp. 470–478. PMLR (2017)
13. Pham Dinh, T., Le Thi, H.A.: Convex analysis approach to dc programming: theory, algorithms and applications. Acta Math. Vietnam **22**(1), 289–355 (1997)
14. Pham Dinh, T., Le Thi, H.A.: A dc optimization algorithm for solving the trust-region subproblem. SIAM J. Optim. **8**(2), 476–505 (1998)
15. Robbins, H., Monro, S.: A stochastic approximation method. Ann. Math. Stat. 400–407 (1951)
16. Xu, Y., Qi, Q., Lin, Q., Jin, R., Yang, T.: Stochastic optimization for dc functions and non-smooth non-convex regularizers with non-asymptotic convergence. In: International Conference on Machine Learning, pp. 6942–6951. PMLR (2019)

An Online DCA and DQN Based Approach for Truck-Drone Delivery Problems

Hoang Viet Nguyen[1], Vinh Thanh Ho[2(✉)], and Hoai Minh Le[1]

[1] Laboratoire de Conception, Optimisation and Modélisation des Systémes (LCOMS), University of Lorraine, Metz, France
minh.le@univ-lorraine.fr

[2] Laboratoire XLIM, Universitè de Limoges, 123 Avenue Albert Thomas, Limoges 87060, France
vinh-thanh.ho@unilim.fr

Abstract. In this paper, we study a hybrid approach for the combined truck-drone delivery problem. This approach integrates DCA (Difference of Convex functions Algorithms) into the Deep Q-learning (DQN) algorithm with the purpose of modifying the Q-values update. A numerical comparison on the performance of DQN and DQN-DCA is given on synthesis datasets to show the effectiveness of the proposed approach for solving the truck-drone delivery problem.

Keywords: Online DCA · DQN · Truck-drone delivery problem

1 Introduction

Logistics and delivery operations play a critical role in many industries, and the introduction of drones has begun to transform this field (see, e.g., [1,7]). Combining drones with traditional truck-based delivery systems offers potential for optimizing processes, reducing costs, and improving efficiency. Drones can bypass traffic and speed up deliveries, but this integration also introduces challenges, including planning and coordinating truck-drone movements, meeting regulatory requirements, and developing the necessary infrastructure.

Various approaches have been proposed to optimize the truck-drone delivery system (see, e.g., [1,7,10,14]). Firstly, heuristic methods use practical strategies to solve optimization problems (see, e.g., [14]). Clustering groups delivery requests by proximity to optimize truck and drone routes, while genetic algorithms refine vehicle assignments through iterative processes. Local search techniques incrementally improve existing routes by adjusting them based on nearby alternatives. Secondly, another approach based on mathematical programming is developed for solving the truck-drone delivery problem (see, e.g., [10]). Mixed integer linear programs focus on optimizing routes considering constraints like capacity and cost as well as handling vehicle assignment. Thirdly,

© The Author(s), under exclusive license to Springer Nature Switzerland AG 2025
H. A. Le Thi et al. (Eds.): ICAMCS 2024, LNNS 1311, pp. 147–158, 2025.
https://doi.org/10.1007/978-3-031-90606-0_13

machine learning techniques are increasingly used to improve delivery planning (see, e.g., [7]). By combining these methods, the truck-drone delivery system can achieve enhanced optimization, adaptability, and precision.

This paper introduces a hybrid approach for optimizing combined truck-drone delivery problems by integrating Deep Q-learning (DQN) with DCA (Difference of Convex Functions Algorithm). Our work aims to assess how this approach enhances combined truck-drone delivery problems. Subsequently, numerical experiments on several synthetic datasets are conducted to observe and compare the performance of DQN and our approach DQN-DCA.

The paper is organized as follows: Sect. 2 describes the truck-drone delivery problem. Section 3 briefly introduces deep reinforcement learning, deep Q-learning, DC programming, and DCA, and discusses their integration. Numerical results are presented in Sect. 4, and the paper concludes in Sect. 5.

2 Problem Description

This section describes the problem of combined truck-drone delivery [7]. The problem can be modelized by utilizing a Markov Decision Process (MDP) with a set of states, actions, transitions, and rewards. States include variables such as truck and drone locations and customer destinations, which change based on actions like moving, waiting, or loading. Actions are rewarded according to the current situation, with the goal of maximizing total rewards. Transition probabilities, influenced by traffic conditions, determine state changes. Bellman optimality equations are used to solve the MDP [18], aiming to find an optimal policy that maximizes expected long-term rewards.

First, we consider the set $\mathcal{V} = \{0, 1, \ldots, n\}$ represents physical locations, where 0 represents delivery hubs and 1 to n represent customer locations. Here we assume that drones can fly directly between two locations without being constrained by physical roads. Trucks and drones must begin deliveries from the delivery hub, visit all customer locations, and finish the delivery at the hub. When deliveries begin, trucks load all customer packages, eliminating the need for trucks or drones to return to the delivery hub to pick up additional packages. A traffic condition variable $\lambda \in \{0, 1, 2\}$ represents road congestion with three levels, where 0 (resp. 1 and 2) represents light (resp. normal and heavy) traffic. The more the value of λ, the more the truck's travel time. The truck's travel time from point i to j is represented by a function $\delta^T_{(i,j)}(\lambda) : \{0, 1, 2\} \to \mathbb{R}^+$. The drone's travel times do not depend on λ and it is denoted by $\delta^D_{(i,j)} \in \mathbb{R}^+$.

State: The state $s = (\alpha_T, \alpha_D, \beta_T, \beta_D, \lambda, \mu, \tau_T, \tau_D, w, d)$ includes the information of system where α_T (resp. α_D) in $\mathcal{V}$ denotes the position of the truck (resp. drone); β_T (resp. β_D) in $\mathcal{V}$ the position of the next destination of the truck and the drone; μ a binary variable indicating whether the drone stays with the truck (μ=0) or not (μ=1); τ_T (resp. τ_D) the remaining time for the truck (resp. the drone) to reach their destination; w the package carried by the drone, $w \in \mathcal{V}\backslash\{0\}$ where $w = 0$ if the drone is empty; d : the customer demands are represented by a vector d of size $|\mathcal{V}|$, where 1 indicates the package is delivered and 0 otherwise.

Action: An action $a = (a_T, a_D)$ consists of the action of the truck and drone. We define $\mathcal{A}_T$ as the set of all available actions for the truck and $\mathcal{A}_D$ as the set of all available actions for the drone. Specifically, $\mathcal{A}_T = \mathcal{V} \cup \{\textbf{wait}\}$ and $\mathcal{A}_D = \mathcal{V} \cup \{\textbf{rest}, \textbf{return}\} \cup \{\textbf{load package } i : i \in \mathcal{V}\backslash\{0\}\}$. The action **wait** lets the truck wait for the drone to complete a delivery and return. The action **rest** allows the drone to stay with the truck without performing any task. The action **return** enables the drone to return to the truck after a delivery. The action **load package** i allows the drone to load a package for customer i. The choice of actions depends on states: let $\mathcal{A}_s$ denote the set of actions available in state s. $\mathcal{A}_{T_s}$ (resp. $\mathcal{A}_{D_s}$) represent the actions available for the truck (resp. drone) in state s, respectively.

For the truck, we have: $\mathcal{A}_{T_s} = \{s(\beta_T)\}$ if $s(\tau_T) \neq 0$; $N(s(\alpha_T))$ if $s(\tau_T) = 0$ and $s(\mu) = 0$; $N(s(\alpha_T)) \cup \{\textbf{wait}\}$ otherwise. When the truck is on the road ($s(\tau_T) \neq 0$) heading to its destination $s(\beta_T)$, the only available action is to continue towards the destination. If the truck has reached its destination or is waiting ($s(\tau_T) = 0$) and the drone is not with the truck ($s(\mu) = 0$), the truck can go to any neighboring location. Additionally, the truck can take the action **wait** to avoid indefinite waiting when the drone is not present. For the drone, the actions $\mathcal{A}_{D_s}$ are determined as follows: $\mathcal{A}_{D_s} = \{s(\beta_D)\}$ if $s(\tau_D) \neq 0$; $\{\textbf{return}\}$ if $s(\tau_D) = 0, s(\mu) = 1$ and $s(w) = 0$; $\{\textbf{rest}\} \cup \{i : i \in \mathcal{V}\backslash\{0\}, i \in s(w)\} \cup$ $\{\textbf{load package } j : j \in \mathcal{V}\backslash\{0\}, s(d)(j) = 0\}$, otherwise.

Transition Probability and Next State: The transition probability can be expressed as the product of the individual transitions of each element $i \in s'$: $p(s'|s,a) = \prod_{i \in s'} p(i|s,a)$ where $p(i|s,a)$ represents the individual transition probability for element $i \in s'$. For example, $p(s'(\alpha_T)|s,a)$ is the probability that the truck's position in the next state is $s'(\alpha_T)$ given the current state s and the action a. Thus, as long as $p(i|s,a) = 0$ for all $i \in s'$, $p(s'|s,a) = 0$. Since the stochasticity of state transitions lies only in the road condition λ, the individual transition probabilities $p(i|s,a)$ of the other elements, i.e., $i \in \{\alpha_T, \alpha_D, \beta_T, \beta_D, \mu, \tau_T, \tau_D\} \cup w \cup d$, are binary. Since the truck makes a decision only after reaching a destination, the road condition is observed only when the arrival time at the destination is 0. So, when $s(\tau_T) \neq 0$, i.e. the truck is still on the road, the road condition remains the same as the one observed by the truck at its last decision, i.e. $p(\lambda'|s,a) = 1$ if $s(\lambda) = \lambda'$; 0 otherwise. When $s(\tau_T) = 0$, i.e. the truck starts moving towards the next customer, the road condition observed by the truck changes according to a Markov process, i.e. $p(\lambda'|s,a) \in [0,1], \quad \forall \lambda' \in \{0,1,2\}$, and $\sum_{\lambda' \in \{0,1,2\}} p(\lambda'|s,a) = 1$.

Reward. A reward $r(s,a)$ is obtained at state s after taking an action a. First at the end of each decision epoch, we reward an action by taking the negative of the time elapsed since the last epoch, so that the cumulative reward can be maximized. When $a_T \in V$ and $a_D \in V$, the reward is defined as: $r(s,a) = -\min\{\delta^T_{s(\alpha_T),a_T}, \delta^D_{s(\alpha_D),a_D}\}$, if $s(\tau_T) = 0, s(\tau_D) = 0$; $-\min\{s(\tau_T), \delta^D_{s(\alpha_D),a_D}\}$ if $s(\tau_T) \neq 0, s(\tau_D) = 0$; $-\min\{\delta^T_{s(\alpha_T),a_T}, s(\tau_D)\}$, otherwise. When $a_T = \textbf{wait}$, we define $r(s,a) = -\delta^D_{s(\alpha_D),a_D}$ if $s(\tau_D) = 0$; $-s(\tau_D)$ otherwise. When $a_D = \textbf{rest}$, the reward is decided only by the truck, i.e. $r(s,a) = -\delta^T_{s(\alpha_T),a_T}$ if $s(\tau_T) = 0$; $-s(\tau_T)$ otherwise. When a_D = **return**, the reward is decided by the longer between the truck and the drone to reach the destination, because for the epoch to end, both the truck and the drone must reach the destination, $r(s,a) = -\max\{s(\tau_T), \delta^D_{s(\alpha_D),s(\beta_T)}\}$ if $s(\tau_T) \neq 0$; $-\max\{\delta^T_{s(\alpha_T),a_T}, \delta^D_{s(\alpha_D),a_T}\}$ otherwise. Finally, when $a_D \in$ {**load package** $i : \forall i \in \mathcal{V}\backslash\{0\}$}, we set $r(s,a) = 0$ and proceed to the next epoch since the truck requires a new action due to the change in the drone's load. When the whole delivery process is completed, a terminal reward R_Δ is granted.

In the sequel, we are going to give solution methods based on reinforcement learning for solving the MDP model just described above.

3 Solution Methods

3.1 Deep Reinforcement Learning and Deep *Q*-Learning

Reinforcement Learning (RL) [18] is a branch of machine learning that focuses on sequential decision-making by an agent through the interaction with an environment (see Fig. 1a). The agent observes the current state of the environment, selects an action based on this state, and the environment transitions to a new state in response to the action. The agent then receives a reward or feedback signal from the environment, which is used to evaluate the quality of the action taken. Over time, the agent's objective is to obtain a policy maximizing cumulative rewards. Unlike other machine learning approaches, RL is characterized by an iterative interaction between the agent and the environment. The agent explores the environment, learns from its experiences, and adjusts its policy accordingly. RL has been successfully applied to various problems, including routing problems, robotics, resource allocation (communication network), etc. (see more in [16]). To achieve its learning objectives, different techniques and algorithms, such as policy gradient methods [19], Monte Carlo methods (see Chap. 5 in [18]), *Q*-learning [20], SARSA algorithms [17] and Deep *Q*-Networks [9], have been developed and utilized to explore the environment, estimate the value of different actions, and update its policy to maximize future rewards. A brief survey on RL algorithms has recently been given in [16].

Among the classical RL algorithms, *Q*-learning is a widely-used RL technique that allows an agent to learn to make optimal decisions in an environment. The agent maintains a function called the *Q*-function, which assigns a value to each state-action pair (s,a). The *Q*-function values are iteratively updated using the Bellman optimality equation [18]:

$$Q(s,a) = r(s,a) + \gamma \cdot \max_{a' \in A} Q(s', a') \tag{1}$$

where $Q(s,a)$ is the *Q*-value for state s and action a, $r(s,a)$ is the immediate reward obtained after taking action a in state s, γ is the discount factor that weighs the importance of future rewards, s' is the next state after taking action a, a' is the next action chosen by the agent. The agent learns by exploring the environment, taking actions, observing rewards, and updating the *Q*-function

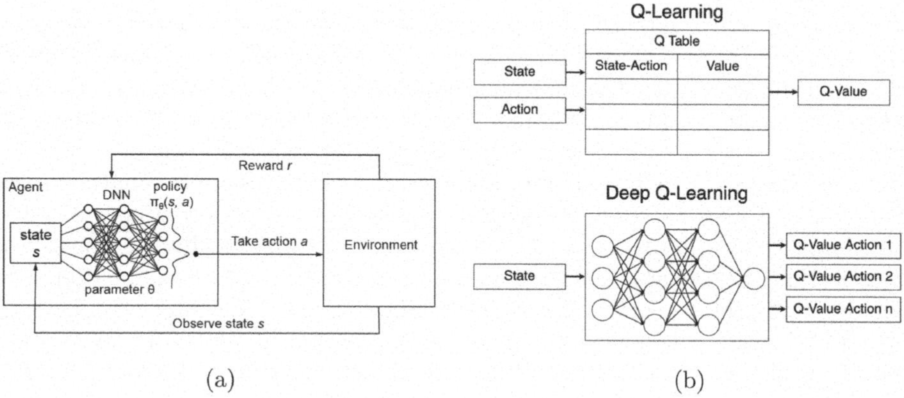

Fig. 1. (a) Deep reinforcement learning [8]; (b) Q-learning versus Deep Q-learning [15]

via the equation (1) to maximize the expected total reward. Through this iterative process, the agent gradually learns the optimal policy corresponding to the optimal Q-function that maximizes cumulative rewards.

Deep Q-learning (DQN) [9] is a significant extension of traditional Q-learning, leveraging deep neural networks as function approximators to estimate Q-values for state-action pairs. In classical Q-learning, a table is maintained to store Q-values for every possible state-action combination, which becomes inefficient or infeasible in complex environments with large or continuous state spaces. DQN overcomes this limitation by using a neural network that takes the current state as input and produces Q-value estimates for all possible actions (see Fig. 1b). The learning process involves adjusting the network's weights and biases to minimize the error between the predicted Q-values and the actual Q-values derived from the Bellman optimality equation (1). This adjustment helps the agent learn more accurate value estimates, thereby improving its decision-making capabilities. To enhance the stability and efficiency of learning, DQN employs techniques such as replay memory. It stores the agent's past experiences, allowing random batches of these experiences to be used during training. This randomization breaks the correlation between consecutive steps, which stabilizes learning and prevents the model from overfitting to recent experiences. Remarkable success of DRL has been marked in complex tasks in logistics and transportation planning to solve complex problems (see, e.g., [11,21,22]).

3.2 Outline of DC Programming, DCA and Online DCA

DC programming and DCA (see, e.g., [12,13]) address DC programs of the form

$$\inf\{f(w) := g(w) - h(w) : w \in \mathbb{R}^d\}, \quad (P_{dc})$$

where $g, h \in \Gamma_0(\mathbb{R}^d)$, the set of all lower semicontinuous proper convex functions on $\mathbb{R}^d$. Such a function f is called DC function, and $g - h$, DC decomposition of f while g and h are DC components of f. The constrained DC program whose feasible set C is convex always can be transformed into the unconstrained DC program by adding the indicator function of C, denoted by χ_C which is defined by $\chi_C(w) = 0$ if $w \in C$, and $+\infty$ otherwise to the first DC component: $\inf\{f(w) := g(w) - h(w) : w \in C\} = \inf\{\chi_C(w) + g(w) - h(w) : w \in \mathbb{R}^d\}$. The main idea of DCA is quite straightforward. It involves approximating a DC (Difference of Convex) problem with a sequence of convex programs. At each iteration l, DCA approximates the concave part $-h$ using its affine majorization (by selecting $z^l \in \partial h(w^l)$) and then minimizes the resulting convex function. The generic DCA scheme can be described in Algorithm 1. DCA has been successfully solved large-scale nonsmooth/nonconvex programs in numerous application areas, especially in machine learning, communication system, biology, finance, etc. (see comprehensible survey papers [5,6] and references therein).

Algorithm 1: Generic DCA scheme

Initialization: Choose an initial point w^0. Set $l \leftarrow 0$.
Repeat
Step 1. Compute $z^l \in \partial h(w^l)$.
Step 2. Compute $w^{l+1} \in \text{argmin } \{g(w) - \langle w, z^l \rangle : w \in \mathbb{R}^d\}$.
Step 3. $l \leftarrow l + 1$
Until stopping condition is satisfied.

The above generic DCA scheme can be adapted to solve online DC problems where the set of predictions is convex and the loss at each step is a DC function as follows (see [2]). At each learning step, a DC loss function F_t is optimized over the convex set of predictions $\mathcal{S}$, resulting in a (standard) DC program. However, since data in an "online" setting arrives sequentially, fully solving this DC problem at each step is not essential. Instead, just one iteration of the DCA is performed. Let T represent the total number of online learning steps. The function F_t can either represent the cumulative loss $\sum_{t=1}^{T} f_t$ or the current loss f_t. In [2], the authors defined F_t simply as the current loss function f_t, say $F_t = f_t := g_t - h_t$ with g_t and h_t being convex functions. The online DCA scheme can be described as follows. The convex subproblem in Online DCA can be solved by any solver for convex optimization problems. In [2], the authors proposed using a subgradient method to completely solve the subproblem or using just one iteration of subgradient method. Online DCA has been applied for solving online learning problems, such as online classification [2] and online prediction with expert advice [3].

Algorithm 2: Online DCA

Initialization: set an initial point w^0, observe a DC loss function $f_0 := g_0 - h_0$.
for $t = 1, 2, \ldots, T$ **do**
1. Predict a vector $w^t \in \mathcal{S}$ by performing one iteration of DCA.
   1.1. Compute a subgradient $z^t \in \partial h_{t-1}(w^{t-1})$.
   1.2. Solve the convex program to get w^t: $\min\{g_{t-1}(w) - \langle z^t, w\rangle : w \in \mathcal{S}\}$.
2. Observe a DC loss function $f_t := g_t - h_t$.
3. Suffer loss $f_t(w^t)$ and update its model.

3.3 Hyrid Approach Based on DRL and Online DCA

Deep Q-Learning for Combined Truck-Drone Delivery Systems. In this section, we describe how DQN is used to learn the delivery policy for the truck and the drone [7]. The DQN algorithm [9] relies on iterative Monte Carlo simulations, where an MDP is repeatedly simulated using a semi-random policy, which is updated until convergence. The MDP for combined truck-drone delivery systems is defined in Sect. 2. Each simulation, called an episode, starts when the truck and drone leave the delivery center and ends when they return after completing deliveries. Neural networks (NNs) in DQN serve as function approximators, estimating the state-action value $Q(s, a)$ defined by $Q_t(s, a) = r_t(s, a) + E\left[\sum_{t'=t}^{T} r_{t+1}(s, a) + r_{t+2}(s, a) + \ldots | s_t, a_t\right], \forall s_t \in \mathcal{S}, a_t \in \mathcal{A}_s$ Moreover DQN stores the simulated episodes of experiences (represented as tuples (s_t, a_t, r_t, s_{t+1})). It then randomly samples mini-batches of past experiences from this memory for training NNs. Because an ϵ-greedy policy is used during the simulation, new experiences added to memory typically yield higher rewards than older ones. Ultimately, the DQN outputs the policy by selecting the action that maximizes the Q-values as follows: $\pi(s) = \arg\max_{a' \in \mathcal{A}_s}\{Q(s, a')\}$. Below is the DQN algorithm (Algorithm 3) [9].

Hybrid Approach of DQN and Online DCA. Now we consider a DCA-based approach in [4] to update the Q-values for each state and action pair (s, a) in Step 3 of DQN instead of using (1). The idea for the Q-values update is to use online DCA for minimizing the ℓ_1-norm of Optimal Bellman Residual (OBR). In particular, let $\mathcal{F}$ denote the space of linear approximation of Q-functions, i.e.

$$\mathcal{F} := \{Q_\theta(s, a) = \langle \theta, \phi(s, a)\rangle \mid \theta \in \mathbb{R}^d\}, \tag{2}$$

where $\phi(\cdot) = (\phi_1(\cdot), \phi_2(\cdot), ..., \phi_d(\cdot)) \in \mathbb{R}^d$ is a vector of basis functions, $\phi_i : S \times A \to \mathbb{R}$, $i = 1, ..., d$ are a set of basis functions defined on $S \times A$. The optimization problem of ℓ_1-norm of Optimal Bellman Residual over the space $\mathcal{F}$ is defined as follows (see details in [4]) : $\min_{\theta \in \mathbb{R}^d} F(\theta) = \|f(\theta)\|_1$ where $f = (f_1, f_2, ..., f_N) : \mathbb{R}^d \to \mathbb{R}^N$, N is the number of transitions in sampled mini-batch (in Step 3.1 of DQN), $N_{\mathcal{A}_s} = |\mathcal{A}_s|$ is the number of possible actions corresponding to states, and each f_i is defined by:

$$f_i(\theta) = \max_{j=1,\ldots,N_{\mathcal{A}_s}} \langle A(i, j), \theta\rangle + b(i).$$

Algorithm 3: DQN: Deep Q-learning with Experience Replay

Initialization : the replay memory $\mathcal{M}$, the Q-network, the maximum no. episodes E.

for *episode* $e = 1,\ldots,E$ **do**

  Initialize the initial state

  for *step* $k = 1,\ldots$ **do**

    Step 1. Action selection:

    1.1. Choose an action using an ϵ-greedy strategy, where with probability ϵ, the agent selects a random action in favor of exploration, and with probability $(1-\epsilon)$, it selects the action with the maximum Q-value predicted by the network Q.

    1.2. Perform the selected action in the environment and observe the next state and the obtained reward.

    Step 2. Experience replay: Store the experience tuple (state, action, reward, next state) in $\mathcal{M}$.

    Step 3. Q-Value update:

    3.1. Sample a mini-batch of transitions from the $\mathcal{M}$.

    3.2. Calculate the target Q-value of sampled transitions using the Bellman optimality equation (1).

    Step 4. Q-Network training: Update the weights of the Q-network by minimizing the mean squared error between the predicted Q-values and the target Q-values. The Q-network is trained using the sampled transitions from $\mathcal{M}$ and their corresponding target Q-values.

    Step 5. Check stopping conditions

$A(i,j) = \gamma\phi(s_i', a_j) - \phi(s_i, a_i)$ and $b(i) = r(s_i, a_i)$, $j = 1, ..., N_{\mathcal{A}_s}$.

To apply online DCA, one considers a DC decomposition $F(\theta) := G(\theta) - H(\theta)$ where $G(\theta) = \|f\|_1 + \sum_{i=1}^N f_i(\theta)$, $H(\theta) = \sum_{i=1}^N f_i(\theta)$. At each iteration k, we calculate $w^k \in \partial H(\theta^k)$ and then solve the resulting convex program:

$$\min_{\theta\in\mathbb{R}^d} \{G(\theta) - \langle w^k, \theta\rangle\} = \min_{\theta\in\mathbb{R}^d} \left\{2\sum_{i=1}^N f_i^+(\theta) - \langle w^k, \theta\rangle\right\}, f_i^+ = \max(f_i, 0)$$

According to [4], we can choose $w^k \in \partial H(\theta^k)$ as follows.

$$w^k = \sum_{i=1}^N \sum_{j_i \in I_i(\theta^k)} \frac{A(i, j_i)}{|I_i(\theta^k)|} \text{ where } I_i(\theta) = \arg\max_{j=1,\ldots,N_{\mathcal{A}_s}} \langle A(i,j), \theta\rangle. \tag{3}$$

Next, we solve the above convex program using one iteration of the subgradient method. Let us define $p(\theta) := 2\sum_{i=1}^N f_i^+(\theta) - \langle w^k, \theta\rangle$. We calculate a subgradient $v^k \in \partial p(\theta^k)$ as follows

$$\partial p(\theta^k) = 2\sum_{i=1}^N \partial f_i^+(\theta^k) - w^k,\ \partial f_i^+(\theta^k) = \begin{cases}\{0\} & \text{if } f_i(\theta^k) \le 0\\ \partial f_i(\theta^k) & \text{otherwise.}\end{cases}$$

$$\partial f_i(\theta) = \partial\left[\max_{j=1,\ldots,N_{\mathcal{A}_s}} \langle A(i,j), \theta\rangle + b(i)\right] = \partial\left[\max_{j=1,\ldots,N_{\mathcal{A}_s}} \langle A(i,j), \theta\rangle\right].$$

Similarly to H, we take $\sum_{j_i \in I_i(\theta^k)} \frac{A(i,j_i)}{|I_i(\theta^k)|} \in \partial f_i(\theta^k)$. Thus, if $f_i(\theta^k) \leq 0$, then we choose $v^k = -w^k$; otherwise, $v^k = 2\sum_{i=1}^{N} \sum_{j_i \in I_i(\theta^k)} \frac{A(i,j_i)}{|I_i(\theta^k)|} - w^k = w^k$. Finally, we obtain the hybrid DQN-DCA algorithm as follows.

Algorithm 4: DQN-DCA

Perform the algorithm DQN in which Step 3.2 is replaced by the following steps with the step size α:

Step 3.2. Calculate the target Q-value of sampled transitions using (2) and its corresponding θ^{k+1} updated as follows:

- **3.2.1.** Calculate w^k using (3).
- **3.2.2.** Calculate $v^k = -w^k$ if $f_i(\theta^k) \leq 0$ and w^k otherwise.
- **3.2.2.** Calculate $\theta^{k+1} = \theta^k - \alpha v^k$.

---

4 Numerical Experiments

In this section, we compare two algorithms for solving the truck-drone delivery problem [7]: DQN and DQN-DCA, using an randomly generated dataset to evaluate their performance. The dataset is simulated as a coordinate set (x, y) of n points, including the warehouse coordinates and the coordinates of $n - 1$ customers. The points are numbered from 0 to $n - 1$, where 0 represents the warehouse, and the numbers $1, 2, \ldots, n - 1$ represent customers. From all pairs of points (i, j) with $i, j \in 0, 1, \ldots, n - 1, i \neq j$, we randomly select 75% of the total number of pairs. These selected pairs represent direct truck routes between two points, meaning there is no need to travel through an intermediate point. The distance between two points is calculated using the Euclidean distance. In this setup, the speed of the drone is assumed to be constant, while the speed of the truck varies depending on traffic conditions. Under normal traffic conditions, the speed of the truck is set to be twice the speed of the drone. The denser the traffic, the slower the truck moves. The random travel time of the trucks is estimated using a 3×3 transition matrix P, which represents the probability distribution of traffic conditions (see details in [7]). The transition matrix with the increasing failure rate is defined as: $P = \begin{pmatrix} 0.3 & 0.5 & 0.2 \\ 0.2 & 0.3 & 0.5 \\ 0.1 & 0.2 & 0.7 \end{pmatrix}$ Its rows (resp. columns) represents light, normal and heavy traffic from top to bottom (resp. from left to right). Under light (resp. heavy) traffic, the truck travel time between nodes is set to be half (reps. double) of the travel time under normal traffic. The agent receives 50 bonus points for each successful delivery, with an additional 1000 points for returning to the warehouse. We test the neural network with different node configurations in $\{64; 128\}$ and we use a discount rate $\gamma = 0.95$, a step size $\alpha = 0.0005$, and an exploration rate ϵ decreasing from 1 to 0.01 after each episode. The basis function ϕ is defined as follows: set $d = \sum_{s \in \mathcal{S}} |\mathcal{A}_s|\}$, and for

$k \in \{1, 2, \dots, d\}$, $\phi_k(s, a) = 1$ if $k = i \times |\mathcal{A}_s| + j$, 0 otherwise, where i (resp. j) is the index of the state (resp. action) element s (resp. a) in $\mathcal{S}$ (resp. $\mathcal{A}_s$). The results of DQN and DQN-DCA in terms of CPU (learning time in seconds) and average reward are presented in Table 1. The performance of two comparative algorithms in the average reward and loss during 1000 episodes are shown in Fig. 2 below.

Table 1. Comparative results of DQN and DQN-DCA with different NN nodes for $n = 10$ and $n = 20$

NN nodes	$n = 10$				$n = 20$			
	DQN		DCA		DQN		DCA	
	CPU	Reward	CPU	Reward	CPU	Reward	CPU	Reward
64	1.238	991.084	4.288	1321.623	2.496	1162.784	18.971	1209.994
128	0.88	1053.4	4.537	1135.9	2.6534	1256.5	22.317	1544.8

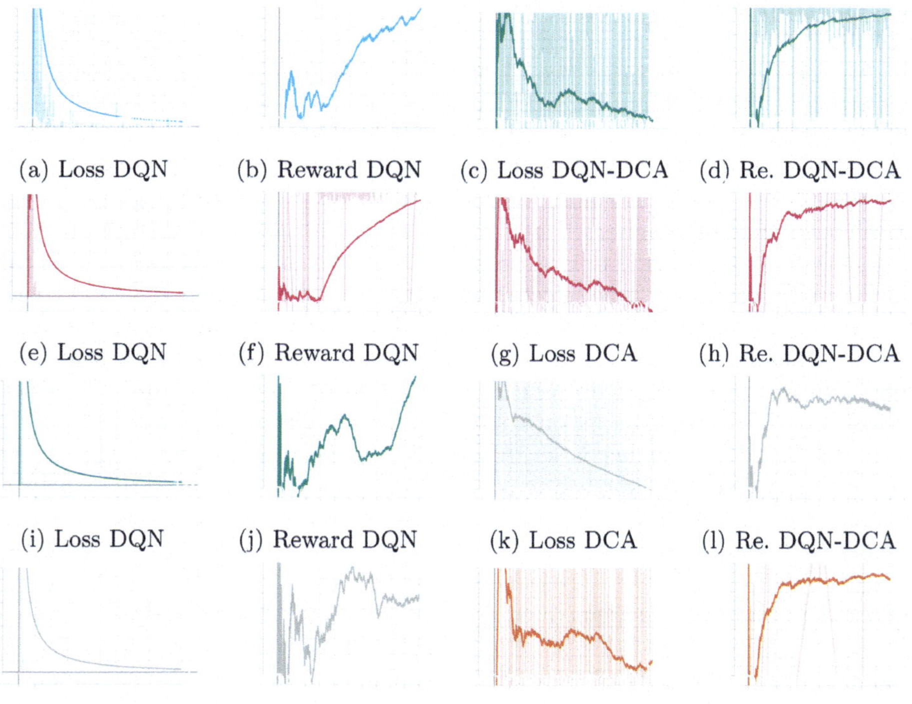

(a) Loss DQN (b) Reward DQN (c) Loss DQN-DCA (d) Re. DQN-DCA

(e) Loss DQN (f) Reward DQN (g) Loss DCA (h) Re. DQN-DCA

(i) Loss DQN (j) Reward DQN (k) Loss DCA (l) Re. DQN-DCA

(m) Loss DQN (n) Reward DQN (o) Loss DCA (p) Re. DQN-DCA

Fig. 2. Performance of DQN and DQN-DCA in the average Reward (or Re.) and loss during 1000 episodes. First row for $n = 10$, 64 nodes. Second row for $n = 10$, 128 nodes. Third row for $n = 20$, 64 nodes. Fourth row for $n = 20$, 128 nodes.

By observing the results in Table 1 and Fig. 2, we see that: first, in terms of rewards, DQN-DCA consistently outperforms the DQN method, indicating superior overall performance and reward accumulation with the online DCA-based approach. Additionally, the reward graphs show that DQN-DCA reaches higher reward levels more quickly than DQN. DQN-DCA also proves to be more efficient, requiring fewer episodes to achieve significant reward improvements. This highlights the effectiveness of our approach's in both exploring and exploiting the state-action space. Regarding CPU time, DQN-DCA may need more time to yield higher average rewards and accelerate the learning process. These key findings suggest that the hybrid DCA and DQN based approach is more efficient and promising for solving the given problem.

5 Conclusions

In this paper, we have developed the RL algorithm DQN-DCA for the combined truck-drone delivery problem where DCA contributes to enhance Q-value updates. Our numerical comparison between DQN and DQN-DCA has demonstrated the effectiveness of the proposed approach over the standard DQN. The results reveal that DQN-DCA not only consistently achieves higher rewards but also reaches these rewards more quickly and efficiently. Future work could explore further enhancements to the DQN-DCA approach in terms of rapidity, such as incorporating adaptive learning rates or expanding the model to handle more complex real-world scenarios and constraints.

References

1. Garg, V., Niranjan, S., Prybutok, V., Pohlen, T., Gligor, D.: Drones in last-mile delivery: a systematic review on efficiency, accessibility, and sustainability. Transp. Res. Part D: Transp. Environ. **123**, 103831 (2023)
2. Le Thi, H.A., Ho, V.T.: Online learning based on online DCA and application to online classification. Neural Comput. **32**(4), 759–793 (2020)
3. Le Thi, H.A., Ho, V.T.: DCA for online prediction with expert advice. Neural Comput. Appl. **33**(15), 9521–9544 (2021). https://doi.org/10.1007/s00521-021-05709-0
4. Le Thi, H.A., Ho, V.T., Pham Dinh, T.: A unified DC programming framework and efficient DCA based approaches for large scale batch reinforcement learning. J. Global Optim. **73**(2), 279–310 (2019)
5. Le Thi, H.A., Pham Dinh, T.: DC programming and DCA: thirty years of developments. Mathematical programming, Special Issue dedicated to 30-th birthday of DC programming and DCA: DC Programming - Theory, Algorithms and Applications **169**(1), 5–68 (2018)
6. Le Thi, H.A., Pham Dinh, T.: Open issues and recent advances in DC programming and DCA. J. Global Optim. **88**(3), 533–590 (2024)
7. Liu, Z., Li, X., Khojandi, A.: The flying sidekick traveling salesman problem with stochastic travel time: a reinforcement learning approach. Transp. Res. Part E: Logist. Transp. Rev. **164**, 102816 (2022)

8. Mao, H., Alizadeh, M., Menache, I., Kandula, S.: Resource management with deep reinforcement learning. In: Proceedings of the 15th ACM Workshop on Hot Topics in Networks, HotNets 2016, pp. 50–56. Association for Computing Machinery, New York, NY, USA (2016)
9. Mnih, V., et al.: Human-level control through deep reinforcement learning. Nature **518**(7540), 529–533 (2015)
10. Moadab, A., Farajzadeh, F., Fatahi Valilai, O.: Drone routing problem model for last-mile delivery using the public transportation capacity as moving charging stations. Sci. Rep. **12**(1), 6361 (2022)
11. Nazari, M., Oroojlooy, A., Snyder, L., Takac, M.: Reinforcement learning for solving the vehicle routing problem. In: Bengio, S., Wallach, H., Larochelle, H., Grauman, K., Cesa-Bianchi, N., Garnett, R. (eds.) Advances in Neural Information Processing Systems, vol. 31. Curran Associates, Inc. (2018)
12. Pham Dinh, T., Le Thi, H.A.: Recent advances in DC programming and DCA. Trans. Comput. Collective Intell. **8342**, 1–37 (2014)
13. Pham Dinh, T., Le Thi, H.A.: Convex analysis approach to d.c. programming: theory, Algorithms and Applications. Acta Mathematica Vietnamica **22**(1), 289–355 (1997), dedicated to Professor Hoang Tuy on the occasion of his 70th birthday
14. Salama, M., Srinivas, S.: Joint optimization of customer location clustering and drone-based routing for last-mile deliveries. Transportation Research Part C: Emerging Technologies **114**, 620–642 (2020)
15. Sebastianelli, A., Tipaldi, M., Ullo, S.L., Glielmo, L.: A deep q-learning based approach applied to the snake game. In: 2021 29th Mediterranean Conference on Control and Automation (MED). pp. 348–353 (2021)
16. Shakya, A.K., Pillai, G., Chakrabarty, S.: Reinforcement learning algorithms: A brief survey. Expert Syst. Appl. **231**, 120495 (2023)
17. Singh, S., Jaakkola, T., Littman, M.L., Szepesvári, C.: Convergence results for single-step on-policy reinforcement-learning algorithms. Mach. Learn. **38**(3), 287–308 (2000)
18. Sutton, R.S., Barto, A.G.: Reinforcement Learning: An Introduction. MIT Press, Cambridge, MA, USA (1998)
19. Sutton, R.S., McAllester, D., Singh, S., Mansour, Y.: Policy gradient methods for reinforcement learning with function approximation. In: Solla, S., Leen, T., Müller, K. (eds.) Advances in Neural Information Processing Systems. vol. 12. MIT Press (1999)
20. Watkins, C.J.C.H.: Learning from Delayed Rewards. Ph.D. thesis, King's College, Cambridge, UK (1989)
21. Zhou, C., Ma, J., Douge, L., Chew, E.P., Lee, L.H.: Reinforcement learning-based approach for dynamic vehicle routing problem with stochastic demand. Computers & Industrial Engineering **182**, 109443 (2023)
22. Zong, Z., Tong, X., Zheng, M., Li, Y.: Reinforcement learning for solving multiple vehicle routing problem with time window. ACM Trans. Intell. Syst. Technol. **15**(2) (2024)

BDCA with Exact Line Search for Symmetric Eigenvalue Complementarity Problems

Yi-Shuai Niu[1(✉)], Hoai An Le Thi[2,3], and Tao Pham Dinh[4]

[1] Beijing Institute of Mathematical Sciences and Applications (BIMSA), Beijing, China
niuyishuai@bimsa.cn

[2] LCOMS, University of Lorraine, Metz, France
hoai-an.le-thi@univ-lorraine.fr

[3] Institut Universitaire de France (IUF), Paris, France

[4] Laboratory of Mathematics (LMI), National Institute of Applied Sciences of Rouen, Saint-Étienne-du-Rouvray, France
pham@insa-rouen.fr

Abstract. We propose an accelerated Difference-of-Convex algorithm (Boosted-DCA, or BDCA) with exact line search for efficiently solving Symmetric Eigenvalue Complementarity Problems (SEiCP). We first reformulate SEiCP to involve only symmetric positive definite matrices and introduce a logarithmic DC formulation. The proposed BDCA enhances the classical DCA with an exact line search based on the real roots of a binomial equation. Numerical experiments demonstrate that BDCA achieves significantly faster convergence and superior solutions compared to DCA, KNITRO, FILTERSD, and MATLAB's FMINCON, particularly for large-scale and ill-conditioned problems.

Keywords: BDCA · Exact Line Search · Symmetric Eigenvalue Complementarity Problems

1 Introduction

The Symmetric Eigenvalue Complementarity Problem (SEiCP) involves finding complementary eigenvectors $x \in \mathbb{R}^n \setminus \{0\}$ and complementary eigenvalues $\lambda \in \mathbb{R}$ such that

$$w = \lambda Bx - Ax, \quad x^\top w = 0, \quad 0 \neq x \geq 0, \quad w \geq 0, \qquad \text{(SEiCP)}$$

where $x^\top$ denotes the transpose of x, A is a real symmetric matrix in $\mathbb{R}^{n\times n}$, and B is a real *symmetric positive definite* (SPD) matrix. The solution set of this problem is denoted by SEiCP(A, B). The SEiCP was originally studied in the context of static equilibrium states of mechanical systems with unilateral friction [9], and has found numerous applications in engineering, as discussed in [8,28]. Regarding the feasibility of (SEiCP), it is established that (SEiCP) always has

© The Author(s), under exclusive license to Springer Nature Switzerland AG 2025
H. A. Le Thi et al. (Eds.): ICAMCS 2024, LNNS 1311, pp. 159–171, 2025.
https://doi.org/10.1007/978-3-031-90606-0_14

a solution [13]. Moreover, the existence of solutions is ensured under the weaker condition that B is *strictly copositive* (SC), meaning that $x^\top Bx > 0$ for all $0 \neq x \geq 0$. The problem (SEiCP) has a positive complementary eigenvalue if and only if there exists some $x \geq 0$ such that $x^\top Ax > 0$ [27]. For instance, when A is an SC matrix, (SEiCP) will have a positive complementary eigenvalue. In general, (SEiCP) has at most $2^n - 1$ positive λ-solutions [27].

In this paper, we demonstrate that any (SEiCP) can be reformulated as an equivalent problem with symmetric positive definite (SPD) matrices $A + \mu B$ for some sufficiently large μ. We then apply the classical DCA [25] and an accelerated DCA (BDCA) with exact line search [22] to solve a logarithmic DC formulation of (SEiCP) involving only SPD matrices. Both DCA and BDCA solve a sequence of convex subproblems with strongly convex objectives over a unit simplex, efficiently addressed using FISTA [4]. BDCA's exact line search is performed inexpensively by finding the real roots of a binomial equation.

DCA has previously been applied to (SEiCP) and its variants [16,19–21]. The key contributions of this paper are twofold: the introduction of a new logarithmic DC formulation based on the SPD reformulation, and the application of BDCA with exact line search to efficiently solve this formulation.

The paper is organized as follows: Sect. 2 establishes the equivalence of any (SEiCP) to one with SPD matrices and introduces the logarithmic DC formulation. Section 3 details the application of BDCA and DCA to solve (LnP). Numerical experiments comparing BDCA and DCA with KNITRO, FILTERSD, and MATLAB's FMINCON on synthetic and Matrix Market datasets are presented in Sect. 4. Finally, we conclude with remarks and potential future research directions.

2 SEiCP Formulations

2.1 Symmetric Positive Definite Formulation

We will prove that any (SEiCP) can be equivalently reformulated as an (SEiCP) involving only symmetric positive definite (SPD) matrices, namely the *Symmetric Positive Definite Formulation.* To demonstrate this, we start with the following equivalent formulation:

$$\lambda x^\top Bx - x^\top Ax = 0, \lambda Bx - Ax \geq 0, x \in \Omega \overset{\Delta}{=} \{x \in \mathbb{R}^n : e^\top x = 1, x \geq 0\}. \quad (1)$$

Clearly, for any solution $(x, \lambda) \in \text{SEiCP}(A, B)$, we have $x^\top Bx > 0$ (since $B \in$ SPD) and the *Rayleigh quotient*

$$\lambda = \frac{x^\top Ax}{x^\top Bx}, \quad (2)$$

is well defined. The following theorem shows that any SEiCP(A, B) is equivalent to SEiCP$(A + \mu B, B)$ for all $\mu \in \mathbb{R}$.

Proposition 1. *For all $\mu \in \mathbb{R}$,*

$$(x, \lambda) \in \mathit{SEiCP}(A, B) \Leftrightarrow (x, \lambda + \mu) \in \mathit{SEiCP}(A + \mu B, B). \quad (3)$$

Proof. For all $\mu \in \mathbb{R}$, we immediately obtain:

$$\begin{cases} (\lambda+\mu)x^\top Bx - x^\top(A+\mu B)x = \lambda x^\top Bx - x^\top Ax, \\ (\lambda+\mu)Bx - (A+\mu B)x = \lambda Bx - Ax, \end{cases}$$

which establishes the desired equivalence. □

The smallest μ guaranteeing $A + \mu B$ to be SPD can be determined by solving the semidefinite program (SDP) $\min\{\mu : A + \mu B \succeq 0\}$, which can be numerically solved by SDP solvers such as MOSEK, SeDuMi, CSDP, DSDP, SDPT3, and SDPA. Alternatively, a sufficiently large μ can be estimated by $|\lambda_{\min}(A)|/\lambda_{\min}(B)$, where $\lambda_{\min}(\cdot)$ denotes the smallest eigenvalue of a matrix.

Thus, without loss of generality, we assume the Assumption 1.

Assumption 1. A and B are SPD matrices for (SEiCP).

Proposition 2. *Under Assumption 1, problem* (SEiCP) *is feasible, and all λ-solutions in SEiCP(A, B) are strictly positive.*

Proof. The feasibility of SEiCP follows from the fact that an SPD matrix A is strictly copositive. As both $x^\top Ax$ and $x^\top Bx$ are strictly positive for all $x \neq 0$, we get from the definition of λ in (2) that all λ-solutions are positive. □

Proposition 1 and Proposition 2 reveal an important fact: by choosing $\mu \geq 0$ sufficiently large so that $A+\mu B \in$ SPD, we can find $(x, \lambda) \in$ SEiCP$(A+\mu B, B)$ with a positive eigenvalue λ, and then $(x, \lambda - \mu) \in$ SEiCP(A, B).

2.2 Logarithmic SPD Formulation

Based on the SPD formulation of (SEiCP), an equivalent formulation, namely the *Logarithmic SPD Formulation*, is given by:

$$\max\{L(x) : x \in \Omega\}, \tag{LnP}$$

where $A, B \in$ SPD and the logarithmic metric function

$$L(x) \triangleq \ln\left(\frac{x^\top Ax}{x^\top Bx}\right) = \ln(x^\top Ax) - \ln(x^\top Bx)$$

is well-defined on Ω. It is easy to show that:

Proposition 3. *Under Assumption 1, any stationary point $\bar{x}$ of* (LnP) *satisfies*

$$(\bar{x}, (\bar{x}^\top A\bar{x})/(\bar{x}^\top B\bar{x})) \in SEiCP(A, B).$$

Proof. It has been shown in [27, Proposition 9] that any stationary point $\bar{x}$ of the *Rayleigh Quotient Formulation* of (SEiCP):

$$\max\left\{\frac{x^\top Ax}{x^\top Bx} : x \in \Omega\right\}, \tag{RP}$$

corresponds to a solution of (SEiCP) such that

$$(\bar{x}, (\bar{x}^\top A\bar{x})/(\bar{x}^\top B\bar{x})) \in \text{SEiCP}(A, B).$$

Given the monotonicity of the logarithmic function, any stationary point of (LnP) also corresponds to a solution of (SEiCP). □

2.3 Logarithmic DC Formulation

The convex constrained Difference-of-Convex (DC) program is defined as:

$$\alpha = \min\{f(x) = g(x) - h(x) : x \in \mathcal{C}\}, \tag{P}$$

where $\mathcal{C}$ is a nonempty closed convex set in $\mathbb{R}^n$, and the objective function f is DC, expressed as the difference of two proper, closed, convex functions g and h.

The problem (LnP) can be reformulated as a DC program:

$$\min\{f(x) = g(x) - h(x) : x \in \Omega\},$$

where $f(x) = \ln(x^\top Bx) - \ln(x^\top Ax)$, and

$$\begin{cases} g(x) = \frac{\eta}{2}\|x\|_2^2 - \ln(x^\top Ax), \\ h(x) = \frac{\eta}{2}\|x\|_2^2 - \ln(x^\top Bx), \\ \nabla h(x) = \eta x - \frac{2Bx}{x^\top Bx}. \end{cases} \tag{4}$$

The term $\varphi_A(x) \triangleq -\ln(x^\top Ax)$ is nonconvex on Ω for SPD matrix A. Therefore, it is necessary to find a parameter η that ensures both g and h are convex over Ω. We will show that there exists a sufficiently large η such that both g and h are strongly convex and have Lipschitz continuous gradients over Ω.

Theorem 1. (Strong Convexity). *Let*

$$\bar{\eta} = 4n \max\{\kappa_A^2, \kappa_B^2\}, \tag{5}$$

where κ_A and κ_B are the condition numbers of A and B, respectively. Then for all $\eta \geq \bar{\eta}$, both g and h defined in (4) are strongly convex over Ω.

Proof. Let $\rho(M)$ denote the spectral radius of M. It follows from the positive definiteness of A that $\forall x \in \Omega$,

$$\rho(\nabla^2\varphi_A(x)) < \rho\left(\nabla^2\varphi_A(x) + \frac{2A}{x^\top Ax}\right) = \rho\left(\frac{4(Ax)(Ax)^\top}{(x^\top Ax)^2}\right),$$

where the strict inequality is due to the renowned monotonicity theorem [12, Corollary 4.3.12]. Then

$$\rho\left(\frac{4(Ax)(Ax)^\top}{(x^\top Ax)^2}\right) = \frac{4\rho((Ax)(Ax)^\top)}{(x^\top Ax)^2} = \frac{4\|Ax\|_2^2}{(x^\top Ax)^2}.$$

Let $A = P^\top \Lambda P$ be the spectral decomposition of A, with $\Lambda = \operatorname{diag}(\lambda_1, \ldots, \lambda_n)$, where $\lambda_1 \leq \cdots \leq \lambda_n$ are the eigenvalues of A, and denote $y = Px$. Then

$$\|Ax\|_2^2 = y^\top \Lambda^2 y = \sum_{i=1}^{n} \lambda_i^2 y_i^2 \leq \lambda_n^2 \|y\|_2^2 = \lambda_n^2 \|x\|_2^2, \tag{6}$$

$$(x^\top Ax)^2 = (y^\top \Lambda y)^2 = \left(\sum_{i=1}^{n} \lambda_i y_i^2\right)^2 \geq \lambda_1^2 \left(\sum_{i=1}^{n} y_i^2\right)^2 = \lambda_1^2 \|y\|_2^4 = \lambda_1^2 \|x\|_2^4. \tag{7}$$

Hence,

$$\rho(\nabla^2 \varphi_A(x)) < \frac{4\|Ax\|_2^2}{(x^\top Ax)^2} \overset{(6)(7)}{\leq} \frac{4\lambda_n^2 \|x\|_2^2}{\lambda_1^2 \|x\|_2^4} = \frac{4\kappa_A^2}{\|x\|_2^2} \leq 4n\kappa_A^2, \quad \forall x \in \Omega.$$

Similarly, we have $\rho(\nabla^2 \varphi_B(x)) < 4n\kappa_B^2$ for all $x \in \Omega$. It follows immediately that for $\eta \geq \bar{\eta} \overset{\Delta}{=} 4n \max\{\kappa_A^2, \kappa_B^2\}$,

$$\nabla^2 g(x) = \eta I - \nabla^2 \varphi_A(x) \succ 0, \quad \nabla^2 h(x) = \eta I - \nabla^2 \varphi_B(x) \succ 0, \quad \forall x \in \Omega.$$

Hence, there exists some $\sigma > 0$ such that $x \mapsto g(x) - \frac{\sigma}{2}\|x\|_2^2$ and $x \mapsto h(x) - \frac{\sigma}{2}\|x\|_2^2$ are convex over Ω, i.e., g and h are σ-strongly convex over Ω for all $\eta \geq \bar{\eta}$. □

Corollary 1 (Smoothness). *For all $\eta \geq \bar{\eta}$, with $\bar{\eta}$ defined in* (5)*, the convex function g (resp. h) defined in* (4) *is L_g-smooth (resp. L_h-smooth) on Ω with*

$$L_g = \eta + 2n\kappa_A \quad \textit{and} \quad L_h = \eta + 2n\kappa_B.$$

Proof. It follows from Theorem 1 that for all $x \in \Omega$,

$$\begin{aligned}
\rho(\nabla^2 g(x)) &= \rho(\eta I - \nabla^2 \varphi_A(x)) = \rho\left(\eta I + \frac{2A}{x^\top Ax} - \frac{4(Ax)(Ax)^\top}{(x^\top Ax)^2}\right) \\
&\leq \rho\left(\eta I + \frac{2A}{x^\top Ax}\right) = \eta + \frac{2\rho(A)}{x^\top Ax} \leq \eta + \frac{2\lambda_n}{\lambda_1/n} = \eta + 2n\kappa_A = L_g,
\end{aligned}$$

where the last inequality is due to $\rho(A) = \lambda_n$ and $x^\top Ax \overset{(7)}{\geq} \lambda_1 \|x\|_2^2 \geq \lambda_1/n$ for all $x \in \Omega$. Hence, $x \mapsto \frac{L_g}{2}\|x\|_2^2 - g(x)$ is convex over Ω, implying that g is L_g-smooth on Ω. A similar result can be obtained for h. □

3 DCA and BDCA for Solving (LnP)

The *DCA*, first introduced by Pham in 1985 and further developed by Le Thi and Pham [17,23–26], is a well-known algorithm for solving the DC program (P). DCA generates a sequence $\{x^k\}$ by solving convex optimization subproblems of the form:

$$\boxed{x^{k+1} \in \operatorname{argmin}\{g(x) - \langle x, y^k\rangle : x \in \mathcal{C}\}, \quad y^k \in \partial h(x^k), \quad k = 0, 1, \ldots} \tag{DCA}$$

where $\partial h(x^k)$ is the subdifferential of h at x^k:

$$\partial h(x^k) \stackrel{\Delta}{=} \{y \in \mathbb{R}^n : h(x) \geq h(x^k) + \langle x - x^k, y\rangle, \forall x \in \mathbb{R}^n\}.$$

The *BDCA* (Boosted DCA), an accelerated variant of DCA, incorporates a line search to enhance convergence. Initially proposed by Artacho et al. in 2018 [1] for unconstrained smooth DC programs, it was later extended by Niu et al. [22,29] to handle convex-constrained nonsmooth and smooth DC programs.

In the DC formulation of (LnP), both DCA and BDCA address the following convex optimization subproblem:

$$\boxed{\min_{x \in \Omega} \left\{ \frac{\eta}{2}\|x\|_2^2 - \ln(x^\top Ax) - \langle x, \nabla h(x^k)\rangle \right\}} \tag{LnPk}$$

which features a strongly convex objective, ensuring a unique solution. FISTA [4], a proximal gradient method with Nesterov's acceleration, is recommended for efficiently solving (LnPk).

FISTA for (LnPk). FISTA is an algorithm for solving composite optimization problems of the form:

$$\min_{x \in \mathbb{R}^n} \phi(x) + \psi(x),$$

where $\phi : \mathbb{R}^n \to \mathbb{R}$ is L_ϕ-smooth and convex, and ψ is convex. Problem (LnPk) can be reformulated as:

$$\min_{x \in \mathbb{R}^n} \underbrace{\frac{\eta}{2}\|x\|_2^2 - \ln(x^\top Ax) - \langle x, \nabla h(x^k)\rangle}_{\phi(x)} + \underbrace{\chi_\Omega(x)}_{\psi(x)},$$

where $\phi(x)$ is L_g-smooth and convex, and $\psi(x) = \chi_\Omega(x)$ is the indicator function[1] of the simplex Ω. Therefore, FISTA can be applied to solve (LnPk), as outlined in Algorithm 1.

[1] $\chi_\Omega(x)$ is the indicator function of Ω, defined as $\chi_\Omega(x) = 0$ if $x \in \Omega$, and $+\infty$ otherwise.

Algorithm 1. FISTA for (LnPk)

INPUT: $y^0 = u^0 = x^k \neq 0$; $t_0 = 1$;
1: **for** $i = 0, 1, \ldots$ **do**
2: $\quad u^{i+1} \leftarrow P_\Omega\left(y^i - \frac{1}{L_g}\nabla\phi(y^i)\right)$;
3: $\quad t_{i+1} \leftarrow \frac{1+\sqrt{1+4t_i^2}}{2}$;
4: $\quad y^{i+1} \leftarrow u^{i+1} + \left(\frac{t_i - 1}{t_{i+1}}\right)(u^{i+1} - u^i)$;
5: **end for**

- In line 2 of Algorithm 1, $P_\Omega(v)$ denotes the *simplex projection* of v, and

$$\nabla\phi(y^i) = \eta(y^i - x^k) - \frac{2Ay^i}{\langle y^i, Ay^i\rangle} + \frac{2Bx^k}{\langle x^k, Bx^k\rangle}.$$

We suggest using Algorithm 2 to compute the simplex projection, which has an $O(n\log(n))$ worst-case complexity [7,11].

Algorithm 2. Simplex Projection

INPUT: $v \in \mathbb{R}^n$;
OUTPUT: $P_\Omega(v)$;
Sort v into z with $z_1 \geq z_2 \geq \cdots \geq z_n$;
$N \leftarrow \max_{1\leq k\leq n}\left\{k : \frac{\sum_{r=1}^k z_r - 1}{k} < z_k\right\}$;
$\lambda^* \leftarrow \frac{\sum_{r=1}^N z_r - 1}{N}$;
return $P_\Omega(v) = [v - \lambda^* e]_+$.

- FISTA achieves an $O(1/i^2)$ convergence rate in function values, where i is the number of iterations [3, Chapter 10.7].

Exact Line Search. Consider the line search problem at z^k along the direction $d^k \neq 0$:

$$\alpha_k = \operatorname{argmin}\{f(z^k + \alpha d^k) : z^k + \alpha d^k \in \Omega, \alpha \geq 0\}.$$

This problem can be efficiently solved using the following proposition:

Proposition 4. *The exact line search in BDCA for* (LnP) *at z^k along d^k is computed as:*

$$\alpha_k = \operatorname{argmin}_\alpha\{q(\alpha) : \alpha \in \{0, \bar{\alpha}_k\} \cup \mathcal{Z}\},$$

where

$$\begin{cases} q(\alpha) = \frac{a_1\alpha^2 + b_1\alpha + c_1}{a_2\alpha^2 + b_2\alpha + c_2}, \\ a_1 = \langle d^k, Bd^k\rangle, \quad b_1 = 2\langle z^k, Bd^k\rangle, \quad c_1 = \langle z^k, Bz^k\rangle, \\ a_2 = \langle d^k, Ad^k\rangle, \quad b_2 = 2\langle z^k, Ad^k\rangle, \quad c_2 = \langle z^k, Az^k\rangle, \\ \bar{\alpha}_k = \min\left\{-\frac{z_i^k}{d_i^k}, i \in \mathcal{I}^k\right\}, \end{cases}$$

with $\mathcal{I}^k = \{i \in \{1, \ldots, n\} : d_i^k < 0\}$*, and* $\mathcal{Z}$ *is the set of all real roots of the quadratic equation:*

$$(a_1 b_2 - a_2 b_1)\alpha^2 + 2(a_1 c_2 - a_2 c_1)\alpha + (b_1 c_2 - b_2 c_1)$$

within the interval $[0, \bar{\alpha}_k]$*. The next iterate* x^{k+1} *is then updated as* $z^k + \alpha_k d^k$*.*

Proof. The proof of this proposition is not difficult and will be provided in the upcoming full-length paper. □

BDCA and DCA for (LnP). We now present the BDCA and DCA algorithms for solving (LnP) in a unified framework, as outlined in Algorithm 3.

Algorithm 3. BDCA/DCA for (LnP)

INPUT: $x^0 \neq 0$; $\eta \geq \bar{\eta}$ with $\bar{\eta}$ defined in (5);
1: **for** $k = 0, 1, \ldots$ **do**
2: Compute $z^k \leftarrow \text{argmin}\left\{\frac{\eta}{2}\|x\|_2^2 - \ln(x^\top Ax) - \langle x, \nabla h(x^k)\rangle : x \in \Omega\right\}$ via FISTA;
3: $d^k \leftarrow z^k - x^k$;
4: Initialize $x^{k+1} \leftarrow z^k$;
5: **if** $\mathcal{A}(z^k) \subset \mathcal{A}(x^k)$ and $\left\langle \frac{Bz^k}{\langle z^k, Bz^k\rangle} - \frac{Az^k}{\langle z^k, Az^k\rangle}, d^k \right\rangle < 0$ **then**
6: Compute $\alpha_k \leftarrow \text{argmin}_\alpha\{q(\alpha) : \alpha \in \{0, \bar{\alpha}_k\} \cup \mathcal{Z}\}$ as described in Proposition 4;
7: Update $x^{k+1} \leftarrow z^k + \alpha_k d^k$;
8: **end if**
9: **end for**

- The DCA algorithm corresponds to Algorithm 3 without lines 5 to 8; including these lines gives BDCA.
- $\mathcal{A}(x^k)$ denotes the active set of Ω at x^k, defined as $\mathcal{A}(x^k) = \{i : x_i^k = 0\}$.
- Both DCA and BDCA satisfy the following convergence theorem:

Theorem 2 (See [15,18,25]**for DCA and** [18,22,29]**for BDCA).** *Let* $\{(x^k, y^k)\}$ *(resp.* $\{(x^k, y^k, z^k)\}$*) be the sequence generated by DCA (resp. BDCA) for* (P) *starting from* $x^0 \in \text{dom}\,\partial h$*. Suppose* g *and* h *are convex over* $\mathcal{C}$ *with moduli* $\rho_g \geq 0$ *and* $\rho_h \geq 0$*. If* $\rho_g + \rho_h > 0$ *and* $\{(x^k, y^k)\}$ *(resp.* $\{(x^k, y^k, z^k)\}$*) is bounded, then:*

- ***(Convergence of*** $\{f(x^k)\}$***)*** *The sequence* $\{f(x^k)\}$ *is non-increasing and convergent.*
- ***(Subsequential convergence of*** $\{x^k\}$***)*** *Any cluster point of* $\{x^k\}$ *is a DC critical point of* (P)[2].

---

[2] A DC critical point x^* is defined as a point where $\partial(g + \chi_{\mathcal{C}})(x^*) \cap \partial h(x^*) \neq \emptyset$.

- ***(Global convergence of*** $\{x^k\}$***)*** *If f is a KL function, $\mathcal{C}$ is semi-algebraic, and h has a locally Lipschitz continuous gradient over $\mathcal{C}$, then $\{x^k\}$ converges to a strongly DC critical point*[3].

The function $f(x) + \chi_{\mathcal{C}}(x)$ is a KL function satisfying the Kurdyka-Łojasiewicz property [18, Definition 3], which is common in optimization, including for semi-algebraic, subanalytic, log, and exp functions [2,5,14].

- Theorem 2 applies to the logarithmic DC formulation because f is KL, g and h are strongly convex on Ω (Theorem 1), h has a locally Lipschitz continuous gradient (Corollary 1), and $\{x^k\}$ and $\{\nabla h(x^k)\}$ are bounded, as $\{x^k\}$ lies within the compact set Ω, and

$$\|\nabla h(x^k)\|_2 = \left\| \eta x^k - \frac{2Bx^k}{\langle x^k, Bx^k \rangle} \right\|_2 \leq \eta + 2\sqrt{n}\kappa_B < \infty, \quad \forall x^k \in \Omega.$$

4 Numerical Simulations

This section presents numerical results for solving (SEiCP) using DCA and BDCA, compared against KNITRO v11.1.0 [6], FILTERSD v1.0 [10], and MATLAB FMINCON on the (LnP) formulation. All experiments were conducted on a 64-bit Windows 10 OS with an Intel i7-10870H 2.20 GHz CPU and 32 GB of RAM, using MATLAB 2021a.

SEiCP Datasets. We tested two sets of problems with B as the identity matrix:

- **Random matrices:** A was generated with elements uniformly distributed in $[-1, 1]$ or $[-10, 10]$, denoted by `RANDEICP(k,m,n)`, where k and m define the range, and n is the matrix order.
- **Matrix Market:** The second set uses matrices from the NEP collection of the *Matrix Market* repository, with 13 matrices of order n ranging from 100 to 800. Asymmetric matrices A were symmetrized as $(A + A^\top)/2$.

To convert A into an SPD matrix, we set $A = A + \mu_{AB}B$, where $\mu_{AB} = \min\{\mu : A + \mu B \succeq 0\} + 1$, computed by solving an SDP using MOSEK.
Setup. The following setups were used:

- **Initialization:** A random initial point $x^0 \neq 0$ uniformly distributed in $[0, 1]^n$ was used for all methods.
- **Termination criteria:** DCA and BDCA stopped if

$$\frac{\|d^k\|_2}{1 + \|z^k\|_2} \leq 10^{-8}.$$

[3] A strongly DC critical point x^* is defined as a point where $\partial h(x^*) \subset \partial(g + \chi_{\mathcal{C}})(x^*)$.

Alternatively, the algorithms terminated after 10,000 iterations. KNITRO and FILTERSD used default settings. For MATLAB FMINCON, `MaxFunEvals` was set to 10^6. FISTA stopped if

$$\frac{\|u^{i+1} - u^i\|_2}{1 + \|u^{i+1}\|_2} \leq 10^{-6}.$$

- **Other settings:** We set $\eta = L_g = n$ for the DC decomposition of (LnP).

Notations. The following notations are used in the numerical results:

- λ: Computed complementary eigenvalue.
- IT: Number of iterations for DCA and BDCA.
- CPU: CPU time in seconds.
- c: Exponent 10^{-c} for the feasibility measure of the computed solution as

$$\|[x]_-\|_2 + \|[w]_-\|_2 + |w^\top x|,$$

  where $[x]_-$ is the vector $[\min\{x_i, 0\}]_{i=1}^n$ and $w = \lambda Bx - Ax$. A larger c indicates better precision.
- avg: Average results for CPU, IT, and c for DCA and BDCA; CPU and c for FMINCON, KNITRO, and FILTERSD.

Numerical Results. The numerical results in Table 1 reveal the following:

- BDCA outperforms DCA, with an 80% reduction in iterations and a 58% reduction in CPU time on average, while also delivering superior solution quality.
- BDCA consistently achieves the best results, with an average CPU time of 0.43 s and the highest solution quality ($c = 6$). DCA follows, with KNITRO, FILTERSD, and FMINCON trailing behind. FMINCON shows the weakest performance in both time and accuracy.
- In some NEP instances, DCA and BDCA exceed the 10000 iteration limit, especially DCA. Despite this, the solutions remain acceptable ($c > 3$). FILTERSD and FMINCON occasionally fail on ill-conditioned NEP instances (e.g., `NEP-olm500` with $\kappa_A = 2.3 \times 10^4$ and `NEP-mhd416a` with $\kappa_A = 2.5 \times 10^3$), but BDCA and DCA successfully solve all test cases.

In conclusion, BDCA not only accelerates the convergence of DCA but also achieves better numerical results, often outperforming other compared solvers. Therefore, BDCA appears to be a highly promising approach for solving (SEiCP), particularly for large-scale and ill-conditioned cases.

Table 1. Comparison of solutions for (SEiCP) obtained by DCA, BDCA, FMINCON, KNITRO, and FILTERSD on RANDEICP and NEP datasets using the (LnP) model. Bold values indicate the minimal CPU time (in seconds) and the best solution precision.

Prob	DCA				BDCA				FMINCON			KNITRO			FILTERSD		
	λ	CPU	IT	c	λ	CPU	IT	c	λ	CPU	c	λ	CPU	c	λ	CPU	c
RANDEICP$(-1, 1, 50)$	3.9518	**0.007**	336	6	3.9518	0.024	89	**8**	3.9518	0.451	5	3.9518	0.145	6	3.9518	0.015	3
RANDEICP$(-1, 1, 100)$	5.5532	0.013	428	6	5.5532	**0.011**	97	**7**	5.5530	0.365	4	5.5532	0.082	6	5.5532	0.026	3
RANDEICP$(-1, 1, 200)$	8.3007	0.038	418	6	8.3007	**0.021**	106	**7**	8.3002	1.826	4	8.3007	0.593	7	8.3007	0.172	3
RANDEICP$(-1, 1, 400)$	11.8781	0.176	1571	6	11.8781	**0.128**	395	**7**	11.8768	14.737	4	11.8781	6.138	7	11.8780	1.893	3
RANDEICP$(-1, 1, 600)$	13.9468	0.385	1114	6	13.9468	**0.206**	274	**8**	13.9765	62.308	3	13.9797	26.038	4	13.9796	4.331	3
RANDEICP$(-1, 1, 800)$	16.3095	1.841	2765	6	16.3095	**0.868**	606	**8**	16.3044	170.292	3	16.3095	86.029	6	16.1180	8.519	4
RANDEICP$(-10, 10, 50)$	32.0015	0.008	604	5	32.0015	**0.004**	84	**6**	32.0009	0.099	3	32.0015	0.022	5	32.0010	0.007	2
RANDEICP$(-10, 10, 100)$	55.0963	0.011	361	5	55.0963	**0.009**	103	**7**	55.0945	0.318	3	55.0963	0.085	5	55.0961	0.035	2
RANDEICP$(-10, 10, 200)$	76.4938	0.074	1134	5	76.4938	**0.037**	251	**6**	76.0301	1.697	3	76.4938	0.545	6	76.4933	0.089	2
RANDEICP$(-10, 10, 400)$	113.2569	0.138	709	5	113.2569	**0.075**	187	**7**	113.2427	13.326	3	113.2569	5.737	5	113.2566	1.420	2
RANDEICP$(-10, 10, 600)$	141.0342	0.309	829	5	141.0342	**0.181**	218	**6**	141.0020	48.917	2	141.0341	24.668	3	141.0340	3.870	2
RANDEICP$(-10, 10, 800)$	159.6700	4.621	8902	5	159.6700	**1.925**	1743	**7**	159.6200	215.517	2	159.6700	80.110	5	159.6698	12.154	2
NEP-bfw398b	−0.0000	0.946	10000	6	−0.0000	**0.105**	426	**7**	−0.0000	11.575	6	−0.0000	10.032	7	−0.0000	0.055	6
NEP-bfw782b	−0.0000	3.573	10000	6	−0.0000	**0.239**	309	6	−0.0000	48.161	6	−0.0000	52.518	**7**	−0.0000	0.170	6
NEP-ck400	4.8132	1.334	10000	2	4.8187	**0.226**	908	**5**	4.8184	6.663	3	4.8187	0.974	5	4.7468	0.124	7
NEP-ck656	4.8048	3.207	10000	2	4.8187	**0.512**	916	**5**	1.0289	7.265	2	4.8187	3.621	4	4.7468	0.156	**7**
NEP-dwa512	0.7722	2.128	10000	4	0.7723	**1.246**	3074	**6**	0.7721	16.891	4	0.7723	14.738	6	0.7723	3.303	6
NEP-lop163	1.1087	0.313	10000	5	1.1087	**0.036**	415	**7**	1.1086	0.605	4	1.1087	0.353	6	1.1087	0.133	5
NEP-mhd416a	1192.3774	1.345	10000	0	1192.8767	**0.077**	294	**4**	1192.7605	3.487	1	1192.8767	1.198	2	256.3736	0.023	0
NEP-olm100	213.6680	0.080	4604	5	213.6680	**0.052**	829	**5**	212.3586	0.575	0	213.6680	0.094	4	212.9640	0.020	0
NEP-olm500	5141.1850	2.371	10000	0	5143.3869	4.042	10000	**2**	4988.8881	121.333	−1	5143.3880	16.240	2	4559.2476	**0.057**	−2
NEP-rbs480a	356.2392	0.182	683	4	356.2392	**0.104**	229	**7**	356.1817	20.428	2	356.2391	8.786	2	356.2388	1.359	1
NEP-rdb200	4.1363	0.068	1025	6	4.1363	**0.026**	183	**9**	4.1351	1.133	4	4.1363	0.488	6	4.1362	0.101	3
NEP-rdb800l	4.3712	2.500	4142	6	4.3712	**0.510**	449	**7**	4.3667	83.324	3	4.3621	74.426	3	4.3711	8.925	3
NEP-tub100	6.2452	0.175	10000	3	6.2452	**0.075**	1230	**5**	6.2151	0.402	2	6.2452	0.089	3	−4.9911	0.008	0
avg.		1.034	4785	5		**0.430**	937	**6**		34.068	3		16.550	5		1.879	3

5 Conclusions

In this paper, we presented a logarithmic DC formulation of (SEiCP) and applied DCA and BDCA to solve this problem. Numerical experiments on both synthetic datasets and the Matrix Market NEP Repository demonstrated that BDCA significantly accelerates DCA, yielding better numerical solutions and outperforming several solvers, including KNITRO, FILTERSD, and MATLAB FMINCON, in terms of average CPU time and solution precision.

Future research directions include:

(i) Applying BDCA to solve asymmetric Eigenvalue Complementarity Problems, symmetric/asymmetric Quadratic Eigenvalue Complementarity Problems, and tensor Eigenvalue Complementarity Problems.

(ii) Improving the estimation of parameters $\bar{\eta}$ and L_g for the logarithmic DC formulation, possibly through an adaptive procedure similar to that used in FISTA, which could lead to better local convex approximations.

Acknowledgments. This work was supported by the Natural Science Foundation of China (Grant No: 11601327) and the Research Grants Council (No. 15304019). Special thanks to Joaquim J. Judice for his encouragement and insightful discussions, and to Defeng Sun and Xun Li for their funding support.

Disclosure of Interests. The authors have no competing interests to declare that are relevant to the content of this article.

References

1. Artacho, F., Fleming, R.M., Vuong, P.T.: Accelerating the dc algorithm for smooth functions. Math. Program. **169**(1), 95–118 (2018)
2. Attouch, H., Bolte, J., Svaiter, B.F.: Convergence of descent methods for semi-algebraic and tame problems: proximal algorithms, forward-backward splitting, and regularized gauss-seidel methods. Math. Program. **137**(1), 91–129 (2013)
3. Beck, A.: First-Order Methods in Optimization. SIAM, Philadelphia (2017)
4. Beck, A., Teboulle, M.: A fast iterative shrinkage-thresholding algorithm for linear inverse problems. SIAM J. Imag. Sci. **2**(1), 183–202 (2009)
5. Bolte, J., Daniilidis, A., Lewis, A.: The Łojasiewicz inequality for nonsmooth subanalytic functions with applications to subgradient dynamical systems. SIAM J. Optim. **17**(4), 1205–1223 (2007)
6. Byrd, R.H., Nocedal, J., Waltz, R.A.: Knitro: an integrated package for nonlinear optimization. In: Large-Scale Nonlinear Optimization, pp. 35–59. Springer, Boston (2006). https://www.artelys.com/docs/knitro
7. Condat, L.: Fast projection onto the simplex and the l1 ball. Math. Program. **158**(1a2), 575–585 (2016)
8. Costa, A., Martins, J., Figueiredo, I.N., Júdice, J.J.: The directional instability problem in systems with frictional contacts. Comput. Methods Appl. Mech. Eng. **193**, 357–384 (2004)
9. Costa, A., Figueiredo, I., Júdice, J., Martins, J.: A complementarity eigenproblem in the stability analysis of finite dimensional elastic systems with frictional contact. In: Complementarity: Applications. Algorithms and Extensions, pp. 67–83. Springer, Boston (2001)
10. Fletcher, R., Curtis, F.E.: Filtersd – a library for nonlinear optimization written in fortran. https://projects.coin-or.org/filterSD
11. Held, M., Wolfe, P., Crowder, H.P.: Validation of subgradient optimization. Math. Program. **6**(1), 62–88 (1974)
12. Horn, R.A., Johnson, C.R.: Matrix Analysis. Cambridge University Press, Cambridge (2012)
13. Júdice, J.J., Sherali, H.D., Ribeiro, I., Rosa, S.: On the asymmetric eigenvalue complementarity problem. Optim. Methods Softw. **24**, 549–586 (2009)
14. Kurdyka, K.: On gradients of functions definable in o-minimal structures. In: Annales de l'institut Fourier, vol. 48, pp. 769–783 (1998)
15. Le Thi, H.A., Huynh, V.N., Pham, D.T.: Convergence analysis of difference-of-convex algorithm with subanalytic data. J. Optim. Theory Appl. **179**(1), 103–126 (2018)
16. Le Thi, H.A., Moeini, M., Pham, D.T., Júdice, J.J.: A dc programming approach for solving the symmetric eigenvalue complementarity problem. Comput. Optim. Appl. **51**, 1097–1117 (2012)
17. Le Thi, H.A., Pham, D.T.: Dc programming and dca: thirty years of developments. Math. Program. **169**(1), 5–68 (2018)
18. Niu, Y.S.: On the convergence analysis of dca. arXiv:2211.10942 (2022)

19. Niu, Y.-S., Júdice, J., Le Thi, H.A., Dinh, T.P.: Solving the quadratic eigenvalue complementarity problem by dc programming. In: Le Thi, H.A., Pham Dinh, T., Nguyen, N.T. (eds.) Modelling, Computation and Optimization in Information Systems and Management Sciences. AISC, vol. 359, pp. 203–214. Springer, Cham (2015). https://doi.org/10.1007/978-3-319-18161-5_18
20. Niu, Y.S., Júdice, J., Le Thi, H.A., Pham, D.T.: Improved dc programming approaches for solving the quadratic eigenvalue complementarity problem. Appl. Math. Comput. **353**, 95–113 (2019)
21. Niu, Y.S., Le Thi, H.A., Pham, D.T., Júdice, J.: Efficient dc programming approaches for the asymmetric eigenvalue complementarity problem. Optim. Methods Softw. **28**, 812–829 (2013)
22. Niu, Y.S., Wang, Y.J., Le Thi, H.A., Pham, D.T.: High-order moment portfolio optimization via an accelerated difference-of-convex programming approach and sums-of-squares. arXiv:1906.01509 (2019)
23. Pham, D.T., Le Thi, H.A.: Dc optimization algorithms for solving the trust region subproblem. SIAM J. Optim. **8**, 476–507 (1998)
24. Pham, D.T., Le Thi, H.A.: The dc programming and dca revisited with dc models of real world nonconvex optimization problems. Ann. Oper. Res. **133**, 23–46 (2005)
25. Pham, D.T., Le Thi, H.A.: Convex analysis approach to d.c. programming: theory, algorithms and applications. Acta Math. Vietnam. **22**(1), 289–355 (1997)
26. Pham, D.T., Souad, E.B.: Algorithms for solving a class of nonconvex optimization problems. methods of subgradients. In: Fermat days 85: Mathematics for Optimization. North-Holland Mathematics Studies, vol. 129, pp. 249–271. Elsevier, B.V. (1986)
27. Queiroz, M., Júdice, J.J., Humes, C.: The symmetric eigenvalue complementarity problem. Math. Comput. **73**, 1849–1863 (2003)
28. Seeger, A.: Eigenvalue analysis of equilibrium processes defined by linear complementarity conditions. Linear Algebra Appl. **294**, 1–14 (1999)
29. Zhang, H., Niu, Y.S.: A boosted-dca with power-sum-dc decomposition for linearly constrained polynomial programs. J. Optim. Theory Appl. **201**(2), 720–759 (2024)

Operations Research and Decision Making

Efficient Heuristic to Solve the Multi-pickup and Delivery Problem with Time Windows

Salma Bouhlla[1,2,3(✉)], Rym Guibadj[1], and Aziz Moukrim[2]

[1] LISIC, Maison de la Recherche Blaise Pascal - BP 719, 50, rue Ferdinand Buisson, 62228 Calais, France
rym.guibadj@univ-littoral.fr
[2] Université de technologie de Compiègne, CNRS, Heudiasyc (Heuristics and Diagnosis of Complex Systems), CS 60319 - 60203, Compiègne Cedex, France
aziz.moukrim@hds.utc.fr
[3] Mobivia Supply Solutions, 2A Bd Van Gogh, 59650 Villeneuve-d'Ascq, France
salma.bouhlla@univ-littoral.fr

Abstract. This paper addresses the Multi-Pickup and Delivery Problem with Time Windows (MPDPTW), an extension of the well-known Pickup and Delivery Problem (PDP). Unlike the PDP, where each request involves one pickup and one delivery, the MPDPTW involves multiple pickups from different locations for a single delivery point. The objective is to minimize logistics costs while respecting time constraints. In this paper, we propose an efficient heuristic that relies on an advanced construction operator that explicitly states precedence constraints to construct high-quality solutions. Experimental results on the MPDPTW benchmark demonstrate that our approach matches all the best-known results from the existing literature and achieves a strict improvement in the relative gap for several instances, with an average enhancement of **0.2%**. These results establish our method as the new state-of-the-art for solving the MPDPTW.

Keywords: Logistics · Multi-pickup and delivery problem · Time windows · Heuristic · Constraint propagation

1 Introduction

In the evolving and highly competitive distribution sector, the demand for faster and more reliable deliveries has been intensifying in recent years. In this context, one of the main challenges in supply management is the coordination of deliveries from multiple collection points, which may represent suppliers or urban storage centers, to various destinations such as stores or final customers. Each destination may receive multiple deliveries throughout the day, resulting in increased logistical complexity and transportation costs.

To address these challenges, we propose a reorganization of deliveries using the Multi-Pickup and Delivery Problem with Time Windows (MPDPTW). The

© The Author(s), under exclusive license to Springer Nature Switzerland AG 2025
H. A. Le Thi et al. (Eds.): ICAMCS 2024, LNNS 1311, pp. 175–187, 2025.
https://doi.org/10.1007/978-3-031-90606-0_15

MPDPTW involves managing requests with multiple pickup nodes and a single delivery node using a fleet of vehicles while respecting time windows and ensuring that each request is handled by one vehicle. The objective is to reduce transportation costs, which might include travel distance or time. This problem formulation is highly relevant in supply chain management, courier services, or any system where multiple collection points feed into a single delivery destination, such as e-commerce fulfillment centers or food delivery services. Solving it efficiently helps reduce operational costs and improve service quality by avoiding disruptions caused by multiple deliveries to the same destination.

The Vehicle Routing Problem (VRP) has been extensively studied due to its importance in logistics and transportation. Solution methods range from exact approaches to various metaheuristics, such as Iterative Local Search (ILS) [6] and Variable Neighborhood Search (VNS) [7], which are effective in handling complex constraints like time windows and vehicle capacities. Metaheuristics like Tabu Search [3] have also shown success in exploring large solution spaces.

One such variant is the Pickup and Delivery Problem with Time Windows (PDPTW), where vehicles are required to pick up and deliver items between specific locations while respecting time window constraints. The PDPTW has also been the subject of considerable attention in the literature. For instance, a comprehensive survey of solution methods for the PDPTW was conducted [9], highlighting various heuristics and metaheuristics that effectively address the time window constraints. More recently, a metaheuristic approach combining local and global search techniques was proposed to effectively solve the PDPTW [6]. Additionally, a unified heuristic approach for various VRP variants, including the PDPTW, was developed using an Adaptive Large Neighborhood Search (ALNS) framework [10].

Building on these advancements, researchers have extended these methods to more complex variants like the Multi-Pickup and Delivery Problem with Time Windows (MPDPTW). This variant adds complexity with multiple pickup locations per request and strict time windows.

The Multi-Pickup and Delivery Problem with Time Windows (MPDPTW) was first introduced in [8]. A mathematical programming formulation was proposed, and the problem was solved exactly via branch-and-bound and heuristically via an ALNS. The proposed ALNS implements five insertion operators and four removal operators. The approaches were evaluated on newly designed benchmark instances inspired by PDPTW benchmark instances.

Two new formulations for the MPDPTW were proposed in [1]: the 2-index formulation and the Asymmetric Representative Formulation (ARF). The problem is solved exactly using a Branch-And-Cut (BC_{ARF}) algorithm, and several families of valid inequalities are introduced to strengthen the LP relaxations of the proposed formulations.

More recently, the effectiveness of employing abstract request-based variables and incorporating additional problem-specific constraints to eliminate infeasible subsets of requests was demonstrated in [4]. An Augmented Two-Index Formulation (ATIF) for the Capacitated Multi-Pickup and Delivery Problems with

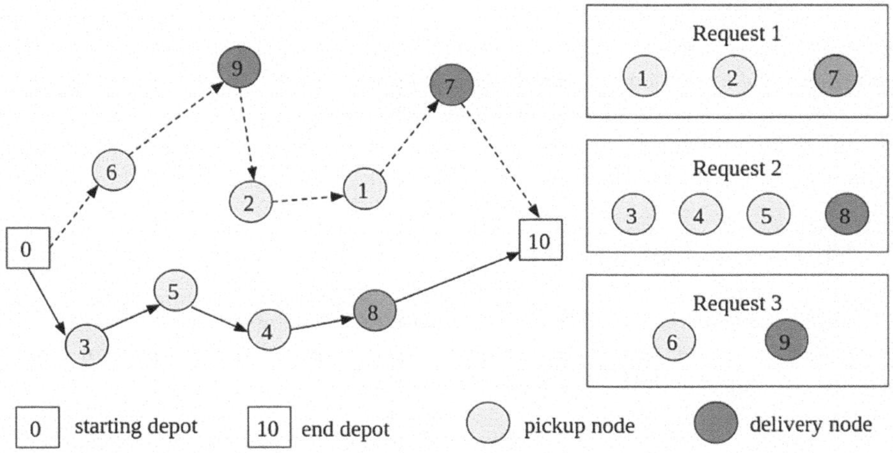

Fig. 1. Example of an MPDPTW solution with 3 requests and 2 vehicles.

Time Windows (CMPDPTW) was introduced, which optimizes vehicle routes by minimizing costs. Using a Branch-And-Cut (BC_{ATIF}) algorithm, this approach optimally solved benchmark instances with up to 50 nodes, many of which remained unsolved by earlier exact methods.

For the CMPDPTW, a new family of Strong Cutting Planes (SCP), known as infeasible request set cuts, has been introduced [5]. These cuts exploit different constraints, and the computational results highlight their effectiveness by achieving a 15% increase in the number of instances solved to optimality.

2 Problem Definition

An instance of the MPDPTW is defined by a set $R = \{r_1, ..., r_n\}$ of n requests to be routed with a fleet of m vehicles. The set $K = \{1, ..., m\}$ represents the available vehicles. The set $P = \{1, ..., p\}$ represents the pickup nodes, while the set $D = \{p+1, ..., p+n\}$ represents the delivery nodes. The number of delivery nodes is equal to n, and the number of pickup nodes is greater than or equal to n. Each request $r_i \in R$ is represented by a set of pickup nodes $P_{r_i} \subset P$ and a single delivery node $d_i \in D$. Each pickup node belongs to only one request, and each request always has at least one pickup node. The set N, which is the union of sets P and D, represents the total set of nodes. The request associated with a node $i \in N$ is noted as $r(i)$. Figure 1 illustrates the definition of the problem.

Let $G = (V, A)$ be the graph where $V = N \cup \{0, p+n+1\}$, with node 0 representing the starting depot and node $p+n+1$ representing the end depot. Each node $i \in V$ is associated with a service time s_i and a time window $[a_i, b_i]$. A vehicle must arrive at node i before the closing of its time window b_i, and if it arrives earlier, it must wait for the opening of the time window a_i. The set of arcs A is defined as $A = V \times V$ minus arcs leading to infeasible solutions. Specifically, the arc (i, j) is excluded if $i \in D$ and $j \in P_{r_i}$ to ensure that a delivery node is not served before its associated pickup nodes. Each arc $(i, j) \in A$ is associated with a non-negative travel cost c_{ij} and a non-negative travel time t_{ij}. Furthermore, $A^+(i)$ represents the set of outgoing arcs from node $i \in V$, and $A^-(i)$ represents the set of incoming arcs to node $i \in V$. Both the cost and time matrices are assumed to satisfy the triangular inequality.

3 Mathematical Model

A feasible solution to the MPDPTW involves creating a set of routes where all requests are assigned to the vehicles. Mathematically, the problem can be formulated using the following decision variables:

- x_{ij}^k: equals 1 if the arc (i, j) is traversed by vehicle k, and 0 otherwise.
- y_{rk}: equals 1 if request r is visited by vehicle k, and 0 otherwise.
- T_i: represents the start time of service at node $i \in V$.

In other words, the solution is obtained by determining which arcs are traversed by each vehicle, whether each request is visited by a vehicle, and the start times of service at each node. The mathematical model is formulated as follows:

$$\min \ f = \sum_{k \in K} \sum_{(i,j) \in A} c_{ij} x_{ij}^k \tag{1}$$

$$\text{s.t.} \quad \sum_{j \in A^+(i)} x_{ij}^k = y_{r(i)k} \qquad \forall k \in K, \forall i \in N \tag{2}$$

$$\sum_{j \in A^-(i)} x_{ij}^k = y_{r(i)k} \qquad \forall k \in K, \forall i \in N \tag{3}$$

$$\sum_{j \in A^+(i)} x_{0j}^k \leq 1 \qquad \forall k \in K \tag{4}$$

$$\sum_{k \in K} y_{rk} = 1 \qquad \forall r \in R \tag{5}$$

$$a_i \leq T_i \leq b_i \qquad \forall i \in V \tag{6}$$

$$T_j \geq T_i + s_i + t_{ij} \qquad \forall (i, j) \in A \tag{7}$$

$$T_{d_r} \geq T_i + s_i + t_{id_r} \qquad i \in P_r, \forall r \in R \tag{8}$$

$$x_{ij}^k \in \{0, 1\} \qquad (i, j) \in A, k \in K \tag{9}$$

$$y_{rk} \in \{0, 1\} \qquad r \in R, k \in K \tag{10}$$

The objective function (1) aims to minimize the total transportation cost associated with the arcs traversed by the vehicles. Constraints (2) and (3) serve as degree constraints, ensuring that all nodes within a request are assigned to the same vehicle while maintaining the pickup and delivery relationships between nodes. Constraint (4) limits the solution to a maximum of K vehicles. Constraint

Algorithm 1. General structure of our approach

```
Input: R, a set of n requests
Output: S_best, the best-found solution
Variables: d_max, diversification degree; S_i, current solution; URequests, unrouted
   requests;
m_si, number of non-empty routes in S_i
d_max ← 3 ; iter ← 0 ; itermax ← α · n
S_i ← guidedConstruction(S_i, R)
while (iter < itermax) do
    d ← random(1, d_max)
    S_i ← randomDestruction(S_i, d)
    URequests ← getUnroutedRequests(S_i)
    S_i ← guidedConstruction(S_i, URequests)          ▷ see Subsection 4.1
    S_i ← localSearch(S_i)                            ▷ see Subsection 4.2
    if Cost(S_i) < Cost(S_best) then
        S_best ← S_i ; iter ← 0 ; d_max ← 3
    else
        d_max ← min(d_max + 1, n/m_si) ; iter ← iter + 1
    end if
end while
```

(5) ensures that each request is served by exactly one vehicle, preventing partial or duplicate fulfillments. To maintain schedule feasibility within the specified time windows, constraints (6) and (7) are enforced. Constraint (8) ensures that the precedence order of pickup and delivery nodes is respected. Finally, constraints (9) and (10) define the binary nature of the decision variables.

4 Solution Method

Destruction and construction operators have been widely used in the literature to solve routing problems [2]. They have proven to be efficient in solving several variants of pickup and delivery problems. The success of these approaches relies on the design of dedicated operators that handle the problem characteristics. Our heuristic relies on a new construction operator specifically designed for the MPDPTW. Unlike existing ALNS approaches [8], which implement several insertion and removal operators, we use only one removal operator as a diversification mechanism and one insertion operator for the intensification mechanism. The key feature of our approach is an efficient construction operator that takes as input a subset of requests and determines the visit order of their associated nodes. It returns a feasible route or detects incompatibilities between nodes requests. This operator either returns a feasible route or detects incompatibilities among the requests.

The general structure of our approach is sketched in Algorithm 1, which starts by generating an initial solution using a guided construction operator. At each iteration, part of the current solution is destroyed by randomly removing d selected requests. The construction operator is then applied to insert the unrouted requests. These destruction and construction phases alternate until reaching the specified limit of consecutive iterations without improvement.

4.1 Precedence Constraints-Based Insertion

The core component of our approach is an efficient Request Insertion Algorithm (RIA), which is used in the *guidedConstruction* function in Algorithm 1. The *guidedConstruction* function aims to progressively build a feasible solution from a partial solution and a subset of unrouted requests. To achieve this, it attempts to insert a set of requests into a partial solution. For evaluating each request's insertion into a route, the function relies on the *Precedence Constraints-Based Insertion* method, as described in Fig. 2.

A significant challenge we encountered is that a single request can consist of multiple pickup nodes, each with its own time window constraints. The order in which these pickup nodes are inserted into a route is critical because an inappropriate insertion order could result in a violation of time windows, even if a feasible insertion exists. To address this, we designed a construction operator that uses precedence constraints to carefully evaluate a request insertion in a given route. This operator is able to identify incompatibilities among a given subset of requests that would prevent a feasible route.

As shown in Fig. 2, the insertion approach evaluates the cost of inserting a request in our RIA. Let $r \in R$ represent a request to be inserted into a specific route k, where $k \in K$. We define a set of nodes $\mathcal{N}_H$, which includes all the pickup nodes of request r as well as the nodes already present in route k. To enforce the necessary visit order, we generate a precedence graph where each node in $\mathcal{N}_H$ is represented as a vertex. An arc (i, j) is added between two nodes i and j if visiting node i must precede visiting node j. This precedence relationship is established in the following scenarios:

1. **Pickup before delivery**: Node j is the delivery point associated with a pickup node i. All pickup nodes for a given request must be visited before its delivery node.
2. **Time window constraints**: Visiting node j before node i would cause a delayed arrival at node i after its time window has closed. In this case, precedence is imposed to ensure that i is visited before j to maintain feasibility within the time windows.
3. **Depot return constraint**: Visiting node j before node i results in a delayed arrival to the end depot. To comply with the depot's closing time, nodes must be visited in an order that prevents delays at the end of the route.

To respect these precedence constraints, the opening time for each node $i \in \mathcal{N}_H$ is adjusted to ensure that it is at least as late as the sum of the opening times of its predecessors plus their respective travel times:

$$a_i \leftarrow \max\left(a_i, \max_{p \in A^-(i)} (a_p + t_{pi})\right)$$

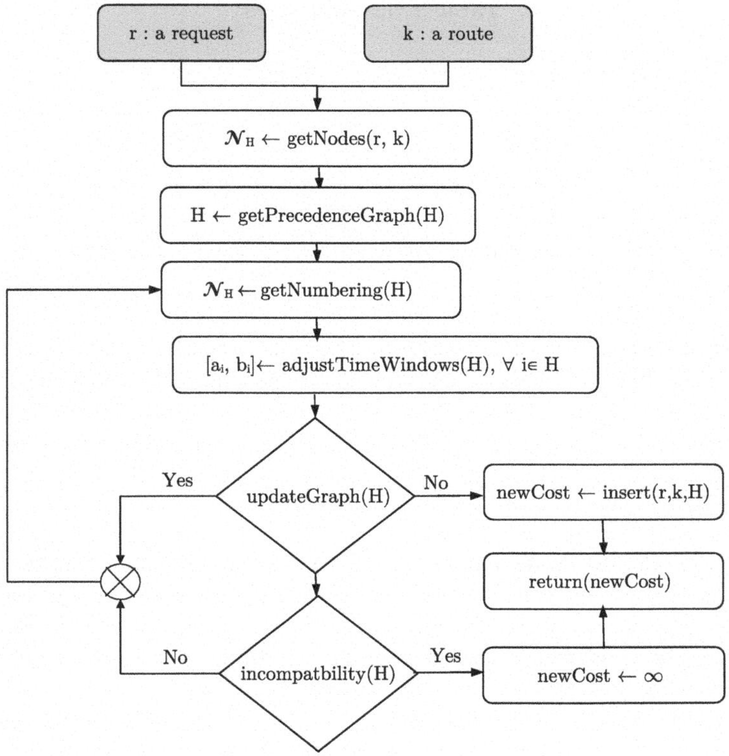

Fig. 2. Illustration showing the evaluation of insertion costs for a request within a specific route during the MPDPTW solution process.

This adjustment ensures that each node is visited within its allowable time window, especially considering that all pickup points must precede the delivery point.

Similarly, the closing time window for each node is adjusted to ensure that it does not exceed the closing time of its successors minus their travel times:

$$b_i \leftarrow \min\left(b_i, \min_{s\in A^+(i)} (b_s - t_{is})\right)$$

These iterative updates continue until no new arcs are added to the precedence graph, ensuring that all constraints are respected. After propagating the precedence constraints, nodes are numbered to define their visiting order, ensuring that for any two nodes i and j, $Number(i) < Number(j)$ if node i must precede node j. This numbering is used to guide the construction of a tour by determining a pre-order for visiting the nodes. The insertion procedure considers different criteria, such as the nearest available node, the earliest time window opening, and others, to determine the order of nodes. If an infeasibility is detected (e.g., conflicting arcs (i, j) and (j, i) in the precedence graph), the insertion is

Table 1. Instance types.

Instance	Without TW		Normal TW		Large TW	
Size	Short	Long	Short	Long	Short	Long
25	w_4_25	w_8_25	n_4_25	n_8_25	l_4_25	l_8_25
35	w_4_35	w_8_35	n_4_35	n_8_35	l_4_35	l_8_35
50	w_4_50	w_8_50	n_4_50	n_8_50	l_4_50	l_8_50
100	w_4_100	w_8_100	n_4_100	n_8_100	l_4_100	l_8_100

deemed impossible, and the algorithm backtracks to find an alternative solution. By leveraging these precedence constraints and dynamically adjusting time windows, our approach effectively reduces the search space and detects infeasibility, ensuring that all requests are handled within their respective constraints.

4.2 Local Search

The local search procedure is composed of six operators that can be classified into two categories: intra-route operators and inter-route operators. Intra-route operators attempt to reduce the travel cost within a single route by rearranging its nodes, while inter-route operators aim to minimize the total travel distance by moving requests between different routes.

At each iteration, the local search procedure randomly applies one of the following operators. For intra-route operations, the *Shift* operator moves a node to a new position within the same route, the *Swap* operator exchanges the positions of two nodes within the same route, and the *Remove and Reinsert* operator removes a node and reinserts it into the best position within the same route. For inter-route operations, the *Relocate* operator moves a request from one route to another, the *Exchange* operator swaps two requests between different routes, and the *Shake and Repair* operator removes three random requests from their current routes and reinserts them into new routes.

The search is performed using a first-improvement strategy, meaning that moves are applied as soon as an improvement is found. This process continues until no further improving move can be identified in the neighborhood.

5 Computational Experiments

This section presents computational experiments carried out to measure the performance of the proposed algorithm. First, we study the performance of the algorithm in terms of its parameters and its components. Then, we provide a comparison with state-of-the-art methods.

5.1 Benchmark Instances

The test instances for the MPDPTW were first introduced in [8]. These instances were derived from the problem instances originally proposed for

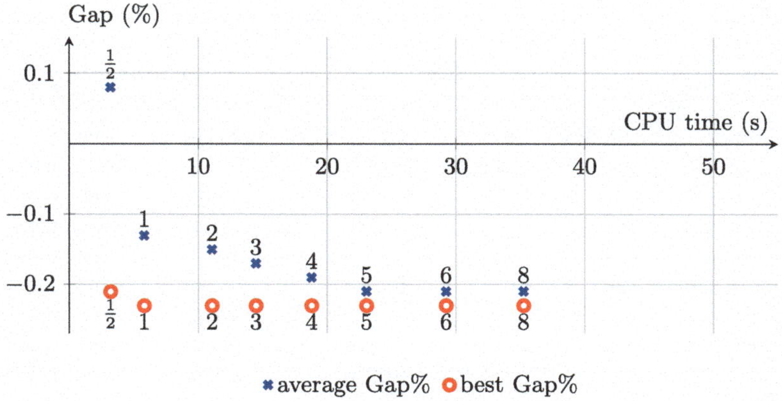

Fig. 3. Performance of our heuristic in terms of the stopping parameter for different values of α.

the PDPTW in [6]. In our study, we used the problem instances introduced in [1], which are publicly accessible at the following link: https://www.leandro-coelho.com/multi-pickup-and-delivery/. The detailed procedure for designing these problem instances can be found in [8]. The details of the instances are provided in Table 1. Each instance is characterized by three key factors: the type of time window (TW), the maximum request length, and the total number of nodes (instance size). The time window type is categorized as follows: "Without," where the TW of the original node is removed; "Normal," where the TW for each node is extended by opening 150 units earlier and closing 150 units later; and "Large," where it is extended by 300 units earlier and 300 units later. Every instance requires a minimum request size of 2 nodes, meaning each request must include at least one pickup and one delivery node. Depending on the instance type, requests may contain up to 4 nodes (Short requests) or 8 nodes (Long requests). The instances vary in size, containing 25, 35, 50, or 100 nodes.

5.2 Stopping Condition Parameter

To determine when our algorithm should stop, we set a limit on the number of iterations, denoted as iter_{max}, during which no improvement in the best solution is observed. To find the most effective stopping condition, we experimented with various combinations of $\text{iter}_{\text{max}} = \alpha \cdot n$. The parameter α was varied from 0.5 to 8. These tests were conducted on the same subset of instances. The results, illustrated in Fig. 3, show how the average gap, the best gap, and the average CPU time in seconds, over ten executions, change depending on this parameter.

The analysis of the results presented in Fig. 3 shows that the average gap decreases as the stopping parameter increases. This suggests that an intermediate level of iterations achieves the best balance between solution quality and computational time. A lower value of the stopping parameter does not allow the

Table 2. Detailed results for our heuristic compared to existing approaches

Instances	BKS	BC_{ARF}		BC_{ATIF}		SCP		ALNS			Notre heuristique			
		solved	cpu	solved	cpu	solved	cpu	*rpe%*	*arpe%*	cpu	improved	*rpe%*	*arpe%*	cpu
n_4_25	4734,8	5	0,2	5	8,8	5	0,7	0	0	2,12	0	0,00	0,41	0,57
n_4_35	6626,0	5	0,4	5	34,8	5	10,8	0	–	–	0	0,00	0,00	0,25
n_4_50	8923	5	56,9	5	171	–	–	0	0	4,30	0	0,00	0,06	0,63
n_4_100	15217,6	0	3600	0	–	–	–	0	0,2	18,4	1	**–0,02**	0,32	5,33
n_8_25	4646,1	5	0,06	5	1,1	5	0,1	0	0	2,3	0	0,00	0,00	0,11
n_8_35	6329,2	5	0,06	5	0,8	5	0,7	0	–	–	0	0,00	0,39	0,21
n_8_50	8521,6	5	0,3	5	5,5	–	–	0	0	3,9	0	0,00	0,00	0,45
n_8_100	16894,2	5	40,1	0	–	–	–	0	0	11,6	0	0,00	0,00	2,15
l_4_25	4117,5	5	88,1	5	56,7	5	94,7	0	0	2,3	0	0,00	0,00	0,18
l_4_35	5655	3	1662,3	5	272	0	3600	0	–	–	0	0,00	0,20	0,48
l_4_50	7213,1	0	3600	1	3117	–	–	0	0	4,6	0	0,00	0,10	1,17
l_4_100	12060,5	0	3600	0	–	–	–	0	0,18	19,1	1	**–0,01**	0,06	12,34
l_8_25	4424	5	0,23	5	5,3	5	0,6	0	0	2,7	0	0,00	0,19	0,21
l_8_35	5907,5	5	1,06	5	7,2	5	55,6	0	–	–	0	0,00	0,35	0,39
l_8_50	7368	4	747,3	5	498	–	–	0	0	4,4	1	**–0,15**	**–0,05**	0,92
l_8_100	13930,9	0	3600	0	–	–	–	0	0,3	13	2	**–0,11**	0,02	4,99
w_4_25	3079,8	0	3600	5	40,9	4	731,7	0	0,2	2,7	0	0,00	0,00	0,65
w_4_35	3974,6	0	3600	0	–	1	2913	0	–	–	2	**–2,72**	**–2,72**	2,70
w_4_50	5108,3	0	3600	0	–	–	–	0	0,1	5,4	2	**–0,11**	**–0,08**	11,02
w_4_100	8432,6	0	3600	0	–	–	–	0	0,9	22,6	3	**–0,09**	0,28	151,33
w_8_25	3047,1	2	2326,6	5	33	5	6	0	0	3,3	0	0,00	0,11	0,96
w_8_35	4333	1	2915,1	0	–	2	2184	0	–	–	3	**–1,20**	**–1,16**	2,92
w_8_50	4970,4	0	3600	0	–	–	–	0	1,35	5,9	2	**–0,13**	**–0,06**	9,99
w_8_100	9221,7	0	3600	0	–	–	–	0	0,58	16,9	2	**–0,27**	0,07	110,48
Average	7280,7	60	1826,6	66	303,7	47	800	0	0,23	8,11	**19**	**–0,20**	**–0,08**	13,35

algorithm to explore the solution space sufficiently, while a higher value results in a significant increase in computational time without a proportional improvement in solution quality. Therefore, we select $\alpha = 5$ as the most suitable value for the stopping parameter in our algorithm for solving the MPDPTW, ensuring an effective balance between computational efficiency and solution quality.

The experiments were conducted on various hardware configurations to assess the performance of different methods. The BC_{ARF} method described in [1] used a 2.1 GHz Intel processor with 40 GB of RAM, while the ALNS approach from [8] ran on a 2.67 GHz Intel processor under Scientific Linux 6.3, with 8 GB of RAM. The BC_{ATIF} method in [4] employed a 1.6 GHz Intel processor laptop with 8 GB of RAM, and the SCP method presented in [5] used a desktop with a 3.20 GHz Intel processor and 8 GB of RAM. Our experiments were performed on an AMD Ryzen 5 PRO 5650U processor with a variable clock speed up to 4.29 GHz, 12 cores, leveraging all available resources for optimal performance evaluation.

The performance of all methods is evaluated using the Relative Percentage Error (rpe) and the Average Relative Percentage Error (arpe) with respect to

Table 3. Improvement Classified According to the Type of TW

TW Type	#	rpe	cpu
Normal TW	1	0	1,21
Large TW	4	–0,03	2,58
Without TW	14	–0,56	36,25
Total	**19**	**–0,20**	13,35

the best-known solution $BestUB$. These two metrics are computed as: $rpe = \frac{Cost_{best} - BestUB}{Cost_{best}} \cdot 100$ and $arpe = \frac{Cost_{avg} - BestUB}{Cost_{avg}} \cdot 100$. $Cost_{best}$ denotes the best cost obtained over 10 runs, and $Cost_{avg}$ the average cost.

Table 2 presents the detailed results of the heuristic compared to existing approaches, namely ALNS [8], BC_{ARF} [1], BC_{ATIF} [4], and SCP [5]. The column *BKS* shows the average of the best-known solutions found in the literature for each block of instances, which serves as a reference point for evaluating the performance of the different methods. The *ALNS* section reports the relative percentage error (*rpe%*), the average relative percentage error (*arpe%*), and the computational time (*cpu*) in seconds. The BC_{ARF}, BC_{ATIF}, and SCP columns provide the number of instances solved to optimality and their corresponding computational times. For *Our heuristic*, we report the number of instances where improvements were achieved, along with the *rpe%*, *arpe%*, and *cpu* time. A dash (-) in any field indicates that the method was not tested on that block of instances by the respective authors. The results demonstrate that our approach consistently improves the best-known solutions, as indicated by the negative *rpe%* and *arpe%* values, and shows competitive computational efficiency, particularly in more complex instances.

The results obtained in Table 2 show that our approach offers significant improvements on several instances. These results highlight the competitive advantage of our heuristic in terms of computational efficiency. While our approach generally exceeds the performance of the existing literature by providing better solutions for several instances, it may require more time in some cases when significant improvements are necessary. However, for some instances as n_8_100, we achieve the best known solution five times faster. Among the 120 instances evaluated, our method improves the best known solution of **19** instances with an average rpe of **–0.2** and an average arpe of **–0.08**, underlining the robustness of our algorithm in providing high-quality solutions.

Table 3 represents the average values associated with different types of TW (Normal, Large, Without) for three different metrics: the number of improved instances (#), rpe, and CPU time (cpu). The results indicate that the absence of TW significantly increases the computational effort, as evidenced by the much higher CPU time (58.48) in the "Without TW" category compared to the other types. The Normal TW configuration shows the lowest rpe (0.00), suggesting that it provides stable performance. However, the Large TW exhibits a slightly negative rpe (-0.03), indicating a notable improvement over the state of the art.

Interestingly, when the time window (TW) is enlarged, there is a gain in terms of rpe, as seen in the "Large TW" case. Overall, the total values reveal that while the Normal TW results in minimal computation time and a neutral rpe, the presence of a larger TW or its absence altogether can both lead to increased computational complexity. However, the expansion of the time window shows potential benefits in performance improvement relative to the benchmark.

6 Conclusion

In this paper, we proposed an efficient heuristic to solve the Multi-Pickup and Delivery Problem with Time Windows (MPDPTW). Our approach relies on a heuristic that integrates an advanced construction operator specifically designed for this problem. Unlike existing methods, our algorithm employs a unique removal and insertion strategy, which has proven effective in handling precedence and time window constraints inherent to the MPDPTW. The computational experiments conducted on benchmark instances demonstrate the superior performance of our heuristic compared to previous approaches. Our method not only achieved improvements in several instances, reducing the relative percentage error (RPE) by an average of **0.2%**, but also reached optimal solutions with significantly lower computational times in some cases.

For future work, we aim to enhance our approach by incorporating additional optimization techniques to further reduce computational time and improve performance in more complex scenarios. We also plan to explore its applicability to other variants of the problem, such as those involving additional constraints, to address emerging needs in logistics and supply chain management.

References

1. Aziez, I., Côté, J.F., Coelho, L.C.: Exact algorithms for the multi-pickup and delivery problem with time windows. Eur. J. Oper. Res. **284**(3), 906–919 (2020)
2. Ben-Said, A., El-Hajj, R., Moukrim, A.: A variable space search heuristic for the capacitated team orienteering problem. J. Heuristics **25**(2), 273–303 (2019)
3. Gendreau, M., Hertz, A., Laporte, G.: A tabu search heuristic for the vehicle routing problem. Manag. Sci. **40**(10), 1276–1290 (1994)
4. Kohar, A., Jakhar, S.K.: A capacitated multi pickup online food delivery problem with time windows: a branch-and-cut algorithm. Ann. Oper. Res. 1–22 (2021). https://doi.org/10.1007/s10479-021-04145-6
5. Kohar, A., Jakhar, S.K., Agarwal, Y.K.: Strong cutting planes for the capacitated multi-pickup and delivery problem with time windows. Transp. Res. Part B: Methodol. **176**, 102806 (2023)
6. Li, H., Lim, A.: A metaheuristic for the pickup and delivery problem with time windows. In: Proceedings of the 13th IEEE International Conference on Tools with Artificial Intelligence (ICTAI 2001), pp. 160–167. Lecture Notes in Computer Science, IEEE, Atlanta, GA, USA (2001)
7. Mladenović, N., Hansen, P.: Variable neighborhood search. Comput. Oper. Res. **24**(11), 1097–1100 (1997)

8. Naccache, S., Côté, J.F., Coelho, L.C.: The multi-pickup and delivery problem with time windows. Eur. J. Oper. Res. **269**(1), 353–362 (2018)
9. Parragh, S., Doerner, K.F., Hartl, R.F.: A survey on pickup and delivery problems. part i: transportation between customers and depot. J. für Betriebswirtschaft **58**(1), 21–51 (2008)
10. Ropke, S., Pisinger, D.: A unified heuristic for a large class of Vehicle Routing Problems with Backhauls. Eur. J. Oper. Res. **171**(3), 750–775 (2006)

A Heuristics for Pickup-Delivery Problem with Cooperative Robots

Chi Thao Nguyen[1], Jean-Philippe Gayon[1], Viet Hung Nguyen[1], and Ta Anh Son[2](✉)

[1] Université Clermont-Auvergne, CNRS, Mines de Saint-Etienne, Clermont Auvergne INP, LIMOS, Clermont-Ferrand, France
{chi_thao.nguyen,j-philippe.gayon,viet_hung.nguyen}@uca.fr
[2] Hanoi University of Science and Technology, Hanoi, Vietnam
son.taanh@hust.edu.vn

Abstract. This paper presents a novel heuristic approach to solving the Pickup and Delivery Problem with Cooperative Robots (PDP-CR), a challenging optimization problem in automated warehouse logistics. The PDP-CR involves coordinating a fleet of robots to complete pickup and delivery tasks, some of which require multiple robots to cooperate. Our heuristic is designed to minimize the makespan by efficiently assigning robots to tasks while adhering to cooperation constraints. We compare the performance of our heuristic with an existing Mixed Integer Linear Programming (MILP) model through computational experiments on both small and large datasets. Results show that the proposed heuristic algorithm significantly reduces computation time while achieving good solutions, making it more suitable for large-scale applications.

Keywords: Pickup and Delivery Problem with Cooperative Robots · Mixed Integer Linear Programming (MILP) · heuristic algorithm

1 Introduction

In the Pickup and Delivery Problem with Cooperative Robots (PDP-CR), we manage a fleet of several identical robots to complete a set of pickup and delivery tasks inside a warehouse. Each task requires a specified number of robots and involves transporting goods from a pick-up point to a destination. For example, a package inside the warehouse may be too large for a single robot to handle. In this case, multiple robots need to cooperate to complete the task. Tasks cannot begin until all required robots are present at the pick-up location. A task is considered done if all robots needed for the task reach its pick-up point and then go to the destination simultaneously. After finishing a task, the robots can be assigned to other ones. All robots start their routes at a starting depot and finish at a return depot. The goal of PDP-CR is to minimize the makespan, i.e. the time at which the last robot returns to the depot.

At a glance, one can find that PDP-CR has some similarities with other problems.

© The Author(s), under exclusive license to Springer Nature Switzerland AG 2025
H. A. Le Thi et al. (Eds.): ICAMCS 2024, LNNS 1311, pp. 188–198, 2025.
https://doi.org/10.1007/978-3-031-90606-0_16

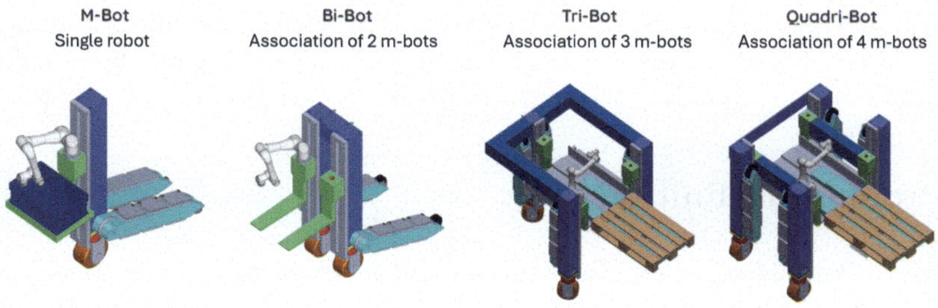

Fig. 1. Reconfiguration of warehouse robots, from https://www.mecabotix.com/

The Vehicle Routing Problem (VRP) [15], one of the most important combinatorial optimization challenges, focuses on efficiently distributing goods or services to customers using a fleet of vehicles, typically starting from a central depot. The Pick-up Delivery Problem (PDP) [1,14] involves transporting goods from an origin to a destination, and its mathematical model is a variant of the VRP with added constraints to ensure certain locations are visited before others. Synchronization in VRP [2,3,5,8] refers to the constraints that require two or more vehicles to complete a single task. For example, the synchronization between multiple homecare staff, they must be simultaneously present at a customer [2] or synchronization between the mailman crews and the vehicles [5]. PDP-CR can be seen as a special case of the Resource Constrained Project Scheduling Problem (RCPSP) with transfer time where there is only one resource and no precedence constraint. We can see the robots as a type of resource, and every task in PDP-CR is a job in RCPSP with transfer time.

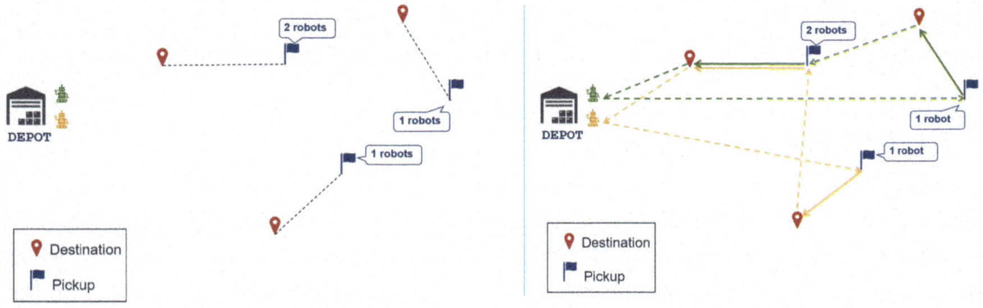

Fig. 2. An example with 2 starting robots. Left: the layout and right: the solution

The primary goal of this paper is to propose a fast heuristic algorithm for solving the Pickup and Delivery Problem with Cooperative Robots (PDP-CR). We also suggest some modifications to an existing Mixed Integer Linear Programming (MILP) formulation and theoretical results for the PDP-CR. The next section will describe the problem in more detail. In Sect. 3, we proposed our modified version of the flow-based MILP model for PDP-CR from [12]. The heuristic will be explained in detail in Sect. 4. Section 5 describe the numerical results and the performance analysis of our heuristic. Finally, the conclusion, some observations, and potential future directions are presented in Sect. 6.

2 Problem Definition

We have a fleet M of m identical robots and a set N of n tasks. Every task i is characterized by its pick-up location p_i, destination d_i, process time c_i, and number of robots required r_i. A task i cannot start until r_i robots arrive at its pickup point. This implies that The early robots that visit the task must wait for all the robots required to come, and then they can start to process the task from its pick-up point to its destination. All the robots start their paths at a starting depot, denoted by 0 and finish their path at a returning depot, denoted by $n+1$. t_{ij} is the time for any robot to move from task i to task j on the graph, corresponding to the time a robot moves from d_i to p_j. The objective of PDP-CR is to minimize the makespan, i.e., when the last robot reaches the return depot.

One can look at Fig. 2 to gain more intuition on the PDP-CR. The left figure illustrates a map with 3 tasks. Every task is denoted by an arc starting from its pickup point and ending at its destination. The task in the middle needs to be served by 2 robots (imagine in a warehouse, this is a large package that does not fit on a single automated guided vehicle, so two robots must come and cooperating carrying this parcel to its destination). The right figure is the optimal paths with 2 starting robots, each tour color corresponding to the path of one robot. One can observe that after two robots finish the "diagonal" tasks, they meet at the pick-up node of the "horizontal" task. If one robot comes first, it must wait for another, and then, they can start processing the task, and return to the depot together. So at some point, the robots may share the same route to perform a task or on the way to another task that needs multiple robots.

3 MILP Formulation

Two formulations for PDP-CR have been proposed in [12]. Namely, the VRP-based formulation and the flow-based formulation. In this section, we will provide a more succinct version of the flow-based formulation in [12]. We consider a robot as a flow unit. Let $x_{ij} = 1$ if there are a robot do task j right after task i, $f_{ij} \in \mathbb{R}$ denotes the number of flow going to task j right after i. $z_i \in \mathbb{R}$ denotes the time robots start to work task i. If $i = n+1$ then z_{n+1} is the time all robots return to the depot. And z_0 is the time robots leave the starting depot. And let $N^+ = \{0\} \cup N$ and $N^- = N \cup \{n+1\}$. The enhanced flow-based model to minimize the makespan of PDP-CR is as follow:

$$\min \quad z_{n+1} \tag{1}$$

$$\text{s.t.} \quad \sum_{j \in N^-} f_{0j} = m, \tag{2}$$

$$\sum_{i \in N^+} f_{i,n+1} = m, \tag{3}$$

$$\sum_{i \in N^+} f_{ij} = r_i \quad \forall j \in N \tag{4}$$

$$\sum_{j \in N^-} f_{ij} = r_i \quad \forall i \in N \tag{5}$$

$$z_0 = 0, \tag{6}$$

$$z_j + M(1 - x_{ij}) \geq z_i + c_i + t_{ij} \qquad \forall i \in N^+, \forall j \in N^-, i \neq j, \tag{7}$$

$$f_{ij} \geq x_{ij} \qquad \forall i \in N^+, \forall j \in N^-, i \neq j, \tag{8}$$

$$r_i x_{ij} \geq f_{ij} \qquad \forall i \in N^+, \forall j \in N^-, i \neq j, \tag{9}$$

$$f_{ij} \in \mathbb{R} \qquad \forall i \in N^+, \forall j \in N^-, i \neq j \tag{10}$$

$$z_i \in \mathbb{R} \qquad \forall i \in N^- \tag{11}$$

$$x_{ij} \in \{0, 1\}, \qquad \forall i \in N^+, \forall j \in N^-, i \neq j \tag{12}$$

In the formulation, the objective function (1) calls for minimizing the time that all robots reach the returning depot z_{n+1}, which is the makespan. Constraints (2) and (3) ensure that there are m flow units leaving the starting depot and m flow units reach the returning depot. (4) and (5) is the flow conversation constraint for each task. The starting time of all flows leaving the source is 0 is guaranteed by constraint (6). The constraint (7) is a Subtour Elimination Constraint (SEC) and is the linearization of the following constraint:

$$x_{ij} = 1 \implies z_j \geq z_i + c_i + t_{ij} \qquad \forall i \in N^+, \forall j \in N^-, i \neq \text{j}$$

The value of M is obtained using Algorithm 1. And finally, constraints (8) and (9) imply the relation between f and x.

Algorithm 1. Choosing Big M

$M \leftarrow 0$
for each node i on G **do**
    $M_i \leftarrow$ max(distance from i to other tasks)
    $M \leftarrow M + c_i + M_i$
end for

4 Greedy Heuristics for PDP-CR

4.1 Constructing a PDP-CR Solution from a Set of Paths

In this subsection, we give an algorithm to construct a PDP-CR solution from a set of paths. Concretely, let α_k be the schedule (or path) of robot k which contains the sequence of executed tasks by robot k. Let also $\alpha = (\alpha_1, \ldots, \alpha_m)$ the vector of all paths. The goal is to find the starting time of every task and the makespan, assuming that they start as soon as possible.

For example, in an instance with 3 starting robots and 6 tasks, tasks 2 and 5 require 2 robots, task 6 requires 3 robots, and the other tasks require one. The path of the first robot is $(0, 1, 6, 2, 5, 4, 7)$, the second robot is $(0, 6, 2, 3, 7)$ and the third is $(0, 6, 5, 7)$. Find the starting time of every task.

Consider the directed graph $\mathcal{F} = (A, V)$ with $V = N \cup \{0, n+1\}$. Arcs in A are constructed as follows. For every path $\alpha_i = (j_1^{\alpha_i}, j_2^{\alpha_i}, \ldots, j_{k_i}^{\alpha_i})$, we have $(j_1^{\alpha_i}, j_2^{\alpha_i}), (j_2^{\alpha_i}, j_3^{\alpha_i}), \ldots, (j_{k_i-1}^{\alpha_i}, j_{k_i}^{\alpha_i}), \in A$. This implies that if $(j, j') \in A$, then there is at least one robot going to task j' after it finishes task j. For the graph $\mathcal{F}$ in the previous example, we have the arcs set $A = \{(0,1), (1,6), (6,2), (2,5), (5,4), (4,7), (0,6), (2,3)\ , (3,7), (6,5), (5,7)\}$ To acquire the solution from the path set α, we need 2 steps:

- **Step 1**: Check if the paths are *valid* and construct the topological sorted sequence of the vertices in $\mathcal{F}$.
- **Step 2**: Apply Algorithm 3.

There are several cases when the paths in α are invalid for a solution. For example, if a path of robot 1 is $(3, \mathbf{2}, \mathbf{5}, 4)$ and robot 2 travels $(1, \mathbf{5}, \mathbf{2})$ then there won't exist a feasible solution. Or if a task requires 2 robots, but is only traversed by one. Proposition 1 gives a condition for the validity of the paths.

Proposition 1. *The path set α are valid for a solution of PDP-CR if and only if:*

- *There exists a topological ordering of $\mathcal{F}$.*
- *task j appears in exact r_j paths in α.*

A topological sort is a graph traversal in which each node j is visited only after all its dependencies are visited. A topological ordering is possible if and only if the graph has no directed cycles, that is, if it is a directed acyclic graph (DAG).

We call L the topological ordering of all the tasks in $\mathcal{F}$. To construct L, we can apply *Kahn' algorithm* (Algorithm 2). Note that a graph may have multiple topological ordering. In the above example, both $(0, 1, 6, 2, 5, 3, 4, 7)$ and $(0, 1, 6, 2, 5, 4, 3, 7)$ are the topological-sorted order of the graph $\mathcal{F}$ of these paths.

After the paths are all valid and L is acquired. We can move to step 2, find the starting time of every task and the makespan by applying Algorithm 3. It's a slightly modified version of finding the longest distance in the DAG algorithm.

Algorithm 2. Kahn's algortihm

Require: a graph G
Ensure: L: a topological sorted order of nodes in G

L: an empty list that will contain the sequence of topological sorted tasks
S: set of all nodes with no incoming edge
while S is not empty **do**
    Remove a node j from S
    add j to L
    for each node j' with an edge from j to j' **do**
        Remove edge (j, j') from the G
        if j' has no other incoming edges **then**
            Insert j' into S
        end if
    end for
end while
if G has edges **then**
    return error (G has at least one cycle)
else
    return L (a topologically sorted order)
end if

Algorithm 3. Construct solution from paths

Require:
A graph G
A topological sorted order L of nodes in G
The transportation time between every pair of tasks and and between every task and the depot $c_{jj'}$
The process time c_j of every task
Ensure:
The starting time of every task z_j and the makespan z_{n+1}

$z_0 \leftarrow 0$;
$z_j \leftarrow -\infty$ for all $j \in N^-$
for each vertex j in L **do**
    for each task j' such that $(j, j') \in G$ **do**
        $z_{cur} \leftarrow z_j + c_j + c_{jj'}$
        if $z_{j'} < z_{cur}$ **then**
            $z_{j'} \leftarrow z_{cur}$
        end if
    end for
end for
return z_j for j and makespan z_{n+1}

4.2 Greedy Algorithm

This section describes a greedy heuristic to acquire a solution for PDP-CR quickly. The idea of Algorithm 4 is to append tasks iteratively to the path of the robots; based on which task, after appending it to its number of required paths, the makespan increases the least.

Algorithm 4. Greedy append

Require:
A set N of n task.
A set M of m robot.
Ensure: Paths of each robot, and the makespan of the solution.
1: $N^u \leftarrow N$ ▷ N^u contains un-inserted tasks
2: $\alpha_k = [0]$ for $k = 1, \ldots, m$ ▷ Initially, every robot start at a depot
3: $l_k = 0$ for $k = 1, \ldots, m$ ▷ Initialize length of each path are 0
4: **while** N^u is not empty **do**
5: $c_{max} \leftarrow +\infty$
6: **for** each task i in N^u **do**
7: $L \leftarrow$ an empty list
8: **for** each robot $k = 1, \ldots, m$ **do**
9: $i_k \leftarrow$ last task visited by k
10: $l'_{ki} \leftarrow l_k + t_{i_k i}$ ▷ l'_{ki} is the length of l if append i to k.
11: Append l'_{ki} to L
12: **end for**
13: $c_{temp} \leftarrow r_i$-th smallest value in L
14: $R \leftarrow$ corresponding tasks of r_i smallest values in L ▷ R is the list of best paths for the current tasks
15: **if** $c_{temp} < c_{max}$ **then**
16: $c_{max} \leftarrow c_{temp}$
17: $i^* \leftarrow i$
18: $R^* \leftarrow R$
19: **end if**
20: **end for**
21: **for** each robot $k \in R^*$ **do**
22: Append i^* to α_k
23: $l_k \leftarrow c_{max} + c_{i^*}$
24: **end for**
25: Remove i^* from N^u
26: **end while**
27: $C_{max} \leftarrow \max_{k=1 \ldots I}(l_k)$
28: **return** α_k for $k = 1, \ldots, m$ and C_{max}

In the following explanation, we use the word "iteration" to refers the loop starting at line 4. Each robot k starts with an initial path containing only the starting depot. For each pair of (robot, task), l'_{ki} save the temporary time of path of robot k. We have the set L to store the data of this temporary length. Since a task i needs r_i robots and the task cannot start before a sufficient number of robots arrive, c_{temp} is the best possible value of the makespan of task i in the current iteration, which is when we append i to r_i paths that have its temporary time smaller or equal to c_{temp} (line 13). At each iteration, we want to append the task that has the smallest value of c_{temp} to the corresponding paths (line'15).

5 Computational Result

We test our algorithms and formulation on a device equipped with Intel(R) Core(TM) i5- 8400 CPU @ 2.80 GHz chipset. The MILP model and the heuristics were implemented using Python 3.9 and the formulation was solved using CPLEX 22.1.0. The small-size data was generated as follows.

- Number of robots $m \in \{1, 2\}$.
- Number of tasks $n \in \{5, 10, 15\}$.

For each pair of chosen (number of tasks, number of robots), we generated 5 random instances according to the following:

- Robot required for each task $r_j \in \{1, 2, \ldots, m\}, \forall j \in N$.
- Coordinate of pickup, destination for each task $(x_j, y_j) \in ([0, 500], [0, 500])$, $\forall j \in N$.

To further test our work, we also acquire large-size instances by adapting data in [10]. The data already have information about the pick-up point and corresponding location of each task. In each instance, we generated a number of robots $r_j \in [1, 5]$ for each task j.

In the tables, the column *obj* is the objective value found by each method, *time* is the solving time in second. *gap* is the optimal gap reported in CPLEX. Table 1 presents results for small-sized, randomly generated datasets where

Table 1. Comparisions between CPLEX and Greedy heuristics on small size, generated data

Data	Robots	Tasks	MILP			Heuristics	
			obj	time	gap	obj	time
g0	1	5	2142	0.06	0.0	2554	0.01
g1	1	5	1668	0.05	0.0	2108	0.01
g2	1	5	2007	0.03	0.0	2603	0.01
g3	1	5	2405	0.05	0.0	2875	0.01
g4	1	10	3990	57.73	0.0	5110	0.03
g5	1	10	3796	48.77	0.0	4654	0.03
g6	1	10	4055	65.94	0.0	4597	0.03
g7	1	10	4511	76.23	0.0	4966	0.03
g12	2	5	2872	0.64	0.0	3161	0.01
g13	1	5	2572	0.38	0.0	2613	0.01
g14	2	5	1533	0.63	0.0	1922	0.01
g15	2	5	2272	0.64	0.0	2424	0.01
g16	2	10	2509	125.22	0.0	2675	0.02
g18	2	10	3072	188.06	0.0	3917	0.03
g19	2	10	3641	6450.69	0.0	4041	0.02

CPLEX can achieve optimal solutions. One can see that the MILP consistently achieves better objective values, indicating optimal solutions. However, it requires significantly more computation time compared to the heuristic, especially as the number of tasks increases. The heuristic provides near-optimal solutions in much less time, making it a practical alternative for larger instances. Table 2 compares performance on larger datasets. Here, CPLEX struggles with higher objective values and reaches the maximum preset solve time (1 h). In contrast, the heuristic yields better objective values (up to 25% smaller) while maintaining low computation times. The heuristics outperform the exact method in both solution quality and solving time, demonstrating the heuristics effectiveness in handling large-scale problems where exact methods are computationally limited.

Table 2. Comparisions between MILP and Greedy heuristics on large data, every instance have approximately 50 tasks

	Data	Robots	MILP		Greedy Heuristics		
			obj	time	obj	time	diff
0	lc101-5	10	444	3600	**421**	0.94	0.05
1	lc102-5	10	476	3600	**419**	0.85	0.12
2	lc103-5	10	438	3600	**418**	0.83	0.05
3	lc104-5	10	438	3600	**407**	0.81	0.07
4	lc105-5	10	461	3600	**425**	0.79	0.08
5	lc106-5	10	454	3600	**449**	0.79	0.01
6	lc107-5	10	**424**	3600	438	0.79	-0.03
7	lc108-5	10	**474**	3600	475	0.80	-0.00
8	lc109-5	10	492	3600	**418**	0.79	0.15
9	lc201-5	10	871	3600	**739**	0.75	0.15
10	lc202-5	10	993	3600	**852**	0.73	0.14
11	lc203-5	10	1112	3600	**958**	0.73	0.14
12	lc204-5	10	890	3600	**671**	0.74	0.25
13	lc205-5	10	832	3600	**762**	0.74	0.08
14	lc206-5	10	873	3600	**751**	0.73	0.14
15	lc207-5	10	719	3600	**632**	0.73	0.12

6 Conclusion

In this paper, we introduced a heuristic to address the Pickup and Delivery Problem with Cooperative Robots (PDP-CR), offering a practical solution to the computational challenges posed by the problem. Our heuristic algorithm outperforms the MILP model in terms of computation time, especially for large problem instances, while maintaining solution quality close to optimal. The results suggest that our approach is effective for real-world applications in automated warehouses, where quick and efficient decision-making is critical. Future work could explore the improvement in the MILP formulations and have metaheuristics to improve the starting solution.

References

1. Berbeglia, G., Cordeau, J.F., Laporte, G.: Dynamic pickup and delivery problems. Eur. J. Oper. Res. **202**(1), 8–15 (2010). https://doi.org/10.1016/j.ejor.2009.04.024
2. Bredström, D., Rönnqvist, M.: Combined vehicle routing and scheduling with temporal precedence and synchronization constraints. Eur. J. Oper. Res. **191**(1), 19–31 (2008). https://doi.org/10.1016/j.ejor.2007.07.033
3. Drexl, M.: Synchronization in vehicle routing-a survey of VRPs with multiple synchronization constraints. Transp. Sci. **46**(3), 297–316 (2012). https://doi.org/10.1287/trsc.1110.0400
4. Haddadene, S.A., Labadie, N., Prodhon, C.: NSGAII enhanced with a local search for the vehicle routing problem with time windows and synchronization constraints. IFAC-PapersOnLine **49**(12), 1198–1203 (2016). https://doi.org/10.1016/j.ifacol.2016.07.671
5. Hollis, B.L., Forbes, M.A., Douglas, B.E.: Vehicle routing and crew scheduling for metropolitan mail distribution at Australia Post. Eur. J. Oper. Res. **173**(1), 133–150 (2006). https://doi.org/10.1016/j.ejor.2005.01.005
6. Hurkens, C.A., Woeginger, G.J.: On the nearest neighbor rule for the traveling salesman problem. Oper. Res. Lett. **32**(1), 1–4 (2004). https://doi.org/10.1016/S0167-6377(03)00093-2
7. Kahn, A.B.: Topological sorting of large networks. Commun. ACM **5**(11), 558–562 (1962). https://doi.org/10.1145/368996.369025
8. Kim, B.I., Koo, J., Park, J.: The combined manpower-vehicle routing problem for multi-staged services. Expert Syst. Appl. **37**(12), 8424–8431 (2010). https://doi.org/10.1016/j.eswa.2010.05.036
9. Laporte, G.: The traveling salesman problem: an overview of exact and approximate algorithms. Eur. J. Oper. Res. **59**(2), 231–247 (1992). https://doi.org/10.1016/0377-2217(92)90138-Y
10. Li, H., Lim, A.: A metaheuristic for the pickup and delivery problem with time windows. In: Proceedings 13th IEEE International Conference on Tools with Artificial Intelligence, ICTAI 2001, pp. 160–167 (2001). https://doi.org/10.1109/ICTAI.2001.974461
11. Liu, R., Tao, Y., Xie, X.: An adaptive large neighborhood search heuristic for the vehicle routing problem with time windows and synchronized visits. Comput. Oper. Res. **101**, 250–262 (2019). https://doi.org/10.1016/j.cor.2018.08.002
12. Nguyen, C.T., Gayon, J.P., Nguyen, V.H., Quilliot, A., Ta, A.S.: Pickup and delivery problem with cooperative robots. In: Proceedings of the 12th International Symposium on Information and Communication Technology, SOICT 2023, pp. 695–700. Association for Computing Machinery, New York, NY, USA, December 2023. https://doi.org/10.1145/3628797.3628988

13. Parragh, S.N., Doerner, K.F., Hartl, R.F.: A survey on pickup and delivery problems. J. für Betriebswirtschaft **58**(1), 21–51 (2008). https://doi.org/10.1007/s11301-008-0033-7
14. Parragh, S.N., Doerner, K.F., Hartl, R.F.: A survey on pickup and delivery problems: part ii: transportation between pickup and delivery locations. J. für Betriebswirtschaft **58**(2), 81–117 (2008). https://doi.org/10.1007/s11301-008-0036-4
15. Toth, P., Vigo, D. (eds.): The vehicle routing problem. SIAM monographs on discrete mathematics and applications, Society for Industrial and Applied Mathematics, Philadelphia, PA (2002)

A Mathematical Model for the Ecosystem of Two Fish Species, Wallago Attu and Pangasius Sanitwongsei, in the Mekong River Basin

Hung Duc Nguyen, Trung Viet Ngo, Anh Duc Nguyen, and Thuy Phuong Nguyen(✉)

Hanoi University of Science and Technology, Hanoi, Vietnam
{hung.nguyenduc,anh.nguyenduc1,thuy.nguyenphuong}@hust.edu.vn,
trung.nv222271m@sis.hust.edu.vn

Abstract. Wallago Attu and Pangasius Sanitwongsei are two types of fish with huge potential for agricultural economic development and commercial value in the freshwater ecosystems of the Mekong Delta region. However, environmental changes and overfishing have threatened these species' population dynamics. Our main concern is to improve the understanding of human impact on the ecosystem's stability. A new model is constructed to represent the food chain involving the two fish species and their common prey. Some notable parameters, such as the harvesting rate of humans and the intraguild predation rate, will be analyzed more closely. Numerical analyses were performed to analyze the existence and stability of five equilibrium points and investigate the impact of varying model parameters. Simulation results and bifurcation diagrams provided insights into how changes in ecological factors influence the long-term population trends of the species. Several complex system behaviors have been observed as the environment changes, causing increased predatory ability of Pangasius Sanitwongsei species. The coexistence of the predators and the prey populations is possible if the harvesting rate falls within the range of 14% to 43%, which requires strict management of fishing activities.

Keywords: Predator-prey Model · Stability Analysis · Hopf Bifurcation · Intraguild Predation

1 Introduction

An ecosystem is a unified entity comprising living organisms and their surrounding physical environment. By studying how ecosystems function, interact, and regulate themselves, we gain a better understanding of the delicate balance of nature and the impact of human activities on the environment. Therefore, research on the ecosystem environment is pivotal in ensuring a sustainable future for humanity and our planet.

© The Author(s), under exclusive license to Springer Nature Switzerland AG 2025
H. A. Le Thi et al. (Eds.): ICAMCS 2024, LNNS 1311, pp. 199–211, 2025.
https://doi.org/10.1007/978-3-031-90606-0_17

To investigate complex and diverse ecosystems, it is advisable to begin by studying simpler ecosystems with fewer influencing factors. In fact, there has been much research describing the ecosystem of predator-prey over the past 40 years. Many studies have taken the first steps in studying ecosystems consisting of one or two predator(s) [7,12], one species will serve as food for the other. Some studies have also shown that competition exists between predators, known as Intraguild Predation [3,4]. Others have also been conducted on control parameters such as human catch coefficient, population density, and environmental carrying capacity [1,2,11]. There has been a fair amount of research done to prove that ecosystems can achieve local stability at equilibrium points, but only a few have clearly stated whether it is globally stable or not [1,2].

In this paper, we introduce a two-predators-one-prey ecosystem in the Mekong River basin, analyze the local/global steady state, and consider and evaluate factors affecting the ecosystem state when environmental capacity changes over time. In recent years, the problem of environmental pollution at the Mekong River has been truly alarming, causing a serious decrease in the number of individual river creatures. Wallago Attu (WA) and Pangasius Sanitwongsei (PS) are two suitable freshwater fish species for research and evaluation. They have been dubbed the "Giant Monsters", relying on giant-sized individuals. PS is slightly better than WA in terms of hunting ability and ecological shape. Due to their beautiful morphology and nutritious fish meat, they are often considered a top hunting target for humans. The IUCN Red List listed PS with the tag name **Critically Endangered** in 2007 [5]. Until 2019, WA was also listed with the tag name **Vulnerable** [6]. Both WA and PS like to eat small fish, worms, shrimp, and other aquatic animals, collectively known as prey. The appearance and distribution of two types of fish are always in the same area in the Mekong River basin [10]. The research in [9] showed that, in an environment with limited food resources, two predator species may kill and feed on each other, which might increase their chances of obtaining sufficient food for survival and maintaining their populations.

The rest of the article is structured as follows: In Sect. 2, we construct the mathematical model of the ecological system. Section 3 provides an analysis of the model using mathematical techniques. We conduct numerical simulations to examine the response of the ecosystem under human impact by analyzing the theoretical results of equilibrium states and bifurcation diagrams based on the consumption rate of WA on prey and catch coefficient in Sect. 4 and the conclusion will be presented in Sect. 5.

2 Model Formulation

This section is dedicated to implementing a mathematical model of predator-prey ecosystems. The food source comprises small fish, worms, shrimp, and other aquatic organisms, collectively represented as a prey species R, which grows as the logistic growth with a growth rate r and the environmental carrying capacity K. Predators that only seek food by hunting prey and their population densities are represented by variable C, in which C_{1A}, C_{1E} are the adult and juvenile population densities of WA, and C_{2A}, C_{2E} are the adult and juvenile ones of PS. If the prey R becomes extinct, the two predator species will decline exponentially $\dot{C}(t) = -mC$ and gradually disappear, so the predator species are completely dependent on the prey species. Adding human-impact factors for the ecosystem of two fish species, the model is formulated in (1), and the biological parameters are listed in Table 1.

$$\begin{cases} \frac{dR}{dt} &= rR\left[1-\frac{R}{K}\right] - a_1RC_{1A} - a_2RC_{2A}, \\ \frac{dC_{1A}}{dt} &= -(h+m_{1A})C_{1A} + \alpha_1 C_{1E}, \\ \frac{dC_{1E}}{dt} &= -m_{1E}C_{1E} - \alpha_1 C_{1E} - b_{21}C_{1E}C_{2A} + a_1e_1RC_{1A} + b_{12}e_1C_{2E}C_{1A}, \\ \frac{dC_{2A}}{dt} &= -(h+m_{2A})C_{2A} + \alpha_2 C_{2E}, \\ \frac{dC_{2E}}{dt} &= -m_{2E}C_{2E} - \alpha_2 C_{2E} - b_{12}C_{2E}C_{1A} + a_2e_2RC_{2A} + b_{21}e_2C_{1E}C_{2A}, \\ \frac{dK}{dt} &= \alpha\left(K-k_1\right)\left(1-\frac{K-k_1}{k_2-k_1}\right). \end{cases} \tag{1}$$

Table 1. Description of parameters's biological meaning.

Parameter	Ecological Meaning
r	Intrinsic growth rate of the prey
h	Catch coefficient of human
k_1	Maximum value of environmental carrying capacity
k_2	Minimum viable restoration of the environment
a_1, a_2	Consumption rate of WA/PS species on prey
b_{12}, b_{21}	Attack rate of WA/PS species on other's juveniles
e_1, e_2	Consumer WA/PS species energy conversion coefficient
α_1, α_2	Proportion of the juveniles of WA/PS species become adults
α	Speed of environmental change
m_{1A}, m_{2A}	Mortality of the adults of WA/PS species
m_{1E}, m_{2E}	Mortality of the juveniles of WA/PS species

As assumed in Sect. 1, we consider the general case of two species coexisting in direct and indirect competition. This directly affects their growth. Using the Holling Type I functional response, we assume that as the population density of predator(s) increases, its consumption rate also increases linearly with the density of the prey species. However, when predator(s) are at very high densities, its consumption rate will decrease. This leads to a balance between the species, and the ecosystem gradually achieves stability. In fact, humans can exploit fish through various methods such as aquaculture, fishing, and selective harvesting, which can be represented by a positive parameter, denoted as h. Due to fishermen are more interested in capturing larger fish rather than juveniles, so the parameter h will mainly affect the adult individuals in both species. An essential factor is the carrying capacity of the environment, denoted as K as described in the last equation of (1), depending on the rainy and dry seasons.

Finally, there are some constraints for parameters at Table 1 to ensure that the model (1) remains within biologically meaningful and feasible ranges, the index $i \in \{1, 2\}$ corresponding to species WA, PS:

- In realistic ecological scenarios, all parameters should have non-negative values.
- The energy conversion coefficient e_i, mortality rates m_{iA}, m_{iE}, proportion of the juveniles α_i, and attack rate b_{12}, b_{21} should all fall in the range of $[0, 1]$.
- For the adult stage, the total number of natural deaths m_{iA} and catch coefficient h must not exceed the total number of species, resulting in $m_{1A} + h \leq 1$ and $m_{2A} + h \leq 1$.
- Since the implementations of these seasons are similar, we will assume that the initial environment corresponds to the beginning of the rainy season, which means $0 < k_2 < k_1$.

3 Mathematical Analysis

3.1 Positivity and Boundedness of Solutions

This subsection aims to demonstrate two fundamental properties of the model's solution: positivity and boundedness. They ensure that our model accurately reflects the dynamics system within ecological meanings.

Theorem 1. *Let's define a positive cone region*

$$H^+ = \{R \geq 0, C_{1A} \geq 0, C_{1E} \geq 0, C_{2A} \geq 0, C_{2E} \geq 0, K \geq 0 \mid k_1 > k_2\}.$$

Then, for every positive initial value

$$(R(0), C_{1A}(0), C_{1E}(0), C_{2A}(0), C_{2E}(0), K(0)) \in H^+,$$

the solution $(R(t), C_{1A}(t), C_{1E}(t), C_{2A}(t), C_{2E}(t), K(t)) \in H^+$ *for all* $t \in \mathbb{R}^+$.

Proof. From the system (1), we have:

$$(R(0), C_{1A}(0), C_{1E}(0), C_{2A}(0), C_{2E}(0), K(0)) \in H^+,$$

We can evaluate the positivity of the solution of the system of equations with the conditions $R \geq 0, C_{1A} \geq 0, C_{1E} \geq 0, C_{2A} \geq 0, C_{2E} \geq 0, K \geq 0$. We also have

$$\begin{aligned}
&f_1(0, C_{1A}, C_{1E}, C_{2A}, C_{2E}, K) = 0 \geq 0\\
&f_2(R, 0, C_{1E}, C_{2A}, C_{2E}, K) = \alpha_1 C_{1E} \geq 0\\
&f_3(R, C_{1A}, 0, C_{2A}, C_{2E}, K) = a_1 e_1 R C_{1A} + b_{12} e_1 C_{2E} C_{1A} \geq 0\\
&f_4(R, C_{1A}, C_{1E}, 0, C_{2E}, K) = \alpha_2 C_{2E} \geq 0\\
&f_5(R, C_{1A}, C_{1E}, C_{2A}, 0, K) = a_2 e_2 R C_{2A} + b_{21} e_2 C_{1E} C_{2A} \geq 0\\
&f_6(R, C_{1A}, C_{1E}, C_{2A}, C_{2E}, 0) = -\alpha \frac{k_1 k_2}{k_2 - k_1} \geq 0.
\end{aligned}$$

Using the result of Nagumo [8], we obtain $R(t) \geq 0, C_{1A}(t) \geq 0, C_{1E}(t) \geq 0, C_{2A}(t) \geq 0, C_{2E}(t) \geq 0, K(t) \geq 0$ with $t \in \mathbb{R}^+$, lead to

$$(R(t), C_{1A}(t), C_{1E}(t), C_{2A}(t), C_{2E}(t), K(t)) \in H^+, \forall t \in \mathbb{R}^+.$$

Hence, the theorem was proven.

Theorem 2. *$(R(t), C_{1A}(t), C_{1E}(t), C_{2A}(t), C_{2E}(t), K(t))$ is bounded for all $t \in [0, +\infty)$ and $k_1 > k_2$.*

Proof. From the last equation of (1), we have

$$\dot{K} = \alpha (K - k_1) \left(1 - \frac{K - k_1}{k_2 - k_1}\right) = \frac{\alpha}{k_2 - k_1}(K - k_1)(k_2 - K).$$

- If $K \in [0, k_2]$, so $\dot{K} \leq 0$. It means $K \leq k_2$.
- If $K \in (k_2, k_1]$, so $\dot{K} \geq 0$. It means $K \leq \max\{k_1, k_2\} = k_1$.
- If $K \in (k_1, +\infty)$, so $\dot{K} \leq 0$. It means $K \leq k_1$.

From that, for all t, K is bounded in $(0, +\infty)$.

We denote that $M = \max\{K, R(0)\}$ and $m = \min\{m_{1A}, m_{1E}, m_{2A}, m_{2E}\}$. It is easy to see that $\dot{R} \leq rR(1 - \frac{R}{K})$, leads to $R(t) \leq M$ for all t, or R is bounded. Consider the Lyapunov function:

$$V_0(R, C_{1A}, C_{1E}, C_{2A}, C_{2E}, K) = R + \frac{C_{1A}}{e_1} + \frac{C_{1E}}{e_1} + \frac{C_{2A}}{e_2} + \frac{C_{2E}}{e_2} + \frac{k_2 - k_1}{\alpha} K. \quad (2)$$

By calculating the derivative of V_0 along the direction of the trajectory of the problem, we obtain that

$$\begin{aligned}
\dot{V}_0 = \; & rR(1 - \frac{R}{K}) - (h + m_{1A})\frac{C_{1A}}{e_1} - m_{1E}\frac{C_{1E}}{e_1} - (h + m_{2A})\frac{C_{2A}}{e_2} - m_{2E}\frac{C_{2E}}{e_2} +\\
& + \frac{b_{12}(e_2 - 1)}{e_2} C_{2E} C_{1A} + \frac{b_{21}(e_1 - 1)}{e_1} C_{1E} C_{2A} + (K - k_1)(k_2 - K).
\end{aligned}$$

Applying the Gronwall inequality, we get

$$\dot{V}_0 + mV_0 \leq (r + m)R + (K - k_1)(k_2 - K),$$

$\Rightarrow V_0(t) \leq V_0(0)e^{-mt} + \dfrac{R(r+m)+(K-k_1)(k_2-K)}{m}(1-e^{-mt})$. Because R and K are bounded as proof above, $V_0(t)$ is bounded when $t \to +\infty$. As a result, $(R(t), C_{1A}(t), C_{1E}(t), C_{2A}(t), C_{2E}(t), K(t))$ is also bounded, and the theorem was proven.

3.2 Equilibria and Their Existence

To achieve equilibrium points, the carrying capacity K reaches the value k_1 or k_2. Since the two cases are similar, we can simplify the calculations by setting K as a constant value k'. The model (1) exhibits five equilibrium points:

- The trivial equilibrium point $E_1 = (0, 0, 0, 0, 0, k')$ always exists, corresponding to the scenario where all species go extinct.
- The equilibrium point $E_2 = (k', 0, 0, 0, 0, k')$ always exists in which both predator species go extinct, and the prey reaches its maximum, which is equal to its carrying capacity.
- The equilibrium point $E_3 = (\bar{R}_2, 0, 0, \bar{C}_{2A}, \bar{C}_{2E}, k')$ corresponding to the case that the first predator species becomes extinct, where $\bar{R}_2 = \frac{(h+m_{2A})(m_{2E}+\alpha_2)}{\alpha_2 a_2 e_2}$, $\bar{C}_{2A} = \frac{r}{a_2}\left(1 - \frac{\bar{R}_2}{k'}\right)$, $\bar{C}_{2E} = \frac{r(h+m_{2A})}{a_2\alpha_2}\left(1 - \frac{\bar{R}_2}{k'}\right)$. Existence of E_3 under conditions: $C_{2A} > 0$ or $1 - \frac{\bar{R}_2}{K}$ or $\bar{R}_2 < k'$.
- The equilibrium point $E_4 = (\bar{R}_1, \bar{C}_{1A}, \bar{C}_{1E}, 0, 0, k')$ corresponding to the case that the second predator species becomes extinct, where $\bar{R}_1 = \frac{(h+m_{1A})(m_{1E}+\alpha_1)}{\alpha_1 a_1 e_1}$, $\bar{C}_{1A} = \frac{r}{a_1}\left(1 - \frac{\bar{R}_1}{k'}\right)$, $\bar{C}_{1E} = \frac{r(h+m_{1A})}{a_1\alpha_1}\left(1 - \frac{\bar{R}_1}{k'}\right)$. Existence of E_4 under conditions: $\bar{C}_{1A} > 0$ or $1 - \frac{\bar{R}_1}{k'}$ or $\bar{R}_1 < k'$.
- The coexist equilibrium point $E_5 = (R^*, C_{1A}^*, C_{1E}^*, C_{2A}^*, C_{2E}^*, k')$ corresponding to the case that all the species co-exist, where

$$\begin{cases} R^* = \dfrac{\dfrac{A_2 b_{12}}{C_2 - D_2} + \dfrac{A_1 b_{21}}{C_1 - D_1} - r}{\dfrac{b_{12}b_{21}e_2}{C_2 - D_2} + \dfrac{b_{12}b_{21}e_1}{C_1 - D_1} - \dfrac{r}{k'}}, \\ C_{1E}^* = \dfrac{A_2 - e_2 b_{21} R^*}{C_2 - D_2}; C_{1A}^* = \dfrac{\alpha_1}{h + m_{1A}} C_{1E}^*, \\ C_{2E}^* = \dfrac{A_1 - e_1 b_{12} R^*}{C_1 - D_1}; C_{2A}^* = \dfrac{\alpha_2}{h + m_{2A}} C_{2E}^*, \end{cases}$$

and

$$\begin{cases} A_1 = m_{1E} + \alpha_1; b_{12} = \dfrac{a_1\alpha_1}{h+m_{1A}}; C_1 = \dfrac{b_{12}e_1\alpha_1}{h+m_{1A}}; D_1 = \dfrac{b_{21}\alpha_2}{h+m_{2A}} \\ = m_{2E} + \alpha_2; b_{21} = \dfrac{a_2\alpha_2}{h+m_{2A}}; C_2 = \dfrac{b_{21}e_2\alpha_2}{h+m_{2A}}; D_2 = \dfrac{b_{12}\alpha_1}{h+m_{1A}}. \end{cases}$$

Existence of E_5 under conditions: $R^* > 0$, $C_{1E}^* > 0$ and $C_{2E}^* > 0$.

3.3 Global Stability Analysis

This section will present the global stability conditions of the system (1) at the equilibrium points. The theorems are built and proven based on the Liapunov's Second Method, which is the construction of the Lyapunov function.

Theorem 3. *The equilibrium point $E_2(k_2, 0, 0, 0, 0, k_2)$ is globally asymptotically stable under following conditions:*

$$\begin{cases} (h + m_{1A})(m_{1E} + \alpha_1) - \alpha_1 a_1 e_1 k_2 & > 0, \\ (h + m_{2A})(m_{2E} + \alpha_2) - \alpha_2 a_2 e_2 k_2 & > 0, \\ k_1 > k_2. \end{cases} \tag{3}$$

Proof. Consider the Lyapunov function:

$$\begin{aligned} V_1(R, C_{1A}, C_{1E}, C_{2A}, C_{2E}, K) = &\left(\frac{R}{k_2} - 1 - \log\frac{R}{k_2}\right) + \frac{a_1}{(h + m_{1A})}C_{1A} + \frac{1}{Ke_1}C_{1E} \\ &+ \frac{a_2}{(h + m_{2A})}C_{2A} + \frac{1}{Ke_2}C_{2E} + \left(\frac{K}{k_2} - 1 - \log\frac{K}{k_2}\right). \end{aligned} \tag{4}$$

It is clear that $V_1 \geq 0$ when $R \geq 0$, $C_{1A} \geq 0$, $C_{1E} \geq 0$, $C_{2A} \geq 0$, $C_{2E} \geq 0$, $K \geq 0$ và $V_1 = 0$ if and only if $R = K = k_2$, $C_{1A} = C_{1E} = C_{2A} = C_{2E} = 0$.
We compute the derivative $\dot{V}_1$ along the solutions of the problem:

$$\begin{aligned} \dot{V}_1 = &\frac{(R - k_2)}{k_2}\frac{\dot{R}}{R} + \frac{a_1}{m_1}\dot{C}_{1A} + \frac{1}{k_2 e_1}\dot{C}_{1E} + \frac{a_2}{m_2}\dot{C}_{2A} + \frac{1}{k_2 e_2}\dot{C}_{2E} + \frac{\dot{K}}{k_2} - \frac{\dot{K}}{K} \\ = &-\frac{r}{k_2^2}(R - k_2)^2 + \frac{b_{21}}{k_2 e_1}(e_1 - 1)C_{1E}C_{2A} + \frac{b_{12}}{k_2 e_2}(e_2 - 1)C_{2E}C_{1A} + \\ &+ \frac{a_1\alpha_1}{(h + m_{1A})k_2}\left[k_2 - \frac{(m_{1E} + \alpha_1)(h + m_{1A})}{a_1 e_1 \alpha_1}\right]C_{1E} \\ &+ \frac{a_2\alpha_2}{(h + m_{2A})k_2}\left[k_2 - \frac{(m_{2E} + \alpha_2)(h + m_{2A})}{a_2 e_2 \alpha_2}\right]C_{2E}. \end{aligned}$$

We can see that $\dot{V}_1 \leq 0$ with condition (3), with equality if only if

$$R = k_2, C_{1A} = 0, C_{1E} = 0, C_{2A} = 0, C_{2E} = 0, K = k_2.$$

While that, the set $M = \{(R, C_{1A}, C_{1E}, C_{2A}, C_{2E}, K) : R = K = k_2\}$ has only one invariant subset $\widehat{M} = (k_2, 0, 0, 0, 0, k_2)$. Applying Lasalle's invariant principle, we can conclude that $(k_2, 0, 0, 0, 0, k_2)$ is globally asymptotically stable on $\mathbb{R}_+^6$.

The global stability condition E_4 is given by the following theorem.

Theorem 4. *The equilibrium point $E_4(\bar{R}_1, \bar{C}_{1A}, \bar{C}_{1E}, 0, 0, k_2)$ is globally asymptotically stable if*

$$\begin{cases} \bar{R}_1 = \dfrac{(h + m_{1A})(m_{1E} + \alpha_1)}{\alpha_1 a_1 e_1} < k_2, \bar{R}_1 < \bar{R}_2 - \dfrac{b_{21}}{e_1 a_2}\bar{C}_{1E}, \\ k_1 > k_2, \underline{R} > k_2, \end{cases} \tag{5}$$

where $\underline{R}$ is the minimum value of R that satisfies $\underline{R} \leq \liminf_{t\to\infty} R(t)$.

Proof. We choose the Lyapunov function as follows:

$$\begin{aligned} V_2(R, C_{1A}, C_{1E}, C_{2A}, C_{2E}, K) &= \bar{R}_1\left(\frac{R}{\bar{R}_1} - 1 - \log\frac{R}{\bar{R}_1}\right) + \\ &+ \frac{\bar{C}_{1A}(m_{1E}+\alpha_1)}{e_1\alpha_1}\left(\frac{C_{1A}}{\bar{C}_{1A}} - 1 - \log\frac{C_{1A}}{\bar{C}_{1A}}\right) + \frac{\bar{C}_{1E}}{e_1}\left(\frac{C_{1E}}{\bar{C}_{1E}} - 1 - \log\frac{C_{1E}}{\bar{C}_{1E}}\right) \\ &+ \frac{m_{2E}+\alpha_2}{e_2\alpha_2}C_{2A} + \frac{1}{e_2}C_{2E} + \left(\frac{K}{k_2} - 1 - \log\frac{K}{k_2}\right). \end{aligned} \tag{6}$$

Obviously, $V_1 \geq 0$ when $R \geq 0$,$C_{1A} \geq 0$,$C_{1E} \geq 0$,$C_{2A} \geq 0$,$C_{2E} \geq 0$,$K \geq 0$ and $V_1 = 0$ if and only if

$$R = \bar{R}_1, C_{1A} = \bar{C}_{1A}, C_{1E} = \bar{C}_{1E}, C_{2A} = 0, C_{2E} = 0, K = k_2.$$

We now compute the derivative of V_2 with respect to t to obtain that:

$$\begin{aligned} \dot{V}_2(R, C_{1A}, C_{1E}, C_{2A}, C_{2E}, K) &= \bar{R}_1\dot{R}\left(\frac{1}{\bar{R}_1} - \frac{1}{R}\right) \\ &+ \frac{\bar{C}_{1A}(m_{1E}+\alpha_1)\dot{C}_{1A}}{e_1\alpha_1}\left(\frac{1}{\bar{C}_{1A}} - \frac{1}{C_{1A}}\right) + \frac{\bar{C}_{1E}\dot{C}_{1E}}{e_1}\left(\frac{1}{\bar{C}_{1E}} - \frac{1}{C_{1E}}\right) \\ &+ \frac{m_{2E}+\alpha_2}{e_2\alpha_2}\dot{C}_{2A} + \frac{\dot{C}_{2E}}{e_2} + \dot{K}\left(\frac{1}{k_2} - \frac{1}{K}\right) \\ &= \left[\left(r - \frac{rR}{K}\right)(R - \bar{R}_1) - \frac{\bar{C}_{1E}(m_{1E}+\alpha_1)}{e_1} + \frac{\bar{C}_{1E}^2(m_{1E}+\alpha_1)^2}{e_1^2 a_1 R\bar{C}_{1A}}\right] \\ &+ a_2\left(\bar{R}_1 + \frac{b_{21}}{e_1 a_2} - \bar{R}_2\right)C_{2A} + \frac{b_{21}}{e_1}(e_1 - 1)C_{1E}C_{2A} \\ &+ \frac{b_{12}}{e_2}(e_2 - 1)C_{1A}C_{2E} - b_{12}\bar{C}_{1E}\frac{C_{2E}C_{1A}}{C_{1E}} - \frac{\alpha(K-k_1)(K-k_2)^2}{k_2(k_2-k_1)K} \\ &- \frac{\bar{C}_{1E}}{e_1}\left[\frac{e_1 a_1 R C_{1A}}{C_{1E}} + \frac{m_{1E}+\alpha_1}{\bar{C}_{1E}}.\frac{C_{1E}}{C_{1A}} + \frac{(m_{1E}+\alpha_1)^2\bar{C}_{1E}}{a_1 e_1 R\bar{C}_{1A}} - 3(m_{1E}+\alpha_1)\right]. \end{aligned}$$

We evaluate each cluster of elements in the formula of $\dot{V}_2$. First, the condition $\underline{R} \leq \liminf_{t\to\infty} R(t)$ can be rewritten as $R(t) > \dfrac{k_2}{2} \quad \forall t \geq t_0$, we can see

$$\begin{aligned} &\left(r - \frac{rR}{K}\right)(R - \bar{R}_1) - \frac{\bar{C}_{1E}(m_{1E}+\alpha_1)}{e_1} + \frac{\bar{C}_{1E}^2(m_{1E}+\alpha_1)^2}{e_1^2 a_1 R\bar{C}_{1A}} \\ \leq\ & r(R - \bar{R}_1)\left[1 - \frac{R}{k_2} - \frac{\bar{R}_1}{R}\left(1 - \frac{\bar{R}_1}{k_2}\right)\right] = \frac{r(R-\bar{R}_1)^2}{R}\frac{k_2 - R - \bar{R}_1}{k_2} \leq 0 \end{aligned}$$

By Cauchy inequality, we have

$$\frac{e_1 a_1 R C_{1A}}{C_{1E}} + \frac{m_{1E}+\alpha_1}{\bar{C}_{1E}}.\frac{C_{1E}}{C_{1A}} + \frac{(m_{1E}+\alpha_1)^2\bar{C}_{1E}}{a_1 e_1 R\bar{C}_{1A}} \geq 3(m_{1E}+\alpha_1),$$

The other clusters that are easily seen are all less than or equal to 0 resulting in $\dot{V}_2 \leq 0$, with equality if and only if $R = \bar{R}_1, \frac{C_{1E}}{C_{1A}} = \bar{C}_{1E} = \frac{(h+m_{1A})}{\alpha_1}, C_{2A} = C_{2E} = 0, K = k_2$. Therefore, the set

$$M = \left\{(R, C_{1A}, C_{1E}, C_{2A}, C_{2E}, K) : R = \bar{R}_1, \frac{C_{1E}}{C_{1A}} = \bar{C}_{1E} = \frac{(h+m_{1A})}{\alpha_1}, C_{2A} = C_{2E} = 0, K = k_2\right\}$$

has a single invariant subset $\widehat{M} = (\bar{R}_1, \bar{C}_{1A}, \bar{C}_{1E}, 0, 0, k_2)$. According to Lasalle's invariant principle, it can be concluded that $E_4(\bar{R}_1, \bar{C}_{1A}, \bar{C}_{1E}, 0, 0, k_2)$ is globally asymptotically stable on $\mathbb{R}^6_+$.

Remark 1. By using similar arguments in Theorem (4), we can prove the globally asymptotically stable E_3 which is as follows:

$$\begin{cases} \bar{R}_2 = \dfrac{(h+m_{2A})(m_{2E}+\alpha_2)}{\alpha_2 a_2 e_2} < k_2, \bar{R}_2 < \bar{R}_1 - \dfrac{b_{12}}{e_2 a_1}\bar{C}_{2E}, \\ k_1 > k_2, \underline{R} > k_2. \end{cases} \tag{7}$$

where $\underline{R}$ is the minimum value of R that satisfies $\underline{R} \leq \liminf_{t\to\infty} R(t)$.

4 Numerical Simulation

In this section, we investigate the effects of the consumption rate of predators and human intervention on our model's behavior through numerical simulations.

4.1 Effects of the Consumption Rate a_1 of WA on Prey

By selecting the suitable parameter values, we present the phase portrait of all equilibrium points in Fig. 1. Then, we aim to investigate the model's behavioral changes as a function of certain factors represented by the parameters.

Consider a_1 as the bifurcation parameter, bifurcation diagrams are shown in Fig. 2. At first, the system undergoes a transcritical bifurcation. In more detail, the system transitions from E_2 to the co-exist stable equilibrium point E_5, and then to the stable equilibrium point E_4. Subsequently, when a_1 is gradually increased towards the threshold 0.355, a Hopf bifurcation occurs, where the appearance of a limit cycle leads to the loss of stability in the system. When a_1 falls within the range $(0.05, 0.25)$, the system attains the optimal equilibrium state E_5, corresponding to the coexistence state of the two predators and their prey. In practice, various aquaculture models have been proposed, such as pond farming, cage culture, and others, which have yielded promising results. These findings may partly help fishermen improve economic efficiency in the context of Wallago Attu farming is increasingly popular, where the need to adjust the feeding regime of this species can contribute to the protection of the ecological environment.

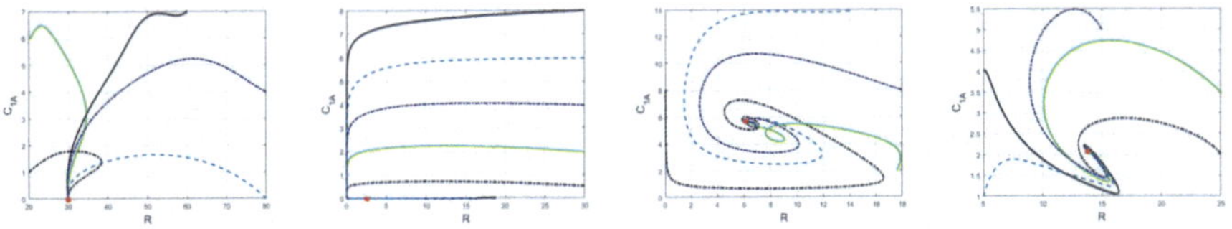

(a) E_2 (b) E_3 (c) E_4 (d) E_5

Fig. 1. The phase portrait of the density of the prey and the species WA in four cases of stable equilibrium points.

4.2 Human Intervention in the System Through Fishing Activities

Human impacts will have a significant influence on the living environment of organisms through fishing activities h. The simulation results are shown in the Fig. 3. Considering the critical threshold $\bar{h}_1 \approx 0.14$, when human interaction is negligible ($h < \bar{h}_1$), the PS species with its better adaptive capacity and predation skills will thrive. The WA species will gradually decline, leading to a rapid decrease in its population and eventual extinction. The system will reach the stable equilibrium state E_3.

Conversely, when human impact on the environment is quite strong, and the catch coefficient h exceeds the threshold $\bar{h}_2 \approx 0.43$, WA will gradually dominate, causing a significant decline in the PS population over time. The system stabilizes at the equilibrium state E_4. In reality, although PS is ecologically stronger, it is more heavily targeted by fishing, leading to a shrinking population, and it is listed in the IUCN Red List.

The system will be destroyed if human influence becomes too severe, with h exceeding the threshold $\bar{h}_3 \approx 0.675$. In this case, both WA and PS species will lose their ability to reproduce and develop, leading to their extinction. Only the prey species, unaffected by predators and humans, will sustainably recover and stabilize at the equilibrium state E_2. It is not uncommon for certain species to become completely extinct due to indiscriminate hunting, and this is when the authorities need to intervene to prevent and mitigate adverse environmental factors.

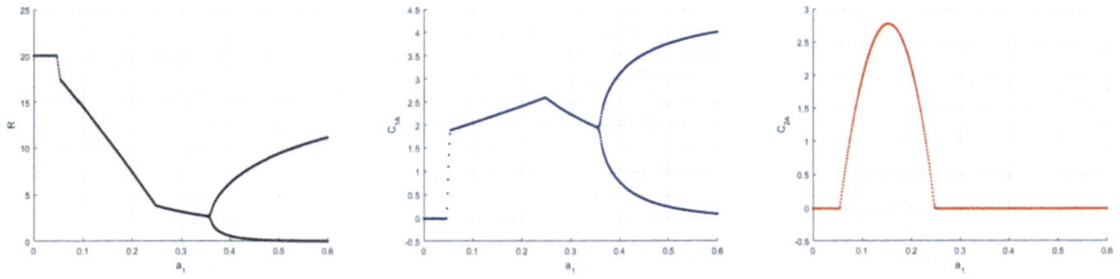

(a) The prey R (b) The adult WA C_{1A} (c) The adult PS C_{2A}

Fig. 2. Bifurcation diagrams of the system with a_1 as the bifurcation parameter.

However, by implementing environmental protection measures and controlling the fishermen's catch h within the range between 14% and 43%, the system can achieve the optimal stable equilibrium state E_5. If human fishing activities are orderly and rational, the environment will be protected, and the economic benefits will be high.

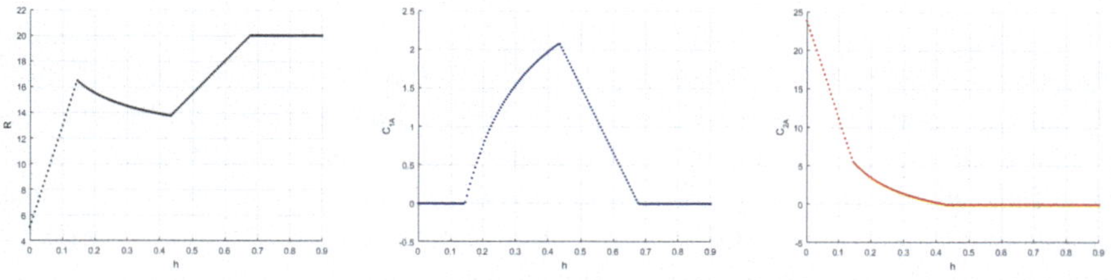

(a) The prey R (b) The adult WA C_{1A} (c) The adult PS C_{2A}

Fig. 3. Bifurcation diagrams of the system with R as the bifurcation parameter.

5 Discussion and Conclusion

Through mathematical analysis and numerical simulations, a mathematical model has established to study the interaction between W. attu, P. Sanitwongsei, and their prey based on the Lotka-Volterra predator-prey model. The system has considered changing environmental carrying capacity and clearly distinguishes between direct and indirect competition between two predator species.

The model analysis has shown the existence of five equilibrium points. Simulation results showed that some of these equilibrium points could be locally or globally stable, depending on the particular conditions. Hopf bifurcation diagrams provided a clear overview of how changes in ecological factors, such as catch coefficient, or internal factors like growth potential for prey species influence the long-term population dynamics of the species. The analysis clarifies the complex interspecies interactions, including direct competition through other's egg predation, as well as indirect competition for shared prey resources within the system. These intricate ecological mechanisms were effectively captured by the developed mathematical model.

We have highlighted the influence of the bifurcation parameter a_1, which represents the predation rate of the WA species. The stable equilibrium states transition from E_2 to optimal equilibrium point E_5 and then to E_4. Continuing to increase a_1 beyond a critical threshold 0.355, Hopf bifurcation occurs, leading to the loss of stability. Fishermen can potentially adjust the parameter a_1 within the range of 5% to 25% through feeding practices, allowing the system to stabilize at the equilibrium state E_5 where all species coexist. These findings may, to some extent, help enhance the economic efficiency of WA fish farming, which is becoming increasingly prevalent.

We also highlight the system's behavior under varying levels of human impact, represented by the parameter h. When the fishing pressure is negligible ($h < \bar{h}_1 \approx 0.14$), the PS species, with its better adaptability and foraging abilities, will predominate, leading to a rapid decline in the WA population and ultimately its extinction. The equilibrium state E_3 is stable, while the others are not. Conversely, when the human impact is quite strong ($h > \bar{h}_2 \approx 0.43$), the PS will become extinct, and the WA species will prevail. This is reasonable, as the PS is currently listed as a critically endangered species. Furthermore, if human influence becomes excessively strong ($h > \bar{h}_2 \approx 0.675$), both species will disappear entirely, and the system will stabilize at E_2. The optimal equilibrium state E_5, where all species coexist stably over time, can be achieved if the harvesting rate h falls within the range of 14% to 43%. This suggests that fisheries management, with clear plans and minimal environmental degradation, is crucial for maintaining a stable balance within the system.

References

1. Al-Moqbali, M.K., Al-Salti, N.S., Elmojtaba, I.M.: Prey-predator models with variable carrying capacity. Mathematics **6**(6), 102 (2018)
2. Al-Salti, N., Al-Musalhi, F., Gandhi, V., Al-Moqbali, M., Elmojtaba, I.: Dynamical analysis of a prey-predator model incorporating a prey refuge with variable carrying capacity. Ecol. Complex. **45**, 100888 (2021). https://doi.org/10.1016/j.ecocom.2020.100888, https://www.sciencedirect.com/science/article/abs/pii/S1476945X20301689
3. Brauer, F., Castillo-Chávez, C.: Mathematical models in population biology and epidemiology, vol. 40, January 2001. https://doi.org/10.1007/978-1-4614-1686-9
4. van der hammen, T., de Roos, A., Sabelis, M., Janssen, A.: Order of invasion affects the spatial distribution of a reciprocal intraguild predator. Oecologia **163**, 79–89 (2010). https://doi.org/10.1007/s00442-010-1575-7
5. IUCN: Pangasius sanitwongsei: Jenkins, a., kullander, F.F., tan, H.H.: March 2007. https://dx.doi.org/10.2305/IUCN.UK.2009-2.RLTS.T15945A5324983.en, title of the publication associated with this dataset: IUCN Red List of Threatened Species
6. IUCN: Wallago attu: Ng, H.H., de alwis goonatilake, s., fernado, m. & kotagama, O.: August 2019, https://dx.doi.org/10.2305/IUCN.UK.2019-3.RLTS.T166468A174784999.en, title of the publication associated with this dataset: IUCN Red List of Threatened Species
7. Lotka, A.J.: The growth of mixed populations: two species competing for a common food supply. J. Washington Acad Sci. **22**(16/17), 461–469 (1932). http://www.jstor.org/stable/24530449
8. Nagumo, N.: Ber die lage der integralkurven gewhnlicher differentialgleichungen. Proc. Phys. Math. Soc. **24**, 551–559 (1942). https://doi.org/10.11429/ppmsj1919.24.0_551
9. Polis, G.A., Myers, C.A., Holt, R.D.: The ecology and evolution of intraguild predation: potential competitors that eat each other. Ann. Rev. Ecol. Evol. System. **20**, 297–330 (1989). https://api.semanticscholar.org/CorpusID:85706047
10. Poulsen, A.F., et al.: Distribution and ecology of some important riverine fish species of the mekong river basin (2004). https://api.semanticscholar.org/CorpusID:247787625

11. Rayungsari, M., Suryanto, A., Muharini, W., Darti, I.: Dynamics analysis of a predator-prey fractional-order model incorporating predator cannibalism and refuge. Front. Appl. Math. Stat. **9** (2023). https://doi.org/10.3389/fams.2023.1122330
12. Wang, M.: Stability and hopf bifurcation for a prey-predator model with prey-stage structure and diffusion. Math. Biosci. **212**(2), 149–160 (2008)

A Genetic Algorithm Approach for Three Level Supply Chain Design

Ta Anh Son[1(✉)] and Tran Thi Thuy An[2]

[1] Faculty of Mathematics and Informatics, Hanoi University of Science and Technology (HUST), Dai Co Viet, Hanoi, Vietnam
son.taanh@hust.edu.vn

[2] Université Clermont-Auvergne, CNRS, Mines de Saint-Etienne, Clermont Auvergne INP, LIMOS, Clermont-Ferrand, France

Abstract. Supply chain problems can have a significant impact on a company's profitability and customer satisfaction. Addressing these challenges often involves a combination of strategic planning, process optimization, and the use of advanced technologies to streamline operations and enhance supply chain resilience. A good supply chain design provides an optimal platform for the efficient supply chain management. This paper analyzes the three-level supply chain problem: goods are transported from suppliers to distribution centers, where goods are repackaged and shipped to customer zones. This problem is formulated in the form of the quadratic binary programming. In this work, we introduce a genetic algorithm to find the optimal solution and demonstrate the effectiveness of our proposed algorithm through a set of different scale problems and compare its results with those obtained by CPLEX.

Keywords: Supply Chain Management · Genetic Algorithm · Meta-heuristics

1 Introduction

Supply chain design provides an optimal platform that ensures the supply of goods to customers. Supply chain design is challenging because supply networks usually have complex structures. The goods from suppliers are transported through many warehouses and agents to the customer. The cost often depends on many factors such as transportation distance, land rental, construction, labor, taxes, etc. A real-world supply chain design problem is usually a large-scale optimization problem with multiple integer variables and many constraints. In this paper, we investigate the three-level supply chain problem: goods provided by suppliers are transported to distribution centers (DCs), then the distribution centers repackage them for the customer zone. The goal was to build DCs that can provide for all customer zones and minimize costs. In [1], the authors present a binary quadratic optimization model for this problem. However, the binary quadratic model cannot be solved in a reasonable time, so the author

© The Author(s), under exclusive license to Springer Nature Switzerland AG 2025
H. A. Le Thi et al. (Eds.): ICAMCS 2024, LNNS 1311, pp. 212–222, 2025.
https://doi.org/10.1007/978-3-031-90606-0_18

has produced a series of model simplifications with linearization, relaxing, and clustering. Clustering and relaxation help to reduce the number of variables in the problem, which reduces the time to solve the problem but can result in a violation of the constraint that each customer zone can only receive from one DC. Therefore, we focus on genetic algorithms - a meta-heuristic approach that proved efficient for many NP problems. Genetic algorithms have been applied to many supply chain optimization problems such as demand prediction, supplier selection, and transportation problems (see [2]). This paper presents a genetic algorithm for the three-level supply chain design problem and compares it with the models given by [1].

2 Mathematics Model

2.1 Three-Level Supply Chain Design

A company has a three-level supply chain: goods are transported from suppliers to distribution centers, where goods are repackaged and shipped to customer zones. The goal was to determine the sizes and locations to build DCs that satisfy to provide all customer zones and also minimize the cost. Only one DC serves each customer zone. To keep it simple, we assume each good comes from only one supplier, but the supplier can have multiple plants. The nearest plants of each supplier will provide DCs. Assume that goods shipped from DCs are in the form of standardized outbound pallets. The demands of each customer zone and the size of DC are presented in terms of pallets per week.

As the problem was the same as in [1], we reused their notations. We index a set of customer zones by

$$r \in R = \{1, 2, \cdots, m\}$$

the set of potential DC locations by

$$d \in D = \{1, 2, \cdots, n\}$$

and the set of DC capacity levels by

$$h \in H = \{1, 2, \cdots \mu\}.$$

For all $d \in D, h \in H$ and $r \in R$, let

$$x_d^h = \begin{cases} 1 & \text{if a DC is built at location d with the capacity level h,} \\ 0 & \text{otherwise,} \end{cases}$$

$$y_{dr} = \begin{cases} 1 & \text{if customer zone r receives goods from DC at location d,} \\ 0 & \text{otherwise,} \end{cases}$$

then $x = \{x_d^h : d \in D, h \in H\}$ and $y = \{y_{dr} : d \in D, r \in R\}$ are the binary variables. The problem has the following constraints:

- Each DC must be assigned a capacity level

$$\sum_{h \in H} x_d^h \leq 1 \quad \forall d \in D.$$

- Every customer zone must be assigned to precisely one distribution center, so we have the following constraints

$$\sum_{d \in D} y_{dr} = 1 \quad \forall r \in R.$$

- Let p_r be the demand from customer zone r and c^h be the capacity of a DC with capacity index h. To ensure each DC could supply for all the customer zones it was assigned, we have the following constraints

$$\sum_{r \in R} p_r y_{dr} \leq \sum_{h \in H} c^h x_d^h \quad \forall d \in D.$$

The cost to be minimized consists of fixed costs, variable costs, and transportation costs. Fixed costs consist of building costs that do not depend on location and land purchasing depends on location. Let b^h be the building fixed costs, l_d^h be the land purchasing for a distribution center to be built at location d with capacity c^h. The total fixed cost is calculated as follows

$$C_1(x) = \sum_{h \in H} b^h \sum_{d \in D} x_d^h + \sum_{h \in H} \sum_{d \in D} l_d^h x_d^h. \tag{1}$$

Variable costs include all the costs that depend on the number of pallets. Due to economies of scale, DCs with larger capacities have cheaper variable costs. Let v^h be the variable costs of operating a DC with capacity c^h. The total variable costs are formulated by the quadratic function as

$$C_2(x, y) = \sum_{h \in H} v^h \sum_{d \in D} x_d^h \sum_{r \in R} p_r y_{dr}. \tag{2}$$

Transportation costs include outbound transportation from DCs to customer zones and inbound from suppliers to DCs. Let k_{dr} be the distance from the DC location d to the customer zone r and ω be the cost of transporting one pallet per kilometer. The total outbound transportation costs are computed as below

$$C_3(y) = \omega \sum_{d \in D} \sum_{r \in R} p_r k_{dr} y_{dr}. \tag{3}$$

Let ω_s be the cost to transport one pallet per kilometer from supplier s to any DC and $\check{k}_{sd}$ be the distance from DC location d to the nearest location of supplier s. Given the varying in-pallets from different suppliers, we can establish a conversion factor under the assumption that the out-bound pallets are consistent in both size and content and ρ_s be the percentage of goods from supplier s on any out-bound pallet. All suppliers' combined inbound transportation costs can be expressed as

$$C_4(y) = \sum_{s\in S}\sum_{d\in D}\sum_{r\in R} \omega_s \rho_s \check{k}_{sd} p_r y_{dr}. \tag{4}$$

Finally, the objective function is the total of (1), (2), (3) and (4)

$$C(x, y) = C_1(x) + C_2(x, y) + C_3(y) + C_4(y).$$

Notice that C_1, C_3, and C_4 are linear, but C_2 is quadratic, so C is quadratic; we have the Binary Quadratic Program (BQP)

BQP

$$\begin{aligned}
\text{Min} \quad & C(x,y) = C_1(x) + C_2(x,y) + C_3(y) + C_4(y) & \\
\text{s.t} \quad & \sum_{h\in H} x_d^h \le 1 & \forall d \in D, \\
& \sum_{d\in D} y_{dr} = 1 & \forall r \in R, \\
& \sum_{r\in R} p_r y_{dr} \le \sum_{h\in H} c^h x_d^h & \forall d \in D, \\
& x_d^h \in \{0,1\} & \forall d \in D, \forall h \in H, \\
& y_{dr} \in \{0,1\} & \forall d \in D, \forall r \in R.
\end{aligned}$$

2.2 Model Simplifications

In this section, we recap the model simplifications in [1] as we will re-implement their models BQP and BLP (Binary Linear Program) to compare the results with our method. The authors of [1] combine three approaches to simplify the model: clustering, variable relaxation, and linearization. The first is clustering the customer zones and relaxing variable y. The set of customer zone R is replaced by the set of clusters and the binary variables y_{dr} were replaced by real-valued variables $y_{dr} \ge 0$ we got the CMBQP (Cluster Mixed Binary Quadratic Program). The CMBQP has fewer variables than the BQP. By assigning to each DC a variable cost based on capacity rather than output by setting they get the BLP. The BLP and BQP is identical except that $C_{2L}(x)$ replaces C_2:

$$C_{2L}(x) = \sum_{h\in H} v^h c^h \sum_{d\in D} x_d^h.$$

The BLP model has the same amount of variables and constraints as the BQP but is more accessible to solve because of the linearity of the cost function. Finally, combining all the simplifications, they got the CMBLP (Cluster Mixed Binary Linear Program), which has linearity and is less variable than the BQP. Unfortunately, in their experiment, the solver failed to solve the BQP model due to the large scale of the problem. However, the BLP gave the best objective compared to other models.

3 Methodology

A genetic algorithm (GA) is a nature-inspired or meta-heuristic algorithm. Based on Charles Darwin's theory, genetic algorithms simulate the natural evolution of animals. With the ability to deal with a wide range of optimization problems, GA has become a conventional and classic method. GA is a gradient-free method, so it can be applied efficiently to solve combinatorial optimization problems where the variables are integers and optimization problems where the objective functions are discontinuous. There are many variants of the genetic algorithm, but most of them follow the "standard" scheme. Firstly, the solutions are encoded into arrays of bits or character strings, i.e., the *chromosomes* (see [3]). This representation is significant; an inappropriate representation may lead to producing invalid offspring, making the algorithm hard to converge. Then, a fitness function must be defined to evaluate the individuals. The simplest way to construct a fitness function is to take the objective function. Now that the problem is formulated, we need to create a population of chromosomes and evaluate them with the fitness function. This initialization can be random or based on some heuristics. Then we repeat the following evolution steps:

1. Crossover: swapping parts of the solutions to produce offspring. The main role is to provide mixing of the solutions and convergence in a subspace.
2. Mutation: changing parts of a solution randomly, mutation provides a mechanism for escaping a local optimum. Usually, the mutation is applied with a small probability, a large mutation rate can prevent the algorithm's convergence.
3. Selection: selecting the individuals to create a new generation. This step should ensure the best individual is passed to the next generation and keep the diversity of the population. Without diversity, the algorithm is prone to premature convergent i.e. converging to a local optimal (see [4] for more detail about the benefits of population diversity in GA).

The formulation of the fitness function, the population size, the choice of the important parameters such as the rate of mutation and crossover, and the selection criteria of new populations are important to the performance of GA. Any inappropriate choice will make it difficult for the algorithm to converge, or it simply produces meaningless results.

The complete algorithm proposed in this paper is described in Algorithm 1.

3.1 Chromosome Representation

Each solution is represented as an array of m integers

$$d_1 d_2 \cdots d_r \cdots d_m$$

where m is the number of customer zones, $d_r \in D$ represents the DC indexes assigned to serve customer zone r. We can easily verify that this encoding is 1–1.

Algorithm 1. GA

Input: Three-level supply chain design problem, $T \in \mathbb{N}$ is the number of generation, $r \in [0, 1]$ is the mutation rate
Output: A solution to the problem
Population initialization
Population evaluation
for $k = 0, \cdots, T$ **do**
  Crossover
  Mutation with rate r
  Select individual for the next generation
end for
Return: Best individual in the last generation.

To decode, we calculate the number of customer zones assigned to each potential DC location and assign the smallest size that satisfies the total demand of its customer zones. DCs that are not assigned any customer zone would not be built.

3.2 Fitness Function

In this paper, we simply use the objective function as the fitness function. As we want to minimize the cost, the smaller the fitness value the better.

3.3 Crossover

We use multi-point crossover (see [5]). First, we pick the two chromosomes as the parents, then exchange some genes of those chromosomes randomly. Chromosomes with bad fitness can produce good fitness chromosomes, so we randomly pick the parent uniformly. For example:
Parent 1: 21|**1**|32|**13**|2221|**1**
Parent 2: 31|**2**|21|**31**|3121|**1**
Offspring 1: 21**2**32**31**2221**1**
Offspring 2: 31**1**21**13**3121**1**.

3.4 Mutation

We perform mutation by randomly picking a chromosome, then picking a random gene in the selected chromosome and changing it to a valid random gene. For example:
Parent: 21231**2**221
Offspring: 21231**1**221.

3.5 Selection

To keep the population diverse, we replace both parents with their offspring if one of the offspring has better fitness than both parents. This way, we also keep the best solution for the next generation.

3.6 Population Initialization

To help the genetic algorithm converge sooner, we propose an algorithm to generate a good population. Our algorithm is based on the idea that in the optimal solution, all customer zones will be assigned to the DC near the customer zone, near supplier plants, or located at a cheap location. We use the following function to approximate the appropriation of DC d to customer zone r:

$$g_r(d) = p_r(\omega k_{dr} y_{dr} + \sum_{s \in S} \omega_s \rho_s \check{k}_{sd} y_{dr} + l_d^0)$$

$$f_r(d) = e^{-\frac{g_r(d)}{T}},$$

where T is a hyper-parameter. $g_r(d)$ is the sum of three costs according to the demand of customer zone r: transportation cost from suppliers to DC d, transportation cost from DC d to customer zone r, and the rental cost at location d. $g_r(d)$ can be seen as an approximation of the contribution of the pair (d, r) to the cost function. As we want to minimize the cost function, the larger this approximation, the lower the probability customer zone r be assigned to DC d. We use the exponential distribution for these probabilities so that in most of the solutions, customer zone r is assigned to the best DC. To demonstrate the

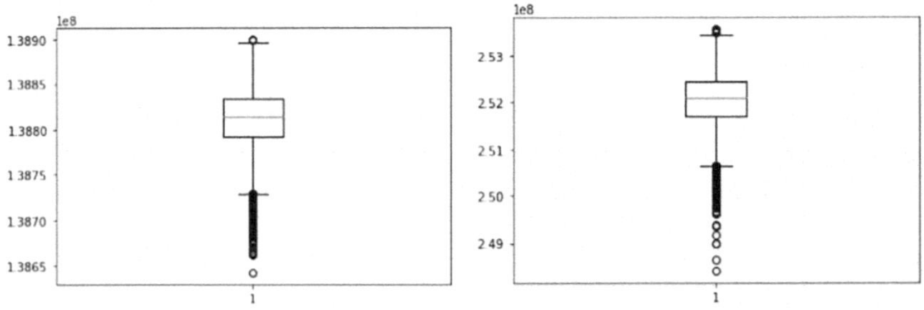

Fig. 1. Comparing the best fitness of the proposed algorithm (left) to random initialization (right)

effectiveness of the proposed algorithm, we run the algorithm 10000 times. We then compare the best fitness values, and the average fitness values of the result populations to the populations generated randomly. Figure 1 and 2 show that our algorithm generates so far better than random initialization.

4 Experiments

To analyze the effectiveness of the proposed algorithm, we generate a series of datasets with different generating strategies and sizes. DC size is selected in [500, 1000, 2000, 5000]. Building costs, renting costs, and variable costs are generated

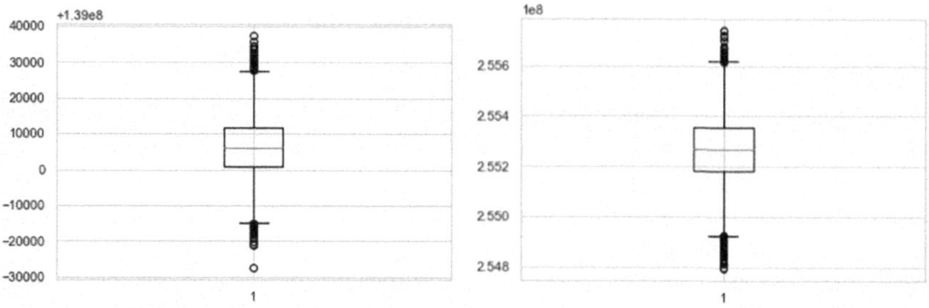

Fig. 2. Comparing the average fitness of the proposed algorithm (left) to random initialization (right)

Table 1. Generating strategies

Strategy	Building cost	Renting cost	Variable cost
A	Strategy 1	Strategy 1	Strategy 1
B	Strategy 1	Strategy 1	Strategy 2
C	Strategy 1	Strategy 2	Strategy 2
D	Strategy 2	Strategy 2	Strategy 2

Table 2. Dataset sizes

Dataset size	Number of suppliers	Number of customer zones	Number of potential DCs	Number of variables	Number of constraints
Size 1	10	800	20	16 080	840
Size 2	15	1000	30	30 120	1060
Size 3	20	1400	40	56 160	1480
Size 4	30	1800	50	90 200	1900
Size 5	40	2300	60	138 240	2420
Size 6	47	2976	83	247 340	3142
Size 7	200	4000	100	400 400	4200
Size 8	200	4800	120	576 480	5040
Size 9	200	6400	160	1 024 640	6720

by two strategies:
Strategy 1: ratio to the size of the DC
Strategy 2: the larger the size of the DC, the cheaper the cost
Combine the strategies in Table 1 with the sizes in 2 we have datasets with different generating strategies and different sizes

Table 3. Experimental results

Data	BQP		BLP			GA		
	Time (min)	Objective	Time (min)	Objective	GAP (%)	Time (min)	Objective	GAP (%)
DataA1	1	**3877034**	**0.06**	3889392	0.319	24	3878943	0.049
DataA2	60	**4614427**	**0.13**	4635526	0.457	24	4618680	0.092
DataA3	60	**6213473**	**0.70**	6235493	0.354	24	6220154	0.107
DataA4	60	**7464768**	**0.65**	7498782	0.456	24	7477132	0.165
DataA5	60	**11790684**	**3**	11814607	0.203	24	11806383	0.133
DataA6	65	23416194	**10**	12296614	0.157	24	**12277382**	0
DataA7	NA	NA	57	17421262	0.044	**24**	**17413614**	0
DataA8	NA	NA	69	21472521	0.077	**24**	**21455925**	0
DataA9	NA	NA	78	27504824	0.104	**24**	**27476170**	0
DataB1	0.42	**3042309**	**0.05**	3055171	0.423	24	3043494	0.039
DataB2	25	**4643796**	**0.09**	4659978	0.348	24	4646686	0.062
DataB3	60	**6804983**	**0.39**	6823589	0.273	24	6814050	0.133
DataB4	60	**7717709**	**0.91**	7736114	0.238	24	7728745	0.143
DataB5	60	**9567551**	**3.38**	9592398	0.26	24	9570628	0.032
DataB6	NA	NA	43	13456680	0.108	**24**	**13442190**	0
DataB7	NA	NA	**23**	18016453	0.123	24	**17994297**	0
DataC1	0.95	**3528184**	**0.06**	3540869	0.36	24	3530466	0.064
DataC2	175	**4470961**	**0.14**	4489823	0.422	24	4477091	0.137
DataC3	70	**6244214**	**0.45**	6263529	0.309	24	6253678	0.151
DataC4	65	**9298675**	**0.73**	9311434	0.137	24	9301551	0.031
DataC5	60	**10284309**	**18**	10312853	0.278	24	10297287	0.126
DataC6	76	25703458	**9**	12772859	0.013	24	**12771252**	0
DataC7	NA	NA	68	17187078	0.116	**24**	**17167166**	0
DataC8	NA	NA	70	21723099	0.028	**24**	**21717040**	0
DataC9	NA	NA	80	27146129	0.154	**24**	**27104273**	0
DataD1	0.5	**3898440**	**0.05**	3911978	0.347	24	3900101	0.042
DataD2	72	**4762217**	**0.13**	4768654	0.135	24	4766941	0.099
DataD3	68	**5971039**	**0.38**	5990804	0.331	24	5971932	0.015
DataD4	68	**7514807**	**0.70**	7540005	0.335	24	7525849	0.146
DataD5	60	**11035884**	**18**	11049677	0.125	24	11037136	0.011
DataD6	91	25540497	**9.4**	12772859	5.747	24	**12078716**	0
DataD7	NA	NA	67	17272447	0.073	**24**	**17259929**	0
DataD8	NA	NA	69	21184675	0.070	**24**	**21169837**	0
DataD9	NA	NA	77	26096077	0.189	**24**	**26046729**	0
				Max GAP(%)	5.747		Max GAP(%)	0.165
				Min GAP(%)	0.013		Min GAP(%)	0.011
				Average GAP(%)	0.393		Average GAP(%)	0.089

We implemented the proposed algorithm in a single thread with Python, fixed the number of generations $T = 1000000$, mutation rate $r = 0.01$, and population size $N = 100$. For the BQP and the BLP model, we let CPLEX utilize all resources on a ThinkCentre server with 32 Gb of RAM and 12 CPUs

3.2 GHz and set a time limit of 1 h. The results are shown in Table 3. The GAPs are calculated as follows:

$$GAP = \frac{\text{Objective value of the algorithm} - \text{The best objective value}}{\text{The best objective value}} \cdot 100\%$$

The BQP and BLP have solving time increase rapidly with the scale of the problem. BQP failed to solve large-scale problems that we marked "NA" in the BQP column. It took the genetic algorithm only 24 min to run 1,000,000 generations to all scales. In all the experiments genetic algorithm gave better solutions than the BLP model. Compared to the BQP model, the genetic algorithm has a GAP of 0.089%, while the GAP of BLP is 0.306%. For large-scale problems, GA is definitely the best choice. For other scale problems, we must note that even when the genetic algorithm has a longer running time, it consumes much less computational resources. As the genetic algorithm was implemented single-threaded, it runs on only one CPU. In contrast, CPLEX utilizes all the CPU and RAM on the servers. Then the genetic algorithm is also a good choice when computational resources are limited.

5 Conclusion

In conclusion, the three-level supply chain design problem presents a significant challenge, particularly in large-scale applications, akin to quadratic binary programming. Through our research, we introduced a genetic algorithm as an effective approach to address this complex issue. The experimental outcomes substantiate the efficacy of our method, demonstrating both stable running times and consistently favorable results. This contribution not only advances the field of supply chain optimization but also underscores the practical applicability of genetic algorithms in real-world logistics scenarios. As we continue to refine and expand upon these methodologies, we pave the way for more efficient and robust supply chain designs that can adapt to the dynamic demands of today's globalized markets.

References

1. Ashtab, S., Caron, R., Selvarajah, E.: A binary quadratic optimization model for three level supply chain design, vol. 17 (2014). https://doi.org/10.1016/j.procir.2014.01.121
2. Min, H.: Genetic algorithm for supply chain modeling: Basic concepts and applications. Int. J. Serv. Oper. Manage. **22**, 143–163 (2015). https://doi.org/10.10007/12345678
3. Yang, X.-S.: Chapter 5 - genetic algorithms. In: Yang, X.-S.(ed.) Nature-Inspired Optimization Algorithms, pp. 77–87. Elsevier, Oxford (2014). https://doi.org/10.1016/B978-0-12-416743-8.00005-1, https://www.sciencedirect.com/science/article/pii/B9780124167438000051

4. Sudholt, D.: The benefits of population diversity in evolutionary algorithms: a survey of rigorous runtime analyses, pp. 359–404 (2020). https://doi.org/10.1007/978-3-030-29414-4_8
5. Kora, P., Yadlapalli, P.: Crossover operators in genetic algorithms: a review. Int. J. Comput. Appl. **162**, 34–36 (2017). https://doi.org/10.5120/ijca2017913370

A New Mathematical Model of the Exam Timetabling Problem at HUST

Ta Anh Son(✉) and Thi Ngan Nguyen

Faculty of Mathematics and Informatics, Hanoi University of Science and Technology, Hanoi, Vietnam
{son.taanh,ngan.nt211310m}@sis.hust.edu.vn

Abstract. Designing an exam schedule for a university is a complex task that poses significant challenges to the person in charge. At Hanoi University of Science and Technology (HUST), this task has become a tedious and time-consuming process due to the various constraints and a large volume of university data involved. To address this, a 0–1 integer programming model is proposed in this paper, which models two sub-problems: (a) the Exam Class Scheduling (ECS), involving the allocation of exams to specific periods and rooms, and (b) the Proctor Exam Assignment (PEA), which deals with assigning proctors to exams scheduled from ECS. The objective of this model is to minimize the number of exam rooms and proctors used. Using the Cplex optimizer 12.0 on a personal computer, we successfully solved the case of the Faculty of Mathematics and Informatics (FAMI) in HUST with moderate-sized data sets. Given that result, it is seen that the proposed model can produce a high-quality exam timetable in an acceptable amount of time.

Keywords: Exam timetabling · Mixed Integer programming · University timetabling · Cplex

1 Introduction

1.1 Exam Timetabling in General

The problem of scheduling exams involves assigning a set of exams to specific timeslots, rooms, and proctors while adhering to a variety of constraints, including both *hard constraints* and *soft constraints*. Some exam scheduling rules and their corresponding constraints are presented below:

1. **Conflict-free schedule:** When scheduling exams, conflicts arise when two or more exams are assigned to the same room, at the same time, and administered by the same proctor. Conflicts also occur when students are scheduled to take multiple exams simultaneously. Therefore, constraints to address these issues must be considered, including:
   + At a time, each student takes at most one exam,
   + At a time, each proctor oversees at most one exam class,
   + At a time, each room can only be used at most once.

© The Author(s), under exclusive license to Springer Nature Switzerland AG 2025
H. A. Le Thi et al. (Eds.): ICAMCS 2024, LNNS 1311, pp. 223–234, 2025.
https://doi.org/10.1007/978-3-031-90606-0_19

2. **Exam schedule must be complete**
   A schedule is considered complete when all exams for each group of students must appear on the schedule.
3. **The exam schedule must ensure the appropriate exam time**
   The scheduling of exams must guarantee that classes with the same exam code (same exam questions) are conducted simultaneously to prevent question leakage. Furthermore, exams of the same subject should be held on the same day.

1.2 Exam Timetabling Problem Is an NP-Complete Problem

The exam timetabling problem is known to be a complex problem that falls under the NP-complete category [2]. As the number of students and the complexity of teaching programs increase, finding reasonable solutions becomes increasingly hard, and the computation time grows exponentially.

1.3 State-of-the-Art Approaches for Solving Exam Timetabling

The exam scheduling problem is not a new problem. In fact, numerous approaches and strategies have been proposed for solving this problem, such as mathematical modeling, the simulated annealing algorithm, genetic algorithm,... Specifically, in 2010, Al-Yakoob et al. [1] introduced two distinct mixed integer programming models to address exam scheduling and proctor (invigilator) assignment challenges. Dener M. et al. (2018) [7] used Generic Algorithm to solve the problem with two steps: assigning courses to sessions and assigning students who are in session to classes. In 2019, Nuno L. et al. [6] proposed a new metaheuristic algorithm using a graph coloring heuristic for solution initialization and the SA metaheuristic for solution optimization. In 2023, Bazari S. et al. [3] proposed integer linear programming with zero-one variables to solve exam scheduling at Ferdowsi University of Mashhad with new constraints on the conflicts of professor's exams and the concurrence of exams.

1.4 Some Exam Scheduling Software in the World

Exam scheduling software is used in educational institutions around the world to manage and organize exams. Some of the popular tools include ExamSoft, CollegeNET, Timetabler, ... However, these tools may not be able to handle complex scheduling constraints; as a result, their ability to tailor the software to specific needs is limited.

1.5 Exam Timetabling at HUST, Our Approach, and Our Contribution

At HUST, the exam schedule is a painstaking and time-consuming task that happens every semester due to large data and complicated requirements, people in charge only use the Excel tool and their experience to build a suitable

schedule. Therefore, finding a way to automate exam scheduling has been of great importance in facilitating operational activities.
Our main contribution is proposing a new and suitable mathematical model to solve the exam scheduling problem at HUST. Below are some features of the exam scheduling problem at HUST:

- **Exam:** Each subject may have multiple exams corresponding to different exam question sets. For instance, *Calculus I* has four exams, each tailored to specific groups of students. In addition, many groups of students may take the same exam, necessitating constraints to be in place to prevent the leakage of exam questions.
- **Facilities:** HUST possesses various rooms that are tailored to meet the specific requirements of exams and have sizes that fit a particular number of students.
- **Students:** Students may be included in multiple groups depending on their class registration and will be required to take a specific set of exams for each group they are part of.
- **Proctors:** Certain specialized exams may necessitate a qualified proctor to oversee the testing process. To ensure efficient scheduling, proctors can indicate their availability by specifying their free/busy schedule.

At HUST, the scheduling task is particularly complex due to the huge volume of data and numerous operational rules and requirements. Therefore, the exam timetabling problem is split into two sub-problems:

- The Exam Class Scheduling (ECS), involving the allocation of exams to specific time slots and rooms to create exam classes.
- The Proctor Exam Assignment (PEA), deals with assigning proctors to these exam classes.

By this division, the scheduling task can be solved efficiently when dealing with large and complex data sets. Building a model based on this approach, it is possible to significantly reduce the size of the problem as well as the computation time.

1.6 About Cplex Optimizer

The CPLEX Optimizer, created by IBM, offers efficient and adaptable mathematical programming solvers for various optimization problems including linear programming, mixed-integer programming, quadratic programming, and quadratically constrained programming. This solver is capable of solving large-scale optimization problems in a relatively short amount of time while still delivering accurate solutions. In fact, it has demonstrated success in solving optimization models containing millions of variables and constraints [8].

In this paper, we especially focus on proposing a new and suitable mathematical model to solve the exam scheduling problem at HUST. Section 2 introduces a new binary integer programming model for this problem. Section 3 shows the numerical simulation of our models and some discussion and conclusion are reported in Sect. 4.

2 Proposed Binary Integer Programming Model for the Exam Timetabling Problem at HUST

In our previous paper [10] course timetabling problem at HUST has been successfully solved. Hence, in this paper, we going on using this approach to address the exam scheduling problem at HUST.

2.1 The First Sub-Problem: Exam Class Scheduling (ECS)

Input: Data sets.
Output: The exam timetabling solution that don't mention proctors.

2.1.1 Input Data: Required data sets is listed below

1. Students groups

+ G: All groups of students, including both single groups and combined groups: $G = \{g_0, g_1, g_2, \ldots, g_{|G|}\}$. For example:
G= {'G0': ['CNTT 01'], 'G1': ['CNTT 01', 'CNTT 02'], 'G2': ['CNTT 02'],...}.

+ $G_{combine}$: Combined groups of students, extracted from G. At HUST, a combined student group is understood as a student group created by combining two or more independent student groups. Creating a combined student group to take the exam together will help reduce the number of exam rooms and proctors. Let us set $G_{combine} = \{b_1, b_2, \ldots, b_j, \ldots, b_{|B|}\}$.

For example: $G_{combine}$ = {'G1': ['CNTT 01', 'CNTT 02'],...}.

+ G_{single}: Single groups of students extracted from G. At HUST, each single student group is understood as an independent student class: $G_{single} = \{s_1, s_2, \ldots, s_i, \ldots, s_{|S|}\}$.

For example: G_{single} = {'G0': ['CNTT 01'], 'G2': ['CNTT 02'],...}.

Note that: $G_{combine} \cap G_{single} = \emptyset, G = G_{combine} \cup G_{single}$.

+ $G_{contain}$ is extracted from $G_{combine}$ vá G_{single}, contains all single groups accompanied by the combined groups containing them: $G_{contain} = \{s_i : b_j | s_i \subset b_j\}$.

2. Exams and subjects

E: Set of all the exams, $E = \{e_0, e_1, e_2, \ldots, e_{|E|}\}$, for example:
E={'E0': 'ĐS-01', 'E1': 'ĐS-02', 'E2': 'ĐS-03', 'E3': 'GTI-01', ...}.

+ S_e: Subjects accompanied by their exams, for example:
S_e={'ĐS': ['ĐS-01', 'ĐS-02', 'ĐS-03'], 'GTI': ['GTI-01', 'GTI-02'], ...}.

+ G_e: Groups of students accompanied by the exams that they need to take in the semester, for example:
G_e = {'G0': ['E0', 'E1', 'E2'], 'G1': ['E0', 'E3', 'E4'], ...}.

+ E_g: Exams accompanied by the student groups taking them, for example:
E_g = {'E0': ['G0', 'G1'], 'E1': ['G0'], ...}.

3. Rooms

+ R: set of all the rooms: $R = \{r_1, r_2, \ldots, r_{|R|}\}$,
for example: R = {'D5-101', 'D5-102', 'TC-101', ...}.

+ R_c: Capacity along with rooms, for example:

$R_c = \{184 : \{\text{'D5-101'},\text{'D5-102'},\text{'D5-103'}\}, 200 : \{\text{'TC-101'}\}...\}$.
Here, there are three rooms having 184 seats: 'D5-101','D5-102','D5-103'; and, there is only one room having 200 seats: 'TC-101'.
+ G_c: Student groups along with their number of students,
for example: G_c = {'G0': '32', 'G1': '80', 'G2': 120, ...}.
+ G_r: is extracted from R_c and G_c contains student groups accompanied by all the rooms that can accommodate them (capacity $(G) \leq$ capacity (R)),
for example: G_r={'G0': ['R0','R1', 'R2'], 'G1': ['R2', 'R3', 'R4'], ...}.
4. D: Exam days of the semester, for example: $D = \{0, 1, 2, 3, 4\}$.
5. P: Exam periods in a day. At HUST, there are five exam periods per day, each of them lasts 90 min, for example: $P = \{0, 1, 2, 3, 4, \}$.

2.1.2 Problem Variables: In order to formulate the problem, we introduced four sets of binary variables below
1. Basic variables:
$x_{e,g,p,d,r}$ where $d \in D, p \in P, g \in G, e \in E, r \in R$
+ $x_{e,g,p,d,r} = 1$ when the exam e for student group g is scheduled at period p on day d in the room r,
+ $x_{e,g,p,d,r} = 0$, otherwise.
2. Auxiliary variables:

- $z_{e,p,d}$ in which $e \in E, p \in P, d \in D$
  - $z_{e,p,d} = 1$ when there is at least one group of students taking exam e at period p on the day d,
  - $z_{e,p,d} = 0$ when there isn't any group of students taking exam e at period p on day d.

- $y_{s,d}$ in which $s \in S, d \in D$
  - $y_{s,d} = 1$ when there is at least one exam of subject s is scheduled on the day d,
  - $y_{s,d} = 0$ when there isn't any exam of subject s is scheduled at period p on day d.

- $h_{d,r}$ in which $d \in D, r \in R$
  - $h_{d,r} = 1$ when room r is used in day d,
  - $h_{d,r} = 0$ when room r is not used on day d.

2.1.3 Constraints: The model incorporates the following constraints

1. Each exam taken by a group of students must be scheduled at only one-time slot in only one room that fits the student group (the constraint ensures completeness)

$$\forall g \in G, \forall e \in G_e[g], \sum_{p \in P} \sum_{d \in D} \sum_{r \in G_r[g]} x_{e,g,p,d,r} = 1. \tag{1}$$

2. At a time, each room must be assigned to at most one exam and at most one group of students (at a time, each room must be used at most once)

$$\forall r \in R, \forall d \in D, \forall p \in P, \sum_{e \in E} \sum_{g \in E_g[e]} x_{e,g,p,d,r} \leq 1. \tag{2}$$

3. The constraint ensures a student takes at most one exam at a time (because a student can be included in both single and combined groups)
$\forall d \in D, \forall p \in P, \forall s_i \in G_{single}$,

$$\sum_{e_{s_i} \in E_g[s_i]} \sum_{r_{s_i} \in G_r[s_i]} x_{e_{s_i}, s_i, p, d, r_{s_i}} + \sum_{b_j \in G_{contain}[s_i]} \sum_{e_{b_j} \in E_g[b_j]} \sum_{r_{b_j} \in G_r[b_j]} x_{e_{b_j}, b_j, p, d, r_{b_j}} \leq 1. \tag{3}$$

4. The constraint about the total number of exam classes (ensure completeness and avoid redundancy)

$$\sum_{e \in E} \sum_{g \in G} \sum_{r \in R} \sum_{p \in P} \sum_{d \in D} x_{e,g,p,d,r} = M, \tag{4}$$

in which M is total number of exam classes, M is calculated from set G.
5. All classes of every exam must be scheduled in the same time slot to prevent leakage of the exam questions
$\forall e \in E, \forall p \in P, \forall d \in D$,

$$\sum_{r \in G_r[g]} \sum_{g \in E_g[e]} x_{e,g,p,d,r} = N_e * z_{e,p,d}, \tag{5}$$

in which N_e is the number of exam classes of exam e. For example, $N_e = 10$ when there are 10 student groups taking exam $e =$'GTI-01'. N_e is calculated from set E_g.
6. Exams of the same subject should be scheduled on the same day
$\forall s \in S_e, \forall d \in D$,

$$\sum_{e \in E} \sum_{g \in G} \sum_{r \in R} \sum_{p \in P} x_{e,g,p,d,r} = K_s * y_{s,d}. \tag{6}$$

in which K_s is the number of exam classes of subject s. For example, subject $s=$ 'GTI' has 2 exams including 'GTI-01' and 'GTI-02', in which 'GTI-01' is taken by 10 student groups and 'GTI-02' is taken by 5 student groups; therefore, $K_s = 10 + 5 = 15$. K_s is calculated from set E_g.
7. Each student should not take more than 2 exams in a day to ensure efficient break time for all $s_i \in G_{single}$, for all $d \in D$:

$$\sum_{e_{s_i} \in E_g[s_i]} \sum_{r_{s_i} \in G_r[s_i]} \sum_{p \in P} x_{e_{s_i}, s_i, p, d, r_{s,j}} + \sum_{b_j \in G_{\text{contain}}[s_i]} \sum_{e_{b_j} \in E_g[b_j]} \sum_{r_{b_j} \in G_r[b_j]} \sum_{p \in P} x_{e_{b_j}, b_j, p, d, r_{b_j}} \leq 2 \tag{7}$$

8. The constraint to define variables $h_{d,r}$:

$$h_{d,r} \leq \sum_{p \in P} \sum_{g \in G} \sum_{e \in E} x_{e,g,p,d,r} \leq P * h_{d,r}, \quad (8)$$

where P is the total number of periods in a day (for example P=5).

2.1.4 Objective Function At HUST, in addition to the purpose of organizing exams, rooms can be used for other purposes such as renting rooms to outside units, lending rooms to students or teachers, etc. with the hope that the rooms reserved for these secondary purposes do not change much between days for ease of registration. Therefore, minimizing the total number of fixed. The proposed objective function is to minimize the number of rooms used in the exam period as follows:

$$\min \sum_{d \in D} \sum_{r \in R} h_{d,r}. \quad (9)$$

2.2 The Second Sub-Problem: Proctor Exam Assignment (PEA)

Input: Data sets.
Output: The complete exam timetabling solution with assigned proctors.

2.2.1 Input Data: Required data sets are listed below
1. T: All timeslots that classes are assigned to. This set is extracted from the output of **ECS**, $T = \{dp\}$ where dp is the pth period on day d, for example: $T = \{00, 01, 02, 03, \dots\}$.
2. C: All exam classes got from the output of the **ECS**, $C = \{c_0, c_1, c_2, \dots, c_{|E|}\}$. For example: C = {'C0':['E1', 'G2', '01'],...}, it presents for Class 'C0': Exam 'E1' taken by student group 'G2' is scheduled at the second period of the first exam day in the semester.
3. L: All proctors, $L = \{l_1, l_2, \dots, l_{|L|}\}$.
For example: L = {L1:'Nguyen Xuan Truong', L2:'Nguyen Canh Nam', ...}.
4. E_l: Exams accompanied by proctors l that are qualified.
For example: E_l = {'E0': ['L0', 'L1], 'E1': ['L2'], ...}.
C_l: All exam classes along with proctors l that are qualified. This set is extracted from E_l and C, for example: C_l = {'C0': ['L0', 'L1], 'C1': ['L2'], ...}.
5. T_c: All timeslots in T accompanied by all the classes that are assigned to each timeslot, for example T_c = {'00': ['C1','C2','C3'], '10': ['C0','C4'], ...}.
6. L_{t1}: Proctors accompanied by the time slots that they are busy,
for example: L_{t1} = {'L0': ['00', '13'], 'L1': ['22', '33'], ...}.
L_t: is extracted from L_{t1} and T, it is a set of proctors accompanied by the real exam time slots that they are busy

$$L_t = \{l : t | l \in L_{t1}, t \in L_{t1}[l] \cap T\},$$

for example: L_t = {'L0': ['13'], 'L1': ['33'], ...}.

2.2.2 Problem Variables: We introduce two sets of binary variables in this model.
+ Basic variables: denoted by $x_{l,t,c}$ where $l \in L, t \in T, c \in C$,
$x_{l,t,c} = 1$ when proctor l is assigned to proctor exam class c at timeslot t.
$x_{l,t,c} = 0$, otherwise.
+ Auxiliary variables: denoted by w_l
$w_l = 1$ when proctor l is assigned to classes in the exam period
$w_l = 0$ when proctor l is not assigned to any classes in the exam period.

2.2.3 Constraints: The model incorporates the following constraints

1. At a time, a teacher is assigned to at most one exam class
   $\forall t \in T, \forall l \in L,$

$$\sum_{c \in T_c[t]} x_{l,t,c} \leq 1 \tag{10}$$

2. At a time, an exam class is assigned to exactly two proctors

$$\forall t \in T, \forall c \in T_c[t], \quad \sum_{l \in L} x_{l,t,c} = 2 \tag{11}$$

3. Do not assign proctors to the exam timeslots that they are busy

$$\forall l \in L_t, \forall t \in L_t[l], \quad \sum_{c \in T_c[t]} x_{l,t,c} = 0 \tag{12}$$

4. The constraint to define variables w_l

$$A * w_l \leq \sum_{t,c} x_{l,t,c} \leq B * w_l \tag{13}$$

   where A is the minimum number of periods that every proctor is required, and B is the maximum number of periods that every proctor is required.

2.2.4 Objective Function In fact, if teachers are scattered across exam classes, it will lead to some difficulties and inconveniences in arrangement. For example, for teachers who are only assigned to proctor a few exams, it will take a lot of effort for them to arrange proctoring work and their own personal plans. The idea here is that we will minimize the number of proctors used in the exam, with the hope that only a certain number of proctors will be used, and the remaining teachers will not have to participate in proctoring the exam, and they can spend the entire exam focusing on their own plans. The proposed objective function is to minimize the number of proctors used as follows:

$$\min \sum_{l \in L} \mathrm{w}_l. \tag{14}$$

3 Model Implementation

The program was developed using Python language and CPLEX 12.8 solver on a personal computer with a Core i5-5200U 2.2 GHz processor and 16G RAM. Initially, the model was tested using small-sized sample data sets, and the results were recorded in Table 1 and Table 2.

Table 1. Result of ECS with small-sized sample data.

Code	Group	Exam	Subject	Room	Period	Day
C0	TT01	GT-01	GT	D3-302	2	4
C1	TT02	GT-02	GT	D3-101	2	4
C2	TT01	ĐS-01	ĐS	D3-101	3	1
C3	TT01, TT02	XSTK	XSTK	D3-302	4	4
C4	HTTTQL	XSTK	XSTK	D3-101	4	4
C5	TT02	ĐS-02	ĐS	D3-101	2	1
C6	HTTTQL	ĐS-02	ĐS	D3-301	2	1
C7	TT01	CSDL	CSDL	D3-302	5	1
C8	TT02	CSDL	CSDL	D3-101	5	1
C9	HTTTQL	CSDL	CSDL	D3-301	5	1

Table 2. Result of PEA with small-sized sample data.

Code	Group	Exam	Proctor 1	Proctor 2	Period	Day
C0	TT01	GT-01	N.D.Tu	N.T.N.Anh	2	4
C1	TT02	GT-02	N.T.Dung	N.X.Truong	2	4
C2	TT01	ĐS-01	N.D.Tu	N.X.Truong	3	1
C3	TT01, TT02	XSTK	N.T.Dung	N.X.Truong	4	4
C4	HTTTQL	XSTK	N.D.Tu	N.T.N.Anh	4	4
C5	TT02	ĐS-02	N.D.Tu	N.T.Dung	2	1
C6	HTTTQL	ĐS-02	N.X.Truong	N.T.T.Huong	2	1
C7	TT01	CSDL	N.D.Tu	N.T.Dung	5	1
C8	TT02	CSDL	L.C.Ngoc	N.X.Truong	5	1
C9	HTTTQL	CSDL	N.T.T.Huong	N.T.N.Anh	5	1

Subsequently, we run the program using medium-sized data sets from Faculty of Mathematics and Informatics (FAMI), with the problem sizes presented in Table 3 and 4.

Computation time and Objective function value of ECS and PEA with FAMI data and MIP gap value equals to 10% and 20% is recorded in Table 5 and 6.

Table 3. Size of ECS with FAMI data

	Exams count	Groups count	Rooms count
Problem size	32	92	44
	Constraints count	Variables count	
Model size	4037	88014	

Table 4. Size of PEA with FAMI data

	Proctors count	Exam classes count
Problem size	58	119
	Constraints count	Variables count
Model size	813	6850

Table 5. Computation time and Objective function value of ECS and PEA with FAMI data and MIP gap value 10%

Sub-problem	GAP	Computation time	Objective function value
ESC	MIP gap = 10%	12555 s	26
PEA	MIP gap = 10%	3379 s	29

Table 6. Computation time and Objective function value of ECS and PEA with FAMI data and MIP gap value 20%

Sub-problem	GAP	Computation time	Objective function value
ESC	MIP gap = 20%	22.89 s	36
PEA	MIP gap = 20%	1.05 s	36

The relative MIP gap tolerance is the difference between the best-known integer solution found so far and the optimal integer solution, expressed as a percentage of the optimal integer solution. Setting a higher relative MIP gap tolerance value will allow the solver to terminate earlier, potentially saving computational resources but resulting in a solution that is further from the optimal solution. Setting a lower relative MIP gap tolerance value will result in a more accurate solution but may require more computational resources to reach.

4 Discussion and Conclusion

The exam schedule obtained from running the model with real data of FAMI has successfully met numerous exam scheduling constraints and regulations at HUST. The model can be enhanced by incorporating soft constraints such as the constraints regarding room switch between consecutive exams of a student in a day, constraints regarding pre-assignments, or optimizing the waiting time between two consecutive exams to allow for effective rest and preparation. In the future, after improvements, this model can be utilized to solve the exam timetabling problem at HUST and other universities.

Cplex represents its computing power when it can solve problems with tens of thousands of constraints and variables, finding near-optimal exam scheduling options. However, when the number of variables and constraints explodes, the solution time of Cplex also increases a lot. This can be seen when comparing problem ECS and problem PEA, problem ECS has high complexity, and the number of variables and constraints is much greater than problem PEA, which is why the time to solve ECS is much longer than PEA. For this reason, when applying the model to real data that is larger than the experimental data, we need to trade-off between execution time and solution quality, sometimes we just need to stop at finding an acceptable solution.

Compared to the semi-manual method used at HUST, which uses Excel and experts' experience, the proposed model has a more optimal solution for using exam rooms and invigilators during exams. However, the model has not brought good results in terms of running time and still needs further improvement optimization.

Compared to the timetable software mentioned in Sect. 1.4, the proposed model has solved some more specific requirements of the exam scheduling problem at HUST such as putting different student groups in the same exam class or scheduling different exams of the same subject, ...

References

1. Salem, M.A., Hanif, D.S., Mona, A.: A mixed-integer mathematical modeling approach, to exam timetabling. CMS **7**, 19–46 (2010). https://doi.org/10.1007/s10287-007-0066-8
2. Michael, W.C., Gilbert, L., Sau Yan, L.: Examination timetabling: algorithmic strategies and applications. J. Oper. Res. Soc. **47**(3), 373–383 (1996). https://doi.org/10.2307/3010580
3. Bazari, S., Pooya, A., Soleimani Fard, O., et al.: Modeling and solving the problem of scheduling university exams in terms of new constraints on the conflicts of professors' exams and the concurrence of exams with common questions. OPSEARCH **60**, 877–915 (2023). https://doi.org/10.1007/s12597-023-00638-z
4. Masri, A., et al.: Intelligent examination timetabling software. Procedia Soc. Behav. Sci. **18**, 600–608 (2011). https://doi.org/10.1016/j.sbspro.2011.05.087
5. Di Gaspero, L., Schaerf, A.: Tabu search techniques for examination timetabling. In: Burke, E., Erben, W. (eds.) PATAT 2000. LNCS, vol. 2079, pp. 104–117. Springer, Heidelberg (2001). https://doi.org/10.1007/3-540-44629-X_7
6. Nuno, L., Fernando, M., Agostinho, C.R.: A fast simulated annealing algorithm for the examination timetabling problem. Expert Syst. Appl. **122**, 137–151 (2019). https://doi.org/10.1016/j.eswa.2018.12.048
7. Dener, M., Calp, M. H.: Solving the exam scheduling problems in central exams with genetic algorithms. Mugla J. Sci. Technol. **4**(1), 102–115 (2018). https://doi.org/10.22531/muglajsci.423185
8. IBM CPLEX Optimizer https://www.ibm.com/analytics/cplex-optimizer

9. Schaerf, A.: A survey of automated timetabling. Artif. Intell. Rev. **13**(2), 87–127 (1999). https://doi.org/10.1023/A:1006576209967
10. Son, T.A., Ngan, N.T.: An integer programming model for course timetabling problem at HUST. In: Tran, D.-T., Jeon, G., Nguyen, T., Lu, J., Xuan, T.-D. (eds.) ICISN 2021. LNNS, vol. 243, pp. 345–354. Springer, Singapore (2021). https://doi.org/10.1007/978-981-16-2094-2_43

Integer Linear Programming Models for Variations of the Discrete Semi-obnoxious Facility Location Problem

Chowhan Tanmay Tushar[1], Vishwanath R. Singireddy[2], and Manjanna Basappa[3(✉)]

[1] BITS Pilani, Hyderabad Campus, Hyderabad 500078, Telangana, India
f20181056@hyderabad.bits-pilani.ac.in

[2] Malla Reddy College of Engineering and Technology, Hyderabad 500100, Telangana, India
vishwanath@mrcet.ac.in

[3] Department of Computer Science and Engineering, NITK Surathkal, Mangalore 575025, Karnataka, India
manjanna@nitk.edu.in

Abstract. In this work, we consider semi-obnoxious facility location problems, motivated by the following scenario: Suppose an international restaurant brand wants to open a chain of branches in a city that is already occupied by several competing brands. We model this as a geometric optimization problem to provide decision-makers with relevant information on how to economically place k new branches among competitors and how to set prices of items in comparison to nearby competitors. We refer to this as the discrete k-semi-obnoxious facility location (DSOFL) problem, classifying it within the context of similar problems in the literature. We introduce two variations of the DSOFL problem and propose integer linear programming (ILP) based exact methods for solving both. Additionally, we discuss the implementation of these models using the Gurobi solver.

Keywords: Semi-obnoxious facility location · Integer linear programming · Red-blue point-set separability

1 Introduction

Most facility location problems were solved by considering two types of facilities. First is the problem of locating desirable facilities such as hospitals, post offices, or fire stations, which must be as close as possible to the demand points (attractive points). Second, the problem of locating undesirable facilities, also known as obnoxious facilities, such as dumping yards, chemical factories, or nuclear plants,

[3]The author was partially supported by SERB TARE Grant TAR/2022/000397.

© The Author(s), under exclusive license to Springer Nature Switzerland AG 2025
H. A. Le Thi et al. (Eds.): ICAMCS 2024, LNNS 1311, pp. 235–246, 2025.
https://doi.org/10.1007/978-3-031-90606-0_20

which need to be located as far away as possible from the demand points (repulsive points) to avoid the nuisance generated by these facilities. However, semi-obnoxious facility location problems aim to locate the facilities not too close to the repulsive points and not too far away from the attractive points. Here, the problem will have two objectives, i.e., it is a bi-objective problem. Consider the following scenario: A government plans to build an airport near a newly developing township. The challenge is to place the airport far enough from residential areas and the city to minimize disturbances, such as noise pollution. However, the airport also needs to be close enough to ensure easy access and convenience for the population, thereby minimizing transportation costs. The location must strike a balance between reducing the negative impact on residents and ensuring the airport remains beneficial and accessible to its users. That is, it has to be away from the city to avoid sound pollution or nuisance generated due to it while it needs to be as near as possible to its customers to minimize the transportation cost. Here, we consider variations by turning one of the two objectives into constraints.

From a geometric point of view, the facility location problems are solved in both continuous (i.e., locating facilities on a continuous plane) and discrete settings (i.e., locating facilities at some of the positions from a given discrete set of candidate locations).

1.1 Problem Statements and Motivation

2D-MinRed-AllBlue: Given a set $B = \{p_{b_1}, p_{b_2}, \dots, p_{b_n}\}$ of n blue points, a set $R = \{p_{r_1}, p_{r_2}, \dots, p_{r_m}\}$ of m red points and a set $F = \{p_{f_1}, p_{f_2}, \dots, p_{f_s}\}$ of s candidate sites within a rectangular region $\mathcal{R}$ and an integer k, pack at most k ($k \leq s$) congruent disks centered at sites in F such that all the blue points are covered, and the number of red points covered is minimized (see Fig. 1).

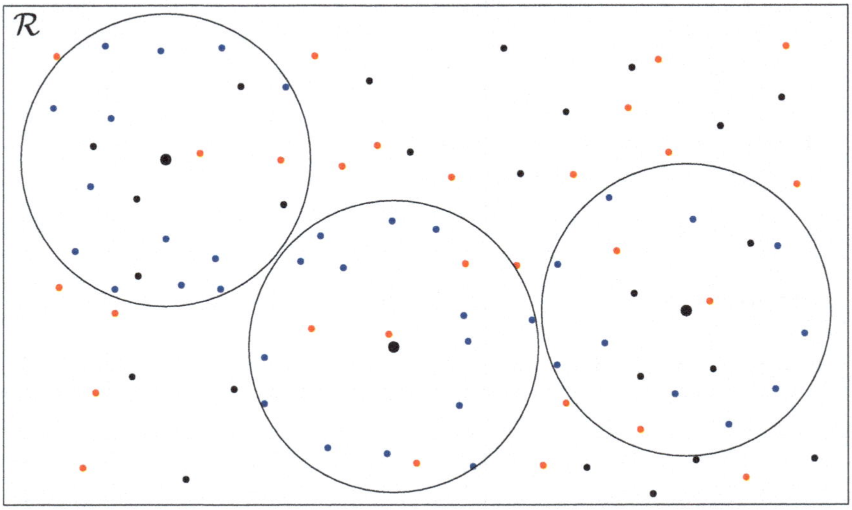

Fig. 1. The optimal solution when $k = 3$

2D-MAXBLUE-NORED: Given a set $B = \{p_{b_1}, p_{b_2}, \ldots, p_{b_n}\}$ of n blue points, a set $R = \{p_{r_1}, p_{r_2}, \ldots, p_{r_m}\}$ of m red points and a set $F = \{p_{f_1}, p_{f_2}, \ldots, p_{f_s}\}$ of s candidate sites within a rectangular region $\mathcal{R}$ and an integer k, pack at most $k > 0$ ($k \leq s$) congruent disks centered at sites in F such that the maximum number of blue points are covered, and none of the red points is covered (see Fig. 2).

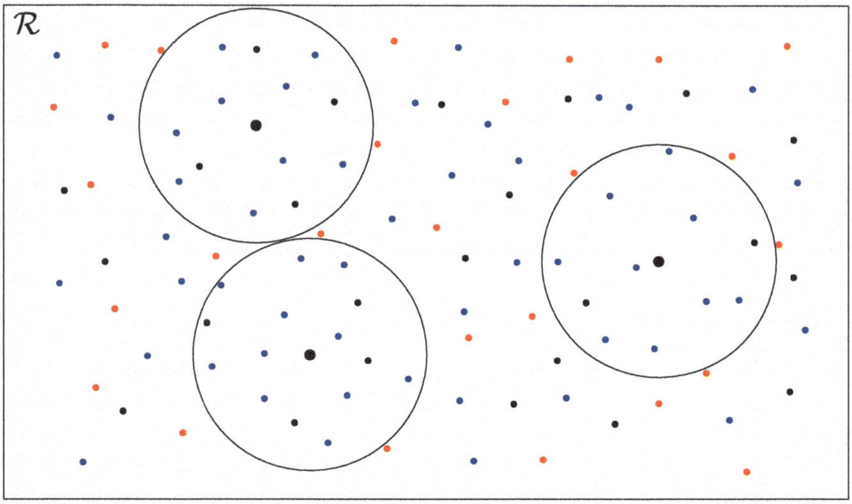

Fig. 2. The optimal solution when $k = 3$

Suppose a pizza company wants to open a certain number of pizza stores in a city with many pizza stores already opened by other companies. Now, this new pizza company has to open stores such that they need to be as close as possible to public areas (e.g., schools, airports, railway stations, and shopping malls) and away from the existing stores to avoid competition. We can solve this problem by considering the existing pizza stores as a set of red points (R), public areas as a set of blue points (B), and the problem is to find k locations among a set of candidate locations (F) that satisfies the distance constraint L (the farthest demand point from where a customer can visit the newly established store or the closest competitor to the newly established store). In this paper, we propose integer linear programming (ILP) models for both the variants of the problem (Sect. 2). Next, we discuss the specific motivations that shape the objective functions in the formulations of both models.

The latter variant of the problem, as the name MAXBLUE-NORED suggests, is avoiding competitors - ideally, no competitor (no red point) at all (i.e., a new pizza store should be located as far away as possible from the existing stores). This may result in fewer customers to the new stores as they have to be placed far from the public areas to avoid existing competitors. However, it is not necessary to reduce the prices of the items as there are no competitors in the vicinity of the new stores. Whoever few customers living close to the new store shall pay and eat. Let l be the number of public areas (demand points) to be covered. Then we give an ILP model with the objective of maximizing the value of l (Subsect. 2.2).

The former variant, the 2D-MinRed-AllBlue, aims to maximize the number of customers, ideally serve or cover all the customers (i.e., a new pizza store should be located in proximity to public areas with a high concentration of potential customers, while minimizing competition from existing stores as much as possible). But to minimize competition, we may need to decrease the prices so that we can attract customers. Let l be the number of competitors affecting the prices in the new stores. We model this variant as an ILP that minimizes l, the number of red points covered (Subsect. 2.1).

For different values of B, R, and L, computing the optimal values l of both variants of the problem above would assist in making the best choices of opening new facilities (stores) by the practitioners for a given city.

1.2 Related Work

Erkut and Neuman [4] define an undesirable (or obnoxious) facility as one that causes harm or inconvenience to nearby individuals while providing a specific product or service. Tralhão et al. [10] proposed a multi-objective approach for locating semi-desirable urban facilities, considering four objectives. The first objective minimizes total investment costs, the second minimizes the average distance between customers and facilities, and the last two objectives address the "pull" and "push" aspects of the decision problem. One objective minimizes the number of individuals too close to any facility, while the other minimizes the number of customers too far from their respective facility. This approach, designed to address semi-obnoxious location problems by separately considering "push" and "pull" objectives, was developed to analyze problems involving heterogeneous demand stakeholders. Coutinho et al. [2] introduced a mixed-integer, bi-objective programming model to determine the optimal locations and capacities of semi-desirable (or semi-obnoxious) facilities. The first objective minimizes total investment costs, while the second minimizes dissatisfaction by combining the "pull" and "push" factors into a single function-where individuals prefer not to live too close to, but also not too far from, the facilities. The model identifies the number of facilities to open, their respective capacities, locations, the share of total demand allocated to each, and the population assigned to each facility. Teran-Somohano et al. [9] studied the problem of locating multiple capacitated semi-obnoxious facilities using evolutionary strategies. These strategies aim to find solutions that effectively balance the complex trade-offs between minimizing the negative impact on nearby populations and meeting the capacity requirements of the facilities. Additional research on the location of multiple semi-obnoxious facilities, particularly using bi-objective models, is available in the literature [3,5,7,11]. In computational geometry, these semi-obnoxious facility location problems are often termed red-blue point separability problems, where the goal is to separate red points from blue points using a hyperplane [8], square, disk, rectangular or circular anulli [1].

2 Formulation of Mathematical Models

Consider the 2D-MinRed-AllBlue parameterized by the radius L, named as $<$2D-MinRed-AllBlue, $L>$. We now formulate the problem as an integer linear program that exactly models $<$2D-MinRed-AllBlue, $L>$. Observe that $<$ 2D-MinRed-AllBlue, $L>$ may be infeasible depending on the parameter's value. However, the original problem has always a feasible solution since we can place one disk (or at most k) of a sufficiently large radius to cover all the blue points. The optimal value for 2D-MinRed-AllBlue of a feasible solution to $<$ 2D-MinRed-AllBlue, $L>$ can be achieved with any candidate value for L. We will show that the number of possible candidate values for the parameter L is finite. It seems that 2D-MinRed-AllBlue is `para-NP-hard` parameterized by the radius: $<$ 2D-MinRed-AllBlue, $L>$ remains `NP-hard` for a constant radius L.

2.1 2D-MinRed-AllBlue Problem

We now give an integer linear programming model for the above problems. The following notations are used:

- B: the set of blue points as attractive demand points (customers).
- R: the set of red points as repulsive demand points (competitors).
- F: the set of candidate locations for the facilities.
- L: the radius of facilities in the optimal packing.
- l: the number of red points covered.
- k: the number of facilities to be located.
- d_{ij}: minimum distance between any two sites $i, j \in (F \cup B \cup R)$.
- $y_i = 1$ if a facility is opened at site $i \in F$, and 0 otherwise.
- $x_{ij} = 1$ for every pair of candidate sites $i, j \in F$, if the two facilities located at i and j respectively are such that $d_{ij} \geq 2L$, and $x_{ij} = 0$ otherwise.
- $z_j = 1$ if a demand point $j \in R$ is covered by some selected facility, and otherwise 0.
- $z_j = 0$, $\forall j \in B$.
- M: the diameter of the point set F.

$$\text{minimize} \quad l \tag{1}$$

$$\text{subject to} \quad \sum_{i \in F} y_i \leq k \tag{2}$$

$$z_j = 0 \qquad \forall j \in B \tag{3}$$

$$z_j + \sum_{i \in F : d_{ij} \leq L} y_i \geq 1 \qquad \forall j \in (B \cup R) \tag{4}$$

$$x_{ij} d_{ij} + (1 - x_{ij}) M \geq 2L \qquad \forall i, j \in F, i \neq j \tag{5}$$

$$y_i + y_j \leq x_{ij} + 1 \qquad \forall i, j \in F, i \neq j \tag{6}$$
$$y_i \geq x_{ij} \qquad \forall i, j \in F, i \neq j \tag{7}$$
$$\sum_{i \in F: d_{ij} \leq L} y_i \leq z_j \qquad j \in R \tag{8}$$
$$\sum_{j \in R} z_j \leq l \tag{9}$$
$$l \geq 0 \qquad l \in \mathbb{Z}^+ \cup \{0\} \tag{10}$$
$$y_i \in \{0, 1\} \qquad i \in F \tag{11}$$
$$z_j \in \{0, 1\} \qquad j \in (B \cup R) \tag{12}$$
$$x_{ij} \in \{0, 1\} \qquad i, j \in F, i \neq j \tag{13}$$

Correctness of the ILP Model: The number of facilities to be established is at most k, which is ensured by constraint (2). A point in $(B \cup R)$ may or may not be covered by a facility in F (4), but all the blue points are to be covered, as ensured by (3) and (4) together and $z_j = 0$, $\forall j \in B$. The inequalities (5, 6, 7 and 8) guarantee that the distance between any two selected facilities is at least $2L$, where M is the diameter of the point set F. The inequalities (6)–(7) make sure that y_i is never strictly less than x_{ij} since the selected facilities i and j can never be less than $2L$ distance apart. The constraints (8–9) guarantee that a red point j is covered by a facility i while $z_j = 1$. Finally, the number of red points covered should be non-negative, as guaranteed by (10).

Optimization Scheme: The above ILP formulation provides an optimal solution for the given radius L. Our objective is to minimize the number of red points covered and ensure coverage of all the blue points. Consequently, the radius of the disks in the optimal solution can be determined by the distances between the blue points and the candidate facility locations. Let $\mathcal{L}$ represent the set of candidate values for L. This set can be defined as follows:

$$\mathcal{L} = \{d_{ij} \mid i \in B \text{ and } j \in F\}$$

Furthermore, the radii of the disks can also be derived from the distances between the candidate facility locations, as the disks must not overlap (i.e., hence form a packing). Therefore, the diameter of a disk could be one of the distances between any pair of candidate facility locations. These distances are subsequently included in the set of candidate radii:

$$\mathcal{L} = \mathcal{L} \cup \{\frac{d_{ij}}{2} \mid i, j \in F \text{ and } i \neq j\}$$

The red points, however, cannot determine the radius of the disks. Let us consider a scenario where a red point determines the radius of the disks, meaning it lies on the boundary of any of the disks. In such a case, we can reduce the radii of the disks until a blue point lies on the boundary. Consequently, the cardinality of the set $\mathcal{L}$ is $O(ns + s^2)$. Hence, we have the following lemma.

Lemma 1. *The set $\mathcal{L}$ of candidate radius is $O(ns + s^2)$.*

The above ILP needs to be solved to optimality using the Gurobi solver $O(ns+s^2)$ times (i.e., for every candidate radius $L \in \mathcal{L}$) to compute an optimal solution to the 2D-MinRed-AllBlue problem. Theoretically, we know that ILP parameterized by the number of variables is FPT [6]. In the above ILP, the number N of variables is $m + n + s^2 + 1$. Let η be the total sum of bits needed to encode the coordinates values of the points in $B \cup R \cup F$ and the real value L and integer value k. Hence, we have the following theorem.

Theorem 1. 2D-MinRed-AllBlue *is solvable in time $O(f(N)\eta \log \eta)$ where $f(N) = 2^{O(N \log N)}(n + s)s$, $N = m + n + s^2 + 1$ is the number of variables and η is the input size.*

2.2 2D-MaxBlue-NoRed Problem

In the second variant of the DSoFL problem, we need to cover none of the red points, but maximize the number of blue points covered. As in the previous section, we consider the 2D-MaxBlue-NoRed problem parameterized by the radius L of the disks, named as <2D-MaxBlue-NoRed, L >. Given an arbitrary L, it is possible that no feasible solution exists for <2D-MaxBlue-NoRed, L > because the disjoint disks of radius L centered at every combination of k points in F might cover at least one red point. However, if we set L to any value within the interval $[0, \min_{i \in F, j \in R} d_{ij})$, the disks of radius L centered at any combination of k points in F will not cover any red points. From there, we can continue shrinking the radius of all the disks until no pair of disks overlaps, resulting in a configuration of pairwise disjoint disks that cover no red points. Hence, we always have a feasible solution for the 2D-MaxBlue-NoRed problem. We now give an integer linear programming model for the <2D-MaxBlue-NoRed, L > problem. The following are the notations that are different from the notations introduced in the previous section. The meaning of the other notations remains the same.

- l: The number of blue points covered.
- $z_j = 0$ if the blue point $j \in B$ is covered, and 1 otherwise.

$$
\begin{aligned}
\text{maximize} \quad & l && && (14)\\
\text{subject to} \quad & \sum_{i \in F} y_i \leq k && && (15)\\
& z_j + \sum_{i \in F : d_{ij} \leq L} y_i \leq 1 && \forall j \in (B \cup R) && (16)\\
& z_j = 1 && \forall j \in R && (17)\\
& x_{ij} d_{ij} + (1 - x_{ij}) M \geq 2L && \forall i, j \in F, i \neq j && (18)\\
& y_i + y_j \leq x_{ij} + 1 && \forall i, j \in F, i \neq j && (19)\\
& y_i \geq x_{ij} && \forall i, j \in F, i \neq j && (20)
\end{aligned}
$$

$$1 - \sum_{i \in F: d_{ij} \leq L} y_i = z_j \qquad \forall j \in B \tag{21}$$

$$\sum_{j \in B} (1 - z_j) \geq l \tag{22}$$

$$l \geq 0 \qquad l \in \mathbb{Z}^+ \cup \{0\} \tag{23}$$

$$y_i \in \{0, 1\} \qquad i \in F \tag{24}$$

$$z_j \in \{0, 1\} \qquad j \in (B \cup R) \tag{25}$$

$$x_{ij} \in \{0, 1\} \qquad i, j \in F, i \neq j \tag{26}$$

Correctness of the ILP Model: The number of facilities to be established is at most k, which is ensured by constraint (15). A point in $(B \cup R)$ may or may not be covered by a facility in F (as allowed by the inequality (16)), but none of the red points must be covered, which is guaranteed by the constraint (16) and (17) together as $z_j = 1$ implies that the point j is not covered by any facility. As in the previous model, the inequalities (18, 19 and 20) guarantee that the distance between any two selected facilities is at least $2L$, where M is the diameter of the point set F. The constraint (21) ensures that if a blue point j is covered by exactly one facility i then $z_j = 0$. This implies that the total number of blue points covered is given by $|B| - \sum_{j \in B} z_j$, which should be non-negative, as ensured by the constraints (22) and (23) together.

Optimization Scheme: The set of candidate radii is the same as in 2D-MinRed-AllBlue problem since, in any optimal solution, there will be at least one blue point lying on the boundary of any of the disks and we can not increase its radius as it may cover a red point. Similarly, all possible pairs of candidate facility locations contribute to the set of all candidate radii. Hence, from Lemma 1, the cardinality of the set of candidate radii is $O(ns + s^2)$. As in the previous section, the above ILP for the <2D-MaxBlue-NoRed, L> problem needs to be solved to optimality for every candidate radius L if the parameterized problem is feasible for that L.

Theorem 2. *2D-MaxBlue-NoRed is solvable in time $O(f(N)\eta \log \eta)$ where $f(N) = 2^{O(N \log N)}(n + s)s$, $N = m + n + s^2 + 1$ is the number of variables and η is the input length.*

2.3 Implementation of the Models Using Gurobi Optimizer

Our implementation was done using Python version 3.11.4 and the Gurobi Solver version 11.0.3. Since we did not find any known benchmark datasets for the DSoFL variants in the literature, we generated instances for both problems by randomly creating points within a predefined bounding box. Below, we explain

Table 1. 2D-MINREDALLBLUE.

s	m	n	k	ILP Average (ms)	ILP Time deviation (ms)
30	20	20	5	2.21	0.81
50	20	20	8	4.59	2.98
50	20	20	10	5.73	1.84
80	40	80	18	17.56	5.35
100	90	100	25	29.12	10.48
150	130	150	40	107.04	39.19
200	130	150	50	304.55	91.97
250	150	200	70	711.39	172.83
300	180	200	75	1261.9	451.03
350	210	250	90	2117.69	833.15
400	250	300	100	2374.12	946.75
450	275	325	120	4680.27	1981.71
500	300	350	125	6069.55	1993.06

our methodology for generating the data and discuss the results observed when testing our models on these generated instances.

2D-MINRED-ALLBLUE: To generate mock data for this study, we employed a systematic approach, as follows. Initially, we positioned k non-overlapping facility locations within a two-dimensional bounding rectangle, ensuring no two facilities intersected. Subsequently, all n blue points were placed such that each point was covered by at least one of these facilities, guaranteeing complete coverage of the blue points. Following this, the remaining $s - k$ facilities were randomly distributed within the two-dimensional space, introducing an element of randomness to the overall facility layout. Finally, m red points were randomly scattered across the space without any constraints regarding their proximity to either the facilities or the blue points. This method ensures the existence of a feasible solution to the problem at all times.

The implementation results are available in Table 1, where the timing versus input size comparisons were made for the ILP approach using Gurobi solver in Python 3.11.4, for the 2D-MINREDALLBLUE problem. Averages were calculated over 20 cases for each s, m, n, and k combination. Note that the ILP times only reflect model optimization in Gurobi, excluding model-building time, which significantly increases computation. The tests were conducted on a machine equipped with an i7-10700 processor and 16 GB RAM.

Table 2. 2D-MaxBlue-NoRed.

s	m	n	k	ILP Average (ms)	ILP Time deviation (ms)
30	20	20	5	1.83	0.46
50	20	20	8	5.15	1.13
50	20	20	10	6.15	3.65
80	40	80	18	30.15	6.00
100	90	100	25	17.08	3.10
150	130	150	40	296.72	62.96
200	130	150	50	257.38	31.18
250	150	200	70	798.59	94.26
300	180	200	75	939.36	198.38
350	210	250	90	3384.30	430.8
400	250	300	100	2049.88	259.77
450	275	325	120	2661.68	463.68
500	300	350	125	1966.00	1313.33

2D-MaxBlue-NoRed: The mock data generation method employed for this problem follows a procedure similar to the previous approach. Initially, k non-overlapping facility locations are placed in a two-dimensional space, ensuring that no two facilities intersect. Subsequently, all m red points are randomly positioned outside the coverage areas of these k facilities. Afterward, n blue points are randomly distributed across the space without any specific restrictions. Finally, the remaining $s - k$ facilities are placed randomly to introduce variability into the facility arrangement. This method ensures the existence of a feasible solution under the given arrangement.

The implementation results are given in Table 2, where the timing versus input size comparisons were made for the ILP approach implemented on the Gurobi solver in Python (version 3.11.4) for the 2D-MaxBlue-NoRed problem. These averages were calculated over 20 generated cases for each s, m, n, and k pair listed. Note that the time listed for the ILP only corresponds to the model optimization times in Gurobi and does not include the model building time that increases the compute time drastically. These results were produced on a machine equipped with an i7-10700 processor and 16 GB RAM.

We also plotted graphs of input sizes versus the time taken by the ILP solver for both problems (see Fig. 3). From the graphs, we observe that the ILP solver's time increases drastically as the input parameters increase for the MinRed-AllBlue problem, whereas for the MaxBlue-NoRed problem, the solver's time initially decreases slightly and then remains moderate. This suggests a difference in the computational complexity of the two problems, which warrants further investigation. It would be interesting to theoretically prove that there exists a parameter in the input for which MaxBlue-NoRed admits an efficient parameterized algorithm.

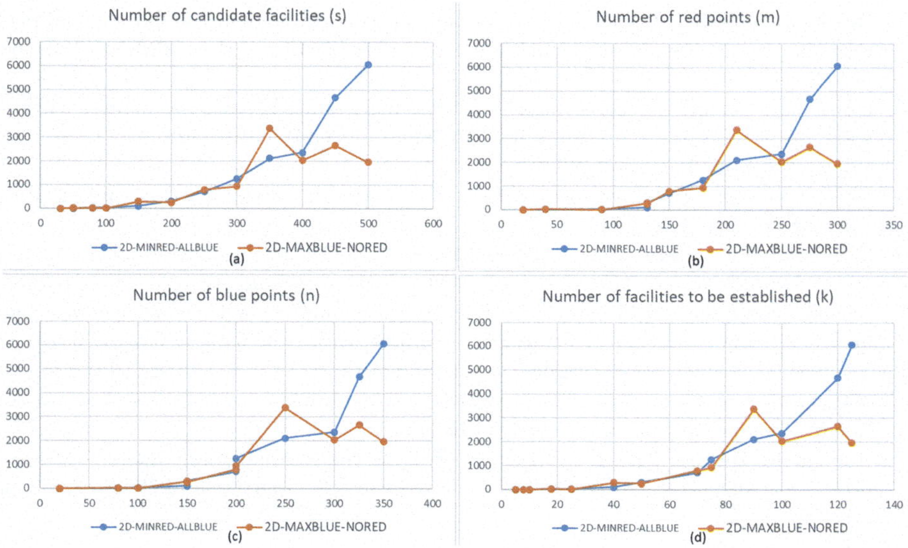

Fig. 3. (a) s Vs Time (ms) (b) m Vs Time (c) n Vs Time (d) k Vs Time

3 Conclusion

In this paper, we introduced two variants of the discrete semi-obnoxious facility location problem and proposed ILP models for each. These problems may have real-world applications in economics and decision-making, particularly in scenarios where businesses must balance maximizing customer reach while minimizing competition, such as selecting optimal locations for new stores in competitive markets. There are two potential directions for future work: (i) Since the problems are NP-hard, their parameterized and approximation complexity could be explored, and (ii) further investigation could focus on developing other exact solution techniques based on integer linear programming models, such as Bender's decomposition, the branch(-and-cut)-and-price approach, or constraint programming.

References

1. Abidha, V., Ashok, P.: Geometric separability using orthogonal objects. Inf. Process. Lett. **176**, 106245 (2022)
2. Coutinho-Rodrigues, J., Tralhão, L., Alçada-Almeida, L.: A bi-objective modeling approach applied to an urban semi-desirable facility location problem. Eur. J. Oper. Res. **223**(1), 203–213 (2012)
3. Eiselt, H.A., Marianov, V.: A bi-objective model for the location of landfills for municipal solid waste. Eur. J. Oper. Res. **235**(1), 187–194 (2014)
4. Erkut, E., Neuman, S.: Analytical models for locating undesirable facilities. Eur. J. Oper. Res. **40**(3), 275–291 (1989)

5. Heydari, R., Melachrinoudis, E.: Location of a semi-obnoxious facility with elliptic maximin and network minisum objectives. Eur. J. Oper. Res. **223**(2), 452–460 (2012)
6. Kannan, R.: Improved algorithms for integer programming and related lattice problems. In: Proceedings of the Fifteenth Annual ACM Symposium on Theory of Computing, pp. 193–206 (1983)
7. Melachrinoudis, E.: Bicriteria location of a semi-obnoxious facility. Comput. Ind. Eng. **37**(3), 581–593 (1999)
8. O'Rourke, J., Rao Kosaraju, S., Megiddo, N.: Computing circular separability. Discrete Comput. Geom. **1**(2), 105–113 (1986). https://doi.org/10.1007/BF02187688
9. Teran-Somohano, A., Smith, A.E.: Locating multiple capacitated semi-obnoxious facilities using evolutionary strategies. Comput. Ind. Eng. **133**, 303–316 (2019)
10. Tralhão, L., Coutinho-Rodrigues, J., Alçada-Almeida, L.: A multiobjective modeling approach to locate multi-compartment containers for urban-sorted waste. Waste Manage. **30**(12), 2418–2429 (2010)
11. Yapicioglu, H., Smith, A.E., Dozier, G.: Solving the semi-desirable facility location problem using bi-objective particle swarm. Eur. J. Oper. Res. **177**(2), 733–749 (2007)

Process Technology Selection in Multi-criteria Group Decision-Making Context: WSAM-OWA-Based TOPSIS Approach

Ryma Zegai[1](✉), Imen Khettabi[2], Lyes Benyoucef[3], and Moncef Abbas[1]

[1] DGRSDT, AMCD&RO Laboratory, USTHB University, Algiers, Algeria
{rzegai,mabbas}@usthb.dz

[2] DGRSDT, LAROMAD Laboratory, USTHB University, Algiers, Algeria
khettabi.imen@usthb.dz

[3] Aix Marseille University, CNRS, LIS UMR 7020, Marseille, France
lyes.benyoucef@lis-lab.fr

Abstract. In today's competitive and interconnected supply chains, decision-making often requires collaboration among multiple decision-makers (DMs) to achieve common goals, such as selecting process technology, identifying suitable partners, adopting new technologies, and establishing operational strategies, to name few. This article addresses the challenge of selecting process technology within the context of multi-criteria group decision-making (MCGDM). The developed approach incorporates the weighted selective aggregated majority-OWA (WSAM-OWA) operator, which accounts for the varying expertise levels of DMs, significantly influencing the aggregation process. To further enhance decision accuracy, a combinative method is used to evaluate the importance of DMs and strengthen preference aggregation. The TOPSIS method is then applied to identify the most suitable alternative, using both subjective and objective weighting methods. Finally, to demonstrate the effectiveness of the proposed approach, an illustrative example dedicated to process technology selection is presented and the obtained numerical results analysed.

Keywords: Group decision making · Multi-criteria · Consensus · WSAM-OWA · Combinative weighting · Prcoess technology selection

1 Context and Motivations

In today's complex decision-making environments, engaging multiple experts is essential for making well informed choices across various domains. This often involves evaluating both qualitative and quantitative criteria, which can complicate the process, especially when striving for consensus among decision makers. Effective aggregation is vital for converting individual opinions into a unified

© The Author(s), under exclusive license to Springer Nature Switzerland AG 2025
H. A. Le Thi et al. (Eds.): ICAMCS 2024, LNNS 1311, pp. 247–259, 2025.
https://doi.org/10.1007/978-3-031-90606-0_21

decision, particularly in group decision-making (GDM), where diverse judgments need to be aligned. This paper presents a noval approach designed to tackle the challenges of multi-expert, multi-criteria decision-making, emphasizing the role of experts and the methods employed to assess their expertise level. Additionally, the importance of criteria weights in comparing alternatives is emphasized.

With the rapid advancement of technologies such as automation, robotics, and AI-systems, selecting the appropriate process technology has become vital for organizations aiming to boost efficiency and competitiveness, especially in advanced manufacturing contexts. It is essential to evaluate various performance measures including cost, productivity, quality, reliability, flexibility, and scalability, since each significantly affects the long-term success of the chosen technology. The complexity and diversity of these criteria often transform the selection process into a multi-criteria decision-making (MCDM) challenge, requiring thorough consideration of factors that influence system functionality and profitability.

In many situations, achieving unanimous agreement among experts is difficult, making majority consensus a more practical solution. Aggregation methods, like the ordered weighted averaging (OWA) operator introduced by Yager in 1988, are key in combining diverse expert judgments into a unified decision. The OWA operator provides flexibility, allowing for a range of consensus building. The OWA operator can be seen as a general class of aggregation functions that fill the gap between these two extremes by weighting and combining inputs in a controlled manner. This operator thus provides a way of balancing different expert opinions while maintaining a structured approach to the decision making process. Numerous studies have explored new ways to apply these operators to decision-making problems, highlighting their adaptability and efficiency in various contexts. Notable contributions to the literature include works by [17,18], and more recent advancements by [8–10]. These studies have extended the theoretical basis of OWA-related operators, while at the same time demonstrating their practical relevance in areas such as multi-criteria decision analysis and group decision making. Furthermore, many published works have employed techniques and methods such as AHP (analytic hierarchy process), TOPSIS (technique for order of preference by similarity to ideal solution), P-SVM (potential support vector machine), PP (preference programming,) integrated a fuzzy multi-criteria group decision-making approach using ranking interval type-2 fuzzy variables [5].

The rest of the paper is organized as follows: Sect. 2 reviews the related works. Section 3 describes the problem under consideration. Section 5 details the proposed approach. Section 6 demonstrates the applicability of the proposed approach using a simple illustrative example. Section 7 concludes the paper with some future research directions.

2 Related Works

In an environment with several decision-makers, the challenge of reaching a collective decision requires careful consideration of the different points of view and interests. Consensus plays a crucial role in this process. Consensus in multi-decision-maker scenarios is a process where all involved decision maker aim to reach a collective agreement, balancing diverse perspectives and preferences. This approach is vital in organizational, political, and collaborative environments, as it ensures that decisions are accepted by all members, reducing conflicts and enhancing the quality of outcomes. Achieving consensus can be challenging due to varying interests, but techniques such as structured communication, negotiation, and iterative feedback loops help to bridge differences [3]. Moreover, consensus-building fosters a sense of shared responsibility, increasing commitment to the final decision [2]. Research suggests that incorporating formal decision support systems can further streamline this process, leading to more effective and equitable outcomes [15].

In the literature, there are some approaches which have been proposed to model the majority concept using OWA operators. [12] proposed the majority additive OWA operator (MA-OWA) to aggregate the argument values that have cardinality greater than one. In the context of GDM, each expert has an associated degree of importance that reflects his/her expertise, knowledge, skill, etc. Motivated by the heterogeneous GDM problems, the inclusion of the reliability of information sources (or degree of importance) is suggested as the extension of SAM-OWA and it is denoted as the weighted SAM-OWA operator.[11] proposed two approaches to deal with this issue. The first is based on the use of the IOWA operator, where the support function is applied to derive a set of order-inducing, scalar-valued variables, i.e., reordered based on the most similar opinions. While the other approach is based on a fuzzy subset that represents the majority opinion under the vague concept. Correspondingly, [1] extended the Pasi-Yager method, specifically based on the IOWA operator, by employing the Minkowski OWA-based similarity measure to obtain the order-inducing variables. Moreover, in their method, instead of synthesizing the consensus on each alternative (classical scheme), they proposed an alternative approach where the consensus measure on each specific criterion (alternative scheme) is implemented. Furthermore, they propose to apply the importance degrees of experts to heterogeneous GDM.

Recently, [4] developed the selective MA-OWA (SMA-OWA) operator to deal with the problem of fast convergence of the associated weights. [13] proposed the selective aggregated majority OWA (SAM-OWA) operator where the cardinality is used to calculate the individual weight for each group of argument values. [7] study in granular computing extended the TOPSIS method by integrating the OWAD (Ordered Weighted Averaging Distance) operator, enhancing the method's capacity to manage uncertainty and fuzzy conditions in multi-criteria decision-making.

3 Problem Description

When organizations face the challenge of selecting the best alternative from a range of options, decision-making becomes a complex process, especially when it involves multiple stakeholders and a mix of qualitative and quantitative factors. To navigate this complexity, a group of decision-makers (DMs) evaluates the alternatives based on a set of predefined criteria. Consider a scenario with K decision-makers, denoted as DM_k, $(k = 1, \ldots, K)$, where each DM evaluates a set of alternatives denoted by A_i $(i = 1, \ldots, m)$, using a set of criteria, namely C_j $(j = 1, \ldots, n)$. The decision-makers evaluations and preferences are recorded in a matrix of m rows and n columns. It is essential to focus on several key criteria to ensure the selected alternative meets the organization's objectives. In our case, the following three criteria are considered:

- **Cost (C_1):** It includes all direct and indirect expenses to implement, maintain, or operate a process or solution, covering both initial investment and recurring operational costs, in the decision-making scenario.
- **Flexibility (C_2):** It refers to a process or system's capacity to adapt changes in demand, production methods, or operational needs without major reconfiguration or delay, accommodating variations in product type, volume, or schedule.
- **Lead time (C_3):** It includes deadlines, processing times, and waiting periods. In production and supply chains, shorter lead times are preferable for greater efficiency, faster customer response, and lower inventory costs.

This research work aims to assist a group of decision-makers (DMs) in ranking and selecting the best process technology from a set of alternatives. To manage the complexity of diverse data types—such as crisp, interval, and linguistic values—we propose a consensus-based method. The approach utilizes the weighted selective aggregated ordered weighted averaging (WSAM-OWA) operator, enabling efficient integration of DMs' inputs and processing of heterogeneous data while considering relevant criteria. An illustrative example is provided to demonstrate the effectiveness of the proposed approach (see Fig. 1).

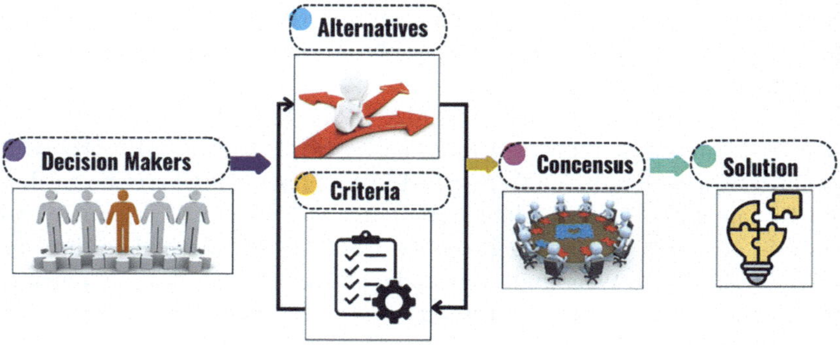

Fig. 1. Problem illustration

4 Theoretical Background

Definition: WSAM-OWA operator is a function $F_{WSAM} : R^n \times N^n \rightarrow R$. that has an associated weighting vector V of dimension n such that $\sum_{i=1}^{n} v_i = 1$ and $v_i \in [0, 1]$, defined as:

$$F_{WSAM}(a_1, a_2, \ldots, a_n) = \sum_{i=1}^{n} w_{i,N} b_{\sigma(i)} \tag{1}$$

where $N = \max_{1 \leq i \leq n} m_i$ and σ denotes a permutation with respect to the cardinality m_i [19]. The associated weights are defined by the recurrent relations:

$$w_{i,1} = \omega_i = \begin{cases} v_i, & \text{if } m_i = 1 \\ \sum_{i=1}^{m_i} v_i, & \text{if } m_i > 1 \end{cases} \tag{2}$$

The cardinal-dependent weights are given as:

$$w_{i,k} = \frac{\omega_i g_{i,k} y_k + w_{i,k-1}}{z_k},$$
$$y_1 = 1, \quad y_k = \begin{cases} 1, & \text{if } \sum_{j=1}^{n} \omega_j g_{j,k} = 0, \\ \frac{\sum_{j=1}^{n} g_{j,k}}{\sum_{j=1}^{n} \omega_j g_{j,k}}, & \text{otherwise,} \end{cases} \tag{3}$$
$$z_1 = 1, \quad z_k = \begin{cases} 1, & \text{if } \sum_{j=1}^{n} \omega_j g_{j,k} = 0, \\ 1 + \sum_{j=1}^{n} g_{j,k}, & \text{otherwise.} \end{cases}$$

where $g_{j,k}$ is:

$$g_{j,k} = \begin{cases} \delta & m_{\sigma(j)} \geq k \\ 1 - \delta & \text{otherwise} \end{cases} \tag{4}$$

The parameter δ is the cardinality relevance factor (CRF) and $1 \leq i \leq n, 2 \leq k \leq N$.

$$\delta = 1 - \left(2 + s^2\left(m_{\sigma(i)}\right)\right)^{-1} \tag{5}$$

where $s^2\left(m_{\sigma(i)}\right)$ is the variance of cardinality value and $\delta \in [0,1]$, such that:

$$E(m_{\sigma_i}) = \sum_{i=1}^{n} w_i m_{\sigma_i}, \quad w_{i,1} = 1/n \quad \text{and} \quad s^2(m_{\sigma_i}) = \sum_{i=1}^{n} \omega_i \left(m_{\sigma_i} - E(m_{\sigma_i})\right)^2$$

Combinative Weighting Method for the Criteria: A subjective and objective combinative clustering (SOC) [14] weighting approach is used with the following calculation process to enhance the comprehensiveness and accuracy of the criterion weights:

$$W_{a.SOC} = \frac{\left[(W_{a.z})^{\alpha_z} (W_{a.k})^{\alpha_k}\right]^{\frac{1}{(\alpha_z + \alpha_k)}}}{\sum_{a=1}^{n} \left[(W_{a.z})^{\alpha_z} (W_{a.k})^{\alpha_k}\right]^{\frac{1}{\alpha_z + \alpha_k)}}} \tag{6}$$

Subjective Weighting by Experienced DMs for a Limited Number of Criteria: The SPD weighting method [14] calculates subjective weights for DMs by having each DM rank the criteria based on importance. The most important criterion receives a weight determined by the DM noted $w_1 = W_{k,1}$, while the remaining criteria are assigned weights proportionally.

$$W_n = W_{kn} * \left(1 - \sum_{m=1}^{n-1} W_m\right) \text{ with: } W_{kn} \leq \frac{W_{n-1}}{1 - \sum_{m=1}^{n-1} W_m} \tag{7}$$

5 Proposed Approach

This section describes the developed approach within the MCGDM context. The approach accounts for both the trust degree (expertise level) of decision-makers and the weights of the criteria (ranking phase), utilizing a combinative weighting method. This method evaluates the trust degree of each DM for every criterion (consensus phase) and then for the overall decision problem (ranking phase), enhancing the aggregation process as proposed by [19]. Furthermore, a combinative weighting method is applied during the ranking phase to improve the accuracy of the criteria weights. By integrating these techniques, our approach, based on the WSAM-OWA operator and TOPSIS with heterogeneous data, becomes more robust, accurately reflecting the significance of DMs and delivering precise criteria weights for refining the selection process. The two main phases of the approach are as follows:

5.1 Consensus Phase

Step 1: Each DM_k should provide his own heterogeneous preferences decision matrix.

Step 2: Convert the heterogeneous preferences decision matrix into homogeneous preferences decision matrix following these steps:

a. On the crisp criteria values, normalize each crisp value in interval of [0,1] using the following two equations:

$$\text{For cost criteria: } r_{ij} = 1 - \frac{x_{ij}}{\sqrt{\sum_{i=1}^{m} x_{ij}^2}}, \quad \text{For benefit criteria: } r_{ij} = \frac{x_{ij}}{\sqrt{\sum_{i=1}^{m} x_{ij}^2}} \tag{8}$$

b. On the interval valued criteria, apply λ-function to get a crisp value from each interval valued using the equation bellow:

$$I_\lambda = (1 - \lambda) \times \alpha_L + \lambda \times \alpha_U \quad \text{with} \quad \lambda \in [0, 1] \tag{9}$$

Then, normalize them using Eq. (8).

c. On each linguistic value criteria, apply the α-cut procedure to defuzzify each linguistic value, resulting in a normalized interval a^α. Subsequently, use the λ-function to get a crisp value from each interval.

$$\alpha\text{-cut procedure:} \quad a^\alpha = [a_L^\alpha, a_U^\alpha] = [(b - a) \times \alpha + a, (c - d) \times \alpha + d] \tag{10}$$

Step 3: Construct the normalized preferences decision matrix which elements are n-tuples as follows:

$$Z = \left(\begin{array}{c|ccc} & C_1 & \cdots & C_n \\ \hline A_1 & \{z_{11}^1, z_{11}^2, \ldots, z_{11}^k\} & \cdots & \{z_{1n}^1, z_{1n}^2, \ldots, z_{1n}^k\} \\ A_2 & \{z_{21}^1, z_{21}^2, \ldots, z_{21}^k\} & \cdots & \{z_{2n}^1, z_{2n}^2, \ldots, z_{2n}^k\} \\ \vdots & \vdots & \ddots & \vdots \\ A_m & \{z_{m1}^1, z_{m1}^2, \ldots, z_{m1}^k\} & \cdots & \{z_{mn}^1, z_{mn}^2, \ldots, z_{mn}^k\} \end{array}\right)$$

Step 4: In this step, we are going to consider the varying levels of skill (trustworthiness) of each DM DM_k for a given criterion, thus each is assigned a value reflecting how much is the influence (weight) of an DM's opinion on each criterion in the decision-making process (DM's Weight), which will be computed by following steps (a), (b) and (c) as detailed bellow:

a) **Determine the DMs opinions evaluations under each criterion by a subjective method:**
Determine the weights of DM's under each criterion by self evaluation which is calculated by The Self-Evaluation Coefficient $\beta_k^{'j}$, based on the degree of conformity between personal traits and the question to be decided. $\beta_k^{'j}$ has values in [0,1], where a value of "0" indicates complete inconsistency between the DM's traits and the criteria, while "1" indicates total consistency between its personal traits and the criteria. To determine the self-evaluation weight under the criterion C_j of decision maker k: $\beta_k^{'j}$, linearly normalize the self-evaluation coefficients $\beta_k^{'j}$ provided by all decision-makers under criterion C_j. The equation is as follows:

$$\beta_k = \frac{\beta'_k j}{\sum_{k=1}^{K} \beta'_k j}, \quad s = 1, 2 \ldots k. \tag{11}$$

b) **Determine the DMs opinions evaluations under each criterion by an objective method:**
By mutual evaluation, ascertain the DMs weights, each DM assesses the other two DMs in turn, allocating relative evaluation weights also known as paired evaluations based on the comparison of the conformance between the two DMs personal qualities and the decision problem. This allows for the determination of each DM relative weights. After adding together all of the evaluations and using linear normalization, the final weights for every DM are determined. The following equation provides the calculation [14]:

$$\gamma_k^j = \frac{\sum_{K=1}^{k} \sum_{b=1}^{K} C_{kab}^j}{\sum_{k=1}^{K} \sum_{a=1}^{K} \sum_{b=1}^{K} C_{kab}^j}, \quad j = 1, 2 \ldots n. \tag{12}$$

c) **Determine the combinative weights of DMs opinions evaluations under each criterion:** In order to provide a more accurate and balanced weighting for the DMs opinions evaluation, the geometric mean and normalization techniques are employed to aggregate the weights from both subjective and objective weighting methods:
At the end of this step we should normalize the obtained values of combinative weights α_k^j such that:

$$\alpha_k^j = \frac{\sqrt{(\beta_k) \cdot (\gamma_k)}}{\sum_{k=1}^{K} \sqrt{\sqrt{(\beta_k) \cdot (\gamma_k)}}}, \quad j = 1, 2 \ldots n. \tag{13}$$

Step 5: Apply the WSAM-OWA operator on each n-tuple of the matrix Z using equations Eqs. (1)–(5).
Step 6: Construct the final normalized collective preferences decision matrix $\tilde{Z}$.

5.2 Ranking Phase

In this phase, TOPSIS method is going to be applied on the collective decision matrix $\tilde{Z}$ to rank the alternatives.
To ensure an increasing comprehensiveness and reliability of the criteria weights, a combinative weighting calculation method is applied to leverage advantages of both subjective and objective weighting calculation methods.
Step 7: Determine the level of expertise of DMs overall the problem by proceeding as detailed bellow:

a) Compute the arithmetic mean of self evaluation coefficients $\beta_k^{'j}$ for each DM_k from **Step 4-a)** to get the mean value of self evaluation coefficients $\beta_k^{'}$ for each DM_k overall the problem.
b) Aggregate the objective opinions evaluations γ_k^j for each C_j for each DM_k obtained in **Step 4-b)** by summing the products of each γ_k^j by the weight W_j of the criterion C_j, at this stage we are going to consider that all criteria have same importance in the decision making process. Consequently, we assess them the same weight as follows:

$$\gamma_k = \sum_{j=1}^{n} W_j * \gamma_k^j, \quad k = 1, 2 \ldots K. \tag{14}$$

c) Calculate the empirical coefficient of each DM_k using the equation:

$$\delta_k = \frac{(\beta_k' - 0.5) + \frac{K}{2}\left(\gamma_k - \frac{1}{K}\right)}{2}, \quad k = 1, 2 \ldots K. \tag{15}$$

If $\delta_k > 0$ the DM_k is considered as experienced, else he's inexperienced.

Step 8: Choose an objective and a subjective weighting methods based on the result of **Step 4-c)**. Then, compute for each criterion C_j its subjective and objective weights.
Step 9: Compute the criteria weights using the subjective and objective clustering combinative weighting method (SOC) using Eq. (6).
Step 10: Apply TOPSIS steps as described in section.

6 Illustrative Numerical Example

Consider a factory aiming to enhance production efficiency and optimize profitability by investing in new process technologies. From a set of four alternatives: A_1, A_2, A_3, A_4, the decision-makers must select the best process technology based on three predefined criteria namely: "Cost" (C_1), represented by crisp values; "Profit Flexibility" (C_2), expressed using linguistic values; and "Lead Time" (C_3), defined with interval values.

Consensus Phase: WSAM-OWA with a combinative weighting method for the DMs expertise level

We start this phase with **Step 1**, by collecting data as shown in Table 1.

Table 1. Heterogeneous preferences decision matrices for DM_1, DM_2, and DM_3.

	A_i	C_1	C_2	C_3		A_i	C_1	C_2	C_3		A_i	C_1	C_2	C_3
	A_1	250,2	M	[3,6]		A_1	240,1	H	[4,7]		A_1	260,4	L	[2,5]
DM_1	A_2	200,9	M	[3,5]	DM_2	A_2	210,5	M	[3,6]	DM_3	A_2	230,3	M	[3,6]
	A_3	325,5	VH	[4,6]		A_3	310,7	VH	[5,8]		A_3	340,9	H	[4,7]
	A_4	500	EH	[6,12]		A_4	480	EH	[7,13]		A_4	490	EH	[6,12]

After normalizing the data as outlined in **Step 2**, we construct the normalized decision matrix Z, as described in **Step 3**, where the elements are 3-tuples (see Table 2).

Table 2. Normalized matrix Z.

	C_1	C_2	C_3
Z_{ij}^s	$\{Z_{i1}^1, Z_{i1}^2, Z_{i1}^3\}$	$\{Z_{i2}^1, Z_{i2}^2, Z_{i2}^3\}$	$\{Z_{i3}^1, Z_{i3}^2, Z_{i3}^3\}$
A_1	{0.37, 0.46, 0.37}	{0.58, 0.43, 0.43}	{0.38, 0.27, 0.39}
A_2	{0.30, 0.31, 0.40}	{0.58, 0.43, 0.58}	{0.33, 0.40, 0.41}
A_3	{0.48, 0.52, 0.38}	{0.88, 0.93, 0.58}	{0.40, 0.56, 0.58}
A_4	{0.74, 0.65, 0.75}	{0.93, 0.93, 0.88}	{0.76, 0.67, 0.57}

Step 4 consists of the evaluation of the trust degree (expertise) of each DM_k on each criterion C_j. To do so, we are going to compute both objective and subjective weights for the DMs over each criterion by proceeding to Eqs. (11) and (12), the results are summarized in Table 3.

Table 3. Subjective DMs weights β_k^j and objective DMs weights y_k^j.

	Objective weighting			Subjective weighting					
	y_k^j			$\beta_k^{'j}$			β_k^j		
	y_1^j	y_2^j	y_3^j	$\beta_1^{'j}$	$\beta_2^{'j}$	$\beta_3^{'j}$	β_1^j	β_2^j	β_3^j
C_1	0.3	0.65	0.55	0.5	0.5	1	0.25	0.25	0.5
C_2	0.575	0.55	0.375	0.4	0.5	0.4	0.307	0.384	0.307
C_3	0.45	0.35	0.7	1	0.8	0.6	0.416	0.333	0.25

The combinative DMs weights calculated by Eq. (13) should be normalized in [0, 1]. Table 4 shows the obtained results.

Table 4. Combinative DMs weights and their normalized values for each criterion C_j.

C_j	a_k^j	DM_1	DM_2	DM_3	C_j	$\hat{a}_k^j$	DM_1	DM_2	DM_3
C_1	a_k^1	0.382	0.563	0.691	C_1	$\hat{a}_k^1$	0,233	0,343	0,422
C_2	a_k^1	0.382	0.563	0.691	C_2	$\hat{a}_k^2$	0,344	0,376	0,278
C_3	a_k^1	0.382	0.563	0.691	C_3	$\hat{a}_k^3$	0,362	0,286	0,350

To demonstrate the calculation steps of **Step 5**, we take as an example the fourth alternative A_4 with criteria C_2 and C_3. After determining $W_{i,1}$ using Eq. 2, the next steps follow accordingly. For more clarity, we start by order each value of each tuple with its corresponding cardinal and the trust degree of each DM_k on each criterion C_j as follows:

A_4 under C_2: $(0.88, 1, 0.278)$ $(0.93, 2, 0.72)$.
A_4 under C_3: $(0.76, 1, 0.362)$ $(0.67, 1, 0.286)$ $(0.57, 1, 0.35)$.

For the values of A_4 with respect to C_2, we got the following results: $E = 1.718$, $\sigma = 0.2005$, and the caridinality relevance factor CRF=0.55. Where, $g_{1,1} = 0.55$, $g_{1,2} = 0.45$, $g_{2,1} = 0.55$, and $g_{2,2} = 0.55$, $y_2 = 1.92650$ and $z_2 = 2.00$.

Finally, we compute $W_{1,2} = 0.260$ and $W_{2,2} = 0.738$. Applying the WSAM-OWA operator, we obtain the aggregated value for the tuple elements of A_4 with respect to C_4:

$$F_{WSAM-OWA}(\{0.93, 0.93, 0.88\}) = 0.91160.$$

By repeating this process for each tuple across all criteria, we obtain the collective preference decision matrix shown in Table 5.

Table 5. Normalized collective preferences decision matrix.

A_i	C_j		
	C_1	C_2	C_3
A_1	0.54	0.477	0.351
A_2	0.345	0.526	0.377
A_3	0.451	0.814	0.508
A_4	0.712	0.916	0.667

Ranking Phase: TOPSIS with combined criteria weighting

At **Step 7**, we start with computing the arithmetic mean of self evaluation coefficients $\beta_k^{'j}$ for each DM_k to get the mean value of self evaluation coefficients for each DM_k overall the problem. We obtain the following values: $\beta_1^{'} = 0.63$, $\beta_2^{'} = 0.6$, $\beta_3^{'} = 0.67$. After that, we aggregate the objective opinions evaluations γ_k^j for each C_j for each DM_k by summing the products of each γ_k^j by the weight W_j of the criterion C_j, by considering $W_j = \left(\frac{1}{3}, \frac{1}{3}, \frac{1}{3}\right)$. At the end of this step, we obtain the objective opinions evaluations for each DM_k overall the problem: $y_1 = 0.4416$, $y_2 = 0.5166$ and $y_3 = 0.5416$.

Finally, we compute the empirical coefficient emp_k to measure the level of expertise of each DM_k using Eq. (15), the findings are presented as follows: $emp_1 = 0.1476$, $emp_2 = 0.1875$, $emp_3 = 0.2375$. Since all the values of $emp_k > 0$, we discern that the three DMs are all with high level of expertise and knowledge in regard to this decision problem.

As all of our DMs possess high expertise, at **Step 8**, they must select appropriate subjective and objective weighting methods to calculate the subjective weight W_j^s and objective weight W_j^o for each criterion. For our example, the DMs chose the SPD method to calculate W_j^s, while the ENTROPY method [6] was used to compute W_j^o. The results are presented in Table 6.

Table 6. Subjective and objective criteria weights.

Criterion	SPD	ENTROPY
C_j	W_j^s	W_j^o
C_1	0.5	0.1499
C_2	0.15	0.7853
C_3	0.07	0.0647

At **Step 9**, since the three DMs have selected one subjective and one objective weighting methods to evaluate the criteria weights, we don't need to set values to α_p and α_q, furthermore, due to the fact that the three DMs are experienced α_s would be prioritized by the DMs in relation to α_o. DMs set α_s and

α_o values respectively equal to 0.6 and 0.4. After applying the W_{soc} formula described in Eq. (6), we obtain the accurate combined criteria weights vector $W_j = (0.462, 0.435, 0.101)$.

Finally at **Step 10**, we apply TOPSIS steps on the normalized collective preferences decision matrix $\tilde{Z}$ on each alternative under all criteria with the vector of combined weights $W_j = (0.462, 0.435, 0.101)$. Since the alternative A_4 has the highest value of closeness coefficient $CC_1 = 1$, followed by A_3 with $CC_3 = 0.5454$, A_1 with $CC_1 = 0.3009$ and finally A_2 with the lowest value $CC_2 = 0.0816$. Hence, the final ranking is $A_4 > A_3 > A_1 > A_2$. It indicates that the best technology for the organization to select is A_4.

7 Conclusion

The challenge of selecting the optimal process technology system is a critical issue for manufacturing companies and operations managers. This research work addressed this challenge within the framework of multi-criteria group decision-making (MCGDM), proposing an approach that integrates the weighted selective aggregated majority-OWA (WSAM-OWA) operator. This operator accounts for the varying expertise levels of decision-makers, which significantly influences the aggregation process. By applying combinative methods, the evaluation of decision-makers' contributions and the importance of criteria were enhanced. An illustrative example demonstrated the practical applicability of the proposed approach, followed by a comprehensive analysis of the obtained numerical results. In the near future, we plan to implement a consensus coupling approach that combines goal programming with a satisfaction function (GPSF) to better address the diverse needs of decision-makers. Additionally, a sensitivity analysis using the Levenshtein distance will help evaluate discrepancies between individual and collective solutions, ensuring more robust and reliable outcomes.

References

1. Bordogna, G., Sterlacchini, S.: Extending the Pasi-Yager method with the Minkowski OWA-based similarity measure for majority opinion modeling. Fuzzy Sets Syst. **253**, 34–48 (2014)
2. Dong, Y., Zhang, Y., Li, J.: Consensus-building and shared responsibility in collaborative decision-making. Group Decis. Negot. **29**(4), 721–736 (2020)
3. Huang, X., Tang, Y., Wang, Y.: Structured communication, negotiation, and decision-making in diverse teams. J. Organ. Behav. **40**(3), 512–527 (2019)
4. Karanik, A., Peláez, J.I., Doña, J.M.: The selective MA-OWA operator for faster convergence in decision-making. Expert Syst. Appl. **64**, 437–447 (2016)
5. Kundu, P., Kar, S., Maiti, M.: A fuzzy multi-criteria group decision making based on ranking interval type-2 fuzzy variables and an application to transportation mode selection problem. Soft Comput. **21**, 3051–3062 (2017)
6. Lotfi, F.H., Fallahnejad, R.: Imprecise Shannon's entropy and multi attribute decision making. Entropy **12**(1), 53–62 (2010)

7. Mardani, A., Nilashi, M., Zavadskas, E.K., Awang, S.R., Zare, H., Jamal, N.M.: Decision-making methods based on fuzzy aggregation operators: three decades review from 1986 to 2017. Int. J. Inf. Technol. Decis. Mak. **17**(2), 391–466 (2018)
8. Merigó, J.M., Gil-Lafuente, A.M.: The induced generalized OWA operator. Inf. Sci. 729–741 (2009)
9. Merigó, J.M., Casanovas, M.: The fuzzy generalized OWA operator and its application in strategic decision making. Cybern. Syst. **41**, 359–370 (2010)
10. Merigó, J.M., Yager, R.R.: Generalized moving averages, distance measures and OWA operators. Int. J. Uncertain. Fuzz. **21**, 533–559 (2013)
11. Pasi, G., Yager, R.R.: Modeling the majority concept using OWA operators. Int. J. Approximate Reasoning **41**(2), 123–145 (2006)
12. Peláez, J.I., Doña, J.M.: Majority additive OWA operator for group decision making. Eur. J. Oper. Res. **145**(3), 632–641 (2003)
13. Peláez, J.I., Doña, J.M., Cruz-Cunha, M.M.: The selective aggregated majority OWA operator for weighted group decision making. Knowl.-Based Syst. **95**, 57–65 (2016)
14. Wan, C., Li, W., Ling, S., Li, C., Yang, G.: Multi-scenario group decision-making based on TOPSIS for deep hole drill parameter optimization. Arab. J. Sci. Eng. **47**(12), 15779–15795 (2022)
15. Xu, Q., Xia, Y.: The impact of formal decision support systems on decision-making outcomes. Decis. Support Syst. **146** (2021)
16. Yager, R.R.: On ordered weighted averaging aggregation operators in multi-criteria decision making. IEEE Trans. Syst. Man Cybern. **18**, 183–190 (1988)
17. Yager, R.R., Kacprzyk, J.: The Ordered Weighted Averaging Operators: Theory and Applications. Kluwer, Norwell, MA, US (1997)
18. Yager, R.R., Filev, D.P.: Induced ordered weighted averaging operators. IEEE Trans. Syst. Man Cybern. **29**(2), 141–150 (1999)
19. Yusoff, B., Merigó, J.M., Ceballos, D., Peláez, J.I.: Weighted-selective aggregated majority-OWA operator and its application in linguistic group decision-making. Int. J. Intell. Syst. **33**(9), 1929–1948 (2018)

Economics and Finance

Graph-Based Features and Refined Diffusion Models for Stock Price Prediction

Cuong Nhat Nguyen[1(✉)], Khang Minh Vuong[1], and Thi Kim Dzung Pham[2]

[1] Swinburne Vietnam, FPT University, Hanoi, Vietnam
{nhatncswh01230,minhvkswh01271}@fpt.edu.vn
[2] Foreign Trade University, Hanoi, Vietnam
dungpk@ftu.edu.vn

Abstract. Predicting stock prices presents significant challenges due to the inherent complexity and chaotic behavior of financial time series data. Traditional deep learning models often struggle to capture long-term dependencies and manage the non-stationary nature of stock prices. To enhance prediction accuracy while effectively addressing these challenges, this paper introduces a framework that employs graph-based feature extraction and Diffusion Variational Autoencoders (Diffusion-VAEs). By transforming stock data into complex graph networks and utilizing feature embeddings alongside a diffusion process, our approach effectively captures both the underlying structure and temporal dependencies, enriching the charac teristics of stock data. Experimental results on real-world stock datasets, spanning both regression and classification tasks, demonstrate that our method significantly outperforms recent state-of-the-art models by substantially lowering MAPE and RMSE while boosting the F1-score by over 20%, proving its accuracy and stability.

Keywords: Stock Price Prediction · Variational Autoencoder · Diffusion Model · Graph-based Technique · Generative Adversarial Network

1 Introduction

Predicting stock prices remains a persistent challenge in the financial industry, where accurate forecasts can result in substantial gains or losses. Traditional prediction methods often struggle with the inherent volatility and noise of financial markets. To tackle these issues, various generative models have been employed, including Generative Adversarial Networks [1] and Wasserstein GANs [2], as well as Variational Autoencoders [3,4]. Recently, the integration of Diffusion-VAEs has emerged as a leading approach, offering improved performance and robustness in stock price prediction [5].

© The Author(s), under exclusive license to Springer Nature Switzerland AG 2025
H. A. Le Thi et al. (Eds.): ICAMCS 2024, LNNS 1311, pp. 263–275, 2025.
https://doi.org/10.1007/978-3-031-90606-0_22

Stock price data is inherently time series data, characterized by its sequential nature and intricate temporal dependencies [6]. Traditional forecasting methods often struggle to capture these complex patterns, resulting in suboptimal predictions. To address these challenges, various traditional feature engineering techniques have been employed, such as the Stochastic Oscillator [7], Moving Average Convergence Divergence [8], Relative Strength Index [9], etc. Although these techniques provide certain improvements, they may not fully leverage the underlying structural information embedded in the data.

To more effectively capture the complexities of stock data, graph-based methods may offer a valuable approach for extracting intricate relationships [10]. Prominent techniques in this domain include Recurrence Networks [11], Visibility Graphs [12], and Transition Networks [13]. Among these, Visibility Graphs are particularly advantageous for extracting the characteristics of stock price due to their simplicity, the absence of the need for parameter optimization, and their ability to effectively capture the essential structural elements of price movements.

In addition to using features extracted through traditional methods, this study proposes a framework that incorporates graph-based embedding features alongside a refined Diffusion model to enhance stock price prediction accuracy. The contributions of the research are as follows:

- **Graph-Based Technique Integration:** We utilize Visibility Graphs to embed stock data, enabling us to extract features that encapsulate the underlying structural information.
- **Enhancement of Diffusion VAEs:** We enhance the computation of the probabilistic distribution within the latent space of the Variational Autoencoder (VAE) used in the Diffusion Model. This optimization improves prediction accuracy and fully leverages the model's capability to capture complex and stochastic patterns.
- **Real-World Testing:** Our research, validated with datasets from five well-known companies (2013–2022), shows superior performance in both regression and classification tasks. The proposed method outperforms baseline models across various metrics, including MAPE, RMSE, Accuracy, Precision, Recall, and F1-score.

The structure of the paper is as follows: Sect. 2 reviews relevant literature. Section 3 describes the methodology in detail, including graph-based feature engineering and the architecture of the diffusion model. Section 4 presents the experimental setup and discusses the results. Finally, Sect. 5 offers conclusions, insights, and suggestions for future research.

2 Related Works

2.1 Feature Engineering

Stochastic Oscillator. As mentioned by Lane in 1985 [7], the Stochastic Oscillator (SO) serves as a representative indicator for stock price data within the market through a technical trading signal. The SO indicator (%k) is calculated by determining the ratio between the difference of the closing price (c) and the lowest price (l), and the difference between the highest price (h) and the lowest price [14].

$$\%k = \frac{c - l}{h - l} \times 100$$

Moving Average Convergence Divergence. The Moving Average Convergence Divergence (MACD) indicator is extensively used to forecast long-term price movements by analyzing the interaction between two exponential moving averages (EMAs) of different periods [8], detailed as follows:

$$EMA = p_i \cdot \frac{2}{n+1} + EMA_{i-1} \cdot \left(1 - \frac{2}{n+1}\right),$$

where p_i represents the asset price for the current period, i, and n denotes the number of data points used in the moving average calculation.

Absolute Returns. The Relative Strength Index (RSI) is to measure the speed and magnitude of price movements [15]. The RSI is computed by comparing the magnitude of recent gains to recent losses over a specified period as:

$$RSI_{i,t} = 100 \times \frac{RS_{i,t}}{1 + RS_{i,t}},$$

where the Relative Strength (RS) is given by:

$$RS_{i,t} = \frac{\sum_{k=0}^{13} \max\left(0, P_{i,t-k} - P_{i,t-1-k}\right)}{\sum_{k=0}^{13} \max\left(0, P_{i,t-1-k} - P_{i,t-k}\right)},$$

with $P_{i,t}$ represents the closing price of stock i on day t.

The combination of different indicators, such as MACD, OS, and RSI, helps improve the model's forecasting accuracy compared to using only the original data by exploring potential trade signals [9].

2.2 Time Series Prediction

Traditional Models. Fundamental problems like binary classification and linear regression have been used in stock price prediction. Traditional classification models predict whether stock prices will rise or fall [16], while regression models forecast their exact numerical values [17]. These models work well for single-step predictions and short-term forecasts but struggle to capture the complex, non-linear relationships and long-term dependencies in financial time series.

Recurrent Neural Networks. To address the limitations of traditional prediction models, Recurrent Neural Networks (RNNs) [18] have been introduced. Among RNN variants, Long Short-Term Memory (LSTM) networks [19] are popular for stock price prediction due to their ability to capture long-term dependencies with memory cells and gating mechanisms. Gated Recurrent Units (GRUs) [20], another RNN variant, streamline this with fewer parameters using update and reset gates [21]. While LSTM and GRU models improve the capture of market dynamics, they often face challenges with predictive accuracy, especially with volatile or complex datasets [22].

Generative Adversarial Networks. Due to their ability to generate realistic data and model complex relationships, Generative Adversarial Networks (GANs) are powerful tools for stock price prediction [1]. Basic GANs typically utilize a 1D-CNN discriminator and a GRU-based generator, optimizing through cross-entropy to minimize the difference between real and generated data. The Wasserstein GAN with Gradient Penalty (WGAN-GP) variant enhances stability by employing Wasserstein distance and adding a gradient penalty for a more stable training process [2]. Despite these advancements, GAN-based models, especially those with RNN architectures, still encounter hyperparameter tuning challenges that can affect predictive accuracy across various datasets [23].

Variational Autoencoder. VAEs are popular generative models in machine learning, with variants like Hierarchical VAE (HVAE) [5], Bidirectional VAE (BVAE) [24], and Nouveau VAE (NVAE) [4], each tailored for specific tasks or datasets. Despite their versatility, VAEs are often unsuitable for stock market prediction due to the high variability from their stochastic latent space, which can result in unpredictable outputs [5]. However, VAEs excel in dimensionality reduction, transforming high-dimensional data into lower-dimensional latent spaces [25]. This capability is valuable for data compression, anomaly detection, and particularly for diffusion models, as discussed in Section Methodology.

2.3 Model Evaluation

Root Mean Squared Error. One of the most commonly used metrics for evaluating the accuracy of predictive models, especially in regression tasks, is the Root Mean Square Error (RMSE) [26]. RMSE measures the average magnitude of the prediction error, providing a clear indication of how closely the predicted values align with the actual outcomes. The RMSE is formally defined as:

$$RMSE = \sqrt{\frac{1}{n}\sum_{i=1}^{n}(y_i - \hat{y}_i)^2}, \tag{1}$$

where n represents the number of observations, y_i is the actual value, and $\hat{y}_i$ is the predicted value for the ith observation. A lower RMSE value indicates a better fit of the model to the data.

Mean Absolute Percentage Error. One widely used metric for evaluating the accuracy of predictive models is the Mean Absolute Percentage Error (MAPE) [27]. This metric is calculated using the following formula:

$$MAPE = \frac{1}{n}\sum_{t=1}^{n}\left|\frac{e_t}{Y_t}\right| \times 100\%, \tag{2}$$

where Y_t represents the actual value, e_t is the error, and n is the number of observations. The MAPE provides an average percentage error, with lower values indicating better model accuracy.

Classification Performance. Evaluating the performance of prediction models for classifying price movements (whether prices will rise or fall) involves using several key metrics, such as accuracy, precision, recall, and F1-score. Using multiple evaluation metrics simultaneously provides insights into various aspects of the model's predictive capabilities, offering a comprehensive view of its performance.

3 Methodology

3.1 Graph-Based Feature Engineering

Recent advances in complex network methods offer new ways to analyze time series data by converting them into graphs, which help uncover hidden structures and better address long-term dependencies and chaotic behaviors [28]. Popular Graph-Based methods include Recurrence Networks (RNs) for statistical similarities [11], Visibility Graphs (VGs) for local convexities [12], and Transition Networks (TNs) for state transitions [13]. We selected VGs due to their parameter-free nature and effectiveness in refining important features from time series data [29]. VGs help identify data characteristics, as shown in Fig. 1, through two main steps:

Time Series to Graph Transformation. The VGs preserve the temporal order and local properties of the original series data. In the graph, each data point in a time series $\{x(t)\}$ (where $t = 1, 2, \ldots, N$) is treated as a node. Two nodes $x(i)$ and $x(j)$ (where $i < j$) are connected if they satisfy the visibility condition:

$$x(k) < x(i) + \frac{x(j) - x(i)}{j - i}(k - i) \text{ for all } k \text{ where } i < k < j.$$

This means that there is a direct line of sight between $x(i)$ and $x(j)$ without any other data points obstructing the view.

The VG algorithm is parameter-free, preserving the temporal structure and local convexities of the original time series, making it straightforward to apply to various types of time series data.

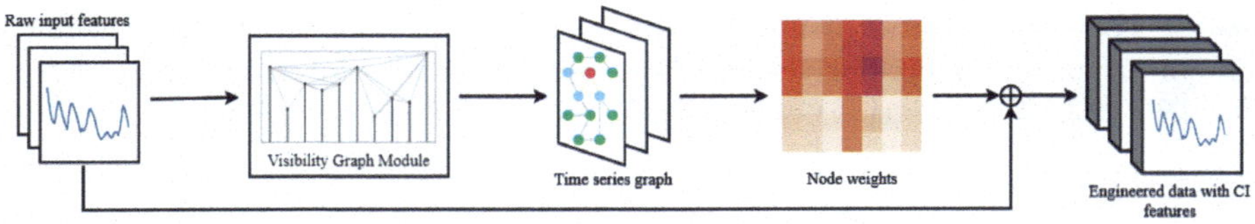

Fig. 1. Graph of stock prices

Node Embeddings and Weights. The Collective Influence (CI) algorithm represents the most influential nodes within the graph [30]. Given a graph $G(V, E)$ where V is the set of nodes and E is the set of edges, the CI algorithm aims to identify nodes whose removal would maximally disrupt the network's connectivity. The influence of a node i^{th} is determined by defining a radius ℓ, which sets the local neighborhood around each node by:

$$CI(i) = (k_i - 1) \sum_{j \in \partial B(i,\ell)} (k_j - 1),$$

where k_i is the degree of the node i, and $\partial B(i, \ell)$ is the boundary of the ball of radius ℓ centered at node i. This measure captures the influence of node i^{th} on the network, taking into account the influence propagated through its neighbors. The nodes are then ranked based on their CI scores, with the top-ranking nodes considered the most influential [31].

3.2 Refined Diffusion Model

The Diffusion-VAEs integrate various VAEs with diffusion probabilistic techniques to tackle the challenges of stochasticity in multi-step regression tasks, particularly in stock price prediction [5]. In this study, we refine the original HVAE [5] to extract latent vector representations from the data while implementing a diffusion model. A detailed implementation of our proposed framework (named DVAE) is illustrated in Fig. 2.

Hierarchical Variational Autoencoder. Formally, let $\mathbf{X}$ denote the input sequence of stock prices and $\mathbf{Y}$ the target sequence we aim to predict. The encoder network $q_\phi(\mathbf{Z}_+|\mathbf{X})$ of the hierarchical VAE learns a latent representation $\mathbf{Z}_+$, which is sampled and fed into the decoder $p_\theta(\mathbf{Y}|\mathbf{Z}_+)$ to generate the predicted sequence: $\mathbf{Z}_+ \sim q_\phi(\mathbf{Z}_+|\mathbf{X}), \quad \hat{\mathbf{Y}} = p_\theta(\mathbf{Y}|\mathbf{Z}_+)$. Here, $\mathbf{Z}_+$ represents a hierarchical stack of latent variables that capture the stochastic and nonlinear relationships within the stock price data. The refined HVAE combines the stochastic probability distributions of latent variables across hierarchical layers using addition, rather than concatenation as in the original HVAE structure. This hierarchical approach enhances the model's ability to manage the inherent randomness of stock prices, which is essential for improving the accuracy and robustness of predictions.

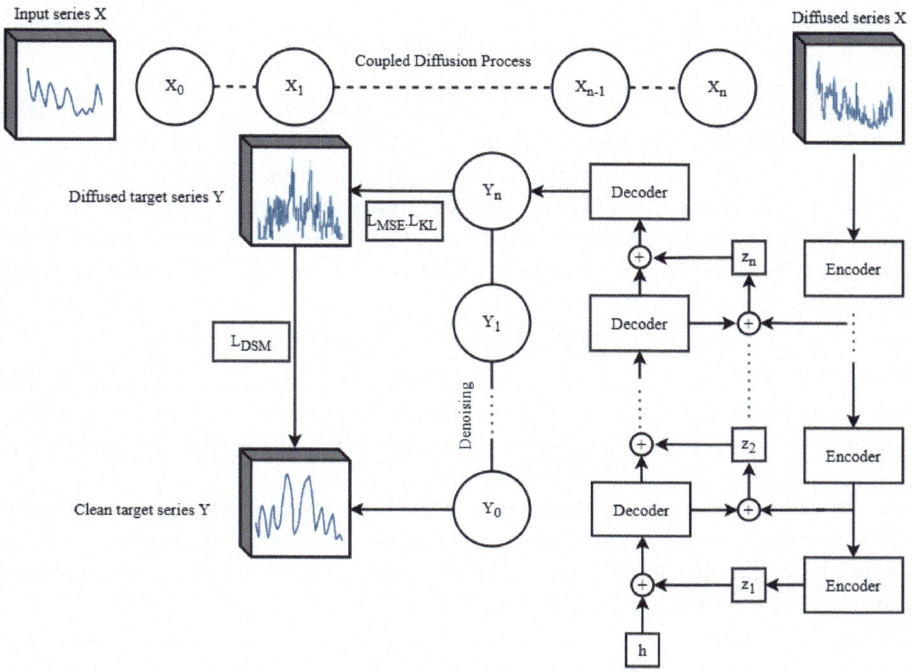

Fig. 2. DVAE architecture

Diffusion Process. To handle the stochastic nature of stock prices more effectively, we employ a diffusion modeling approach. This method introduces Gaussian noise into the input data X and target data Y through a Markov chain of diffusion steps, producing progressively noisier samples $[X_1, X_2, \ldots, X_n]$ and $[Y_1, Y_2, \ldots, Y_n]$.

At each diffusion step t, the noise variance is controlled by a variance schedule $\beta_t \in [0, 1]$. Additionally, a scaling factor λ is applied to modify the variance schedule specifically for the target data, allowing for a separate noise level adjustment between the input and target sequences. The reparameterization trick is used to obtain tractable closed-form samples X_t and Y_t at arbitrary steps. The model is trained to predict target sequences under varying noise levels, enhancing its robustness and generalizability across market conditions.

Total Loss Function. The total loss function for our proposed diffusion model is meticulously designed to optimize the performance of the model by combining the following components.

The **Mean Squared Error (MSE)** component, denoted as L_{MSE}, is used to minimize the difference between the predicted and actual stock prices, that is calculated as:

$$L_{\text{MSE}} = \frac{1}{n}\sum_{i=1}^{n}(y_i - \hat{y}_i)^2,$$

where y_i represents the actual stock prices, $\hat{y}_i$ the predicted prices, and n the number of samples. This component ensures that the model's predictions are as close as possible to the actual values, which is crucial for accurate forecasting.

The **Kullback-Leibler (KL) divergence** component, L_{KL}, measures the difference between the learned latent variable distribution and a prior distribution, typically a standard normal distribution. The KL divergence is given by:

$$L_{\text{KL}} = \mathbb{E}_{q_\phi(\mathbf{Z}_+|\mathbf{X})}\left[\log p_\theta(\mathbf{Y}|\mathbf{Z}_+)\right] - D_{KL}\left(q_\phi(\mathbf{Z}_+|\mathbf{X}) \parallel p(\mathbf{Z}_+)\right),$$

where D_{KL} denotes the KL divergence between the learned posterior $q_\phi(\mathbf{Z}_+|\mathbf{X})$ and the prior $p(\mathbf{Z}_+)$. This component ensures that the latent space learned by the encoder is regularized and follows a known distribution, aiding in stable and interpretable latent variable representations.

The **Denoising Score Matching (DSM)** component, L_{DSM}, adapts principles from diffusion probabilistic models. This component is vital for training the model to handle noisy data and effectively recover the true sequences from noisy inputs. The DSM term can be expressed as:

$$L_{\text{DSM}} = \mathbb{E}_{p(x_t|x_0)}\left[\|s_\theta(x_t, t) - \nabla_{x_t}\log p(x_t|x_0)\|^2\right],$$

where s_θ is the score function approximated by the model, x_t represents the noisy data at time step t, and x_0 is the original data. This term trains the model to match the gradients, learning an energy function that helps in denoising the sequences and capturing the true underlying patterns in the data.

The total loss function L_{total} combines three above components with respective weighting factors ζ and η to balance their contributions:

$$L_{\text{total}} = L_{\text{MSE}} + \zeta \cdot L_{\text{KL}} + \eta \cdot L_{\text{DSM}},$$

where ζ and η are hyperparameters that control the influence of the KL divergence and DSM components, respectively. This composite loss function ensures that the model not only accurately predicts the stock prices but also maintains a well-regularized latent space and effectively denoises the sequences, thereby improving the overall robustness and generalizability of the diffusion model.

4 Experiment

4.1 Datasets and Baseline Models

Datasets. For our experiments, we utilized a stock dataset[1] comprising 11,420 samples from five major technology companies—Apple, Amazon, Google, Meta, and Netflix—spanning the years 2011 to 2020. We meticulously partitioned the

[1] https://www.kaggle.com/datasets/soumendraprasad/stock.

dataset into training, validation, and test sets with a ratio of 8:1:1. This division is essential for ensuring the reliability and generalizability of the predictive models.

Baseline Models. To prove the effectiveness of the proposed DVAE, we selected advanced baseline models for comparison. GANs - strong performance in simulating intricate financial dynamics and capturing complex patterns in financial time series [2]; WGANs - ability to address stability issues during training and offer robust, reliable predictions through the use of the Wasserstein distance metric [2]; DARNN-CAAN - combines the effectiveness of capturing long-term dependencies as seen in DARNN [32], with the ability to enhance prediction accuracy by leveraging interrelationships among time series through self-attention mechanisms in price graphs (namely VGs), as demonstrated by CAAN [29].

4.2 Parameters Setting

In our model configuration, the input sequence length (T) was consistently set to 12 days with 2 cases of output prediction horizons, specifically 5 days and 10 days. The Adam optimizer, initialized with a learning rate of 3×10^{-3}, and a fixed batch size of 16 was employed for training, and each model underwent 20 epochs for individual stocks, selecting the iteration with the most favorable validation results. The diffusion process entailed 10 steps with a β schedule ranging from 0.01 to 0.1 and λ set to 0.5. The DSM network featured 16 channels, a dimension of 16, and a size of 5, considering 5 diffusion samples. Hyperparameter trade-offs underwent a grid search with a step size of 0.1 within the range [0.1, 1], establishing ζ at 0.5 and η at 1. Each experiment was repeated five times per stock, with the reported results including the overall MAPE, RMSE, and classification performance metrics.

4.3 Experimental Result

Experiment 1: Effectiveness of Regression Prediction Models. This experiment assesses the stock price prediction performance of three models: GAN, WGAN, and DVAE under different conditions: (1) using raw data for 5-step predictions, (2) using data with graph-based features extracted from VGs for 5-step predictions, and (3) using data with graph-based features extracted from VGs for 10-step predictions.

Table 1. Stock close price prediction with baseline models and DVAE

		Apple		Amazon		Google		Netflix		Meta	
		RMSE	MAPE	RMSE	MAPE	RMSE	MAPE	RMSE	MAPE	RMSE	MAPE
raw	GAN	9.76	19.85	10.42	16.24	10.67	19.39	39.09	17.92	20.94	18.67
	WGAN	8.57	17.43	10.25	15.95	9.16	16.64	33.24	15.24	19.69	17.56
	DVAE	**2.13**	**4.33**	**3.68**	**5.73**	**2.72**	**4.94**	**13.35**	**6.12**	**5.8**	**5.17**
5-step	GAN	5.39	10.97	9.55	14.87	6.14	11.16	34.07	15.62	11.60	10.34
	WGAN	5.24	10.67	7.59	11.83	5.55	10.09	29.84	13.68	9.84	8.78
	DVAE	**1.97**	**4.02**	**2.83**	**4.41**	**1.95**	**3.55**	**10.16**	**4.66**	**4.88**	**4.35**
10-step	GAN	6.02	12.25	10.17	15.84	7.82	14.22	37.47	17.18	14.98	13.36
	WGAN	6.75	13.74	9.36	14.58	7.51	13.65	31.69	14.53	14.46	12.89
	DVAE	**2.74**	**5.58**	**4.65**	**7.24**	**3.04**	**5.53**	**16.18**	**7.42**	**7.45**	**6.64**

The results displayed in Table 1 highlight the superior performance of our proposed DVAE model compared to GAN and WGAN. The DVAE model consistently delivers the lowest RMSE and MAPE values across all datasets and time horizons, clearly demonstrating its effectiveness in capturing intricate stock price patterns. For instance, in the 5-step predictions, DVAE achieves an RMSE of 1.97 and MAPE of 4.02 for Apple, significantly outperforming GAN (5.39, 10.97) and WGAN (5.24, 10.67). Similarly, in the 10-step predictions, DVAE continues to excel with an RMSE of 2.74 and MAPE of 5.58 for Apple, again surpassing GAN and WGAN models.

These findings illustrate the DVAE model's robustness and precision, making it a highly effective tool for stock price prediction, especially in scenarios requiring longer-term forecasting. The model's consistent outperformance of traditional GAN-based approaches establishes DVAE as a more effective method, particularly when enhanced with advanced graph-based feature engineering techniques.

Experiment 2: Stock Trend Classification. This experiment demonstrates the effectiveness of DVAE compared to DARNN-CAAN. To ensure a fair comparison, we converted the problem into a classification task and utilized metrics such as accuracy, precision, recall, and F1-score to evaluate both models. The results of Experiment 2 represent predicted values, focusing on the trends indicating whether stock prices will rise or fall, informed by the data and graph-based features extracted from VGs.

Table 2. Stock trend classification with DARNN-CAAN and D-VAE models

Model	DARNN-CAAN				DVAE			
	Accuracy	Precision	Recall	F1-score	Accuracy	Precision	Recall	F1-score
Apple	0.55	0.74	0.49	0.59	**0.77**	**0.78**	**0.74**	**0.76**
Amazon	0.56	0.75	0.51	0.58	**0.78**	**0.76**	**0.78**	**0.77**
Google	0.57	0.74	0.53	0.60	**0.78**	0.74	**0.78**	**0.76**
Meta	0.53	0.77	0.51	0.58	**0.83**	**0.85**	**0.77**	**0.81**
Netflix	0.54	0.76	0.54	0.57	**0.87**	**0.86**	**0.88**	**0.87**

The results of stock trend classification presented in Table 2 demonstrate that the DVAE model consistently outperforms the DARNN-CAAN model. The DVAE model achieves an F1-score exceeding 0.76, reflecting approximately a 20% improvement over the DARNN-CAAN model. With DARNN-CAAN, although the precision results of 0.74–0.77 are high, the recall scores ranging from 0.49–0.54 are considerably lower. This indicates that this model misses many positive instances, resulting in a higher number of false negatives, and consequently lower accuracy and F1-score. This issue is addressed by DVAE, which demonstrates improved and balanced performance with both precision and recall scores ranging from 0.74 to 0.88. This demonstrates that the DVAE model performs effectively even with classification tasks and provides more reliable and balanced predictions for future stock trends.

5 Conclusion

The research demonstrates the superior performance of the proposed framework with the enhanced Diffusion model for stock price and trend prediction. Experiments comparing it with recent models like GAN and WGAN show that DVAE consistently offers better accuracy and reliability. By using a graph-based technique to convert time series data into a complex network and employing graph embeddings, our model captures intricate structural details often missed by traditional models. Integrating these graph-based features into the Diffusion model provides robust support for investors and financial analysts. Future work should explore further refinements and applications to other financial domains to develop reliable predictive models in complex financial markets.

References

1. Diqi, M., Hiswati, M.E., Nur, A.S.: StockGAN: robust stock price prediction using GAN algorithm. Int. J. Inf. Technol. **14**(5), 2309–2315 (2022)
2. Lin, H., Chen, C., Huang, G., Jafari, A.: Stock price prediction using generative adversarial networks. J. Comput. Sci. **17**(3), 188–196 (2021)

3. Gunduz, H.: An efficient stock market prediction model using hybrid feature reduction method based on variational autoencoders and recursive feature elimination. Financ. Innov. **7**(1), 1–24 (2021). https://doi.org/10.1186/s40854-021-00243-3
4. Vahdat, A., Kautz, J.: NVAE: a deep hierarchical variational autoencoder. Adv. Neural. Inf. Process. Syst. **33**, 19667–19679 (2020)
5. Koa, K.J., Ma, Y., Ng, R., Chua, T.-S.: Diffusion variational autoencoder for tackling stochasticity in multi-step regression stock price prediction. In: Proceedings of the 32nd ACM International Conference on Information and Knowledge Management, pp. 1087–1096 (2023)
6. Tran, D.T., Iosifidis, A., Kanniainen, J., Gabbouj, M.: Temporal attention-augmented bilinear network for financial time-series data analysis. IEEE Trans. Neural Netw. Learn. Syst. **30**(5), 1407–1418 (2018)
7. Lane, G.C.: Lane's stochastics: the ultimate oscillator. J. Tech. Anal. **21**, 37–42 (1985)
8. Aguirre, A., Medina, R., Méndez, N.: Machine learning applied in the stock market through the moving average convergence divergence (MACD) indicator. Invest. Manage. Financ. Innov. **17**(4), 44 (2020)
9. Phuong, L.: Investor sentiment by relative strength index and stock return: empirical evidence on Vietnam's stock market. Accounting **7**(2), 451–456 (2021)
10. Yang, Y., Wang, J., Yang, H., Mang, J.: Visibility graph approach to exchange rate series. Phys. A **388**(20), 4431–4437 (2009)
11. Donner, R.V., Zou, Y., Donges, J.F., Marwan, N., Kurths, J.: Recurrence networks-a novel paradigm for nonlinear time series analysis. New J. Phys. **12**(3), 033025 (2010)
12. Lacasa, L., Luque, B., Ballesteros, F., Luque, J., Nuno, J.C.: From time series to complex networks: the visibility graph. Proc. Natl. Acad. Sci. **105**(13), 4972–4975 (2008)
13. Nicolis, G., Cantu, A.G., Nicolis, C.: Dynamical aspects of interaction networks. Int. J. Bifurcat. Chaos **15**(11), 3467–3480 (2005)
14. Paik, C.K., Choi, J., Vaquero, I.U.: Algorithm-based low-frequency trading using a stochastic oscillator and William% R: a case study on the us and korean indices. J. Risk Financ. Manage. **17**(3), 92 (2024)
15. Sami, H.M., Ahshan, K.A., Rozario, P.N., Ashrafi, N.: Evaluating the prediction accuracy of MACD and RSI for different stocks in terms of standard market suggestions. Can. J. Bus. Inf. Stud. **7820**, 137–143 (2022)
16. Ding, Q., Wu, S., Sun, H., Guo, J., Guo, J.: Hierarchical multi-scale gaussian transformer for stock movement prediction. In: IJCAI, pp. 4640–4646 (2020)
17. Cakra, Y.E., Trisedya, B.D.: Stock price prediction using linear regression based on sentiment analysis. In: 2015 International Conference on Advanced Computer Science and Information Systems (ICACSIS), pp. 147–154. IEEE (2015)
18. Selvin, S., Vinayakumar, R., Gopalakrishnan, E., Menon, V.K., Soman, K.: Stock price prediction using LSTM, RNN and CNN-sliding window model. In: 2017 International Conference on Advances in Computing, Communications And Informatics (ICACCI), pp. 1643–1647. IEEE (2017)
19. Sunny, M.A.I., Maswood, M.M.S., Alharbi, A.G.: Deep learning-based stock price prediction using LSTM and bi-directional LSTM model. In: 2020 2nd novel intelligent and leading emerging sciences conference (NILES), pp. 87–92. IEEE (2020)
20. Gao, Y., Wang, R., Zhou, E.: Stock prediction based on optimized LSTM and GRU models. Sci. Program. **2021**(1), 4055281 (2021)

21. Sethia, A., Raut, P.: Application of LSTM, GRU and ICA for stock price prediction. In: Information and Communication Technology for Intelligent Systems: Proceedings of ICTIS 2018, vol. 2, pp. 479–487. Springer (2019)
22. Mateus, B.C., Mendes, M., Farinha, J.T., Assis, R., Cardoso, A.M.: Comparing LSTM and GRU models to predict the condition of a pulp paper press. Energies **14**(21), 6958 (2021)
23. Roy, S., Mehera, R., Pal, R.K., Bandyopadhyay, S.K.: Hyperparameter optimization for deep neural network models: a comprehensive study on methods and techniques. Innov. Syst. Softw. Eng. 1–12 (2023)
24. Lee, Y., Shin, J., Jung, K.: Bidirectional variational inference for non-autoregressive text-to-speech. In: International Conference on Learning Representations (2021)
25. Todo, W., Laurent, B., Loubes, J.-M., Selmani, M.: Dimension reduction for time series with variational autoencoders (2022)
26. Hodson, T.O.: Root mean square error (RMSE) or mean absolute error (MAE): when to use them or not. Geosci. Model Dev. Discuss. **2022**, 1–10 (2022)
27. Amar, S., Sudiarso, A., Herliansyah, M.K.: The accuracy measurement of stock price numerical prediction. In: Journal of Physics: Conference Series, vol. 1569, p. 032027. IOP Publishing (2020)
28. Donner, R.V., Small, M., Donges, J.F., Marwan, N., Zou, Y., Xiang, R., Kurths, J.: Recurrence-based time series analysis by means of complex network methods. Int. J. Bifurcat. Chaos **21**, 1019–1046 (2011)
29. Wu, J., Xu, K., Chen, X., Li, S., Zhao, J.: Price graphs: utilizing the structural information of financial time series for stock prediction (2021)
30. Zhu, F.: Improved collective influence of finding most influential nodes based on disjoint-set reinsertion. Sci. Rep. **8**(1), 14503 (2018)
31. Kong, D., Fu, C., Yang, J., Xu, D., Han, L.: The impact of the collective influence of search engines on social networks. IEEE Access **5**, 24920–24930 (2017)
32. Wu, J., Xu, K., Chen, X., Li, S., Zhao, J.: Price graphs: utilizing the structural information of financial time series for stock prediction. Inf. Sci. **588**, 405–424 (2022)

Potential Customer Detection for Decentralized Finance Applications Based on Wallet Transactions

Huy Hai Nguyen[1], Viet-Bang Pham[2], Lam Hoang[2], and Binh Minh Nguyen[1,3](✉)

[1] School of Information and Communication Technology, Hanoi University of Science and Technology, Hanoi, Vietnam
minh.nguyenbinh@hust.edu.vn
[2] Centic.io, Inc., Hanoi, Vietnam
[3] Institute for Digital Technology and Economy, Hanoi University of Science and Technology, Hanoi, Vietnam

Abstract. On blockchain platforms, individuals can engage in transactions and authentication processes without undergoing identity verification. This characteristic presents challenges for third-party service providers, such as Decentralized Applications, in accurately identifying and targeting their customer base. Consequently, the demand for detecting potential users has surged recently. However, most solutions today rely on off-chain data such as those on public forums and social media. Although these methods are straightforward, they demand significant time and effort and exhibit limited scalability and flexibility. In this research, we propose a novel methodology to identify potential users for a given lending decentralized application. The proposed approach involves developing a machine learning-based solution that leverages on-chain transaction data, rather than solely depending on the rule-base. Specifically, we employ blockchain mechanisms to gather and analyze user data from DApps. Based on the identified patterns, we formulate hypotheses and apply machine learning techniques to train the model. The focus of this research is on predicting potential users for DApps operating on EVM (Ethereum Virtual Machine) blockchain networks, with evaluations conducted on the seven largest chains: Ethereum, BNB Chain, Polygon, Fantom, Avalanche, Optimism, and Arbitrum. The proposed detection method demonstrates promising results, achieving an accuracy rate and F-2 score of approximately 89% and 72% respectively. Furthermore, the customer acquisition cost (CAC) is estimated to be around $3,133, which is superior to the strategies employed by DApps like 1inch and three times more cost-effective than Uniswap.

Keywords: Customer Detection · Recommender system · Web3 · Blockchain data · Wallet transaction · Decentralized finance

© The Author(s), under exclusive license to Springer Nature Switzerland AG 2025
H. A. Le Thi et al. (Eds.): ICAMCS 2024, LNNS 1311, pp. 276–287, 2025.
https://doi.org/10.1007/978-3-031-90606-0_23

1 Introduction

Blockchain technology nowadays has revolutionized various finance industries by offering decentralized and secure solutions for transactions and data management. One of its fundamental appeals is providing anonymity during authentication and transactions, which complicates identifying transaction participants. While this anonymity enhances privacy and security, it presents challenges for third-party service providers in the blockchain ecosystem. These providers struggle to identify and target their customer base effectively, which is essential for maintaining profitability and ensuring sustained revenue streams. Simultaneously, the development of Decentralized Applications (DApps)—an ecosystem of financial applications built on blockchain technology—across diverse sectors such as finance, gaming, and supply chain management, lending, is rapidly advancing. This growth underscores the pressing need to understand DApp's users and target the right customer on the whole blockchain. Such identification is crucial for attracting and growing the user base necessary for the development of these applications. Organizations can boost profits by almost 100% by engaging just 5% more of their customers [10].

Currently, most research on identifying potential entities in the blockchain domain focuses on security aspects, particularly on classical blockchain platforms such as Bitcoin and Litecoin [9,11]. These studies often propose methods related to tracking potential illegal wallets or scam transactions. However, when it comes to identifying the potential customers for organizations on advanced networks utilizing the Ethereum Virtual Machine (EVM), there is a noticeable lack of studies addressing this challenge. Instead, the primary approaches currently used for DApps predominantly employ rule-based and off-chain data to detect potential users. Although these methods can be effective in generating revenue, they lack flexibility and scalability. For example, strategies leveraging gamified challenges [1,2] to identify potential users who earn the most rewards may only attract participants primarily motivated by rewards rather than genuine engagement, and they require significant costs and time investments to create campaigns. Similarly, heuristic methods that rely on wallet participation in centralized exchanges (CEX) or airdrop events [3] are inapplicable to wallets that do not engage in these activities.

In this paper, we pursue machine learning approaches to detect potential entities on DApps, offering several advantages over previous methods. Initially, we developed a comprehensive dataset and identified wallet transaction patterns, proposing two detection strategies: an unsupervised method based on Collaborative Filtering and a supervised method using Machine Classification. Our methodology demonstrated an accuracy rate of approximately 89% and an F-2 score of 72%, with a customer acquisition cost (CAC) of around \$3,133, outperforming leading DApps like 1inch and Uniswap. Our findings are applicable to any EVM-based DApp blockchain network.

The rest of this paper is organized as follows. Related work is reviewed in Sect. 2. Our design methodology is described in Sect. 3. In Sect. 3.1, we introduce ETL module which delves into the details of data processing and wallet embedding. In Sect. 3.2, we introduce the Machine Learning Module which states hypotheses and develops different Machine Learning approaches to detect potential users. Section 3.4 presents the results of our experiment. The paper is concluded in Sect. 4 with pointers to future research.

2 Related Work

Today's techniques to detect potential entities of wallets can be categorized into two groups: off-chain rule-based and Machine Learning based. Off-chain rule-based is currently more popular due to its simplicity and efficacy. Recent strategies for pinpointing potential users within blockchain ecosystems integrate heuristic rule-based methods with gamification and advanced analytics. Platforms like QuestN [1] and Zealy [2] employ gamified challenges and campaigns to engage users, where DApps formulate tasks that users complete to earn rewards or tokens. This approach not only boosts user interaction and structured onboarding but also fosters community cohesion. However, it may attract participants primarily motivated by rewards rather than genuine engagement, necessitating continuous investment in new challenges and rewards to sustain interest. Additionally, the effectiveness of these campaigns in detecting potential users depends on the duration and resources required to execute them, which may incur significant costs and time investments for DApps.

Apart from heuristic methods, clustering has also emerged as a Machine Learning approach for potential user detection. Pioneering research in the domain of Bitcoin transaction analysis has illustrated the efficacy of clustering algorithms in identifying potential entities of interest. The study Bitcoin Transaction Graph Analysis [7] introduced a clustering approach for Bitcoin wallets based on transaction volumes and temporal patterns, coupled with the Pagerank algorithm to identify prominent wallets. This work has significantly contributed to enhancing the attribution capabilities in blockchain analytics and cybersecurity by linking wallet addresses to real-world entities. Further advancements in the clustering of Bitcoin wallets are seen in the work by [6], which clusters wallets by geographical regions using transaction timezone data, achieving 72% accuracy in identifying wallet groups. In the context of detecting illicit activities, Kamuhanda et al. (2023)[9] employed community detection and clustering algorithms to identify illegal wallet clusters within Bitcoin networks. The study used Leiden, Louvain [15], and Label Propagation algorithms, alongside Spectral Clustering [16] and DBSCAN [12], to analyze Bitcoin transaction networks from 2010 to 2020, revealing that clusters identified by Leiden and Louvain had the strongest connectivity. However, these clustering methods have limitations in identifying potential users in DApps. The reliance on clustering algorithms to group entities with similar attributes may overlook variations in user behaviors and engagement patterns, highlighting the challenge of accurately modeling user diversity in such systems.

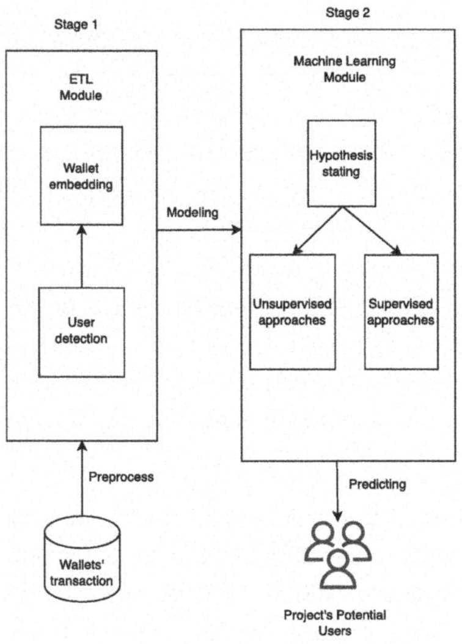

Fig. 1. System architecture

3 System Design

The structural framework of our entity detection system is illustrated in Fig. 1. It comprises two primary components: the ETL (Extract, Transform, Load) module and the Machine Learning module. The ETL module is responsible for the initial stage of our framework, processing blockchain transactions and events data to construct wallet profiles. The output of this module is the embedding of wallets, which facilitates the modeling process in the second stage, aimed at identifying potential users of blockchain projects.

3.1 ETL Module

User Detection. To identify potential entities of DApps, it is essential to analyze the behavior of their users, making user detection a crucial step. This process involves examining the interactions between wallets and the project's smart contracts. Smart contracts, which are self-executing agreements with terms directly encoded, operate on blockchain platforms. They facilitate, verify, and enforce agreements autonomously, eliminating the need for intermediaries and ensuring trust and automation in DApps. In other words, the majority of services within DApps are facilitated through interactions between users' wallets and the DApp's smart contracts. The addresses of a DApp's smart contracts are publicly known. For example, Aave's smart contracts include Aave V2 Lending Pool[1], Aave V3 Lending Pool[2]. So to discover smart contracts is straightforward. In

[1] Aave V2 Lending Pool: 0x7d2768de32b0b80b7a3454c06bdac94a69ddc7a9.
[2] Aave V3 Lending Pool: 0x87870bca3f3fd6335c3f4ce8392d69350b4fa4e2.

this context, we define Dapps' users as those whose wallets either transfer assets to or receive assets from the Dapps' smart contracts.

In the Dapps' user exploration, we try to figure out which transactional features could tell the characteristics of the Dapps' users. Therefore, we denote that feature as the contribution of users to the Dapps. Drawing inspiration from Yang's research [17] on the evaluation of lending DApps, which highlights that the primary revenue indicators for a lending project are derived from the assets that wallets interact with through its smart contracts, we propose a contribution score to quantify user contributions to DApps. This score aggregates the total assets transferred to and received from the DApps' smart contracts, with a higher score indicating a greater user contribution to the DApps.

Wallet Embedding. We then embed wallets into a vector that represents their transactional behaviors. Each wallet i is represented as a feature vector $w_i = (a, b, f)$ that amalgamates three groups of features: Asset, Balance, and Frequency. Here, a is a vector representing asset features, b representing the balance features, and f representing the frequency features.

Asset Features. Given that the asset aspect has a highly positive correlation to the contribution score, we consider three specific types of asset features which are: the total money in USD that a wallet has transferred, and the average money in USD of each transaction a wallet transfers.

Balance Features. The balance features also provide insights into the transactional patterns of wallets, as evidenced by the correlation between them and the contribution score. Then, we choose features related to balance, including the total balance in USD of a wallet, and the impurity of balance distribution to each token that a wallet holds which represents the way that a wallet manages its token's balance.

Frequency Features. The data exploration also reveals a stable pattern in frequency features, such as the total number of tokens interacted with and the total number of addresses that wallets interact with. Therefore, we propose a list of features related to these wallet habits, including the number of transactions, wallets, and tokens a wallet interacts with, as well as features representing how a wallet manages its transaction timing and transaction tokens.

3.2 Machine Learning Module

After obtaining insights from the data exploration steps and gaining the wallet embedding vector, we posit hypotheses and begin to formulate the problems for developing Machine Learning algorithms in the Machine Learning module.

The First Hypothesis and Collaborative Filtering Approach. In the first hypothesis, we assert that there exist smart contracts that are analogous to the smart contracts of DApps in terms of their interactions with surrounding wallets, specifically, with the assets engaged by these wallets. If such smart contracts exist, they will likely attract the same wallets both presently and in the future.

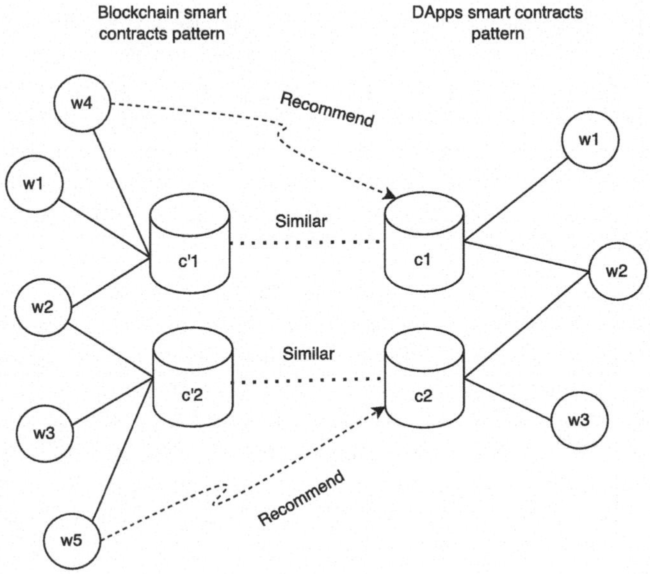

Fig. 2. Potential wallets recommendation with collaborative filtering approach

Leveraging this hypothesis, we propose to treat potential user detection as a recommendation problem using the Collaborative Filtering approach. Collaborative Filtering [14] is an unsupervised method commonly used in recommender systems to predict a user's item preferences by aggregating preferences from many users. In Collaborative Filtering, the relationship between users and items is determined by rating scores, recommending items similar to those a user has liked in the past. For the potential user recommendation problem, instead of considering the User-Items-Rating relationship, we consider the Contract-Wallet-Rating relationship with the following definition:

- Contracts: The smart contracts of lending DApps and the other smart contracts that DApps users engaged in.
- Wallets: The embedding of blockchain wallets.
- Rating: The total assets that a wallet interacts with a smart contract.

The idea is to detect contracts similar to those used in lending DApps and then recommend users of detected contracts to those in lending DApps. The detailed concept is illustrated in Fig. 2.

When implementing filtering algorithms, traditional methods like content-based and item-based recommendations face challenges in DApps, particularly the "grey sheep problem"[3] and the "cold-start problem"[4]. To address these,

[3] The grey sheep problem: The issue when the system faced with users who possess unique or unconventional preferences.

[4] The cold start problem: The issue when the system needs help to make accurate recommendations for new users or items with limited interaction history.

deep learning approaches, such as Transformer-based Collaborative Filtering, are employed, including GlobalLocal Kernel-based matrix completion (Glocal-K) and Graph Convolutional Matrix Completion (GCMC). Glocal-K integrates both global patterns and local details to improve accuracy, making it effective for cold-start scenarios. In contrast, GCMC uses graph convolutional networks to model data as a bipartite graph, enabling it to handle the grey sheep problem by leveraging relational structures. Both methods enhance matrix completion by addressing different issues.

The Second Hypothesis and Classification Approaches. We also posit an additional hypothesis that the transactional behaviors of wallets remain relatively stable after they join DApps. This consistency allows for the identification of potential users by analyzing the patterns of current DApp users. In other words, we assume that the patterns of potential users and a specific group of existing DApp users are the same. By detecting wallets exhibiting similar transactional patterns, these wallets can be classified as potential users. We thus treat the potential user detection problem as a Binary Classification problem that must determine whether or not wallets are of the potential users.

Labeling Method. In the context of binary classification, a prevalent supervised learning challenge in Machine Learning, it is imperative to establish a precise labeling methodology. To determine the ground truth for this problem, we identify potential users of DApps as wallets likely to engage with DApp services in the near future. We define the near future as a period of approximately one month, aligning with the average duration of contemporary blockchain marketing strategies. Thus, we consider new users of DApps as potential users from the past and utilize their behavior for the labeling process.

We specifically analyze data from the past month to construct the training features. At the current moment, we determine the contribution score range for new users joining the DApps and label the training data accordingly. At time (t), we define a contribution score range, (*potential_range*). If any wallets in the training data, constructed at the time (t-1), have a contribution score within (*potential_range*), they are labeled as True. Otherwise, they are labeled as False. Wallets without a contribution score (as they have not yet joined the DApps) are also labeled as False. This labeling method is validated by data exploration, which indicates that nearly all new DApp users share a specific range of contribution scores.

Feature Selection. After generating the labeled data, we now focus on selecting features to build a binary classification model. Firstly, we propose groups of features coming from the embedded wallet as the statistic features. As defined in the data exploration, these feature groups include asset, balance, and frequency. In addition, we assume that sequential features significantly influence labeled data, particularly the amount of assets a wallet spends in each transaction. To capture this pattern, we employ a Long Short-Term Memory (LSTM) [18] sequence model alongside natural language processing (NLP) preprocessing techniques. In particular, we leverage labeled data in the past and use LSTM to carry out a

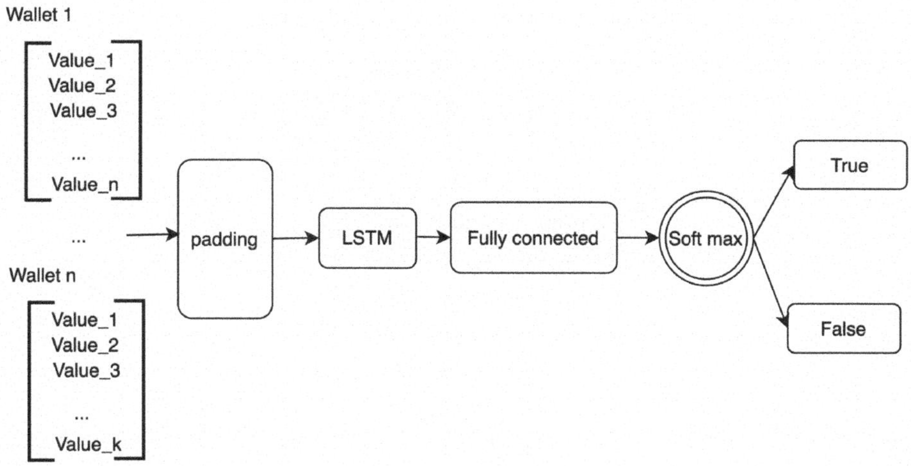

Fig. 3. Sequential model structure

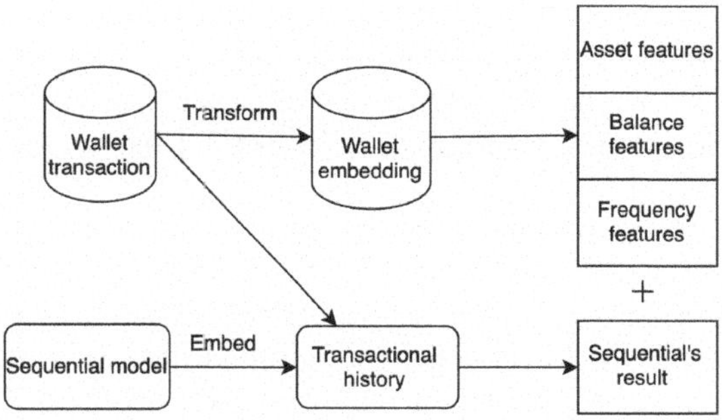

Fig. 4. Feature selection: statistic features and sequential features

binary sequential classification with the input of the wallet's transaction history. The pipeline of the sequential model is shown in Fig. 3.

Here, each wallet is represented as a sequential vector w_i with n dimension (n is the total number of transactions that a wallet transfers) including all the wallet's transaction history. These inputs undergo the padding layer to ensure uniform sequence length and then feed into the LSTM network, which maintains hidden states with the sigmoid activation function and cell states. The LSTM output then passes through a fully connected layer, learning relevant features. Finally, a softmax function assigns probabilities to different classes ("True" or "False"), resulting in the classification of the input sequence. Therefore, we leverage the output of the sequential model and consider it as a sequential feature for the overall classification. In particular, the embedded prediction from the sequential model at (t-1) is considered a sequential feature and is combined with statistical features from $t-1$ to construct a classification model for predict-

Table 1. Chain networks information

Chain	ChainId	entities
BNB Chain	0x38	5,723,972
Ethereum	0x1	4,443,970
Polygon	0x89	1,837,251
Fantom	0xfa	617,997
Arbitrum	0xa4b1	287,823
Optimism	0xa	163,467
Avalanche	0xa86a	121,542

ing potential outcomes at the current time (t), the detailed idea of the training process is described in Fig. 4.

In terms of learning models, due to the dynamic characteristics inherent in transactional data, the study prefers models that exhibit rapid training. This is the reason standard Machine Learning models are prioritized over Deep Learning. Therefore, we experiment with the following methods: Support Vector Machines (SVM) [8] as the underlying classification model; Random Forest [4] as the Bagging Ensemble model, and XGBoost [5] as the Boosting Ensemble model.

We conducted the experiment on the data we obtained and embedded the wallet described in the previous section. Particularly, we have collected more than twenty million entities over seven chains: BNB, Ethereum, Polygon, Fantom, Arbitrum, Optimism, and Avalanche, the detailed data is shown in Table 1. In this study, we experimented to detect potential users on Aave, Compound, Lido, Rocket Pool, and KyberSwap as the top projects in three different categories: Lending, Liquidity Staking, and Decentralized Exchanges (DEXes).

3.3 Evaluation Metrics

Besides the model's performance, another concern has also risen stating the effect of the practical model's result on the marketing strategy.

Customer Acquision Cost (CAC) is a key metric for evaluating the efficiency of marketing strategies, calculated by dividing the total cost of acquiring new customers by the number of customers gained. A lower CAC indicates more effective marketing efforts, essential for profitability and long-term growth.

In the blockchain domain, CAC is important for DApps, especially in assessing the effectiveness of Web3 marketing strategies like airdrop campaigns. Research from Sixdegree indicates that CAC in Web3 services ranges from \$50 to \$1,500, significantly higher than in traditional financial institutions. To analyze CAC further, we track predicted wallet addresses that interact with DApps within one month of an airdrop. The "airdropped token value" is calculated by multiplying the number of predicted wallets by the average expenditure per wallet, typically around \$1. The revised CAC formula is

Table 2. Models'result in marketing backtest

Model	Recall	CAC
SVM + sequential +statistical feat	0.65	$3,830
Random Forest + sequential + statistical feat	0.61	$4,502
XG Boost + sequential + statistical feat	**0.67**	**$3,133**
Glocal K	0.54	$10,120
GCMC	0.51	$11,141

$$\mathrm{CAC} = \frac{\mathrm{num\_predicted} \times \mathrm{airdrop\_value}}{\mathrm{num\_predicted\_join\_dapps}} (\mathrm{USD})$$

- *num_predicted* is the number of wallets predicted by the model.
- *airdrop_value* is the average value of the token that DApps spend for each wallet (here is considered as $1).
- *num_predicted_join_dapps* is the number of wallets predicted by the model that truly join the DApps.

Recall is another metric to evaluate the practical effectiveness of the model is recall. Recall in this context represents the coverage of the set of predicted wallets compared to the number of wallets that participate in DApps each month. In other words, it is the proportion between the number of wallets joining DApps predicted by the model and the total number of wallets joining DApps within a month. The higher the Recall, the better coverage the predicted wallets over the participant joining DApps each month. Recall is represented as the formula:

$$\mathrm{Recall} = \frac{\mathrm{num\_join}}{\mathrm{total\_join}}$$

- *num_join* is the number of wallets joining DApps predicted by the model.
- *total_join* is the total number of wallets joining DApps within a month.

3.4 Experiment Results

The comparative analysis of various models for recommending potential users for DApps, evaluated using Recall and Customer Acquisition Cost (CAC) metrics shown in Table 2, reveals noteworthy insights into the effectiveness of Classification and Collaborative Filtering approaches. Classification models, particularly XGBoost and SVM, exhibit superior performance compared to collaborative filtering models such as Glocal K and GCMC. Specifically, XGBoost, enhanced with sequential and statistical features, achieves the highest Recall of 0.67 and the lowest CAC of $3,133. This indicates that XGBoost not only covers a larger proportion of potential wallets but also does so with greater cost efficiency. These models' ability to handle diverse and detailed data inputs allows them to better

Table 3. Models'result vs other DApps' strategies in marketing backtest

Model	CAC
ParaSwap	$1,017
Best model from our research	**$3,133**
1inch	$4,505
Uniswap	$11,208

understand and predict user behaviors, making them highly effective for this task.

In contrast, collaborative filtering models such as Glocal K and GCMC underperform in comparison. Glocal K achieves a Recall of 0.54 and a CAC of $10,120, while GCMC has a Recall of 0.51 and a CAC of $11,141. These higher CAC values and lower Recall rates suggest that these models are less efficient in identifying and engaging potential wallets for DApps. The collaborative filtering approach might struggle with issues such as the cold-start problem and sparse data, which can limit its effectiveness in this context.

We also compared the CAC value of the best performance model to other DApps' marketing strategies (with the data provided from the research of Sixdegreelab [13]). The detailed comparison is shown in Table 3. The table compares the Customer Acquisition Cost (CAC) across different marketing strategies employed by DApps, highlighting the performance of the best model alongside ParaSwap, 1inch, and Uniswap. ParaSwap achieves the lowest CAC at $1,017, indicating its efficiency in acquiring new customers at minimal cost. The best model follows with a CAC of $3,133, demonstrating competitive cost-effectiveness in customer acquisition compared to traditional DApps' strategies. In contrast, 1inch shows a higher CAC of $4,505, while Uniswap exhibits the highest CAC at $11,208. These results underscore the effectiveness of the best model in optimizing marketing strategies for DApps, offering a compelling alternative to existing approaches by achieving a favorable balance between acquisition cost and efficiency. This analysis positions the best modelas a promising solution for enhancing customer acquisition efforts within the DApps ecosystem.

4 Conclusion and Future Work

In conclusion, we have developed an innovative system designed to automatically identify potential users for DApps. Our system comprises two main components: the ETL module and the Machine Learning module. The ETL module generates an embedding of a wallet, capturing its transactional behavior. The Machine Learning module employs various approaches, including unsupervised methods like Collaborative Filtering and supervised methods using Classification models and sequential features. This combination has resulted in excellent prediction performance, addressing the limitations of previous methods and achieving a Customer Acquisition Cost (CAC) of $3,133.

Our system leverages the strengths of recommender systems and machine learning to create robust hypotheses based on actual events, enabling the construction of predictive models without the need for strict labeling rules or prior assumptions. Additionally, it offers significant time and resource savings by identifying potential users without the necessity of building and maintaining a community presence on social platforms for DApps.

Future work will focus on evaluating the effectiveness of various learning techniques for our predictive model and developing a real-time system that can be seamlessly integrated into practical applications.

References

1. Questn. https://questn.com/
2. Zealy. https://zealy.io/
3. Allen, D.W., Berg, C., Lane, A.M.: Why airdrop cryptocurrency tokens? J. Bus. Res. **163**, 113945 (2023)
4. Breiman, L.: Random forests. Mach. Learn. **45**, 5–32 (2001)
5. Chen, T., Guestrin, C.: XGBoost: a scalable tree boosting system. In: Proceedings of the 22nd ACM SIGKDD International Conference on Knowledge Discovery and Data Mining, pp. 785–794 (2016)
6. DuPont, J., Squicciarini, A.C.: Toward de-anonymizing bitcoin by mapping users location. In: CODASPY 2015, pp. 139–141. Association for Computing Machinery, New York (2015)
7. Fleder, M., Kester, M.S., Pillai, S.: Bitcoin transaction graph analysis. arXiv preprint arXiv:1502.01657 (2015)
8. Hearst, M.A., Dumais, S.T., Osuna, E., Platt, J., Scholkopf, B.: Support vector machines. IEEE Intell. Syst. Their Appl. **13**(4), 18–28 (1998)
9. Kamuhanda, D., Cui, M., Tessone, C.J.: Illegal community detection in bitcoin transaction networks. Entropy **25**(7), 1069 (2023)
10. Madhani, P.M.: Effective marketing strategy with blockchain implementation: enhancing customer value propositions. IUP J. Bus. Strategy **19**(1), 7–35 (2022)
11. Newman, M., Girvan, M.: Finding and evaluating community structure in networks. Phys. Rev. E **69**, 026113 (2004)
12. Schubert, E., Sander, J., Ester, M., Kriegel, H.P., Xu, X.: DBSCAN revisited, revisited: why and how you should (still) use DBSCAN. ACM Trans. Database Syst. (TODS) **42**(3), 1–21 (2017)
13. Sixdegreelab: Airdrop and liquidity mining - CAC. https://blog.sixdegree.xyz/airdrop-liquidity-mining-02-cac/
14. Su, X., Khoshgoftaar, T.M.: A survey of collaborative filtering techniques. Adv. Artif. Intell. **2009**(1), 421425 (2009)
15. Traag, V.A., Waltman, L., Van Eck, N.J.: From Louvain to Leiden: guaranteeing well-connected communities. Sci. Rep. **9**(1), 1–12 (2019)
16. Von Luxburg, U.: A tutorial on spectral clustering. Stat. Comput. **17**, 395–416 (2007)
17. Yang, S., Cui, W.: An evaluation system for DeFi lending protocols. In: 2023 42nd Chinese Control Conference (CCC), pp. 8888–8893. IEEE (2023)
18. Yu, Y., Si, X., Hu, C., Zhang, J.: A review of recurrent neural networks: LSTM cells and network architectures. Neural Comput. **31**(7), 1235–1270 (2019)

Evaluating the Influence of Diversification Strategies on Portfolio Optimization: A Case Study in Vietnam

Nguyen Thi Lan Anh[1], Tuan Viet Trinh[1], Nguyen Ngoc Dai[2], and Quang Thuan Nguyen[1(✉)]

[1] International School, Vietnam National University, Nguy Nhu Kon Tum Street, Thanh Xuan District, Hanoi, Vietnam
nguyenquangthuan@vnu.edu.vn

[2] Université de Montréal, 2900 Edouard Montpetit Blvd, Montréal, QC H3T 1J4, Canada

Abstract. This research addresses the gap in understanding the Vietnamese stock market's diversification strategies and financial indicators. We examine how diversification constraints impact portfolio optimization, employing Value at Risk (VaR) as the risk measure. Our two-stage approach involves developing a robust model that maximizes investment levels while minimizing risk, and then integrating various diversification constraints (VaR and MILP). We prioritize measurable financial metrics (ROE, ROA, P/E, ROS) over industry benchmarks, aligning with Factor Investing to identify stocks with strong short-term potential. Our research addresses the lack of recent studies on portfolio optimization that consider both economic cycles and simulation techniques (Monte Carlo Stimulation), which are vital for navigating market uncertainties. We emphasize the importance of monitoring model stability using the Population Stability Index (PSI) across different economic phases. By integrating PSI with traditional metrics like Sharpe Ratio (SR), Calmar Ratio (CR), and Out-of-sample Mean Return (oMR), our study promotes a more thorough evaluation of portfolio optimization models for stronger investment strategies. Lastly, we found that Portfolio Optimization is not a one-size-fits-all solution. Our analysis underscores the importance of aligning your risk tolerance with the current economic phase and tailoring your model selection accordingly. A dynamic, diversified approach and expert financial advice will yield the best results.

Keywords: Portfolio Optimization · Diversification Strategies · Constraints Impact

1 Introduction

Portfolio optimization is the process of selecting the best mix of assets to maximize returns and minimize risk, based on an investor's goals. Introduced by Harry Markowitz in the 1950s, Modern Portfolio Theory (MPT) provided a foundation

© The Author(s), under exclusive license to Springer Nature Switzerland AG 2025
H. A. Le Thi et al. (Eds.): ICAMCS 2024, LNNS 1311, pp. 288–299, 2025.
https://doi.org/10.1007/978-3-031-90606-0_24

for diversification and optimization by balancing expected return and risk. Over time, the concept has evolved, incorporating additional factors such as transaction costs, liquidity constraints, and business cycles to address the increasing complexity of financial markets [11].

Today, rapid data processing and technological advancements have introduced new dimensions to portfolio management [2]. The integration of machine learning, programming languages like Python and R [12], and advanced methodologies such as Monte Carlo Simulation and Neural Networks [16] has revolutionized portfolio optimization. These tools enable dynamic adjustments to portfolio strategies, allowing investors to adapt to changing market conditions.

While global financial research has expanded significantly, in Vietnam, research on diversification strategies for individual investors and fund managers remains limited, particularly concerning the impact of business cycles, investor preferences, and economic uncertainties. This gap makes it difficult for investors to make informed decisions and manage risks effectively. Our study aims to address this gap by exploring diversification strategies and financial indicators that influence portfolio optimization in the Vietnamese stock market. We consider economic cycles to offer insights that can benefit both investors and fund managers.

The study leverages established tools like Value at Risk (VaR) and Monte Carlo Simulation, along with innovative techniques, to examine diversification constraints thoroughly. We develop and assess portfolio optimization models under various constraints, evaluating their performance across different phases of the business cycle. Our detailed analysis of key performance metrics seeks to identify the best strategies for building robust and efficient portfolios, providing actionable insights for managing risk and enhancing returns in Vietnam's evolving financial landscape.

Historically, portfolio optimization has undergone a significant transformation, evolving from the early frameworks of Modern Portfolio Theory to the incorporation of complex risk models like the Black-Litterman Model and Bayesian approaches [13]. The 21st century, marked by advancements in computational power, has enabled deeper integration of quantitative approaches, expanding beyond traditional measures to include dynamic risk factors and adaptive strategies [8].

As the financial landscape continues to evolve, this research combines historical insights with cutting-edge technological applications, contributing to both the academic field of portfolio optimization and providing practical guidance for market participants. By examining the interaction between diversification strategies, financial indicators, and economic cycles, our study empowers investors and fund managers to build resilient portfolios that can withstand the uncertainties of financial markets, with a particular focus on Vietnam.

2 Methodology

This study uses a two-stage approach to assess diversification constraints' impact on portfolio optimization in the Vietnamese market. The first stage develops a base model using Value at Risk (VaR) to manage risk and limit asset exposure. The second stage evaluates the model's performance across different economic phases by adjusting expected returns, using metrics like Mean Return (MR), Sharpe Ratio (SR), Calmar Ratio (CR), and Stability Index (SI). Tableau and CPLEX solver support data visualization and optimization.

2.1 Model Development

Diversification Constraints Selection. The selection of diversification constraints is key to portfolio optimization, focusing on three types: Lower and Upper Bound Constraints to prevent over-concentration, Sector or Industry Constraints to limit capital allocation and reduce risk, and the Base Model Framework, which uses Value at Risk (VaR) to assess downside risk under adverse market conditions.

Base Model Formulation and Diversification Constraints Integration. The base model applies linear and mixed-integer programming to optimize portfolio allocation while minimizing Value at Risk (VaR). It uses binary variables to capture worst-case scenarios aligned with the selected confidence level. VaR's worst realization assesses potential losses under adverse market conditions, offering insights into downside risk and forming the basis for evaluating investment risks. The base model applies linear and mixed-integer programming to optimize portfolio allocation while minimizing Value at Risk (VaR). It uses binary variables to capture worst-case scenarios aligned with the selected confidence level. VaR's worst realization assesses potential losses under adverse market conditions, offering insights into downside risk and forming the basis for evaluating investment risks.

$$M(x) = y_s = \sum_{i=1}^{n} r_{is} x_i. \tag{1}$$

Let $r_{is}, i = 1, 2, \cdots, n$, denote the rate of return of the asset in scenario s. Additionally, let $x_i, i = 1, 2, \cdots, n$, represent the proportion of the total investment allocated to the asset i. The portfolio optimization model with the worst realization as a safety measure (the MinMax model) can be formulated as follows:

$$\max \quad f \tag{2}$$

$$\sum_{i=1}^{\mathrm{n}} r_{it} x_i \geq f \quad t = 1, \ldots, T, \quad \mu = \sum_{i=1}^{\mathrm{n}} \mu_i x_i, \quad \mu \geq \mu_0, \quad x \in Q$$

In finance and banking literature, quantile is commonly referred to as the Value-at-Risk (VaR) measure, which is essentially the expected rate of return (worst realization) not worse than specified confidence interval for a given portfolio. To formulate the VaR optimization problem, we introduce the following notations.

Parameters. The following parameters are used in the model:

- S: A set of scenarios.
- p_s: The probability that scenario s will happen.
- I: The set of assets.
- K: The set of sectors.
- r_{is}: The return rate of asset i in scenario s.
- μ_i: The mean return of asset i.
- LB, UB: The lower and upper bounds of the allocation in assets.
- N: The number of sectors.
- B: A very large constant.
- C: Expected return.
- e_{ik}, $i \in I$, $k \in K$: Binary indicators, equal to 1 if stock i belongs to sector k, and 0 otherwise.
- α: Confidence level.

Variables. The following variables are involved in the optimization process:

- z_s: Binary variable, representing if scenario s is recognized as one of the worst scenarios.
- x_i: A continuous variable representing the proportion of money allocated to asset i.
- y_k: Binary variable, representing if sector k is selected for investing.
- g_i: Indicator variable, representing whether asset i is chosen.

Base Model . Other models will add more constraints compared with Model 1.

$$\max \quad f$$

$$\text{s.t} \quad f \geq \sum_{i \in I} r_{is} x_i - M z_s, \quad s \in S \tag{3}$$

$$\sum_{s \in S} p_s z_s = \alpha, \tag{4}$$

$$\sum_{i \in I} x_i = 1, \tag{5}$$

$$\sum_{i \in I} \mu_i x_i \geq C, \tag{6}$$

$$0 \leq x_i \leq 1, \quad i \in I, \tag{7}$$

$$z_s \in \{0, 1\}, \quad s \in S. \tag{8}$$

The objective function minimizes the Value at Risk. Constraints (3) enforce that binary variable z_s takes the value 1 whenever variable f is greater than the portfolio returns under scenario s ($f \geq f_s = \sum_{i \in I} r_{is} x_i$). This constraint defines the return of the portfolio at the α percentile. Constraint (4) guarantees that the probability of all scenarios satisfying $f \geq f_s$ is equal to α. Constraint (8) ensures that the total proportion of money allocation in the portfolio equals 1. Constraint (5) imposes a lower bound on the total expected return. Finally, constraints (6) and (7) define the domains for variables x_i and z_i for all $i \in I$, respectively.

Lower and Upper Bound Constraints (LUBC Model). This model integrates constraints that impose minimum (LB) and maximum (UB) investment proportions for all assets [5]. The model is formulated as:

$$\begin{aligned} \max \quad & f \\ \text{s.t.} \quad & (3) - (8), \\ & x_i \leq B\, g_i, \quad i \in I, && (9) \\ & x_i \geq LB\, g_i, \quad i \in I, && (10) \\ & x_i \leq UB, \quad i \in I, && (11) \\ & g_i \in \{0,1\}, \quad i \in I. && (12) \end{aligned}$$

Constraints (9) and (10) together ensure that if an asset is selected for investment, the corresponding allocation must be greater than LB. Constraint (11) enforces an upper bound on the allocation of investor funds. Constraints (12) impose the integrality property on variables g_i, for all $i \in I$.

Sector Constraints Integrated (ST Model). This model incorporates a constraint that restricts the number of chosen assets to a specific number N of pre-defined sectors or industries. Portfolio optimization using MAD. The problem is formulated as:

$$\begin{aligned} \max \quad & f \\ \text{s.t.} \quad & (3) - (8), \\ & \sum_{k \in K} y_k \geq N, && (13) \\ & e_{ik} x_i \leq y_k, \quad i \in I, k \in K && (14) \\ & \sum_{i \in I} e_{ik} x_i \geq LB_k y_k, \quad k \in K && (15) \\ & y_k \in \{0,1\}, \quad k \in K. && (16) \end{aligned}$$

This model incorporates a constraint that restricts the number of chosen assets to a specific number, N, of pre-defined sectors or industries. Portfolio optimization using MAD. Constraint (13) is designed to uphold the condition where the number of selected sectors is fixed at N. Constraints (14) and (15) together enforce that y_k equals 1 if and only if asset i is chosen and belongs to sector k.

Comprehensive Constraints (Full Model). This model combines both lower/upper bound and sector constraints, ensuring a diversified portfolio with controlled individual asset exposure and industry representation. The problem is formulated as:

$$
\begin{aligned}
\max \quad & f && (17)\\
\text{s.t.} \quad & (3) - (8), \\
& x_i \leq B \cdot g_i, \quad i \in I, \\
& x_i \geq LB \cdot g_i, \quad i \in I, && (18)\\
& x_i \leq UB, \quad i \in I, && (19)\\
& g_i \in \{0,1\}, \quad i \in I, && (20)\\
& \sum_{k \in K} y_k \geq N, && (21)\\
& e_{ik} x_i \leq y_k, \quad i \in I,\, k \in K, && (22)\\
& \sum_{i \in I} e_{ik} x_i \geq LB_k y_k, \quad k \in K, && (23)\\
& y_k \in \{0,1\}, \quad k \in K. && (24)
\end{aligned}
$$

By systematically integrating these constraints, we create a series of portfolio optimization models that not only optimize risk (VaR) but also enforce desired levels of diversification within the portfolio.

2.2 Model Evaluation and Performance Metrics

The second stage rigorously evaluates the models using out-of-sample validation with a rolling window technique, incorporating business cycle phases from the Vietnamese economy and stock market to ensure practical relevance. The out-of-sample validation process uses data specific to these phases, closely resembling real-world investment scenarios [10]. We adjust expected returns for each model according to the identified business cycle phase, capturing the dynamic nature of investor expectations. The models are efficiently solved using the CPLEX solver, and performance is evaluated using various metrics.

Out-of-Sample Mean Return (MR). The out-of-sample mean return represents the average return an investor can expect to earn on a portfolio using data not employed in its construction. This metric is defined as:

$$\text{Mean Return} = \frac{1}{S} \sum_{s=1}^{S} r_s, \tag{25}$$

where S is the number of scenarios s, and r_s is the return in scenario s.

Calmar Ratio (CR) and Sharpe Ratio (SR). These ratios offer insights into risk-adjusted performance.

Calmar Ratio is defined as:

$$\text{Calmar Ratio} = \frac{\text{Return over}[0,T]}{\text{MDD over}[0,T]}, \tag{26}$$

where the Maximum Drawdown (MDD) is calculated as:

$$\text{MDD} = \left[\max_{t \in [0,T]} \left(\sum_{t=1}^{T} (R_t) - R_T \right) \right]. \tag{27}$$

The Sharpe Ratio is defined as:

$$\text{Sharpe Ratio} = \frac{R_p - R_f}{\sigma(R_p)}, \tag{28}$$

where

- R_p: The multi-period return over period T;
- R_f: The base currency risk-free return over period T;
- $\sigma(R_p)$: The volatility of the portfolio return over period T.

Population Stability Index (SI). The Population Stability Index (SI) emphasizes the stability of the models over time. It tracks the stability of the portfolios by measuring the change in allocation weights across periods. The SI is defined as:

$$\text{SI} = \frac{1}{T-1} \frac{1}{N} \sum_{i=1}^{N} \sum_{t=1}^{T-1} \Delta w_{i,j}, \tag{29}$$

where:

- Let N be the total number of unique tickers in the dataset;
- Let T be the total number of iterations;
- Let $w_{i,j}$ be the weight of ticker j at iteration i;
- $\sum_{i=1}^{T-1} w_{i,j}$: Sum over all iterations (from 1 to $T-1$);
- $\sum_{i=1}^{N} w_{i,j}$: Sum over this unique ticker;
- $\Delta w_{i,j}$: The absolute change in weight for ticker j from iteration i to iteration $i+1$.

2.3 Portfolio Optimization Simulation and Validation Framework

In our study, we employed Geometric Brownian Motion (GBM) to simulate thousands of potential market scenarios, providing comprehensive insights into portfolio performance under uncertain market conditions. As suggested by [9], this method models continuous asset price movements by incorporating drift and volatility. We further enhanced these simulations with Gradient Boosting

Machines (GBMs) to better adapt to complex market dynamics, combining both techniques for more realistic risk analysis.

We also integrated Vietnam's business cycle data from 2020 to 2023 [15], allowing the model to reflect real-world investment scenarios [1]. This included phases such as Peak (2nd quarter 2020 - 1st quarter 2021), Contraction (2nd quarter 2021), Expansion (4th quarter 2021 - 2nd quarter 2022), and Trough (4th quarter 2022 - 3rd quarter 2023), each characterized by varying GDP growth rates, inflation, and unemployment. For instance, during the expansion phase, GDP growth was 2–3%, inflation held steady at 2%, and unemployment ranged between 3.5–4.5%.

To validate the model, we applied a rolling window approach [3,4] with window lengths of 60, 120, and 240 months, generating 36 rolling windows of out-of-sample returns for statistical significance (Fig. 1). This ensured the model's robustness across diverse market environments.

Lastly, we implemented a 7-element experimental framework that included data training, scenario generation, portfolio optimization, and performance evaluation using metrics like Mean Return (MR), Sharpe Ratio (SR), Calmar Ratio (CR), and Stability Index (SI). These metrics were applied to both in-sample and out-of-sample datasets and benchmarked against existing portfolios to ensure the model's robustness and effectiveness, as supported by [6] and [14].

3 Results

3.1 Impacts of Diversification Constraints

The study evaluated various portfolio models using Mean Return (MR), Sharpe Ratio (SR), Calmar Ratio (CR), and Population Stability Index (PSI). All models had positive returns, with medians from 0.083 to 0.107 (Fig. 1). Fig. 1 also illustrates the following: (i) The LBUB model provided the highest potential return, as indicated by the upper whisker of 0.710, but also carried the greatest risk, reflected by the lower whisker of −0.213. (ii) The Base and ST models achieved a balance between strong returns, with upper whiskers exceeding 0.64, and moderate risk.

The Full model had the lowest risk (IQR: 0.20 - Fig. 1) but lower returns (upper whisker: 0.496 - Fig. 1), while the Base model led in risk-adjusted performance (Sharpe Ratio median: 0.355 - Fig. 1). Base and ST models had significant downside risks. The Calmar Ratio showed favorable return-to-drawdown ratios for Base, ST, and LBUB, with the ST model appealing to aggressive investors. All models were stable with low PSI values, though Base had slightly higher instability. Overall, LBUB achieved the highest returns with more risk, while the Full model was more suitable for risk-averse investors.

3.2 Impacts of Business Cycle Models

This section explores how portfolio optimization metrics-Mean Return (MR), Sharpe Ratio (SR), Calmar Ratio (CR), and Population Stability Index (PSI)-

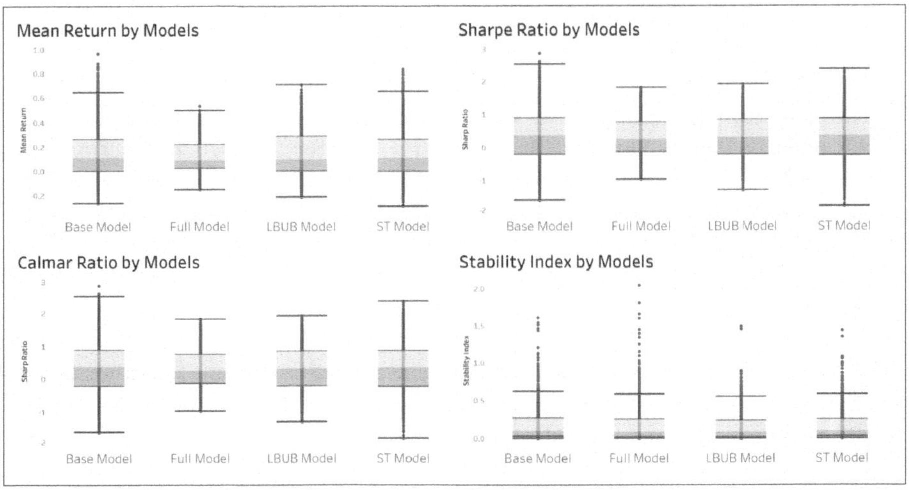

Fig. 1. The Results From Models Without Integration.

change across economic phases, stressing the need to align portfolios with business cycles. During contraction, portfolios showed modest returns (median MR of 0.060 - Fig. 2) with some risk of loss (lower whisker of −0.059 - Fig. 2).

3.3 Suggested Models Across Business Cycle Phases

This section examines how portfolio optimization models align with different phases of the business cycle, focusing on key performance indicators: Mean Return (MR), Sharpe Ratio (SR), Calmar Ratio (CR), and Population Stability Index (PSI), to assess performance and stability across contraction, expansion, peak, and trough phases [7]. During contractions, the ST Model stands out for risk-seeking investors with the highest mean return (0.0750 - Fig. 3), while the LBUB Model offers a balanced option (mean return: 0.0658 - Fig. 3). The Full Model, with the lowest negative return (−0.0026 - Fig. 3), is ideal for minimizing losses. In expansions, the Full Model continues to limit losses (−0.0026), and the ST Model shows resilience with the least negative return (−0.0485 - Fig. 3). At market peaks, the Base Model (mean return: 0.3197) and LBUB Model (mean return: 0.3140 - Fig. 3) both demonstrate their ability to leverage market momentum.

The Sharpe Ratio is explored to identify models with optimal risk-adjusted returns. The ST Model consistently performs well across contraction, trough, and peak phases, making it an ideal choice for investors seeking both diversification and a favorable risk-return balance.

The Full Model underperforms in most phases, indicating a need for adjustment or that it's less suited for current market conditions. During expansions, all models show negative Sharpe Ratios, but the ST Model suffers the least (−0.417 - Fig. 3), demonstrating its ability to limit losses. The Calmar Ratio highlights

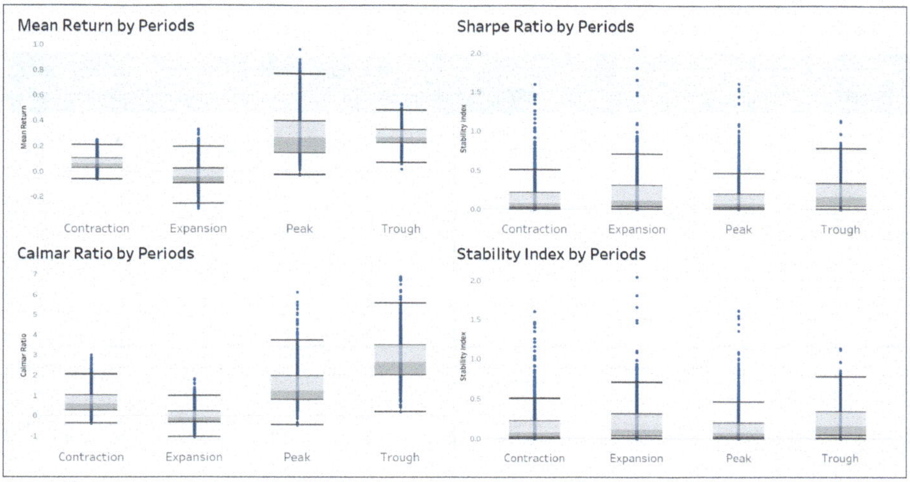

Fig. 2. The Results With Indicators (4 Business Cycle Phases Integrated).

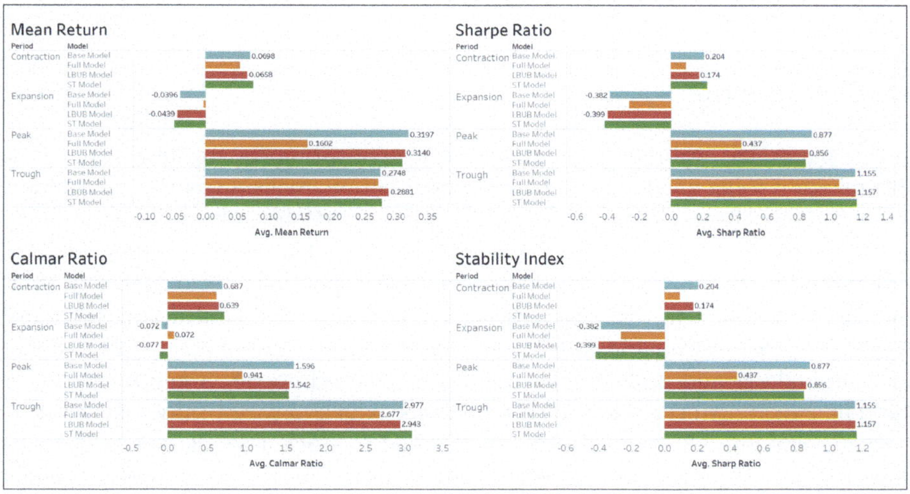

Fig. 3. The Results With Indicators (4 Business Cycles Integrated).

the trough phase as the most favorable for risk-adjusted returns, with the ST Model excelling (CR: 3.095 - Fig. 3) (Table 1) .

Even in contractions, models deliver positive returns over drawdowns, with the ST Model holding a CR of 0.714 (Fig. 3). In contrast, the peak phase poses significant risks, with most models showing negative CRs, except the Base Model, which shows resilience (CR: 1.596 - Fig. 3). PSI values reveal the LBUB Model has superior stability during contractions (PSI: 0.1072 - Fig. 3), while the Full Model unexpectedly excels during trough phases (PSI: 0.1213 - Fig. 3), suggesting strategic adjustments during declines.

Table 1. Recommendations and suggested models for investors.

Economic Phase	Model	Calmar Ratio (CR)	Performance & Risk Characteristics	Suitable Investor Type	Economic Phase (Right Side)	Model	Calmar Ratio (CR)	Performance & Risk Characteristics	Suitable Investor Type
Trough (Recovery)	Base	2.917	Highest CR, strong risk-adjusted returns	Conservative, Balanced (Minimizes potential losses, good balance between risk & reward)	Peak (Overheated Market)	Base	1.596	Positive CR, resilience during challenging phases	Conservative (Maintains some growth potential)
	Full	2.677	High CR, good risk-adjusted returns	Conservative, Balanced (Prioritizes capital preservation, good balance)		Full	-0.054	Negative CR, losses outweigh returns	Not Recommended (High risk of capital loss)
	LBUB	2.8	High CR, good risk-adjusted returns	Conservative, Balanced (Minimizes potential losses)		LBUB	-0.077	Negative CR, losses outweigh returns	Not Recommended (High risk of capital loss)
	ST	3.095	Highest CR, exceptional risk-adjusted returns	Balanced (Prioritizes both capital preservation and growth potential)		ST	-0.099	Negative CR, losses outweigh returns	Not Recommended (High risk of capital loss)
Contraction (Downturn)	Base	0.688	Positive CR, moderate risk-adjusted returns	Conservative (Minimizes potential losses)	Expansion (Growth)	Base	-0.072	Positive CR, potential for increased drawdowns	Not Recommended (May experience capital erosion)
	Full	0.614	Positive CR, moderate risk-adjusted returns	Conservative (Minimizes potential losses)		Full	-0.072	Negative CR, potential for increased drawdowns	Not Recommended (May experience capital erosion)
	LBUB	0.634	Positive CR, moderate risk-adjusted returns	Conservative (Minimizes potential losses)		LBUB	0.077	Positive CR, lower risk-adjusted returns	Balanced (May offer some growth with moderate risk)
	ST	0.714	Positive CR, moderate risk-adjusted returns	Conservative (Minimizes potential losses)		ST	0.099	Positive CR, lower risk-adjusted returns	Balanced (May offer some growth with moderate risk)

4 Conclusion

In this paper, we evaluated the impact of diversification constraints on portfolio optimization using a value-at-risk objective function, focusing on the Vietnamese stock market. Diversification effectively reduces portfolio risk while maintaining returns, making it essential in portfolio construction. Our study addresses gaps in understanding how diversification strategies affect portfolio performance in Vietnam. By examining various diversification constraints and considering economic cycles, we provide valuable insights for investors and fund managers to enhance decision-making and manage risk more efficiently. However, a key limitation is the reliance on historical data for simulations, which may not fully reflect future market behavior. The analysis was also limited to stock asset classes, suggesting that further research could explore diversification constraints across a broader range of investment types.

References

1. Aldcroft, D.H., Richardson, H.W.: The business cycle. The British Economy 1870–1939, pp. 23–61 (1969)
2. Beccalli, E., Elliot, V., Virili, F.: Artificial intelligence and ethics in portfolio management. In: Digital Business Transformation: Organizing, Managing and Controlling in the Information Age, pp. 19–30. Springer (2020)
3. Bergmeir, C., Benítez, J.M.: On the use of cross-validation for time series predictor evaluation. Inf. Sci. **191**, 192–213 (2012)
4. Bessler, W., Drobetz, W., Seim, M., Zimmermann, J.: Equity issues and stock repurchases of initial public offerings. Eur. Financ. Manag. **22**(1), 31–62 (2016)

5. Chekhlov, A., Uryasev, S., Zabarankin, M.: Drawdown measure in portfolio optimization. Int. J. Theor. Appl. Finance **8**(01), 13–58 (2005)
6. Cogneau, P., Hübner, G.: The 101 ways to measure portfolio performance. Available at SSRN 1326076 (2009)
7. DeMiguel, V., Garlappi, L., Uppal, R.: Optimal versus naive diversification: how inefficient is the 1/N portfolio strategy? Rev. Financ. Stud. **22**(5), 1915–1953 (2009)
8. Fama, E.F., French, K.R.: Size and book-to-market factors in earnings and returns. J. Financ. **50**(1), 131–155 (1995)
9. Fathi Vajargah, K., Shoghi, M.: Simulation of stochastic differential equation of geometric Brownian motion by quasi-Monte Carlo method and its application in prediction of total index of stock market and value at risk. Math. Sci. **9**, 115–125 (2015)
10. Gerrard, R., Kyriakou, I., Nielsen, J.P., Vodička, P.: On optimal constrained investment strategies for long-term savers in stochastic environments and probability hedging. Eur. J. Oper. Res. **307**(2), 948–962 (2023)
11. Kolm, P.N., Tütüncü, R., Fabozzi, F.J.: 60 years of portfolio optimization: practical challenges and current trends. Eur. J. Oper. Res. **234**(2), 356–371 (2014)
12. Mansini, R., Ogryczak, W., Speranza, M.G.: Twenty years of linear programming based portfolio optimization. Eur. J. Oper. Res. **234**(2), 518–535 (2014)
13. Martínez-Nieto, L., Fernández-Navarro, F., Carbonero-Ruz, M., Montero-Romero, T.: An experimental study on diversification in portfolio optimization. Expert Syst. Appl. **181**, 115203 (2021)
14. Shper, V., et al.: Investigation of the real meaning of the stability index and its empirical analysis. Processes (2023)
15. Statista: GDP growth rate in Vietnam 2020-2023, by quarter (2024). https://www.statista.com/statistics/1195301/vietnam-gdp-growth-rate-by-quarter/
16. Zhu, Y., Escobar-Anel, M.: A neural network Monte Carlo approximation for expected utility theory. J. Risk Financ. Manag. **14**(7), 322 (2021)

Leveraging Graph Neural Networks and Optimization Algorithms to Enhance Anti-money Laundering Systems

Thi Thu Trang Phan(✉)

Hanoi University of Science, VNU, Hanoi, Vietnam
21001593@vnu.edu.vn

Abstract. Anti-money laundering (AML) is garnering increasing attention from governments and media, both domestically and globally. While several cases have been uncovered, the number still does not correspond to the full risk posed by this type of crime in Vietnam, highlighting the necessity for the development of more effective anti-money laundering (AML) systems. Traditional rule-based methods face challenges in terms of adaptability and accuracy when flagging suspicious transactions. Graph Neural Networks (GNNs) have emerged as a highly regarded model due to their ability to learn from graph-structured data, such as society networks, and financial transaction networks. This study experimented with the application of rewiring algorithms such as Batch Ollivier Ricci Flow (BORF) and First-order Spectral Rewiring (FoSR) to address over-smoothing and overs-quashing problems, enhancing the accuracy of GNNs for multi-layer cases. To the best of my knowledge, this is the first study to evaluate BORF on a financial dataset. The results demonstrated that Graph-SAGE, utilizing the Elliptic imbalanced dataset combining both global and local features and employing FoSR optimization, achieved the highest accuracy in most experiments, with F1 score and Recall score outperforming statistic machine learning models. These findings indicate the strong potential for GNNs to be implemented in AML systems.

Keywords: Batch Ollivier Ricci Flow · First-order Spectral Rewiring · Anti-Money Laundering · Graph Neural Networks

1 Introduction

Money laundering is the process through which criminals conceal the illegal origins of their money. Criminals use sophisticated methods to transform their illegally gained assets into seemingly legitimate wealth. This presents challenges for law enforcement and financial investigators to trace the money back to its criminal source. According to Vietnam Law E-Journal [1], most money laundering schemes follow three stages: Placement, Layering, and Integrating.

© The Author(s), under exclusive license to Springer Nature Switzerland AG 2025
H. A. Le Thi et al. (Eds.): ICAMCS 2024, LNNS 1311, pp. 300–311, 2025.
https://doi.org/10.1007/978-3-031-90606-0_25

The first stage, Placement, involves the introduction of illegally obtained funds into the legitimate financial system. The goal is to convert large amounts of cash from criminal activities into more manageable and less suspicious forms within legitimate financial channels. For instance, criminals may deposit cash into bank accounts below reporting thresholds, purchase financial instruments, exchange currency, or employ similar methods to avoid detection.

After Placement, the process moves to the layering stage, where criminals conduct a series of complex financial transactions to obscure the origin of the illicit funds. This creates a complex trail of transactions, making it difficult for law enforcement and financial investigators to trace the money back to its illegal source. Common techniques include setting up multiple shell companies to transfer and control funds or falsifying invoices to facilitate cross-border transfers.

The final stage is integration, in which illicit money is reintroduced (to the market), appearing completely "clean" and legal. Criminals may invest illicit money in companies that appear legitimate or lend the funds, receiving legal repayments from borrowers later on.

In Vietnam, the Anti-Money Laundering Law was introduced in 2013. However, it was not until the 2015 Criminal Code, amended and supplemented in 2017 [2], article 324, that the legal framework to address the legalization of money and assets obtained through criminal activities was fully established. This provided the investigative police agency the grounds to initiate and investigate money laundering cases such as those involving Giang Kim Dat (2016), Phan Sao Nam (2018), and, most recently, Truong My Lan (2024), who is involved in laundering 445.6 billion VND. Despite that, Vietnam Law E-Journal [1] noted that the number of prosecuted money laundering cases remains low compared to the potential risks of this crime in Vietnam.

Overall, compared to international standards, Vietnam's legal system for combating money laundering is still in the early stages of development and implementation. The Anti-Money Laundering Law No. 14/2022/QH15 of Vietnam [3] clearly outlines the groups that must pay special attention to anti-money laundering measures in article 2, ranked as follows:

1. Financial institutions.
2. Relevant non-financial entities.
3. Vietnamese and foreign individuals, organizations, and entities interacting with financial institutions.
4. Other entities involved in anti-money laundering.

Currently, banks in Vietnam are the primary and leading groups implementing comprehensive AML measures and regulations. Others, however, remained at the stage of research and observation until they were required to adopt AML solutions and review their internal activities by the 2022 investigative law.

There are two research methods to detect money laundering transactions: the rule-based method [10,11] and the AI-based method. The rule-based method, which is currently adopted by most financial institutions, is quick and simple to implement. However, this method faces a very high false positive rate (FPR) of 95–98% according to HOVSTADIUS, David [6]. An alternative is to deploy AI for anti-money laundering systems. This requires high-quality and clean datasets for statistical machine-learning models. However, in practice, datasets often contain noise and imbalance (e.g., a low proportion of illicit transactions and enormous amounts of unlabeled data), which can lead to poor classification performance in machine learning models.

As a result, researchers are in search of methods that are independent of labels and can leverage connections between accounts to identify money laundering activities. In recent years, Graph Neural Networks (GNNs) have emerged as powerful tools for analyzing non-Euclidean networks such as social and communication networks. Financial transaction networks, a non-Euclidean network, can be effectively analyzed using GNNs through link or node classification tasks.

This study applies GNNs with a focus on node classification to identify specific accounts involved in illegal activities. Nonetheless, one downside of applying GNN architectures with fewer layers is the limited scope of node dependencies, leading to poor long-range dependency learning. To enhance this and detect correlations among distantly connected money laundering transactions, increasing the depth of the GNNs is necessary. However, deeper GNN architectures face issues such as oversmoothing and oversquashing, which result in the degradation of information as it propagates through the layers.

To address these challenges, this study applied state-of-the-art rewiring algorithms, containing Batch Ollivier Ricci Flow (BORF) [4] and First-order spectral rewiring (FoSR), for optimizing large graph-based networks. Due to the lack of research demonstrating the performance of BORF on large financial datasets, it is hypothesized that implementing BORF on raw large data will improve the classification metrics of GNNs in the context of financial transaction networks, outperforming FoSR and demonstrating the advantages of GNNs over other methods. The goal is to evaluate the effectiveness of GNNs in constructing AML solutions.

This study is divided into the following sections: Sect. 2 explores the current approaches to anti-money laundering in the financial industry. Section 3 describes Batch Ollivier Ricci Flow theory and Graph Neural Network disciplines. Section 4 introduces the training data and explains the training procedure. Section 5 highlights the experimental setup and results of the study. Next, Sect. 6 analyses the results from all the experiments. Finally, Sect. 7 gives a summary of the study.

2 Related Work

2.1 Rule-Based Approach

The traditional approach is constructing rule-based systems to detect and flag suspicious transactions based on a set of predefined rules established by financial experts. Technically, these systems operate through a series of 'if-else' conditions.

Bellomarini, Luigi, et al. [11] applied Knowledge Graph theory to implement a rule-based system in real-world applications. They identified challenges in processing uncertain data and suggested using machine learning methods to address these issues. Butgereit, Laurie [10] proposed a rule-based algorithm based on the

12 Red Flags of the US Immigration and Customs Enforcement agency. They recognized its potential for tagging data, which could be used in machine learning models.

While rule-based systems are fast and straightforward to implement in financial systems, they come with significant drawbacks. Hovstadius, David [6] pointed out that rule-based systems result in high False Positive Rates, typically around 95–98%. Additionally, rule-based approaches struggle to adapt when criminals employ sophisticated money laundering strategies.

In conclusion, while rule-based systems are quick and simple to deploy, they rely heavily on the expertise of domain specialists and are less effective in detecting complex money laundering schemes, especially when criminals are constantly developing more advanced techniques.

2.2 AI-Based Approach

In the AI-based approach, there are two popular trends in previous research: statistical machine learning methods and graph-based methods.

When it comes to statistical machine learning models, previous research focused on supervised and unsupervised learning approaches. Supervised models, such as Support Vector Machines (SVM) [16], XGBoost, Logistic Regression [15], Random Forest [18] and unsupervised models, such as Kmeans [19] and Gaussian Mixture Model [20], have demonstrated notable accuracy.

Moreover, these models perform best when applied to high-quality, clean datasets. In particular, unsupervised learning requires datasets that contain features indicative of money laundering, while supervised learning relies on well-labeled data.

However, according to HOVSTADIUS, David [6], real-world transaction data from banks often lacks reliable labels, presenting problems as the models could not fully reflect actual money laundering patterns. As a result, statistical machine learning models perform poorly in real-world settings.

In contrast, graph-based methods have been shown to be effective in handling unreliable data cases by HOVSTADIUS, David [6] and some other authors. HOVSTADIUS, David [6] compared the performance of AML using GCN, GAT, and GraphSAGE against statistical machine learning methods. The results showed that GAT outperformed machine learning approaches, particularly in cases where data tags were mislabeled.

In general, machine learning methods perform well when reliable, high-quality datasets are available. Currently, these methods are still applied in AML systems. However, banks often struggle to accurately tag reliable data, making it challenging to achieve high classification results with these methods.

On the other hand, graph-based methods excel at identifying suspicious transactions, especially in cases involving unreliable data. The main drawback of these methods, however, is their black-box nature, which makes it extremely difficult to explain their output.

2.3 Hybric Approach

Combining rule-based systems with machine learning approaches has been a popular suggestion by previous authors such as Bellomarini, Luigi, et al. [11], and Butgereit, Laurie [10]. However, this hybrid approach faces a significant limitation: the high false positive rate of rule-based systems makes it difficult to prepare reliable labeled training data for supervised learning models. For unsupervised learning models, the challenges include data imbalance and noise.

Nevertheless, the combination of GNNs and rule-based systems shows promise for deployment in real-world financial systems. As mentioned earlier, GNNs are black-box models, making it challenging to explain their output. In this hybrid approach, GNNs play a crucial role in identifying suspicious transactions, while rule-based systems provide explanations for flagged transactions. By comparing these transactions against known rules, the system can offer specific reasons for flagging a transaction as suspicious, thus addressing the issue of explainability.

3 Methodology

This section introduces two preprocessing methods to address oversquashing and oversmoothing in GNNs, followed by a brief overview of GNNs, including their definition and core principles.

3.1 Batch Ollivier Ricci Flow

Batch Ollivier Ricci Flow (BORF) is a graph rewiring algorithm proposed by Nguyen, Khang, et al. [4], based on the Ollivier-Ricci curvature theory. The study demonstrated the effectiveness of BORF in reducing oversmooting and oversquashing problems, outperforming other methods such as, First-order spectral rewiring [14], on the specified datasets.

Ollivier-Ricci curvature measures the "curvature" of edges in a graph. The formula for the Ricci curvature of an edge (u, v) is:

$$\kappa(u, v) = 1 - \frac{W(m_u, m_v)}{d(u, v)} \tag{1}$$

where $\kappa(u, v)$ is the Ollivier-Ricci curvature between u and v.

$W(m_u, m_v)$ is the Wasserstein distance between the distributions of the neighbors of u and v.

$d(u, v)$ is the shortest distance between u and v.

This algorithm works by dividing the data into N batches. For each batch, it identifies A edges with the lowest Ollivier-Ricci curvature, which contribute to oversquashing in GNNs, and B edges with the highest Ollivier-Ricci curvature, which lead to oversmoothing. For each low-curvature edge (uj, vj), the algorithm adds a new edge (h, k) to the graph, where h and k are bottleneck nodes from the optimal transport plan. Finally, it removes A edges with the highest curvature to mitigate over-smoothing.

3.2 First-Order Spectral Rewiring

First-order spectral rewiring (FoSR), proposed by Karhadkar, Baner-Jee, et al. [14], is a rewiring algorithm that relies on the spectral gap and is designed to address the over-squashing problem in GNNs.

The spectral gap is calculated as the difference between the second-largest eigenvalue and the largest eigenvalue of the graph's normalized Laplacian matrix. A large spectral gap indicates information passing through layers is well-preserved, while a small spectral gap poses a bottleneck in the architecture. FoSR optimizes the spectral gap by strategically filtering and adding edges that maximize it. By increasing the spectral gap, the method helps preserve the flow of information, thereby reducing oversquashing.

3.3 Graph Neural Networks

To date, various versions of GNNs have been developed. When considering homogeneous graphs, there are three common types: GAT, GCN, and GraphSAGE. GCN is the simplest of the three GNN architectures, extending convolutional neural networks to graph data through a straightforward message-passing algorithm. In contrast, GraphSAGE optimizes the process by sampling and aggregating features from a node's local neighborhood. Unlike the other networks, GAT leverages an attention mechanism to weigh neighboring nodes during feature aggregation, enabling adaptive learning of node representations in graphs.

In each layer, GNNs employ a neural message-passing algorithm that facilitates information exchange between nodes in a graph and updates node representations using neural networks. In this framework, each node is initially represented by a tensor. Messages, which are tensor representations derived from node features and/or edge attributes, propagate information along the edges of the graph.

In GNNs, the message-passing algorithm involves two main steps:

1. Aggregation: Detecting and combining the information from the neighborhood nodes of the target node.
2. Update: Combining the information of neighborhood nodes of the target node and target node to update tensor presentation for the target node.

4 Implementation

This section provides a comprehensive overview of the dataset and the specific configuration used for the training process.

4.1 Dataset

This study utilizes the Elliptic Dataset [12], a public Bitcoin graph-structured dataset from the MIT-IBM Watson AI Laboratory [13]. It consists of 203,769 nodes and 234,355 edges, representing licit, illicit, and unknown financial transactions. The dataset contains three main files: a 'classes' file which indicates the transaction status; a 'features' (or nodes) file, which corresponds to individual transactions; and an 'edges' file which represents the transaction flows or connections between transactions. For example, if user A transfers money to user B, users A and B are represented as nodes, the transaction flow between them is the edge, and the status of this transaction is stored in the "classes" file.

In the 'classes' file, there are 3 labels: illicit (2%), licit (21%), and Unknown (77%). The 'features' file contains 166 features for each node, which are as follows:

1. 94 features about local information from nodes: time step, input/output numbers, fees, and volume
2. 72 features about global information from neighboring nodes: statistical measures like maximum, minimum, standard deviation, and correlation coefficients

Additionally, the data is structured into 49 time steps, each representing approximately two weeks of transaction data. Notably, there are no connections between transactions from different time steps, implying each time step operates independently.

4.2 Training

This study implements two strategies for processing the dataset to make it suitable for machine learning models and GNN models.

Since the initial 'features' file lacks a header, the first step is to assign appropriate labels. The first column represents the user_id (denoting the node), and the second column indicates the time step. The following 94 columns contain local information about the node, while the last 72 columns provide global information.

For training the machine learning models, this study uses the 'classes' and 'features' files (skipping the edges file). After merging the two tables, the column named 'class' is treated as the label, while the remaining columns are considered attributes. To efficiently classify illicit and licit transactions, this study focuses on two clear labels: illicit and licit, skipping the label named 'Unknown'. The dataset is split into training and testing sets at a 7:3 ratio. Then, four statistical machine learning models-Logistic Regression (with class_weight = 'balanced'), SVM (with class_weight = 'balanced'), XGBoost, and Balanced Random Forest-are employed as they are well-suited for handling imbalanced datasets. Finally, the best model is tested on the holdout test set to evaluate its performance.

For training the GNN models, this study uses the 'features', 'edges' and 'classes' files. It follows a similar approach as in machine learning approach, using the labeled dataset while removing any entries labeled as 'Unknown'. The experiment considers two scenarios: one using data with all local and global features, and another using data with only the local features of the node. In both cases, the dataset is split into a 65% training set, a 15% validation set, and a 20% test set. Training involves three types of GNN models, including GCN, GAT, and GraphSAGE, with 13,000 epochs (early stopping is applied when the accuracy and loss curves converge). The loss function is cross-entropy, with a learning rate of 9×10^{-3} and a weight decay of 5×10^{-4}.

5 Experiment Setup and Results

In this section, several experiments are conducted to evaluate the performance of the rewiring algorithm in GNN models, including GAT, GCN, and GraphSAGE, compared to the standard GNN case. The evaluation considers variations in the number of layers and both dataset cases. Additionally, the study compares the performance of traditional machine learning models with GNN models to demonstrate the advantages of using GNN models over traditional machine learning approaches.

5.1 Training Setup

For training the GNNs, this study sets up 13,000 epochs (with early stopping), uses cross-entropy loss, and employs the Adam optimization method.

5.2 Metrics

$$\text{Precision} = \frac{\text{True Positives}}{\text{True Positives} + \text{False Positives}} \tag{2}$$

$$\text{Recall} = \frac{\text{True Positives}}{\text{True Positives} + \text{False Negatives}} \tag{3}$$

$$\text{F1 Score} = 2 \times \frac{\text{Precision} \times \text{Recall}}{\text{Precision} + \text{Recall}} \tag{4}$$

5.3 Results

Tables 1, 2, and 3 present metrics for illicit transactions using GNN models with graph-based convolution layers varying in number from 2 to 5 to 7. The results are provided for two datasets: one containing all features and the other containing only local features.

From each of the previously considered experiments, this study selected the best model to compare the performance of state-of-the-art machine learning models.

Table 1. Performance metrics of illicit transactions for 2-layer GNN Models for 2 data preparation cases.

Model	Manual			BORF			FOSR		
	Precision	Recall	F1	Precision	Recall	F1	Precision	Recall	F1
GCN Conv (local)	0.843	0.448	0.585	0.844	0.450	0.587	0.836	0.449	0.584
GCN Conv (local + global)	0.797	0.498	0.613	0.800	0.503	0.618	0.796	0.509	0.621
GAT Conv (local)	0.811	0.667	0.732	0.825	0.664	0.735	0.818	0.663	0.732
GAT Conv (local + global)	0.826	0.729	0.774	0.841	0.722	0.777	0.825	0.728	0.773
SAGE Conv (local)	0.947	0.766	0.847	0.951	0.760	0.844	0.951	0.766	0.849
SAGE Conv (local+global)	0.945	**0.834**	**0.886**	0.943	**0.837**	**0.887**	0.940	**0.845**	**0.890**

Table 2. Performance metrics of illicit transaction for 5-layer GNN Models for 2 data preparation cases.

Model	Manual			BORF			FOSR		
	Precision	Recall	F1	Precision	Recall	F1	Precision	Recall	F1
GCN Conv (local)	0.809	0.446	0.575	0.815	0.457	0.586	0.82	0.451	0.582
GCN Conv (local + global)	0.685	0.517	0.589	0.694	0.517	0.593	0.658	0.537	0.591
GAT Conv (local)	0.810	0.588	0.681	0.830	0.611	0.704	0.850	0.602	0.705
GAT Conv (local + global)	0.632	0.642	0.637	0.750	0.423	0.510	0.640	0.630	0.635
SAGE Conv (local)	0.930	0.784	0.851	0.930	0.790	0.854	0.942	0.783	0.855
SAGE Conv (local+global)	0.941	**0.820**	**0.876**	0.945	**0.812**	**0.873**	0.946	**0.820**	**0.878**

Table 3. Performance metrics of illicit transaction for 7-layer GNN Models for 2 data preparation cases.

Model	Manual			BORF			FOSR		
	Precision	Recall	F1	Precision	Recall	F1	Precision	Recall	F1
GCN Conv (local)	0.766	0.455	0.571	0.758	0.460	0.573	0.771	0.455	0.572
GCN Conv (local + global)	0.589	0.537	0.562	0.590	0.541	0.565	0.631	0.533	0.578
GAT Conv (local)	0.717	0.480	0.575	0.699	0.484	0.572	0.709	0.481	0.573
GAT Conv (local + global)	0.526	0.576	0.550	0.548	0.561	0.554	0.469	0.601	0.527
SAGE Conv (local)	0.924	0.791	0.852	0.947	0.789	0.861	0.947	0.780	0.855
SAGE Conv (local+global)	0.938	**0.817**	**0.874**	0.943	**0.827**	**0.881**	0.956	**0.813**	**0.879**

Table 4. Summary table of metrics of illicit transaction between machine learning models and the best GNN models with different layer structures.

Model	Precision	Recall	F1_Score	F1 Micro AVG
Logistic Regression	0.222	0.792	0.346	0.812
SVM	0.353	0.685	0.466	0.901
XGBoost	**0.989**	0.593	0.741	0.974
Balanced Random Forest	0.756	0.680	0.716	0.966
SAGEConv (local + glocal) (2 layers + FoSR)	0.94	**0.845**	**0.89**	**0.98**
SAGEConv (local + glocal) (5 layers + FoSR)	0.946	0.82	0.878	0.978
SAGEConv (local + glocal) (7 layers + BORF)	0.943	0.827	0.881	0.979
SAGEConv (fully features) (7 layers + FoSR)	0.956	0.813	0.879	0.979

6 Discussion

The results, as presented in the first three tables, show that all metrics for illicit transactions across the considered GNN models tend to decrease as the number of layers increases. Among the GNN models, the one utilizing SAGE Convolution with the dataset containing all features demonstrated the best performance. Notably, when using the FoSR method, this model outperformed the BORF method in both the 2-layer and 5-layer configurations. However, in the 7-layer architecture, the same model with the BORF method emerged as the best, exhibiting superior classification performance for illicit transactions.

Despite the fact that the 2-layer architecture outperformed other multi-layer models, it is not suitable for real-world banking systems. Real datasets often contain millions of entities, and focusing on a small scope may not be optimized for detecting suspicious transactions. A multi-layer architecture is necessary to capture long-range dependencies more effectively in real-world scenarios.

Referring to Table 4, the best models from the three previous experiments outperform all other machine learning models. While most machine learning models showed low recall and F1 scores, the top-performing GNN models exhibited high performance across all metrics, achieving the highest Micro Average F1 score of 98%. Notably, GNN models are independent of data labels, whereas statistical machine learning models depend on them. This characteristic makes GNN models a potential solution for real-world AML systems, which often encounter transactions with unreliable labels.

Another significant finding is the computational efficiency of the FoSR algorithm, which required less than 20 GB of RAM to perform best with models having fewer than 7 layers. In contrast, the BORF algorithm requires substantially higher computational resources (a minimum of 100 GB for the 2-layer architecture, and higher resources when the number of layers grows). While the algorithm demonstrates its best performance in the 7-layer architecture experiment, further experimentation to fully evaluate BORF's performance with large graph-structured datasets is required for future work.

7 Conclusions

In conclusion, this study demonstrated the outstanding performance of GNN models compared to statistic machine learning models and highlighted their high potential for deployment in AML systems. The limitation of this study lies in the absence of a real-world banking transaction dataset, which prevents a thorough evaluation of the actual efficiency of GNNs and the complexity of interpreting non-Euclidean networks.

Furthermore, this research evaluated the performance of two rewiring algorithms, FoSR and BORF, as the depth of the network increases. For large datasets, all experiments showed that FoSR provided more stable results while requiring lower computational resources. Despite the fact that Nguyen, Khang, et al. [4] had demonstrated that BORF outperformed FoSR in every specific dataset used by the authors, BORF demanded very high computational resources and performed best only in the 7-layer architecture experiment. More experiments are needed to make a clear decision about the true efficiency of this algorithm.

Acknowledgments. I would like to extend our heartfelt gratitude to our supervisors, Dr. Nguyen Thi Bich Thuy from Hanoi University of Science, VNU, and Do Xuan Hanh from Smartlog Company, as well as FSS Company, for their invaluable feedback, guidance, and support.

Disclosure of Interests. The paper was conceived and completed by student from Hanoi University of Science (HUS), Phan Thi Thu Trang.

References

1. Vietnam law E-Journal. https://phaply.net.vn/phuong-thuc-thu-doan-cua-toi-pham-rua-tien-tai-viet-nam-va-giai-phap-nang-cao-hieu-qua-phong-chong-a255175.html. Accessed 08 July 2024
2. Vietnam criminal code 2015. https://thuvienphapluat.vn/van-ban/EN/Bo-may-hanh-chinh/Law-12-2017-QH14-amendments-100-2015-QH13/364731/tieng-anh.aspx. Accessed 20 June 2017
3. Law No. 14/2022/QH15 anti-money Laundering. https://thuvienphapluat.vn/van-ban/EN/Tien-te-Ngan-hang/Law-14-2022-QH15-Anti-Money-Laundering/547270/tieng-anh.aspx. Accessed 15 Nov 2022
4. Nguyen, K., et al.: Revisiting over-smoothing and over-squashing using Ollivier-Ricci curvature. In: International Conference on Machine Learning, pp. 25956–25979. PMLR (2023)

5. Chen, Z., Van Khoa, L.D., Teoh, E.N., Nazir, A., Karuppiah, E.K., Lam, K.S.: Machine learning techniques for anti-money laundering (AML) solutions in suspicious transaction detection: a review. Knowl. Inf. Syst. **57**(2), 245–285 (2018). https://doi.org/10.1007/s10115-017-1144-z
6. Hovstadius, D.: Anti-Money laundering with unreliable labels (2024)
7. Author, A.-B.: Contribution title. In: 9th International Proceedings on Proceedings, pp. 1–2. Publisher, Location (2010)
8. Usman, A., Naveed, N., Munawar, S.: Intelligent anti-money laundering fraud control using graph-based machine learning model for the financial domain. J. Cases Inf. Technol. **25**(1), 1–20 (2023)
9. Bakry, A.N., et al.: Automatic suppression of false positive alerts in anti-money laundering systems using machine learning. J. Supercomput. **80**(5), 6264–6284 (2024)
10. Butgereit, L.: Anti money laundering: rule-based methods to identify funnel accounts. In: 2021 Conference on Information Communications Technology and Society (ICTAS). IEEE (2021)
11. Bellomarini, L., et al.: Rule-based anti-money laundering in financial intelligence units: experience and vision. RuleML+ RR (Suppl.) **2644**, 133–144 (2020)
12. Elliptic. www.elliptic.co
13. Weber, M., et al.: Anti-money laundering in bitcoin: experimenting with graph convolutional networks for financial forensics. In: KDD 2019 Workshop on Anomaly Detection in Finance, Anchorage, AK, USA (2019)
14. Karhadkar, K., Banerjee, P. K., Montufar, G.: FoSR: first-order spectral rewiring for addressing oversquashing in GNNs. In: International Conference on Learning Representations (2023)
15. Friedman, J., Hastie, T., Tibshirani, R.: Regularization paths for generalized linear models via coordinate descent. J. Stat. Softw. **33**, 1–22 (2010)
16. Cortes, C., Vapnik, V.: Support-Vector networks. Mach. Learn. **20**, 273–297 (1995)
17. Chen, T., Guestrin, C.: XGBoost: a scalable tree boosting system. In: Proceedings of the 22nd ACM SIGKDD International Conference on Knowledge Discovery and Data Mining (2016)
18. Rigatti, S.J.: Random forest. J. Insur. Med. **47**(1), 31–39 (2017)
19. Chen, Z., Xia, S.: K-means clustering algorithm with improved initial center. In: 2009 Second International Workshop on Knowledge Discovery and Data Mining. IEEE (2009)
20. Reynolds, D.A.: Gaussian mixture models. Encycl. Biometrics **741**, 659–663 (2009)

Engineering Systems

Salinity Optical Sensor Based on $Si\text{-}Si_3N_4$ Composite Waveguide Structure

Thanh Tung Bui, H. Y. Pham, and T. A. Nguyen(✉)

International School, Vietnam National University Hanoi, Cau Giay Dist., 144 Xuan Thuy Street, Hanoi, Vietnam

nguyentheanh@vnu.edu.vn

Abstract. This study presents a salinity optical sensor based on a composite $Si\text{-}Si_3N_4$ waveguide structure. The sensor leverages the hybridization of Si and Si_3N_4 waveguide modes to enhance light-matter interactions, critical for high sensitiity optical sensing. This has been achieved by incorporating the waveguide structure, in a Mach-Zehnder Interferometer (MZI), into the sensor to detect changes refractive index arising from different concentrations of salt. Simulation results show that the sensor exhibits a high sensitivity with 88.74 nm/ RIU wavelength shift which is equivalent to 55 pm/% salt concentration. This method is a reliable low-cost option for real time salinity monitoring in multiple aquatic environments.

Keywords: Salinity Optical Sensor · Mach Zehnder Interferometer · $Si\text{-}Si_3N_4$ Composite Waveguide

1 Introduction

Salinity is a central factor in the analysis of aquatic systems, it contributes to controlling biological mechanisms [1], influencing water chemistry [2] and changes at physical level of waters bodies [3]. Accurate and reliable salinity measurements are essential for a wide range of applications, including oceanography [4], climate research [5], water quality monitoring [6] and aquaculture management [7], conductivity-based sensors [8, 9] are used for traditionally measuring salinity using the global network; however, variation in temperature, biofouling and need of frequent calibration become limitations to it [10, 11].

In response to these challenges, the development of optical salinity sensors has emerged as a promising alternative. These sensors operate based on the principle of light interaction with water [12], offering a non-invasive [13], high-resolution [14] method for salinity measurement. Optical sensors offer the benefit of longer lasting and more accurate readings due to their reduced susceptibility to interferences from the environment. Furthermore, because they can provide continuous, real-time data, they are especially useful in dynamic environments where salinity levels can change quickly.

This work presents a salinity optical sensor based on hybrid $Si\text{-}Si_3N_4$ waveguide structure. The waveguide core consists of Si_3N_4 strip underneath thin Si slab, separated by a thin SiO_2 layer. The hybridization of the Si planar waveguide mode with the

© The Author(s), under exclusive license to Springer Nature Switzerland AG 2025
H. A. Le Thi et al. (Eds.): ICAMCS 2024, LNNS 1311, pp. 315–323, 2025.
https://doi.org/10.1007/978-3-031-90606-0_26

Si_3N_4 strip waveguide mode confines the optical field within both the Si and Si3N4 layers. Since the Si slab waveguide is on top, the hybrid waveguide mode can achieve strong light-matter interaction, which is essential for high-sensitivity optical sensing. For instance, in an interferometer sensor, the waveguide can be utilized to form a highly sensitive Mach-Zehnder interferometer, enabling the detection of small changes in the refractive index due to environmental factors with high precision.

2 Si-Si_3N_4 Composite Waveguide Structure

The waveguide cross-section is illustrated in Fig. 1(a), where a 30 nm thin Si slab is positioned top a 220 nm thick Si_3N_4 strip, underneath of Si slab is clad in SiO2, the air cap Si on top of waveveguide. A 200 nm gap separates the Si slab from the Si3N4 strip. Figure 1(b) shows the mode profile of the fundamental TE mode at a wavelength

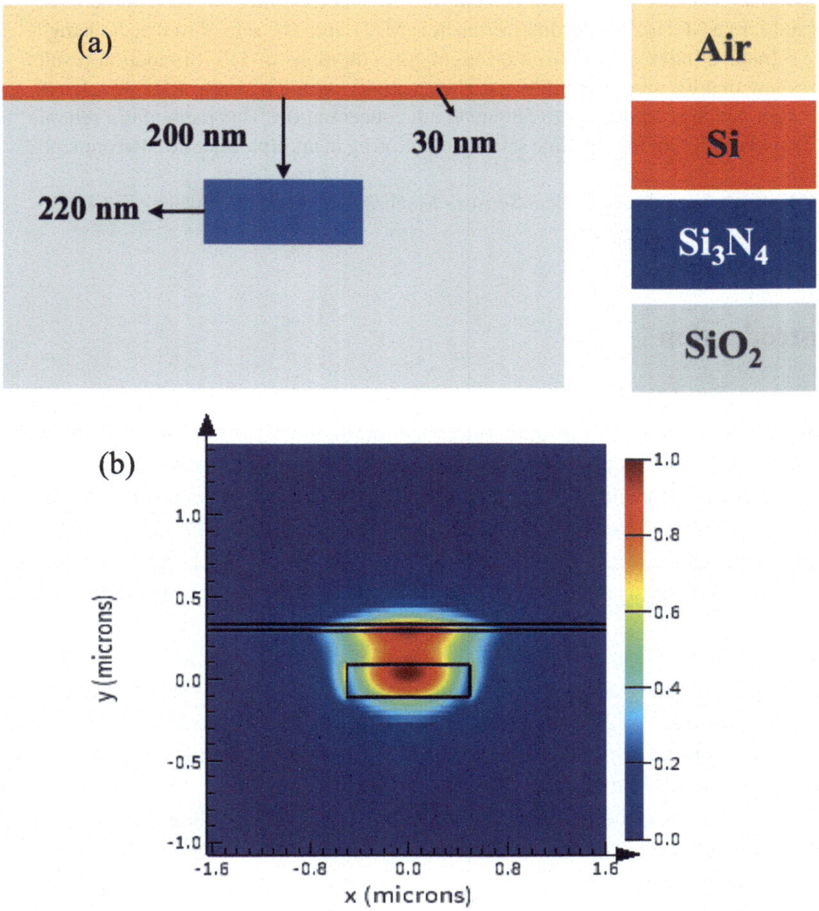

Fig. 1. (a) Cross section of the Si-Si_3N_4 composite waveguide structure. (b) Fundamental TE mode profile of the waveguide.

of 1550 nm. The optical field is confined not only within the Si_3N_4 strip but also within the Si slab.

The waveguide mode can be engineering by adjusting the width of the Si_3N_4 strip. As shown in Fig. 2, Finite Diffirence Eigenmode (FDE) simulations depict the cut-off condition for single-mode guidance at a wavelength of 1550 nm for air cladding, where only modes with an effective index greater than 1.52 are guided. To ensure single-mode operation, the width of the Si_3N_4 strip (with a fixed height of 220 nm) should be kept below 1.35 μm. If the waveguide width exceeds this threshold, a second TE mode (TE1) will appear.

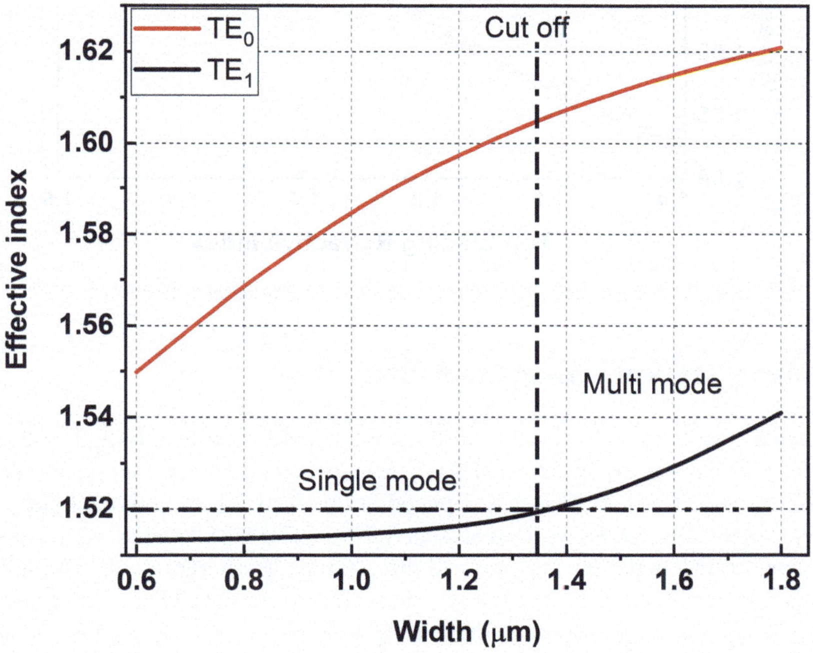

Fig. 2. Cut-off condition of the single-mode waveguide.

The waveguide structure is well-suited optical sensing. By changing the refractive index of top clading, we can alter the effective index of the waveguide mode is shown in Fig. 3, the refractive index of the top cladding significantly impacts the waveguide effective index, particularly as it increases from 1 (the refractive index of air) to 1.46. By integrating this waveguide structure into a Mach-Zehnder Interferometer (MZI), we can develop an optical sensor capable of determining material concentrations in solutions.

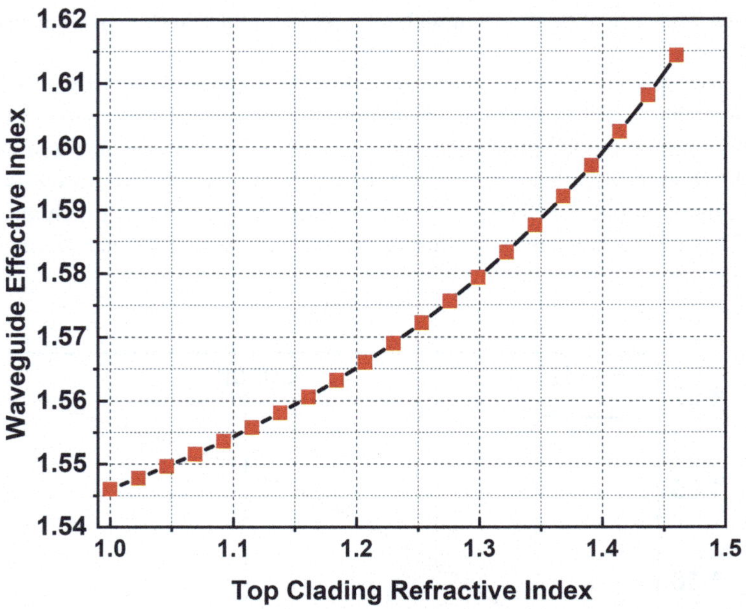

Fig. 3. Waveguide effective index versus top clading refractive index.

3 Salinity Optical Sensor Configure

In this study, we utilize a Si-Si_3N_4 composite waveguide stucture to form the two arms of a Mach-Zehnder Interferometer (MZI) for salinity detection. The configuration of the MZI-based optical salinity sensor is depicted in Fig. 4(a), where both the light source and detector are Optical Network Analyzers (ONAs). The MZI structure comprises two arms: one serving as the sensing arm and the other as the reference arm. Lasing light, within a wavelength range of 1500–1600 nm, is emitted from the ONA source and split into two paths: one passing through the sensing arm and the other through the reference arm, both of equal length $L = 130\ \mu m$

In the sensing arm, the Si slab on top of the waveguide is exposed to the environment or the material under investigation. Changes in the top cladding conditions, such as variations in salt concentration in water, cause alterations in the waveguide's effective index Δn within the sensing arm. This leads to a phase shift, $\Delta\emptyset = \frac{2\pi}{\lambda}L\Delta n$, in the light propagating through this arm. Meanwhile, the reference arm remains isolated from external influences, providing a stable baseline for comparison. This ensures that any detected changes in the interference pattern are solely due to variations in the sensing arm.

After passing through both arms, the light is recombined, producing an interference pattern with intensity described by $I = 2I_0 + 2\sqrt{I_0}cos\Delta\emptyset$. The ONA detector captures this interference pattern, which shifts based on the phase difference between the two arms, as illustrated in Fig. 4(b). This phase difference directly corresponds to changes in the sensing arm, allowing the sensor to detect and accurately quantify the variations.

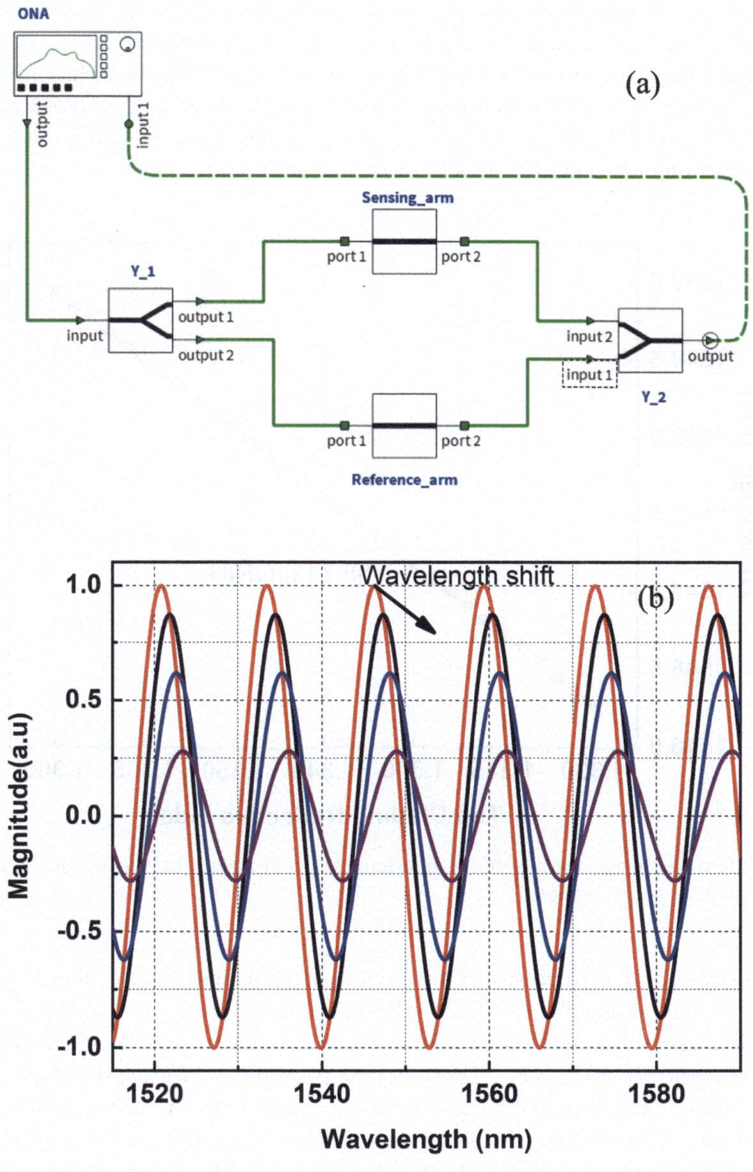

Fig. 4. (a) Salinity optical sensor configure, (b) Shifted interference pattern

To validate the optical sensor's performance in detecting salt concentration, the interconnect model of Ansys Lumerical was utilized, as illustrated in Fig. 4. Initially, the refractive index of the top cladding in the waveguide structure was set to 1.33, corresponding to the refractive index of pure water [15, 16] for the reference arm (see Fig. 1). Subsequently, the refractive index of the top cladding in the sensing arm was adjusted from 1.33 to 1.36, represeting saltwater with concentrations ranging from 0% to 50% [17–19]. This adjustment resulted in an increase in the effective index of the waveguide from 1.5814 to 1.59045, as shown in Fig. 3, leading to a phase difference between the two arms. The sensor demonstrated a high sensitivity, with a wavelength shift of 88.74 nm/RIU corresponding to 55 pm/ % of salt concentration, as indicated in Fig. 5, confirming its responsiveness to variations in refractive index.

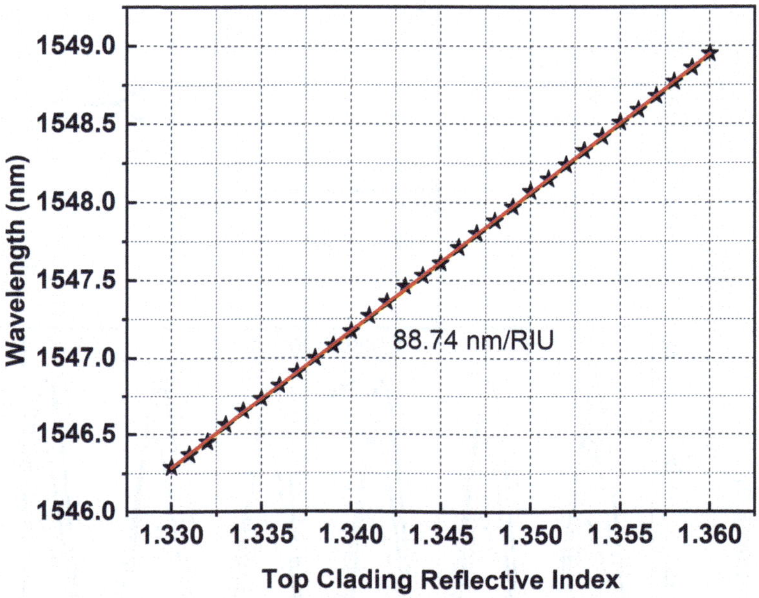

Fig. 5. Shifted resonance wavelength of interference pattern with the various of top cladding refractive index, correspondingly.

Additionally, Fig. 4(b) illustrates a reduction in the intensity of the interference pattern due to the increasing phase difference between the two arms. For a waveguide length of $L = 130\ \mu m$, the phase difference reaches π within the effective index range of 1.58417 to 1.5905, as shown in Fig. 6(a). This outcome is promising for the development of a low-cost salinity optical sensor based on the waveguide structure depicted in Fig. 6(b). In this sensor configuration, the ONA used in Fig. 4(a) is replaced by a single-mode laser (CWL) with a wavelength of 1549.5 nm, aligned with the resonance wavelength of the interference pattern in Fig. 4(b), and an Avalanche Photodetector (APD) is used for output detection. The output power decreases from −10 dBm to −60 dBm as the effective index of the sensing arm increases (Fig. 6(c)), indicating that the measurement results can be digitized based on the photodetector output.

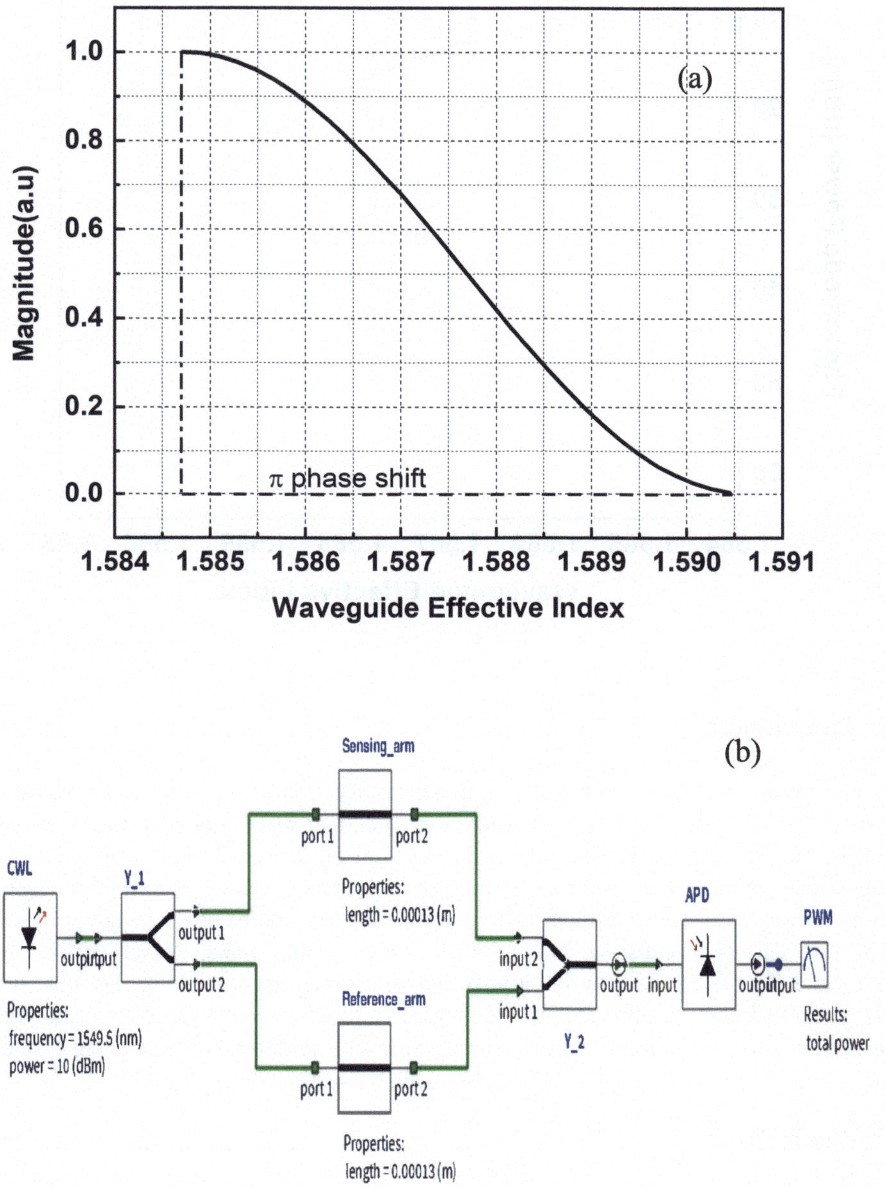

Fig. 6. (a) The π phase shift in a 130 μm waveguide results in a change of the effective index from 1.58417 to 1.59, (b) Low-cost salinity optical sensor configure for real time measurement, (c) Output power on APD versus effective index of waveguide.

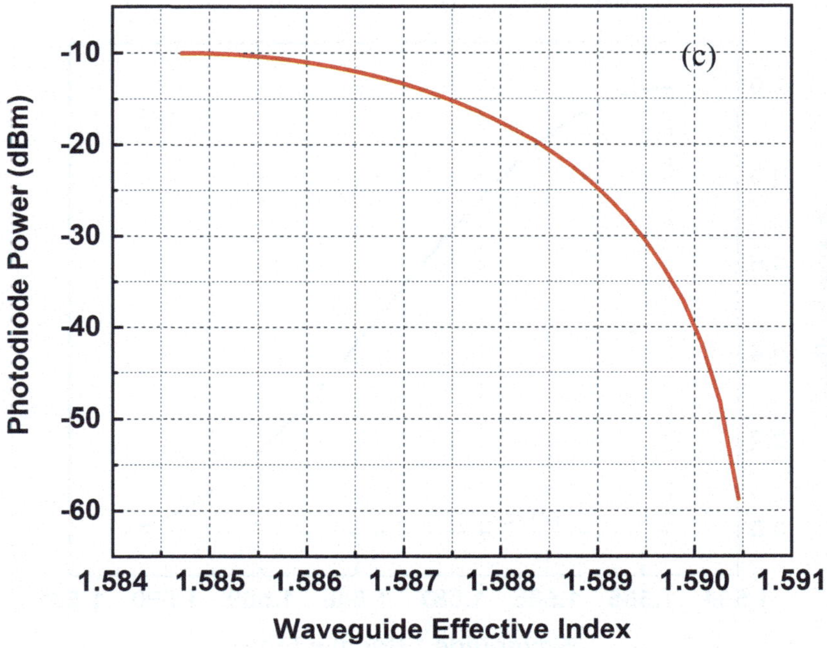

Fig. 6. (*continued*)

4 Conclusion

In this work, we have developed a high-sensitivity salinity optical sensor based on composite Si-Si_3N_4 waveguide structure integrated in Mach-Zehnder Interferometer (MZI). The hybrid waveguide design enhances light-matter interaction, enabling precise detection of salinity changes through variations in the refractive index. Simulation results show a significant wavelength shift per refractive index unit to prove its suitability for real-time salinity monitoring. These sensor features could make it essential for many fields including oceanography, environmental monitoring or aquaculture. The next steps in our research are the optimization of the sensor system for wider range of environmental conditions and further investigation on integration with autonomous monitoring systems.

References

1. Cañedo-Argüelles, M., Kefford, B., Schäfer, R.: Salt in freshwaters: causes, effects and prospects-introduction to the theme issue. Philosophical Transactions of the Royal Society B **374**(1764), 20180002 (2019)
2. Lintern, A., et al.: Key factors influencing differences in stream water quality across space. Wiley Interdiscip. Rev. Water **5**(1), e1260 (2018)
3. Karne, Avinash, V., Prabhakar, Kulkarni, D.: Studies on physico-chemical characteristics of freshwater bodies in Khatav Tahsil, Maharashtra. Nature, Environment and Pollution Technology **8**(2), 247–251 (2009)

4. Vinogradova, N., et al.: Satellite salinity observing system: Recent discoveries and the way forward. Frontiers in Marine Science **6,** 243 (2019)
5. Gu, L., et al.: Advances in the technologies for marine salinity measurement. J. Marine Sci. Eng. **10**(12), 2024 (2022)
6. Adjovu, G.E., et al.: Overview of the application of remote sensing in effective monitoring of water quality parameters. Remote Sensing **15**(7), 1938 (2023)
7. Hu, Z., et al.: A method overview in smart aquaculture. Environmental Monitoring and Assessment **192,** 1–25 (2020)
8. Benjankar, R., Kafle, R.: Salt concentration measurement using re-usable electric conductivity–based sensors. Water, Air, & Soil Pollution **232**, 1–16 (2021)
9. Thirstrup, C., Deleebeeck, L.: Review on electrolytic conductivity sensors. IEEE Transactions on Instrumentation and Measurement **70,** 1-22 (2021)
10. Heydari, F., Javad, M., Tabatabaei, N., Rezai, P.: Low-cost resistive microfluidic salinity sensor for high-precision detection of drinking water salt levels. ACS Omega **7**(18), 15529–15539 (2022)
11. Fitridge, I., Dempster, T., Guenther, J., De Nys, R.: The impact and control of biofouling in marine aquaculture: a review. Biofouling **28**(7), 649–669 (2012)
12. Elsherif, M., et al.: Optical fiber sensors: Working principle, applications, and limitations. Advanced Photonics Research **3**(11), 2100371 (2022)
13. Nag, A., Mukhopadhyay, S.C., Kosel, J.: Sensing system for salinity testing using laser-induced graphene sensors. Sensors and Actuators A: Physical **264**, 107–116 (2017)
14. Sayed, H., Aly, A.H.: Salinity optical sensor by using two-dimensional photonic crystals: computational study. Materials Science and Engineering: B **269**, 115169 (2021)
15. Price, Chelsea, L., et al.: Simultaneous retrieval of the size and refractive index of suspended droplets in a linear quadrupole electrodynamic balance. The Journal of Physical Chemistry A **124**(9), 1811–1820 (2020)
16. Le, T.H.H., Morita, A., Tanaka, T.: Refractive index of nanoconfined water reveals its anomalous physical properties. Nanoscale horizons **5**(6), 1016–1024 (2020)
17. Lin, Z., et al.: High-sensitivity salinity sensor based on etched C-type micro-structured fiber sensing structure. Sensors and actuators A: physical **339**, 113518 (2022)
18. Li, H., et al.: Theoretical and experimental characterization of a salinity and temperature sensor employing optical fiber surface plasmon resonance (SPR). Instrumentation Science & Technology **48**(6), 601–615 (2020)
19. Wang, F., Wei, Y., Han, Y.: High sensitivity and wide range refractive index sensor based on surface plasmon resonance photonic crystal fiber. Sensors **23**(14), 6617 (2023)

Novel of Frequency Stabilizing Strategy for Delayed Smart Grids Incorporated Wind Turbines via Memorable Estimator

Hung Quoc Dao[1], Ngoc Thanh Pham[1(✉)], Minh Duc Vu[1], Viet Tung Nguyen[1], Duc Duong Nguyen[2], Doan Dong Nguyen[1], and Nham Do Thi[3]

[1] International School, Vietnam National University Hanoi, Hanoi, Vietnam
{21070370,pnthanh,22070119,21070348,dongnd}@vnu.edu.vn
[2] University of Economics - Technology for Industries, Hanoi, Vietnam
ndduong86.ddt@uneti.edu.vn
[3] Vietnam National University of Agriculture, Hanoi, Vietnam
dtnham@vnua.edu.vn

Abstract. This paper concerns a frequency stabilizing strategy to reinstate the stability of delayed smart grids incorporated wind turbines fluctuated due to unexpected change in load demand. Because of the use of networked communication, time delays happen in the control-loop of both thermal and renewable generating units. We first consider a new mathematical representation of a smart grid comprising time delays related to thermal power plan and wind turbine. We derive new stabilizable conditions for synthesizing frequency controller to guarantee performance of the grid. The validation of the proposed scheme is carried out with a grid incorporating wind turbine.

Keywords: Time-delay Power System · Memorable Observer · Load Frequency Stabilisation

1 Introduction

The disparity between power provided and consumed take place suddenly owing to unexpected variation in the power demands. If the dilemma is sufficiently dealt, the inequality will results in dangerous sequence in the operation of power grids, for instance the frequencies of system may largely fluctuate as well as diverge far from the desirable positions [1–3]. The primary goal of frequecy regulation is to retain the system frequency around planned values by appropriate control strategy [4–7]. Recently, there have been research attempt on the mixing of wind power into the grids [3,8,9]. Wind turbines own friendly environmental characteristics, for instance lower greenhouse pollution, wind turbines can contribute into frequency control to backing power sources in order to diminish the departures of system frequency.

Relating to the implementation, the combination of Wind turbine into a smart grid involves networked communication to renovate data information of

© The Author(s), under exclusive license to Springer Nature Switzerland AG 2025
H. A. Le Thi et al. (Eds.): ICAMCS 2024, LNNS 1311, pp. 324–335, 2025.
https://doi.org/10.1007/978-3-031-90606-0_27

wind generating units. Thus, the happening of delays is obvious by the reason of quick jam in communication channels, reducing quality of packets of data [10,11]. In connection with the frequency adjustment of power grid, the incidence of delays leads to the deviation in systems frequency. To accomplish the harmful disadvantages of delays, control regimes for LFC of the grids have been developed [12–15]. In [12,13], technique of Lyapunov-Krakowski functional with Linear Matrix Inequality was utilize controllers to assure that the asymptotic stability of systems. Full state feedback regime was deployed with an assumption of the accessibility and measurable potential of state information. Nonetheless, there is limitation of measurement is available for the implementation of controllers in hand-on application. For that reason, the improvement of a new frequency control strategy is valuable to overcome the drawback of delays in managing intelligent grids.

In this manuscript, frequency stabilization of smart grids with wind turbines via memorable estimator is developed. We derive a mathematical expression of a power grid in the collaboration of wind turbine. To establish our proposed control structure, we select a power grid with a coal consumed power unit and a unit wind turbine having sufficient reserve power. To manage with matter of delayed signal, the famous Jensen inequality is exploited to generate synthesis criteria in linear representation. The controller gains can be acquired by an effective procedure. The rest of this manuscript includes Sect. 2 shows a mathematical representation of a power grid incorporated renewable section with the existence of delays. Section 3 provides stability and stabilization criterion for the research systems With numerical scenario of simulations, the usefulness our offered control strategy will be validated in Sect. 4. As a last point, the conclusion of the research is given at Sect. 5.

2 System Description

Figure 1 illustrates a digram of power grid with wind turbine arising from [8,9]. Due to the unexpected variation in load demand $\Delta P_l(t)$, the frequency deviation of power grid,$\Delta f(t)$, becomes fluctuated. To handle this issue the control center send a power command, ΔP_c, and voltage request, $\Delta V_{qr}(t)(t)$, to thermal and winds turbines. Due to the use of open communication infrastructure to deliver the input signals, the time delay, ζ, exist within the power grids which can degrade the performance of the system.

In Fig. 1, $\Delta P_m(t)$, $\Delta P_v(t)$, $\Delta P_{ld}(t)$, $\Delta P_e(t)$ are the change in the turbine output mechanical power, opening of the valve degree, the load demand and wind turbine's output power. $\Delta i_q(t)$, $\omega(t)$ are current deviation of the rotor and speed deviation. M_w, M_m are wind turbine and constant inertia R constant inertia. $\Delta \mathcal{T}_w$ is the mechanical wind torque, D is load deviation ratio, R is the coefficient of speed drop, $\mathcal{W}_{wo}$, $\mathcal{I}_{wo}$ are selected appropriate values. T_t and T_g is turbine and governor time constant. Other notations and corresponding parameters can be found at the Appendix.

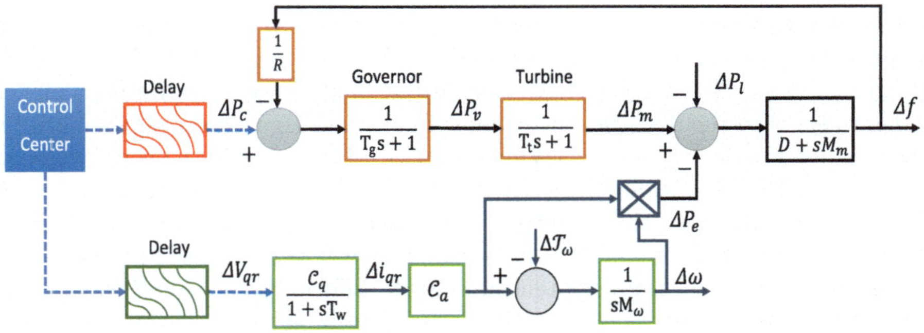

Fig. 1. Block diagram of transfer function of power systems with wind turbines

The system dynamic are

$$\begin{aligned}
\Delta \dot{P}_v(t) &= \frac{1}{T_g}\Delta P_c(t-\zeta) - \frac{\Delta f(t)}{RT_g} - \frac{\Delta P_v(t)}{T_g},\\
\Delta \dot{P}_m(t) &= \frac{\Delta P_v(t)}{T_t} - \frac{\Delta P_m(t)}{T_t};\ \ \Delta \dot{P}_e(t) = \mathcal{C}_w \mathcal{W}_{wo}\Delta i_{qr}(t) + \mathcal{C}_w \mathcal{I}_{wo}\omega(t),\\
\Delta \dot{f}(t) &= -\frac{D}{M_m}\Delta f(t) + \frac{1}{M_m}\Delta P_m(t) - \frac{1}{M_m}\Delta P_l(t) - \frac{1}{M_m}\Delta P_e(t)\\
\Delta \dot{i}_q(t) &= \frac{\mathcal{C}_q}{T_w}\Delta V_{qr}(t-\zeta) - \frac{\Delta i_{qr}(t)}{T_w};\ \dot{\omega}(t) = \frac{\mathcal{C}_w}{M_w}\Delta i_{qr}(t) - \frac{\Delta \mathcal{T}_w(t)}{M_w}.
\end{aligned} \tag{1}$$

In frequency alteration of single power system, a virtual variable, $\delta(t) = \int \beta \Delta f(t)dt$ is included for preserving zero value of frequency deviation. Let we define $\vartheta(t) \in \mathbb{R}^n$, $u(t) \in \mathbb{R}^m$ $\varpi(t)\mathbb{R}^r$ and $y(t) \in \mathbb{R}^k$ are control signal, disturbance vector, vector of measurements, where

$$\begin{aligned}
\vartheta(t) &= [\Delta f(t)\ \Delta P_m(t)\ \Delta P_v(t)\ \delta(t)\ \Delta i_q(t)\ \omega(t)]^T,\\
u(t) &= [\Delta P_c\ \Delta V_q]^T, d(t) = [\Delta P_l\ \Delta \mathcal{T}_w]^T;\ \ y(t) = [\Delta f(t)\ \delta(t)]^T
\end{aligned} \tag{2}$$

Accordingly, from (1)–(2), a state-space expression of system in Fig. 1 is

$$\begin{aligned}
\dot{\vartheta}(t) &= \mathcal{A}\vartheta(t) + \mathcal{F}u(t-\zeta) + \mathcal{F}_d\varpi(t),\\
y(t) &= \mathcal{M}\vartheta(t).
\end{aligned} \tag{3}$$

In (3), $\mathcal{A} \in \mathbb{R}^{n\times n}, \mathcal{F} \in \mathbb{R}^{n\times m}, \mathcal{F}_d \in \mathbb{R}^{n\times p}, \mathcal{M} \in \mathbb{R}^{k\times n}$ are known matrices given at the Appendix. The control system of studied grid is driven by a structure of full feedback strategy (see Fig. 2) as follows

$$u(t) = \Sigma\vartheta(t) \tag{4}$$

System (4) becomes

$$\dot{\vartheta}(t) = \mathcal{A}\vartheta(t) + \mathcal{A}_h\vartheta(t-\zeta) + \mathcal{F}_d\varpi(t); \mathcal{A} = \mathcal{F}\Sigma \tag{5}$$

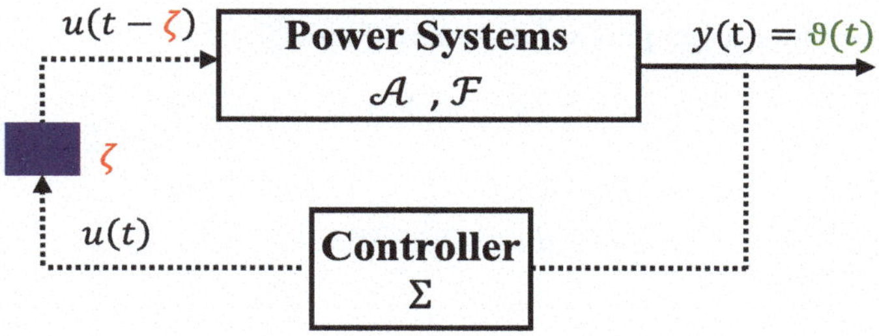

Fig. 2. A block of full state feedback control scheme

Problem Statement: As can be seen that, with a given mathematical representation of the power systems in (3), any load frequency full state feedback gain, Σ can be constructed to stabilize the studied system. However, as mentioned in the introduction, a full feedback strategy, $u(t) = \Sigma\vartheta(t)$, require all knowledge of state variables, $\vartheta(t)$, which means all signals need be measurable. This assumption is invalid in the practical operation of power systems due to only several signals, $y(t)$ can be measured. Hence, the purpose of this paper is to develop an appropriate feedback control scheme to stabilize the system while remaining the advantages of full state feedback strategy.

3 Control Strategy

In this research, we consider impulsive change in system disturbance including load demand and wind torque. By considering that all impact of disturbances be considered as initial value of system and can be ignored during the controller design. Hence, a new representation of system (5) is

$$\dot{\vartheta}(t) = \mathcal{A}\vartheta(t) + \mathcal{A}_h\vartheta(t-\zeta); \mathcal{A}_h = \mathcal{F}\Sigma \tag{6}$$

3.1 Stability Criteria

First, we develop the stability condition in Theorem 1 representing a sufficient condition ensuring the stability of system (6) with a given controller Σ. Then, we derive a stabilization condition in Theorem 2 to establish the controller gain Σ to warrant the system (6) is asymptotically stable.

Theorem 1. *For prescribe value $\zeta > 0$, asymptotic stability of system (6) is certified if matrices $\Pi, \Psi, \Upsilon \in \mathbb{R}^n > 0$ are available in symmetric definite form and satisfies the following linear condition*

$$\Theta = \begin{bmatrix} \Omega_1 & \Omega_2 \\ \Omega_3 & \Omega_4 \end{bmatrix} < 0 \tag{7}$$

where $\Omega_1 = \mathcal{A}^T\Pi + \Pi\mathcal{A} + \Psi + \zeta^2\mathcal{A}^T\Upsilon\mathcal{A} - \Upsilon$, $\Omega_2 = \Pi\mathcal{A}_h + \zeta^2\mathcal{A}^T\Upsilon\mathcal{A}_h + \Upsilon$, $\Omega_3 = \mathcal{A}_h^T\Pi + \zeta^2\mathcal{A}_h^T\Upsilon\mathcal{A} + \Upsilon$, $\Omega_4 = \zeta^2\mathcal{A}_h^T\Upsilon\mathcal{A}_h - \Psi$.

Proof. We consider a Luapunov function as follow

$$\Lambda(\vartheta,t)=\Lambda_1(\vartheta,t)+\Lambda_2(\vartheta,t)+\Lambda_3(\vartheta,t) \tag{8}$$

where $\Lambda_1(\vartheta,t),=\vartheta^T(t)\Pi\vartheta(t)$; $\Lambda_2(\vartheta,t)=\int_{t-\zeta}^{t}\vartheta^T(s)\Psi\vartheta(s)ds$, $\Lambda_3(\vartheta,t)=\zeta\int_{-\zeta}^{0}\int_{t+l}^{t}\dot{\vartheta}^T(u)\Upsilon\dot{\vartheta}(u)dudl$. Derivation of $\Lambda(\vartheta,t)$ is

$$\dot{\Lambda}(\vartheta_t,t)=\sum_{i=1}^{3}\dot{\Lambda}_i(\vartheta_t,t), \tag{9}$$

where $\dot{\Lambda}_1(\vartheta_t,t)=\dot{\vartheta}^T(t)\Pi\vartheta(t)+\vartheta^T(t)\Pi\dot{\vartheta}(t)$, $\dot{\Lambda}_2(\vartheta_t,t)=\vartheta^T(t)\Psi\vartheta(t)-\vartheta^T(t-\zeta)\Psi\vartheta(t-\zeta)$, $\dot{\Lambda}_3(\vartheta_t,t)=\zeta^2\dot{\vartheta}^T(t)\Upsilon\dot{\vartheta}(t)-\zeta\int_{-\zeta}^{0}(\dot{\vartheta}^T(t+l)\Upsilon\dot{\vartheta}(t+l))dl$

By using Lemma 1 at the Appendix, we obtain the following inequality

$$-\zeta\int_{-\zeta}^{0}(\dot{\vartheta}^T(t+l)\Upsilon\dot{\vartheta}(t+l))dl\leqslant-\Big(\vartheta^T(t)-\vartheta^T(t-\zeta)\Big)\Upsilon\Big(\vartheta(t)-\vartheta(t-\zeta)\Big) \tag{10}$$

Finally, based on the results in (9)–(10), we have

$$\dot{\Lambda}(\vartheta,t)\leqslant\left[\vartheta^T(t)\ \vartheta^T(t-\zeta)\right]\Theta\begin{bmatrix}\vartheta(t)\\ \vartheta(t-\zeta)\end{bmatrix}. \tag{11}$$

Hence, for $\Theta<0$, $\dot{\Lambda}(\vartheta_t,t)<0$, the stability of system (6) is confirmed.

Theorem 2. *For pre-arranged scalars $\zeta,\epsilon>0$, asymptotic stability of system (6) is guaranteed if symmetric definite matrices $\mathcal{P}\in\mathbb{R}^n>0,\mathcal{Q}\in\mathbb{R}^n>0$ and $\mathcal{Y}\in\mathbb{R}^n$ are available with appropriate dimension which satisfies the following condition.*

$$\Theta_1=\begin{bmatrix}\mathcal{P}^T\mathcal{A}^T+\mathcal{A}\mathcal{P}+\mathcal{Q}-\frac{\mathcal{P}}{\zeta^2\epsilon} & \mathcal{P}^T\mathcal{A}^T & \mathcal{F}\mathcal{Y}+\frac{\mathcal{P}}{\zeta^2\epsilon} & \mathcal{P}^T\mathcal{A}^T\\ \mathcal{A}\mathcal{P} & -\epsilon\mathcal{P} & \mathcal{F}\mathcal{Y} & -\epsilon\mathcal{P}\\ \mathcal{Y}^T\mathcal{F}^T+\frac{\mathcal{P}}{\zeta^2\epsilon} & \mathcal{Y}^T\mathcal{F}^T & -\mathcal{Q}-\frac{\mathcal{P}}{\zeta^2\epsilon} & \mathcal{Y}^T\mathcal{F}^T\\ \mathcal{A}\mathcal{P} & -\epsilon\mathcal{P} & \mathcal{F}Y & -\epsilon\mathcal{P}\end{bmatrix}<0 \tag{12}$$

and full state feedback controller gain Σ can be obtained as $\Sigma=\mathcal{Y}(\mathcal{P})^{-1}$

Proof. From the Theorem 1, we have $\Theta<0$. Now we apply schur complement to obtain $\Theta^*<0$

$$\Theta^*=\begin{bmatrix}\mathcal{A}^T\Pi+\Pi\mathcal{A}+\Psi-\Upsilon & \mathcal{A}^T & \Pi\mathcal{A}_h+\Upsilon & \mathcal{A}^T\\ \mathcal{A} & -\Upsilon_0 & \mathcal{A}_h & -\Upsilon_0\\ \mathcal{A}_h^T\Pi+\Upsilon & \mathcal{A}_h^T & -\Psi-\Upsilon & \mathcal{A}_h^T\\ \mathcal{A} & -\Upsilon_0 & \mathcal{A}_h & -\Upsilon_0\end{bmatrix} \tag{13}$$

We denote $\mathcal{Q}=\Pi^{-T}\Psi\Pi^{-1}$, $\mathcal{P}=\Pi^{-1}$,$\mathcal{P}^T=\mathcal{P}$; $\mathcal{Y}=\Sigma\mathcal{P}$, $\mathcal{P}^T\Sigma^T=\mathcal{Y}^T$,$\zeta^2\Upsilon=\frac{1}{\epsilon}\Pi$; $\Upsilon_0=\zeta^{-2}\Upsilon^{-1}=\epsilon\Pi^{-1}=\epsilon\mathcal{P}$, $\Gamma=diag(\Pi^{-1},I,\Pi^{-1},I)$, $\bar{\Upsilon}=\Pi^{-T}\Upsilon\Pi^{-1}=\mathcal{P}^T\Upsilon\mathcal{P}=\frac{\mathcal{P}^T\mathcal{P}^{-1}}{\zeta^2\epsilon}\mathcal{P}=\frac{\mathcal{P}}{\zeta^2\epsilon}$, $\Sigma=\mathcal{Y}\mathcal{P}^{-1}$.

By Pre and post multiplying Θ^* to Γ and Γ^T we obtain

$$\Theta_1 = \Gamma \Theta^* \Gamma^T. \tag{14}$$

Hence, it is very clear that if $\Theta_1 < 0$ (condition (12)) holds, the condition (condition (7)) and system (6) is asymptotically stable.

3.2 Feedback Scheme Implementation with Memorable Observer

We develop memory estimator with the following expression

$$\dot{\hat{\vartheta}}(t) = \mathcal{A}\hat{\vartheta}(t) + \mathcal{F}\hat{u}(t-\zeta) + \mathcal{L}(y(t) - \hat{y}(t)), \tag{15}$$

where $\hat{\vartheta} \in R^n, \hat{y}(t) \in R^k$ denote an approximation of state variables $\vartheta(t), y(t)$. $\hat{u}(t-\zeta)$ is the memorable value of the estimation, $\hat{u}(t-\zeta)$ of $u(t-\zeta)$. It can be seen that, in our observer structure the observer signals are reconstructed by memorable $\hat{u}(t-\zeta)$ and output measurements $y(t)$, which is computed within the system loop of observer, rather than instant value of control signals as conventional observer. Which indicate the effectiveness of our implementation due to less information is require for synthezing observer.

It is very clear, when the state information is well reconstructed, $\hat{\vartheta}(t) \rightarrow \vartheta(t)$, hence $\Sigma\hat{\vartheta}(t-\zeta) \rightarrow \Sigma\vartheta(t-\zeta)$, Σ is controller gain from (4). By using the memorable observer to establish control signal $\hat{u}(t-\zeta) = \Sigma\hat{\vartheta}(t-\zeta)$, system (6) becomes

$$\dot{\vartheta}(t) = \mathcal{A}\vartheta(t) + \mathcal{A}_h\hat{\vartheta}(t-\zeta), \mathcal{A}_h = \mathcal{F}\Sigma \tag{16}$$

We define observer error as $\varsigma(t) = \vartheta(t) - \hat{\vartheta}(t)$, which is the error between state variables and the reconstructed signals. The error dynamics of the estimator is

$$\dot{\varsigma}(t) = \dot{\vartheta}(t) - \dot{\hat{\vartheta}}(t) = (\mathcal{A} - \mathcal{L}\mathcal{M})\varsigma(t) \tag{17}$$

It is noted that $\varsigma(t)$ need convert to 0 after some settling time to ensure the performance of observer resulting in the accurate operation of the whole systems. In order to achieve that, $\mathcal{L}$ will be obtained such system (17) be asymptotically stable. In the following, Theorem 3 is provided to derive observer gain, $\mathcal{L}$.

Theorem 3. *System* (17) *become stable asymptotically if the existence of symmetric definite matrices $\Pi_1 \in \mathbb{R}^{n\times n} > 0, \mathcal{Y}_1 \in \mathbb{R}^{n\times n} > 0$ is ensured and they satisfies the following linear criteria*

$$\Theta_\varsigma = (\mathcal{A}^T\Pi_1 + \Pi_1\mathcal{A}) - (\mathcal{M}^T\mathcal{Y}_1^T + \mathcal{Y}_1\mathcal{M}) \tag{18}$$

And $\mathcal{L}$ can be obtained $\mathcal{L} = \Pi_1^{-1}\mathcal{Y}_1$.

Proof. We consider a Luyaponov candidature of observer error

$$V_\varsigma(\varsigma, t) = \varsigma^T(t)\Pi_1\varsigma(t) \tag{19}$$

Obtaining derivation of $V(\varsigma(t), t)$, we achieve the following result

$$\dot{V}_\varsigma(\varsigma, t) = \varsigma^T(t)\Big((\mathcal{A}^T \Pi_1 + \Pi_1 \mathcal{A}) - (\mathcal{M}^T \mathcal{Y}_1^T + \mathcal{Y}_1 \mathcal{M})\Big)\varsigma(t) \tag{20}$$

where $\mathcal{Y}_1 = \Pi_1 \mathcal{L}$. Therefore, if $\Theta_\varsigma < 0$, we will have $\dot{V}_\varsigma(\varsigma(t), t) < 0$, then system (15) is stable.

With the availability full state feedback controller Σ and the observer $\mathcal{L}$, the augmented system can be obtain as follow

$$\dot{\vartheta}_g(t) = \mathcal{A}_g \vartheta_g(t) + \mathcal{A}_{hg} \vartheta_g(t - \zeta) \tag{21}$$

$$\dot{\vartheta}_g(t) = \begin{bmatrix} \dot{\vartheta}(t) \\ \dot{\hat{\vartheta}}(t) \end{bmatrix}, \mathcal{A}_g = \begin{bmatrix} \mathcal{A} & 0 \\ \mathcal{L}\mathcal{M} & \mathcal{A} - \mathcal{L}\mathcal{M} \end{bmatrix}, \vartheta_g(t-\zeta) = \begin{bmatrix} \vartheta(t-\zeta) \\ \hat{\vartheta}(t-\zeta) \end{bmatrix}, \mathcal{A}_{hg} = \begin{bmatrix} 0 & \mathcal{F}\Sigma \\ 0 & \mathcal{F}\Sigma \end{bmatrix}.$$

We apply *Corollary* 1 that follows *Theorem* 1 to test the stability of (21)

Corollary 1. *For prescribe value $\zeta > 0$, asymptotic stability of system (6) is certified if matrices $\Pi, \Psi, \Upsilon \in \mathbb{R}^{2n} > 0$ are available in symmetric definite form and satisfies the following linear condition*

$$\Theta_g = \begin{bmatrix} \Omega_{1g} & \Omega_{2g} \\ \Omega_{3g} & \Omega_{4g} \end{bmatrix} < 0 \tag{22}$$

where $\Omega_{1g} = \mathcal{A}_g^T \Pi + \Pi \mathcal{A}_g + \Psi + \zeta^2 \mathcal{A}_g^T \Upsilon \mathcal{A}_g - \Upsilon$, $\Omega_{2g} = \Pi \mathcal{A}_{hg} + \mathcal{A}_g^T \Upsilon \mathcal{A}_{hg} + \Upsilon$, $\Omega_{3g} = \mathcal{A}_{hg}^T \Pi + \mathcal{A}_{hg}^T \Upsilon \mathcal{A}_g + \Upsilon$, $\Omega_{4g} = \zeta^2 \mathcal{A}_{hg}^T \Upsilon \mathcal{A}_{hg} - \Psi$.

Algorithm to obtain memorable observer based feedback controller is shown in Fig. 3 and Table 1.

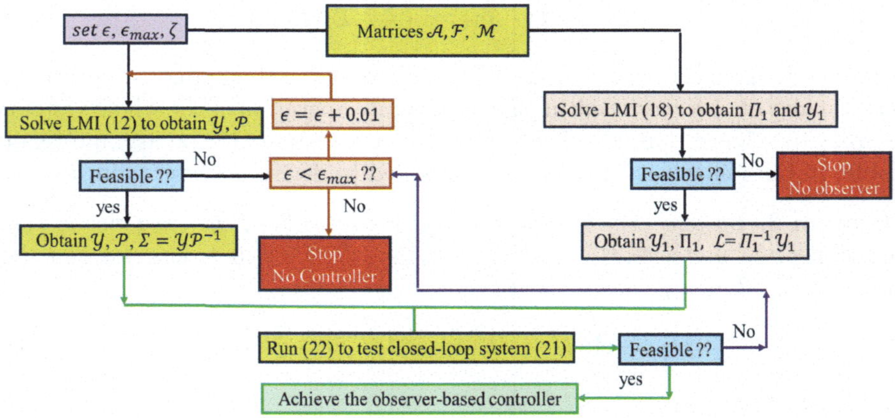

Fig. 3. A diagram presentation of proposed algoritm

Table 1. Control Algorithm

Step 1	Obtain $\mathcal{A}$, $\mathcal{F}$, $\mathcal{M}$, in (2) from the Appendix
Step 2	Solve condition (12) by using Robust control toolbox MATLAB with
Step 3	Solve condition (18) by using Robust control toolbox MATLAB to obtain $\mathcal{Y}_1$ optimisation process of ϵ to obtain $\mathcal{Y}$ and $\mathcal{P}$. Hence, controller gain: $\Sigma = \mathcal{Y}\mathcal{P}^{-1}$ and Π_1. Hence, the memorable estimator gain $\mathcal{L} = \Pi_1^{-1}\mathcal{Y}_1$
Step 4	Run test condition (21) to validate the performance of Power Systems

4 System Verification

4.1 Effectiveness of Control Strategy

In this section, we will take various scenarios of simulation to test the effectiveness of our proposed control regime. To do that, the performance of closed-loop power systems quipped full state feedback controller Σ and the performance of closed-loop power systems embedded a memorable observer based feedback controller gains $(\Sigma, \mathcal{L})$ are considered. Simulation of Scenarios I_1-I_4 are conducted by using MATLAB software and C-programming language. In this scenario, we consider a topology of a power system consisting a thermal power plant and a wind turbine as in Fig. 1. By applying Theorem 2 in (12) and algorithm in Sect. 3, we obtain controller Σ with given value of time delay, $\zeta = 0.094$s and given searching factor, $\epsilon = 71.43$. The controller gains are

$$\Sigma = \begin{bmatrix} 2.6111 & -0.0206 & -0.1403 & -0.2172 & 0.0027 & 0.3003 \\ 51.9992 & 1.4502 & -0.4262 & -0.8493 & -11.9024 & 5.5063 \end{bmatrix} \tag{23}$$

By applying Theorem 3, we obtain the memorable estimator gain, $\mathcal{L}$ as

$$\mathcal{L} = \begin{bmatrix} 0.5380 & 0.2709 & -199.8845 & 0.2115 & 0.3292 & -1.6011 \\ 0.2197 & -0.0077 & 0.0024 & 0.5000 & 0.1027 & -0.4046 \end{bmatrix}^T \tag{24}$$

For demonstrative purpose, Fig. 4 shows the trajectories of frequency deviation, $\Delta f(t)$ of closed-loop system with full-state feedback controller, Σ and of closed-loop system with our proposed memorable estimator based feedback controller, $(\Sigma,\ \mathcal{L})$. It is noted that the initial values of system's state vector and their estimation are $\vartheta^0 = [0.01\ \ 0.012\ \ -0.01\ \ 0.005\ \ -0.005\ \ -0.008]$ and $\hat{\vartheta}^0 = [-0.005\ \ -0.01\ \ -0.003\ \ 0.002\ \ -0.005\ \ 0.02]$.

As can be clearly seen that, the system frequency, $\Delta f(t)$, is pushed back to desirable values after some seconds of settling times by applying both controller. Hence, our proposed controller structure can achieve the requirement of frequency control of power systems as well as similar performance to full state feedback control systems. On the other hand, we undertake to observe the trajectories of frequency deviation, $\Delta f(t)$ and its estimation, $\hat{f}(t)$, with controller,

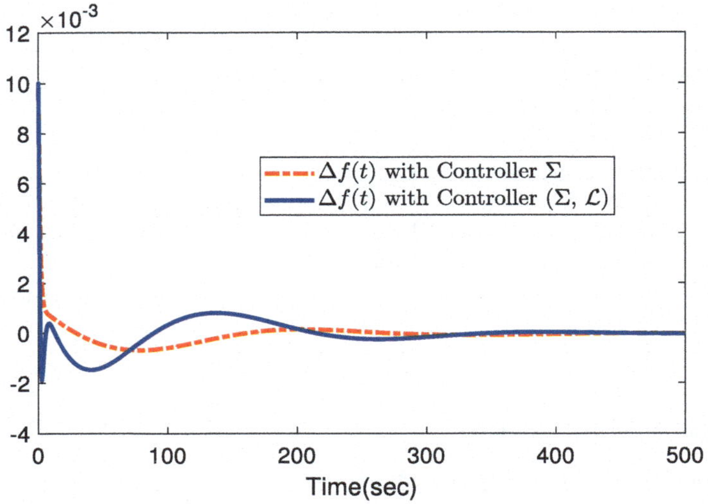

Fig. 4. Response of $\Delta f(t)$ of the closed-loop systems with controllers Σ and $(\Sigma, \mathcal{L})$

$\Sigma, \mathcal{L}$ (see Fig. 5). It is obvious that from different initial position, the estimation value of frequency deviation, $\Delta \hat{f}(t)$ can convert to the real value $\Delta f(t)$ after a duration of settling-time. This result verify the effectiveness of our estimator as well as control strategy.

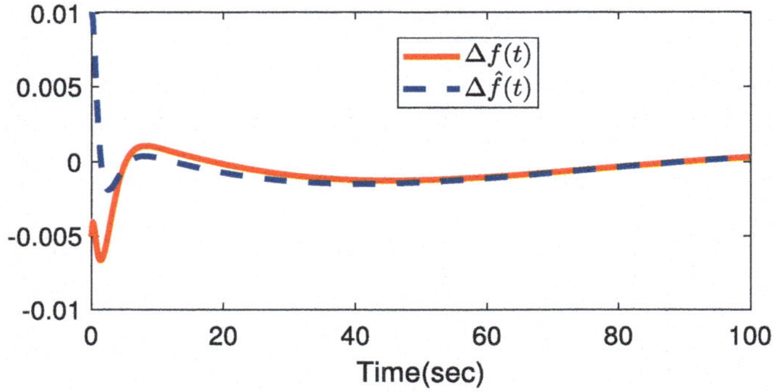

Fig. 5. Trajectories of $\Delta f(t)$ and $\Delta \hat{f}(t)$ with controllers $(\Sigma, \mathcal{L})$

4.2 The Impacts of Communication Delays

In this subsection, we will consider the inverse impact of communication of time delays on the operation of the studied system. The structure of system without time-delay is

$$\dot{\vartheta}(t) = \mathcal{A}\vartheta(t) + \mathcal{F}\Sigma_1\vartheta(t) \tag{25}$$

By using apply Theorem 2 for case $\zeta = 0$, we obtain

$$\Sigma_1 = \begin{bmatrix} 16.0782 & -3.1852 & 0.5475 & -0.5780 & -0.8573 & 6.2208 \\ 83.8627 & 23.4362 & 46.7674 & 17.5680 & 1.7538 & -105.3917 \end{bmatrix} \tag{26}$$

Now we obtain the controller Σ_1 and subject into system (22) when $\zeta = 0.04\ s$.

$$\dot{\vartheta}(t) = \mathcal{A}\vartheta(t) + \mathcal{F}\Sigma_1\vartheta(t-\zeta) \tag{27}$$

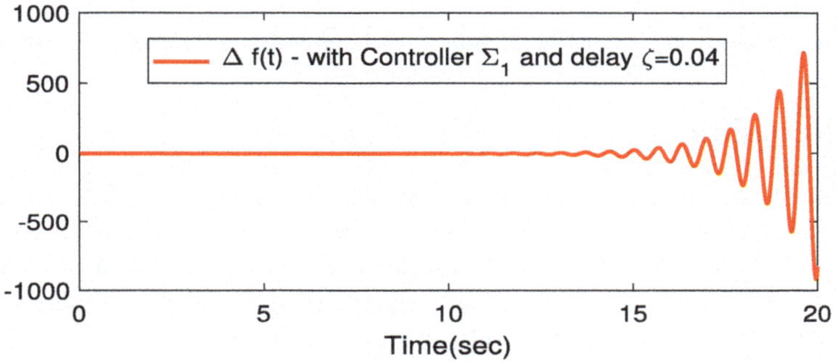

Fig. 6. Trajectories of $\Delta f(t)$ of the systems with controllers (Σ_1, delay $\zeta = 0.04$)

Figure 6 shows the trajectories of frequency deviation, $\Delta f(t)$ of closed-loop system with (Σ_1 and time delay, $\zeta = 0.04$. As can be seen that the system becomes unstable due to the existence of time delay ζ. Hence, it is very important to develop suitable control strategy to restore system's stability. This, again, verify the effectiveness of our control method.

4.3 Robustness to Sudden Demand Changes

In this subsection, we will consider the robustness of our control scheme subjects to sudden change in load demand $\Delta P_l(t)$. To do that, we run simulation of our systems with a time-varying change in power demand $\Delta P_l(t) = 0.01 + 0.01\cos(\sqrt{2}t)$. Note that we use $\Delta T_w^T = 0.01$. Now, we re-establish augmented system (21) with a consideration of load demand change.

$$\dot{\vartheta}_g(t) = \mathcal{A}_g\vartheta_1(t) + \mathcal{A}_{hg}\vartheta_g(t-\zeta) + \mathcal{F}_g\varpi(t), \tag{28}$$

where $\mathcal{F}_g = [\mathcal{F}_d^T \ \ [0]]^T$, $\varpi(t) = [\Delta P_l^T(t) \ \ \Delta T_w^T]^T$ and other notations are given in (21). Figure 7 shows the trajectories of the augmented systems under influence of time-varying power demand. As can be seen that our proposed controller can achieve the main requirement of LFC such bringing frequency deviation around zeros after some seconds.

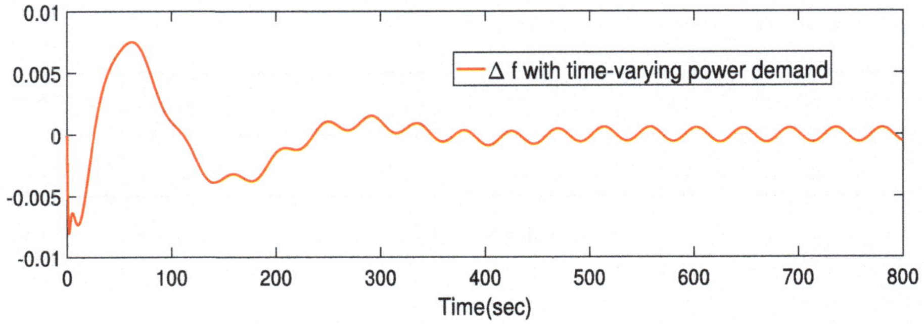

Fig. 7. Response of $\Delta f(t)$ of the system with time-varying change in power demand

5 Conclusion

This paper have considered a frequency stabilizing strategy for reinstate the stability of Smart grids incorporated wind turbines fluctuated due to unexpected change in load demand. We have developed a new mathematical representation of a smart grid comprising time delays related to thermal power plan and wind turbine. We have derived new stabilizable conditions for synthesizing frequency controller to guarantee performance of the grid. The validation of the proposed scheme is carried out with a grid incorporating wind turbine. In this paper, for simplicity, we merely consider the system model with a linear expression of wind turbines and the latency in the communication channels are a constant value. This is the limitation of our research. In the future, we will take a consideration of complex and non-linear model of wind turbines as well as the communication latency is a time-varying functions and further with bounded interval/or random delays.

A Appendix

Lemma 1. *(Jensen Inequality): For a given matrix* $R^{(}n \times n) > 0$ *and for all continuous functions* $\omega \in [a, b]$, *the following inequality holds:*
$I(\omega) = \int_a^b \left(\omega^T(u)\right) R\left(\omega(u)\right) du \geqslant \frac{1}{b-a}(\int_a^b \left(\omega^T(u)\right) du\right) R\left(\int_a^b(\omega(u))du\right)$

System Data: System data is given as follows $D = 1.0$, $M_m = 1.0$, $T_t = 0.3$ s, $R_s = 0.05$, $T_g = 0.1$ s, $T_w = 1382.18$ s, $M_w = 9$, $\mathcal{I}_{wo} = 0.8$, $\mathcal{W}_{wo} = 1.17$, $\mathcal{C}_q = 181.16$, $\mathcal{C}_w = 0.98$.

$$\mathcal{A} = \begin{bmatrix} -\frac{D_d}{M_m} & \frac{1}{M_m} & 0 & 0 & -\frac{\mathcal{C}_w \mathcal{W}_o}{M_m} & \frac{\mathcal{C}_w \mathcal{I}_o}{M_m} \\ 0 & -\frac{1}{T_t} & \frac{1}{T_t} & 0 & 0 & 0 \\ -\frac{1}{RT_g} & 0 & -\frac{1}{T_g} & 0 & 0 & 0 \\ 1 & 0 & 0 & 0 & 0 & 0 \\ 0 & 0 & 0 & 0 & -\frac{1}{T_w} & 0 \\ 0 & 0 & 0 & 0 & \frac{X_3}{M_w} & 0 \end{bmatrix} \mathcal{F} = \begin{bmatrix} 0 & 0 \\ 0 & 0 \\ \frac{1}{T_g} & 0 \\ 0 & 0 \\ 0 & \frac{\mathcal{C}_q}{T_w} \\ 0 & 0 \end{bmatrix} \mathcal{F}_d = \begin{bmatrix} -\frac{1}{M_m} & 0 \\ 0 & 0 \\ 0 & 0 \\ 0 & 0 \\ 0 & 0 \\ 0 & -\frac{1}{M_w} \end{bmatrix}$$

$$\mathcal{M} = \begin{bmatrix} 1\,0\,0\,0\,0\,0 \\ 0\,0\,0\,1\,0\,0 \end{bmatrix}$$

References

1. Kundur, P.: Power System Stability and Control. McGraw-Hill (1994)
2. Ibraheem, I., Kumar, P., Kothari, D.P.: Recent philosophies of automatic generation control strategies in power systems. IEEE Trans. Power Syst. **20**(1), 346–357 (2005)
3. Pandey, S.K., Mohanty, S.R., Kishor, N.: A literature survey on load-frequency control for conventional and distribution generation power systems. Renew. Sustain. Energy Rev. **25**, 318–334 (2013)
4. Trinh, H., Fernando, T., Iu, H., Wong, K.P.: Quasi-decentralized functional observers for the LFC of interconnected power systems. IEEE Trans. Power Syst. **28**(3), 3513–3514 (2013)
5. Bevrani, H., Ghosh, A., Ledwich, G.: Renewable energy sources and frequency regulation: survey and new perspectives. IET Renew. Power Gener. **4**(5), 438–457 (2010)
6. Aldeen, M., Trinh, H.: Load frequency control of interconnected power systems via constrained feedback control schemes. Int. J. Comput. Elect. Eng. **20**(1), 71–88 (1994)
7. Pham, T.N., Trinh, H., Hien, L.V.: Load frequency control of power systems with electric vehicles and diverse transmission links using distributed functional observers. IEEE Trans. Smart Grid **7**(1), 238–252 (2015)
8. Liu, F., Ma, J.: Equivalent input disturbance-based robust LFC strategy for power system with wind farms. IET Gener. Transm. Distrib. **12**(20), 4582–4588 (2018)
9. Venkateswaran, R., Joo, Y.H.: Retarded sampled-data control design for interconnected power system with DFIG-based wind farm: LMI approach. IEEE Trans. Cybern. **52**(7), 5767–5777 (2021)
10. Bhowmik, S., Tomsovic, K., Bose, A.: Communication models for third party load frequency control. IEEE Trans. Power Syst. **19**(1), 543–548 (2004)
11. Tomsovic, K., Bakken, D.E., Venkatasubramanian, V., Bose, A.: Designing the next generation of real-time control, communication, and computations for large power systems. Proc. IEEE **93**(5), 965–979 (2005)
12. Yu, X., Tomsovic, K.: Application of linear matrix inequalities for load frequency control with communication delays. IEEE Trans. Power Syst. **19**(3), 1508–1515 (2004)
13. Dey, R., Ghosh, S., Ray, G., Rakshit, A.: H_∞ load frequency control of interconnected power system with communication delays. Int. J. Elect. Power Energy Syst **42**(1), 672–684 (2012)
14. Zhang, C.K., Jiang, L., Wu, Q.H., He, Y., Wu, M.: Delay-dependent robust load frequency control for time delay power system. IEEE Trans. Power Syst. **28**(3), 2192–2201 (2013)
15. Bevrani, H., Hiyama, T.: On load frequency regulation with time delays: design and real time implementation. IEEE Trans. Energy Convers. **24**(1), 292–300 (2009)

Sliding Mode Control for The Two-Wheels Inverted Pendulum Mobile Robot via Optimisation Bounded-Output Observer

Dzung Manh Do[1], Xuan Minh Dinh[1], Tuan Nguyen Van[1(✉)], and Hai Xuan Le[2]

[1] Faculty of Mechanical Engineering and Mechatronics, Phenikaa University, Yen Nghia, Ha Dong, Hanoi 12116, Vietnam
tuan.nguyenvan@phenikaa-uni.edu.vn

[2] Faculty of Applied Sciences, International School, Vietnam National University, Hanoi 100000, Vietnam

Abstract. The two-wheel inverted pendulum mobile robot (TWIPMR) has high nonlinearities, uncertain parameters, and input disturbance. Based on the sliding mode control (SMC) and the disturbance observer, this paper proposes an adaptive controller for the TWIPMR with three independent motions. The novel disturbance observer with the saturation of the hardware is proposed to estimate the input disturbance of the TWIPMR system and reduce the computational cost of the existing observers. The low-pass regulated filters are integrated into the design procedure to eliminate the high-order harmonic and radio frequency interference of the TWIPMR system. The stability of the closed-loop system under the effect of the AFTSMC controller is ensured systematically through Lyapunov's second law. The simulation results demonstrate the effectiveness of the proposed controller and disturbance observer.

Keywords: Two-Wheels Inverted Pendulum Mobile Robot (TWIPMR) · Sliding Mode Control (SMC) · Disturbance observer · Low-pass regulated filter

1 Introduction

In recent years, the Two-Wheels Invert Pendulum Mobile Robot has received much attention from scientists and engineers around the world owing to its potential in the transportation fields [3]. One of the significant capabilities of the TWIPMR system is that it can be moved flexibly in many different topographics, especially in the metropolis area. The balancing archived of this robot when it moves is important in the development of the more complex types of mobile robots. Therefore, it is also necessary to continuously update and supplement research works on the TWIPMR system.

© The Author(s), under exclusive license to Springer Nature Switzerland AG 2025
H. A. Le Thi et al. (Eds.): ICAMCS 2024, LNNS 1311, pp. 336–347, 2025.
https://doi.org/10.1007/978-3-031-90606-0_28

According to paper [16], the structure of the TWIPMR system consists of two wheels with an inverted pendulum pattern linked in the middle. Therefore, the equilibrium plane degenerates to a straight line, leading to the TWIPMR system itself being an unstable system with strong nonlinearities from the dynamic perspective. It is a challenge in the TWIPMR control topic since the designed controller not only makes it balanced but also keeps its balance over a long time interval. Recently, there have been numerous controllers that can control the TWIPMR, from the classic ones [5] to advanced ones such as the H_∞ switching T-S fuzzy control [8], the model reference adaptive control [6], the neural-based adaptive control [16], the nonlinear model predictive control [12], controller via invariance and immersion method [1], time-delay control [4]. For the TWIPMR system, sliding mode control (SMC) [7,10] is the most effective method for designing the controller due to its robustify and simplicity. Therefore, this paper also used the SMC controller for the TWIPMR system.

Theoretically, to successfully design the SMC for a nonlinear system with environmental disturbances, the design procedure needs to know the information of the disturbances for combining into the input signal. Unfortunately, the desired quality of the SMC controller is only guaranteed with match disturbances [17], and the environmental disturbances themselves are unpredictable. To overcome this disadvantage, the equipment called the disturbance observer has been solved to estimate all unpredictable environmental disturbances. In [11], authors proposed the full-order data-driven adaptive disturbance observers (DADOs) for estimating unknown input gains, as well as the match and mismatch external disturbances via data-driven adaption. The stability of the closed-loop system when using DADOs and the backstepping controller was proven systematically. Similarly, authors in paper [9] proposed an unmatched disturbance estimator using the time-receding optimisation principle for a class of multi-input multi-output nonlinear non-autonomous systems with unmatched disturbance in both continuous-time and discrete-time domains. Additionally, the disturbance observer is also used to improve the quality of the controller such as the observer studied in the article [14]. Although these mentioned observers give good estimated quality with a small enough truncation error, authors when designing ones do not the saturation of the actuator. Therefore, this paper also proposes a new disturbance observer that not only can estimate the unpredictable disturbances with small enough truncation error but also satisfies the boundedness of the TWIPMR system's actuated through by *Projection* discontinuous mapping. Another new approach of the proposed observer is utilising the least-mean-square optimisation to construct the observation laws. The stability of the closed-loop system, including SMC controller and proposed optimisation observer is impressively proven through theorems. That is the main contribution of this manuscript.

The remaining part of this article is: Sect. 2 presents the problem formulation. Section 3 presents the primary design, as well as the design of the proposed observer. All the theorems for the stability of the closed-loop system are also demonstrated in Sect. 3. All simulation results are given in Sect. 4. Conclusions and future works are presented in the final section.

In this paper, $\mathbf{x}$ is symbolized for the column vector of real numbers $x_1, x_2, ..., x_n$ with n dimensions. $\mathbf{x}^T$ is the transpose vector of vector $\mathbf{x}$ and $\dot{\mathbf{x}}$ is the time derivation of vector $\mathbf{x}$. $\|.\|$ is the notation for 2-norm of vector $\mathbf{x}$. Matrix $\Theta_{m\times n}$ is a matrix with m rows and n columns including all zero numbers. The matrix Θ_n is a short presentation of matrix $\Theta_{n\times n}$. R^n is the space of all n-dimensional vectors and $R^{m\times n}$ is the space of all matrices which have m rows and n columns. In addition, the *Projection* discontinuous mapping is defined as:

$$h(x) = Proj\left(z(x), z_{\min}, z_{\max}\right) = \begin{cases} z(x), \text{ if } z_{\min} < x < z_{\max} \\ z_{\max}, \text{ if } x \geq z_{\max} \\ z_{\min}, \text{ if } x \leq z_{\min} \end{cases} \tag{1}$$

where $x \in R$, $h(.)$ and $z(.)$ are both continuous functions over x and $z_{\min}, z_{\max} \in R$, $z_{\min} < z_{\max}$.

2 Problem Formulation

The mathematical model of the TWIPMR nonlinear system, which is given in the article [12,15] as the form of an affine nonlinear system:

$$\ddot{\mathbf{x}} = \mathbf{F}(\mathbf{x}) + \mathbf{G}(\mathbf{x})\mathbf{u} + \mathbf{d} \tag{2}$$

where $\mathbf{x} = \left[x\ \theta\ \psi\right]^T$ is the state vector of the TWIPMR in which θ and ψ are the rotational angles of the system in horizontal and vertical, respectively. $\mathbf{u} \in R^m$ is an input vector that is generated by the motor of two wheels. $\mathbf{d} \in R^n$ is the unknown vector of the match and mismatch environmental disturbances affected in the system in the working process. Matrix $\mathbf{g}(\mathbf{x}) \in R^{2n\times n}$ is the matrix of input gain where $\mathbf{G}(\mathbf{x})$ is an invertible and continuous matrix. Vector $\mathbf{F}(\mathbf{x})$ is the vector including all kinematic and dynamical properties of TWIPMR system such as gravitational component, holonomic and nonholonomic constraints, and $\mathbf{F}(\mathbf{x})$ is a continuous vector. Matrix $\mathbf{G}(\mathbf{x})$ and vector $\mathbf{F}(\mathbf{x})$ are both calculated in the article [16].

The objectives of this paper are:

- Determine the state feedback controller $\mathbf{u} = \mathbf{u}(\mathbf{x})$ in order to drive the TWIPMR's states to track the reference signal $\mathbf{x}_d$ in a short of a time interval.
- Design an equipment $\hat{\mathbf{d}} = \hat{\mathbf{d}}(\mathbf{x}, \dot{\mathbf{x}}, t)$ satisfying $\lim_{t\to+\infty} \hat{\mathbf{d}} = \mathbf{d}$ based on bounded least-mean-square method. This equipment is called the optimisation output-bounded observer.

To perform these tasks, two assumptions below are held:

Assumption 1: *All states of the TWIPMR system, their first-order derivations, and the input disturbances are bounded in the convex domain $\mathcal{D}$.*

Assumption 2: *rank* $[\mathbf{G}(\mathbf{x})] = 3, \forall \mathbf{x}, t \geq 0$.

These assumptions are reasonable because, in reality, the TWIPMR's movement is limited by the hardware structure and the control signal created by the actuator cannot be arbitrarily large to generate an infinite velocity for each state.

3 Main Design

3.1 Design Sliding Mode Controller

Definite the positional tracking error as a form $\mathbf{e} = \mathbf{x}_d - \mathbf{x}$. And then, the sliding surface for the positional controller is calculated by:

$$\mathbf{S} = \dot{\mathbf{e}} + \mathbf{F}(\mathbf{e}) \tag{3}$$

where $\mathbf{F}(\mathbf{e})$ is a differentiable and continuous function. Take the derivative in both sides of the Eq. (3) gives:

$$\dot{\mathbf{S}} = \ddot{\mathbf{x}}_d - \mathbf{F}(\mathbf{x}) - \mathbf{G}(\mathbf{x})\mathbf{u} - \mathbf{d} + \frac{\partial \mathbf{F}(\mathbf{e})}{\partial \mathbf{e}}\dot{\mathbf{e}} \tag{4}$$

To guarantee the stability of the closed-loop system under the positional controller, set the Eq. (4) to be equal to:

$$\ddot{\mathbf{x}}_d - \mathbf{F}(\mathbf{x}) + \frac{\partial \mathbf{F}(\mathbf{e})}{\partial \mathbf{e}}\dot{\mathbf{e}} - \mathbf{G}(\mathbf{x})\mathbf{u} - \mathbf{d} = -\lambda \mathbf{S} - \beta \tanh(\mathbf{S}) \tag{5}$$

Denote $\mathcal{F} = \ddot{\mathbf{x}}_d - \mathbf{F}(\mathbf{x}) + \frac{\partial \mathbf{F}(\mathbf{e})}{\partial \mathbf{e}}\dot{\mathbf{e}}$. λ and β are two positive real numbers. From (5), we have the ideal finite-time terminal sliding mode positional controller:

$$\mathbf{G}(\mathbf{x})\mathbf{u} = \mathcal{F} - \mathbf{d} + \lambda \mathbf{S} + \beta \tanh(\mathbf{S}) \tag{6}$$

Due to satisfying Assumption 2, the SMC controller is determined by:

$$\mathbf{u} = \left[\mathbf{G}(\mathbf{x})^T \mathbf{G}(\mathbf{x})\right]^{-1} \mathbf{G}(\mathbf{x})^T \left(\mathcal{F} + \hat{\mathbf{d}} + \lambda \mathbf{S} + \beta \tanh(\mathbf{S})\right) \tag{7}$$

In Eq. (7), $\hat{\mathbf{d}}$ is the estimated disturbances of $\mathbf{d}$.

3.2 Design the Optimisation Bounded-Output Observer

A novel step-wise mode disturbance observer is proposed in this section to estimate the real lumped disturbances. This observer is designed to satisfy the limitations of the actuators. Substituting Eq. (7) into Eq. (2) gives:

$$\ddot{\mathbf{x}} = \mathbf{F}(\mathbf{x}) + \hat{\mathcal{F}} + \lambda_f \mathbf{S} + \beta_j \tanh(\mathbf{S}) - \hat{\mathbf{d}}_j + \mathbf{d}_j \tag{8}$$

where $\mathbf{x} = x, \theta, \psi$ and $j = 1, 2, 3$. Replacing $\ddot{\mathbf{x}}$ on the left side with its estimated value by low-pass filter $\hat{f}$ we have:

$$\hat{\mathbf{x}} \approx \mathbf{F}(\mathbf{x}) + \hat{\mathcal{F}} + \lambda_f \mathbf{S} + \beta_j \tanh(\mathbf{S}) - \hat{\mathbf{d}}_j + \mathbf{d}_j \tag{9}$$

Replacing $\mathbf{d}_j$ by $\hat{\mathbf{d}}_j^{k+1}$ which is the value to be estimated at time instance t_{k+1}, Eq. (9) is converted to the discrete form:

$$\hat{\mathbf{x}}_k \approx \mathbf{F}^k(\mathbf{x}) + \hat{\mathcal{F}}^k + \lambda_f \mathbf{S}^k + \beta_j \tanh(\mathbf{S}^k) - \hat{\mathbf{d}}_j^k + \hat{\mathbf{d}}_j^{k+1} \tag{10}$$

Note that $\hat{\mathbf{d}}_j^{k+1}$ need to be satisfied the bounded condition: $\mathbf{d}_{\text{min}} \leq \hat{\mathbf{d}}_j^{k+1} \leq \mathbf{d}_{\text{max}}$. In Eq. (10), $\hat{f}_k$, $\mathbf{F}^k(\mathbf{x})$, $\hat{\mathcal{F}}^k$, $\mathbf{S}^k$ are the short presentation of $\hat{f}(t_k)$, $\mathbf{F}(\mathbf{x}, t_k)$, $\hat{\mathcal{F}}(t_k)$, $\mathbf{S}(t_k)$ at time instance $t = t_k$, respectively. $\hat{\mathbf{d}}_j^k$ is the estimated value of the environmental disturbances at $t = t_k$. From Eq. (10), the adaptive formula for observing the lumped disturbance is determined as:

$$\begin{cases} \hat{\mathbf{d}}_j^{k+1} \approx \hat{\mathbf{d}}_j^k - \mathbf{H}(\mathbf{x}) \\ \hat{\mathbf{d}}_j^{k+1} \leq \mathbf{d}_{\text{max}} \\ \hat{\mathbf{d}}_j^{k+1} \geq \mathbf{d}_{\text{min}} \end{cases} \tag{11}$$

Therefore,

$$\hat{\mathbf{d}}_j^{k+1} = Proj\left(\hat{\mathbf{d}}_j^k - \mathbf{H}(\mathbf{x}), \mathbf{d}_{\text{min}}, \mathbf{d}_{\text{max}}\right) \tag{12}$$

Where $\mathbf{H}(\mathbf{x}) = F_j^k(\mathbf{x}) + \hat{\mathcal{F}}_f^k + \lambda_f \mathbf{S}^k + \beta_j \tanh(\mathbf{S}^k) - \hat{f}_k$, $f = x, \theta, \psi$ and $j = 1, 2, 3$.

3.3 Stability Analysis

Theorem 1: The learning-based estimator $\mathbf{d}_j^{k+1}$ given in Eq. (12) minimises the approximation (10) if $\hat{\mathbf{d}}_j^{k+1} \in [\mathbf{d}_{\text{min}}, \mathbf{d}_{\text{max}}]$. Moreover, in case $\hat{\mathbf{d}}_j^{k+1} \notin [\mathbf{d}_{\text{min}}, \mathbf{d}_{\text{max}}]$, the truncation error of approximation (10) is always bounded by a positive constant α.

Proof: Denote the truncation error of approximation (10) at $t = t_k$ as:

$$\varepsilon_k = \mathbf{H}(\mathbf{x}) - \hat{\mathbf{d}}_j^k + \mathbf{d}_j^k \tag{13}$$

And the cost function for minimizing ε_k is defined with a quadratic form:

$$J_k(\mathbf{d}_j^k) = \frac{1}{2}\varepsilon_k^T \varepsilon_k \rightarrow \min \tag{14}$$

Substituting Eq. (13) into Eq. (14) gives:

$$J_k(\mathbf{d}_j^k) = \frac{1}{2}\left[\left(\mathbf{H}(\mathbf{x}) - \hat{\mathbf{d}}_j^k\right) + \mathbf{d}_j^k\right]^T \left[\left(\mathbf{H}(\mathbf{x}) - \hat{\mathbf{d}}_j^k\right) + \mathbf{d}_j^k\right] \tag{15}$$

$$\Leftrightarrow J_k(\mathbf{d}_j^k) = \frac{1}{2}\left[J_H + \mathbf{d}_j^{kT}\mathbf{d}_j^k + 2\left(\mathbf{H}(\mathbf{x}) - \hat{\mathbf{d}}_j^k\right)^T \mathbf{d}_j^k\right] \tag{16}$$

Where $J_H = \left(\mathbf{H}(\mathbf{x}) - \hat{\mathbf{d}}_j^k\right)^T \left(\mathbf{H}(\mathbf{x}) - \hat{\mathbf{d}}_j^k\right)$. The optimal solution $\mathbf{d}_j^{*k}$ of minimizing $J \rightarrow \min$ must be satisfied:

$$\mathbf{d}_j^{*k} = \underset{\mathbf{d}_{\min} \leq \mathbf{d}_j^k \leq \mathbf{d}_{\max}}{\operatorname{argmin}} \quad J_k(\mathbf{d}_j^k) \tag{17}$$

The constraint problem (17) gives solution by the gradient descent method [13]:

$$\mathbf{d}_j^{*k} = Proj\left(\hat{\mathbf{d}}_j^k - \mathbf{H}(\mathbf{x}), \mathbf{d}_{\min}, \mathbf{d}_{\max}\right) \tag{18}$$

which coincides with $\hat{\mathbf{d}}_j^{k+1}$ given in (12). In case $\hat{\mathbf{d}}_j^{k+1} \notin [\mathbf{d}_{\min}, \mathbf{d}_{\max}]$, it can be seen from Eq. (13) that:

$$\|\varepsilon_k\| = \left\|\mathbf{H}(\mathbf{x}) - \hat{\mathbf{d}}_j^k + \mathbf{d}_j^k\right\| \leq \left\|\mathbf{H}(\mathbf{x}) - \hat{\mathbf{d}}_j^k\right\| + \left\|\mathbf{d}_j^k\right\| \tag{19}$$

Note that $\mathbf{H}(\mathbf{x}) - \hat{\mathbf{d}}_j^k \leq \max\limits_{\mathbf{x} \in \mathcal{D}} \mathbf{H}(\mathbf{x}) - \mathbf{d}_{\min}$. Therefore,

$$\left\|\mathbf{H}(\mathbf{x}) - \hat{\mathbf{d}}_j^k\right\| \leq \left\|\max_{\mathbf{x} \in \mathcal{D}} \mathbf{H}(\mathbf{x}) - \mathbf{d}_{\min}\right\| \leq \alpha_1 \tag{20}$$

Where $\alpha_1 > 0$. Therefore,

$$\|\varepsilon_k\| \leq \alpha_1 + \max_{\mathbf{d} \in \mathcal{D}} \left\|\mathbf{d}_j^k\right\| = \alpha = \text{const} \tag{21}$$

This proof is complete.

Remark 3: *Due to the dependence of* $\mathbf{H}(\mathbf{x})$ *on the sliding function* $F_f(e_f)$, $f = x, \theta, \psi$, *we can reduce* α *by changing the sliding function* $F_f(e_f)$. *The effectiveness of this method is demonstrated in the simulation section.*

Remark 4: *The novelty of the proposed disturbance observer (12) is that the right-hand side of the estimator (12) satisfies the bounded condition* $\mathbf{d}_{\min} \leq \mathbf{d}_j^{k+1} \leq \mathbf{d}_{\max}$ *of the actuators. This problem has not been considered in these previous works [9, 15]. Furthermore, by utilizing the low-pass filter for* $\ddot{f}, \ddot{f}_d$ *estimation, we can adjust the filter's time parameter to reduce truncation approximation. It is impossible when using the backward differentiation formula in the paper [15].*

Theorem 2: The sliding mode controller (7) including the positional controller, the rotational controller, and the anti-swing controller will make the closed-loop system that to be stable at the neighbourhood of the origin $\mathcal{O}$ which is determined by:

$$\mathcal{O} = \left\{\mathbf{S}V(\mathbf{S}) \leq \min\left\{\left(\frac{\mathbf{d}}{(1-q)\varphi_1}\right)^{\frac{1}{\phi_2}}, \left(\frac{\mathbf{d}}{(1-q)\varphi_2}\right)^{\frac{1}{\phi_1}}\right\}\right\} \tag{22}$$

Furthermore, the closed-loop system is stable at the attractive domain $\mathcal{O}$ in a fixed-time: $T \leq T_1 = \frac{1}{\varphi_1 q(\phi_2 - 1)} + \frac{1}{\varphi_2 q(1-\phi_1)}$.

Proof: To prove the stability of the closed-loop system, a candidate Lyapunov function is defined as the quadratic form:

$$V(\mathbf{S}) = \frac{1}{2}\sum_{f=1}^{3} S_f^T P S_f + \frac{1}{2}\sum_{f=1}^{3}\left(\int_{t_o}^{t} S_f(\tau)d\tau\right)^T Q\left(\int_{t_o}^{t} S_f(\tau)d\tau\right) \tag{23}$$

where P, Q are two positively defined matrices. Take the time derivative on both sides of the Eq. (23) gives:

$$\dot{V}(\mathbf{S}) = \sum_{f=1}^{3} S_f^T P \dot{S}_f + \frac{1}{2}\sum_{f=1}^{3}\left(\int_{t_o}^{t} S(\tau)d\tau\right)^T QS \tag{24}$$

From Eq. (4), we have:

$$\dot{\mathbf{S}}_f = \hat{x}_d - \mathbf{F}_f(\mathbf{x}) - \hat{\tau}_j - \mathbf{d}_j + \frac{\partial F_f(e_f)}{\partial e_f}\dot{e}_f \tag{25}$$

where $f = x, \theta, \psi$; $\hat{f}_d$ is the approximation of $\ddot{f}_d$ by using the low-pass filter. Substituting the Eq. (7) into the Eq. (25) we have:

$$\dot{\mathbf{S}}_f = (\hat{x}_d - \ddot{x}_d) + \left(\hat{\mathbf{d}}_j - \mathbf{d}_j\right) - \lambda_f \mathbf{S} - \beta_j \tanh(\mathbf{S}) \tag{26}$$

Denote that $\tilde{x}_d = \hat{x}_d - \ddot{x}_d$ and $\tilde{\mathbf{d}}_j = \hat{\mathbf{d}}_j - \mathbf{d}_j$. From these Eqs. (26), (24), we have:

$$\begin{aligned}\dot{V}(\mathbf{S}) = &-\sum_{f=1}^{3} S_j^T P\lambda_f S_j - \sum_{f=1}^{3} S_j^T P\tanh(S_j) - \sum_{f=1}^{3} Q\beta_j\left(\int_{t_o}^{t}\mathbf{S}d\tau\right)^2 \\ &-\sum_{f=1}^{3}\left(\int_{t_o}^{t}\mathbf{S}(\tau)d\tau\right)^T Q\beta_j\left(\int_{t_o}^{t}\tanh(\mathbf{S})(\tau)d\tau\right) + \mathcal{R}\left(\tilde{x}_d, \tilde{\mathbf{d}}_j\right)\end{aligned} \tag{27}$$

where the function $\mathcal{R}\left(\tilde{x}_d, \tilde{\mathbf{d}}_j\right) = -\sum_{f=1}^{3}\left(\int_{t_o}^{t}\mathbf{S}(\tau)d\tau\right)^T Q\left(\int_{t_o}^{t}\left(\tilde{x}_d + \tilde{\mathbf{d}}_j\right)d\tau\right) - \sum_{f=1}^{3} S_j^T P\tilde{x}_d - \sum_{f=1}^{3} S_j^T P\tilde{\mathbf{d}}_j$. If denote $\mathbf{d}t > 0$ is a small time interval and

$$\mathbf{d}_1 = \inf_{[t_o;t_o+\mathbf{d}t)}\ \inf_{[t_o+\mathbf{d}t;t_o+2\mathbf{d}t)} \cdots \inf_{[t_o+n\mathbf{d}t;+\infty)}\ \inf_{\tilde{f}_d,\tilde{\mathbf{d}}_j,j=1,2,3}\left\|\mathcal{R}\left(\tilde{f}_d, \tilde{\mathbf{d}}_j\right)\right\| > 0 \tag{28}$$

$$\mathbf{d}_2 = \sup_{[t_o;t_o+\mathbf{d}t)}\ \sup_{[t_o+\mathbf{d}t;t_o+2\mathbf{d}t)} \cdots \sup_{[t_o+n\mathbf{d}t;+\infty)}\ \sup_{\tilde{f}_d,\tilde{\mathbf{d}}_j,j=1,2,3}\left\|\mathcal{R}\left(\tilde{f}_d, \tilde{\mathbf{d}}_j\right)\right\| > 0 \tag{29}$$

On the one hand, at $t = t_k \in [t_o + m\mathbf{d}t; t_o + (m+1)\mathbf{d}t), 1 \le m \le n$ and $\mathbf{d}_{\min} \le \mathbf{d}_j^k \le \mathbf{d}_{\max}$, we can deduce that $\lim_{t\to+\infty}\left\|Proj\left(\hat{\mathbf{d}}_j^k - \mathbf{H}(\mathbf{x}), \mathbf{d}_{\min}, \mathbf{d}_{\max}\right) - \mathbf{d}_j^k\right\| \le \varepsilon, \forall\varepsilon > 0$ leading to $\mathbf{d}_1 \le \left\|\mathcal{R}\left(\tilde{f}_d, \tilde{\mathbf{d}}_j\right)\right\| \le \mathbf{d}_2, \forall t \ge t_o$. On the other hand, due to the boundedness of tanh(.) function, these exist four real positive numbers

$0 < \phi_1 < 1$, $\phi_2 > 1$, $\varphi_1 = \min(\lambda_f, \beta_f), \varphi_2 > 0$ that make the candidate Lyapunov function and its first-order derivation in (23), (27) to satisfy:

$$-\varphi_1 V^{1-\phi_1}(\mathbf{S}) - \varphi_2 V^{\phi_2} + \mathbf{d}_2 \leq \dot{V}(\mathbf{S}) \leq -\varphi_1 V^{\phi_1}(\mathbf{S}) - \varphi_2 V^{\phi_2}(\mathbf{S}) + \mathbf{d}_1 \quad (30)$$

According to the paper [2], there exists a fixed-time T satisfying $t_o < T \leq \frac{1}{\varphi_1 q(\phi_2 - 1)} + \frac{1}{\varphi_2 q(1-\phi_1)}$ such as $\forall t \geq T$, the stability of the closed-loop system at the neighbourhood of the origin is guaranteed in which the attractive domain is $\mathcal{O}$ which is determined by:

$$\mathcal{O} = \left\{ \mathbf{S} V(\mathbf{S}) \leq \min \left\{ \left(\frac{\mathbf{d}}{(1-q)\varphi_1} \right)^{\frac{1}{\phi_2}}, \left(\frac{\mathbf{d}}{(1-q)\varphi_2} \right)^{\frac{1}{\phi_1}} \right\} \right\} \quad (31)$$

Furthermore, the candidate Lyapunov function V satisfying

$$\exists \hat{\gamma} \in [0,1], \hat{\gamma} V(S_{1f}) + (1-\hat{\gamma}) V(S_{2f}) \leq V(\hat{\gamma} S_{1f} + (1-\hat{\gamma}) S_{2f}) \quad (32)$$

Where $S_{1f} = \mathbf{S}(t_1)$, $S_{2f} = \mathbf{S}(t_2)$, $t_1 \neq t_2; t_1, t_2 \geq T > t_o$, so the Lyapunov function cannot have a local maximum, thus, $\mathbf{S} \in \mathcal{O}$ either $\dot{V}(\mathbf{S}) < 0$ or $\dot{V}(\mathbf{S}) = 0$. Therefore, $\lim_{t \to +\infty} \|\mathbf{S}(t)\| = 0$ is ensured. This is the second proof. The proof for Theorem 2 is complete.

4 Simulation Results

In the simulation section, we will simulate the effectiveness of the AFTSMC controller and the proposed disturbance observer with four different sliding functions which are $f_1 = F_f(e_f) = \lambda_j e_f^{\gamma_j}$, $f_2 = F_f(e_f) = \lambda_j \tanh^{\gamma_j}(e_f)$, $f_3 = F_f(e_f) = \lambda_j \text{arcsinh}^{\gamma_j}(e_f)$, $f_4 = F_f(e_f) = \lambda_j \text{arcsinh}^{\gamma_j}(e_f) \sqrt[\chi]{\|e_f\|}$. Where $f = x, \theta, \psi$ and $j = 1, 2, 3$, and f_1, f_2, f_3 and f_4 satisfies the condition mentioned in Remark 1. The effectiveness of the proposed control strategy will be verified through the simulation process on Matlab/Simulink. First, the parameters of the TWIPMR system and the AFTSMC controller are presented in Table 1. Besides, the quality of the AFTSMC controller is demonstrated by considering five different time parameters T_s that is $0.001(s)$, $0.005(s)$, $0.01(s)$, $0.0008(s)$ and $0.05(s)$.

Table 1. The parameters of the TWIPMR system

$l = 0.08(m)$	$r = 0.033(m)$	$d = 0.16(m)$	$g = 9.81(m/s^2)$
$m = 0.04(kg)$	$M = 0.5(kg)$	$C = 5.10^{-4}(Ns/m)$	$K_m = 10000 Nm/A$
$J = \frac{1}{2}mr^2$	$K_r = \frac{1}{4}mr^2$	$I_1 = I_3 = \frac{1}{2}md^2$	$I_2 = \frac{1}{2}md^2, \chi = 7$
$\lambda_x = \lambda_\theta = \lambda_\psi = 50$	$\beta_x = \beta_\theta = \beta_\psi = 50$	$\lambda_1 = \lambda_2 = \lambda_3 = 50$	$\gamma_1 = \gamma_2 = \gamma_3 = 1.3$

In this section, we assumed that the environmental disturbances which affect the closed system as to be a combination of the sinusoid signals:

$$\mathbf{d}(t) = \begin{bmatrix} 5\left(\sin(3\pi t) - \sin(6\pi t + \pi/4)\right) \\ 5\left(\sin(\pi t + \pi/4) + \cos(0.5\pi t + \pi/6)\right) \\ 5\left(\cos(5\pi t) + \sin(\pi t + \pi)\right) \end{bmatrix} \tag{33}$$

And the boundaries of the estimated disturbance that we select are $\mathbf{d}_{\max} = 10$ and $\mathbf{d}_{\min} = -10$. The initial conditions for x, θ, ψ are chosen as $x(t_o) = 2(m)$, $\theta(t_o) = \pi/6(rad)$, $\psi(t_o) = -\pi/3(rad)$. All velocities are equal to zeros in the initial time. The reference signals are selected to be equal to zero.

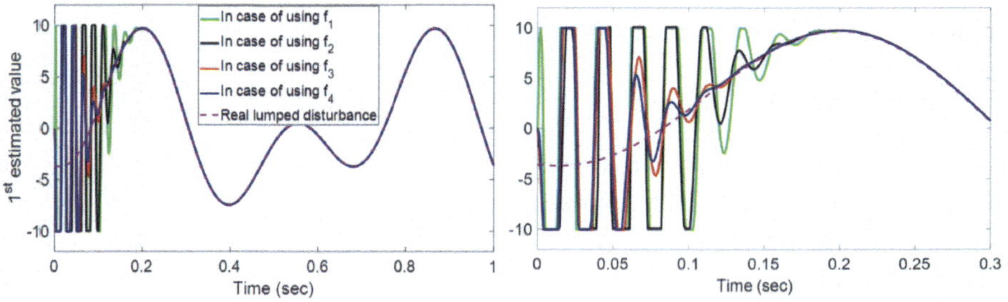

Fig. 1. The first result of the disturbance estimation within 1 s and 0.3 s.

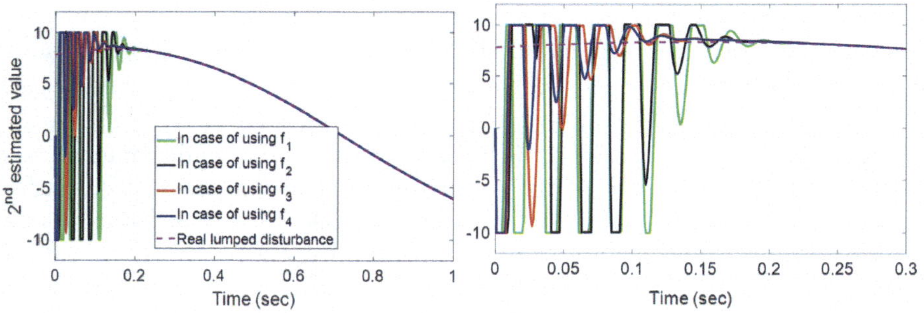

Fig. 2. The second result of the disturbance estimation within 1 s and 0.3 s.

The results of the disturbance approximation are shown from the Figs. 1, 2 and 3. The effectiveness of the proposed optimisation observer is demonstrated. On the one hand, this observer can estimate unpredictable environmental disturbances with small accuracy in a short time interval despite the match or mismatch disturbance. Moreover, the estimated results do not overcome the restriction of the actuators by applying the Gradient descent method for the constraint optimal issue. On the other hand, by changing the sliding function, the accuracy of the proposed observer can be improved significantly as mentioned

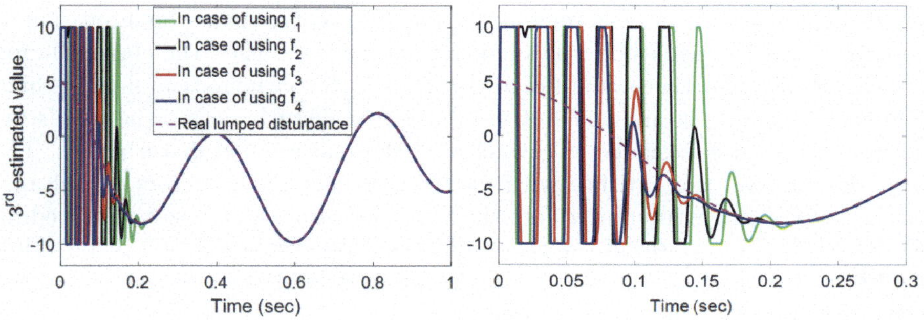

Fig. 3. The third result of the disturbance estimation within 1 s and 0.3 s.

in Remark 3. In detail, the estimated results with the sliding function f_3 or f_4 (red line and blue line) give a higher control quality than the other functions. Simultaneously, the time interval for approximation processes is shorter (about from 0.1 s to 0.15 s).

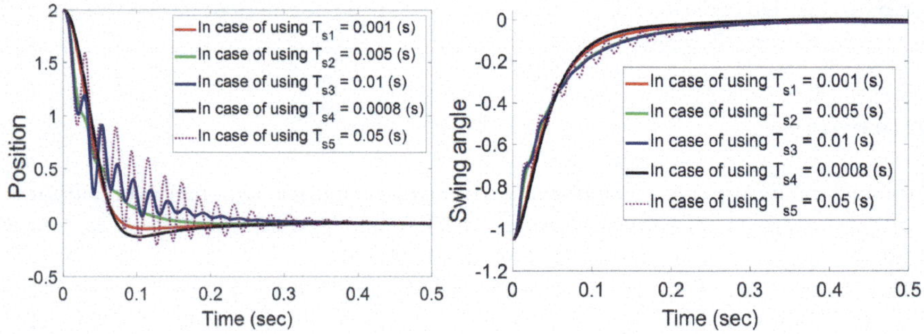

Fig. 4. The positional response and the anti-swing response with different low-pass filter's time parameters within 0.5 s.

The results shown in Fig. 4 are the numerical simulation demonstrations for the proposition mentioned in the Theorem 3 that the closed system's stability does not depend on the time parameter of the low-pass filter. Á can be seen from this figure, the smaller the low-pass filter's time parameter is, the higher the control performance is. With the five different values of T_s, it also can conclude that whether T_s is still small enough to satisfy these conditions $\mathbf{d}_1 \leq \left\| \mathcal{R}\left(\tilde{f}_d, \tilde{\mathbf{d}}_j, T_s\right) \right\| \leq \mathbf{d}_2$, the stability of the closed TWIPMR system is not affected.

5 Conclusion

This work proposed a sliding mode controller (SMC) to drive the states of the TWIPMR system to track the desired states in a short time interval. The proposed controller with the different sliding functions could make the input signals smoother, and its amplitude was smaller than itself by using the classic sliding mode controller. Furthermore, due to the TWIPMR mathematical model's high nonlinearities and the influence of unpredictable disturbances, a novel disturbance observer was designed to estimate all the real lumped disturbances. The proposed observer was satisfied with the restriction of the actuators. Simultaneously, low-pass filters integrated into the SMC procedure and observer procedure allowed the engineers to adjust the truncation approximation. All results simulated via the Matlab-Simulink platform demonstrated the effectiveness of the SMC controller and the proposed observer.

Acknowledgments. The authors would like to hugely thank our colleagues in the Faculty of Mechanical Engineering and Mechatronics, Phenikaa University for their constructive and valuable support in building the best submission. Additionally, Dzung Manh Do is also a Master's student in the School of Electrical and Electronic Engineering, Hanoi University of Science and Technology. Dzung Manh Do was funded by the Master Scholarship Programme of Vingroup Innovation Foundation (VINIF), code VINIF.2024.ThS.14.

Disclosure of Interests. The authors declare that no conflicts of interest or financial benefits can affect this paper.

References

1. Astolfi, A., Ortega, R.: Immersion and invariance: a new tool for stabilization and adaptive control of nonlinear systems. IEEE Trans. Autom. Control **48**(4), 590–606 (2003). https://doi.org/10.1109/TAC.2003.809820
2. Ba, D., Li, Y.X., Tong, S.: Fixed-time adaptive neural tracking control for a class of uncertain nonstrict nonlinear systems. Neurocomputing **363**, 273–280 (2019). https://doi.org/10.1016/j.neucom.2019.06.063
3. Chang, C.L., Chang, S.Y.: Using reinforcement learning to achieve two-wheeled self balancing control. In: 2016 International Computer Symposium (ICS), pp. 104–107 (2016). https://doi.org/10.1109/ICS.2016.0029
4. Cho, S.T., Jung, S.: A time-delayed control method for robust balancing performance of a two-wheel mobile robot. In: 2013 13th International Conference on Control, Automation and Systems (ICCAS 2013), pp. 1769–1772 (2013). https://doi.org/10.1109/ICCAS.2013.6704224
5. Gandarilla, I., Santibáñez, V., Sandoval, J., Romero, J.G.: PID passivity-based control laws for joint position regulation of a self-balancing robot. Control. Eng. Pract. **116**, 104927 (2021). https://doi.org/10.1016/j.conengprac.2021.104927
6. Jleilaty, H.A., Asmar, D., Daher, N.: Model reference adaptive control of a two-wheeled mobile robot. In: 2019 International Conference on Robotics and Automation (ICRA), vol. 21, pp. 1155–1161 (2019). https://doi.org/10.1109/ICRA.2019.8793633
7. Kharola, A., Raiwani, S., Kharola, S.: Soft-computing-based real-time control of two wheel mobile robot (TWMR). In: Handbook of Research on Modeling, Analysis, and Control of Complex Systems, pp. 434–449 (2021). https://doi.org/10.4018/978-1-7998-5788-4.ch017

8. Koopaee, M.A.J., Majd, V.J.: An h_∞ switching T-S fuzzy controller design for position control of two-wheeled mobile robots. In: 2015 3rd RSI International Conference on Robotics and Mechatronics (ICROM), pp. 049–054 (2015). https://doi.org/10.1109/ICRoM.2015.7367759
9. Nguyen, P.D., Nguyen, N.H.: A simple approach to estimate unmatched disturbances for nonlinear nonautonomous systems. Int. J. Robust Nonlinear Control **32**(17), 9160–9173 (2022). https://doi.org/10.1002/rnc.6324
10. Pang, H., Liu, M., Hu, C., Zhang, F.: Adaptive sliding mode attitude control of two-wheel mobile robot with an integrated learning-based RBFNN approach. Neural Comput. Appl. **34**(17), 14959–14969 (2022). https://doi.org/10.1007/s00521-022-07304-3
11. Peng, Z., Wang, D., Wang, J.: Data-driven adaptive disturbance observers for model-free trajectory tracking control of maritime autonomous surface ships. IEEE Trans. Neural Netw. Learn. Syst. **32**(12), 5584–5594 (2021). https://doi.org/10.1109/TNNLS.2021.3093330
12. Shen, J., Hong, D.: Model predictive control using dynamic model decomposition applied to two-wheeled inverted pendulum mobile robot. In: 2022 19th International Conference on Ubiquitous Robots (UR), pp. 332–337 (2022). https://doi.org/10.1109/UR55393.2022.9826244
13. Stephen Boyd, L.V.: Convex Optimization. Cambridge University Press (2019)
14. Sun, X., Zhang, Y., Lei, G., Guo, Y., Zhu, J.: An improved deadbeat predictive stator flux control with reduced-order disturbance observer for in-wheel PMSMS. IEEE/ASME Trans. Mechatron. **27**(2), 690–700 (2022). https://doi.org/10.1109/TMECH.2021.3068973
15. Tran, K.G., Nguyen, N.H., Nguyen, P.D.: Observer-based controllers for two-wheeled inverted robots with unknown input disturbance and model uncertainty. J. Control Sci. Eng. **2020**, 1–12 (2020). https://doi.org/10.1155/2020/7205737
16. Van Nguyen, T., Le, H.X., Tran, H.V., Nguyen, D.A., Nguyen, M.N., Nguyen, L.: Adaptive dynamic programming based control scheme for uncertain two-wheel robots. In: 2021 IEEE International Conference on Autonomous Robot Systems and Competitions (ICARSC), pp. 111–116 (2021). https://doi.org/10.1109/ICARSC52212.2021.9429805
17. Vu, Q.V., et al.: An adaptive hierarchical sliding mode controller for autonomous underwater vehicles. Electronics **10**(18), 2316 (2021). https://doi.org/10.3390/electronics10182316

Statistical Approach to Finding Design Solutions for a Dynamic Control System that Are Robust to Multifactorial Uncertainty

Thi Thanh Van Do[1](✉), Thi Vui Le[1], and Manh Cuong Nguyen[2]

[1] Vietnam Women's Academy, Hanoi, Vietnam
{vandtt,vuilt}@hvpnvn.edu.vn
[2] Air Defense Technical Institute, Hanoi, Vietnam

Abstract. The article presents a statistical approach to finding design solutions for a dynamic controlled system that are robust to multifactorial uncertainty. The robustness of the design decision-making task is determined with respect to a multitude of uncontrolled factors. Some solutions to design problems are shown using the method of finding design solutions for a dynamic controlled system that are robust to multifactorial uncertainty.

Keywords: Statistical approach · stability · multifactor uncertainty · design solution · dynamic controlled system

1 Introduction

When developing dynamic controlled systems (DCS) under conditions of multifactor uncertainty, one of the main problems is the correct representation of the relationship between the modes of disturbed motion of the DCS and the uncontrolled factors causing such disturbed motion. An important factor in this issue is the formation of conditions for stable motion of the DCS in relation to initial disturbances, without establishing a connection between these initial disturbances and the real characteristics of the acting uncontrolled factors. In this case, the stability conditions establish the nature of the behavior of the trajectory of the DUS, leaving open the question of the optimal properties of the trajectory. Here, for the formation of optimal control of the DUS, the conditions of design stability are necessary.

The theory and methods of statistical synthesis in the most general case are built on the following basis: $<V, P_M, \Delta>$, where V is the initial set of statistical samples; P_M is the set of mathematical models that are formed as a result of statistical synthesis; Δ - is the set of external criteria.

The main task of the theory and methods of statistical synthesis is to find a model of optimal complexity, in which the adopted external criterion reaches its lower limit [2]:

$$\Delta^* = \inf_{V, P_M} \Delta(V, P_M). \tag{1}$$

© The Author(s), under exclusive license to Springer Nature Switzerland AG 2025
H. A. Le Thi et al. (Eds.): ICAMCS 2024, LNNS 1311, pp. 348–355, 2025.
https://doi.org/10.1007/978-3-031-90606-0_29

In contrast to (1), in the classical theory of self-organization of complex systems [2] a slightly different problem is solved, namely:

$$\Delta^* = \inf_{P_M} \Delta(P_M), \tag{2}$$

which means that in (2) the object of choice is only the mathematical model of the system under study, while in (1) the statistical sample also varies in some way.

Let us consider a statistical approach to finding design solutions for a dynamic controlled system that are robust to multifactorial uncertainty.

2 Statistical Approach to Finding Design Solutions for a Dynamic Control System that Are Robust to Multifactorial Uncertainty

When studying the design stability of a dynamic controlled system, uncontrolled factors of the external environment and principles of optimality are considered.

If the criterion functions $J_i(d)$ are concave and continuous, and the region of admissible solutions is a convex, closed set, then any optimality principle can be represented in additive form [1]:

$$J = \sum_{i=1}^{l} \alpha_i J_i, \tag{3}$$

where J_i - partial optimality criteria; α_i - weight coefficients,

$$\alpha \in \Gamma^+ = \left\{ \alpha : \sum_{i=1}^{l} \alpha_i = 1, \alpha_i \geq 0, i = \overline{1,l} \right\}.$$

Depending on the specific combination of coefficients α_i, $i = 1, l$, the corresponding principle of optimality and the corresponding design solution are formed.

We will call the stability of design solutions to the principles of optimality (to the set Γ^+) and to the set of uncontrolled factors (to the set W, where W - is the set of possible states of the environment) functional stability.

Let the designed dynamic controlled system be modeled under the action of uncontrolled factors from a given set W_U. The quality of the system is estimated by l partial optimality criteria $J_i(d)$, $i = \overline{1,l}$. Then the system has functional stability to the factors of situational and multicriterial uncertainty if, during its operation, its quality indicators $J_i(d)$, $i = \overline{1,l}$ take values no less than the specified levels (assuming that all functionals must be maximized). The decision-making problem consists of solving the following nonlinear operator equation:

$$Cd = J, \text{где } J = \begin{pmatrix} J_1 \\ J_2 \\ \cdots \\ J_l \end{pmatrix}. \tag{4}$$

The solution to problem (4) can be reduced to the following equivalent problem of minimizing the functional:

$$f[d] = \|Cd - J\|. \tag{5}$$

In this case, the expression $F[C, J] = \underset{d \epsilon D}{inf} f[d]$, is called the measure of incompatibility on the set D of the operator Eq. (4).

Perturbations from the set W (situational uncertainty) lead to a perturbation of the operator C, which we denote by $\tilde{C}$, and perturbations from the set Γ^+ (multi-criteria uncertainty) correspond to perturbations of the right-hand sides $\tilde{J}$. Then the problem of making design decisions under conditions of mixed uncertainty has the form

$$\tilde{C}d = \tilde{J} \tag{6}$$

Let instead of exact input data $\omega = (C, J) \epsilon W$ approximate input data is known $\omega_\eta = (C_h, J_\delta)$, that satisfy the inequalities: $\|\tilde{J} - J\| \leq \delta, \delta \geq 0$; $\|\tilde{C} - C\| \leq h, h \geq 0$, и $\eta = (\delta, h)$ and specifies the error of the input data.

Problem (4) will be stable if the condition [1] is met:

$$\lim_{\eta \to 0} F[\tilde{C}, \tilde{J}] = F[C, J] \tag{7}$$

Condition (7) means that as the perturbations tend to zero, the measure of incompatibility of the perturbed Eq. (6) tends to the measure of incompatibility of the exact Eq. (4), т.е. i.e. an equation in which the right-hand sides J and the operator C are precisely specified. With regard to design problems, precisely specified values of J and C correspond to precisely specified values of uncontrolled factors and to precisely specified optimality principle, which, as already noted, is impossible. Therefore, any values can be taken as precisely specified values $\omega \in W$ and $\alpha \in \Gamma^+$. In the limiting case, this situation corresponds to the case

$$F\left[\tilde{C}, \tilde{J}\right] = F[C, J] \tag{8}$$

under any $\eta \in W \times \Gamma^+$.

Condition (8) can be interpreted as follows. Let one of the partial optimality criteria J_i, $i = \overline{1, l}$, determining the quality of a dynamic controlled system, correspond to the probability of task completion (or the probability $P_{\pi op}$ of hitting a target for aircraft (AC)). Stability with respect to uncontrolled factors by virtue of (8) means, that the criterion $P_{\pi op}$ does not depend on their impact and, therefore, with a rational choice of the aircraft's appearance characteristics, this criterion should be equal to one. The stability of the design solution to the principles of optimality means that the relative values of all particular optimality criteria should be equal to each other and, therefore, by virtue of the condition $P_{\pi op} = 1$, they are equal to one [4].

To determine the stability of a decision-making problem under uncertainty (situational stability), we define a parametric family of operators C_n^h that approximates the unperturbed operator C_0 in the sense of [10]

$$\left|C_n^h d - C_0 d\right| \leq h\|d\|.$$

Let for arbitrary sequences of operators from the approximating family $\{C_n^h\}$, $h_n \to 0$ and $\{\varepsilon_n\}$, $\varepsilon_n > 0$, $\varepsilon_n \to 0$ sequences of solutions satisfying the inequalities $\{d_n\}$, $d_n \in D$ are constructed from numbers: $C_n^h d_n - inf C_n^h d \le \varepsilon_n$.

Definition 1. A decision-making problem $f_h[d] = \|C^h d - J_0\|$ on $D_Y \times W, D_Y \subset \tilde{D}$ is called stable with respect to a set of uncontrollable factors W, if for any sequences of uncontrollable factors $\{\omega_n\}$, $\omega_n \epsilon W$, for which $\|C_n^h - C_0\| \to 0$, and for any sequence of numbers $\{\varepsilon_n\}$, $\varepsilon_n > 0$, $\varepsilon_n \to 0$ any sequence of decisions $\{d_n\}$, $d_n \in D$, satisfying the inequalities.

$$\|C_n^h d_n - J_0\| - \inf_{d \epsilon D} \|C_n^h d_n - J_0\| \le \varepsilon_n,$$

will be the minimizing sequence of the original functional $f_0[d]$:

$$\|C_0 d_n - J_0\| \to \inf_{d \epsilon D} \|C_0 d_n - J_0\| = \mu_D[C_0, J_0].$$

The process of finding design solutions that are stable to small changes in uncertainty factors (in this case, the principles of optimality and uncontrolled factors) generally comes down to the following operations:

- finding regularizing operators;
- determining the regularization parameter based on additional information.

The selection of possible solutions is carried out on the basis of the variational principle, where a non-negative functional $\Omega[d]$, called the stabilizing functional, is used. In the work [5] it is shown that the regularizing operator $R(J_\delta, \alpha)$ can be found from the solution of the problem

$$\inf_{d \in D_\delta} \Omega[d] \text{ при } \rho_F(Cd, J_\delta) = \delta \tag{9}$$

Here $D_\delta = D \cap Q_\delta$, Q_δ - is the set of solutions $d \in D$, for which equality $\rho_F(Cd, J_\delta) = \delta$ is satisfied.

Problem (9) is a problem of variational calculus for a conditional extremum, which is solved by the method of indefinite Lagrange multipliers, i.e. the minimum of the functional is determined

$$M^\alpha[d, J_\delta] = \rho_F^2(Cd, J_\delta) + \alpha\Omega[d], \tag{10}$$

where parameter α is determined from the condition: $\rho_F(Cd_\alpha, J_\delta) = \delta$.

So, α - A.N. Tikhonov's regularizer for the operator equation $Cd_\alpha = J_\delta$ has the form [6]:

$$R : \alpha \to d = \arg\min_{d \in D} \left\{ \rho_F^2(Cd, J_\delta) + \alpha\Omega[d] \right\} \tag{11}$$

When D, F - are Hilbert spaces, the functionals in (11) can be represented as

$$R : \alpha \to d = \arg\min_{d \in D} \left\{ u\|Cd - J_\delta\|_F^2 + \|d\|_D^2 \right\} \tag{12}$$

where $u = 1/\alpha$.

The essence of the regularizer is that the original operator equation, which in this case describes the decision-making problem, is replaced by a new, regularized equation

$$CRu = J_\delta \tag{13}$$

i.e. the mathematical model of the system is corrected in some way.

In computational schemes, the parameter of the unified integral principle of optimality is selected as follows:

Let δ - be the error of the right-hand side of Eq. (13), corresponding to the spread of the applied optimality principles. A finite segment of a monotonic sequence of numbers $h_0, h_1, \ldots, h_n$, s considered, which can be specified, for example, in the form of a geometric progression $h_k = h_0 q^k, k = 0,1, 2, \ldots, q > 0$.

For each, a dependence d_{h_k} is found that minimizes the functional $M^{h_k}[d, J_\delta]$ and the discrepancy $\rho_F(Cd_{h_k}, J_\delta)$ is calculated. As the desired value h, we take such h_{k_0} for which the equality is satisfied with the required accuracy

$$\rho_F\left(Cd_{h_{k_0}}, J_\delta\right) = \delta. \tag{14}$$

The solution of equation $\rho_F(Cd_h, J_\delta) = \varphi(h) = \delta$ with respect to parameter h is significantly complicated for the case of a nonlinear operator equation, which is what the decision-making problem generally represents. In this case, the residual $\varphi(h) = \rho_F^2(Cd_h, J_\delta)$, is generally not a continuous function, and it may not be strictly monotone.

3 Some Solutions of Design Problems Using the Method of Searching for Design Solutions of a Dynamic Controlled System that Are Stable to Multifactor Uncertainty

3.1 The Design Problem Can Be Represented as an Operator Equation $Ad = J_\delta$

Where A - is a linear compact operator, the right-hand side is specified with an error, which corresponds to multicriterial uncertainty. The variational problem has the form

$$R : u \to d = \arg\min_{d \in D}\left\{u\|Ad - J_\delta\|_F + \|d\|_D^2\right\} \tag{15}$$

Euler's equation for this problem:

$$d + uA^*Ad = uA^*J_\delta \tag{16}$$

where A^* is the adjoint operator to A. The formula of the method is:

$$d = Ru = \left(E + uA^*A\right)^{-1}A^*(uJ_\delta) \tag{17}$$

where E is the unit operator.

The calculation is much simpler using the Lagrangian regularizer

$$R : u \to d = \arg\min_{d \in D}\left\{-\langle Ad, J_\delta\rangle_F + \|d\|_D^2\right\} \tag{18}$$

Here a stable solution can be found in the following explicit form:

$$d = Ru = A^*u \tag{19}$$

3.2 The Design Problem is Represented as a Linear Operator Equation $A^h d = J_\delta$.

This case corresponds to the adoption of a design solution based on vector criterion (J_δ) under the action of uncontrolled factors (A^h).

The variational problem in the form of a Tikhonov regularizer has the form

$$R : u \to d = \arg\min_{d \in D}\left\{u\|Ad - J_\delta\|_F + \|d\|_D^2\right\}$$

The Euler equation here is the same as in case I, the regularized solution is written as follows:

$$d = Ru = \left(E + uA^{*h}A^h\right)^{-1} A^{*h}(uJ_\delta)$$

Minimization of the residual $\|Ad - J_\delta\|$ is carried out until the condition $\left\|A^h Ru - J_\delta\right\| \le \eta$ where η is the specified accuracy, $\eta = \delta + h$.

Solution of the problem in the form of a Lagrangian regularizer

$$R : u \to d = \arg\min_{d \in D}\left\{\mathrm{Re}\left\langle A^h d, J_\delta\right\rangle + \|d\|_D^2\right\}$$

is determined by the corresponding formula of the method

$$d = Ru = A^{*h}u \tag{20}$$

where A^{*h} - transposed and complex conjugate matrix to A^{*h}.

Algorithm of the Method [7]

- Calculate the transposed matrix A^{*h}.
- Calculate the matrix $C^h = A^h A^{*h}$.
- Calculate the initial value of the regularization parameter $\|u\| \le u_0$:

$$u_0 = \frac{\max_{i \in [1,l]} |J_i|}{\max_{i \in [1,l]} \sum_{j=1}^{l} \left|c_{ij}\right|}, \tag{21}$$

where J_i - private optimality criteria, $i = \overline{1, l}$; c_{ij} – matrix elements C^h

- Determine the regularization parameter $u_{h\delta}$ from the system of inequalities:

$$\begin{cases} \|u\| \le u_0; \\ \left\|C^h u - J_\delta\right\| - \inf\limits_{\|u\| \le u_0} \left\|C^h u - J_\delta\right\| \le \varepsilon. \end{cases} \tag{22}$$

- Using the regularization parameter found in point 4, the condition is checked

$$\left\|A^h A^{*h} u_{h\delta} - J_\delta\right\| \le h + \delta \tag{23}$$

If condition (23) is not satisfied, then the regularization parameter u_0 is increased by some value $\Delta u_0 : u_1 = u_0 + \Delta u_0$ and a return to step 4 is made with a new parameter u_1.

The regularization parameter is increased with a step of Δu_0 until condition (23) is satisfied. When condition (23) is satisfied, a transition to step 6 is made.

A stable, regularized solution is determined: $d_{h\delta} = A^{*h} u_{h\delta}$.

3.3 A Design Decision Making Problem Described by a Nonlinear Operator Equation

If the problem of making design decisions is unstable in the sense of any of the definitions 1–3, then it is necessary to construct regularizing algorithms.

The conditions of functional uncertainty and its special cases of multicriterial and situational uncertainty lead to the corresponding stability conditions, which, as was shown above, are expressed in the form of complex nonlinear operator equations. This most important case in the problem of stability of decision-making problems is also the most complex in terms of computation.

In smoothing functionals, stabilizers are written in explicit analytical form, while the use of statistical synthesis allows us to obtain so-called algorithmic stabilizers. For this, the left and right parts of statistical samples (input and output information) must reflect the stability conditions. Thus, it is possible to "construct" a sample so that the solution obtained from it will satisfy the condition of multi-criteria or situational stability.

Algorithmic construction of stability conditions and corresponding solutions presupposes a fundamentally different construction of the region of admissible solutions.

Let us consider the standard definition of the regularizer R from the condition of minimization of the smoothing functional:

$$R : u \to d = \arg\min_{d \in D}\left\{u\|Cd - J_\delta\|_F + \|d\|_D^2\right\} \tag{24}$$

with stabilizing functionality $\Omega = \|d\|_D^2$.

As can be seen, here an approximate solution $\tilde{d} \in D$ is selected, which, according to the construction of the smoothing functional, is stable, it is selected in the form

$$\tilde{d} = Ru, \tag{25}$$

where u - regularization parameter. For example, in the case where C - is a linear compact operator, solution (25) is written as:

$$\tilde{d} = C^* u, \tag{26}$$

where C^* - adjoint operator to C.

Thus, the structure of the regularizer (26) is completely determined by the solution of the variational problem. When constructing a regularizer based on statistical synthesis, its structure corresponds to the limit element of the minimizing sequence of the functional $f_0[d] = \|C_0 - J_0\|$, which must be stable for any sequences $w_n = \{C_n, J_n\}$ of the initial data from the sets $\left\{C : \left\|C^h - C\right\| \le h\right\}$ and $\{\|J_0 - J_\delta\| \le \delta\}$.

Conclusion

The obtained results of this work allow us to draw the following conclusions:

1) The statistical approach to finding design solutions for a dynamic controlled system that are stable to multifactorial uncertainty determines the stability of the problem of making design decisions for a dynamic controlled system in relation to a set of uncontrolled factors.

2) The use of statistical synthesis allows us to obtain algorithmic stabilizers in solving design problems that are stable to multifactorial uncertainty for a dynamic controlled system.

References

1. Balyk, V.M.: Statistical Synthesis of Design Solutions in the Development of Complex Systems, 280 p. MAI Publishing House (2011)
2. Ivakhnenko, A.G.: Inductive Method of Self-organization of Models of Complex Systems, 296 p. Nauk. Dumka, Kiev (1981)
3. Mironov, V.V., Svertsev, N.A.: Method of Analysis of System Stability and Motion Controllability, 166 p. RUDN University Publishing House (2002)
4. Tarasov, E.V., Valyk, V.M.: Methods of Designing Aircraft: A Tutorial, 412 p. MAI-PRINT Publishing House (2008)
5. Tikhonov, A.N., Arsenin, V.Ya.: Methods for Solving Ill-Posed Problems, 288 p. Nauka (1986)
6. Tarasov, E.V., Balyk, V.M.: Methods of Decision-Making in the Design of Technical Systems. Study Guide, M MAI Publishing House (1993)
7. Tarasov, E.V., Yufa, D.I.: Methods of decision-making with multifactor uncertainty in design studies of aircraft. MAI Bull. **14**(1) (2007)

Design Smart Factory Using RTDS and 800xA SCADA Based on IEC 60870-5-104

Duc Duong Nguyen[1(✉)], Manh Hung Do[2], Ngoc Thanh Pham[3], Van Dung Ngo[4], and Viet Phuong Pham[2]

[1] University of Economics – Technology for Industries, Hanoi, Vietnam
nddduong86.ddt@uneti.edu.vn
[2] Hanoi University of Science and Technology, Hanoi, Vietnam
{hung.domanh,phuong.phamviet}@hust.edu.vn
[3] International School, Vietnam National University, Hanoi, Vietnam
pnthanh@vnu.edu.vn
[4] Air Defense-Air Force Academy, Hanoi, Vietnam

Abstract. Nowadays, in the fourth industrial revolution, the transformation of traditional factories into smart factories is being executed extremely strongly. This paper presents the applied ability of Real Time Digital Simulator (RTDS) to digitize the brine filtration process of a chemical factory or any smart factories based real-time simulations of the level and concentration process controls. In addition, it can be connected to a SCADA system (ABB 800xA) through IEC 60870-5-104 protocol (IEC 104) and OLE for Process Control (OPC). The connection is realized by Triangle MicroWorks SCADA Data Gateway (SDG triangle). RTDS is a powerful system which can optimize supervisory in any smart factories to improve efficiency, reduce costs, and provide real-time visibility into control all levels of smart factories.

Keywords: RTDS · Smart Factory · SDG triangle · OPC · Object Linking and Embedding (OLE)

1 Introduction

Today, the transformation of traditional factories into smart factories is an inevitable trend in the fourth industrial revolution. The design rules of a smart factory includes modularity, interoperability, decentralization, virtualization, service orientation and real-time capability [1]. Digitalization is the most important matter to design a smart factory. Although it brings many practical benefits, implementing it is not easy and causing many managers to worry. RTDS Technologies is the world leader in real-time digital simulator system, used in many fields, such as: microgrid testing, distribution automation [2], inverter testing [3], wind turbine generator [4, 5], cyber security [6–9], solar technology [10], … The RTDS system is an effective solution for the digitization of smart factories.

This paper presents the process to model the brine filtration process of a chemical factory, applying ratio control strategy combined with feedback control strategy

© The Author(s), under exclusive license to Springer Nature Switzerland AG 2025
H. A. Le Thi et al. (Eds.): ICAMCS 2024, LNNS 1311, pp. 356–369, 2025.
https://doi.org/10.1007/978-3-031-90606-0_30

and advanced control algorithms to resolve the effects of coupler between brine level and concentration and upgrade the quality and stability of the control system, therefore upgrades the quality of caustic soda, using RTDS/RSCAD. Otherwise, it can be connected to a ABB SCADA system through IEC 104 and OPC. IEC 104 is a widely used automation protocol with a server/client architecture. Data may be exchanged between the RTDS system and external equipment over a LAN/WAN using IEC 104 standard. The specific simulation consequences and analysis will be discussed in the following sections.

2 Model of a Brine Filtration Process and Control Method

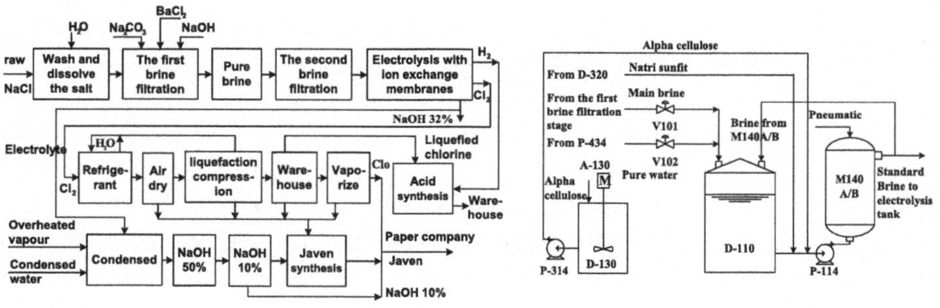

(a) (b)

Fig. 1. Block diagram of caustic soda - chlorine production line (a); P&ID of second brine filtration process (b)

Bai Bang Chemical Factory has a caustic soda production line using the ion exchange membrane method with a design capacity of 7000 tons/year. This production line has many stages, including: salt washing and dissolution stage, the first brine filtration stage, the second brine filtration stage, the electrolysis stage, and the treatment of electrolysis products. In which, the second brine purification step with the main object being the D110 preliminary brine tank plays a very important role, creating brine that reaches standard concentration (305 ± 5 kg/m^3) to put into the electrolysis tank (Fig. 1) [11].

Simplifying the components of the D110 tank, we have the tank diagram illustrated in Fig. 2.

In this model, F_1, F_2, F_3 are volume flow rate of preliminary brine, pure water, brine from M140A/B filter tank, output brine (m^3/h)(m^3/h); x_1, x_2, x are concentration of preliminary brine, pure water, output brine (%); h is water level in D110 tank, with H_{max} = 3 (m); ρ is water's specific weight, with $\rho = 1000$ (m^3/h)(kg/m^3); ρ_1 is preliminary brine's specific weight, with $\rho_1 = 1197$ (kg/m3); ρ_2 is output brine's specific weight, with = 1194 (kg/m3); V is water volume in D110 tank; A is D110 bottom tank; R is radius of D110 tank, with $R = 1.4$ (m) [11].

According to the conservation law of mass and chemical balance [12], we have (1):

$$\frac{dh}{dt} = \frac{1}{A}\left(\frac{\rho_1}{\rho}F_1 + \frac{\rho_2}{\rho}F_2 + F_3 - F\right) = f;\ \frac{dx}{dt} = \frac{1}{V}\left[\frac{\rho_1}{\rho}F_1x_1 - x\left(\frac{\rho_1}{\rho}F_1 + \frac{\rho_2}{\rho}F_2\right)\right] = g \quad (1)$$

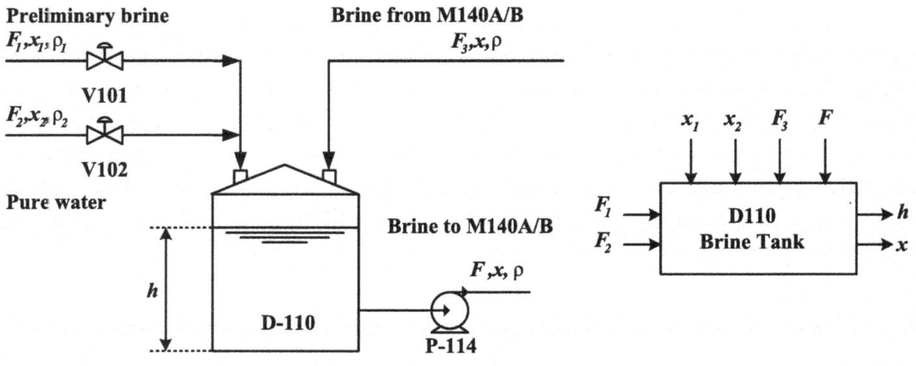

Fig. 2. P&ID of D-110 tank

From (1), we have laplace equations in (2):

$$\Delta h(s) = \frac{1}{As}\left(\frac{\rho_1}{\rho}\Delta F_1 + \frac{\rho_2}{\rho}\Delta F_2 + \Delta F_3 - \Delta F\right);\ \Delta x(s) = \frac{K_1\Delta F_1}{1+T_x s} + \frac{K_2\Delta F_2}{1+T_x s} + \frac{K_3\Delta x_1}{1+T_x s} \quad (2)$$

with $T_x = \frac{\rho\overline{V}}{a}$; $a = \rho_1\overline{F_1} + \rho_2\overline{F_2}$; $K_1 = \frac{\rho_1(\overline{x_1}-\overline{x})}{a}$; $K_2 = -\frac{\rho_2\overline{x}}{a}$; $K_3 = \frac{\rho_1\overline{F_1}}{a}$

To facilitate in analyzing and synthesizing controllers, we consider the equations in percentage form, from (2), we have (3):

$$\Delta h_p(s) = \frac{\Delta F_{1p}}{sT_{h1}} + \frac{\Delta F_{2p}}{sT_{h2}} + \frac{\Delta F_{3p} - \Delta F_p}{sT};\ \Delta x_p(s) = \frac{L_1\Delta F_{1p}}{1+T_x s} + \frac{L_2\Delta F_{2p}}{1+T_x s} + \frac{L_3\Delta x_{1p}}{1+T_x s} \quad (3)$$

with $\Delta h_p = \frac{\Delta h}{h_{\max}}$; $\Delta F_{1p} = \frac{\Delta F_1}{F_{\max}}$; $\Delta F_{2p} = \frac{\Delta F_2}{F_{\max}}$; $\Delta F_{3p} = \frac{\Delta F_3}{F_{\max}}$; $\Delta F_p = \frac{\Delta F}{F_{\max}}$; $T = \frac{Ah_{\max}}{F_{\max}}$;

$$T_{h1} = \frac{\rho}{\rho_1}T;\ T_{h2} = \frac{\rho}{\rho_2}T; L_1 = K_1F_{\max}; L_2 = K_2F_{\max}; L_3 = K_3\rho_1$$

From (3), we have the block diagram of the D110 tank model with two valves illustrated in Fig. 3.

The object transfer functions of D-110 tank are illustrated in (4):

$$G_{11}(s) = \frac{1}{sT_{h1}};\ G_{12}(s) = \frac{1}{sT_{h2}};\ G_{21}(s) = \frac{L_1}{1+T_x s};\ G_{22}(s) = \frac{L_2}{1+T_x s} \quad (4)$$

The noise transfer functions of D-110 tank are illustrated in (5)

$$G_{d1}(s) = \frac{\Delta h_p}{\Delta F_{3p} - \Delta F_p} = \frac{1}{sT};\ G_{d2}(s) = \frac{L_3}{1+T_x s} \quad (5)$$

After the recent supervisory, current status of the second brine filtration stage has some problems. A very important issue for this stage is to maintain the brine put into the electrolysis tank at the concentration of 305 ± 5 kg/m^3 or 25.2% ÷ 25.9% because at

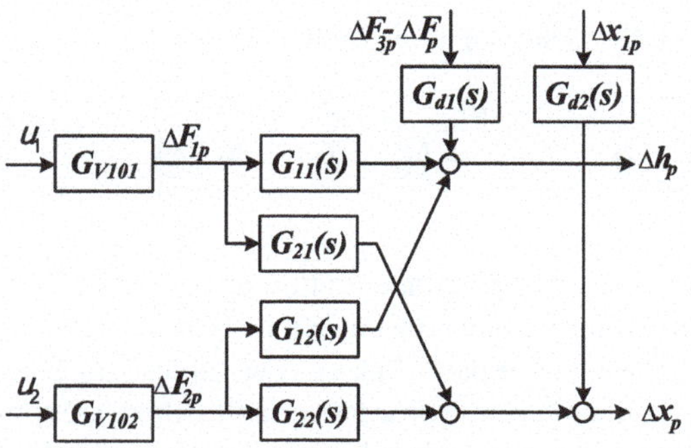

Fig. 3. Block diagram of D-110 tank

this concentration the ion exchange membrane electrolysis process reaches the highest efficiency. The entire electrolysis line for caustic soda-chlorine of this chemical plant is provided by China. Based on the actual operation of this line for more than twenty years, the quality of caustic soda product depends a lot on the quality of the control system. But in fact, the current control part is not complete. For example, for the brine mixing process in tank D110, the plant still uses the decentralized control method with the structure illustrated in Fig. 4a [12]. Appling Matlab-Simulink, some results are shown in Fig. 4b.

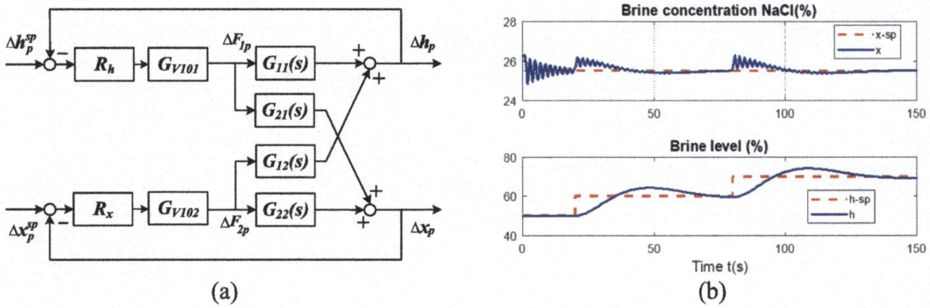

Fig. 4. Decentralized control method (a), concentration and level response in this method (b)

Thanks to the decentralized control, the brine concentration value can closely follows its set point value. However, from twentieth second, when the level set point value changes from 50% to 60% and to 70%, it also affects the concentration, causing the concentration to change, the overcorrection is larger and the transient time is longer. Although the concentration also returns to the equilibrium value, it proves that the decentralized control method has not resolved the effects of coupler between brine level and concentration.

According to [12], the relative gain is caculated in (6):

$$\lambda = \lim_{s \to 0} \frac{G_{11}}{G_{11} - \frac{G_{12}G_{21}}{G_{22}}} = \frac{\rho_1}{\rho_1 - \rho_2 \frac{K_1}{K_2}} = \frac{\overline{x}}{\overline{x_1}} = \frac{25.54\%}{26.32\%} = 0.971 \Rightarrow 0.5 < \lambda < 1 \tag{6}$$

with working point value of brine concentration at the outlet $\overline{x} = 25.54\%$, working point value of the preliminary brine concentration: $\overline{x_1} = 26.32\%$.

From (6) and [12], we have a pair of process variables $\Delta h_p - \Delta F_{1p}$; $\Delta x_p - \Delta F_{2p}$.

To eliminate interactions between process variables, there are some methods, such as: ratio control, decoupler control and Model Predictive Control (MPC). The design basis of an MPC controller is based on a fixed sampling cycle, while each process variable has different fast and slow changing characteristics. To achieve the common requirements of many variables, we must choose the smallest control cycle, thus increasing the load on the CPU. In addition, adjusting the parameters of the MPC controller in practice is very difficult and takes a lot of time [12]. The decoupler control method is rather simple. Therefore, we will choose the ratio control strategy combined with feedback control strategy, shown in Fig. 5.

According to [12], ratio control strategy is to maintain the ratio between two variables F_1 and F_2 at a set value to indirectly control the concentration of brine with FFIC-102 and LIC-101 controllers.

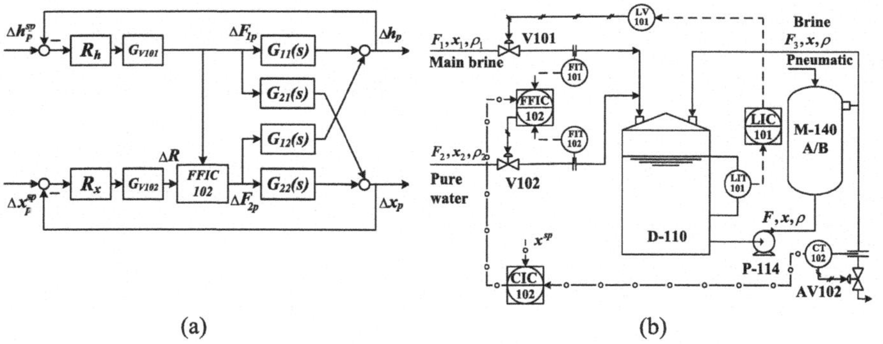

Fig. 5. Ratio control strategy (a), P& ID of brine filtration process in ratio control strategy (b)

Assume

$$R = \frac{\rho_2 F_2}{\rho_1 F_1 + \rho_2 F_2} \Rightarrow \frac{\rho_1 F_1}{\rho_1 F_1 + \rho_2 F_2} = 1 - R \tag{7}$$

From (2) and (7), we have Eq. (8).

$$\frac{dx}{dt} = \frac{\rho_1 F_1 + \rho_2 F_2}{\rho V}[(1 - R)x_1 - x] = k;\ \Delta x_p(s) = \frac{N_1}{1 + sT_x}\Delta R + \frac{N_2}{1 + sT_x}\Delta x_{1p} \tag{8}$$

with $N_1 = \frac{-\overline{x_1}}{\rho}$; $N_2 = \frac{\rho_1}{\rho}\frac{\overline{x}}{\overline{x_1}}$; $T_x = \frac{\rho \overline{V}}{\rho_1 \overline{F_1} + \rho_2 \overline{F_2}}$

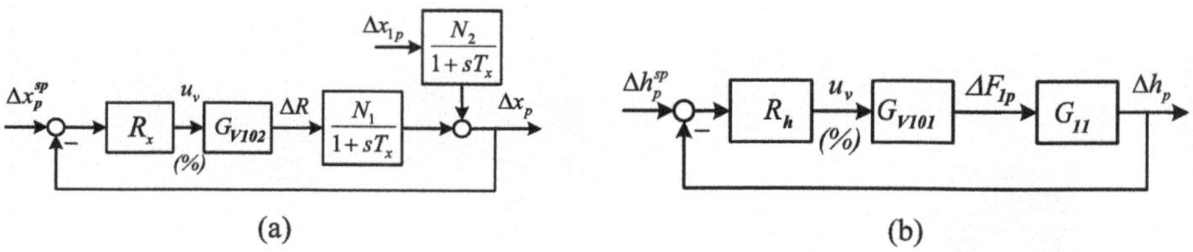

Fig. 6. Concentration control loop (a), level control loop (b)

According to [12] and (8), we have control loops shown in Fig. 6.

According to the IMC-PID (Internal Model Control) regulation of Chien& Fruehauf [12], we have parameters of the concentration controller, shown in (9).

$$\begin{aligned} R_x &= K_{Px}\left(1+\frac{1}{s\tau_{ix}}+s\tau_{dx}\right);\ K_{px}=\frac{T_x+T_{v2}}{\tau_c K_{S2}};\\ \tau_{ix} &= T_x+T_{v2};\ \tau_{dx}=\frac{T_xT_{v2}}{T_x+T_{v2}};\ K_{S2}=N_1K_{v2} \end{aligned} \tag{9}$$

From (7), we have (10)

$$\Delta F_{2p}=M_1\Delta F_1+M_2\Delta R;\ M_1=\frac{\rho_1}{\rho_2}\left(\frac{1}{1-\overline{R}}-1\right);\ M_2=\frac{\rho_1}{\rho_2}\frac{\overline{F_1}}{\left(1-\overline{R}\right)^2F_{max}} \tag{10}$$

According to (10) and the IMC-PID regulation of Chien& Fruehauf [12], we have parameters of the level controller, shown in (11):

$$R_h=K_{Ph}\left(1+\frac{1}{s\tau_{ih}}+s\tau_{dh}\right);\ K_{ph}=\frac{2\tau_c+T_{v1}}{K_{S1}\tau_c^2};\ \tau_{ih}=2\tau_c+T_{v1};\ \tau_{dh}=\frac{2\tau_cT_{v1}}{2\tau_c+T_{v1}} \tag{11}$$

3 System Configuration Analysis

Nowadays, there are many solutions to digitze a smart factory. After many years of research, we propose a simulation system to digitze a smart factory, illustrated in Fig. 7.

The system includes two components: RTDS and SCADA system. In RTDS, there are a NovaCor CPU (10 licensed cores), a GTNETx2 card, a GTSYNC card (External Time Synchronization), a GNSS clock, a network Switch and the first workstation. The GTSYNC card synchronizes RTDS timestep to external time reference (e.g. GPS clock) and synchronizes devices under test. The function of GTNETx2 card is communication with external devices over some protocols (IEC 61850; DNP3 and IEC 60870-5-104; PMU IEEE C37.118; Mobus TCP, RTU over TCP, ASCII over TCP). The NovaCor CPU is the IEC 104 Server. The first workstation is installed RSCAD FX V2.1 packet, played the role of as the IEC 104 client. The SCADA system includes at least two workstations, the second workstation and the third workstation. The second workstation is installed with Industrial^{IT} 800xA (ABB) V5.1 software, playing the role as an engineering station

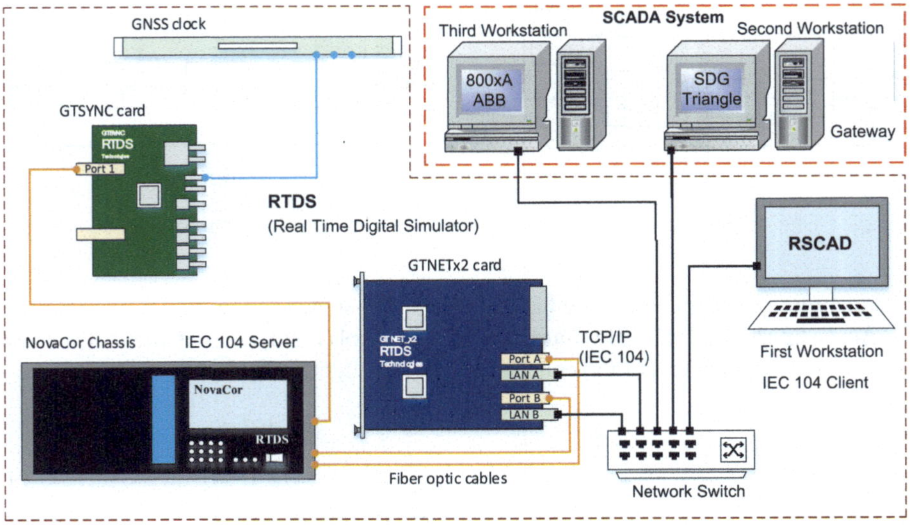

Fig. 7. System configuration of smart factory simulation

(client), and SDG triangle (Triangle MicroWorks SCADA Data Gateway) which is communication protocol driver. The values and databases that the protocol transmits over the wire are provided by SDG. It plays the role as a gateway. It aggregates data from IEC 60870-5-101, IEC 60870-5-104, DNP3, IEC 61850, TASE.2/ICCP, and Modbus slave devices, or OPC Server, and then supplies this data to other control systems supporting IEC 60870-5-101, IEC 60870-5-104 (IEC104), DNP3, IEC 61850, TASE.2/ICCP, and Modbus communication protocols or OPC Client [13]. The third workstation is installed with Industrial^{IT} 800xA (ABB) V5.1 software, playing the role as the SCADA server (aspect server and connectivity server). In some cases, the second workstation and the third workstation can be merged into one server.

Establishing a connection according to TCP/IP with the above structure, the link layer parameters of devices are setup as Table 1.

Table 1. Table link layer parameters

Device	IP Address	Device	IP Address
NovarCor CPU	172.24.9.14	GTNETx2 card	172.24.9.111 (port 1)
RSCAD	172.24.9.100		172.24.9.112 (port 2)
SDG Triangle	172.24.9.150		IP Port Number: 2404
800xA ABB	172.24.9.160		Common address: 3

In the application layer of IEC104, the systems often establish the data type, address as the Table 2.

Table 2. Table signal classification used in IEC104

Data Type	Variable Type	Address
MV (Measured value)	T13	6004,6005,6006, …
SP (Single Point Information)	T1	1002, 1001, 1000, …
DP (Double Point Information)	T3	9000,9001,9002, …
SC (Single Command)	T45	4000, …
SetC (Set point Command)	T50	8000,8001,8002

3.1 First Workstation's Configuration

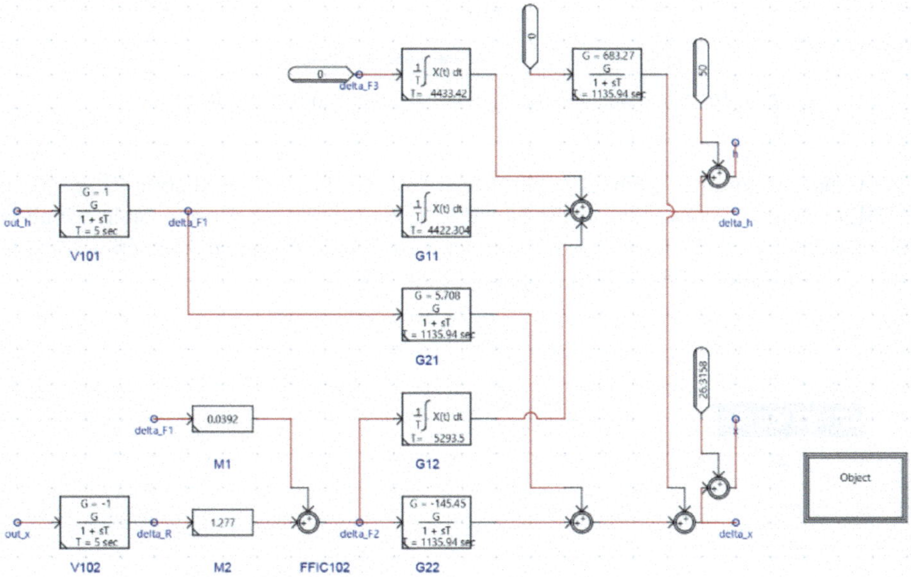

Fig. 8. Block diagram of brine filtration process in RSCAD

In the first workstation (RSCAD/RTDS), we use real pole function (REALPL), integrator component (INTGL), summing function (SUM3), GAIN, schedular component (SCHED), constant … to build the block diagram of brine filtration process (Fig. 8), the block diagram of level controller (Fig. 9) and the block diagram of the concentration controller (Fig. 10).

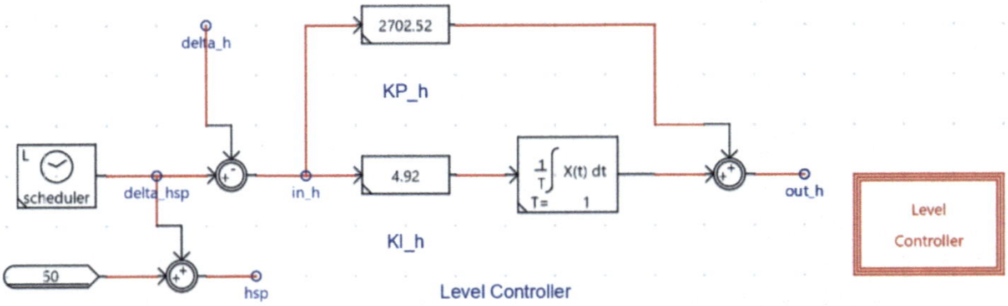

Fig. 9. Block diagram of the level controller in RSCAD

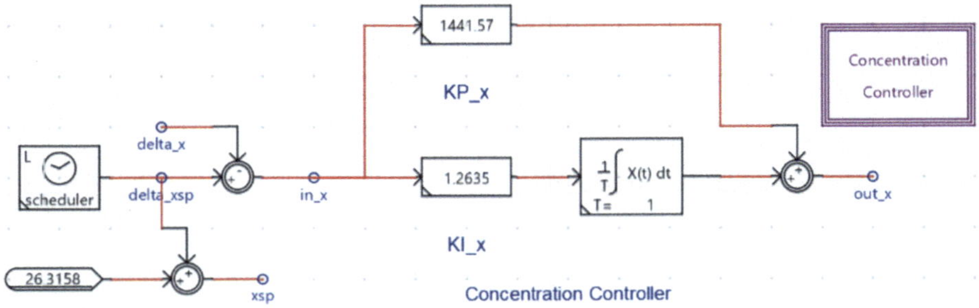

Fig. 10. Block diagram of the concentration controller in RSCAD

In the proposed system configuration of smart factory simulation (Fig. 5), a GTNETx2 card is connected to the NovaCor CPU using a fiber optic cable which is connected between GT Ports on each hardware unit's rear. A typical connection for IEC 104 communication in the RTDS is shown in Fig. 11.

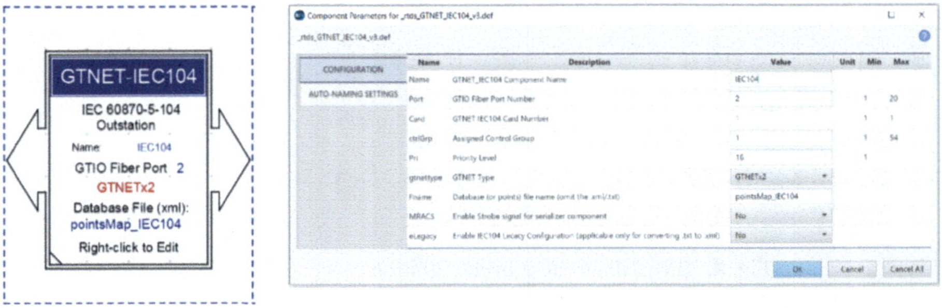

Fig. 11. Configuration menu of GTNET-IEC104-v3 Component

A database file is required to map the IEC 104 data references to signal names in RSCAD FX Draft. It is generated by "right clicking on GTNET- IEC104-v3 component → Edit → IEC 60870-5-104 Database File" as shown in Fig. 12 with Analog Status, Analog Control, Binary Status, Binary Control, Outstation …

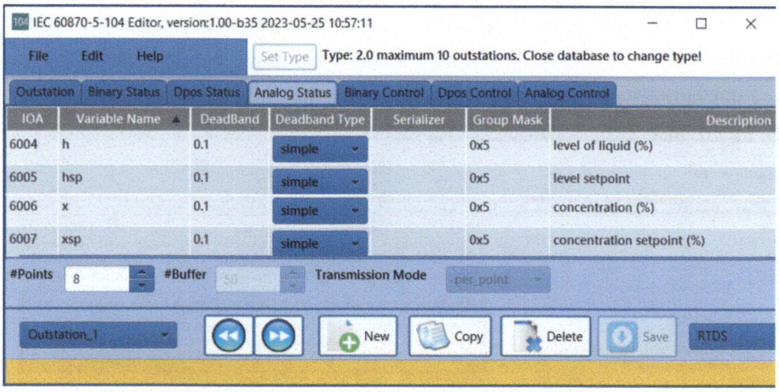

Fig. 12. Signal configuration in IEC 104

3.2 Second Workstation's Configuration

The second workstation (SDG) is a workstation installed SDG triangle, played the role of a gateway that convert signals from RTDS (IEC 104) to the workstation of the SCADA (OPC DA). SDG software structure is shown in Fig. 13 [13].

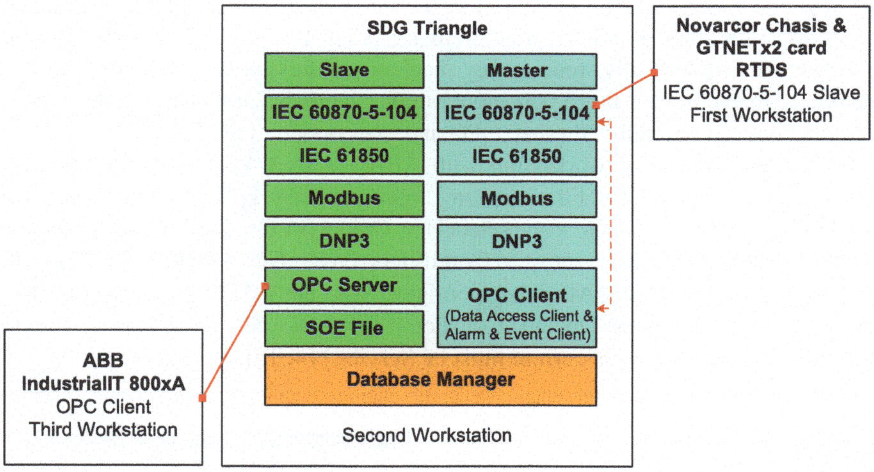

Fig. 13. SDG software structure in the second workstation

In addition, the second workstation is also installed ABB's 800xA packet as client. It is connected to the third workstation (ABB's aspect server and connectivity server) as client – server topology [14]. At first, the second worksation is registered as SDG GUI OPC Server. Master Data Objects (MDOs) are points that obtain their value from an external Slave device (RTDS simulator) to IEC104 master (SDG gateway). MDOs include measured values T13 (h, hsp, x, xsp), single Point Information T1, set point Comand T50. After creating a TCP/IP Channel (M104) named 104M, we need add sessions and MDOs shown in Fig. 14, and then save them to file tmwgtway.ini. After that, the second worksation is registered as SDG Service OPC Server to configure the third worksation [13].

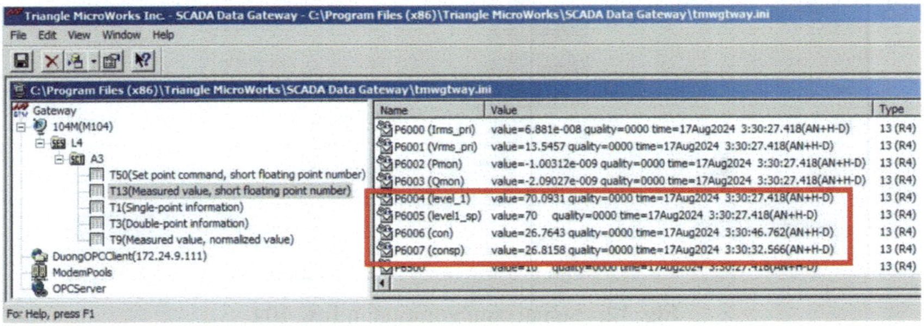

Fig. 14. Organization of Point Mapping File Data

3.3 Third Workstation's Configuration

Industrial[IT] 800xA is the most modern SCADA system from ABB, integrated for process control applications, suitable for chemical, thermal power, cement plants… Compared to systems from other brands such as Siemens, Allen Bradley, it has many outstanding advantages, specifically: redundancy mechanism, flexibility, connecting up to 48 controllers, managing up to 15,000 input/output channels, supporting many types of communications, especially ethernet, OPC and fiber optic …[14, 15].

To configure the third worksation with ABB's 800xA V5.1, we can conduct as following steps sequentially. First step, in Engineering Workplace\ Control Structure\ Root, Domain, we can right-click and select New Object (PLC Generic Control Network) and name it as SDG. At here, move to Controllers, create New Object (PLC Controller Type) and name it as RTDS. After then configure SDG and RTDS following Fig. 15. In Service Structure\ Services\ OpcDA_Connector\ SD_SDG\ SP_TTG\ Service Provider Definition\ Configuration, its current must be Service [14, 15].

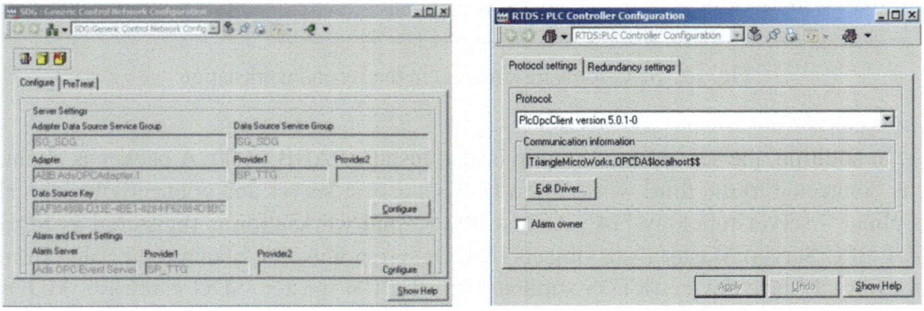

Fig. 15. Configuration Steps in ABB's 800xA

The next step, we can move to RTDS\ PLC Uploader \Advanced tab, push to Append Objects button and move to Standard tab, push to Start Button [15]. So RTDS's variables in SDG triangle of second workstation have been updated into ABB's 800xA SCADA of third workstation, shown in Fig. 16a.

4 Simulation Result and Discussions

In ABB's SCADA, the graphic display, faceplates, log configurtations, trend displays of the brine filtration process are created by the Process Portal A software [15], illustrated in Fig. 16b.

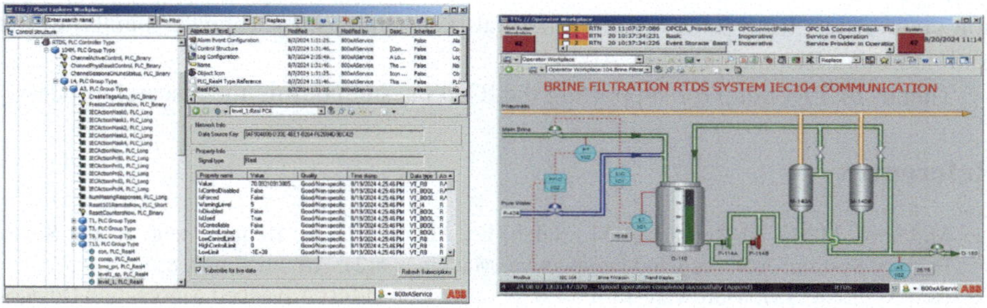

(a) (b)

Fig. 16. RTDS's variable in ABB's 800xA (a) and main graphic display (b)

Running the simulation on RTDS in the same condition as Fig. 4, we have the trend display of brine level and concentration in RTDS/RSCAD (Fig. 17a); trend display in ABB SCADA (Fig. 17b). The initial brine is brought to tank D110, so the initial concentration in the tank is 26.32%. When we set the concentration values to 25.8% respectively, thanks to the FFIC-102 controller, the brine concentration has decreased and followed the set point value. When the level set point value changes from 50% to 60%, and to 70%, the level gradually increases and follows the set point value. At that time, the concentration response is not affected. It proves that the FFIC-102 controller and the control loops can work well because of the real-time function of two systems.

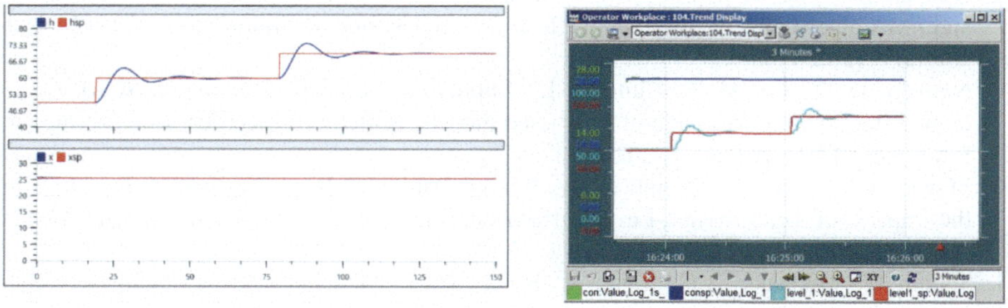

(a) (b)

Fig. 17. Trend displays in RTDS system (a) and in ABB's 800xA (b)

5 Conclusions

In this paper, we use RTDS to digitize the brine filtration process of a chemical factory or any smart factories and successfully connect between RTDS and ABB SCADA through IEC104 protocol based on GTNETx2 card and SDG triangle. It is applied by ratio control strategy combined with feedback control strategy to resolve the effects of coupler between brine level and concentration and upgrade the quality and stability of the control system. Simulation results are real-time and close to reality and can be trusted. RTDS system can be used as a pilot to test solutions to troubleshoot problems in the smart factories. It brings outstanding benefits to businesses, so the production efficiency is increasingly improved, and the product quality is synchronized. In addition, RTDS system can be one of the effective solutions to test and calibrate IEC 104 communication connections from power plants (thermal power plants, hydropower plants) to the National Load Dispatch Centers A0, A1.

Thanks to TT-Group for supporting the RTDS system during the research process.

References

1. Mabkhot, M.M., Al-Ahmari, A.M., Salah, B., Alkhalefah, H.: Requirements of the Smart Factory System: A Survey and Perspective. Machines **6**(2), Art. no. 2 (2018)
2. Reiz, C., Leite, J.B.: Hardware-In-the-Loop Simulation to Test Advanced Automation Devices in Power Distribution Networks. IEEE Transactions on Power Delivery **36**(4), 2194–2203 (2021)
3. Bountouris, P., Tzelepis, D., Abdulhadi, I., Coffele, F.: Dynamic Model of Commercially Available Inverters With Validation Against Hardware Testing. IEEE Transactions on Power Systems **38**(3), 2007–2017 (2023)
4. Kim, G.-H., Kim, Y.-J., Park, M., Yu, I.-K., Song, B.-M.: RTDS-based real time simulations of grid-connected wind turbine generator systems. In: 2010 Twenty-Fifth Annual IEEE Applied Power Electronics Conference and Exposition (APEC), pp. 2085–2090 (2010)
5. Do, M.-H., Nguyen, D.-D.: Effects of actuator faults on large-scale wind turbine dynamics with an industry-standard controller. E3S Web Conf. **514**, 03007 (2024)
6. Devanarayana, C., Zhang, Y., Kuffel, R.: Testing Cyber Security of Power Systems on a Real Time Digital Simulator. In: 2019 IEEE 8th International Conference on Advanced Power System Automation and Protection (APAP), pp. 1166–1170 (2019)
7. Nguyen, D.-D., Do, M.-H.: Two Methods for Detecting the Linear Attack on SCADA Systems, In: Advances in Engineering Research and Application, Cham, pp. 929–940 (2022)
8. Nguyen, D.D., Le, M.T., Cung, T.L.: The ability to detect the linear attack of WL-CUSUM and FMA algorithms. In: Bulletin of Electrical Engineering and Informatics, vol. 12, no. 1, Art. no. 1 (2023)
9. Nguyen, D.-D., Le, M.-T., Cung, T.-L.: Improving intrusion detection in SCADA systems using stacking ensemble of tree-based models. Bulletin of Electrical Engineering and Informatics **11**(1), Art. no. 1 (2022)
10. Maharjan, M., Ekic, A., Strombeck, B., Wu, D.: An RTDS-Based Testbed for Investigating the Impacts of Transmission-Level Disturbances on Solar PV Operation. Energies **14**(13), Art. no. 13 (2021)
11. Tauqir, W., Maduako, A.C., Yang, X., Bollas, G.M.: Modeling aqueous multi-electrolyte systems for the simulation and optimization of concentrated brine processes. In: Kokossis, A.C., Georgiadis, M.C., Pistikopoulos, E. (eds.) Computer Aided Chemical Engineering, vol. 52, pp. 733–738. Elsevier (2023)

12. Skogestad, S., Postlethwaite, I.: Multivariable Feedback Control: Analysis and Design. John Wiley & Sons (2005)
13. Triangle Microworks, Inc.: SCADA Data Gateway Implementation Guide. Scribd (2020). https://www.scribd.com/document/523594387/SDG-Implementers-Guide
14. ABB.: System 800xA 5.1 Network Configuration (2010)
15. ABB.: System 800xA, PLC Connect, Configuration (2010)

A Boundary Element Method for Indentation of a Rigid Indenter and a Layer-Auxetic BCS Substrate

Xuan Thanh Nguyen[1,2], Dinh Duc Nguyen[2], and Van Thuong Nguyen[3](✉)

[1] Faculty of Civil and Environment, Thai Nguyen University of Technology, Thai Nguyen 250000, Vietnam

[2] Faculty of Civil Engineering, VNU University of Engineering and Technology, Hanoi 10000, Vietnam

[3] School of Aerospace Engineering, VNU University of Engineering and Technology, Hanoi 10000, Vietnam

thuong.nv@vnu.edu.vn

Abstract. We use the boundary element method to study numerically the contact behaviors of an auxetic layer-substrate structure indented by a rigid indenter. The indenter considered in this paper is selected as a parabolic indenter. The layer-substate comprises the PLA for the layer and the bone-inspired cellular structure (BCS) for the substrate. In developing the boundary element method, the special fundamental solution for elastic bimaterials is used. Thus, no meshes are required on the contact surface between the layer and substrate, which is beneficial in computational and modeling the problem. With the boundary element method, numerical examples are provided, and we have shown that the computed solutions agree with the available solutions in the literature. Parametric studies have also been performed to study the influence of the BCS substrate and the layer on the indentation resistance of the structure. It was found that the BCS auxetic substrate and layer's thickness significantly influenced the structure's indentation stress and the indentation resistance of the structures.

Keywords: auxetic · indentation · boundary element method · bone-inspired cellular structure

1 Introduction

Layer-substrate structures are widely used in engineering applications [1–8]. Depending on the application, the layer can be a thin film or a protective surface of the substrate [1–10]. If the substrate is made of cellular structure material, it can be used as an impact, indentation protection, or energy absorption component of the structure [9, 10]. Due to the importance of the layer-substrate structure, especially when subjected to indenting loads, several different indentation problems can be found in the literature [1–5, 11, 12]. In solving the indentation problems with the layered substrate, numerous solution methods have been proposed, such as analytical approaches [4] and numerical approaches [11–13]. Among different methods, analytical methods are applied only

© The Author(s), under exclusive license to Springer Nature Switzerland AG 2025
H. A. Le Thi et al. (Eds.): ICAMCS 2024, LNNS 1311, pp. 370–380, 2025.
https://doi.org/10.1007/978-3-031-90606-0_31

to simple problems. Numerical methods such as finite and boundary elements can be applied to complicated practical engineering problems. When observing the published works with finite and boundary element methods, we found that the finite element method (FEM) required to mesh the entire calculated domains. For the layer-substrate problems, a finer mesh is usually required for the layer. FEM is, therefore, time-consuming. The boundary element method (BEM) can be considered promising in replacing the finite element method in solving the contact problem. This method requires mesh only on the boundary, which reduces the problem dimensions and provides more accurate results. It has attracted the attention of many researchers in recent years [14–20]. In these studies, the anisotropic elastic materials, piezoelectric and MEE materials are considered. To the authors' knowledge, however, no published works using BEM can be found in the literature, especially for the layer-substrate structure with cellular structure material.

With this understanding, in this paper, we present a BEM for the contact analysis of a layer substrate with bone-inspired cellular structure material (BCS). In this BEM, the special fundamental solution for the bimaterial of anisotropic elastic material is employed [21]. With this special fundamental solution, no mesh is required on the bimaterial surface, which reduces the mesh and computational time and enhances the accuracy of the solution at the same time. To demonstrate the correctness of our proposed method, numerical results are provided, which further allow us to perform parametric studies of the BCS substrate and the layer on the indentation responses of the structure.

2 Indentation of a Rigid Indenter on a Layer-BCS Substrate

Consider the frictionless contact of a rigid indenter on a layer-BCS substrate, as shown in Fig. 1. Here, the indenter head is parabolic.ssis 3.3 GPa and Poisson ratio is 0.35. The layer is made of PLA material, whose Young's modulus is 3.3 GPa and Poisson ratio is 0.35. The BCS material is artificially designed, and its geometry mimics the cancellus bone [9, 10, 22]. It was found that the BCS materials consisted of 9 independent elastic constants, which vary according to the base material and thickness and the angles of the sub-cell. Suppose the PLA material is used as the base material, and a uniform thickness of 1 mm is used for the cell walls. In that case, the elastic material of these materials can be obtained as listed in Table 1 below using Castigliano's theorem and the shear stress/strain deformation mechanisms [22].

Table 1. The material properties of BCS materials [22]

β	E_1	E_2	E_3	G_{31}	G_{32}	G_{12}	υ_{13}	υ_{23}	υ_{12}
550	21.02	59.48	680.50	142.14	109.89	0.34	0.011	0.031	0.13
82.5^0	21.51	83.04	692.12	142.73	113.62	0.24	0.011	0.042	0.00
130^0	20.06	52.09	753.80	172.13	107.06	0.13	0.009	0.024	−0.19

This table shows that depending on the angle β, the BCS material may exhibit positive, zero, and negative Poison ratios. If the negative Poison is negative, the material is so-called auxetic material. In this table, the subscripts 1, 2, and 3 denote the x, y, and z directions. The unit of Young's moduli and shear moduli are MPa. To see the mathematical model development used to evaluate these material properties, one can refer to [22] for details.

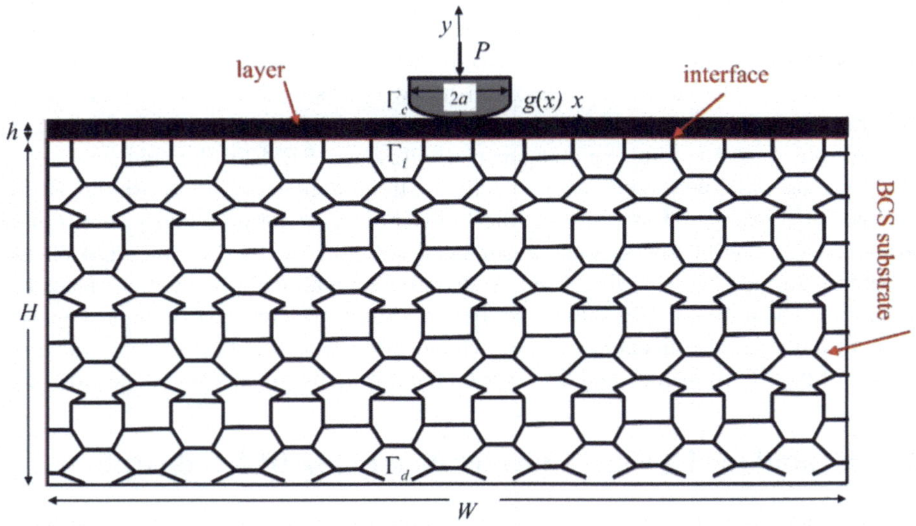

Fig. 1. Indentation of a rigid indenter and a layer-BCS substrate. Here, (x,y) is the global coordinate. $g(x)$ is the indenter profile, $g(x) = 0$ for the flat-ended indenter; $g(x) = x^2/2R$ for the parabolic indenter; $2a$ is the width of the indenter; h is the layer's thickness; H and W is the width of the BCS substrate.

If we let $g(x)$ and v_0 be the profile and the normal translation of the indenter, and P be the applied load on the indenter, the boundary conditions for this problem can be expressed as Eqs. (1)–(6), in which Eqs. (1–2) denotes the frictionless boundary conditions at the top surface of the layer, Eq. (3) denotes the traction and displacement continuity along the interface of the layer and the substrate; Eq. (4) denotes the prescribed displacement at the lower surface of the substrate; Eq. (5) denotes the traction free condition, and Eq. (6) denotes the equilibrium condition of the indenter.

$$t_1^{(l)}(x,0) = 0,\ u_2^{(l)}(x,0) = g(x) + v_0,\ u_3^{(l)}(x,0) = 0,\ x \in \Gamma_c, \tag{1}$$

$$t_i^{(l)}(x,0) = 0,\ x \notin \Gamma_c \tag{2}$$

$$u_i^{(l)}(x,-h)=u_i^{(b)}(x,-h),\ t_i^{(l)}(x,-h)=-t_i^{(b)}(x,-h),\ x\in\Gamma_i \tag{3}$$

$$u_b^{(l)}(x,-H,-h)=0,\ x\in\Gamma_d, \tag{4}$$

$$\begin{aligned}&t_i^{(l)}(-W/2,y)=0,\ t_i^{(l)}(W/2,y)=0,\ -h<y\le 0,\\&t_i^{(b)}(-W/2,y)=0,\ t_i^{(b)}(W/2,y)=0,\ h-h-h<y<=,\end{aligned} \tag{5}$$

$$\int_{\Gamma_c} t_2^{(l)}(x,0)dx=P \tag{6}$$

where Γ_c, Γ_i and Γ_d denote, respectively, the contact region, the interface between the layer and substrate and the prescribed displacement boundary conditions; $u_i^{(l)}$ and $t_i^{(l)}$, are, respectively, the layer's displacements and tractions; $u_i^{(b)}$ and $t_i^{(b)}$, are, respectively, the substrate's displacements and tractions; $y=-h$ denotes the location of the interface between the layer and substrate; H and W are, respectively, the height and width of the foundation; $g(x)$ is the profile of the indenter. For the parabolic indenter, we have $g(x)=x^2/2R$, where R is the radius of curvature. To solve the boundary conditions stated above, the boundary element method stated in the next section will be employed.

3 Boundary Element Method for Layer-Substrate Structure

Following the derivation procedure of BEM, the boundary integral equations can be expressed in the following system of algebraic equations as [14–16, 21]

$$\sum_{n=1}^{N}\mathbf{Y}_{in}\mathbf{u}_n=\sum_{n=1}^{N}\mathbf{G}_{in}\mathbf{t}_n,\ i=1,2,...,N. \tag{7}$$

where $\mathbf{u}_n$ and $\mathbf{t}_n$ are the displacement and traction at the boundary nodes; N is the number of the boundary node; $\mathbf{Y}_{in}$ and $\mathbf{G}_{in}$ are matrices of influence coefficients of nodes i and n. These influence matrices can be evaluated by using a proper fundamental solution. This study selects the fundamental solution based on the bimaterial Green's function of two-dimensional anisotropic elastic material.

The bimaterial Green's function [21]

$$\begin{aligned}&\left[u_{ij}^{*(l)}(\mathbf{x},\boldsymbol{\xi})\right]=2\mathrm{Re}\left\{[\mathbf{A}_l\mathbf{F}_l(p^{(l)})]^T\right\},\ \left[t_{ij}^{*(l)}(\mathbf{x},\boldsymbol{\xi})\right]=2\mathrm{Re}\left\{[\mathbf{B}_l\mathbf{F}_{l,s}(p^{(l)})]^T\right\},\\&\left[u_{ij}^{*(b)}(\mathbf{x},\boldsymbol{\xi})\right]=2\mathrm{Re}\left\{[\mathbf{A}_b\mathbf{F}_b(p^{(b)})]^T\right\},\ \left[t_{ij}^{*b)}(\mathbf{x},\boldsymbol{\xi})\right]=2\mathrm{Re}\left\{[\mathbf{B}_b\mathbf{F}_{b,s}(p^{(b)})]^T\right\},\end{aligned} \tag{8}$$

where

$$\mathbf{F}_l(p^{(l)})=\frac{1}{2\pi i}\left\{<\ln(p_\alpha^{(l)}-\hat{p}_\alpha^{(l)})>\mathbf{A}_l^T+\sum_{k=1}^{3}<\ln(p_\alpha^{(l)}-\overline{\hat{p}}_k^{(l)})>\tilde{\mathbf{A}}_l\right\}, \tag{9}$$

$$\mathbf{F}_b(p^{(b)})=\frac{1}{2\pi i}\sum_{k=1}^{3}<\ln(p_\alpha^{(b)}-\hat{p}_k^{(b)})>\tilde{\mathbf{A}}_b, \tag{10}$$

and

$$
\begin{aligned}
& p_\alpha^{(l)} = x + \mu_\alpha^{(l)} y,\ \hat{p}_\alpha^{(l)} = \hat{x} + \mu_\alpha^{(l)} \hat{y}, \\
& p_\alpha^{(b)} = x + \mu_\alpha^{(b)} y,\ \hat{p}_\alpha^{(b)} = \hat{x} + \mu_\alpha^{(b)} \hat{y}, \\
& \tilde{\mathbf{A}}_l = \mathbf{A}_l^{-1}(\overline{\mathbf{M}}_b + \mathbf{M}_l)^{-1}(\overline{\mathbf{M}}_b - \overline{\mathbf{M}}_l)^{-1}\overline{\mathbf{A}}_l \mathbf{I}_j \overline{\mathbf{A}}_l^T, \\
& \tilde{\mathbf{A}}_b = \mathbf{A}_b^{-1}(\mathbf{M}_b + \overline{\mathbf{M}}_l)^{-1}\overline{\mathbf{A}}_l^T \mathbf{I}_j \mathbf{A}_l^T,\ \mathbf{M}_l = -i\mathbf{B}_l \mathbf{A}_l^{-1}, \mathbf{M}_b = -i\mathbf{B}_b \mathbf{A}_b^{-1}.
\end{aligned}
\tag{11}
$$

where (x,y) and $(\hat{x}, \hat{y})$ are the location of the field point and source point; the superscript (l) and l imply the parameters and values related to the layer, while the superscript (b) and b denote the parameters and values related to the BCS substrate; μ_α is the material eigenvalue, and **A** and **B** are the material matrices [21]. The above fundamental solution automatically satisfies the boundary conditions (3) set for the common interface between the layer and substrate Γ_i. Thus, we only need the boundary nodes on the boundaries outside the layer and substrate. When all the boundary nodes are considered, and the boundary conditions stated in Sect. 2 are taken into consideration, Eq. (7) produces a $(3N + 1) \times (3N + 1)$ system of linear equations in which all the variables are uniquely determined. The variables of this linear system of linear equations include the displacement vector $\mathbf{u}$ and the traction vector $\mathbf{t}$ assembled by all the nodal displacements and tractions u_n and t_n and the rigid translation of the indenter v_0. It should be noted that the contact region may be unknown in advance for the indentation problem with a parabolic indenter. To get the $(3N + 1) \times (3N + 1)$ system of linear equations, the contact region and its associated boundary conditions for the contact nodes must be initially assumed. The solutions of $\mathbf{u}_n$, $\mathbf{t}_n$ and v_0 of the system of linear equations, therefore, need to be further checked with the contact criteria set for all contact nodes. For the current problem, the contact criteria include the condition that the normal stress of all contact nodes should be negative, and the gap between the indenter and the layer should be greater than or equal to zero. If positive normal stress is found for a certain contact node, its associated boundary condition should be changed from (1) to (2). If the gap is negative, the boundary condition of the associated contact node should reverse from (2) to (1). This process is repeated until all contact nodes pass the contact criteria. The iterative procedure applied for the contact problems can be found in many published works [13–16, 21] and is omitted here for brevity.

4 Numerical Examples

4.1 Solution Verification

In this section, numerical results will be presented to demonstrate the accuracy of the BEM. For this purpose, we consider the indentation of a parabolic indenter on a PLA foundation. To simulate the PLA foundation in BEM, both the layer and substrate are now made of PLA, the material properties of which are given in Sect. 2. The indenter width is selected as $a = 0.02$ m, and the foundation has a width of $W = 2$ m and a height of $H = 1$ m. The layer thickness is $h = 0.05$ m. Since the dimensions of the foundation are very large compared to the dimensions of the indenter, we can also apply the analytical solutions to this problem. Figures 2 and 3 show the surface deformation and contact

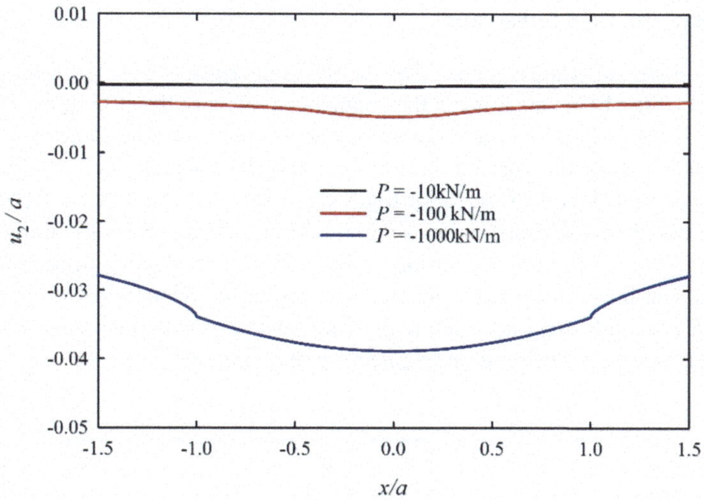

Fig. 2. Surface displacements along the layer's upper surface for different loads when the materials of the layer and BCS substrate are both PLA.

pressure under the indenter for different values of applied load P. These figures show that our solutions are well-matched with those obtained using the analytical method [21]. In addition, when the applied load increases, the surface deformation and contact pressure under the indenter also increase, which is reasonable and consistent with our engineer's intuition.

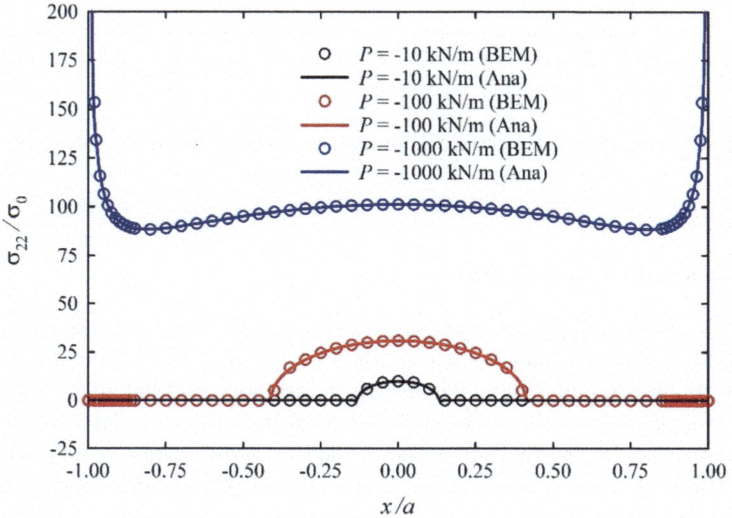

Fig. 3. Stress along the layer's upper surface for different loads when the materials of the layer and BCS substrate are both PLA. Here, $\sigma_0 = P_0/2a,\ P_0 = -10$ kN/m.

4.2 Effect of the BCS Substrate

In this section, the substrate is replaced by the BCS materials, as given in Table 1, to study the influence of the BCS substrate on the contact responses of the structures. The applied load is $P = -100$ kN/m. The other conditions are chosen to be the same as in Sect. 4.1. Figures 4 and 5 show the surface deformation and the contact stress along the upper surface of the PLA layer. From these figures, we see that the auxetic BCS substrate (with a negative Poisson ratio) reduces the maximum contact pressure along the layer. At the same time, it increases the surface deformation of the layer's upper surface. The reduction of the indentation stress means that auxetic BCS materials provide a better ability to prevent high stress in structure than the substrates with positive or zero Poisson ratios, which is always desirable in engineering applications.

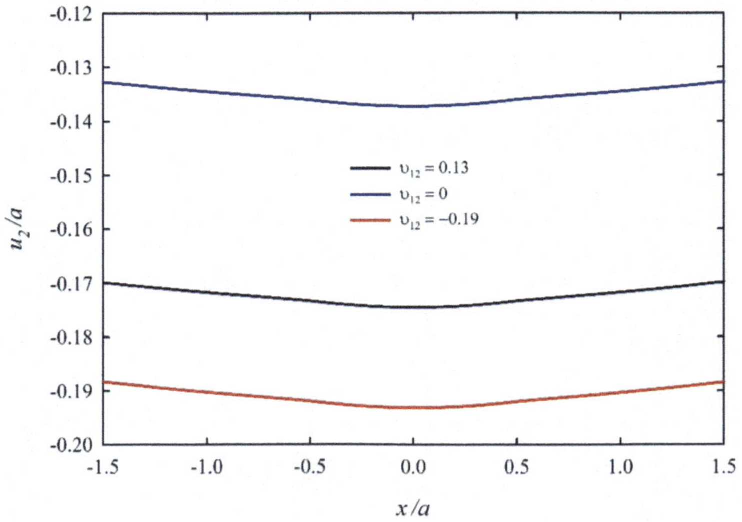

Fig. 4. The layer's upper surface deformation with different BCS substrate types.

4.3 Influence of Layer Thickness

To study the effect of the layer thickness, we consider the contact of the rigid indenter, as described in Sect. 4.1, and the auxetic BCS foundation. The applied load is $P = -100$ kN/m. Figures 6 and 7 show the surface deformation and indentation stress along the layer's upper surface. It can be deduced from these figures that when the layer's thickness increases, the surface deformation and the contact region reduce. Meanwhile, the contact region reduces, and the maximum contact pressure increases. The contact is complete when the thickness is zero and switches to incomplete when the layer's thickness increases. These observations are reasonable since the layer is stiffer than the BCS substrate, and the increase in the layer's thickness makes the indentation resistance of the structure stronger.

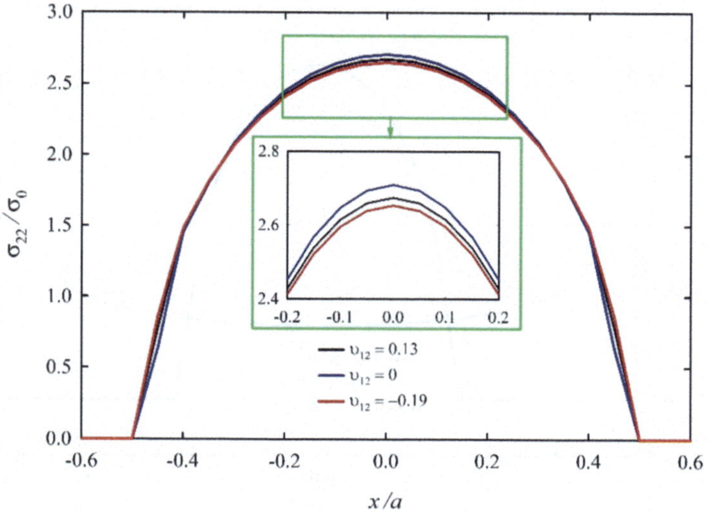

Fig. 5. The contact pressure along the layer's upper surface with different BCS substrate types. Here, $\sigma_0 = P_0/2a$, $P_0 = -100$ kN/m.

5 Conclusions

In this paper, the boundary element method was used to solve the indentation problem of a rigid indentation on a layer-BCS substrate structure. By using the special fundamental solutions of bimaterials, no meshes were needed along the interface, which benefited in

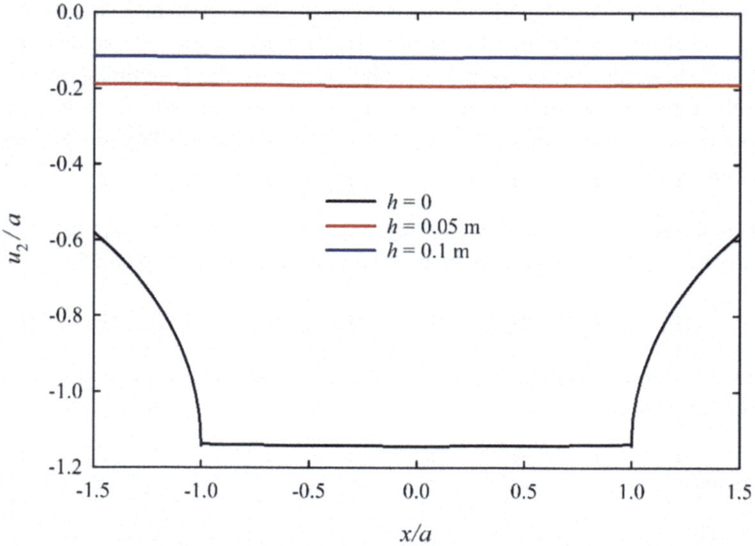

Fig. 6. The surface deformation along the layer's upper surface with the layer's thickness.

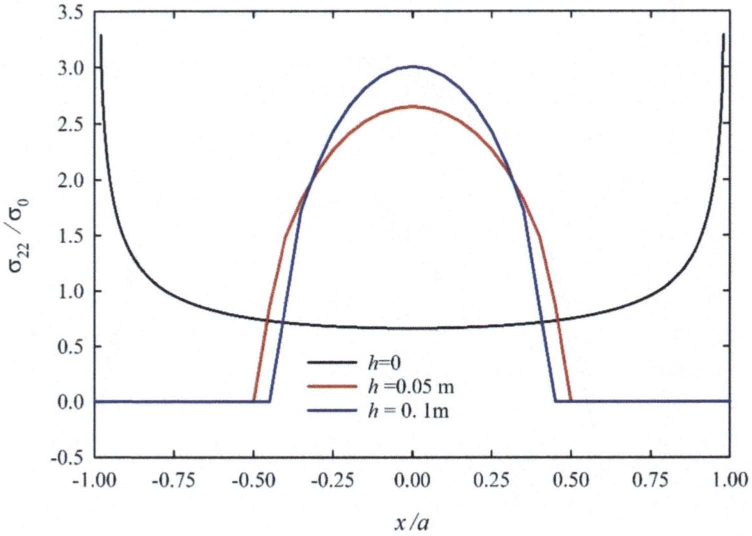

Fig. 7. The contact pressure along the layer's upper surface with the layer's thickness. Here, $\sigma_0 = P_0/2a$, $P_0 = -100$ kN/m.

terms of modeling, accuracy and computational resources. The numerical results demonstrated the correctness and applicability of the developed BEM. The influence of the BCS substrate and the layer's thickness on the structure's contact behavior was analyzed. It is found that the BCS auxetic substrate can effectively reduce the contact pressure along the layer's upper surface. The increase of the layer's thickness can significantly improve the indentation resistance. The BEM and the investigation in this paper are useful and can be further used in the design and optimization of the layer-substrate structures made of BCS materials. Finally, although the present BEM possesses the above advantages and demonstrates its applicability for solving and analyzing the structures with the layer-auxetic substrate, in this study, only static indentation problems are considered. The authors plan to extend this method to consider the dynamic indentation problems such as impact in the future.

Acknowledgments. Nguyen Van Thuong was funded by the Postdoctoral Scholarship Programme of Vingroup Innovation Foundation (VINIF), code VINIF.2024.STS.25.

Disclosure of Interests. The authors have no competing interests to declare that are relevant to the content of this article.

References

1. Ciprari, S., Ripard, V., Saulot, A., Massi, F.: Investigation of third body role in dry contacts: experimental procedure to dissociate the effects of substrate and interface layer on the contact pair frictional response. Tribol. Int. **190**, 109047 (2023)

2. Ding, S., Hu, Y., Jian, B., Zhang, Y., Xia, R., Hu, G.: A semi-analytical solution to incipient sliding contact on viscoelastic layer-elastic substrate with imperfectly bonded interface. Appl. Math. Modell. **125**, 35–58 (2024)
3. Cao, R., Li, L., Li, X., Mi, C.: On the frictional receding contact between a graded layer and an orthotropic substrate indented by a rigid flat-ended stamp. Mech. Mater. **158**, 103847 (2021)
4. Zhang, W.H.: Study on parameter measurement of elastic thin films with substrate effect based on contact mechanics. Int. J. Solids Struct. **271–272**, 112254 (2023)
5. Argatov, I.I., Borodich, F.M., Epshtein, S.A., Kossovich, A.L.: Contact stiffness depth-sensing indentation: understanding of material properties of thin films attached to substrates. Mech. Mater. **114**, 172–179 (2017)
6. Pham, H.C., Nguyen, V.H., Truong, T.T.T., Duc, N.D.: Nonlocal strain gradient-based geometrically nonlinear vibration analysis of double curved shallow nanoshell containing functionally graded layers. Aerospace Sci. Technol. **151**, 109310 (2024)
7. Nguyen, H.V., Phung, V.M., Duc, N.D.: Finite-element modeling for static bending and free vibration analyses of double-layer non-uniform thickness FG plates taking into account sliding interactions. Archiv. Civ. Mech. Eng. **24**(2), 10 (2024)
8. Van Quyen, N., Duc, N.D.: Vibration and nonlinear dynamic response of nanocomposite multi-layer solar panel resting on elastic foundations. Thin-Walled Struct. **177**, 109412 (2022)
9. Ghorbani, F., Gharehbaghi, H., Farrokhabadi, A., Bolouri, A.: Investigation of the equivalent mechanical properties of the bone-inspired composite cellular structure: analytical, numerical and experimental approaches. Compos. Struct. **309**, 116720 (2023)
10. Ghazlan, A., Ngo, T., Nguyen, T., Linforth, S., Le, T.V.: Uncovering a high-performance bio-mimetic cellular structure from trabecular bone. Sci. Rep. **10**, 14247 (2020)
11. Wu, J., Ru, C.Q.: Spherical indentation of an elastic layer on a rigid substrate revisited. Thin Solid Films **669**, 500–508 (2019)
12. Vieira, V.F., Shigaki, Y., Martins, P.S., Ba, E.C.T., Dias, C.A.R.: Nanoindentation test of a DLC coated high-speed steel substrate using a two-dimensional axisymmetric finite element method. Diam. Relat. Mater. **134**, 109792 (2023)
13. Wu, C., Zhang, L., Singhatanadgid, P., Zhang, D.: Thermoelastic analysis of a bi-layered system with the single domain inclusion-based boundary element method. Eng. Anal. Bound. Elem. **157**, 374–389 (2023)
14. Nguyen, V.T., Chen, G.T., Hwu., C.: Multibody contact of two-dimensional anisotropic elastic/piezoelectric/magneto-electro-elastic solids. Eng. Anal. Bound. Elem. **146**, 767–785 (2023).
15. Nguyen, V.T., Hwu, C.: Boundary element method for contact between multiple rigid punches and anisotropic viscoelastic foundation. Eng. Anal. Bound. Elem, **118**, 295–305 (2020). https://doi.org/10.1016/j.enganabound.2020.07.001
16. Nguyen, V.T., Hwu, C.: Indentation by multiple rigid punches on two-dimensional anisotropic elastic or viscoelastic solids. Int. J. Mech. Sci. **178**, 105595 (2020)
17. Nguyen, V.T., Chen, G.T., Hwu, C.: Multibody contact of two-dimensional anisotropic elastic/piezoelectric/magneto-electro-elastic solids. Eng. Anal. Boundary Elem. **146**, 767–785 (2023)
18. Rodríguez-Tembleque, L., Buroni, F.C., Sáez, A., Aliabadi, M.H.: 3D coupled multifield magneto-electro-elastic contact modelling. Int. J. Mech. Sci. **114**, 35–51 (2016)
19. Rodríguez-Tembleque, L., Sáez, A., Aliabadi, M.H.: Indentation response of piezoelectric films under frictional contact. Int. J. Eng. Sci. **107**, 36–53 (2016)
20. Rodríguez-Tembleque, L., Buroni, F.C., Sáez, A.: Boundary element analysis of the frictionless indentation of piezoelectric films. Europ, J. Comput. Mech. **25**(1–2), 24–37 (2016)

21. Hwu, C.: Anisotropic Elastic Plates. Springer, New York (2010)
22. Nguyen, V.T., Khoa, N.D., Ngo, T., Duc, N.D.: Analytical investigation of in-plane and out-of-plane elastic properties of bone-inspired cellular structures. Aerosp. Sci. Technol. **147**, 109012 (2024)

The Effect of Leading-Edge Serration on Aerodynamic Characteristics of an Ishii Airfoil

The Nguyen Ngoc[1], Anh Le Dinh[1], The Hung Tran[2], and Minh Do Van[3(✉)]

[1] School of Aerospace Engineering, VNU University of Engineering and Technology, Hanoi, Vietnam
[2] Faculty of Aerospace Engineering, Le Quy Don Technical University, Hanoi, Vietnam
[3] Faculty of Special Equipments, Le Quy Don Technical University, Hanoi, Vietnam
minhdv100@lqdtu.edu.vn

Abstract. In this study, the effect of the serrate leading edge of the wing used software commercial ANSYS was investigated. An airfoil profile named Ishii airfoil, which has been proven to have optimal aerodynamic quality under low Reynolds conditions, is used in this study to examine the impact of a serrated edge on the wing. A Reynolds averaged Navier-Stokes (RANS) simulation is applied for the simulation. The numerical results are then verified by experimental data at the same conditions. The results show that the accuracy of the numerical simulation at angles of attack below 6 degrees is sufficiently reliable (less than 0.5%). Additionally, the results indicate that the serrated model at the leading edge of the wing helps delay the formation of stall angle. However, in the serrated model, a decrease in aircraft performance is also observed before the stall angle occurs. Detailed pressure fields, lift and drag are presented in this study for different leading-edge configurations.

Keywords: Serrations · Ishii airfoil · CFD

1 Introduction

The development of science and technology today is increasingly advancing, with continuous new research that has practical applications in life as well as for humanity, and even in the military field. Prominent among them are micro aerial vehicles (MAVs), unmanned aerial vehicles (UAVs), and various types of space exploration aircraft like those for Mars… Which have numerous applications in life and science. These types of aircraft have many outstanding features such as: not being limited by human physiology, flexible design, long battery life, wide speed range, and extensive airspace [1]. Due to their high applicability and diversity, ranging from civilian to military uses, the improvement of aerodynamic characteristics for these types of vehicles is currently a focus of research for the scientific and industrial communities. These types of ultra-small, unmanned flying devices typically operate in low Reynolds number environments, with sizes and shapes very different from conventional aircraft. Therefore, it has attracted a

© The Author(s), under exclusive license to Springer Nature Switzerland AG 2025
H. A. Le Thi et al. (Eds.): ICAMCS 2024, LNNS 1311, pp. 381–391, 2025.
https://doi.org/10.1007/978-3-031-90606-0_32

lot of research aimed at improving the aerodynamic characteristics of this type of aircraft wing.

One of the methods to improve aerodynamic quality is controlling flow separation, which occurs when the airflow separates from the pressure suction of the wing on an aircraft. There are two active and passive ways to help control airflow separation on an aircraft wing. The proactive approach is to install complex and expensive control systems on aircraft, while the passive approach is to alter the design shape of various components of the flying device to optimize aerodynamic quality and airflow over the flying object. One of the passive methods that has been researched recently is changing the shape of the leading edge of the aircraft wing, This is a new passive method that has recently attracted researchers. This idea is inspired by nature, specifically a humpback whale.

There are two types of leading-edge shapes of aircraft wings that are currently being researched: leading-edge protuberances (LEPs) and serrated leading edge (SLEs) [2–6]. These studies present some similar and different conclusions regarding aircraft performance, such as delaying the onset of the stall angle, and some findings about the performance of the wings after stalling. In research on airfoil shapes (LEPs), it is concluded that [2–4], wings with a serrate shape at the leading edge of the wing help delay the onset of stall angle, while also reducing the increase in drag coefficient after stall. The lift coefficient can increase by up to 16% and the drag coefficient can decrease by up to 17% [3]. Meanwhile, previous research on sawtooth wing shapes (SLEs) has often focused on studying flat plates with serration shapes at the leading edge of the wing [5, 6]. These studies show that the aerodynamic characteristics of a flat plate with serrated edges compared to a regular flat plate are almost identical, and there are also slight improvements in the performance of the stall angle [6]. In addition, some research used serrate trailing egde to reduce the noise [8–11].

In this work, we investigate the influence of aerodynamic characteristics and flow turbulence when altering the shape of the leading edge of an aircraft wing under low Reynolds number conditions. The shape used in the research is a serration shape, and the airfoil profile in the paper is inspired by Anyoji et al. [7], which is called the Ishii airfoil. This type of wing design has been proven to optimize aerodynamic quality at low Reynolds numbers. The RANS simulation model is applied to predict the flow around the wing of the aircraft, particularly the behavior of the flow around the profile of the leading edge of the wing. The wavelength and amplitude of the serration at the leading edge of the wing have also been altered to gain a deeper understanding of the aerodynamic characteristics of this type of serrated wing.

2 Model and Numerical Simulation Method

2.1 Model Geometry and Domain

In this study, the wing shape used is called an Ishii airfoil, which was researched by Anyoji et al. at Tohoku University, Japan. The shape of the Ishii airfoil is shown in Fig. 1. The Ishii wing has a maximum wing thickness ratio of 7.1% at $x/c = 0.25$ and a maximum camber of 2.3% at $x/c = 0.62$. The Mach number M was set to 0.2 and the Reynolds number $Re = 23000$ [7].

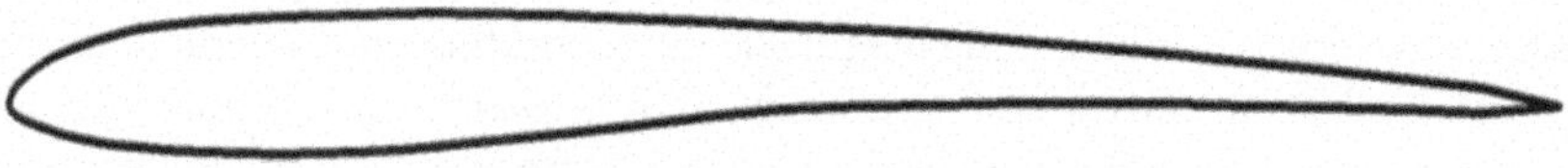

Fig. 1. The shape of the Ishii airfoil.

The wing was used in this research and compared with the previous results of Anyoji et al., Table 1 shows details of the geometry parameters of the Ishii airfoil. The standard PP1 model is like the model in the previous research by Anyoji [7], with a wingspan of 100 mm and a chord length of 50 mm. In addition, the study was conducted to examine the aerodynamic effects associated with changing the leading edge of the wings with the PP2 and PP3 wing types.

Table 1. Details table of chord-coordinates of Ishii airfoil [7].

x/c	y/c (suction side)	y/c (pressure side)	x/c	y/c (suction side)	y/c (pressure side)
1	0.005	−0.005	0.275	0.0476	−0.0219
0.95	0.0098	−0.0031	0.25	0.0473	−0.0236
0.9	0.0142	−0.0016	0.225	0.0468	−0.0248
0.175	0.0453	−0.0258	0.2	0.046	−0.0254
0.15	0.0443	−0.0254	0.1	0.041	−0.0239
0.125	0.043	−0.0248	0.08	0.0384	−0.0222
0.85	0.0188	−0.0002	0.07	0.0366	−0.0212
0.8	0.0227	0.0004	0.06	0.0342	−0.02
0.75	0.0268	0.0011	0.05	0.0314	−0.0188
0.7	0.0303	0.0013	0.04	0.0287	−0.0169
0.65	0.0338	0.0011	0.03	0.0248	−0.0145
0.6	0.0368	0.0003	0.025	0.0227	−0.0131
0.55	0.0392	−0.0006	0.02	0.0201	−0.0119
0.5	0.0411	−0.0023	0.015	0.0173	−0.0102
0.45	0.0429	−0.006	0.01	0.014	−0.0085
0.4	0.0446	−0.0111	0.005	0.0098	−0.0061
0.35	0.0463	−0.0158	0.002	0.0058	−0.0037
0.3	0.0473	−0.0198	0.001	0.0036	−0.0026
			0	0	0

In this probe, research was conducted on three types of wings, as illustrated in Fig. 2 a). The two types of wings are PP1, PP2, and PP3. With PP1 is the standard wing model implemented for comparison and validation with the previous research results of Anyoji [7]. Meanwhile, the PP2 and PP3 wings are a type of serrated wing at the leading edge of the wing, consisting of the base wing length and the protruding wing length as illustrated in Fig. 2 b). Each tooth corresponds to a wavelength λ, where $\lambda = 6.25$ mm. The heights of the serrate of the PP2 and PP3 wings are $A1 = 2.5$ mm and $A2 = 6$ mm, respectively.

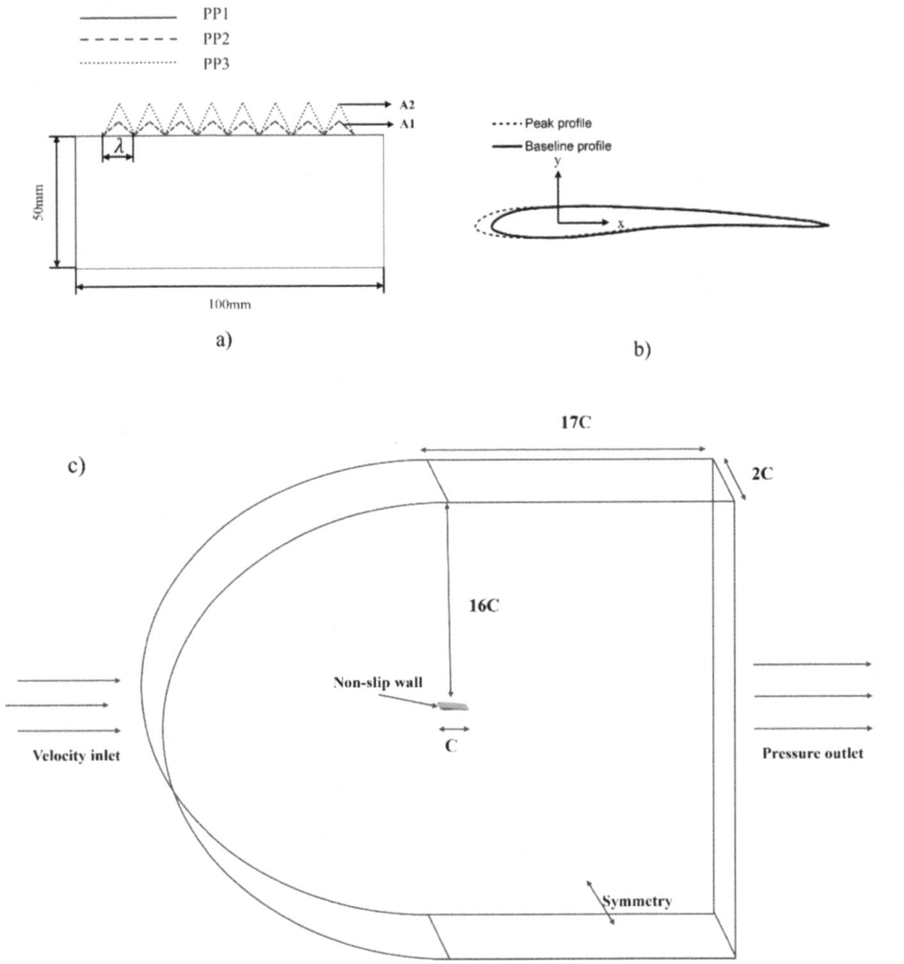

Fig. 2. Models and Computational domain, a) illustrate three types of models, b) illustrate the protrusion profile of the model, c) Domain area.

The domain area is illustrated in Fig. 2 c, with a height of the circular arc dimension $R = 16C$, a length behind the circular arc of $17C$, and a width of $2C$. Where C is the chord line of the wing ($C = 50$ mm). In this computational domain, the input velocity domain is the circular supply area in front of the aircraft wing, above and below the wing, with the output pressure set to free flow. Both walls of the computational domain were set as a symmetric surface, and the wing was set as the wall function.

2.2 Grid and Computational Method

The structured grid is applied around the airfoil profile and the computational domain, with images captured from various angles as shown in Fig. 3. The thickness of the first grid layer around the wing boundary is approximately 4.10^{-5} m, corresponding to $y+ \approx 1$, suitable for ensuring sufficiently accurate calculations. The length from the wing region to the boundary layer is wide enough for the numerical simulation to be minimally affected by the boundary region, as previously shown in Fig. 2 c). In addition, the calculated boundary area has a curved shape at the leading edge of the wing, which helps to adjust the wind direction and facilitates the examination of various angles of attack.

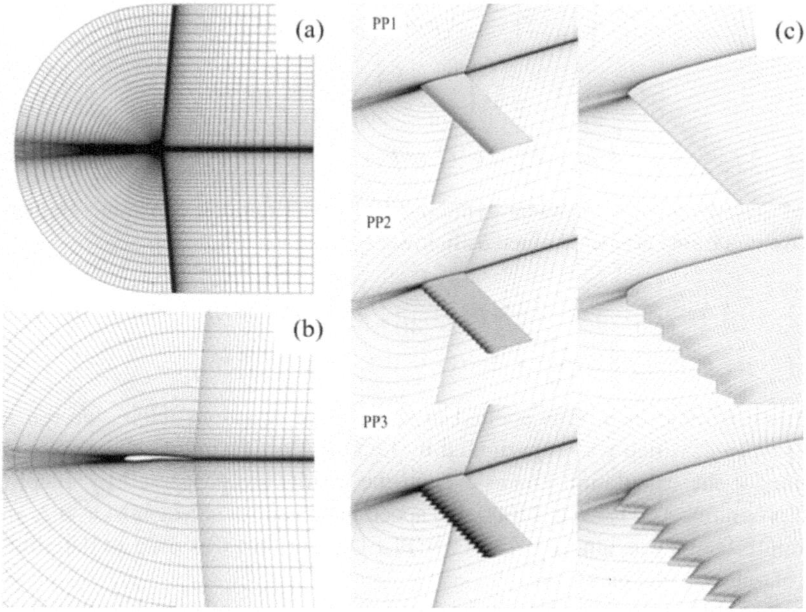

Fig. 3. Structured grid diagrams: (a) overall flow field grid; (b) grid near wing surface; (c) airfoil and surrounding grid (PP1,PP2 and PP3).

The theme of the Shear Stress Transfer k-ω computational model is applied for calculations in this domain. The SST k-ω model has the advantage of fast computation and effectively handles flow separation problems around the boundary layer. The transmission equation and the turbulent kinetic energy equation are defined by the following equations:

$$\frac{\partial(\rho k)}{\partial t} + \frac{\partial(\rho u_j k)}{\partial x_j} = P + \beta^* \rho \omega k + \frac{\partial}{\partial x_j}\left[(\mu + \sigma_k \mu_t)\frac{\partial k}{\partial x_j}\right] \tag{1}$$

$$\frac{\partial(\rho \omega)}{\partial t} + \frac{\partial(\rho u_j \omega)}{\partial x_j} = \frac{\gamma}{\nu_t} P - \beta \rho \omega^2 + \frac{\partial}{\partial x_j}\left[(\mu + \sigma_k \mu_t)\frac{\partial \omega}{\partial x_j}\right] + 2(1 - F_1)\frac{\rho \sigma_{\omega 2}}{\omega}\frac{\partial k}{\partial x_j}\frac{\partial \omega}{\partial x_j} \tag{2}$$

$$\nu_t = \frac{a_1 k}{\max(a_1 \omega; \Omega F_2)} \tag{3}$$

where:

$$F_2 = \tanh\left[\left[\max\left(\frac{2\sqrt{k}}{\beta\omega y}, \frac{500\nu}{y^2\omega}\right)\right]^2\right] \tag{4}$$

$$P_k = \min\left(\tau_{ij}\frac{\partial U_i}{\partial x_j}, 10\beta k\omega\right) \tag{5}$$

$$F_1 = \tanh\left\{\left\{\min\left[\max\left(\frac{\sqrt{k}}{\beta\omega y}, \frac{500\nu}{y^2\omega}\right), \frac{4\sigma_{\omega 2}k}{CD_{k\omega}y^2}\right]\right\}^4\right\} \tag{6}$$

$$\begin{aligned} CD_{k\omega} &= \max\left(2\rho\sigma_{\omega 2}\frac{1}{\omega}\frac{\partial k}{\partial x_i}\frac{\partial \omega}{\partial x_i}, 10^{-10}\right) \\ \phi &= \phi_1 F_1 + \phi_2(1 - F_1) \end{aligned} \tag{7}$$

Additionally, μ is the turbulent dynamic viscosity, The other constant parameters in the above equations are determined as below:

$$\begin{aligned} &\alpha_1 = \frac{5}{9}, \ \alpha_2 = 0.44, \beta_1 = \frac{3}{40}, \ \beta_2 = 0.0828, \beta^* = 0.09, \\ &\sigma_{k1} = 0.85, \ \sigma_{k2} = 1, \sigma_{\omega 1} = 0.5, \ \sigma_{\omega 2} = 0.856 \end{aligned}$$

A finite volume method is applied to solve these equations. Here, we use Ansys Fluent software, which was copyrighted by Le Quy Don Technical University, Hanoi, Vietnam, for the calculation process. Then the results are processed by Tecplot and Matlab programs. The Re was 23000, the speed had been set at Mach M = 0.2, and the temperature factor was ignored. The COUPLED algorithm was used to determine the values of velocity and pressure.

3 Results and Discussions

3.1 Validation of Previous Data Results

The change in the position of the separation point and the reattachment point is illustrated in Fig. 4 a). The line separation and reattachment of the PP1 wing were extracted at the main cross-section of the wing and were determined from the friction coefficient C_f. These points were compared with the previous research of Anyoji et al. [7] from an angle of 0° to 10°. The graph shows that the flow separation point of the PP1 wing is like Anyojj et al. results from attack angles of 6° to 8°. Meanwhile, the phenomenon of Anyoji alignment appeared early at an angle of 4°. The difference between PP1 and Anyoji is the separation of lines at angles of less than 6°. The phenomenon of flow reattachment shows a significant discrepancy at the angle of 6°, where the reattachment point of Anyoji is located at *x/c* ≈ 0.658, while the position of the reattachment point of the PP1 wing is at $x/c \approx 0.315$. The reason for this discrepancy is due to the differences in simulation methods. In research by Anyoji et al., a large eddy simulation (LES) model was used, thus predicting the turbulent flow around the surface is more detailed compared to the *k-ω SST* model. However, this also requires higher computational resources and time.

To clarify the position discrepancy at an angle of attack of 6°, Fig. 4 b) shows a comparison of the pressure distribution around the wing surface with the experimental results of Anyoji at an angle of attack of 6°. The results show that the lower surface pressure of the wing is quite like the experimental results. However, the pressure distribution on the upper surface of the wing shows a clear difference, as at the position from $0 \leq x/c \leq 0.23$ of PP1 it is lower than the experimental data, while from the position $0.23 \leq x/c \leq 0.6$ the pressure distribution is higher than the experimental results. This is also considered as the main influence on the lift curve of the wing at this angle of attack.

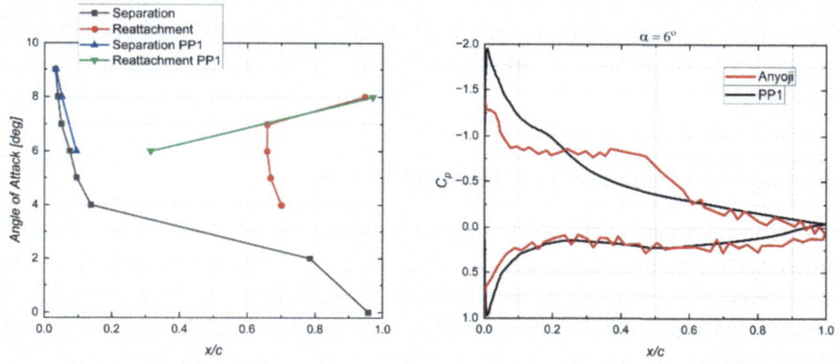

a) Comparison of separation and reattachment

b) Comparision of pressure distribution

Fig. 4. a) Separation and Reattachment location defined by coefficients friction C_f compared with Anyoji's data; b) Validation of pressure distribution.

3.2 Aerodynamic Performance

Figure 5 shows the aerodynamic performance of three types of wings PP1-PP3 and compares it with the experimental results of Anyoji et al. [7] from an angle of attack of 0° to 20°. The aerodynamic quality graph shows that both the lift coefficient and drag coefficient curves for all three types of wings follow the same trend as the experimental results. However, the numerical simulation results for the PP1 wing show the early stall angle at (AOA = 6°) compared to the experimental results at (AOA = 9°). The lift value of PP1 also demonstrates high accuracy at angles of attack of 4° and 6°, with an error margin of less than 0.5%. The wing with a serrated profile on the leading edge of the PP2 and PP3 wings shows a stall angle (AOA = 8°), and the stall angle of this profile occurs later compared to the PP1 wing. This is consistent with previous studies on wings with the protrusion profile at the leading edge of the wing.

Comparing the aerodynamic quality of wing PP1 with PP2 and PP3 before and after the stall angle (AOA = 8°), it can be observed that before the stall phenomenon occurs,

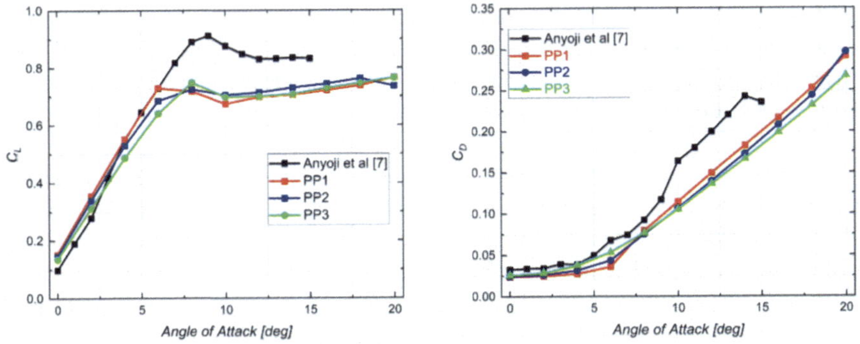

Fig. 5. Comparison of aerodynamic characteristics.

both the lift and drag graphs of wings PP2 and PP3 are not more optimal than those of wing PP1. After the stall angle, the performance of PP2 and PP3 is better than that of PP1. The performance of the PP2 and PP3 wings presented in this study also aligns with previous research [2–4].

3.3 Comparison of Surface Flow Between Models

In the previous section on the aerodynamic performance of wings, some conclusions were drawn regarding the performance of three types of wings before and after the angle of attack stall occurs. In this section, the surface flow at angles of attack from 0° to 20° is represented, clarifying the relationship between surface flow and the shape at the leading edge of the wing in relation drag force. The surface flow of the angle of attack on three types of wings before stall (from 0° to 8°) has been illustrated in Fig. 6. As can be seen, the flow on the PP1 wing from 0° to 4° is relatively stable from the beginning to the end of the wing. However, with the PP2 and PP3 wing types, there is a slight formation of vortices between the grooves at the serrated leading edge, and the pressure at the serrated wing tip shows a gradual decrease towards the grooves. At the angle of attack 8 °, the chaos and vortex are clearly forming. The flow of all three types of blades reverses from the trailing edge to the leading edge, creating areas of low-pressure vortices.

The spiral bundles at the wing with serrated ends PP2 and PP3 are more clearly visible in the grooves. The relationship between the amplitude of the sawtooth shape and the vortex core is also evident as the width of the vortex core appears on the PP2 wing near the leading edge of the wing, while the vortex core of the PP3 wing type extends from the tip to the middle of the wing body. The formation of these vortex bundles at an angle of attack of 4° to 8° directly affects aerodynamic performance as it contributes to increased drag. This is consistent with the drag graph shown in Fig. 5, where the drag force of the serrated wings PP2 and PP3 is highest from 0° to 8°. In contrast to that aerodynamic performance, Fig. 7 shows the surface flow over the wing from an angle of attack of 12° to 20°. The surface flow of the three types of wings continues to trend from the trailing edge of the wing to the leading edge of the wing. However, the flow in the grooves of the serrated wing no longer appears. The explanation for this phenomenon is that vortices of equal size moving in opposite directions meet in the trough, disrupting

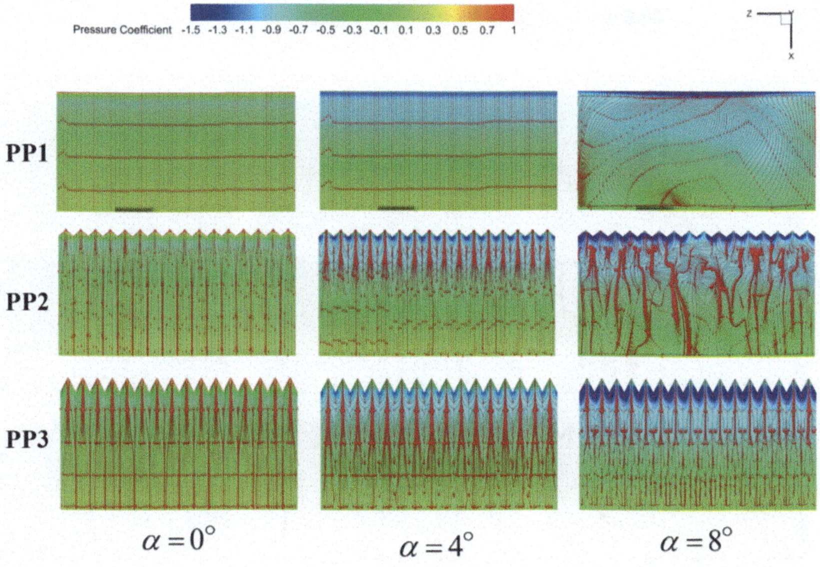

Fig. 6. Surface flow of three types of wings from 0 ° to 8°.

the structure of the swirling bubble, thereby reducing some of the drag force caused by the vortex region. This contributes to improving the increase in drag when the angle of attack is raised, while also demonstrating the relationship between wave amplitude and the increase in drag, as shown in Fig. 5.

3.4 Pressure Contribution

Figure 8 shows the comparison of pressure profiles from 0° to 20 °C for two types of wings, PP2 and PP3, at the peak and trough positions (illustrated in Fig. 2b). The pressure distribution values at the trough cross-section are taken at the midpoint of the aircraft wingspan, x = 0 mm, and the peak cross-section is taken at the top right next to the trough, x = +3.125 mm. The close relationship between the lift force of a wing and the pressure is formed due to the difference between the upper and lower surfaces. The graph comparing the pressure distribution around the peak and trough positions of the two types of wings shows that most of the difference in pressure distribution is concentrated at the leading edge of the wing. In most cases, the pressure distribution at the peak section of the PP2 wing on the pressure side is greater than that of the PP3 wing, and the suction side pressure of PP2 is also lower than that of PP3. This difference in the pressure distribution of PP2 being greater than that of PP3 also demonstrates that the lift force graph in Fig. 5 for PP2 is better than that of PP3 in most cases.

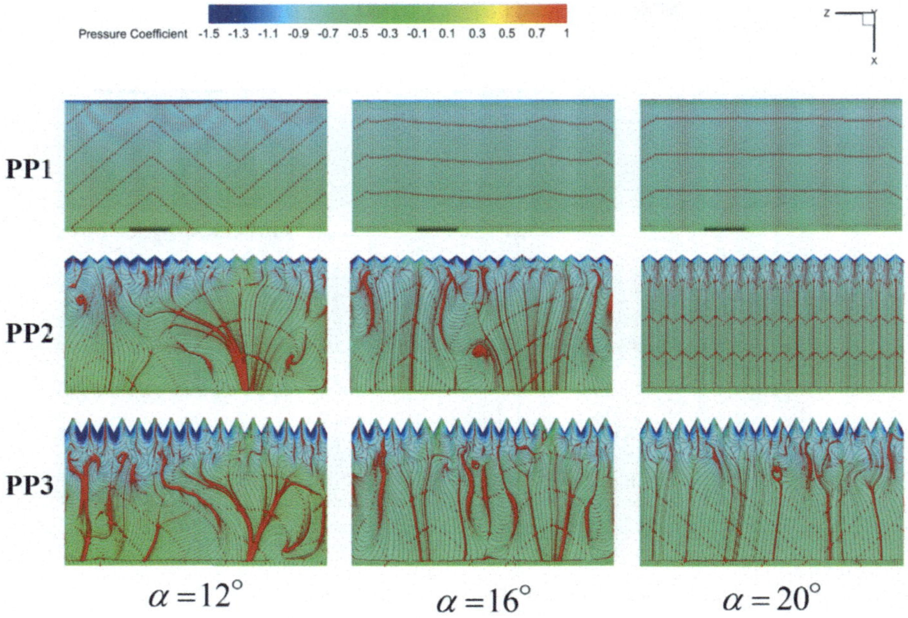

Fig. 7. Surface flow of three types of wings from 12° to 20°.

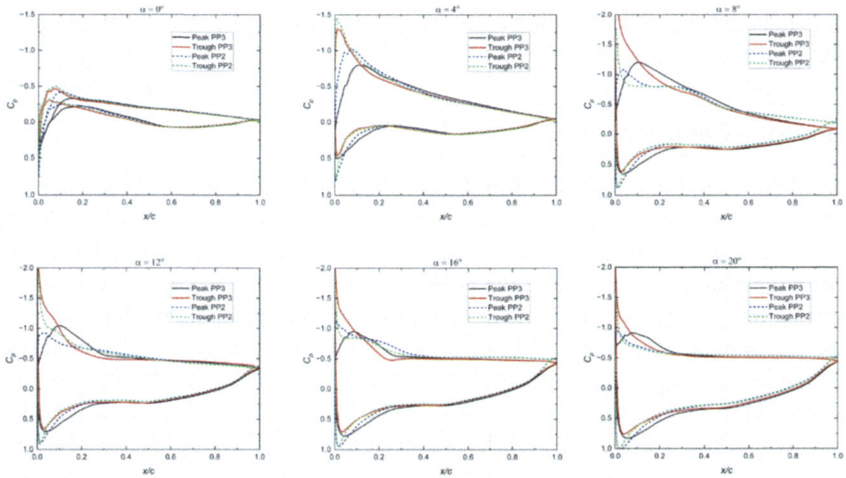

Fig. 8. Comparison of pressure distribution between PP2 and PP3.

4 Conclusions

In this investigation, the *k-ω* SST model was applied to predict the flow, as well as to calculate the performance of three types of aircraft wings PP1, PP2, and PP3 for angles of attack from 0° to 20°, and compared with previous research by Anyoji et al., leading to the following conclusions:

For the PP1 wing type, numerical simulation yields accurate results for angles of attack less than or equal to 6° with an error of less than 0.5%. The separation flow occurs at an angle of around 6°. However, the lift coefficient remains nearly constant for higher angle of attack, which is different from previous airfoil models.

The serrated wing profiles PP2 and PP3 have also demonstrated optimization in delaying the onset of stall angle, significantly improving the performance of the aircraft wing after the stall condition. At the same time, it also shows the relationship between amplitude and resistance by demonstrating the delay in the growth of drag.

In future study, the unsteady flow should be analyzed to understand the unsteady characteristics and dominant modes of the flow. However, further numerical schemes should be developed for this purpose.

References

1. Bangkui, F., Yun, L., Ruiyu, Z., Qiqi, F.: Review on the Technological Development and Application of UAV Systems. Chinese J. Electron. **29**(2), (2020)
2. Cai, C., Zuo, Z., Liu, S., Wu, Y.: Effect of a single leading-edge protuberance on NACA 634- 021 airfoil performance. In: 16th International Symposium on Transport Phenomena and Dynamics of Rotating Machinery, Conference (2016)
3. Wei, X., Li, D., Li, S., Chang, H., Fu, X., Zuo, Z., Wang, H.: Effect of leading-edge protuberances on swept wing aircraft performance. Int. J. Fluid Eng. **1** (2024)
4. Zhang, M.M., Wang, G.F., Xu, J.Z.: Experimental study of flow separation control on a low-Re airfoil using leading-edge protuberance method. Experim. Fluids **55**(4) (2014)
5. Cranston, B., Laux, C., Altman, A.: Leading edge serrations on flat plates at low reynolds number. In: 50th AIAA Aerospace Sciences Meeting including the New Horizons Forum and Aerospace Exposition, Conference (2012)
6. Mangeol, E., Ishiwaki, D., Wallisky, N., Asai, K., Nonomura, T.: Compressibility effects on Flat-Plates with Serrated Leading-Edges at a Low Reynolds Number. Experim. Fluids **58**(11), (2017)
7. Anyoji, et al.: Computational and experimental analysis of a high-performance airfoil under low-reynolds-number flow condition. J. Aircraft **51**(6) (2014)
8. Smith, T.A., Klettner, C.A.: Airfoil trailing-edge noise and drag reduction at a moderate Reynolds number using wavy geometries. Phys. Fluids **34**, 117107 (2022)
9. Avallone, F., Van Der Velden, W., Ragni, D., Casalino, D.: Noise reduction mechanisms of sawtooth and combed-sawtooth trailing-edge serrations. J. Fluids Mech. **848**, 560–591 (2018)
10. Hu, Y.S., Wan, Z.-H., Ye, C.-C., Sun, D.-J., Lu, X.-Y.: Noise reduction mechanisms for insert-type serrations of the NACA-0012 airfoil. J. Fluids Mech. **941**, A57 (2022)
11. Moreau, D.J., Doolan, C.J.: Noise-reduction mechanism of a flat-plate serrated trailing edge. AIAA J. **51**(10), 2513–2522 (2013)

Identify and Control the Speed of DC Motor Using Fuzzy-PI on Microcontroller Platforms

Tuan Anh Than Ngoc[1], Xuan The Nguyen[1], Dang Dang Xuan[2], Giang Le Bui Truong[3], Duc Le Huy[1], Van Dung Le[4], and Hai Xuan Le[1(✉)]

[1] International School, Vietnam National University, Hanoi 100000, Vietnam
{21070240,21070106,18071517,hailx}@vnu.edu.vn
[2] University of Engineering and Technology, Vietnam National University, Hanoi 100000, Vietnam
[3] Faculty of International Education, University of Transport and Communications, Hanoi 100000, Vietnam
giang882126013@lms.utc.edu.vn
[4] Hung Vuong University, Ho Chi Minh City 35000, Vietnam
levandung@hvu.edu.vn

Abstract. A DC (Direct Current) electric motor system, also known as a DC motor, is a type of motor that operates using direct current power and is propelled in a specific direction by electric current. In this article, the research team describes how they identified DC motor system models on the Arduino circuit platform by utilizing PI (Proportional Integral) and fuzzy PI algorithms in conjunction with the MATLAB Identification tool to control the motor speed. The algorithm evaluates the relationship between the input and output signals using experimental techniques. Subsequently, a PI and fuzzy PI controller for the motor speed is created. The results demonstrate how effectively the system's quality responds to the specified values while maintaining setup time.

Keywords: MATLAB Identification · PI and fuzzy PI controller · Arduino

1 Introduction

DC electric motor systems are widely used in constructions, agriculture and in vari-ous industrial settings. The control of engine speed is crucial for maintaining safety, preventing overloading, and averting hazards such as explosions and fires. However, through careful selection of system components, these issues has been effectively addressed. Therefore, automatic speed stabilization is crucial when operating larger systems that require precise control.

The biggest challenge in working with DC electric motor systems lies in determining the system model, as many motors available in the market lack accurate parameters due to label loss or other reasons. Finding the engine speed

© The Author(s), under exclusive license to Springer Nature Switzerland AG 2025
H. A. Le Thi et al. (Eds.): ICAMCS 2024, LNNS 1311, pp. 392–401, 2025.
https://doi.org/10.1007/978-3-031-90606-0_33

system was mentioned in [3] in order to determine the appropriate controller. The development of a control and monitoring system based on a microcontroller is discussed in [2].

In industrial applications, PI and fuzzy PI controllers are widely used for controlling DC motors. In [4], the author constructed a PI controller specifically for a DC motor, a fuzzy PI intelligent algorithm was employed in [1] to control the same DC motor. Both [4] and [1] produced favorable results with the systems employing PI and fuzzy PI algorithms. The research team then made the decision to construct a fuzzy PI and PI controller in order to regulate the speed of the DC motor model.

In order to evaluate the efficacy of intelligent control algorithms in stabilizing speed in DC motor systems, we will adopt an experimental system modeling approach in this article and proceed with a construction of a physical system.

2 System Model

We need to build the speed transfer function in order to derive the DC motor system model. In this process, the MATLAB identification toolbox will be employed to aid in obtaining the model of the complete DC system. An H-bridge circuit will be utilized to control the direct current motor. We plan to install a speed sensor to monitor the motor's speed. The open model of the system will serve as the foundation for developing the control law. The block diagram of the system open model, which includes the Arduino microcontroller, encoder sensor, DC electric system, and H-bridge circuit, is shown in Fig. 1.

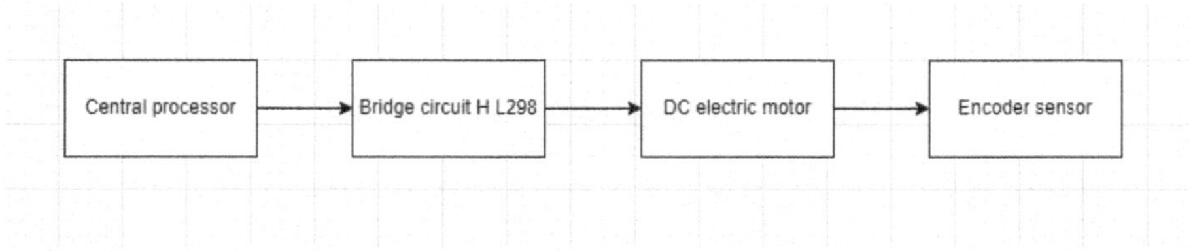

Fig. 1. System model diagram.

We will identify the system model using the MATLAB identification toolbox after sampling the data. The input and output data are carefully documented and saved in an Excel file after the test is finished, see Fig. 2. This well-organized documentation is necessary for a thorough examination of the test results. By serving as a centralized repository, the Excel file facilitates quick reference, comparison, and pattern recognition for subsequent analysis tasks.

First, we will export data in the Excel file to MATLAB's Workspace by using the XLSREAD() command as shown in Fig. 3. The input and output data, labeled as U1 and Y1, respectively, will be extracted from the Excel file and imported into MATLAB for further analysis and modeling.

	A	B	C	D	E	F	G
1			10ms	10	0.0129		
2				10	1.4468		
3				10	2.0854		
4				10	0.1542		
5				10	0.4694		
6				10	1.1129		
7				10	0.431		
8				10	0.4645		
9				10	2.1322		
10				10	1.3497		
11				10	1.8258		
12				10	0.4433		
13				10	0.7097		
14				10	0.0779		
15				10	0.6848		
16				10	1.2315		
17				10	0.2933		
18				10	0.5952		
19				10	1.3111		
20				10	0.0751		
21				10	0.777		
22				10	1.6535		
23				10	1.2363		
24				10	1.7817		
25				10	1.4749		
26				10	1.662		
27				10	2.3725		
28				10	1.6016		
29				10	1.0046		
30				10	3.0779		

Sheet1

Enter

Fig. 2. Data input and output in Excel.

```
>> u1=xlsread('C:\Users\le huy duc\Desktop\data',1,'D1:D700')

>> y1=xlsread('C:\Users\le huy duc\Desktop\data',1,'E1:E700')

>> save ident u1 y1
```

Fig. 3. Export data from Excel to MATLAB in the correct order.

We proceed by selecting the Time Domain Data option and providing the necessary input components, such as the name and sample time. In this case, the sample time is set to 0.001 s or 10 ms. As the configuration complete, we can now observe the visualized imported data in the "Time Plot" Window (Figs. 4 and 5).

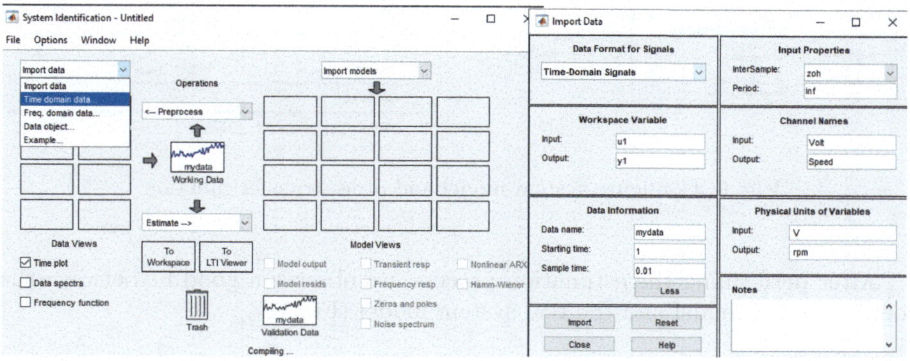

Fig. 4. Configuration of input and output data.

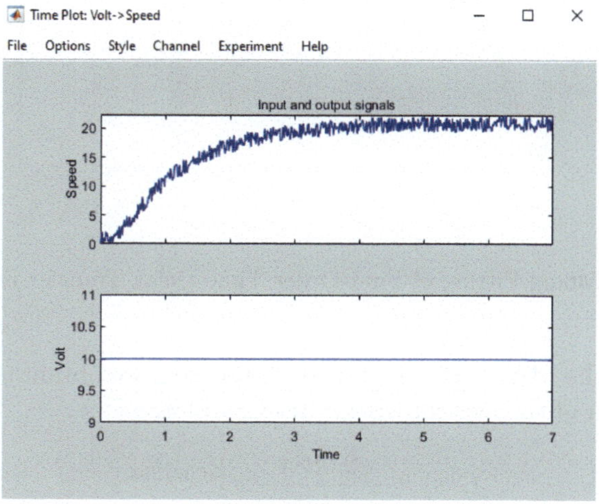

Fig. 5. Input U_1 (volt) and output Y_1 (speed) in time domain.

The next step is to determine the system transfer function by choosing "Process Model" and "Estimate" in the SIT System Identification interface. We will select the first-order transfer function model (Fig. 6).

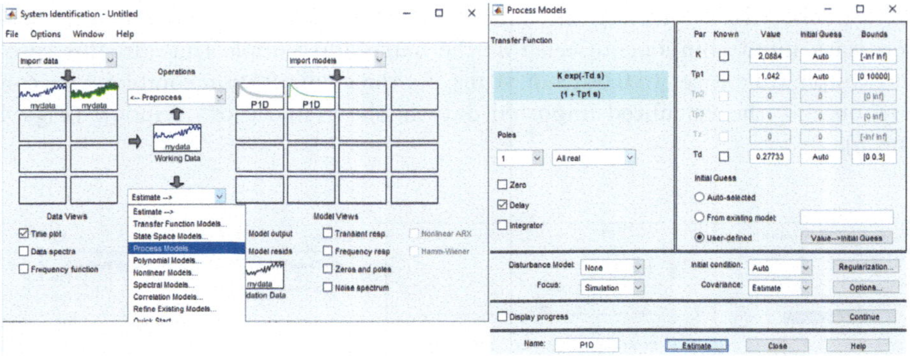

Fig. 6. Configure system model and order transfer function.

After performing the estimation process, we observe a good fit between the computational model and the real system model (Fig. 7).

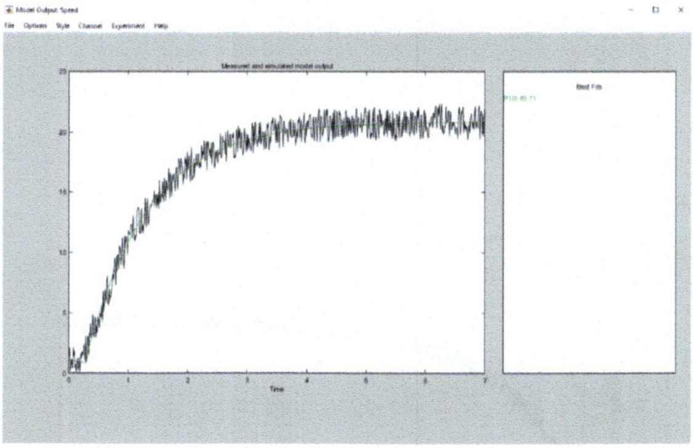

Fig. 7. Model Fitting of First-Order Time-Delay Transfer Function

After using the MATLAB Identification toolbox, we obtain the open system transfer function of the system with a fit of 85.71%:

$$G(s) = \frac{2.0884}{1 + 1.042s}.e^{-0.27733s} \tag{1}$$

Once the system model has been established, we can proceed with designing the PI controller as discussed in Part 3.

3 Design a DC Motor System Using PI and Fuzzy PI

3.1 Design the PI Control Algorithm

In order to use the PI algorithm to design a control system, we approximate the transfer function (1) to a second-order transfer function and calculate the parameters as below:

$$G(s) = \frac{2.0884}{1 + 1.042s}.e^{-0.27733s} = \frac{2.0884}{(1 + 1.042s)(1 + 0.27733s)} \tag{2}$$

The results after calculate the PI parameters is shown in Table 1

Table 1. PI parameter values

Method	Parametric Equation	Parameter value
Ziegler-Nichols 1	$K_p = \frac{1.2T}{kL} \approx 1.619$	$k = 2.0884$
	$T_I = 3.333L = 12.019$	$T = 1.042$
	$T_I = T_1 = 1.042$	$L = 0.27733$
Module Optimization	$K_P = \frac{T_1}{2kT_2} = 0.8995$	$T_1 = 1.042, T_2 = 0.27733$
	$K_I = \frac{1}{2kT_2} = 0.8632$	$k = 2.0884$

3.2 Design of a Fuzzy PI Control Algorithm

Fuzzy PI must be used in the control system design process. The control rules are defined using an "If Then" structure, and the input error (e) is measured as the "Difference between the output speed and the set-point variable". The computer applies rules and computes control signals to ascertain the armature voltage output after being converted into an input for a fuzzy controller. Using conventional techniques, analog controllers are created to implement the fuzzy control system. The basic rules controller is designed to be easily understood and maintained by non-expert users. The rules specify language variables as follows [5]: *NB*: Negative Big; *MN*: Medium Negative; *NS*: Negative Small; *ZE*: Zero; *PS*: Positive Small; *PM*: Positive Medium; *PB*: Positive Big.

We will perform defuzzification using the following two methods [6]: Maximum Method and Weighted Average Method.

Maximum Method. Firstly, we determine the domain containing the value y' with y' is the value at which $\mu_{B'}(y)$ reaches the max value:

$$G = \{y \in Y | \mu_{B'}(y) = H\} \tag{3}$$

Table 2. Fuzzy rules to control DC motor speed

E\CE	NB	MN	NS	ZE	PS	PM	PB
NB	NB	NB	NB	NB	NM	NS	ZE
MN	NB	NB	NB	NM	NS	ZE	PS
NS	NB	NB	NM	NS	ZE	PS	PM
ZE	NB	NM	NS	ZE	PS	PM	PB
PS	NM	NS	ZE	PS	PM	PB	PB
PM	NS	ZE	PS	PM	PB	PB	PB
PB	ZE	PS	PM	PB	PB	PB	PB

Then, we determine y' in one of three ways:
Average Principle:

$$y' = \frac{y_1 + y_2}{2} \tag{4}$$

Left-Bound Principle:

$$y' = y_1 \tag{5}$$

Right-Bound Principle:

$$y' = y_2 \tag{6}$$

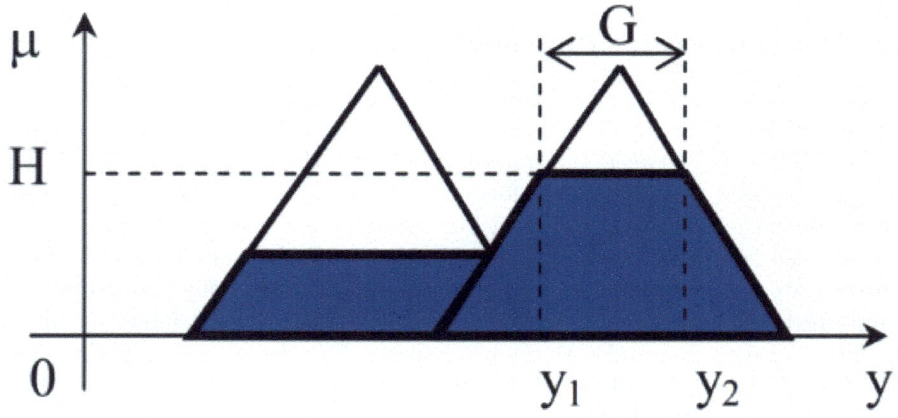

Fig. 8. Fuzzy composition rules.

Weighted Average Method. The defuzzification process also has an impact on the system's overall operational state and complexity. Since all control laws (Ri) participate equally and accurately in the defuzzification process, the centroid point technique is preferred. However, this approach is computationally intensive and ignores the compliance of the control clauses. Another limitation of this approach is the possibility that the key point we identify has very little or no dependency on the output. The domain of the output fuzzy values must be a linked function when establishing the membership function to prevent the issue mentioned above. The value y' is determined according to (3), where S is the domain of the fuzzy set B':

$$y' = \frac{\int_S y\mu(y)dy}{\int_S \mu(y)dy} \tag{7}$$

3.3 Software Design for Two Systems: PI and Fuzzy PI

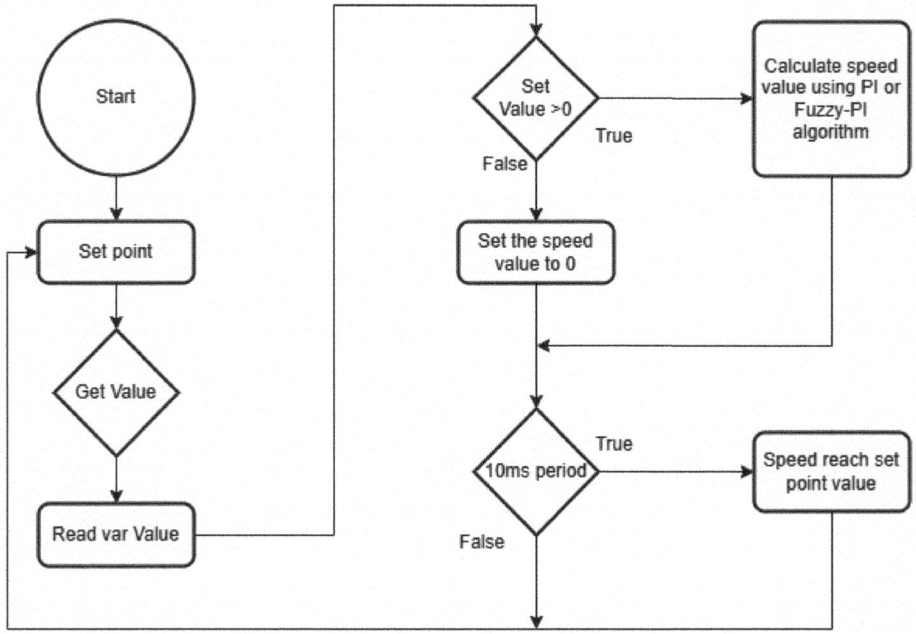

Fig. 9. PI or fuzzy PI controller structure diagram.

The implementation of the algorithms shown in Fig. 8 will utilize the Arduino microcontroller. By using the composition law from Table 2 or the parameters of the Ziegler-Nichols method 1, the coefficients ki and kp are computed from Table 1 and defuzzified using the maximum method. The central processing unit (CPU) of the Arduino is responsible for three primary tasks. The Arduino first reads and processes the value of the "`set_value`" system speed setting. To stabilize the motor speed, Arduino will recalculate and transmit the running speed value to the H-bridge circuit, which is based on the PI control algorithm and the variation between the value and the system's set value (`set_value`). This calculation will use the actual measurement (`rpw_value`) as its foundation. Every 10 ms, these three tasks will be completed. Finally, the serial monitor receives the values that the Arduino's register encoder reads (Fig. 9).

3.4 Experimental Outcomes

The achieved speed value of 150 rpm will be entered into the code (Set point). The speed values obtained through the Arduino serial communication can be used to generate a graph using MATLAB software. The experimental results will be made available through the following graphs (Figs. 10 and 11):

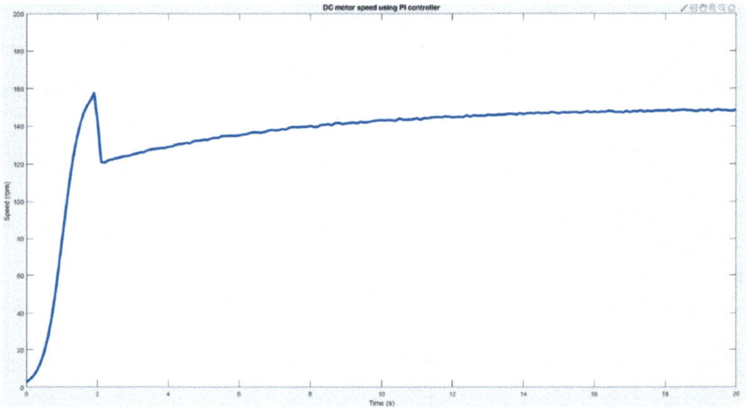

Fig. 10. The PI algorithm is used in the system response.

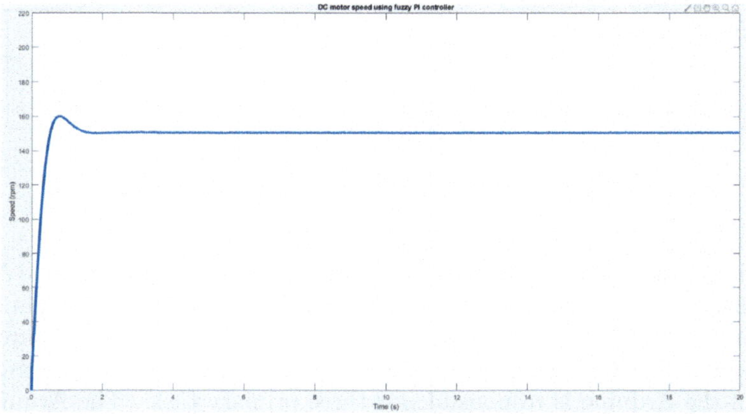

Fig. 11. The fuzzy PI algorithm is used in the system response.

Table 3 provides a summary of the two experimental findings utilizing the fuzzy PI and conventional PI algorithms.

Table 3. Comparison between PI and Fuzzy-PI control on the same system

Control System	F-PI	PI
Rise time	Fast, approximately equal 1.5 s	Quite slow, approximately equal 2 s
Overshoot	Small, approximately equal 4.5%	Small, approximately equal 4.7%
Static error	Negligible	Negligible
Stabilization	Quite stable to process disturbances	Quite Stable to process disturbances

4 Conclusion

The results obtained from simulations and experimental techniques unequivocally demonstrate the superior performance of the Fuzzy-PI controller when compared to the standard PI controller. Implementing intelligent controllers such as Fuzzy-PI is remarkably effective in achieving precise control for large capacity motors. This research adopts a comprehensive methodology that incorporates rigorous testing to tackle the challenge of DC electric motor identification. The simulation results are meticulously validated using MATLAB/Simulink software, confirming the robustness and effectiveness of the proposed approach. The diverse and multidisciplinary outputs of the project play a crucial role in facilitating the development of practical and impactful application drivers.

Acknowledgments. The author would like to expresses gratitude to Dr. Hai Xuan Le for their assistance and collaboration. The research group provided all the necessary funding for this research.

Disclosure of Interests. The author declares no conflicts of interest.

References

1. Almatheel, Y.A., Abdelrahman, A.: Speed control of DC motor using fuzzy logic controller. In: 2017 International Conference on Communication, Control, Computing and Electronics Engineering (ICCCCEE), pp. 1–8. IEEE (2017)
2. Javed, A.: Building Arduino projects for the Internet of Things. In: Experiments with Real-World Applications, pp. 15–34. Apress Media, LLC, USA (2016)
3. Wu, W.: DC motor parameter identification using speed step responses. Model. Simul. Eng. (1), 189757 (2012)
4. Zhao, X.: DC motor speed control system based on PI controller. In: IOP Conference Series: Materials Science and Engineering, vol. 677, p. 052007. IOP Publishing (2019)

Integrating Hand Tracking Technology for Enhanced Hand Rehabilitation: A Flappy Bird Game Approach

Tran Anh Vu[1], Nguyen Phan Kien[1], Mai Hoang Anh[1], Phan Tung Duong[1], Nguyen Minh Duc[1], Hoang Quang Huy[1], and Pham Thi Viet Huong[2](✉)

[1] School of Electrical and Electronic Engineering, Hanoi University of Science and Technology, Hanoi, Vietnam
{Vu.trananh,Kien.nguyenphan,Huy.hoangquang}@hust.edu.vn,
{Anh.mh213652,Duong.pt213661,Duc.nm213665}@sis.hust.edu.vn

[2] International School, Vietnam National University, Hanoi, Vietnam
huongptv@vnuis.edu.vn

Abstract. The hand is a part of the body that performs many important functions for humans. It is involved in most tasks and activities, especially in gripping and holding objects. At the same time, the hand is also a part that is easily injured due to frequent movement, collisions, and contact with various objects. Therefore, the purpose of this research is to establish a hand rehabilitation tool that helps patients increase their motivation for exercise and improve outcomes after rehabilitation treatment by using OpenCV to recognize hand gestures and convert them into actions for playing the game Flappy Bird. The popular designs such as the MediaPipe, and OpenCV are used to design hand movements, which in turn are used to develop the application named Flappy Bird game. The result of this program will be the gaming data of patients to assess the level of hand recovery, thereby supporting experts in making accurate evaluations of the patient's recovery capability and designing more intensive exercises.

Keywords: Game · Rehabilitation · Hand tracking · OpenCV · MediaPipe

1 Introduction

In recent years, the intersection of technology and healthcare has led to innovative solutions that revolutionize traditional rehabilitation methods. One of the most popular methods today is to use robotic arms to support exercise and rehabilitation exercises according to each stage of the patient's recovery [1]. One such promising avenue involves the integration of hand-tracking technology into therapeutic gaming applications. In the world, there is some research about combining play games in hand-rehabilitation. Some outstanding research like playing puzzles, controlling the falling blocks by moving up, down, left, and right according to your hand movements [2] or catch the falling object in VE. Using one hand to catch rabbit appear in one hole [3] or using EMG like quick contraction, prolonged contraction, the simultaneous contraction of flexor and extensor to control car in racing and control the direction of movement [4].

© The Author(s), under exclusive license to Springer Nature Switzerland AG 2025
H. A. Le Thi et al. (Eds.): ICAMCS 2024, LNNS 1311, pp. 402–417, 2025.
https://doi.org/10.1007/978-3-031-90606-0_34

In this study, our focus lies in investigating hand function recovery, specifically targeting hand grasping and opening. Building upon medical insights regarding the pivotal role of hand grasping and opening in the rehabilitation of hand function, especially is palm recovery. The rehabilitation of hand function represents a cornerstone in the multidisciplinary approach to stroke recovery, constituting a pivotal component in the holistic restoration of individuals' functional capabilities. Recognizing the intricate interplay between neurological plasticity and rehabilitative interventions, targeted exercises focusing on hand grasping and opening are meticulously designed to exploit the brain's remarkable capacity for adaptive change following injury. Through systematic and sustained practice, these interventions induce a cascade of neurophysiological processes, orchestrating structural and functional reorganization within the central nervous system [5]. Moreover, these targeted interventions serve to fortify hand functionality, bolster muscle strength, and notably augment both grip strength and dexterity. Such concerted efforts not only contribute to motor recovery but also play a critical role in fostering independence and ultimately elevating the overall quality of life for stroke survivors or Parkinson's patients. Moreover, to augment the efficacy of our rehabilitation model, real-time motion tracking capabilities are implemented to capture and analyze participants' hand movements during gameplay. This data is then securely stored and managed using Firebase, a robust cloud-based platform, facilitating seamless access and remote monitoring by healthcare professionals. Firebase's scalability and real-time synchronization features are leveraged to streamline the assessment of rehabilitation progress, allowing for timely adjustments to the treatment plan as needed. The synergy of gamification, advanced technology, and data-driven insights is harnessed in the innovative rehabilitation model, aiming to revolutionize the approach to hand function recovery. By enhancing the engagement, accessibility, and personalization of therapy, individuals undergoing rehabilitation are empowered to achieve optimal outcomes and regain independence in their daily lives.

2 Related Works

2.1 MediaPipe and OpenCV

In this study, MediaPipe and OpenCV are used to support the extraction of images and coordinates from hands.

MediaPipe is a collection of cross-platform, intervention-friendly, and extremely lightweight machine learning solutions. The advantages of this solution include providing fast inference, ease of installation and deployment, and being open-source and free. Some popular applications include face detection, face mesh, hand detection, and human pose estimation. In short, MediaPipe is a very powerful tool that is easy to use and works across platforms. Many individuals and organizations have used this toolkit to optimize their computer vision-related tasks due to its strengths.

OpenCV is the leading open-source library for computer vision and machine learning, and it now features GPU acceleration for real-time operations. It is widely used in many fields, such as street view imaging, automated inspection and monitoring, robotics and self-driving cars, medical image analysis, image/video search and retrieval, movies (3D structure from motion), and interactive installation art. OpenCV also includes many features and applications; this library can be divided into groups of features and modules such as image/video processing and I/O (core, imgproc, highgui), object detection (objdetect, features2d, nonfree), geometry-based monocular or stereo computer vision (calib3d, stitching, videostab), computational photography (photo, video, superres), machine learning and clustering (ml, flann), and CUDA acceleration (gpu).

2.2 Pygame

Pygame, a cross-platform set of Python modules, is utilized to create the Flappy Bird game tailored for hand rehabilitation. The game requires users to control the flapping motion of a bird through hand gestures, thereby promoting various hand movements and exercises. Pygame's versatility allows for customization of game mechanics and visual elements. The Flappy Bird game screen is not just a conventional entertainment game; it serves as a clear testament to the versatility and robustness of the Pygame library in creating unique and feature-rich gaming experiences. From managing entities such as the bird, pipes, and score, to handling player interactions, each aspect of the game is meticulously and effectively constructed. Controlling the bird entails more than arbitrary mouse clicks; it demands player concentration and swift reflexes to avoid collisions with pipes. The player holds the hand to click the mouse cursor on the screen to control the flappy bird to go up and open hand to control bird go down. Flexibility and acumen become decisive factors in navigating through obstacles or facing defeat. The scoring system transcends mere abstract numbers displayed on the screen. Rather, it constitutes a clear reflection of the player's performance and skill level in the game. Each achieved score represents an accomplishment, an expression of effort and the ability to overcome challenges. When a collision occurs, it signifies not just the end of the game but also serves as a reminder of the caution and patience required to meet expectations. Players learn from each failure, enhancing their skills and strategies to strive for higher scores in subsequent attempts. This unequivocally underscores the competitive and engaging nature of the game, as each player engages in a race to achieve their best performance.

2.3 Tkinter

The Tkinter user interface is not merely a simplistic smiley window; rather, it serves as a sophisticated platform with robust interaction capabilities and integrated features. This facilitates the convenient and efficient collection of user personal information, transcending mere data input to become an integral part of the user experience. Every element on the interface is meticulously designed to optimize user experience. From arranging input fields logically to selecting appropriate colors and icons, every detail is considered to create a user-friendly interface. More importantly, the combination of intuitive interface and input functionality not only provides flexible access for users but also demonstrates the flexibility and utility of the application. Users can effortlessly input their personal information, such as name, gender, and date of birth, without encountering obstacles or barriers in the process. Additionally, storing user information is not merely about recording data but also entails care and respect for their privacy rights. From inputting phone numbers to storing input timestamps, each step is designed to provide convenience and reliability for users. In this manner, users can rest assured that their personal information is protected and handled safely and reliably.

2.4 Distance and Angle in the Kinematic

For the evaluation of finger flexion to calculate hand recovery levels, we propose a formula based on the coordinates of landmarks provided by Mediapipe to calculate the angle between the vectors connecting the joints. Each finger has key landmarks at the joints: MCP (the joint between the finger and the hand), PIP (the joint between the proximal and middle phalanges of the finger), DIP (the joint between the middle and distal phalanges), CMC (the joint between the wrist bone and the metacarpal bone), and finally the Tip (the fingertip). To calculate the flexion angle of the finger or hand, we need to use the points: MCP, PIP, DIP, and finally CMC. The formula for calculating the angle between the fingers uses the coordinates of the landmarks to compute the angle between the vectors connecting the joints. We use the formula for the dot product of vectors. The formula is expressed as follows:

$$\theta = \arccos(\frac{v1 \cdot v2}{||v1||||v2||})$$

In which, vector v1 is calculated from the coordinates of the first landmark to the middle landmark containing the coordinates of 3 points in space, for example, from MCP to PIP, and vector v2 is calculated from the middle landmark to the last landmark, also containing the coordinates of 3 points in space, for example, from PIP to DIP. This will allow us to calculate the flexion angle of the finger at the PIP joint.

For the thumb, it is a special case, as it cannot be calculated in the usual way and requires the use of more specific formulas [6-10]. In research, we find this theory that is utilized to detect the position of hand marks and the angles of fingers when the palm is open or closed, with a particular focus on the movement and angle of the thumb and index finger. The model described in [8] and [9] is implemented in this research, with some modification. Forward Kinematics: The kinematics of the hand is described by using the modified Denavit–Hartenberg (MDH) parameters introduced by Craig [7]. CMC abduction angles ϑ^{I}CMC, a for fingers i = II–V are fixed; ϑ^{I}CMC, a is the angle between the finger axis and the base of the hand; and ϑ^{III}CMC, a = $\pi/2$ by definition.; θ^{III}CMC, f = 0 by definition since CMC flexion is measured with respect to the direction of the third metacarpal bone and, θ^{I}MCP, f and θ^{I}IP,f, f for the thumb are not shown since they are analogous to θ^{I}MCP, f and θ^{I}DIP,f. The torsion of the thumb around its axis is measured by the twist angle α^{I}MCP. Thumb twist is not a fixed angle. The torsion of the thumb is observed to change as it is flexed toward the palm; however, a quantitative expression for this phenomenon has not been found in the literature. Good results were obtained by using the following approach:

$$\alpha^{l}_{MCP} = \alpha\left(\theta^{l}_{CMC,f}\right) = \frac{\alpha_0}{2}\left(2 - \frac{\theta^{l}_{CMC,f} - \theta^{l}_{CMC,f,min}}{\theta^{l}_{CMC,f,max} - \theta^{l}_{CMC,f,min}}\right)$$

where α_0 is the torsion angle measured when the thumb is extended.

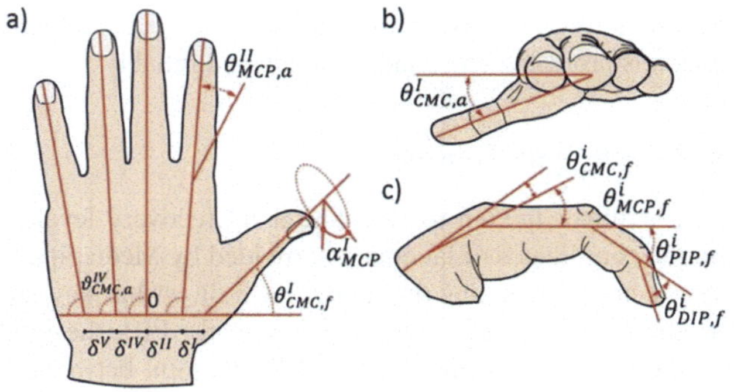

Fig. 1. Angles are defined in the kinematic model of the hand.

Figure 1 shows the hand angles used for calculations during gameplay.

Figure 1(a) Fixed abduction angles ϑ^{I}CMC, a and mobile abduction angles θ^{I}MCP,a for fingers II–V (the same Greek letter, but different symbols have been used to distinguish fixed and mobile angles). Angles are shown only for some of the fingers, the others being others being analogous. α^{I} MCP is the twist angle of the thumb. The flexion angle of the thumb at the CMC joint is indicated with θ^{I} CMC, f. The distance between finger bases is indicated with δ^{I}. Figure 1 (b) Thumb CMC abduction angle θ^{I}CMC, a seen from fingertips. Figure 1 (c) Flexion angles for fingers i = II–V. For the thumb, θ^{I}MCP, f and θ^{I}DIP,f are analogous to θ^{I}MCP, f and θ^{I}DIP,f; hence, they are not shown. θ^{I}CMC, f is measured with respect to the direction of the third metacarpal bone, chosen as indicator of the global orientation of the hand; hence θ^{III}CMC, f = 0 always by definition.

When the hand is closed, twist angle α^{I} MCP depends on the flexion angle (θ^{I}CMC, f) at the CMC (Carpometacarpal) joint of thumb. As the thumb towards the palm like θ^{I}CMC, increases from minimum value to maximum value the twist angle α^{I} MCP changes according to the formular in the part before. The hand is intended to be tracked when it opens, as shown in Fig. 1 (a), and the distances between the joints of the fingers were measured for analysis. Angles denoted as (θ^{I}MCP, a) when I equal II to V. It represents the movement of angles at the joints between the finger bones and the palm (metacarpophalangeal joint-CMP and carpometacarpal joint – CMC). The distances ϑ represent for distances between the bases of the fingers. Because of angles of fingers when the palm open or close is different so that, the angle of thumb and other fingers are calculated when palm close, the angle (θ^{I}CMC, c) represents for the rotational and flexion movements of the fingers when the hand is closed, and all is represented by Fig. 1(b). However, detecting whether the palm is open or closed is not sufficient. It was determined that calculating the angles of all the fingers, particularly the flexion of the thumb and index finger, is necessary when the palm is closed. Angles: θ^{I}CMC, f, θ^{I}MCP, f, θ^{I}PIP, f and θ^{I}DIP, f indicate the flexion at different points of the thumb and index finger when the fingers bend inward toward the palm and Fig. 1(c) represents for this.

2.5 Data Storage

Firebase is a cloud platform developed by Google that provides many services for building web and mobile applications, offering various tools such as storage, real-time databases, cloud functions, and authentication. Firebase allows developers to focus on building applications without worrying about the underlying infrastructure. With Firebase, we can easily build scalable, secure, and high-performance applications that can reach a global audience. It is commonly used to build real-time data-driven applications and is especially popular for developing mobile and web apps with real-time data synchronization features.

3 Methodology

Figure 2 shows the block diagram of the proposed game's system.

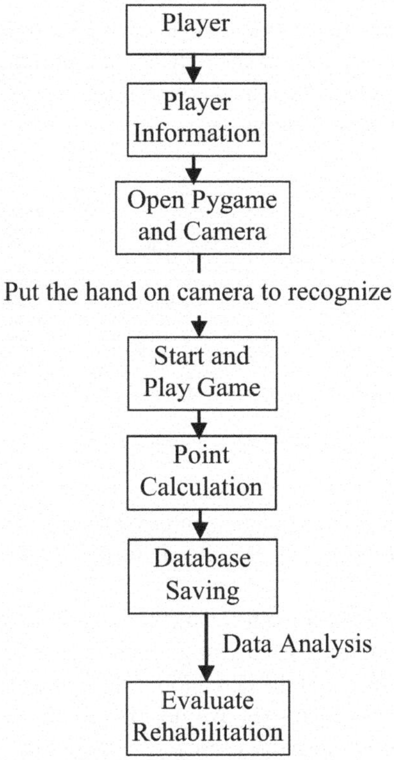

Fig. 2. The block diagram

First, to begin, the player will enter some basic information for future sessions, including phone number, name, age, etc., as a step to distinguish between different players. After entering the information, the game interface will appear, and the camera will be activated. Then, the player will hold their hand in front of the camera at a distance of about 40 cm and start playing. Clenching the fist will make the bird fly up, while opening the hand will make the bird fall freely. When the bird hits an obstacle, the game will pause to calculate the player's score for that round, while the data will be saved and sent to the server for doctors to analyze and assess the patient's recovery progress through different sessions. The more the game is played, the more results are collected, making it easier and more objective to evaluate the patient's recovery level. This study leans towards supporting hand rehabilitation as it provides a more visual way to track recovery progress. Patients can directly assess and observe their own improvement. Therefore, to optimize the rehabilitation process, we recommend combining traditional methods with these new studies, which will make the rehabilitation process less monotonous.

3.1 Detect Hand and Hand-Landmarks on Camera

The patients will sit in front of the camera and raise their hands so that the camera can capture their hands and display them on the screen. At this stage, the OpenCV and Media-pipe libraries will assist us in identifying each joint point of the fingers and hand in the image and tracking every hand movement. In addition, the coordinates of each pixel point of the hand can be determined whether the hand is clenched or open, even without the camera capturing it clearly.

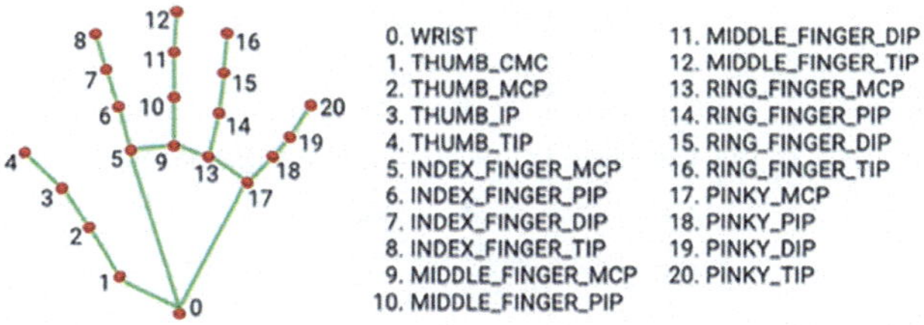

Fig. 3. 21 hand landmarks

Twenty-one 3D landmarks of a hand are inferred from just a single frame of machine learning (ML) shown in Fig. 3 [11]. Details about the detection technique are deeper presented in [12].

In this study we used Media-pipe to identify the hand land-marks. Twenty-one 3D landmarks of a hand are inferred from just a single frame of machine learning (ML) shown in Fig. 3 [11]. Details about the detection technique are deeper presented in [12]. After consulting some research [6–10], We decided to choose land marks 6,10,14,19 and 8,12,16,20 to determine the palm closes or opens. The reason for this choice is because points 8,12,16,20 represent the fingertips, when the hand is clenched, the above points will change position, specifically they will point towards the palm of the hand and points 6,10,14,19 when the hand is clenched, their rotational moment will also change, changing the twist angle.

3.2 Calculate the Distance Between Landmarks, Angle of Fingers and Detect Hand Grip

The input data consists solely of hand images and movements, which is insufficient to determine the speed of hand recovery. Therefore, a technique is introduced to compute the distance between each finger, the distance from the top-most pixel point of the finger to the lowest point of the hand, and the angles formed by the fingers during hand-gripping activity. We develop a formula based on input data derived from the pixel positions captured by Mediapipe and then use vector calculation formulas to determine the distance between the fingers and the wrist point, helping to identify whether the action is gripping or opening the hand. Additionally, we use the vectors produced and the formula proposed in Sect. 2.4 to calculate the bending angles of the fingers. In the angle computation step, particular emphasis is placed on finding the angle between the thumb and index finger, as illustrated in Fig. 3. By analyzing these distances and angles, it becomes possible to assess whether a patient is regaining the ability to fully grip their hand. The assessment will be based on theoretical indicators of hand grip. Subsequently, the distance from the highest point of the finger to the lowest point will be calculated. This data aids in determining whether the hand grip is tight, moderate, or extended, and whether the handshake is tight or moderate, or the hand is extended.

3.3 Using Hand Movements to Control Game, Extract Data and Evaluate Data

In a typical game, players usually use a mouse to click or the spacebar on the keyboard to control the bird to jump. However, in this study, we use a hand-closing and opening mechanism to control the bird as it flies through obstacles. Closing the hand mimics a mouse click. The player places their hand in front of the camera at a distance of around 40 cm, with an allowable error margin of about 15 cm to ensure flexibility and high accuracy. This approach enables the system to function as a rehabilitation aid for patients needing hand rehabilitation, making the process less monotonous and dull compared to traditional methods, which can often be somewhat dry and repetitive. This method not only helps maintain the player's motivation but also enhances the effectiveness of the rehabilitation process by requiring dexterity and focus.

3.4 Personal Information Management

We did not focus too much on building personal information management, so we only created a GUI window using Tkinter that allows users to input basic, simple information such as name, age, and date of birth. After the player enters the basic information and starts the game, at the end of their play session, the gameplay data is transferred to a JSON file and directly saved to Firebase. This data includes the player's initially entered personal information, the game results, extracted hand data, and an assessment of the patient's recovery progress.

Calibration

Central to ensuring the precision and accuracy of the acquired hand data was the meticulous implementation of a calibrated process. This calibration regimen, comprising multiple steps, was meticulously devised to mitigate potential sources of error and uncertainty inherent in the imaging and measurement processes.

- Placement of Reference Marker: In the palm of the hand, a subject's, that is, there is a small 2 cm x 2 cm black square decal. No more, no less. This just acts as a marker of some type for references in subsequent measurements and comparisons.
- Positioning of the Hand: The subject's hand, the decal is already in place, is put 40 cm in front of the camera. Such standardization ensures that there would be uniformity in the process of collecting data by having the same point of reference all along.
- Image Capture and Initial Analysis: It records an image of the hand with the decal in place. The captured image is processed by a computer to determine what fraction of pixels within the image value the black decal and how many depict dimensions of the entire hand, generally.
- Calculation of Actual Hand Size: Using the known size of the black decal (2 cm x 2 cm), the calculated pixel ratio is applied to derive the actual hand size data. This step ensures accurate calibration and normalization of the acquired image data.
- Recalibration for Distance Variations: If there are variations in the distance between the camera and the subject's hand during data collection, a recalibration process is initiated. This involves reassessing the pixel ratio between the decal at the new distance and the original reference image. The recalculated ratio is then used to compute the precise dimensions of the hand at each respective distance.
- Ensuring Consistency and Accuracy: By adhering to this meticulous calibration process, consistency and accuracy in the hand measurements are maintained across various experimental conditions. This framework ensures robust and reliable acquisition of hand data, supporting the integrity and validity of subsequent analyses and findings.

The experimental protocol entailed the participation of a cohort comprising 5 individuals, each serving as subjects for meticulous assessment aimed at scrutinizing the fidelity of hand data obtained vis-à-vis their authentic physical characteristics, encompassing size, orientation, and spatial positioning. To comprehensively explore the robustness of the proposed methodology, experiments were conducted under varying conditions, including controlled room illumination and natural lighting within an expansive environment, at three discrete distances separating the camera from the subject's hand: 30 cm, 40 cm, and 50 cm. This deliberate variation in environmental factors and spatial configurations aimed to account for real-world scenarios, thereby enhancing the generalizability and practical applicability of the findings.

Table 1 shows the calibration's results.

Table 1. The length measurement results from wrist to middle finger in the extension condition in the experiment.

Volunteer	Condition	Results in distance 30 cm (cm)	Results in distance 40 cm (cm)	Results in distance 50 cm (cm)	Actual size (cm)
01	Room light	18.4	18.7	18.6	18.7
	Natural light	18.5	18.8	18.5	
02	Room light	18.2	18.1	18.3	18.0
	Natural light	18.4	18.2	18.0	
03	Room light	17.9	17.5	17.8	17.6
	Natural light	17.6	17.4	17.4	
04	Room light	14.0	14.3	14.3	14.2
	Natural light	14.1	14.2	14.3	
05	Room light	13.9	13.8	13.9	13.9
	Natural light	14.0	14.1	13.9	

4 Experiment Results and Discussion

Experiment

There are some main points that participants should notice. We give technical requirements and explain how to play. Those requirements are for patients in rehabilitation upper limbs.

Technical requirements: the laptop with Windows 10 or higher is required, which has a webcam and the game be installed on it. Each player stays about 100 cm apart. Players will sit straight in front of the laptop and hold their hands up so that the webcam can detect the image of their hands. We decided to let the patients play in a room with white, fluorescent light that was not affected by sunlight. The instructors explained how to play the game to the patients, how to score, the conditions needed to participate in the game and conducted a trial game for the patients.

How to play:

Holding the hand is equivalent to controlling the bird to fly up or releasing the hand is equivalent to controlling the bird to descend during the process of controlling the bird's flight path to avoid the bird from crashing into the pipe.

When the bird passes 1 pipe you get a point. If you control the bird to hit the pipe the game will over, and the score will be displayed on the screen. Then, you can press the screen again to play again and the score will be saved to the system.

Figure 4 shows the webcam screen and corresponding Flappy Bird game controls.

Participants: We conducted the game with the participation of 5 healthy people without any medical conditions and 5 people in the process of rehabilitation with each stage of recovery. We will give each person 10 times and will take the average result of 10 measurements to evaluate.

Fig. 4. Movement of bird after griping a hand

Results

The results obtained through the gaming process will include videos of the player's finger and hand movements, providing suggestions for assessing hand recovery levels based on the player's score. For videos tracking the player's movements, patients can carry out rehabilitation at home or anywhere else, allowing their movements to be recorded and sent to doctors in different locations for monitoring. Doctors can fully observe the patient's movements remotely to assess their recovery level and design subsequent exercises based on their recovery progress.

Regarding recovery results based on game scores, games are designed with levels that progressively increase in difficulty. This approach allows for a more structured and incremental assessment of the patient's recovery, providing a clear indication of improvement as they advance through the levels. At easier levels, the game allows patients to move slowly and gently, suitable for beginners in the recovery process. At more challenging levels, the game requires higher demands on finger and hand movements, closer to normal movement levels to pass. As a result, the game outcome will relatively accurately reflect the hand's recovery level based on the player's progress through the levels.

The provided data will be the results of players over 10 sessions per day in 3 days, after which we will calculate the average result to compare the recovery level of patients in the trial. A game score below 10 points will correspond to a recovery rate of approximately 40%-50%, with the required hand movement increasing progressively at scores of under 20, under 30, and over 30, which correspond to recovery rates of 50%-70%, above 70%, and full recovery (normal). At these levels, the player's score clearly reflects the recovery stage of the patient. The subjects, who are hand rehabilitation patients, were assessed for their recovery level by doctors. When compared with the results during the trial, we can see that the game results relatively estimated the patients' recovery level, thereby assisting doctors in more easily evaluating the recovery progress (Table 2).

Table 2. Results of average game score between 5 normal individuals and 5 patients undergoing rehabilitation in 3 days.

	Mean score(day 1)	Mean score(day 2)	Mean score(day 3)
Volunteer 1(Normal)	33.5	35.5	35.6
Volunteer 2(Normal)	34.1	36.5	37
Volunteer 3(Normal)	32.2	35.2	36
Volunteer 4(Normal)	35.6	37.1	38.5
Volunteer 5(Normal)	34.9	37.3	37.6
Volunteer 6 (80% Recovery)	26	28.4	30.5
Volunteer 7 (70% Recovery)	24.2	26.2	28.1
Volunteer 8 (60% Recovery)	13.2	14.8	16.2
Volunteer 9 (50% Recovery)	8.1	9.4	10.7
Volunteer 10 (40% Recovery)	6	7.1	8.1

The game data results show that normal individuals can easily surpass the 30-point mark, while others will find it more challenging. This is because at this score level, the game requires more flexible movements, something that patients may struggle to achieve. Similarly, at the 20-point mark, the game also requires a certain degree of flexibility in movement, meaning only those with a recovery level of 70% or higher can meet this requirement. Scoring 10 points or below also provides a relatively accurate assessment for participants as described above.

For results obtained by tracking pixels extracted from the palm, these pixels are monitored and recorded to assess the flexion of the fingers. This data provides valuable insights into the patient's finger movements and helps evaluate their recovery progress. This result becomes a useful parameter for doctors to provide comments on the recovery outcomes. The parameters will include pixel data from the fingers, which will be used to calculate the degree of flexion, extension, abduction, adduction, and rotation of the fingers. These results will serve as a support for experts and doctors to make a final assessment of the patient's recovery level (Table 3).

The results are obtained using pixel data on the hand generated by MediaPipe to calculate the flexion of a finger. The angles between the segments connecting the MCP-PIP (metacarpophalangeal-proximal interphalangeal) joints and the PIP-DIP (proximal interphalangeal-distal interphalangeal) joints of each finger are calculated. After computing the angles for each joint, the average of these angles is taken to determine the overall flexion of the hand. This method provides a comprehensive measure of hand flexion during movement.

Discussion

The table contains three columns that representing the performance of five normal people and five patients with different progress of rehabilitation (50%, 40%, 30%, 20%, 10%) in three days play this game. For normal people, the results achieved are quite high and

Table 3. Results of comparison of finger flexion and extension movements between 5 normal individuals and 5 patients undergoing rehabilitation.

Number order	Flexion (Degrees)	Extension (Degrees)
Volunteer 1(Normal)	60	25
Volunteer 2(Normal)	62	27
Volunteer 3(Normal)	58	26
Volunteer 4(Normal)	61	28
Volunteer 5(Normal)	59	25
Volunteer 6 (80% Recovery)	48	20
Volunteer 7 (70% Recovery)	42	18
Volunteer 8 (60% Recovery)	36	15
Volunteer 9 (50% Recovery)	30	12
Volunteer 10 (40% Recovery)	24	10

much higher than those who are in the process of rehabilitation. It can be easily seen that the average score of the highest normal person will be 4.5 times higher than that of the person who has recovered 10% of hand function (6–8 points), 3.8 times higher than that of the person who has recovered 20% (10–12 points) of hand function and 2.5 times higher than that of the person who has recovered 30% (13–16 points) of hand function and nearly 1.5 times higher than that of the person who has recovered 40% and 50% (22–28 points) of hand function. The reason why people with 10% and 20% hand function recovery have low results as shown in the table above is because we have set the difficulty level of the game to increase gradually with every 10 points achieved, so with the easiest level, most players can achieve 10 points, but it is quite difficult for those who have just recovered 10% and 20%. However, by participating in our game, it is easy to see that the patient's score has increased compared to the first day. Specifically, even patients with 10% and 20% recovery achieved at least 7 points on the first day and their scores improved significantly on the second day (from 7 to 10 points) and increased slightly on the third day (10 to 11 points). Applying our game to hand rehabilitation helps improve the condition and speed of hand recovery of patients and that has been shown through the increase in patient scores day by day. However, the increase in scores of patients and normal people may be due to their gradual familiarity with our game, which makes them score higher than before (Fig. 5).

As for patient outcomes over the days, we can see that for most patients there is growth in terms of outcomes. This shows that the patient is gradually regaining his hand movement skills, reflecting that the recovery process is progressing positively over the days. For the group of patients with a high degree of recovery (80% and 70%), we can clearly see through the mean value of day 1 compared to day 3 is 26.and 30.5, 24.2 and 28.1. Patients with high levels of recovery tend to achieve a greater increase in average outcomes, indicating that they make rapid progress in recovery. For the group of patients with low recovery (50% and 40%), the values were, respectively 8.1 and 6 on day 1 and

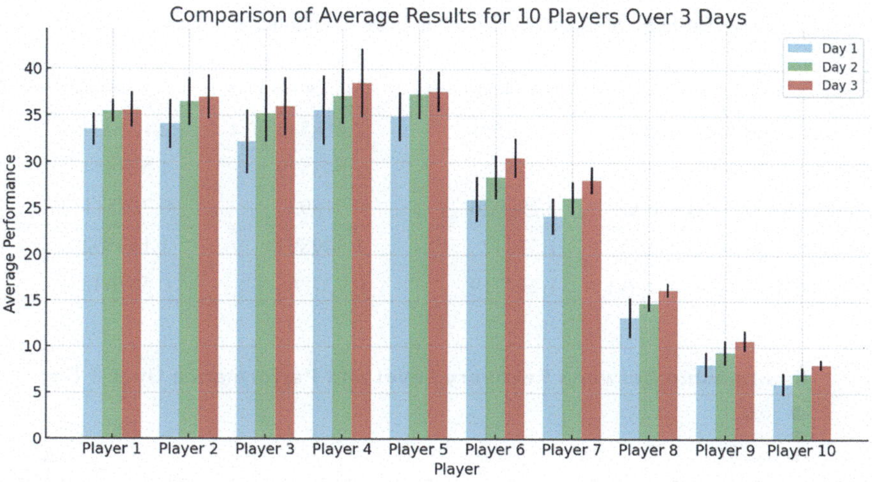

Fig. 5. Illustrate the average score and Standard deviation of 10 volunteers in 3 days

10.7 and 8.1 on day 3. These patients may still see improvement but may be at a slower rate, reflecting limited progression due to low levels of recovery (Table 4).

Table 4. Average score of patients in 3 days

	Mean score(day 1)	Mean score(day 2)	Mean score(day 3)
Volunteer 6 (80% Recovery)	26	28.4	30.5
Volunteer 7 (70% Recovery)	24.2	26.2	28.1
Volunteer 8 (60% Recovery)	13.2	14.8	16.2
Volunteer 9 (50% Recovery)	8.1	9.4	10.7
Volunteer 10 (40% Recovery)	6	7.1	8.1

For the standard deviation of the values over 3 days, the decrease in the standard deviation over the days indicates that the patient's results are gradually becoming more stable. This reflects the increased control and coordination of their hand movements. When the results are less volatile, it can be concluded that the patient is gaining better hand control and gradually regains stability in operation. With a high recovery group (80% and 70%), these patients were able to achieve stability sooner, indicating faster hand control and recovery. The Standard deviation of patients with low recovery (50% and 40%) decreases but may be slower, reflecting that they are on the road to recovery, but still have difficulty maintaining stable results (Table 5).

Overall, for the group of patients with a high level of recovery (80% and 70%), we saw an increase in mean outcomes and a rapid reduction in standard deviation, indicating that they had made significant progress in restoring hand function. These patients were able to complete the exercises well and improve stability in performance.

Table 5. Standard deviation of score of patients in 3 days

	Std Dev(day 1)	Std Dev(day 2)	Std Dev (day 3)
Volunteer 6 (80% Recovery)	2.538591	2.458545	2.173067
Volunteer 7 (70% Recovery)	2.043961	1.813529	1.523884
Volunteer 8 (60% Recovery)	2.250926	0.918937	0.788811
Volunteer 9 (50% Recovery)	1.37032	1.349897	1.159502
Volunteer 10 (40% Recovery)	1.247219	0.737865	0.567646

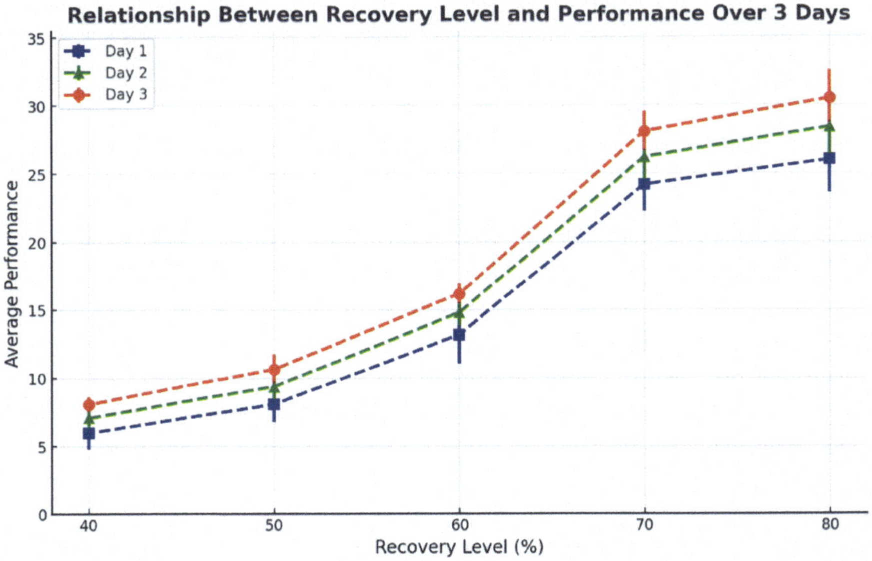

Fig. 6. The relation between recovery level and performance of patient over 3 days

For the intermediate group (60% and 50%), they have an improvement, but the pace may be slower, especially in the early days. The gradual decline of the standard deviation suggests that they are gradually regaining better hand control, but the recovery may take longer. The last group is the low recovery group, which may see a small increase in mean outcomes, but a slower decrease in standard deviation indicates a slow recovery and unstable outcomes (see Fig. 6).

5 Conclusion

Limitation of our study is that when comparing the assessment results with experts, there is still an error when the hand recovery level is only 30% or less, but when the hand is lightly clenched or stretched at 30% or more, our system can accurately assess when comparing. The reason for this challenge is that when the patient's grip strength is at

30% or less, difficulty in gripping is experienced, and the libraries being used, such as OpenCV and Media Pipe, often struggle to accurately assess the specific grip level, with the hand frequently being misinterpreted as fully straight.

In the future, these limitations will be addressed by proposing an automatic system through which appropriate exercises will be identified and provided, exercise results will be tracked, and hand rehabilitation progress will be monitored remotely. This system would eliminate the requirement for both doctors and patients to be physically present at medical facilities.

In conclusion, in our study, by allowing patients with different levels of hand function to play and normal people to play, after three days of playing, the scores of both normal people and rehabilitation patients increased. This shows that our game system is very easy to access and use, and for patients, their average scores gradually increased over the days of playing, contributing to proving the usefulness of this game in hand function rehabilitation. Although there are still limitations that we have mentioned above, in the future we will improve this product and overcome the above limitations as well as implement this model more widely.

References

1. Yue, Z., Zhang, X., Wang, J.: Hand rehabilitation robotics on Poststroke motor (2017)
2. Pastor, I., Hayes, H.A., Bamberg, S.J.M.: A feasibility study of an upper limb rehabilitation system using Kinect and computer games (2012)
3. Burke, J.W., McNeill, M., Charles, D., Morrow, P., Crosbie, J., McDonough, S.: Serious games for upper limb rehabilitation following stroke. In: 2009 Conference in Games and Virtual Worlds for Serious Applications (2009)
4. Cosima Prahm 1 Author Orcid Image ; Ivan Vujaklija 2, 3 Author Orcid Image ; Fares Kayali 4 Author Orcid Image ; Peter Purgathofer 4 Author Orcid Image ; Oskar C Aszmann 1 Author Orcid Image, Game-Based Rehabilitation for Myoelectric Prosthesis Control
5. Martinez, R., Garcia, M., Lopez, A.: Neural mechanisms underlying the effects of hand rehabilitation exercises on motor recovery post-stroke: a neuroimaging study. Neurorehabil. Neural Repair **38**(3), 210–220 (2023)
6. Yasumuro,Y., Chen, Q., Chihara, K.: 3D modeling of human hand with motion constraints. In: Proceedings of International Conference Recent Adv. 3-D Digit. Imag. Modeling, pp. 275–282 (1997)
7. Kumar Parida,P., Bhusan Biswal, B.: Design and analysis of a multifingered robot hand. IAES Int. J. Robot. Autom., **1**(2), 69–77 (2012)
8. Cobos,S., Ferre, M., Sanchez Uran, M.A., Ortego, J., Pena, C.: Efficient human hand kinematics for manipulation tasks. In: Proceedings of the IEEE/RSJ International Conference on Intelligent Robots and Systems, pp. 2246–2251 (2008)
9. Zhou, J., Yi, J., Chen, X., Liu, Z., Wang, Z.: BCL-13: A 13-DOF soft robotic hand for dexterous grasping and in-hand manipulation. IEEE Robot. Autom. Lett. **3**(4), 3379–3386 (2018)
10. Craig,J.J.: Introduction to Robotics, Mechanics and Control, 3rd ed. Pearson, London (1986)
11. MediaPipe. Hand landmarks detection guide. https://developers.google.com/mediapipe/solutions/vision/hand_landmarker
12. Zhang, F., et al.: Mediapipe hands: on device real-time hand tracking (2020). https://arxiv.org/abs/2006.10214

Autonomous Systems

Enhanced Trajectory Tracking for Car-Like Mobile Robots

Son Tung Dang[1], Xuan Minh Dinh[2], Tuan Nguyen Van[2], Trung Nam Pham[2], Ngoc Linh Nguyen[1], and Hai Xuan Le[1](✉)

[1] International School, National University Hanoi, Hanoi, Vietnam
hailx@vnu.edu.vn

[2] Faculty of Mechanical Engineering and Mechatronics, Phenikaa University, Yen Nghia, Ha Dong, Hanoi 12116, Vietnam
tuan.nguyenvan@phenikaa-uni.edu.vn

Abstract. This study proposes a Fast Integral Terminal Sliding Mode Control (FITSMC) method to enhance the trajectory tracking capability of car-like mobile robots. This method combines the superior advantages of existing Sliding Mode Control (SMC) methods, including the fast finite-time convergence of Fast Terminal Sliding Mode Control and the reduction of steady-state error of Integral Sliding Mode Control. The effectiveness of the proposed method is verified through simulations, demonstrating that the robot can quickly and stably achieve the desired reference trajectory in finite time, outperforming traditional sliding mode controllers.

Keywords: Car-Like Mobile Robots · Fast Integral Terminal Sliding Mode Control · Trajectory Tracking Control · Sliding Mode Control

1 Introduction

The world is witnessing a rapid and continuous technological revolution. In this context, mobile robots have drawn significant attention from the scientific and engineering communities, especially in the fields of automation and robotics. Not only do they bring economic benefits, but innovations in this field also open up numerous opportunities for development in both industry and defense. Notably, with the robust growth of the global automotive industry, four-wheeled mobile robots, such as car-like mobile robots, are being widely studied because they provide an ideal platform for researching, developing, and testing algorithms before applying them in real-world scenarios. One of the prominent research directions today is enhancing the trajectory tracking capability of these robots.

A recent study [1] proposed an Adaptive Backstepping Hierarchical Sliding Mode Controller (ABHSMC) using a Radial Basis Function (RBF) neural network for trajectory tracking control of mobile robots. The RBF neural network is utilized to estimate the unknown nonlinear components of the model. Although the simulation and experimental results demonstrate the effectiveness of the ABHSMC algorithm, this study did

© The Author(s), under exclusive license to Springer Nature Switzerland AG 2025
H. A. Le Thi et al. (Eds.): ICAMCS 2024, LNNS 1311, pp. 421–432, 2025.
https://doi.org/10.1007/978-3-031-90606-0_35

not thoroughly analyze the finite-time convergence of the system. In another study [2], a PID controller was used to control the robot to meet the requirements for trajectory tracking. While the results achieved the proposed objectives, the PID controller exhibited relatively slow response compared to the reference signal and did not ensure finite-time convergence, potentially reducing feasibility, accuracy, and stability in more complex and realistic conditions. Another study [3] proposed an adaptive trajectory tracking control scheme for car-like mobile robots, with input constraints such as velocity and steering angle. The study employed a Backstepping technique with control gains automatically adjusted based on tracking error and desired vehicle speed, but it only considered the kinematic model to reduce computational burden. The effectiveness of this method was demonstrated through simulations on Matlab/Simulink and experiments on the Quanser Qcar autonomous vehicle model, showing precise trajectory tracking without violating constraints on speed and steering angle. However, given the current advances in powerful and cost-effective embedded devices, considering the full dynamic model of the robot is necessary to more accurately reflect real-world conditions. Some other studies in the field of Adaptive Nonlinear Control, such as studies [4–6], combined the use of fuzzy logic and achieved impressive results. However, the performance and stability of this approach depend heavily on the expertise involved in designing appropriate fuzzy rules.

Recently, Model Predictive Control (MPC) methods have also been widely applied to mobile robots and autonomous vehicles [7–10]. In general, MPC is a flexible optimization method, particularly effective in handling system constraints to achieve control objectives. However, MPC methods require complex calculations, consuming significant computational resources and having a long convergence time [1], especially for nonlinear systems. Additionally, MPC relies on a system model for predictions, and if the model is inaccurate, control performance can be compromised. In study [8], the authors developed a Linear Time-Varying Model Predictive Control (LTVMPC) to address the trajectory tracking control of autonomous vehicles, particularly during cornering, where control accuracy is often low and real-time performance is reduced. Simulation results of the LTVMPC controller demonstrated superior robustness and noise resistance, while significantly improving tracking accuracy in complex driving environments. Moreover, the real-time performance of LTVMPC is better due to reduced computations compared to earlier MPC techniques [9]. In study [10], an MPC controller with Lyapunov-based input constraints (LMPC) was proposed to address the trajectory tracking problem for mobile robots. This method overcame the computational overhead of terminal constraints in trajectory tracking and reduced the required computational resources. However, both studies [9, 10] have not been tested in practical scenarios. In reality, implementing MPC techniques is very complex, requiring high expertise, a sufficiently accurate system model, and significant computational resources on hardware devices.

Inspired by the studies above, this research proposes using the Fast Integral Terminal Sliding Mode Control (FITSMC) method to address the finite-time stability problem of mobile robot systems, a challenge not tackled in the previously analyzed studies. This method combines the superior advantages of existing SMC methods, including the fast finite-time convergence of Fast Terminal Sliding Mode Control and the reduction of steady-state error of Integral Sliding Mode Control. The effectiveness of the proposed method is verified through simulations, demonstrating that the robot can quickly and stably achieve the desired reference trajectory in finite time, outperforming traditional sliding mode controllers.

The remainder of the paper is organized as follows: Sect. 2 provides a detailed description of the mathematical model of the system. Section 3 introduces the proposed Fast Integral Terminal Sliding Mode Control method. Section 4 presents simulation results along with analyses and evaluations. Finally, Sect. 5 provides conclusions and suggests future research directions.

2 Mathematical Model of the Car-Like Mobile Robot System

The mathematical model of the car-like mobile robot system has been established and presented in various studies [11, 12]. This model employs the Ackerman steering principle to create an instantaneous center of rotation $O\prime$ along the rear axle. The kinematic model can be simplified into a bicycle model by merging the two wheels on the rear axle and the two wheels on the front axle into a single equivalent wheel for each. The front wheel is designed for steering, while the rear wheel remains fixed in direction. As depicted in Fig. 1, L (m) represents the distance between the front and rear axles, l (m) denotes half the distance between the two rear wheels, and b is the distance from the center of mass to the midpoint of the rear axle. The linear velocity and steering velocity are denoted as v (m/s) and ω (rad/s), respectively. Additionally, (x, y) represents the position, θ (rad) is the orientation angle, and δ (rad) is the steering angle of the robot in the OXY coordinate system. R is the distance from the center of rotation O' to the midpoint C of the rear axle.

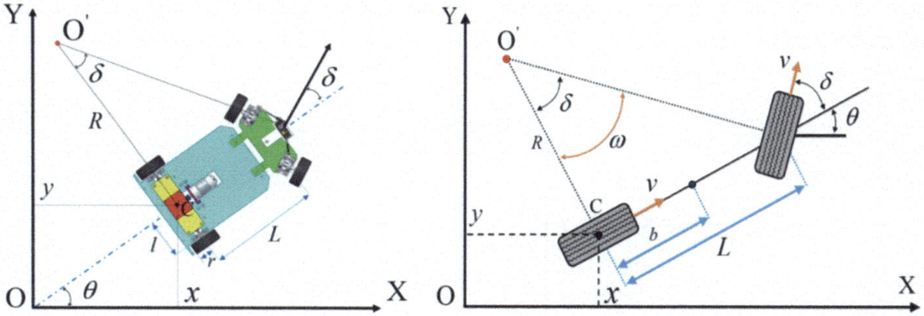

Fig. 1. Mathematical Model of a Car-Like Mobile Robot System

The kinematic model of the car-like mobile robot system is constructed based on geometric relationships in the OXY coordinate system, as expressed by the following equations:

$$\begin{cases} \dot{x} = v\cos(\theta); \; \dot{\theta} = v/R = (v/L)\tan(\delta) \\ \dot{y} = v\sin(\theta); \qquad\qquad \dot{\delta} = \omega \end{cases} \tag{1}$$

The dynamic equations of the car-like mobile robot are represented by the following equation:

$$\mathbf{M}(\mathbf{q})\ddot{\mathbf{q}} + \mathbf{V}(\mathbf{q}, \dot{\mathbf{q}})\dot{\mathbf{q}} = \mathbf{B}(\mathbf{q})\mathbf{T} - \mathbf{A}^T(\mathbf{q})\boldsymbol{\lambda} \tag{2}$$

where, $\mathbf{q} \in \mathbb{R}^{4\times1}$ is the generalized coordinate of the robot, $\mathbf{M}(\mathbf{q}) \in \mathbb{R}^{4\times4}$ is a positive definite symmetric matrix, $\mathbf{V}(\mathbf{q}, \dot{\mathbf{q}}) \in \mathbb{R}^{4\times4}$ is the Coriolis and centrifugal matrix, $\mathbf{T} \in \mathbb{R}^{2\times1}$ is the input control torque vector, $\mathbf{B}(\mathbf{q}) \in \mathbb{R}^{4\times2}$ is the transformation matrix, $\mathbf{A}(\mathbf{q}) \in \mathbb{R}^{2\times4}$ is the matrix associated with constraints, and $\boldsymbol{\lambda} \in \mathbb{R}^{2\times1}$is the Lagrange vector.

Specifically, these matrices and vectors are calculated as follows [11]:

$$\mathbf{V}(\mathbf{q}, \dot{\mathbf{q}}) = \begin{bmatrix} 0\ 0 & -I_C\dot{\theta}\cos(\theta) & 0 \\ 0\ 0 & -I_C\dot{\theta}\sin(\theta) & 0 \\ 0\ 0 & I_b & I_w \\ 0\ 0 & I_w & I_w \end{bmatrix}, \ \mathbf{B}(\mathbf{q}) = \begin{bmatrix} \cos(\theta) & 0 \\ \sin(\theta) & 0 \\ L\sin(\delta)\cos(\delta) & 0 \\ 0 & 1 \end{bmatrix}$$
$$\mathbf{M}(\mathbf{q}) = \begin{bmatrix} m & 0 & I_C\sin(\theta) & 0 \\ 0 & m & -I_C\cos(\theta) & 0 \\ I_C\sin(\theta) & -I_C\cos(\theta) & I_b & I_w \\ 0 & 0 & I_w & I_w \end{bmatrix}, \ \mathbf{A}^T(\mathbf{q})\boldsymbol{\lambda} = \mathbf{A}^T(\mathbf{q})\begin{bmatrix} \lambda_1 \\ \lambda_2 \end{bmatrix} \tag{3}$$

where, $I_b = m_C(L-b)^2 + 4l^2m_{\mathrm{w}} + I_{bzz} + 4I_{\mathrm{wzz}}$, $I_{\mathrm{w}} = \frac{1}{2}m_{\mathrm{w}}r^2$, $m = m_C + 4m_{\mathrm{w}}$, $I_C = (L-b)m_C + 2Lm_{\mathrm{w}}$ with m_C and m_{w} representing the mass of the mobile robot and the mass of each wheel, respectively. I_{w} is the moment of inertia of each wheel about the wheel axis, while I_{bzz} and I_{wzz} are the moments of inertia of the mobile robot and each wheel about the z-axis, respectively. Note that the z-axis is the axis around which the mobile robot rotates.

From Eq. (1), we can rewrite it as follows:

$$\dot{\mathbf{q}} = \mathbf{S}\omega \tag{4}$$

where:

$$\mathbf{S} = \begin{bmatrix} \cos(\theta) & 0 \\ \sin(\theta) & 0 \\ \frac{\tan(\delta)}{L} & 0 \\ 0 & 1 \end{bmatrix} \in \mathbb{R}^{4\times2}, \ \boldsymbol{\omega} = \begin{bmatrix} v & \omega \end{bmatrix}^T \in \mathbb{R}^{2\times1} \tag{5}$$

Substituting Eq. (4) into (2), noting that $\mathbf{S}^T\mathbf{A}^T(\mathbf{q}) = 0$, we obtain the dynamic equation of the car-like mobile robot as follows:

$$\overline{\mathbf{M}}(\mathbf{q})\dot{\omega} + \overline{\mathbf{V}}(\mathbf{q}, \dot{\mathbf{q}})\boldsymbol{\omega} = \overline{\mathbf{B}}(\mathbf{q})\mathbf{T} \tag{6}$$

The matrices and vectors in Eq. (6) are defined as follows:

$$\begin{aligned} \overline{\mathbf{M}}(\mathbf{q}) =& \mathbf{S}^T\mathbf{M}(\mathbf{q})\mathbf{S} \in \mathbb{R}^{2\times2}, \overline{\mathbf{B}}(\mathbf{q}) = \mathbf{S}^T\mathbf{B}(\mathbf{q}) \in \mathbb{R}^{2\times2}, \\ \overline{\mathbf{V}}(\mathbf{q}, \dot{\mathbf{q}}) =& \mathbf{S}^T\left(\mathbf{M}(\mathbf{q})\dot{\mathbf{S}} + \mathbf{V}(\mathbf{q}, \dot{\mathbf{q}})\mathbf{S}\right) \in \mathbb{R}^{2\times2} \end{aligned} \tag{7}$$

At this point, Eq. (6) can be rewritten for the purpose of controller design as follows:

$$\dot{\omega} = \overline{\mathbf{M}}^{-1}(\mathbf{q})\left[-\overline{\mathbf{V}}(\mathbf{q}, \dot{\mathbf{q}})\boldsymbol{\omega} + \overline{\mathbf{B}}(\mathbf{q})\mathbf{T}\right] = \overline{\mathbf{M}}^{-1}(\mathbf{q})\boldsymbol{\tau} + \overline{\mathbf{D}}(\mathbf{q}, \dot{\mathbf{q}}) \tag{8}$$

In this context, the vector $\overline{\mathbf{D}}(\mathbf{q}, \dot{\mathbf{q}}) = -\overline{\mathbf{M}}^{-1}(\mathbf{q})\overline{\mathbf{V}}(\mathbf{q}, \dot{\mathbf{q}})\boldsymbol{\omega}$ represents the nominal dynamic components that have been predetermined, and $\boldsymbol{\tau} = \overline{\mathbf{B}}(\mathbf{q})\mathbf{T}$ is the control torque.

3 Design of Fast Integral Terminal Sliding Mode Control for Car-Like Mobile Robots

Over the past 60 years since its inception, Sliding Mode Control (SMC) has gained popularity due to its simplicity and robust capability when systems face uncertainties in modeling and unknown external disturbances [13]. This method, based on Lyapunov stability theory, allows systems to achieve asymptotic stability and has been widely applied in various fields. However, the asymptotic convergence property of SMC results in slower convergence speeds as the system approaches equilibrium, making it impossible to reach this state in a finite time. This limitation poses challenges in practical applications that require fast convergence, especially in the control and navigation of mobile robots.

Based on these analyses, this study proposes a Fast Integral Terminal Sliding Mode Control (FITSMC) method for car-like mobile robots to enable them to quickly achieve the desired trajectory, maintain high stability, and ensure that system errors converge to zero in finite time. This approach combines the strongest advantages of current SMC methods, including the fast finite-time convergence of Fast Terminal Sliding Mode Control (FTSMC) [14] and the ability to minimize steady-state error inherent to Integral Sliding Mode Control (ISMC) [15]. The structural diagram of the FITSMC controller is shown in Fig. 2.

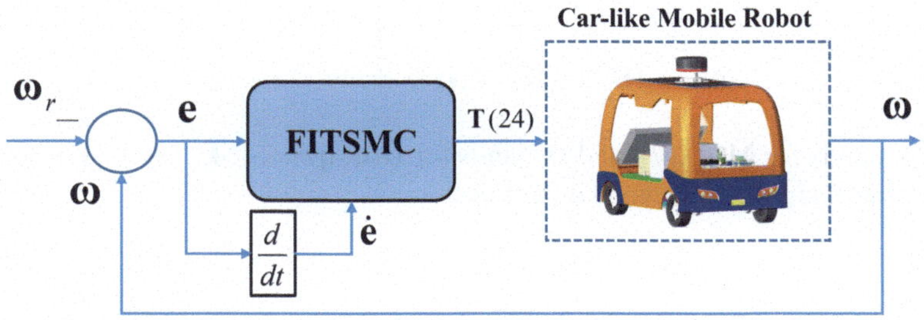

Fig. 2. Structural Diagram of the FITSMC Controller.

The steps for designing the FITSMC controller are detailed as follows:

Setting the system error as in Eq. (9), we adopt this error formulation to facilitate integration with future navigation structures.

$$\mathbf{e} = \begin{bmatrix} e_1 \\ e_2 \end{bmatrix} = \begin{bmatrix} v - v_r \\ \omega - \omega_r \end{bmatrix} = \boldsymbol{\omega} - \boldsymbol{\omega}_r \tag{9}$$

The sliding surface is designed as follows:

$$\mathbf{s} = \mathbf{e} + \boldsymbol{\alpha} \int_0^t \mathbf{e} dt + \boldsymbol{\lambda} sig^{\gamma}(\mathbf{e}) \tag{10}$$

The derivative of the sliding surface in Eq. (10) is obtained as:

$$\dot{\mathbf{s}} = \dot{\mathbf{e}} + \boldsymbol{\alpha}\mathbf{e} + \gamma \boldsymbol{\lambda} |\mathbf{e}|^{\gamma-1} \dot{\mathbf{e}} \tag{11}$$

where: $\boldsymbol{\alpha} = diag[\alpha_1, \alpha_2]$ is a diagonal matrix with $\alpha_1 > 0, \alpha_2 > 0$ being constants; $\boldsymbol{\lambda} = diag[\lambda_1, \lambda_2]$ is a diagonal matrix with $\lambda_1 > 0, \lambda_2 > 0$ being constants, $|\mathbf{e}|^{\gamma-1} = diag\{|e_1|^{\gamma-1}, |e_2|^{\gamma-1}\}$, $sig^{\gamma}(\mathbf{e}) = \left[\, |e_1|^{\gamma} sign(e_1)\ |e_2|^{\gamma} sign(e_2) \right]^T$, $\gamma \in (0, 1)$. Besides, we selected the sliding surface in Eq. (10) to achieve a balance between rapid convergence, precision, and robustness.

Select the Lyapunov function satisfying:

$$V = \frac{1}{2}\mathbf{s}^T\mathbf{s} = \frac{1}{2}s_1^2 + \frac{1}{2}s_2^2 \tag{12}$$

Then, the time derivative of (12) is obtained as:

$$\dot{V} = \mathbf{s}^T\dot{\mathbf{s}} = \mathbf{s}^T\left(\dot{\mathbf{e}} + \boldsymbol{\alpha}\mathbf{e} + \gamma\boldsymbol{\lambda}|\mathbf{e}|^{\gamma-1}\dot{\mathbf{e}}\right) \tag{13}$$

At this point, we define:

$$\begin{aligned} \boldsymbol{\psi} &= \gamma\boldsymbol{\lambda}|\mathbf{e}|^{\gamma-1};\ \boldsymbol{\chi} = \left[\overline{\mathbf{M}}^{-1} + \psi\overline{\mathbf{M}}^{-1}\right]^{-1} \\ \boldsymbol{\zeta} &= -\left(\overline{\mathbf{D}} + \psi\overline{\mathbf{D}}\right) + (\mathbf{I}_2 + \boldsymbol{\psi})\dot{\omega}_r \end{aligned} \tag{14}$$

where: $\overline{\mathbf{M}}$ stands for $\overline{\mathbf{M}}(\mathbf{q})$ (7), and $\overline{\mathbf{D}}$ stands for $\overline{\mathbf{D}}(\mathbf{q}, \dot{\mathbf{q}}) = -\overline{\mathbf{M}}^{-1}(\mathbf{q})\overline{\mathbf{V}}(\mathbf{q}, \dot{\mathbf{q}})\boldsymbol{\omega}$.

The control signal is designed as follows:

$$\boldsymbol{\tau} = \boldsymbol{\chi}\left[-\mathbf{K}_1\mathbf{s} - \left[\mathbf{K}_2 + (\mathbf{I}_2 + \boldsymbol{\psi})\varepsilon\right] sign(\mathbf{s}) + \boldsymbol{\zeta} - \boldsymbol{\alpha}\mathbf{e}\right] \tag{15}$$

In which: $\mathbf{K}_1 = diag[k_1, k_1]$ is a diagonal matrix with $k_1 > 0$ as constants, $\mathbf{K}_2 = diag[k_2, k_2]$ is a diagonal matrix with $k_2 > 0$ as constants.

Remark 1. This control signal (15) is effective when the object dynamics are known, and there are no disturbances. However, in practice, it is often difficult to determine exact dynamics, and disturbances may occur. Therefore, additional compensation or adaptive control strategies might be necessary to maintain stability and performance under uncertain conditions.

Then, substituting the control signal (15) into (13) and combining it with Eq. (9), we obtain:

$$\begin{aligned}\dot{V} &= -\mathbf{s}^T\mathbf{K}_1\mathbf{s} - \mathbf{s}^T\left[\mathbf{K}_2 + (\mathbf{I}_2 + \boldsymbol{\psi})\varepsilon\right]sign(\mathbf{s}) \\ &= -\mathbf{s}^T\mathbf{K}_1\mathbf{s} - \mathbf{s}^T\mathbf{K}_2 sign(\mathbf{s}) - \mathbf{s}^T(\mathbf{I}_2 + \boldsymbol{\psi})\varepsilon sign(\mathbf{s})\end{aligned} \tag{16}$$

From (16), with ε being a positive constant, we can easily deduce that:

$$\dot{V} \leq -\mathbf{s}^T\mathbf{K}_1\mathbf{s} - \mathbf{s}^T\mathbf{K}_2 sign(\mathbf{s}) \tag{17}$$

It is evident that: $\left(|s_1|^2 + |s_2|^2\right)^2 \geq s_1^2 + s_2^2$. Therefore, we obtain:

$$|s_1| + |s_2| \geq \sqrt{s_1^2 + s_2^2} \tag{18}$$

Therefore, from (17) and (18), we have:

$$\begin{aligned}\dot{V} &\leq - k_1\left(s_1^2 + s_2^2\right) - k_2(|s_1| + |s_2|) \\ &\leq - k_1\left(s_1^2 + s_2^2\right) - k_2\sqrt{s_1^2 + s_2^2} \leq -2k_1V - \sqrt{2}k_2V^{\frac{1}{2}}\end{aligned} \tag{19}$$

Based on Eq. (19), it is easy to deduce that:

$$dt \leq \frac{dV}{-2k_1V - \sqrt{2}k_2V^{\frac{1}{2}}} = \frac{2dV^{\frac{1}{2}}}{-2k_1V^{\frac{1}{2}} - \sqrt{2}k_2} \tag{20}$$

Assuming the time to reach the reference signal is t_r, with the Lyapunov function at that moment satisfying $V_{t_r} = 0$. Then, integrating both sides of Eq. (20), we obtain:

$$\int_0^{t_r} dt \leq 2\int_0^{t_r} \frac{dV^{\frac{1}{2}}}{-2k_1V^{\frac{1}{2}} - \sqrt{2}k_2} = \frac{1}{k_1}\ln\left(\frac{2k_1V_0^{\frac{1}{2}} + \sqrt{2}k_2}{\sqrt{2}k_2}\right) \tag{21}$$

Then, from Eq. (21), we obtain:

$$t_r \leq \frac{1}{k_1}\ln\left(\frac{2k_1V_0^{\frac{1}{2}} + \sqrt{2}k_2}{\sqrt{2}k_2}\right) \tag{22}$$

Remark 2. Based on Eq. (17), it can be observed that the derivative of the system's Lyapunov function satisfies the condition $\dot{V} \leq -\mathbf{s}^T\mathbf{K}_1\mathbf{s} - \mathbf{s}^T\mathbf{K}_2 sign(\mathbf{s}) \leq 0$, ensuring that the system achieves stability according to Lyapunov's stability theory. Moreover, from Eq. (22), it is evident that the system converges in finite time.

Additionally, to reduce the "chattering" phenomenon on the sliding surface, the function $sign(\mathbf{s})$ in Eq. (15) is replaced by the function $sat(\mathbf{s})$. Thus, Eq. (15) becomes:

$$\boldsymbol{\tau} = \boldsymbol{\chi}\left[-\mathbf{K}_1\mathbf{s} - \left[\mathbf{K}_2 + (\mathbf{I}_2 + \boldsymbol{\psi})\varepsilon\right]sat(\mathbf{s}) + \boldsymbol{\zeta} - \alpha\mathbf{e}\right] \tag{23}$$

Remark 3. By introducing the saturation function, we strike a balance between robustness and smooth control, mitigating actuator wear caused by chattering while maintaining the performance of the sliding mode controller.

Thus, the input torque vector of the robot is calculated by:

$$\mathbf{T} = \overline{\mathbf{B}}^{-1}(\mathbf{q})\boldsymbol{\tau} \tag{24}$$

4 Simulation Results and Discussion

This section conducts simulation verification using Matlab & Simulink software to confirm the effectiveness of the Fast Integral Terminal Sliding Mode Control (FTISMC) algorithm. The proposed FTISMC method is compared with the Sliding Mode Control (SMC). The simulation scenarios are designed to closely resemble real-world problems, facilitating future experimental implementation. To ensure objectivity in the simulation, evaluation, and comparison of the FITSMC and SMC control methods, the parameters in the mathematical model of the mobile robot are selected from study [11]. Specifically: $m_C = 2.3(\text{kg})$, $m_\text{w} = 0.23(\text{kg})$, $L = 0.2(\text{m})$, $l = 0.05(\text{m})$, $b = 0.1(\text{m})$, $I_{bzz} = 0.015(\text{kg} \cdot \text{m}^2)$, $I_{\text{wzz}} = 0.002(\text{kg} \cdot \text{m}^2)$.

To ensure the practicality of the research, the authors conduct simulations with two types of input reference signals: regular and irregular. Specifically, regular reference signals typically represent pre-programmed trajectories, such as straight-line or circular motion. On the other hand, irregular reference signals represent random, unpredictable situations, reflecting the complex real-world operating environment of robots. The two types of input reference signals are represented by mathematical Eqs. (25) and (26). Equation (25) represents a regular reference signal, while Eq. (26) represents a non-regular, varying reference signal:

$$\boldsymbol{\omega}_r = \left[\cos(0.5t)\ \cos(0.5t)\right]^T \tag{25}$$

$$\boldsymbol{\omega}_r = rand * \left[\cos(0.5t)\ \cos(0.5t)\right]^T \tag{26}$$

The parameters of the FITSMC controller are chosen as follows: $\boldsymbol{\alpha} = diag[3.2, 3.2]$, $\boldsymbol{\lambda} = diag[0.8, 0.8]$, $\varepsilon = 0.08$, $\mathbf{K}_1 = diag[15, 15]$, $\mathbf{K}_2 = diag[35, 35]$, $\gamma = 0.2$.

4.1 Case of Regular Reference Signal

Based on the simulation results shown in Fig. 3, it can be observed that in the case of a regular reference signal, the output response for both the linear velocity and steering velocity of the mobile robot system using the FITSMC method (depicted by the blue line) reaches stability in less than 0.1 s. For the SMC method (depicted by the green line), the system's output response reaches the reference signal after approximately 0.15 s. Regarding the system's steady-state error, Figs. 4 and 5 demonstrate that the steady-state error for the FITSMC method completely converges to zero, whereas with the SMC method, there is still a residual steady-state error even after the system has stabilized. Therefore, in the case of a regular reference signal and no noise, the advantages of FITSMC in terms of response time and reference tracking accuracy are evident compared to the SMC control method.

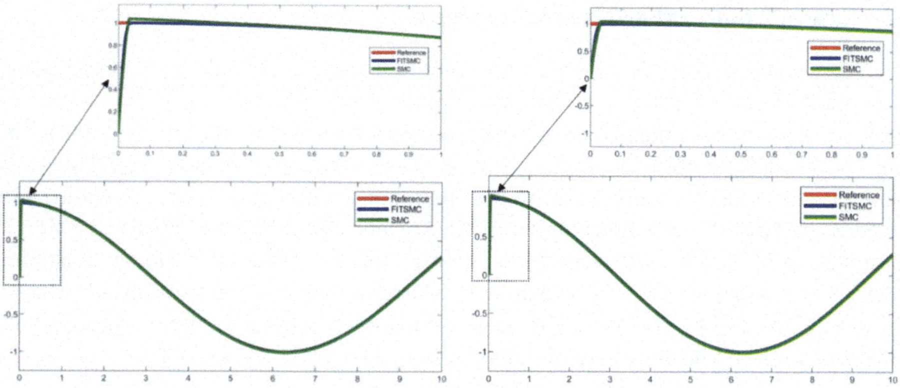

Fig. 3. Output response of linear velocity (left) and steering velocity (right).

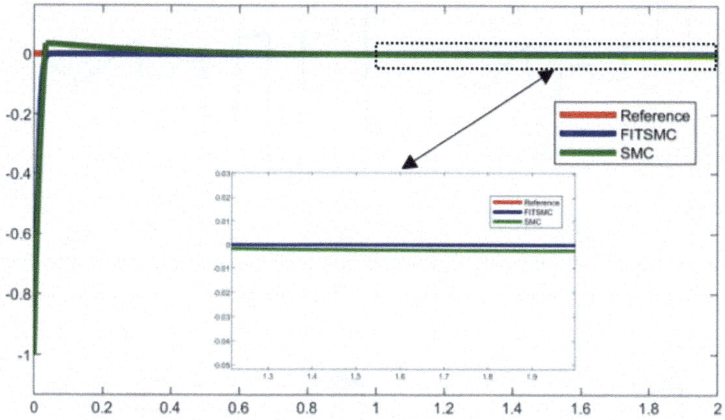

Fig. 4. Linear velocity error of the robot.

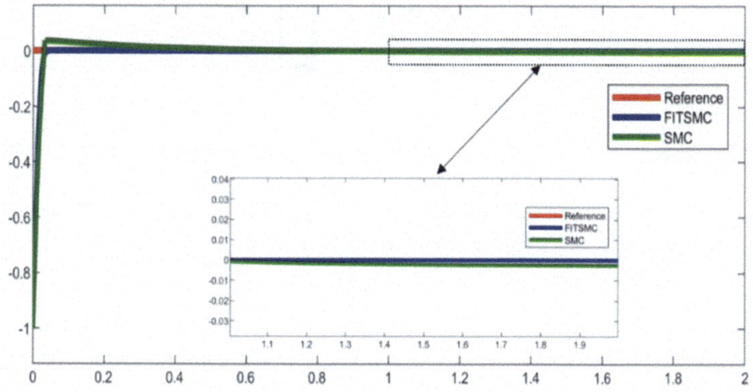

Fig. 5. Steering velocity error of the robot.

4.2 Case of Non-regulated Reference Signal

From the simulation results shown in Figs. 6 and 7, it is evident that the system's response using the FITSMC method performs relatively well even in the case of non-regulated reference signals. Specifically, the car-like mobile robot system quickly responds to the reference signals for both linear velocity and steering velocity at each moment. However, some overshoot is unavoidable due to the continuously changing reference nature. For the SMC method, the system's output response reaches the reference signal more slowly compared to FITSMC. Furthermore, the steady-state error of the system, as illustrated in Fig. 8, is smaller with FITSMC than with SMC. This further confirms the advantages of FITSMC in terms of response time and accuracy when tracking the reference signal compared to the SMC control method.

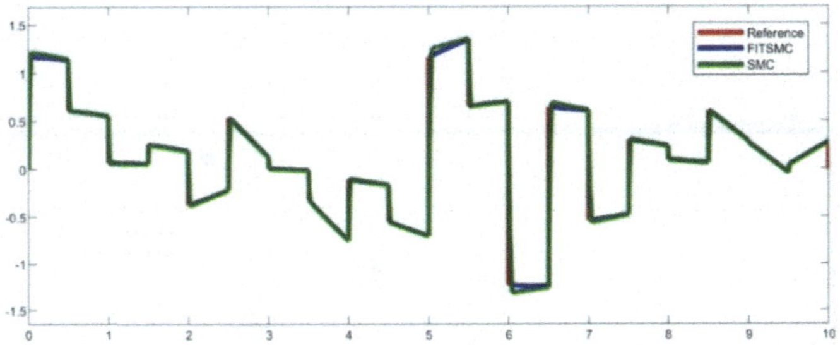

Fig. 6. Output response of the robot's linear velocity with a non-regulated reference signal.

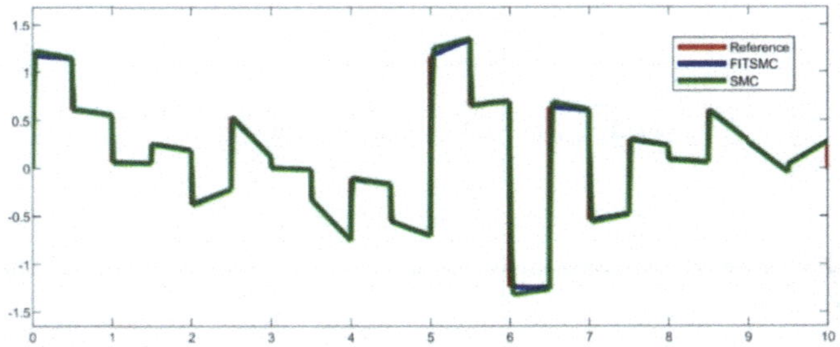

Fig. 7. Output response of the robot's steering velocity with a non-regulated reference signal.

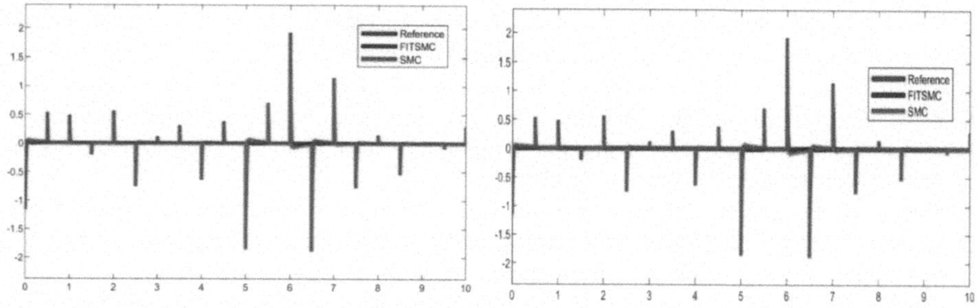

Fig. 8. Linear velocity error (left) and steering velocity error (right) of the robot.

5 Conclusions and Future Directions

In this study, we proposed a Fast Integral Terminal Sliding Mode Control (FITSMC) method to enhance the trajectory tracking capability of car-like mobile robots. The FITSMC method combines the outstanding advantages of current sliding mode control methods, including the fast finite-time convergence of Fast Terminal Sliding Mode Control and the minimization of steady-state errors offered by Integral Sliding Mode Control. Simulation results demonstrate that the FITSMC method enables the robot to quickly and stably reach the desired reference trajectory in finite time, proving to be more effective than traditional sliding mode controllers. In the future, the FITSMC method will be validated on a real mobile robot platform.

Acknowledgments. This research is funded by the International School, Vietnam National University, Hanoi (VNU-IS), and is also supported by Phenikaa University.

References

1. Dang, S.T., Dinh, X.M., Kim, T.D., Xuan, H.L., Ha, M.H.: Adaptive backstepping hierarchical sliding mode control for 3-wheeled mobile robots based on RBF neural networks. Electronics **12**(11), 2345 (2023)
2. Cáceres Flórez, C.A., Rosário, J.M., Amaya, D.: Control structure for a car-like robot using artificial neural networks and genetic algorithms. Neural Comput. Appl. **32**(20), 15771–15784 (2020)
3. Hu, J., Zhang, Y., Rakheja, S.: Adaptive trajectory tracking for car-like vehicles with input constraints. IEEE Trans. Industr. Electron. **69**(3), 2801–2810 (2021)
4. Shui, Y., Zhao, T., Dian, S., Hu, Y., Guo, R., Li, S.: Data-driven generalized predictive control for car-like mobile robots using interval type-2 T-S fuzzy neural network. Asian J. Control **24**(3), 1391–1405 (2022)
5. Ben Jabeur, C., Seddik, H.: Design of a PID optimized neural networks and PD fuzzy logic controllers for a two-wheeled mobile robot. Asian J. Control **23**(1), 23–41 (2021)
6. Bouhamatou, Z., Abedssemed, F.: Fuzzy synergetic control for dynamic car-like mobile robot. Acta Mechanica et Automatica **16**(1), 48–57 (2022)
7. Yeom, K.: Kinematic and dynamic controller design for autonomous driving of car-like mobile robot. Int. J. Mech. Eng. Robot. Res. **9**(7), 1058–1064 (2020)

8. Cao, J., Song, C., Peng, S., Song, S., Zhang, X., Xiao, F.: Trajectory tracking control algorithm for autonomous vehicle considering cornering characteristics. IEEE Access **8**, 59470–59484 (2020)
9. Wang, Z., Li, G., Jiang, H., Chen, Q., Zhang, H.: Collision-free navigation of autonomous vehicles using convex quadratic programming-based model predictive control. IEEE/ASME Trans. Mechatron. **23**(3), 1103–1113 (2018)
10. Zhang, J.J., Fang, Z.L., Zhang, Z.Q., Gao, R.Z., Zhang, S.B.: Trajectory tracking control of nonholonomic wheeled mobile robots using model predictive control subjected to lyapunov-based input constraints. Int. J. Control. Autom. Syst. **20**(5), 1640–1651 (2022)
11. Hussein, M., Zhang, Y., Liu, Z.: Distributed fault-tolerant formation control design via high-order sliding mode for a team of car-like vehicles. IEEE Trans. Intell. Veh. **9**, 4221 (2023)
12. Wang, Z., Li, P., Li, Q., Wang, Z., Li, Z.: Motion planning method for car-like autonomous mobile robots in dynamic obstacle environments. IEEE Access **11**, 137387–137400 (2023)
13. Yu, X., Feng, Y., Man, Z.: Terminal sliding mode control–an overview. IEEE Open J. Ind. Electron. Soc. **2**, 36–52 (2020)
14. Yu, X., Zhihong, M.: Fast terminal sliding-mode control design for nonlinear dynamical systems. IEEE Trans. Circuits Syst. I Fundam. Theory Appl. **49**(2), 261–264 (2002)
15. Hamayun, M.T., Edwards, C., Alwi, H.: Integral Sliding Mode Control, pp. 17–37. Springer International Publishing, Singapore (2016)

Trajectory Tracking Control for Quadrotor UAV with Suspended Load Using Sliding Mode Control via Extended States Observer

Duc Manh Do[1], Chi Nhan Nguyen[1], Thai Duong Tran[1], Duc Thinh Le[2], Van Dung Le[3], and Tung Lam Nguyen[1](✉)

[1] Hanoi University of Science and Technology, Hanoi 100000, Vietnam
lam.nguyentung@hust.edu.vn

[2] Thuyloi University, Hanoi 100000, Vietnam
thinhld@tlu.edu.vn

[3] Hung Vuong University, Phutho 290000, Vietnam
levandung@hvu.edu.vn

Abstract. In this paper, to stabilize the quadrotor unmanned aerial vehicles (QUAV) with suspended load against external disturbances and uncertain parameters, the sliding mode control (SMC) is utilized to control the QUAV follow the desired trajectory, while the extended states observer (ESO) is used to estimate states of the plant and disturbances for the position and altitude control loops. The dynamic model of the QUAV is established based on Euler angles under the influence of aerodynamic moments and force. To control the UAV along the reference trajectory, a controller with two loops is utilized: an outer loop is position control loop, and an inner loop is control-altitude loop. Both of them are designed utilizing SMC with ESO to ensure the sustainability of the controller under disturbances. These disturbances are environmental external forces, the effect of slung payload, and uncertain model parameters. Finally, Matlab/Simulink is used to verify effectiveness and feasibility of the proposed SMC-ESO control system for QUAV.

Keywords: Unmanned Aerial Vehicles · trajectory tracking control · Sliding Mode Control · Extended states observer

1 Introduction

In recent years, drones have been increasingly popular in both military and civilian applications [1, 2]. Among the types of drones available, quadrotors have seen significant improvements in terms of size, weight, and performance. With the ability to take off, land vertically, and navigate through all terrain, drones are being utilized in various industries, including agriculture, search and rescue, and performing dangerous tasks. A QUAV can be considered as a strong coupling nonlinear system. It is a complex system with six degrees of freedom (DoF) outputs, but it has only four control inputs. Moreover, QUAVs have an average of good interference prevention capabilities. However, the outside environment can readily alter their aerodynamic properties. Due to their usefulness and versatility, scientists are researching ways to enhance the stability and efficiency of the quadrotor.

© The Author(s), under exclusive license to Springer Nature Switzerland AG 2025
H. A. Le Thi et al. (Eds.): ICAMCS 2024, LNNS 1311, pp. 433–442, 2025.
https://doi.org/10.1007/978-3-031-90606-0_36

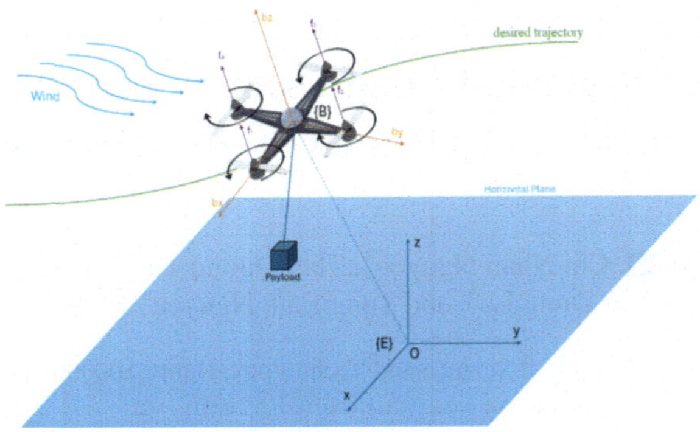

Fig. 1. The quadrotor and the slung payload suspended by a cable in windy environment

A mathematical model of the drone must be created in order to guarantee that drone control strategies operate. The Newton-Euler method has been used in various studies to construct mathematical models of drones. In study [3], authors had considered that air resistance and aerodynamic friction affect to the plant, which resulted in a more complete and realistic model. In [4], the impact of the suspended slung-payload has been regarded as a periodic disturbance that can be considered as an external system. After finish contracting the mathematical model, then a control system is built. The most commonly used control strategies are Linear and Nonlinear controllers, for which there are numerous studies proposing different approaches like PID [5, 6], SMC [7, 8], and BSP [9, 10]. In [6], author suggested designing six nonlinear PID controllers for each DoF subsystems then compare it with PID linear controller. The results showed that linear PID was less effective than nonlinear PID, but nonlinear PID had slower response times and suffered from overshooting. Study [8] proposed the Adaptive SMC which demonstrated its effectiveness using Lyapunov Theory and compared it with LQR (Linear Quadratic Regulator) control and ADRC (Active Disturbance Rejection Control). However, this study didn't mention oject's position control problem. Study [10] compared the advantages and disadvantages of using BSP, Integral Terminal SMC, and Adaptive BSP controllers by simulating the strategy of reducing the object's load while flying. It showed that the Adaptive BSP controller was most effective. For higher control quality, some studies propose using observers to estimate plant's states [11, 12]. In [12], High Gain observer is proposed to compensate for wind disturbance, therefore increasing control performance. Study [13] mentioned an extended states observer capable of observing disturbances affecting the system.

This paper gives a mathematical model of the drone that takes into account unknown disruptions as well as external factors including wind impact, aerodynamic friction, air resistance, and suspension load changes. To guid QUAV follow the desired path, a control system based on cascade control structure is proposed with control loops for position control and rotation angle control. The position control loop will output a virtual control signal to provide a desired signal to the rotation angle control loop and to ensure the stability of the rotation angle control loop. The article proposes to use an SMC for both position and rotation control loops. And then, the paper will propose using ESO to estimate QUAV's states and compensate for disturbances. The article will compare the system response when using ESO and without obsevers to demonstrate control quality.

The paper presents two main contributions. Firstly, it proposes the use of an extended observer set to counteract the impact of the environment and uncertain, and the estimation of status parameters through ESO can help reduce the number of sensors required. The second contribution is the SMC-ESO strategy control is proposed and demonstrated by the Lyapunov theory and simulated in Matlab\simulink.

2 Mathematical Model of QUAV

The QUAV model includes 4 motors attached to 4 propellers placed diagonally across each other. The movement of QUAV is controlled by the thrust generated by each motor when rotating. In this paper, the QUAV model is considered to be attached with an additional load and moving in a windy environment, as illustrated in Fig. 1. By changing the force and moment of the motors, you can create various movements. These movements can include translational movement in the three directions of the Cartesian coordinate system $\mathbf{P} = \left[x\, y\, z\right]^T$ or by the Euler angle rotations $\mathbf{Q} = [\phi\, \theta\, \psi]^T$ (roll, pitch, and yaw). Consider the UAV in two frames: {E} is the inertial frame attached to the earth and {B} is fixed to the vehicle and constrained to move with it. Therefore, the rotation matrix is given by:

$$\mathbf{R}_{\mathrm{B}}^{\mathrm{E}} = \left(\mathbf{R}_{\mathrm{E}}^{\mathrm{B}}\right)^{\mathrm{T}} = \begin{bmatrix} c_\theta c_\psi & s_\phi s_\theta c_\psi - c_\phi s_\psi & c_\phi s_\theta c_\psi + s_\phi s_\psi \\ c_\theta s_\psi & s_\phi s_\theta s_\psi + c_\phi c_\psi & c_\phi s_\theta s_\psi - s_\phi c_\psi \\ -s_\theta & s_\phi c_\theta & c_\phi c_\theta \end{bmatrix} \tag{1}$$

where $c_\alpha = \cos(\alpha)$, $s_\alpha = \sin(\alpha)$, and $\mathbf{W}_B = \left[p\, q\, r\right]^T$ are the rotation speeds in body-fixed frame {B} can be written according to velocities in inertial frame as:

$$\begin{bmatrix} p \\ q \\ r \end{bmatrix} = \begin{bmatrix} 1 & 0 & -s\theta \\ 0 & c\theta & s\phi c\theta \\ 0 & -s\phi & c\phi c\theta \end{bmatrix} \begin{bmatrix} \dot{\phi} \\ \dot{\theta} \\ \dot{\psi} \end{bmatrix} \tag{2}$$

2.1 Translational Motions of the QUAV in the Inertial Frame {E}

Based on Newton's second law, the position dynamics can be formulated as [3]:

$$\mathbf{F} - \mathbf{F_g} - \mathbf{F_c} + \mathbf{D_1} = m\ddot{\mathbf{P}} \tag{3}$$

With $U_1 = k_t\left(w_1^2 + w_2^2 + w_3^2 + w_4^2\right)$ is total thurst from four rotors, k_t is rotor coefficient, w_i is the rotation speed of each rotor. The rotors' thurst impact on the QUAV is defined as:

$$\mathbf{F} = \begin{bmatrix} F_x \\ F_y \\ F_z \end{bmatrix} = \mathbf{R}_{\mathrm{B}}^{\mathrm{E}} \begin{bmatrix} 0 \\ 0 \\ U_1 \end{bmatrix} = \begin{bmatrix} s\phi s\psi + c\phi s\theta c\psi \\ -s\phi c\psi + c\phi s\theta s\psi \\ c\phi c\theta \end{bmatrix} U_1 \tag{4}$$

The gravity $\mathbf{F_g} = \left[0\,0\,mg\right]^{\mathrm{T}}$, where is the total mass, and is the gravitational acceleration. With k_{cx}, k_{cy}, k_{cz} are the translational drag coefficients, $\mathbf{F_c} = \left[k_{cx}\dot{x}\; k_{cy}\dot{y}\; k_{cz}\dot{z}\right]^{\mathrm{T}}$ are the drag forces. The disturbance $\mathbf{D_1}$ will be defined in Sect. 2.3. Where $\mathbf{f_1}(\dot{\mathbf{P}}, \mathbf{Q}) = \frac{1}{m}(-\mathbf{F_g} - \mathbf{F_c})$, (3) can be written as:

$$\ddot{\mathbf{P}} = \frac{\mathbf{1}}{m}\left(\mathbf{F} - \mathbf{F_g} - \mathbf{F_c} + \mathbf{D_1}\right) = \frac{\mathbf{F}}{m} + \mathbf{f_1} + \frac{\mathbf{D_1}}{m} \tag{5}$$

2.2 Rotational Motion of the QUAV

In body frame $\{B\}$, the rotational motion of the QUAV can be described as [3]:

$$\mathbf{M} + \mathbf{M_{gry}} - \mathbf{M_c} + \mathbf{D_2} = \mathbf{I}\left(\dot{\mathbf{W}}_\mathbf{B} + \mathbf{W_B} \times \mathbf{W_B}\right) \tag{6}$$

where k_d is the rotor coefficient, l is the length of QUAV's arm. The rotors' moment:

$$\mathbf{M} = \begin{bmatrix} lU_2 \\ lU_3 \\ U_4 \end{bmatrix} = \begin{bmatrix} lk_t\left(w_2^2 - w_4^2\right) \\ lk_t\left(w_3^2 - w_1^2\right) \\ k_d\left(w_1^2 - w_2^2 + w_3^2 - w_4^2\right) \end{bmatrix} \tag{7}$$

The gyroscopic moment $\mathbf{M_{gry}} = J\left[p\Omega\; -q\Omega\; 0\right]^{\mathrm{T}}$, where J is the inertia of a rotor, $\Omega = w_1 - w_2 + w_3 - w_4$. Drag moment $\mathbf{M_c} = \left[k_{c\phi}\dot{\phi}^2\; k_{c\theta}\dot{\theta}^2\; k_{c\psi}\dot{\psi}^2\right]^{\mathrm{T}}$, where $k_{c\phi}, k_{c\theta}, k_{c\psi}$ are positive orientation drag coefficients. $\mathbf{I} = \mathrm{diag}\left(I_{xx}, I_{yy}, I_{zz}\right)$ is the moment of inertia of the QUAV. The disturbance $\mathbf{D_2}$ will be defined in Sect. 2.3. Equation (6) can be rewritten as follows:

$$\ddot{\mathbf{Q}} = \mathbf{I}^{-1}\mathbf{M} + \mathbf{f_2} + \mathbf{I}^{-1}\mathbf{D_2} \tag{8}$$

where $\mathbf{f_2}(\dot{\mathbf{Q}}) = \begin{bmatrix} \frac{I_{yy} - I_{zz}}{I_{xx}}\dot{\phi}\dot{\theta} - \frac{J}{I_{xx}}\dot{\theta}\Omega - \frac{k_{c\phi}\dot{\phi}^2}{I_{xx}} \\ \frac{I_{zz} - I_{xx}}{I_{yy}}\dot{\psi}\dot{\phi} + \frac{J}{I_{yy}}\dot{\phi}\Omega - \frac{k_{c\theta}\dot{\theta}^2}{I_{yy}} \\ \frac{I_{xx} - I_{yy}}{I_{zz}}\dot{\phi}\dot{\theta} - \frac{k_{cy}\dot{\psi}^2}{I_{zz}} \end{bmatrix}$

2.3 Disturbance Impact on QUAV

In this paper, the disturbances considered include forces from the slung payload, the impact of wind, and uncertain parameters. In standard practical cases, when accounting for suspended loads in the model, they can be considered as disturbances with partially known information. For instance, if the payload oscillates or the quadrotor follows a periodic reference trajectory, the load can be represented as a cyclic disturbance using an exogenous system [4].

$$\begin{cases} \dot{\delta} = \mathbf{A}\delta \\ \mathbf{d_m} = \mathbf{B}\delta \end{cases} \tag{9}$$

where $\boldsymbol{\delta}$ represents the state of the external system, $\mathbf{d_m}$ are the forces impacting to QUAV, σ_i is the disturbance frequency along the i-axis,

$$\mathbf{A} = \begin{bmatrix} \mathbf{A}_x & \mathbf{0}_{2\times2} & \mathbf{0}_{2\times2} \\ \mathbf{0}_{2\times2} & \mathbf{A}_y & \mathbf{0}_{2\times2} \\ \mathbf{0}_{2\times2} & \mathbf{0}_{2\times2} & \mathbf{A}_z \end{bmatrix}, \mathbf{A}_i = \begin{bmatrix} 0 & \sigma_i \\ -\sigma_i & 0 \end{bmatrix}_{i=x,y,z}, \mathbf{B} = \begin{bmatrix} 1\,0\,0\,0\,0\,0 \\ 0\,0\,1\,0\,0\,0 \\ 0\,0\,0\,0\,1\,0 \end{bmatrix}$$

Suppose that the cyclic perturbation $\mathbf{d_m}$ due to the slung payload mainly affecting the position of the quadrotor. Uncertain parameters and model errors affecting forces and moments are represented as $\mathbf{d_1}, \mathbf{d_2}$. It is challenging to precisely forecast the speed and direction of winds. It affects both position and orientation In this paper, the impact of wind is modeled as a combination of force $\mathbf{d_{wf}}$ and moment $\mathbf{d_{wt}}$, which are limited to the derivative component $\dot{\mathbf{d}}_{\mathbf{wf}} < \overline{\mathbf{d_{wf}}}, \dot{\mathbf{d}}_{\mathbf{wt}} < \overline{\mathbf{d_{wt}}}$, where $\overline{\mathbf{d_{wf}}}, \overline{\mathbf{d_{wt}}}$ are positive. Therefore, the disturbances in Eqs. (3) and (6) are defined as follows:

$$\begin{cases} \mathbf{D_1} = \mathbf{d_m} + \mathbf{d_{wf}} + \mathbf{d_1} \\ \mathbf{D_2} = \mathbf{d_{wt}} + \mathbf{d_2} \end{cases} \tag{10}$$

3 Controller Design

3.1 Extended States Observer

The role of the ESO is to assess the state of a QUAV and the effects of disturbances on it, and then adjust control signal to compensate for those interferences, thus ensuring the stability of the QUAV. Via ESO, $\hat{\mathbf{X}}_1 = \left[\hat{\mathbf{P}}\ \hat{\mathbf{Q}}\right]^{\mathrm{T}}; \hat{\mathbf{X}}_2 = \left[\widehat{\dot{\mathbf{P}}}\ \widehat{\dot{\mathbf{Q}}}\right]^{\mathrm{T}}; \hat{\mathbf{X}}_3 = \left[\hat{\mathbf{D}}_1\ \hat{\mathbf{D}}_2\right]^{\mathrm{T}}; \hat{\mathbf{f}} = \left[\mathbf{f}_1\left(\widehat{\dot{\mathbf{P}}}, \hat{\mathbf{Q}}\right)\ \mathbf{f}_2(\hat{\mathbf{Q}})\right]^{\mathrm{T}}$ are estimated signals from real states $\mathbf{X}_1 = \left[\mathbf{P}\ \mathbf{Q}\right]^{T}$; $\mathbf{X}_2 = \left[\dot{\mathbf{P}}\ \dot{\mathbf{Q}}\right]^{\mathrm{T}}\mathbf{X}_3 = \left[\frac{\mathbf{D}_1}{\mathbf{m}}\ \mathbf{I}^{-1}\mathbf{D}_2\right]^{\mathrm{T}}; \mathbf{f} = \left[\mathbf{f}_1\ \mathbf{f}_2\right]^{\mathrm{T}}$. $\mathbf{a} = \left[\frac{\mathbf{F}}{\mathbf{m}}\ \mathbf{I}^{-1}\mathbf{M}\right]^{\mathrm{T}}$ and $g(t) = \dot{X}_3$, Eqs. (5) and (8) can be written as:

$$\begin{cases} \dot{\mathbf{X}}_1 = \mathbf{X}_2 \\ \dot{\mathbf{X}}_2 = \mathbf{a} + \mathbf{f} + \mathbf{X}_3 \\ \dot{\mathbf{X}}_3 = \mathbf{g}(t) \end{cases} \tag{11}$$

The extended states observer is designed as [13]:

$$\begin{cases} \dot{\hat{\mathbf{X}}}_1 = \hat{\mathbf{X}}_2 + \frac{\mathrm{a}_1}{\varepsilon}\left(\mathbf{X}_1 - \hat{\mathbf{X}}_1\right) \\ \dot{\hat{\mathbf{X}}}_2 = \mathbf{a} + \hat{\mathbf{f}} + \hat{\mathbf{X}}_3 + \frac{\mathrm{a}_2}{\varepsilon^2}\left(\mathbf{X}_1 - \hat{\mathbf{X}}_1\right) \\ \dot{\hat{\mathbf{X}}}_3 = \frac{\mathrm{a}_3}{\varepsilon^3}\left(\mathbf{X}_1 - \hat{\mathbf{X}}_1\right) \end{cases} \tag{12}$$

The observer error is difined as $\boldsymbol{\eta} = \left[\eta_1\ \eta_2\ \eta_3\right]^{\mathrm{T}}$, where $\eta_1 = \frac{\mathbf{X}_1 - \hat{\mathbf{X}}_1}{\varepsilon^2}$; $\eta_2 = \frac{\mathbf{X}_2 - \hat{\mathbf{X}}_2}{\varepsilon}$; $\eta_3 = \mathbf{X}_3 - \hat{\mathbf{X}}_3$. Then

$$\varepsilon\dot{\boldsymbol{\eta}} = A\boldsymbol{\eta} + \varepsilon\mathbf{b}\mathbf{g}(t) \tag{13}$$

where $\mathbf{A} = \begin{bmatrix} -a_1\mathbf{I}_{6\times6} & \mathbf{I}_{6\times6} & \mathbf{0}_{6\times6} \\ -a_2\mathbf{I}_{6\times6} & \mathbf{0}_{6\times6} & \mathbf{I}_{6\times6} \\ -a_3\mathbf{I}_{6\times6} & \mathbf{0}_{6\times6} & \mathbf{0}_{6\times6} \end{bmatrix}$; $\mathbf{b} = \begin{bmatrix} \mathbf{0}_{6\times6} \\ \mathbf{0}_{6\times6} \\ \mathbf{I}_{6\times6} \end{bmatrix}$, and a_1, a_2, a_3 are the control parameters chosen so that matrix **A** is Hurwitz. Based on stability theory, for any given symmetric positive definite matrix $\mathbf{K} \in R^{18}$, there exists an unique symmetric positive definite matrix **H** satisfying the Lyapunov equation as follows:

$$\mathbf{A}^{\mathrm{T}}\mathbf{H} + \mathbf{H}\mathbf{A} + \mathbf{K} = \mathbf{0} \tag{14}$$

Define the Lyapunov function as $V{=}\varepsilon\boldsymbol{\eta}^{\mathrm{T}}\mathbf{H}\boldsymbol{\eta}$, then

$$\begin{aligned} \dot{V} &= \varepsilon\dot{\boldsymbol{\eta}}^{\mathrm{T}}\mathbf{H}\boldsymbol{\eta} + \varepsilon\boldsymbol{\eta}^{\mathrm{T}}\mathbf{h}\dot{\boldsymbol{\eta}} = (\mathbf{A}\boldsymbol{\eta} + \varepsilon\mathbf{b}\mathbf{g}(t))^{\mathrm{T}}\mathbf{H}\boldsymbol{\eta} + \boldsymbol{\eta}^{\mathrm{T}}\mathbf{H}(\mathbf{A}\boldsymbol{\eta} + \varepsilon\mathbf{b}\mathbf{g}(t)) \\ &= \boldsymbol{\eta}^{\mathrm{T}}\left(\mathbf{A}^{\mathrm{T}}\mathbf{H} + \mathbf{H}\mathbf{A}\right)\mathbf{A} + 2\varepsilon\boldsymbol{\eta}^{\mathrm{T}}\mathbf{H}\mathbf{b}\mathbf{g}(t) \le -\boldsymbol{\eta}^{\mathrm{T}}\mathbf{K}\boldsymbol{\eta} + 2\varepsilon\|\mathbf{H}\mathbf{b}\| \cdot \|\boldsymbol{\eta}\| \cdot \|\mathbf{g}(t)\| \end{aligned}$$

When $\mathbf{g}(t)$ is limited by a value L, $\dot{V} \le -\lambda_{min}(\mathbf{K})\|\boldsymbol{\eta}\|^2 + 2\varepsilon L\|\mathbf{H}\mathbf{b}\| \cdot \|\boldsymbol{\eta}\|$, where $\lambda_{min}(\mathbf{K})$ is the minimum eigenvalue of **K**. With $\dot{V} \le 0$, observer error is limited as $\|\boldsymbol{\eta}\| \le \frac{2\varepsilon L\|\mathbf{H}\mathbf{b}\|}{\lambda_{\min}(\mathbf{K})}$. So that, the smaller positive ε is, the smaller observer error is.

3.2 Sliding Mode Controller for Position and Orientation.

To control the QUAV follow a desired trajectory $\mathbf{P_d} = \left[x_d\ y_d\ z_d\right]^{\mathrm{T}}$ and ψ_d, the tracking error is defined as $\mathbf{e}_1 = \mathbf{P_d} - \mathbf{P}$; $\dot{\mathbf{e}}_1 = \dot{\mathbf{P}}_\mathbf{d} - \dot{\mathbf{P}}$. The sliding mode function as:

$$\mathbf{s}_1 = \mathbf{c}_1\mathbf{e}_1 + \dot{\mathbf{e}}_1 \tag{15}$$

Then $\dot{\mathbf{s}}_1 = \mathbf{c}_1\dot{\mathbf{e}}_1 + \ddot{\mathbf{e}}_1 = \mathbf{c}_1\dot{\mathbf{e}}_1 + \ddot{\mathbf{P}}_\mathbf{d} - \frac{1}{m}(\mathbf{F} + m\mathbf{f}_1 + \mathbf{D}_1)$. Based on exponential reaching law and in order to restrain the chattering phenomenon, the saturated function sat() is adopted instead of sign() [13]

$$\dot{\mathbf{s}}_1 = -\,\mathbf{n}_1 \cdot \mathrm{sat}(\mathbf{s}_1) - \mathbf{k}_1\mathbf{s}_1 \tag{16}$$

where $\mathbf{c}_1 = \mathrm{diag}(c_x, c_y, c_z)$, $\mathbf{n}_1 = \mathrm{diag}(n_x, n_y, n_z)$, $\mathbf{k}_1 = \mathrm{diag}(k_x, k_y, k_z)$ are positive control parameters. With estimated states in Eq. (12), the control signal is designed as:

$$\mathbf{F} = m\left(-\hat{\mathbf{f}}_1 - \frac{\hat{\mathbf{D}}_1}{m} + \ddot{\mathbf{P}}_\mathbf{d} + \mathbf{c}_1\hat{\dot{\mathbf{e}}}_1 + \mathbf{n}_1.\mathrm{sat}(\hat{\mathbf{s}}_1) + \mathbf{k}_1\hat{\mathbf{s}}_1\right) \tag{17}$$

From Eq. (4), the desired signal ϕ_d, θ_d are:

$$\begin{aligned} U_1 &= \sqrt{F_x^2 + F_y^2 + F_z^2} \\ \theta_d &= \arctan\left(\frac{F_x\cos\psi + F_y\sin\psi}{F_z}\right) \\ \phi_d &= \arctan\left(\frac{F_x\sin\psi - F_y\cos\psi}{F_z}\cos\theta_d\right) \end{aligned} \tag{18}$$

Select the Lyapunov function as $V_1 = \frac{1}{2}\mathbf{s}_1^{\mathrm{T}}\mathbf{s}_1$. Therefore, have

$$\begin{aligned}
\dot{V}_1 &= \mathbf{s}_1^{\mathrm{T}}\dot{\mathbf{s}}_1 = \mathbf{s}_1^{\mathrm{T}}\left(\mathbf{c}_1\dot{\mathbf{e}}_1 + \ddot{\mathbf{P}}_{\mathbf{d}} - \frac{1}{m}(\mathbf{F} + m\mathbf{f}_1 + \mathbf{D}_1)\right)\\
&= (\hat{\mathbf{s}}_1 - \mathbf{s}_1 - \hat{\mathbf{s}}_1)^{\mathrm{T}}\left(\mathbf{n}_1\mathrm{sat}(\hat{\mathbf{s}}_1) + \mathbf{k}_1\hat{\mathbf{s}}_1 + \hat{\mathbf{D}}_1 - \mathbf{D}_1 + \mathbf{c}_1\left(\hat{\dot{\mathbf{e}}}_1 - \dot{\mathbf{e}}_1\right)\right)\\
&\le -\mathbf{n}_1\|\hat{\mathbf{s}}_1\| - \mathbf{k}_1\hat{\mathbf{s}}_1^{\mathrm{T}}\hat{\mathbf{s}}_1 + \mathbf{n}_1\|\hat{\mathbf{s}}_1 - \mathbf{s}_1\| + \|\hat{\mathbf{s}}_1 - \mathbf{s}_1\|\|\mathbf{k}_1\|\|\hat{\mathbf{s}}_1\| + \left(\|\hat{\mathbf{D}}_1 - \mathbf{D}_1\| + \|\hat{\mathbf{f}}_1 - \mathbf{f}_1\| + \mathbf{c}_1\left\|\hat{\dot{\mathbf{P}}} - \dot{\mathbf{P}}\right\|\right)\|\hat{\mathbf{s}}_1\|\\
&\quad + \|\hat{\mathbf{s}}_1 - \mathbf{s}_1\|\left(\|\hat{\mathbf{D}}_1 - \mathbf{D}_1\| + \|\hat{\mathbf{f}}_1 - \mathbf{f}_1\| + \mathbf{c}_1\left\|\hat{\dot{\mathbf{P}}} - \dot{\mathbf{P}}\right\|\right)
\end{aligned}$$

Because estimated values from extended states observer are limitted, this term is limited, so with appropriate positive control parameters $\mathbf{n}_1$, $\mathbf{k}_1$ then $\dot{V}_1 \le 0$. In oder to control the rotational of the quadrotor, the inner control loop designed similarly to the position control loop.

$$\mathbf{M} = \mathbf{I}\left(-\hat{\mathbf{f}}_2 - \mathbf{I}^{-1}\hat{\mathbf{D}}_2 + \ddot{\mathbf{Q}}_{\mathbf{d}} + \mathbf{c}_2\hat{\dot{\mathbf{e}}}_2 + \mathbf{n}_2 \cdot \mathrm{sat}(\hat{\mathbf{s}}_2) + \mathbf{k}_2\hat{\mathbf{s}}_2\right) \tag{19}$$

where $\mathbf{c}_2 = \mathrm{diag}(c_\phi, c_\theta, c_\psi)$, $\mathbf{n}_2 = \mathrm{diag}(n_\phi, n_\theta, n_\psi)$, $\mathbf{k}_2 = \mathrm{diag}(k_\phi, k_\theta, k_\psi)$ are positive control parameters.

4 Simulation Results

The simulation is conducted using Matlab/Simulink. To compare the results, simulations using 3 controllers: proposed controller 1 is SMC combined with ESO, controller 2 is regular SMC, and MPC (controller 3) proposed in [14]. Consider the error of the model parameters to be about 10% different from the actual value, and the QUAV is flying with the slung payload suspended by a cable in windy environment. The parameters of QUAV and motor are taken from [15]. Figures 2, 3, 4 and 5 show the response trajectory over time of these controllers. In these figures, with controller 3, sometimes have larger tracking errors because of impacts from environment, load, and uncertain parameters by observers. Both controller 1 and controller 2 can solve this problem. In controller 2, with high **n** control parameter, the tracking errors are reduced, but it makes high "percent of overshoot" at the initial time. Via ESO, controller 1 solves these problems. By estimating the disturbances, this controller can resolve the impacts of the environment and uncertain parameters without high control parameters as controller 2. Figure 6 shows more clearly the stability of controllers 1 and 2 in space.

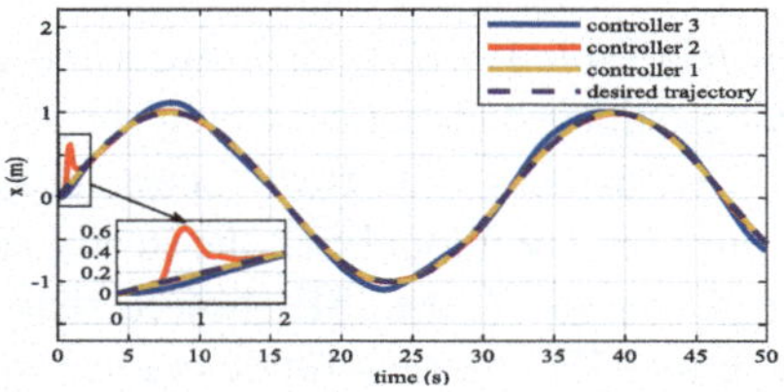

Fig. 2. Response x

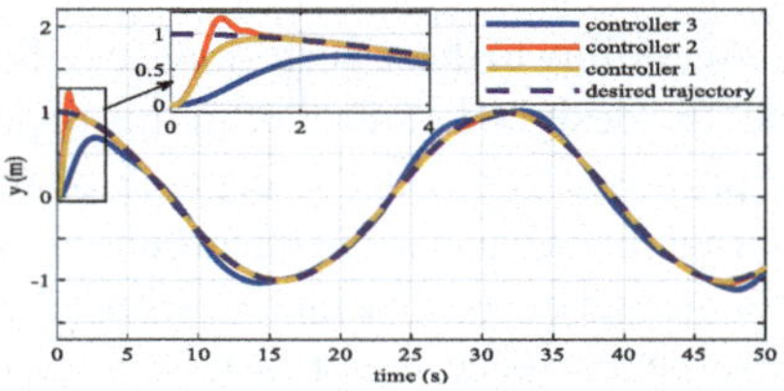

Fig. 3. Response y

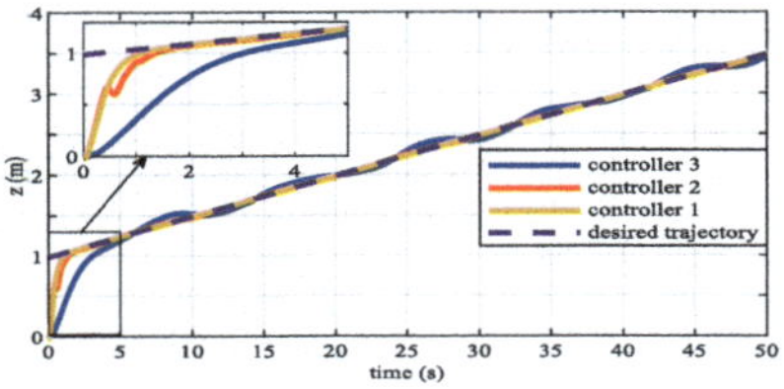

Fig. 4. Response z

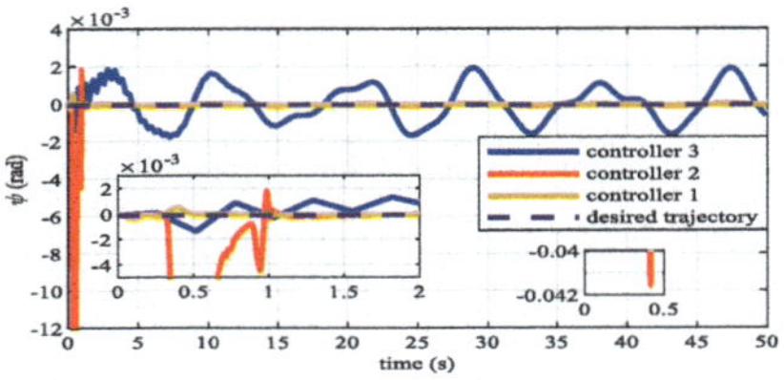

Fig. 5. Response ψ

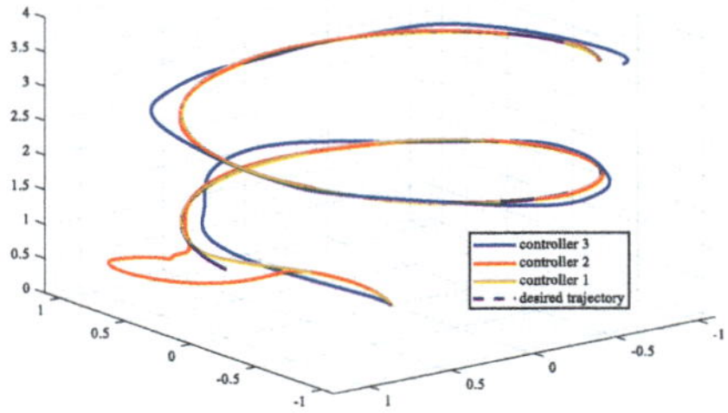

Fig. 6. Tracking trajectory in space

Aiming to evaluate the tracking outcomes in quantitation, three performance indices ISE, IAE, and ITAE described in [16] are used in this study. The ISE_p and ISE_a indices are integral squared errors of position and attitude. The lower these indices are, the faster convergence speed of the controller is IAE_p and IAE_a are integral absolute errors of position and attitude. The smaller the IAE indice, the less oscillation the system experiences. However, this also means that the convergence speed is slower. Integral time-multiplied absolute position and attitude errors are ITAEp and ITAEa, which prioritize steady-state error over initial response.

$$ISA_p = \int_0^t \left(e_x^2(t) + e_y^2(t) + e_z^2(t)\right)dt \qquad ISA_a = \int_0^t \left(e_\phi^2(t) + e_\theta^2(t) + e_\psi^2(t)\right)dt$$

$$IEA_p = \int_0^t \left(|e_x(t)| + |e_y(t)| + |e_z(t)|\right)dt \quad IEA_a = \int_0^t \left(|e_\phi(t)| + |e_\theta(t)| + |e_\psi(t)|\right)dt$$

$$ITEA_p = \int_0^t t\left(|e_x(t)| + |e_y(t)| + |e_z(t)|\right)dt \; ITEA_p = \int_0^t t\left(|e_\phi(t)| + |e_\theta(t)| + |e_\psi(t)|\right)dt$$

According to these performance indices shown in Table 1, with both position and attitude, the designed controller in this paper provides smallest statistics, which showsthat the proposed controller is capable of remain its robustness and effectiveness, despite external disturbances and uncertainties. With controller 2, without ESO, the steady-state error is higher, and because of high "percent of overshot", the ISE showed that the convergence speed is not fast. With controller 3, by MPC in the attitude control loop, although has a fast convergence speed, there are many tracking errors.

Table 1. The performance comparisons of different controllers

Indices	ISEp	ISEa	IAEp	IAEa	ITAEp	ITAEa
Proposed controller 1	**0.638**	**0.001263**	**1.128**	**0.03305**	**4.487**	**0.2292**
Controller 2	0.7135	1.958	2.242	1.099	26.31	0.8829
Controller 3	0.7405	0.0034	2.6616	0.6167	47.8913	12.7748

5 Conclusions

With the combination of sliding mode control and extended state observer, this paper has proposed SMC-ESO controller and demonstrated its stability against impacts from environment, load, and uncertain parameters by simulation on Mathlab\Simulink. This solves the stability problem of traditional SMC against large external disturbances and also reduces the number of sensors on the QUAV because the observer has estimated the object's states. When reducing the number of sensors on the object, not only the total weight but also the cost to create the QUAV will decrease. This controller has partly solved the trajectory tracking problem of the QUAV. However, because of the high speed and using many derivative stages, when the desired signal changes suddenly, there can appear overshot, chattering phenomenon and can cause instability to the system. To solve this problem, in the future, filters will be considered to reduce the sudden changes in the desired signals.

References

1. Ahirwar, S., et al.: Application of drone in agriculure. Int. J. Curr. Microbiol. App. Sci. **8**(01), 2500–2505 (2019)
2. Kocer, B.B., Tjahjowidodo, T., Seet, G.G.L.: Centralized predictive ceiling interaction control of quadrotor VTOL UAV. Aerosp. Sci. Technol. **76**, 455–465 (2018)
3. Labbadi, M., Boukal, Y., Cherkaoui, M.: QUAV modeling. In: Kacprzyk, J. (eds.) Advanced Robust Nonlinear Control Approaches for Quadrotor Unmanned Aerial Vehicle, pp 19–47 (2022)
4. Guo, K., Jia, J., Yu, X., Guo, L., Xie, L.: Multiple observers based anti-disturbance control for a quadrotor UAV against slung-payload and wind disturbances. Control. Eng. Pract. **102**, 104560 (2020)
5. He, Z., Zhao, L.: A simple attitude control of quadrotor helicopter based on ziegler-nichols rules for tuning pd parameters. Sci. World J. **2014**, 1 (2014)
6. Najm, A.A., Ibraheem, I.K.: Nonlinear PID controller design for a 6-DOF UAV quadrotor system. Eng. Sci. Technol. Int. J. **22**(4), 1087–1097 (2019)
7. Labbadi, M., Cherkaoui, M., Guisser, M.: Modeling and robust integral sliding mode control for a quadrotor unmanned aerial vehicle. In: 2018 6th International Renewable and Sustainable Energy Conference (IRSEC). IEEE (2018)
8. Huang, T., et al.: Robust tracking control of a quadrotor UAV based on adaptive sliding mode controller. Complexity **2019**, 1–15 (2019)

9. Hassani, H., Mansouri, A., Ahaitouf, A.: Control system of a quadrotor UAV with an optimized backstepping controller. In: 2019 International Conference on Intelligent Systems and Advanced Computing Sciences (ISACS). IEEE (2019)
10. Dalwadi, N., Deb, D., Muyeen, S.M.: Adaptive backstepping controller design of quadrotor biplane for slung-payload delivery. IET Intell. Transp. Syst. **16**(12), 1738–1752 (2022)
11. Anwar, J., et al.: High gain and super-twisting sliding mode observers for the adaptive control of quadrotor UAV. In: 2016 13th International Bhurban Conference on Applied Sciences and Technology (IBCAST). IEEE (2016)
12. Rodriguez-Mata, A.E., Farza, M., M'Saad, M.: Altitude control of quadrator uavs using high gain observer-based output feedback high gain regulator. In: 2019 8th International Conference on Systems and Control (ICSC). IEEE (2019)
13. Liu, J., et al.: Advanced Sliding Mode Control, pp. 174–191. Springer Berlin Heidelberg (2011)
14. Iyer, C.A.: Model Predictive Control (MPC) of quadrotors using LPV techniques (Master's thesis, Universitat Politècnica de Catalunya) (2020)
15. Nguyen, H.S., et al.: Advanced motion control of a quadrotor unmanned aerial vehicle based on extended state observer. In: International Conference on System Science and Engineering. IEEE (2023)
16. Liu, K., Wang, R., Zheng, S., Dong, S., Sun, G.: Fixed-time disturbance observer-based robust fault-tolerant tracking control for uncertain quadrotor UAV subject to input delay. Nonlinear Dyn. **107**(3), 2363–2390 (2022)

Data-Based Optimal Tracking Control for Underactuated Quadrotor System with Unknown Dynamic and External Disturbance

Gia Khiem Dinh[1], Thanh Tung Bui[2], Van Nam Tran[3], and Phuong Nam Dao[1(✉)]

[1] School of Electrical and Electronic Engineering, Hanoi University of Science and Technology, Hanoi, Vietnam
nam.daophuong@hust.edu.vn
[2] International School, Vietnam National University, Hanoi, Vietnam
[3] Phenikaa University, Hanoi, Vietnam

Abstract. This paper is invested to address the optimal tracking problem for the nonlinear quadrotor with unknown dynamics and external disturbance. First, the quadrotor is an underactuated system with six degrees of freedom and four inputs. Moreover, the quadrotor is considered under the influence of the parameter uncertainties due to the unknown payloads and the external disturbance. To address this problem, a robust control structure is given, which consists of the position and attitude controllers, the position controller achieves the position tracking and generates the attitude reference, and the attitude controller archives the stability of the quadrotor. By using the RL method, the optimal control policies can be learned without the dynamic information. Finally, a simulation built in Matlab software is given to verify the effectiveness of the proposed robust controller.

Keywords: Data-based control · Reinforcement learning · Quadrotor · Optimal tracking control

1 Introduction

With increasing global attention to applications across various industries, quadrotors have attracted much attention in many control scenarios such as finite-time fault-tolerant control [5], attitude regulation [1]. Due to the complex nonlinear dynamic of the quadrotor, appropriate control designs are needed to obtain the control objective. Unlike classical control procedures such as backstepping technique in [5], optimal control can handle with complex constraints, minimizing cost function [1]. However, due to the challenges of directly solving optimal control design, Reinforcement learning (RL) control has been studied to develop for robotics control systems [2,3]. The disadvantage in RL control

© The Author(s), under exclusive license to Springer Nature Switzerland AG 2025
H. A. Le Thi et al. (Eds.): ICAMCS 2024, LNNS 1311, pp. 443–451, 2025.
https://doi.org/10.1007/978-3-031-90606-0_37

[3] is to require the knowledge of model, which is addressed by Off-Policy technique to obtain RL control under the complete complete dynamic uncertainties [2,4,6]. Furthermore, it is necessary to emphasize that the multivariable problem in quadrotor systems makes developing RL control difficult with cascade control structure. On the other hand, due to external disturbance in quadrotors, it requires the consideration of handling disturbance in optimal control problem. Motivated by above discussions, we intend to enhance the Model-Free RL control using Off-Policy technique for quadrotor systems while ensuring control effectiveness for tracking control problems. In this article, we devote to coping with optimal control problem of quadrotor systems by using a cascade control structure with data-based RL control scheme for each sub-system.

2 Preliminaries

2.1 Quadrotor System

The quadrotor aircraft is a rigid machine system with four rotors that can be independently controlled to generate the thrust and moments needed for its movement. In the light of [1], let $\mathbb{E} = [E_x, E_y, E_z]$ define the Earth fixed frame and $\mathbb{B} = [B_x, B_y, B_z]$ is the body fixed frame. Moreover, let $p = [x, y, z]^T \in \mathbb{R}^3$ be the position of the quadrotor in Earth fixed frame and $\xi = [\phi, \theta, \psi]^T \in \mathbb{R}^3$ is the Euler angles following the pitch angle ϕ, the roll angle θ, and the yaw angle ψ satisfying $0 \leq |\phi| < \pi/2$ and $0 \leq |\theta| < \pi/2$. According to [1], the full of quadrotor dynamics contains two subsystems, including the position subsystem and the attitude subsystem which are described by

$$\ddot{p} = \frac{T_p}{m} R_{B2E} c_z - g c_z + d_p \tag{1}$$

$$\ddot{\xi} = J^{-1} C(\xi, \dot{\xi})\dot{\xi} + J^{-1}\tau + d_\xi \tag{2}$$

where d_p and d_ξ are the external disturbance, m is mass of the quadrotor, $J = \text{diag}(J_x, J_y, J_z) \in \mathbb{R}^3$ is the moment matrix, $c_z = [0, 0, 1]^T$, and g is gravitational acceleration constant. The following rotation matrix $R_{B2E} \in \text{RO}(3)$ transfers the position of the quadrotor from the body fixed frame to the Earth fixed frame.

$$R_{\mathbb{B}2\mathbb{E},i} = \begin{bmatrix} c_\theta s_\psi & c_\psi s_\phi s_\theta - c_\phi s_\psi & s_\phi s_\psi + c_\phi c_\psi s_\theta \\ c_\theta s_\psi & c_\phi c_\psi + s_\phi s_\theta s_\psi & c_\phi s_\theta s_\psi - c_\psi s_\phi \\ -s_\theta & c_\theta s_\phi & c_\phi c_\theta \end{bmatrix} \tag{3}$$

where $s_i = \sin(i)$ and $c_i = \cos(i)$. The Coriolis matrix $C(\zeta_i, \dot{\zeta}_i) \in \mathbb{R}^{3\times3}$ is shown in [5]. The total thrust $T_p = k_f u_z$ and torque $\tau = [lk_f u_\phi, lk_f u_\theta, k_t u_\psi]^T \in \mathbb{R}^3$. In practical application, the desired control input $[u_z, u_\phi, u_\theta, u_\psi]$ can be generated by adjusting the speed of rotors ($\omega_i, i = 1, 2, 3, 4$). The main objective of this paper is to design a data-based RL controller that achieves optimal tracking with a predefined trajectory for the underactuated quadrotor subjected to unknown dynamics and external disturbance. Both position and attitude controllers in Fig. 1 are applied data-based RL algorithm (see Sect. 2.2). By using the RL algorithm, the optimal control policies can be learned without the knowledge of the quadrotor dynamic information.

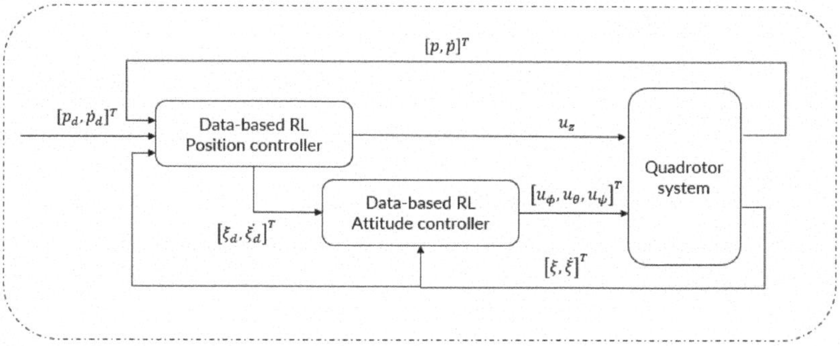

Fig. 1. The control structure of the quadrotor.

2.2 Data-Based RL Optimal Tracking Control

First, consider a nonlinear affine system being defined as

$$\dot{x} = f(x) + g(x)u(x) + d \tag{4}$$

where $x = [x_1, ..., x_n]^T \in \mathbb{R}^n$ is state vector, $u(x) = [u_1(x), ..., u_m(x)] \in \mathbb{R}^m$ is feedback control input vector, and $d = [d_1, ..., d_q] \in \mathbb{R}^q$ is external disturbance vector. $f(x) \in \mathbb{R}^n$ is the drift dynamics and $g(x) \in \mathbb{R}^{n \times m}$ is input dynamics.

Assumption 1. *It is assumed that $f(x)$ and $g(x)$ are unknown and satisfy the Lipchitz continuous condition with $f(0) = 0$. The external disturbance d is unknown and bounded.*

Assumption 2. *The desired trajectory $r(t) \in \mathbb{R}^n$ is generated from a command generator system following as*

$$\dot{r}(t) = f_d(r(t)) \tag{5}$$

where $f_d(r(t)) \in \mathbb{R}^n$ satisfy Lipschitz continuous condition with $f_d(0) = 0$.

Let $e(t) = x(t) - r(t) \in \mathbb{R}^n$ as tracking error, the tracking error dynamics is given as

$$\dot{e} = f(x) + g(x)u(x) + d - f_d(r) \tag{6}$$

The optimal control policy $u^*(x)$ is designed to attenuate the influence of the external disturbance d, the disturbance attenuation condition is defined as

$$\int_t^{\infty} e^{-\alpha(\tau - t)}(e^T Q e + u^T R u) d\tau \leq \gamma^2 \int_t^{\infty} e^{-\alpha(\tau - t)}(d^T d) d\tau \tag{7}$$

where α is positive discounted factor, $\gamma > 0$ is the amount of disturbance attenuation, $Q \in \mathbb{R}^{n \times n}$ and $R = R^T \in \mathbb{R}^{m \times m}$ are positive define matrix. It can be considered that the effects of the external disturbance can be attenuated by a degree at least equal to γ. Once, the discounted performance function can be defined as

$$V(e, u, d) = \int_t^{\infty} e^{-\alpha(\tau - t)}(e^T Q e + u^T R u - \gamma^2 d^T d) d\tau \tag{8}$$

3 Control Design

The control structure is proposed in Fig. 1 with the RL control design to be presented in Sects. 3.1 and 3.2.

3.1 Position Controller Design

Let $z_p = [p^T, \dot{p}^T] \in \mathbb{R}^6$ as position state vector, the position subsystem in Eq. (1) can be rewritten as

$$\dot{z}_p = A_p z_p + B_p u_p + D_p d_p \tag{9}$$

where $A_p = \begin{bmatrix} I_3 & 0_3 \\ 0_3 & 0_3 \end{bmatrix}$, $B_p = \begin{bmatrix} 0_3 \\ m^{-1} k_f I_3 \end{bmatrix}$, and $D_p = \begin{bmatrix} 0_3 \\ I_3 \end{bmatrix}$. The virtual control input $u_p = [u_{px}, u_{py}, u_{pz}]^T$ is defined as

$$\begin{aligned} u_{px} &= u_z(sin\phi sin\psi + cos\phi cos\psi sin\theta) \\ u_{py} &= u_z(cos\phi sin\theta sin\psi - cos\psi sin\phi) \\ u_{pz} &= u_z(cos\phi cos\theta) - g \end{aligned} \tag{10}$$

It is assumed that the desired position state $z_{pd} = [p_d^T, \dot{p}_d^T]^T \in \mathbb{R}^6$ (with $p_d = [x_d, y_d, z_d]^T \in \mathbb{R}^3$) is generated by the generator system following as

$$\dot{z}_{pd} = A_{pd} z_{pd} \tag{11}$$

where $A_{pd} \in \mathbb{R}^{6\times 6}$ is system matrix. Let $Z_p = [z_p^T, z_{pd}^T]$ as the augmented position state, the augmented position subsystem is given as

$$\begin{aligned} \dot{Z}_p &= AZ_p + Bu_p + Dd_p \\ e_p &= C_p Z_p \end{aligned} \tag{12}$$

where $e_p = [e_x, e_y, e_z] \in \mathbb{R}^3$ donates the position tracking error, $A = \text{diag}(A_{pd}, A_p)$, $B = [0_{3\times 6}, B_p^T]^T$, $D = [0_{3\times 6}, D_p^T]^T$, and $C = [I_3, 0_3, -I_3, 0_3]$. To achieve the optimal tracking objective, similar to the performance function in Eq. (8), a discounted performance function is defined as

$$V_p(Z_p, u_p, d_p) = \int_t^{\infty} e^{-\alpha_p(\tau - t)} (Z_p^T \overline{Q}_p Z_p + u_p^T R_p u_p - \gamma_p^2 d_p^T d_p) d\tau \tag{13}$$

where $\overline{Q}_p = C_p^T Q_p C_p$, α_p is positive discounted factor, $\gamma_p > 0$ is the amount of disturbance attenuation, $Q_p \in \mathbb{R}^{3\times 3}$ and $R_p = R_p^T \in \mathbb{R}^{3\times 3}$ are positive define matrix. To find the optimal position policies, a RL Algorithm is applied with three neural networks (NNs) to approximate the optimal policies

$$\begin{aligned} \hat{V}_p^i(Z_p) &= [W_{vp}^i]^T \varphi_v(Z_p) \\ \hat{u}_p^i(Z_p) &= [W_{up}^i]^T \varphi_u(Z_p) \\ \hat{d}_p^i &= [W_{dp}^i]^T \varphi_d(Z_p) \end{aligned} \tag{14}$$

where $W_{vp}^{i} \in \mathbb{R}^{l_v}$, $W_{up}^{i} \in \mathbb{R}^{3\times l_u}$, and $W_{dp}^{i} \in \mathbb{R}^{3\times l_d}$ are weight matrixes, and $\varphi_v(Z_p) \in \mathbb{R}^{l_v}$, $\varphi_u(Z_p) \in \mathbb{R}^{l_u}$, and $\varphi_d(Z_p) \in \mathbb{R}^{l_d}$ are action functions with l_v, l_u, l_d are the number of hide layer neuron. By the Least-square method, the optimal position policies can be learned without the quadrotor dynamic information. According to Eq. (10), once the optimal virtual position control policy is determined, the desired Euler angles $\xi_d = [\phi_d, \theta_d, \psi_d]^T \in \mathbb{R}^6$ and the total force u_z are computed by setting the desired yaw angle equal to zero.

Algorithm 1. Data-based RL Algorithm for Position controller

1: **Initialization** with $u_p = u_p^s + u_p^e$ with u_p^s is the stabilizing control input and u_p^e is the exploring noise input satisfying the persistently exciting (PE) condition, collect the quadrotor position data X_p, u_p, and d_p.

2: Solving HJI equation to obtain V_p^i and u_p^{i+1}.

$$\begin{aligned} V_p^i(Z_p(t+T)) - V_p^i(Z_p(t)) = &-\int_t^{t+T} (Z_p^T \overline{Q}_p Z_p) d\tau - \int_t^{t+T} ([u_p^i]^T R u_p^i) d\tau \\ &+ \gamma_p^2 \int_t^{t+T} ([d_p^i]^T d_p^i) d\tau + \alpha_p \int_t^{t+T} (V_p^i(Z_p)) d\tau \\ &- 2\int_t^{t+T} ([u_p^{i+1}]^T R_p (u_p - u_p^i)) d\tau \\ &- 2\gamma_p^2 \int_t^{t+T} ([d_p^{i+1}]^T (d_p - d_p^i)) d\tau \end{aligned} \tag{15}$$

3: Set $i = i + 1$ and go to 2 until $||u_p^i - u_p^{i+1}|| \leq \epsilon_p$

$$\begin{aligned} u_z &= (u_{pz} + m k_f^{-1} g)/\cos(\phi)/\cos(\theta) \\ \phi_d &= \arcsin((\cos(\phi)\sin(\theta)\sin(\psi) - u_{py}/u_z)/\cos(\psi)) \\ \theta_d &= \arcsin((u_{px}/u_z - \sin(\phi)\sin(\psi))/\cos(\phi)/\cos(\psi)) \end{aligned} \tag{16}$$

3.2 Attitude Controller Design

Let $z_e = [\xi^T, \dot{\xi}^T] \in \mathbb{R}^6$ as attitude state vector, the attitude subsystem in Eq. (2) can be rewritten as

$$\dot{z}_e = f_e(z_p) + g_e u_e + H_e d_e \tag{17}$$

where $f_e = \begin{bmatrix} \dot{\xi} \\ J^{-1} C(\xi, \dot{\xi})\dot{\xi} \end{bmatrix}$, $g_e = \begin{bmatrix} 0_3 \\ \mathrm{diag}(lk_f, lk_f, k_t) \end{bmatrix}$, and $H_e = \begin{bmatrix} 0_3 \\ I_3 \end{bmatrix}$, $u_e = [u_\phi, u_\theta, u_\psi]^T$ is the desired attitude control input. Assumed that the desired attitude reference $z_{ed} = [\xi_d^T, \dot{\xi}_d^T]^T \in \mathbb{R}^6$ can be generated by the system using Eq. (16)

$$\dot{z}_{ed} = f_{ed}(z_{ed}) \tag{18}$$

where $f_{ed}(z_{ed})$ is unknown smooth function. Let $Z_e = [z_e^T, z_{ed}^T]$ as the augmented attitude state, the augmented attitude subsystem is given as

$$\begin{aligned}\dot{Z}_e &= F(Z_e) + Gu_e + Hd_e \\ e_e &= C_e Z_e\end{aligned} \tag{19}$$

where $e_e = [e_\phi, e_\theta, e_\psi] \in \mathbb{R}^3$ donates the attitude tracking error, $F = [f_{ed}(z_{ed}), f_e(z_e)), G = [0_{3\times 6}, g_e^T]^T$, $h = [0_{3\times 6}, H_e^T]^T$, and $C_e = [I_3, 0_3, -I_3, 0_3]$. To achieve the optimal tracking objective, similar to the performance function in Eq. (8), a discounted performance function is defined as

$$V_e(Z_e, u_e, d_e) = \int_t^\infty e^{-\alpha_e(\tau - t)}(Z_e^T \overline{Q}_e Z_e + u_e^T R_e u_e - \gamma_e^2 d_e^T d_e)d\tau \tag{20}$$

where $\overline{Q}_e = C_e^T Q_e C_e$, α_e is positive discounted factor, $\gamma_e > 0$ is the amount of disturbance attenuation, $Q_e \in \mathbb{R}^{3\times 3}$ and $R_e = R_e^T \in \mathbb{R}^{3\times 3}$ are positive define matrix. Similar to the Sect. 3.1, three neural networks (NNs) are employed to approximate the optimal policies

Algorithm 2. Data-based RL Algorithm for Attitude controller

1: **Initialization** with $u_e = u_e^s + u_e^e$ with u_e^s is the stabilizing control input and u_e^e is the exploring noise input satisfying the persistently exciting (PE) condition, collect the quadrotor position data Z_e, u_e, and d_e.

2: Solving HJI equation to obtain V_e^i and u_e^{i+1}.

$$\begin{aligned}V_e^i(Z_e(t+T)) - V_e^i(Z_e(t)) = &- \int_t^{t+T} (Z_e^T \overline{Q}_e Z_e)d\tau - \int_t^{t+T} ([u_e^i]^T R u_e^i)d\tau \\ &+ \gamma_e^2 \int_t^{t+T} ([d_e^i]^T d_e^i)d\tau + \alpha_e \int_t^{t+T} (V_e^i(Z_e))d\tau \\ &- 2\int_t^{t+T} ([u_e^{i+1}]^T R_e(u_e - u_e^i)d\tau \\ &- 2\gamma_e^2 \int_t^{t+T} ([d_e^{i+1}]^T (d_e - d_e^i))d\tau\end{aligned} \tag{21}$$

3: Set $i = i + 1$ and go to 2 until $||u_e^i - u_e^{i+1}|| \leq \epsilon_e$.

$$\begin{aligned}\hat{V}_e^i(Z_p) &= [W_{ve}^i]^T \varphi_v(Z_e) \\ \hat{u}_e^i(Z_p) &= [W_{ue}^i]^T \varphi_u(Z_e) \\ \hat{d}_e^i &= [W_{de}^i]^T \varphi_d(Z_e)\end{aligned} \tag{22}$$

where $W_{ve}^i \in \mathbb{R}^{l_v}$, $W_{ue}^i \in \mathbb{R}^{3\times l_u}$, and $W_{de}^i \in \mathbb{R}^{3\times l_d}$ are weight matrixes, and $\varphi_v(Z_e) \in \mathbb{R}^{l_v}$, $\varphi_u(Z_e) \in \mathbb{R}^{l_u}$, and $\varphi_d(Z_e) \in \mathbb{R}^{l_d}$ are action functions with l_v, l_u, l_d are the number of hide layer neuron. By the Least-square method, the optimal position policies can be learned without the quadrotor dynamic information.

4 Simulation Result

In this section, a simulation result is built in Matlab software, which is given to verify the optimal tracking performance of the proposed controller. The quadrotor parameters as $m = 1.7\,(\mathrm{kg})$, $k_f = 1$, $k_t = 1$, $l = 0.2\,(\mathrm{m})$, $J = 10^{-3}\mathrm{diag}([4.9, 4.9, 9.8])\,(\mathrm{Kgm}^2)$, and $g = 9.8\,(\mathrm{m/s}^2)$. The desired position is generated by the system with

$$A_{pd} = \begin{bmatrix} 0 & 1 & 0 & 0 & 0 & 0 \\ -0.04 & 0 & 0 & 0 & 0 & 0 \\ 0 & 0 & 0 & 0 & 1 & 0 \\ 0 & 0 & -0.04 & 0 & 0 & 0 \\ 0 & 0 & 0 & 0 & 1 & 0 \\ 0 & 0 & 0 & 0 & 0 & 1 \end{bmatrix} \tag{23}$$

The initial position of three quadrotors are $p_d = [0, 5, 0]^T$, $\dot{p}_d = [1, 0, 1]^T$, $p = [2, 4, 1.5]^T$, $\dot{p} = [0, 0, 0]^T$, $\xi = [0, 0, 0.0]^T$, and $\dot{\xi} = [0, 0, 0]^T$. The external disturbance is added in the quadrotor dynamics as

$$\begin{aligned} d_p &= \begin{bmatrix} \sin(0.2)^2 \cos(0.1t)\,(\sin(x) + \sin(x)\cos(\dot{x})) \\ \sin(0.1t)^2 \sin(0.2t)\,(\sin(y) + \sin(y)\cos(\dot{y})) \\ \sin(0.15t)^2\,(\sin(z) + \sin(z)\cos(\dot{z})) \end{bmatrix} \\ d_e &= \begin{bmatrix} 0.5\sin(0.2)\left(\sin(\phi) + \sin(\dot{\phi})\right) \\ 0.5\sin(0.1t)\left(\sin(\theta) + \sin(\dot{\theta})\right) \\ 0.5\sin(0.15t)\left(\sin(\psi) + \sin(\dot{\psi})\right) \end{bmatrix} \end{aligned} \tag{24}$$

First, a PD controller is used as the initial stabilizing for three quadrotors to generate data with the exploring noises chosen as the sum of sin function with random frequencies. The optimal term in RL algorithms for Position and Attitude control is chosen as $\gamma_p = \gamma_e = 3$, $\alpha_p = \alpha_e = 0.05$, $Q_p = 50I_3$, $R_p = I_3$, $Q_e = 100I_3$, $R_e = I_3$.

Figure 2a and 2b show the convergence of the RL algorithm of the position and attitude control, respectively. After 5 iterations, the weight of NNs converge. Figure 3a and 3b show the position and attitude tracking errors of the quadrotor, respectively. After 2 s, the position tracking error converges to zero and after 0,5 s the attitude tracking error converges to zero. To verify the trajectory tracking effectiveness, the trajectory tracking of the quadrotor is shown in Fig. 4.

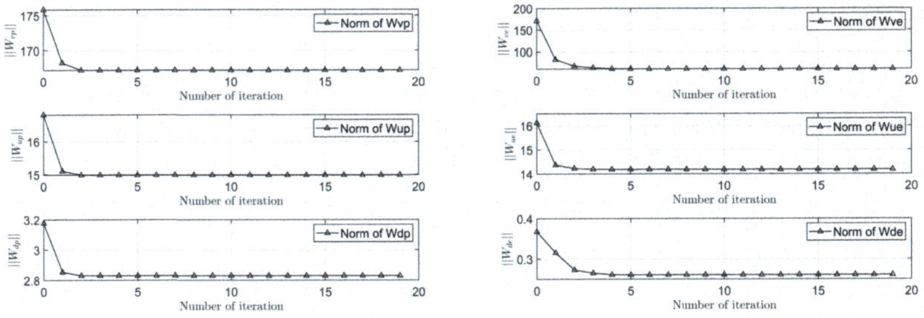

(a) Convergence of Algorithm 3 (b) Convergence of Algorithm 4

Fig. 2. The convergence of RL algorithm.

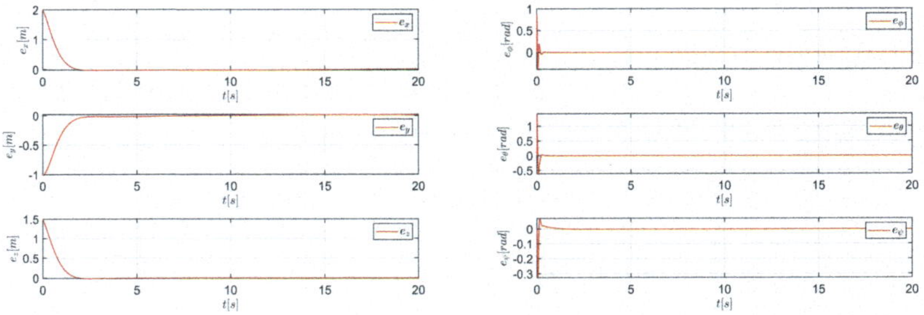

(a) The position tracking error (b) The attitude tracking error

Fig. 3. The quadrotor tracking errors.

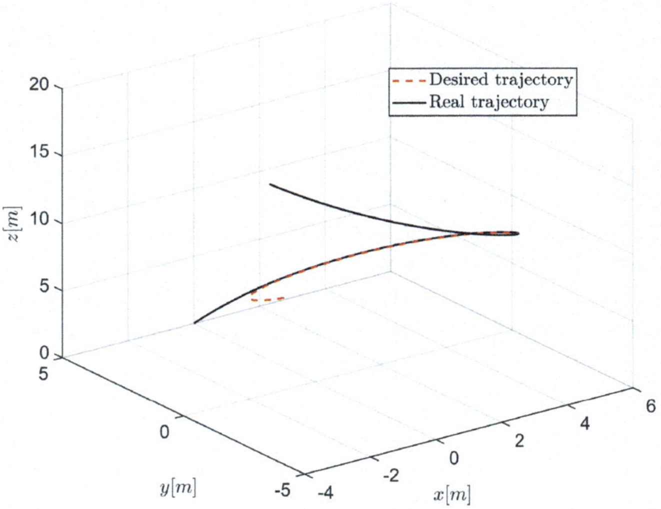

Fig. 4. The 3D trajectory.

5 Conclusion

This paper aims to design a robust control law to achieve optimal trajectory tracking for the underactuated quadrotor system subject to high nonlinearity, strong coupling, parameter uncertainties, and external disturbance. By using the data-based RL method for both position and attitude controllers, the optimal control policies can be learned without the knowledge of the quadrotor dynamic information. The position controller achieves the optimal trajectory tracking and generates the desired attitude reference, while the attitude controller achieve the stability of quadrotor system. Finally, a simulation result is given to demonstrate the effectiveness of the proposed controller.

References

1. Chen, Z., Zhong, W., Xie, S., Zhang, Y., Yuen, C.: Observer-based robust integral reinforcement learning for attitude regulation of quadrotors. Knowl.-Based Syst. 112360 (2024)
2. Dao, P.N., Duc, H., et al.: Nonlinear rise based integral reinforcement learning algorithms for perturbed bilateral teleoperators with variable time delay. Neurocomputing **605**, 128355 (2024)
3. Nguyen, K., Dang, V.T., Pham, D.D., Dao, P.N.: Formation control scheme with reinforcement learning strategy for a group of multiple surface vehicles. Int. J. Robust Nonlinear Control **34**(3), 2252–2279 (2024)
4. Tian, Y., Wan, H., Karimi, H.R., Luan, X., Liu, F.: Model-free aperiodic tracking for discrete-time systems using hierarchical reinforcement learning. Neurocomputing 128486 (2024)
5. Wu, Y., Liang, H., Xuan, S., Zhang, X.: Extended state observer based finite-time fault-tolerant formation control for multi-UAVs. J. Franklin Inst. 107158 (2024)
6. Zhu, L., Wei, Q., Guo, P.: Online off-policy reinforcement learning for optimal control of unknown nonlinear systems using neural networks. IEEE Trans. Syst. Man Cybern. Syst. (2024)

Focal Hanning Loss: Revisiting the Heatmap Classification for UAV Self-localization

Van Quan Ngo[1], Nguyen Hai Long[3], Phan Huy Anh[2], Thi Thanh Tam Bui[2], and Chi Thanh Nguyen[1](✉)

[1] Institute of Information Technology, AMST, Hanoi, Vietnam
thanhnc@ioit.ai.vn
[2] Electronic Institute, AMST, Hanoi, Vietnam
[3] Hanoi University of Science and Technology, Hanoi, Vietnam

Abstract. This paper introduces the Focal Hanning Loss, a novel approach to improve UAV self-localization using heatmap classification. Building upon recent advancements in vision-based localization methods, we address limitations in existing techniques, particularly the Hanning Loss used in WAMF-FPI. Focal Hanning Loss incorporates a modulation factor that dynamically adjusts the importance of samples based on their difficulty, effectively mitigating issues such as red clusters in heatmaps and imbalanced weighting between positive and negative samples. Additionally, we enhance the model's robustness through data augmentation techniques, which include random cropping and scaling of satellite images to simulate multi-scale variations. To validate the effectiveness of Focal Hanning Loss combined with data augmentation, we trained the WAMF-FPI model on a custom dataset derived from DenseUAV, achieving a 3.4-point improvement in the Relative Distance Score metric compared to original Hanning Loss.

Keywords: UAV Localization · Heatmap · Vision Transformer

1 Introduction

Unmanned Aerial Vehicles (UAVs) have gained significant popularity in recent years, finding applications in various fields such as surveillance, agriculture, and mapping. UAVs heavily rely on GPS for navigation and localization, which can be unreliable in areas with weak signals. This limitation underscores the need for alternative localization methods.

Traditionally, feature matching methods such as SIFT (Scale-Invariant Feature Transform) were employed for localization [1]. However, these methods have several weaknesses. They are sensitive to variations in lighting and viewpoint, leading to reduced performance in cluttered or dynamic environments. Additionally, handling large-scale data or real-time requirements poses challenges.

© The Author(s), under exclusive license to Springer Nature Switzerland AG 2025
H. A. Le Thi et al. (Eds.): ICAMCS 2024, LNNS 1311, pp. 452–461, 2025.
https://doi.org/10.1007/978-3-031-90606-0_38

Another approach that gained traction is image retrieval [2–4]. In this method, the satellite image is divided into multiple overlapping patches. These patches are then embedded into feature vectors and stored in a vector database for similarity-based querying. Despite its utility, this approach has notable weaknesses. The accuracy is highly dependent on how the satellite images are split, requiring extensive pre-processing. Furthermore, it is computationally expensive in terms of memory and processing time.

In the past two years, a new approach has been introduced. These methods [5–9] involve using the UAV query image and the satellite search image as inputs to produce a heatmap, where each pixel's value represents the probability of that pixel being part of the foreground. WAMF-FPI introduces the Hanning Loss (HL), which assigns different weights to positive examples to emphasize the center of the region. Despite its advancements, there is still room for improvement due to certain weaknesses, which will be discussed in Sect. 3.2.

To address these shortcomings, we propose a novel loss function, Focal Hanning Loss (FHL). Our method incorporates a modulation factor that dynamically adjusts the importance of samples based on their difficulty, effectively mitigating issues caused by small negative weights. Additionally, we enhance the model's robustness through data augmentation techniques, which include random cropping and scaling of satellite images to simulate multi-scale variations. To validate the effectiveness of FHL combined with data augmentation, we trained the WAMF-FPI model on a custom dataset derived from DenseUAV, achieving a 3.4-point improvement in the Relative Distance Score (RDS) metric compared to HL.

The paper is structured as follows: Sect. 2 reviews related work in UAV localization. Section 3 introduces our proposed Focal Hanning Loss (FHL) and details its methodology. Section 4 outlines the experimental setup, including datasets and evaluation metrics. Section 4.3 presents the results and discussion. Finally, Sect. 5 concludes with a summary of our contributions and future research directions.

2 Related Work

2.1 Overview of Heatmap Classification UAV Localization Methods

After a long period dominated by image retrieval methods, FPI [5] pioneered the heatmap classification approach by introducing a new benchmark dataset called UL14 and adopting a Siamese-like architecture for their model, utilizing convolutional or vision transformer networks as the backbone feature extractor and the correlation operation from SiamFC [10].

WAMF-FPI [6] explicitly uses the Twins PCPVT-S [11] as the backbone model and integrates a weighted fusion module for feature correlation at different scales. This approach incorporates both low-channel spatial information and high-channel semantic information to enhance localization accuracy. Additionally, WAMF-FPI introduces the Hanning Loss, which assigns larger weights to the central region of positive samples. However, while improving positioning accuracy, it suffers from the occurrence of false positives when *negative_ weight* is set small, and blindly increasing this hyperparameter is not the right practice.

Unlike the two-stream transformer-based approach for UAV and satellite images in WAMF-FPI, [7–9] employed a one-stream transformer as the backbone extractor, arguing that this setup facilitates information interactions between the query and the search map in the early stages of the model, thereby increasing performance.

2.2 Focal Loss

To address the issue of extreme class imbalance in one-stage object detectors, [12] introduced the Focal Loss, a modification of the standard cross-entropy loss. This loss function incorporates a modulating factor that reduces the weight of well-classified examples, allowing the model to focus more on harder, misclassified samples. By doing so, Focal Loss effectively mitigates the overwhelming influence of easy-to-classify negatives, which tend to dominate the learning process in scenarios with a high imbalance between foreground and background classes. The approach has proven particularly useful in one-stage object detectors, where the model predicts across a vast number of potential object locations and must overcome the challenge of detecting sparse foreground objects against a majority of background regions.

Despite its significant success in the domain of object detection, the application of Focal Loss in heatmap classification, especially for UAV localization tasks, remains relatively underexplored. In UAV localization, where heatmaps are used to represent the probability of a UAV's location within a satellite image, the class imbalance issue persists, making effective loss functions crucial for accurate predictions. Our proposed Focal Hanning Loss (FHL) builds on the key strengths of both Focal Loss and Hanning Loss to address this challenge. By combining the hard example mining capabilities of Focal Loss with the spatial weighting properties of Hanning Loss, FHL offers a more powerful and precise solution for heatmap classification in UAV localization, ensuring better performance in identifying the UAV's position in complex environments.

3 Methodology

3.1 Problem Formulation and Hanning Loss

Formulation. Let $\mathbf{u}$ denote a UAV image of size $m \times m \times 3$, and let $\mathbf{s}$ denote a satellite image of size $n \times n \times 3$. The goal is to predict a heatmap $\mathbf{p}$ of size $n \times n$, where each pixel represents a probabilistic value that is sigmoid-bounded between 0 and 1. This prediction can be approximated by a function

$$F_\theta : (\mathbf{u}, \mathbf{s}) \rightarrow \mathbf{p}, \tag{1}$$

where θ denotes the parameters of the model. Suppose $\mathbf{h}$ denotes the ground truth heatmap. For a pixel i, define

$$\mathbf{h}_i = \begin{cases} 1, & \text{if } i \in \mathcal{R} \\ 0, & \text{otherwise,} \end{cases} \tag{2}$$

where $\mathcal{R}$ is a square region of size $R \times R$ centered at the pixel closest to the true UAV position. The hyperparameter R determines the boundary between positive and negative samples. A smaller R makes the model more vulnerable to noise and outliers, leading to overfitting and reduced generalization ability [8]. In practice, we chose R of size 33. The predicted UAV position is determined by the pixel corresponding to the maximum value in the predicted heatmap, given by

$$\hat{i} = \arg\max_{i} \mathbf{p}_i. \tag{3}$$

We can easily calculate the real-world coordinate of predicted position using left-top and right-bottom coordinates of the satellite image.

$$x_{\hat{i}} = x_{\text{lt}} + \left(\frac{\hat{i}_x}{n}\right) \cdot (x_{\text{rb}} - x_{\text{lt}}), \tag{4}$$

$$y_{\hat{i}} = y_{\text{lt}} + \left(\frac{\hat{i}_y}{n}\right) \cdot (y_{\text{rb}} - y_{\text{lt}}). \tag{5}$$

Hanning Loss. The Hanning Loss (HL) is defined as

$$HL(\mathbf{p}, \mathbf{h}) = \mathbf{w}BCE(\mathbf{p}, \mathbf{h}), \tag{6}$$

where BCE is the Binary Cross-Entropy Loss, given by

$$BCE(\mathbf{p}, \mathbf{h}) = -\log(\mathbf{q}), \tag{7}$$

where

$$\mathbf{q}_i = \begin{cases} \mathbf{p}_i, & \text{if } \mathbf{h}_i = 1, \\ 1 - \mathbf{p}_i, & \text{otherwise,} \end{cases} \tag{8}$$

and $\mathbf{w}$ is the Hanning weight tensor defined as

$$\mathbf{w}_i = \begin{cases} \frac{normalized\_hanning\_window(i)}{negative\_weight+1}, & \text{if } i \in \mathcal{R}, \\ \frac{negative\_weight}{no\_of\_negative\_samples \times (negative\_weight+1)}, & \text{otherwise.} \end{cases} \tag{9}$$

The normalized Hanning window is visualized in Fig. 1 and is defined as

$$normalized\_hanning\_window(i) = \frac{\sin^2\left(\frac{\pi x}{R-1}\right)\sin^2\left(\frac{\pi y}{R-1}\right)}{\sum_{x=0}^{R-1}\sum_{y=0}^{R-1}\sin^2\left(\frac{\pi x}{R-1}\right)\sin^2\left(\frac{\pi y}{R-1}\right)}, \tag{10}$$

where $(x, y) \in [0, R-1]^2$ represents the coordinates of pixel i within the region $\mathcal{R}$.

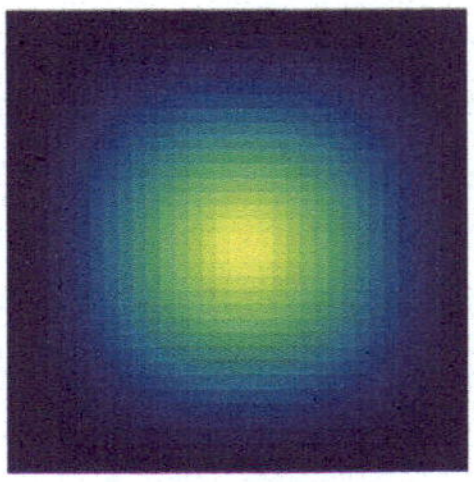

Fig. 1. The normalized 2D hanning window.

3.2 Focal Hanning Loss

Without the *negative_weight* hyperparameter, the weight of negative samples is significantly smaller than that of their positive counterparts, leading to the appearance of many red clusters in the heatmap, as shown in Fig. 2. [6] proposed a solution by introducing this hyperparameter. However, we believe this approach is not ideal, as it reduces the weight of positive samples by a factor of *negative_weight*, which can lower the intensity of these regions and potentially result in a predominantly blue heatmap.

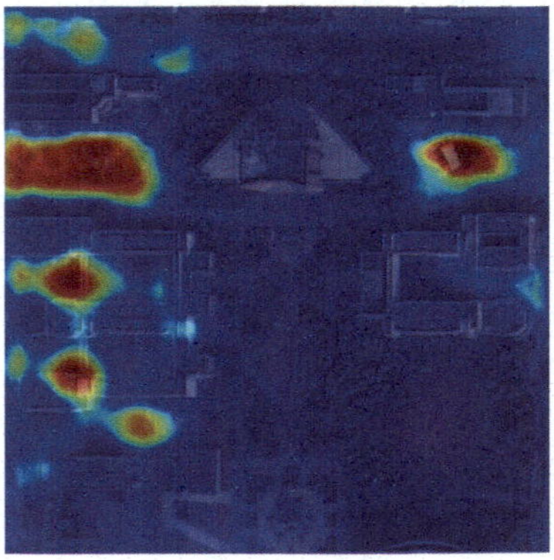

Fig. 2. Red clusters in the heatmap caused by small negative weights.

To address this issue, we propose a new loss function called the Focal Hanning Loss (FHL), defined as

$$FHL(\mathbf{p}, \mathbf{h}) = -\mathbf{w}(\mathbf{1} - \mathbf{q})^{\gamma} \log(\mathbf{q}). \tag{11}$$

The modulation factor $(1-\mathbf{q})^{\gamma}$ was first introduced in [12]. This factor reduces the importance of the loss for easy samples and increases it for hard samples. For example, when $\gamma = 2$, a sample with $\mathbf{q}_i = 0.1$ will be weighted 81 times more than a sample with $\mathbf{q}_i = 0.9$. The red clusters mentioned above are, in fact, candidate regions detected by the model. With this new loss function, the model is guided to distinguish among candidate regions at a finer level of detail. In practice we use a γ value of size 1.

3.3 WAMF-FPI

To experiment with the effectiveness of our proposed loss function, we will use the WAMF-FPI [6] model. Specifically, we utilize Spagnolo's implementation [13] with slight modifications of our own (see Fig. 3).

The model consists of a backbone module and a WAMF (Weight-Adaptive Multi-Feature Fusion) module. The backbone module is a two-stream network that adopts the Twins PCPVT-S [11] architecture to extract features at different scales. These feature tensors are later fused together using the correlation operation from SiamFC [10]. After that, we acquire three heatmap logits from three different scale levels, each assigned a weight. These weighted logits are summed and passed through a sigmoid function to produce the final heatmap output.

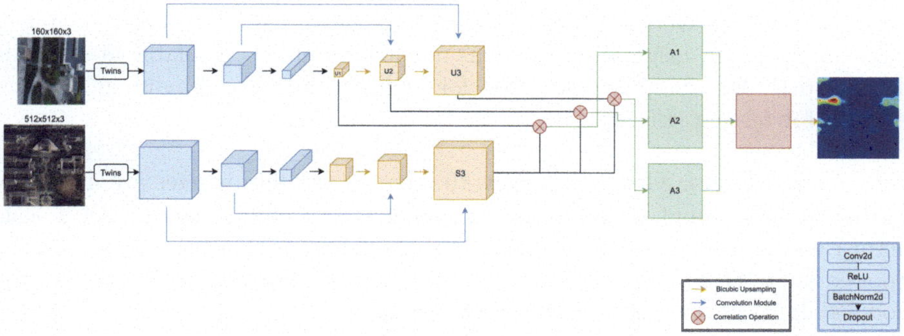

Fig. 3. The architecture of WAMF-FPI.

We move the Batch Normalization layer after the ReLU. While the order of these layers in the Convolution module is still debatable, we believe it is better to filter out negative signals before applying batch normalization. Moreover, Softmax is used for normalizing the weights of A_1, A_2, and A_3 to avoid complexity associated with negative weights when using sum normalization.

4 Experiments

4.1 Implementation Details

The size of the UAV and satellite images are 160×160 and 512×512, respectively. The training process is conducted on a single RTX 4090 GPU with a batch size of 8. The optimizer used is AdamW, with a learning rate of 0.004 for the fusion module and 0.001 for the pretrained backbone extractor. The learning rate scheduler is a multi-step scheduler with a decay factor of 0.2.

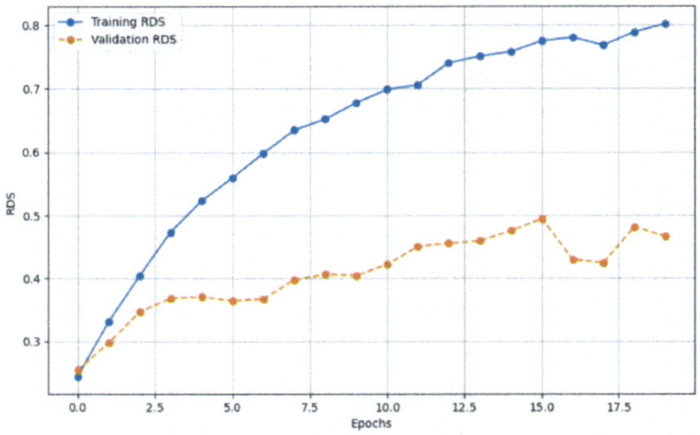

Fig. 4. RDS metric on training and validation sets over epochs.

4.2 Dataset and Evaluation Metrics

Dataset. For the dataset, we use 9,099 UAV images from 14 universities in the DenseUAV dataset [2]. The corresponding satellite images of these 14 campuses are retrieved from Google Maps. This setup replicates the UL14 dataset introduced in [5], as the original dataset was not made publicly available at the time of this work. Ten campuses are used for training, and the remaining four are reserved for testing (Fig. 4).

Data Augmentation. Given that only 10 training satellite images are available, we diversify the input by randomly cropping the satellite images with a scale factor ranging from 1 to 1.5, ensuring that the true position remains within the cropped area. This augmentation strategy reduces overfitting and enhances the model's ability to handle multi-scale variations.

Metrics. We employed the Relative Distance Score (RDS) as the main metric for evaluating performance. The formula is shown in Eq. 12.

$$RDS = e^{-k\sqrt{\frac{\frac{dx}{w}^2 + \frac{dy}{w}^2}{2}}}, \qquad (12)$$

where dx and dy are the distances between the predicted pixel and the ground truth pixel in the x and y directions, respectively; w and h are the width and height of the satellite image. The parameter k is an adjustable constant, which is set to 10 in this paper. We also use the Meter-level Accuracy (MA@K), which indicates the ratio of samples with a distance error of less than K meters. Although this metric is intuitive, it is not entirely reliable because it is heavily affected by the scale of the satellite image.

4.3 Result

Table 1. The impact of FHL and data augmentation (DA). * denotes Spagnolo's implementation, and ** denotes ours.

Model	DA	FHL	RDS (%)	MA@5 (%)	MA@10 (%)	MA@20 (%)	MA@50 (%)
WAMF-FPI*			43.21	-	-	-	-
WAMF-FPI**			44.51	-	-	-	-
WAMF-FPI**		✓	45.96	-	-	-	-
WAMF-FPI**	✓		46.06	3.22	**13.17**	31.83	49.94
WAMF-FPI**	✓	✓	**49.44**	**3.96**	12.44	**33.03**	**54.27**

The results of our experiments are summarized in Table 1, which compares the performance of Spagnolo's implementation and our modified WAMF-FPI model using both the original HL and the proposed FHL. The key performance metrics include the RDS and MA@K across various thresholds (5 m, 10 m, 20 m, 50 m).

Our FHL-enhanced model demonstrates a clear improvement over the original implementation, particularly when combined with data augmentation techniques. Specifically, we observe a 3.44% increase in RDS with data augmentation and 1.45% increase in RDS without data augmentation, indicating that FHL enhances the model's ability to accurately localize UAVs. This suggests that FHL effectively mitigates the issues of red clusters in heatmaps and imbalanced sample weighting, leading to a more precise localization of the UAV's position. Moreover, the MA@K metrics further validate the superiority of FHL. For instance, the MA@50 improves by approximately 4.33% when FHL is applied, demonstrating that the model can maintain high accuracy even at larger error thresholds. This is particularly important in real-world applications where slight deviations in UAV positioning may still be acceptable, but large errors can significantly impact operational effectiveness.

The qualitative results depicted in Fig. 5 provide visual confirmation of these improvements. The heatmaps generated using FHL show a notable reduction in false positives, especially in areas that were previously prone to red clusters. The cyan star, indicating the predicted UAV position, is consistently closer to the white star, which represents the ground truth, thereby confirming the quantitative improvements observed. In conclusion, our results demonstrate that FHL is a significant advancement over existing methods, providing improved accuracy and robustness in UAV self-localization. The incorporation of data augmentation further enhances these benefits, making FHL a viable option for real-world applications.

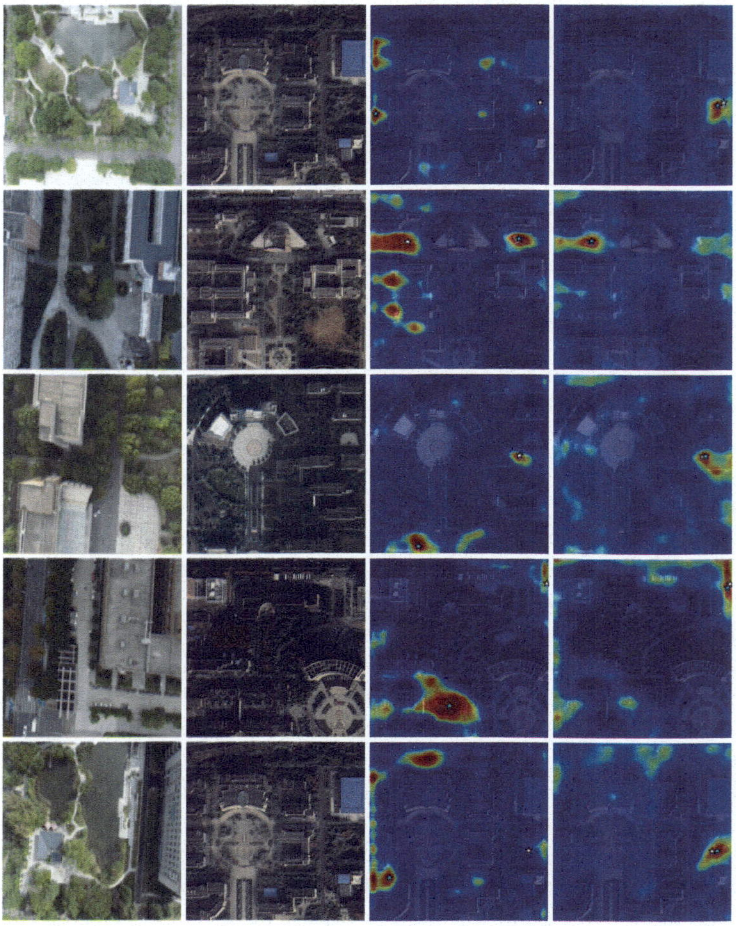

Fig. 5. The third column is the output produced when using HL in training, and the fourth column is the output produced when using FHL in training. The white star indicates the ground truth position and the cyan star indicates the predicted position.

5 Conclusion

Our proposed Focal Hanning Loss (FHL) significantly improves UAV localization accuracy compared to the standard Hanning Loss. The 3.4-point increase in RDS demonstrates FHL's effectiveness in addressing imbalanced sample weighting and reducing false positives in heatmap classification. This improvement is particularly notable given the challenging nature of UAV localization, where accuracy can significantly impact navigation and decision-making processes.

In future work, we aim to develop more robust data augmentation strategies to better handle real-world scenarios, where environmental conditions may vary drastically. A generative module could be introduced [8] to simulate diverse environmental changes, such as varying weather, lighting, or landscape alterations, enhancing the model's ability to generalize across different conditions. This would improve the robustness of UAV localization in dynamic, real-life applications.

Disclosure of Interests. Authors have no conflict of interest to declare.

References

1. Quan, N.V., et al.: Speeding up SIFT for faster feature-based UAV visual localization. In: International Conference on Advances in Information and Communication Technology, pp. 83–90. Springer, Cham (2023)
2. Dai, M., et al.: Vision-based UAV self-positioning in low-altitude urban environments. IEEE Trans. Image Process. **33**, 493–508 (2024). https://doi.org/10.1109/TIP.2023.3346279
3. Dai, M., et al.: A transformer-based feature segmentation and region alignment method for UAV-view geo-localization. IEEE Trans. Circuits Syst. Video Technol. **32**(7), 4376–4389 (2021)
4. Li, Q., et al.: GeoFormer: an effective transformer-based Siamese network for UAV geo-localization. IEEE J. Sel. Top. Appl. Earth Obs. Remote Sens. (2024)
5. Dai, M., et al.: Finding point with image: an end-to-end benchmark for vision-based UAV localization. arXiv preprint arXiv:2208.06561 (2022)
6. Wang, G., et al.: WAMF-FPI: a weight-adaptive multi-feature fusion network for UAV localization. Remote Sens. **15**(4), 910 (2023)
7. Chen, J., et al.: OS-FPI: a coarse-to-fine one-stream network for UAV geo-localization. IEEE J. Sel. Top. Appl. Earth Obs. Remote Sens. (2024)
8. Fan, J., et al.: A cross-view geo-localization algorithm using UAV image and satellite image. Sensors **24**(12), 3719 (2024)
9. He, Y., et al.: DCD-FPI: a deformable convolution-based fusion network for unmanned aerial vehicle localization. IEEE Access (2024)
10. Bertinetto, L., Valmadre, J., Henriques, J.F., Vedaldi, A., Torr, P.: Fully-convolutional Siamese networks for object tracking. In: Hua, G., Jégou, H. (eds.) ECCV 2016. LNCS, vol. 9914, pp. 850–865. Springer, Cham (2016). https://doi.org/10.1007/978-3-319-48881-3_56
11. Chu, X., et al.: Twins: revisiting the design of spatial attention in vision transformers. In: Advances in Neural Information Processing Systems, vol. 34, pp. 9355–9366 (2021)
12. Lin, T.-Y., et al.: Focal loss for dense object detection. In: Proceedings of the IEEE International Conference on Computer Vision, pp. 2980–2988 (2017)
13. Spagnolo, G.: Custom Twins transformer for Multi-View Image Fusion. https://github.com/spagnoloG/uav-localization-2023. Accessed 31 Mar 2024

Output Constrained Control Based Barrier Lyapunov Functions for Lateral Dynamics of Autonomous Vehicles

Van Trung Nguyen[1], Ngoc Chau Tran[1], Tien Dung Nguyen[1], Nhu Toan Nguyen[1], Thanh Ha Vo[2], Tung Lam Nguyen[1(✉)], and Duc Thinh Le[3(✉)]

[1] Hanoi University of Science and Technology, Hanoi 100000, Vietnam
lam.nguyentung@hust.edu.vn
[2] University of Transport and Communications, Hanoi 100000, Vietnam
[3] Thuyloi University, Hanoi 100000, Vietnam
thinhld@tlu.edu.vn

Abstract. This article studies control systems for an automatic lane change system for autonomous vehicles. In this study, a control law is based on a backstepping techniques (BSP) using barrier Lyapunov function (BLF) to constrain the tracking error to be limited within a predefined range and avoid the chattering phenomenon. The slow and fast dynamics of the system are controlled separately by the proposed control technique following the model reduction approach. A conventional sliding mode controller is used as a standard controller to evaluate the proposed controller. Co-simulation based on CarSim and Matlab/Simulink are used to demonstrate the effectiveness and robustness. Performance index of autonomous vehicles such as IWPEI, ITAE, IAPE, and IASE is expressed to quantitatively evaluate the proposed control system.

Keywords: Autonomous vehicle · Lateral dynamics · Lane changing · CarSim · Barrier Lyapunov Function

1 Introduction

The studies on autonomous vehicles have been widely conducted by many universities, research centers and car companies [1]. Because it has many advantages, such as improving car preeminent caused by driver error. However, there are several challenges associated with it, the most important is path tracking control which allows the vehicle to automatically follow the desired trajectory by adjusting longitudinal and lateral dynamics. The longitudinal control is designed to handle the challenges of the desired acceleration and the operations of speeding up and braking the vehicle. While lateral control plays a crucial role in lane keeping, lane changing, and achieving the desired lateral position [2]. In small road curvature conditions to simplify the dynamic model and separate controller design, the longitudinal and lateral motions are often considered separate [3]. In

© The Author(s), under exclusive license to Springer Nature Switzerland AG 2025
H. A. Le Thi et al. (Eds.): ICAMCS 2024, LNNS 1311, pp. 462–474, 2025.
https://doi.org/10.1007/978-3-031-90606-0_39

this paper, the lateral control of autonomous vehicles for lane change maneuvers is presented.

The problem of lateral control is a complicated field, there are many control strategies have been proposed to design different control topologies including PID control [4], Sliding Mode Control [5], Backstepping Control [6]. In [7], the authors introduce a combination of backstepping and sliding mode controller, and particle swarm optimization algorithm (PSO) is used to find optima values for some parameters affecting controller performance. So, the simulation results of the integrated controller are better than those of the backstepping controller, especially with low friction conditions. In [8], a hybrid approach with the use of an adaptive waypoint generator is developed, the authors have provided the background information, such as the timing and constraint conditions of the longitudinal distance along the road parametrized and the lateral distance orthogonal to the curve so the reference trajectory is generated. Model predictive control is a commonly used control strategy in the lateral dynamics for autonomous vehicles because of its capacity to resolve parametric uncertainties and time-varying problems [9,10]. However, the main problem in using MPC for practical vehicles is still the relatively high computational complexity. In [11], the authors proposed a novel approach, the model of the vehicle is reduced to a slow and a fast dynamic with the yaw is fast dynamics and the lateral position is slow dynamics. Then, this enables to reduce control design complexity by designing two separate controllers by inducing an auxiliary control input in slow dynamics. The authors developed a sliding mode controller using the Barrier Lyapunov Function to ensure the robustness of the system, but the control signal faced to chattering phenomenon then actuators could not respond, so this controller could not be used in practice.

In this study, backstepping control utilizing the Barrier Lyapunov Function was proposed to ensure constraints of the lateral error and avoid the chattering phenomenon. The co-simulation tests for validation of the proposed method are performed using the high-precision dynamic simulator CarSim and Matlab/Simulink. The results show that the efficiency of the proposed controller has outperformed the sliding mode control with lower errors. The main contribution of the article is to propose a controller that uses the Barrier Lyapunov Function for the Backstepping control method, helping us to constrain the vehicle's trajectory tracking error. Moreover, it is proposed to use the IWPEI evaluation index to show the performance of the proposed controller when implemented on CarSim.

The organization of this paper is outlined as follows: Sect. 2 delineates the vehicle dynamics model and the dynamic model in terms of error with respect to road. Section 3 introduces the backstepping controller with the barrier Lyapunov function. Section 4 evaluates the performance of the control system through simulation combining Matlab/Simulink and CarSim. Finally, Sect. 5 presents the conclusions.

2 Vehicle Dynamics Model

In this section, a math model for the lateral dynamics of the autonomous vehicle is presented. A two-degree-of-freedom vehicle model is considered, as shown in Fig. 1. Two degrees of freedom are represented by the vehicle's yaw angle ψ and lateral position y.

Assumption 1. Forces on front and rear vehicle wheels satisfy $F_f^l = F_f^r$, $F_r^l = F_r^r$ where F_f^l, F_f^r, F_r^l, F_r^r are the forces on the left-fronts, right-front, left-rear, and right-rear wheels. The longitudinal speed of the vehicle is assumed to be constant.

Remark 1. According to Assumption 1, the combined force on the rear tires of the vehicle F_r and the front tires of the vehicle F_r becomes double the force acting on the wheel such that $F_f = 2F_f^l = 2F_f^r$ and $F_r = 2F_r^l = 2F_r^r$. Therefore, the vehicle in Fig. 1 can be modeled with a bicycle model while ignoring the pitch and the roll motions.

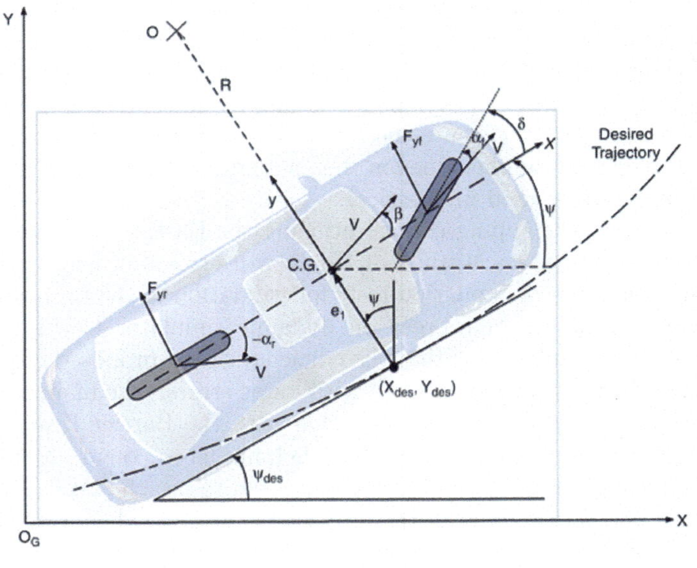

Fig. 1. Lateral dynamics of the vehicle.

According to Assumption 1, the full 3-DOF vehicle model is reduced to the 2-DOF model of the lateral position y and the yaw angle ψ. Motion along axis in Fig. 1 using Newton's second law:

$$ma_y = F_{yf} + F_{yr} \tag{1}$$

where a_y is the lateral acceleration at the center of gravity, m is the vehicle's mass, F_{yf} and F_{yr} are forced a acting on the front and rear wheels, respectively. Lateral acceleration includes the acceleration a_y due to y axis motion and the centripetal acceleration $V_x\dot{\psi}$ [12]. The vehicle lateral motion is described as follows:

$$m(\ddot{y} + V_x\dot{y}) = F_{yf} + F_{yr} \tag{2}$$

Considering the torque around the z axis, the moment balance about the z axis is given by:

$$I_z\ddot{\psi} = l_f F_{yf} - l_r F_{yr} \tag{3}$$

where I_z is the initial moment, l_f and l_r are the distance from the ear and front wheels to the vehicle's center of gravity, respectively. The tire slip angle is the angle between the direction of the tire and the direction of the wheel velocity vector. The slip angle of front and rear tire can be written as:

$$\alpha_f = \delta - \theta_f; \alpha_r = -\theta_r \tag{4}$$

where θ_f is the angle between the velocity vector and the vehicle's longitudinal axis; θ_r is the rear tire velocity angle and δ is the steering angle. Experimental results show that when the slip angle is mall, the lateral force of the tire is proportional to the slip angle [12]. Hence, the front tire and rear tire forces are defined as:

$$\begin{aligned} F_{yf} &= 2C_f\alpha_f = 2C_f(\delta - \theta_f) \\ F_{yr} &= 2C_r\alpha_r = 2C_r(-\theta_r) \end{aligned} \tag{5}$$

where C_f and C_r front and rear cornering stiffness. θ_f and θ_r approximated by:

$$\theta_f = \frac{\dot{y} + l_f\dot{\psi}}{V_x}; \theta_r = \frac{\dot{y} - l_r\dot{\psi}}{V_x} \tag{6}$$

Substituting from (5), (6) into (2), (3), the vehicle lateral dynamics model is obtained:

$$\ddot{y} = -2\frac{C_f + C_r}{mV_x}\dot{y} - (V_x + 2\frac{C_f - C_r}{mV_x})\dot{\psi} + \frac{2C_f}{m}\delta \tag{7}$$

$$\ddot{\psi} = -2\frac{l_f C_f - l_r C_r}{I_z V_x}\dot{y} - 2\frac{l_f^2 C_f + l_r^2 C_r}{I_z V_x}\dot{\psi} + \frac{2l_f C_f}{I_z}\delta \tag{8}$$

For the lateral dynamics control to guarantee good tracking performance, both the lateral error and yaw angle should be considered at the same time, which are given by:

$$e_\psi = \psi - \psi_d; \dot{e}_y = \dot{y} + V_x e_\psi \tag{9}$$

$$\ddot{e}_y = -\frac{2C_f+2C_r}{mV_x}\dot{e}_y + \frac{2C_f+2C_r}{I_z}e_\psi - \frac{2l_fC_f-2l_rC_r}{mV_x}\dot{e}_\psi + (-V_x - \frac{2l_fC_f-2l_rC_r}{mV_x})\dot{\psi}_d + \frac{2C_f}{m}\delta \tag{10}$$

$$\ddot{e}_\psi = -\frac{2l_f^2C_f+2l_r^2C_r}{I_zV_x}\dot{e}_\psi - \frac{2l_fC_f-2l_rC_r}{I_zV_x}\dot{e}_y + \frac{2l_fC_f-2l_rC_r}{I_z}e_\psi - \frac{2l_f^2C_f+2l_r^2C_r}{I_zV_x}\dot{\psi}_d + \frac{2l_fC_f}{I_z}\delta \tag{11}$$

Furthermore, $x_1 = e_y$, $x_2 = \dot{e}_y$, $x_3 = e_\psi$, $x_4 = \dot{e}_\psi$ are defined:

$$\begin{cases} \dot{x}_1 = x_2 \\ \dot{x}_2 = k_1x_2 + k_2x_3 + k_3x_4 + \gamma_1\delta \end{cases} \tag{12}$$

$$\begin{cases} \dot{x}_3 = x_4 \\ \dot{x}_4 = k_4x_2 + k_5x_3 + k_6x_4 + \gamma_2\delta \end{cases} \tag{13}$$

where $k_1 = -\frac{2C_f+2C_r}{mV_x}$, $k_2 = -V_xk_1$, $k_3 = -\frac{2C_fl_f-2C_rl_r}{mV_x}$, $\gamma_1 = \frac{2C_f}{m}$, $k_4 = -\frac{2l_fC_f-2l_rC_r}{I_zV_x}$, $k_5 = -V_xk_4$, $k_6 = -\frac{2l_f^2C_f+2l_r^2C_r}{I_zV_x}$, $\gamma_2 = \frac{2l_fC_f}{I_z}$.

3 Backstepping Control Using Barrier Lyapunov Function

We assume that the dynamics of (13) is much faster than the dynamics of (12) [11], so with a short time, approach its "quasi-steady-state" while x_4 and δ approach to 0 so that the slow dynamic will be:

$$\begin{cases} \dot{x}_1 = x_2 \\ \dot{x}_2 = k_1x_2 + k_2\bar{x}_3 \end{cases} \tag{14}$$

where $\bar{x}_3$ is a quasi-steady state of x_3 and can be used as an auxiliary input for slow dynamics. In this article, the control law is designed with (13) and (14) to act as a fast system and a slow system. The control input of fast dynamics δ follows the auxiliary input $\bar{x}_3$ then the slow dynamics will reach steady state. This subsection designs the controller to satisfy the lateral position and yaw constraint.

Theorem 1. *Considering the lateral dynamic with $x_1(0) \in X_p, x_3(0) \in X_y$ where $X_p = x_1 : |x_1| < k_p$, $X_y = x_3 : |x_3| < k_y$ and k_p, k_y is the positive constant constraint of x_1, x_3. It supposes that the control laws are given by:*

$$\bar{x}_3 = \frac{1}{k_2}(-\frac{x_1}{k_p^2 - x_1^2} + \dot{x}_{2d} - k_1x_2 - l_2\tanh(z_2)) \tag{15}$$

$$\delta = \frac{1}{\gamma_2}(-\frac{x_3}{k_y^2 - x_3^2} + \dot{x}_{4d} - k_4x_2 - k_5\bar{x}_3 - k_6x_4 - l_4\tanh(z_4)) \tag{16}$$

Proof of Theorem 1: The stability of slow dynamic. Let us define V_1 and the derivative of V_1 with respect to time is given by the formula:

$$\begin{aligned} V_1 &= \tfrac{1}{2}\log\left(\tfrac{k_p^2}{k_p^2 - x_1^2}\right); \\ \dot{V}_1 &= \tfrac{x_1\dot{x}_1}{k_p^2 - x_1^2} = \tfrac{x_1 x_2}{k_p^2 - x_1^2} \end{aligned} \tag{17}$$

x_{2d} and z_2 are defined as:

$$\begin{aligned} x_{2d} &= -l_1 x_1 (k_p^2 - x_1^2) \\ z_2 &= x_2 - x_{2d} \end{aligned} \tag{18}$$

where l_1 is the positive constant. Substituting the values of x_{2d} and z_2, $\dot{V}_1$ becomes:

$$\dot{V}_1 = -l_1^2 x_1^2 + \frac{x_1 z_2}{k_p^2 - x_1^2} \tag{19}$$

The derivative of z_2 with respect to time is given by the formula:

$$\dot{z}_2 = k_1 x_2 + k_2 \bar{x}_3 - \dot{x}_{2d} \tag{20}$$

V_2 is defined and the derivative of V_2 with respect to time is given by the formula:

$$V_2 = V_1 + \frac{1}{2} z_2^2 \tag{21}$$

$$\dot{V}_2 = \dot{V}_1 + z_2 \dot{z}_2 = -l_1 x_1^2 + \frac{x_1 z_2}{k_p^2 - x_1^2} + z_2 (k_1 z_2 + k_2 \bar{x}_3 - \dot{x}_{2d}) \tag{22}$$

with the auxiliary control input $\bar{x}_3$ in Eq. (15), $\dot{V}_2$ becomes:

$$\dot{V}_2 = -l_1 x_1^2 - z_2 l_2 \tanh(z_2) \tag{23}$$

where l_2 is the positive constant. Because $x_1(t) \in X_p$ for $t > 0$ so $V_1 \geq 0$ and $V_2 \geq 0$. Moreover $\dot{V}_2 \leq 0$ in Eq. (23) so V_2 reaches to zero. After V_2 converges to zero, $\ln(\frac{k_p^2}{k_p^2 - x_1^2})$ and z_2 converge to zero, resulting in $x_1 = 0$ and $x_2 = 0$.

The stability of fast dynamic. Let us define V_3 and the derivative of V_3 with respect to time is given by the formula:

$$\begin{aligned} V_3 &= \tfrac{1}{2}\log\left(\tfrac{k_y^2}{k_y^2 - x_3^2}\right); \\ \dot{V}_3 &= \tfrac{x_3\dot{x}_3}{k_y^2 - x_3^2} = \tfrac{x_3 x_4}{k_y^2 - x_3^2} \end{aligned} \tag{24}$$

To obtain the stability of system, x_{4d} and z_4 are defined as:

$$\begin{aligned} x_{4d} &= -l_3 x_3 (k_y^2 - x_3^2) \\ z_4 &= x_4 - x_{4d} \end{aligned} \tag{25}$$

where l_3 is the positive constant. Substituting the values of x_{4d} and z_4, $\dot{V}_3$ becomes:

$$\dot{V}_3 = -l_3^2 x_3^2 + \frac{x_3 z_4}{k_y^2 - x_3^2} \tag{26}$$

The derivative of z_4 with respect to time is given by the formula:

$$\dot{z}_4 = k_4 x_2 + k_5 \bar{x}_3 + k_6 x_4 + \gamma_2 \delta - \dot{x}_{4d} \tag{27}$$

V_4 is defined and the derivative of V_4 with respect to time is given by the formula:

$$V_4 = V_3 + \frac{1}{2} z_4^2 \tag{28}$$

$$\dot{V}_4 = \dot{V}_3 + z_4 \dot{z}_4 = -l_3 x_3^2 + \frac{x_3 z_4}{k_y^2 - x_3^2} + z_4(k_4 x_2 + k_5 \bar{x}_3 + k_6 x_4 + \gamma_2 \delta - \dot{x}_{4d}) \tag{29}$$

with the control input δ in Eq. (16), $\dot{V}_4$ becomes:

$$\dot{V}_4 = -l_3 x_3^2 - z_4 l_4 \tanh(z_4) \tag{30}$$

where l_4 is the positive constant. Because $\bar{x}_3(t) \in X_y$ for $t > 0$ so $V_3 \geq 0$ and $V_4 \geq 0$. Moreover $\dot{V}_4 \leq 0$ in Eq. (30) so V_4 reaches to zero. After V_4 converges to zero, $\ln(\frac{k_y^2}{k_y^2 - x_3^2})$ and z_4 converge to zero, resulting in $x_3 = \bar{x}_3 = 0$ and $x_4 = 0$.

4 Simulation and Results

Co-simulation is performed using Matlab/Simulink environment and CarSim to demonstrate two proposed controllers. The Simulink and the CarrSim can illustrate realistic cars with roll, pitch, and yaw motions. In CarSim, the road environment and vehicle parameters such as friction coefficient and speed can be set. The output of CarSim such as lateral position, lateral velocity, and yaw angle are used for control design. We used the B-class Hatchback and changed two parameters of front and rear cornering stiffness. To link Carsim and Matlab/Simulink, set Simulink sampling time to 0.0001 s, Carsim output sampling time to 0.0001 s, frequency 10000 Hz. The mass of the vehicle is 1500 kg. The yaw inertia is 1350 kgm^2. The front and rear tires concerning stiffness are 55000 N/rad and 120000 N/rad, respectively. The front wheels and rear wheels distance from CG are 1.5 m and 2 m, respectively. The control structure is shown in Fig. 2. To verify the stability of the proposed control system, the autonomous vehicle was simulated performing lane changes at a speed of 20 m/s, the Double Lane Change (DLC) and Single Lane Change (SLC) simulations were conducted in accordance with ISO 14791:2000 and ISO 3888-1:2018 standards.

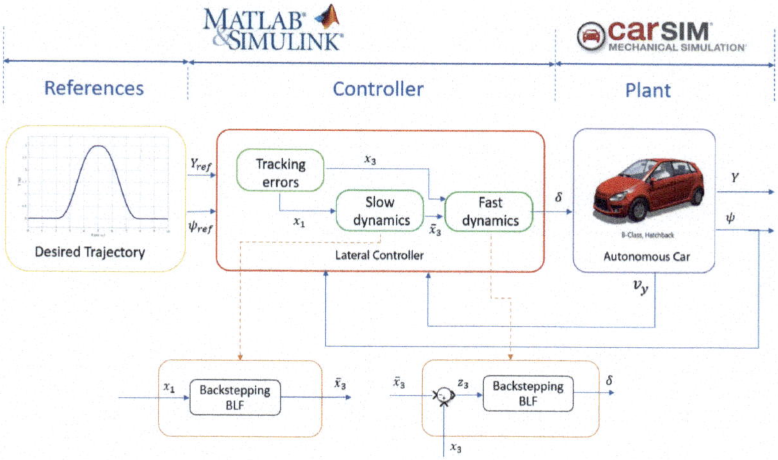

Fig. 2. The control structure of lane changing system.

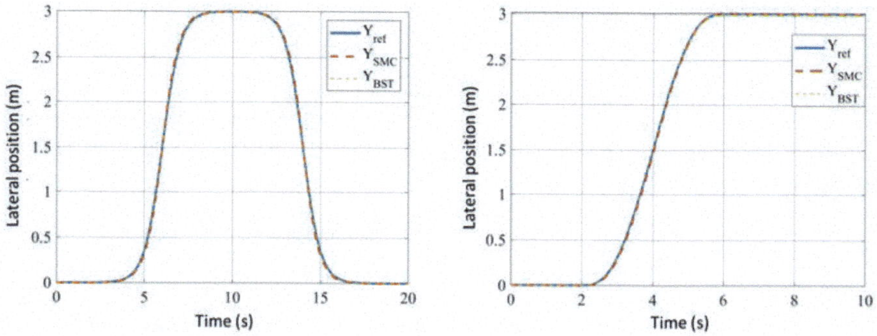

(a) (b)

Fig. 3. Lateral position tracking in DLC and SLC Maneuver.

Figure 3 and Fig. 4 show that both the SMC controller and the BLF-based BSP controller can keep the vehicle on the desired trajectory - DLC and SLC. The result of Fig. 5 shows that SMC shows relatively higher error and violates the lateral position error constraint. Figure 6 shows that BSP using BLF outperforms the SMC.

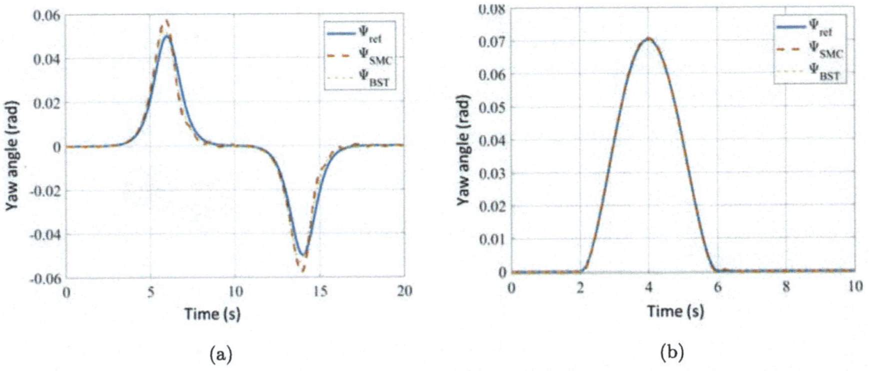

Fig. 4. Yaw angle tracking in DLC and SLC Maneuver.

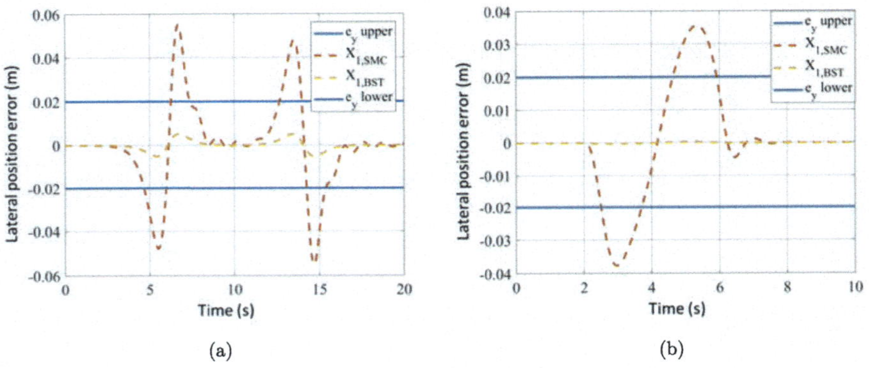

Fig. 5. Lateral position error tracking in DLC and SLC Maneuver.

The SMC displays a relatively higher error and violates the yaw error constraint. Conversely, BSP+BLF satisfies the desired output constraints and hence demonstrates improved performance compared to SMC. The control input for both controllers is compared in Fig. 7. From the results, it can be deduced that BSP+BLF results in less chattering as compared to SMC.

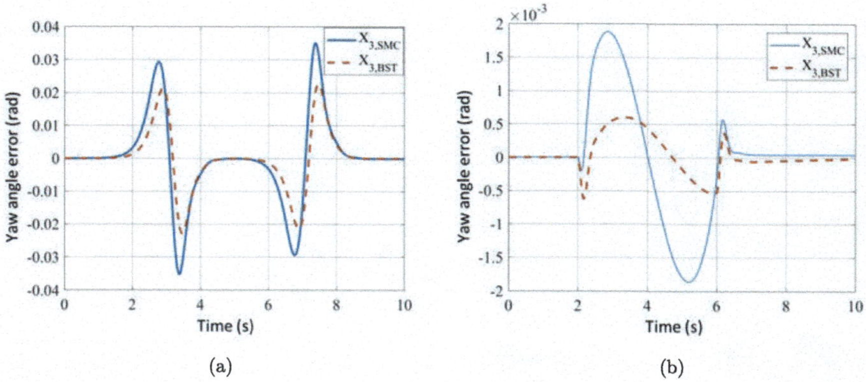

Fig. 6. Yaw angle error tracking in DLC and SLC Maneuver.

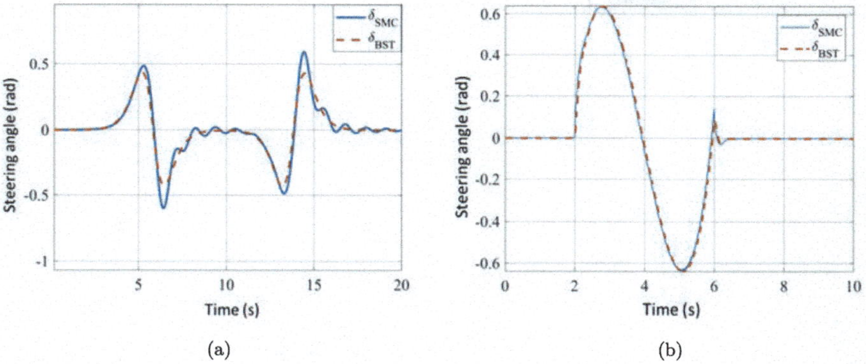

Fig. 7. Control input in DLC and SLC Maneuver.

Figure 8 shows autonomous vehicle responses on CarSim. The results show that the output controller works well for the autonomous vehicle model on CarSim. In addition to using images to evaluate the performance of the proposed controller, the IWPEI index is also used to evaluate the control quality. The IWPEI [13] index is built based on 3 factors: The integral absolute lateral tracking and yaw rate errors (IATE), The integral time product of the absolute lateral tracking error and yaw rate error (IAPE), and The integral of the absolute value of the front steering angle (IASA).

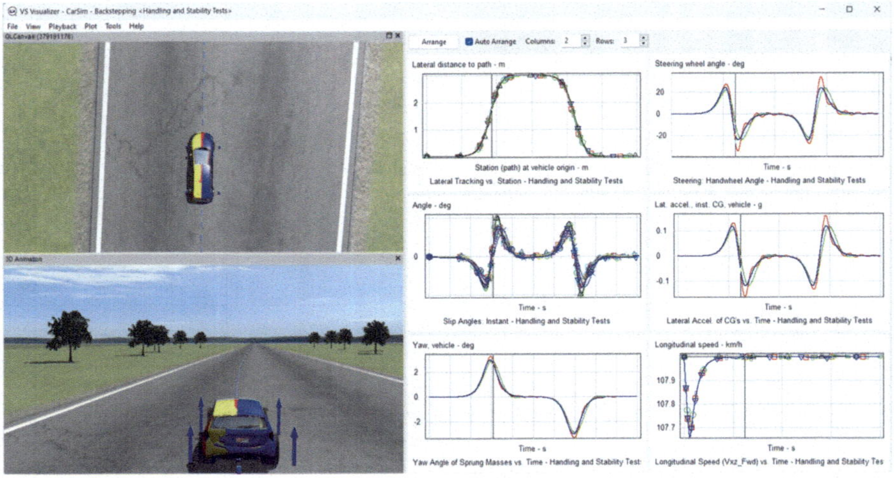

(a)

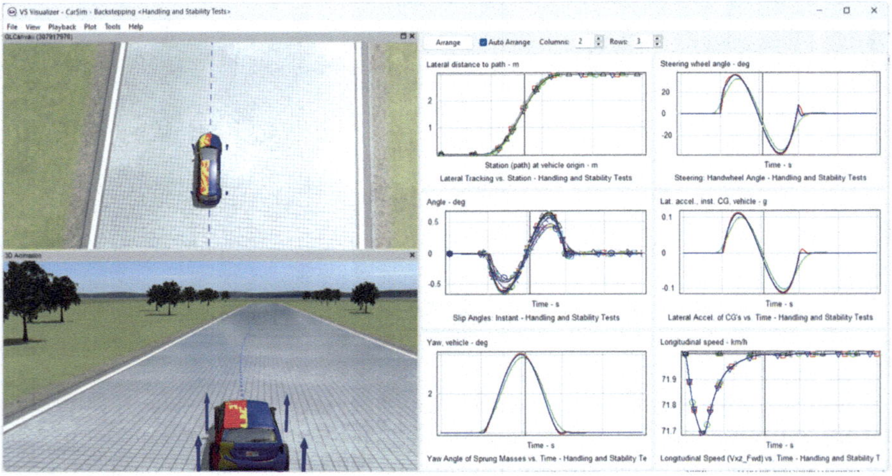

(b)

Fig. 8. Autonomous Vehicle's Response in CarSim in DLC and SLC Maneuver.

The results in Table 1 show that the proposed method achieves the lowest values among the control performance metrics across different simulation scenarios, indicating that the proposed controller is more robust and accurate than the SMC controller. Moreover, the proposed controller ensures that the state deviations remain within a predefined region, making it more suitable for practical applications.

Table 1. Evaluation index comparison in DLC Maneuver and SLC Maneuver.

	DLC Maneuver		SLC Maneuver	
	SMC	BSP+BLF	SMC	BSP+BLF
ITAE	0.49	0.08	0.15	0.012
IAPE	3.06	0.72	0.62	0.05
IASA	3.24	2.64	1.87	1.65
IWPEI	27.02%	16.58%	25.7%	15.03%

5 Conclusion

In this paper, the backstepping controller with the barrier Lyapunov function is designed without considering the path disturbance of the autonomous vehicle. The autonomous car model is reduced to slow and fast dynamics and BSP with BLF is used to track the desired output of the system while constraining the output within certain limits. The control performance is evaluated based on the newly proposed IWPEI, including control input, lateral position error, and yaw rate error. Comparing conventional SMC, the simulation results show that the proposed BSP+BLF generates continuous control input without the chattering phenomenon, resulting in a smaller and smoother lateral position error. In summary, the simulation results show that the BSP using BLF controller generates a control input signal without the chattering phenomenon, lateral position error and yaw angle error are smaller. In the future, derivative restrictions in higher-order systems will be handled by control improvements. Disturbance elements, such as measurement noise and uncertainty noise, will be incorporated to demonstrate the controller's effectiveness. In addition, this control system may be extended to include other autonomous vehicle components, such as lane change control and torque distribution to all four wheels, to allow for stable movement while preventing rollover.

References

1. Badue, C., et al.: Self-driving cars: a survey. Expert Syst. Appl. **165**, 113816 (2021)
2. Filho, C.M., Wolf, D.F., Grassi, V., Osório, F.S.: Longitudinal and lateral control for autonomous ground vehicles. In: 2014 IEEE Intelligent Vehicles Symposium Proceedings, pp. 588–593. IEEE (2014)
3. Attia, R., Orjuela, R., Basset, M.: Combined longitudinal and lateral control for automated vehicle guidance. Veh. Syst. Dyn. **52**(2), 261–279 (2014)
4. Marino, R., Scalzi, S., Netto, M.: Nested PID steering control for lane keeping in autonomous vehicles. Control. Eng. Pract. **19**(12), 1459–1467 (2011)
5. Tagne, G., Talj, R., Charara, A.: Higher-order sliding mode control for lateral dynamics of autonomous vehicles, with experimental validation. In: 2013 IEEE Intelligent Vehicles Symposium (IV), pp. 678–683. IEEE (2013)
6. Kang, C.M., Kim, W., Chung, C.C.: Observer-based backstepping control method using reduced lateral dynamics for autonomous lane-keeping system. ISA Trans. **83**, 214–226 (2018)

7. Norouzi, A., Masoumi, M., Barari, A., Sani, S.F.: Lateral control of an autonomous vehicle using integrated backstepping and sliding mode controller. Proc. Inst. Mech. Eng. Part K: J. Multi-body Dyn. **233**(1), 141–151 (2019)
8. Yang, Q., Shi, Q., Saraoğlu, M., Janschek, K.: A hybrid approach using an adaptive waypoint generator for lane-changing maneuver on curved roads. In: Bargende, M., Reuss, H.-C., Wagner, A. (eds.) 21. Internationales Stuttgarter Symposium. P, pp. 127–141. Springer, Wiesbaden (2021). https://doi.org/10.1007/978-3-658-33521-2_10
9. Cheng, S., Li, L., Chen, X., Jian, W., et al.: Model-predictive-control-based path tracking controller of autonomous vehicle considering parametric uncertainties and velocity-varying. IEEE Trans. Ind. Electron. **68**(9), 8698–8707 (2020)
10. Wang, Y., Liu, Z., Zuo, Z., Li, Z., Wang, L., Luo, X.: Trajectory planning and safety assessment of autonomous vehicles based on motion prediction and model predictive control. IEEE Trans. Veh. Technol. **68**(9), 8546–8556 (2019)
11. Khan, R., Malik, F.M., Mazhar, N., Raza, A., Azim, R.A., Ullah, H.: Robust control framework for lateral dynamics of autonomous vehicle using barrier Lyapunov function. IEEE Access **9**, 50513–50522 (2021)
12. Rajamani, R.: Vehicle Dynamics and Control. Springer, Cham (2011)
13. Ao, D., Huang, W., Wong, P.K., Li, J.: Robust backstepping super-twisting sliding mode control for autonomous vehicle path following. IEEE Access **9**, 123165–123177 (2021)

Development of Autonomous Landing System for Reconnaissance UAVs

Minh Tri Nguyen Sy[1,2], Xuan Minh Dinh[1(✉)], Quang Quan Do[1], Ngoc Thanh Ta[1,3], Van Duy Pham[1], Ngoc Dinh Ta[3], and Xuan Manh Bui[1]

[1] HTI Technological Solutions One Member Limited Company (HTSC), HTI Group, Hanoi City, Vietnam
minh.dx@htigroup.vn
[2] The Dewey Schools, Hanoi City, Vietnam
[3] HTI Digital Services One Member Limited Company (HTI Services), HTI Group, Hanoi City, Vietnam

Abstract. This study outlines a process for developing an autonomous landing system for Quadrotor reconnaissance UAVs. We begin with an overview of the system, including mathematical models for the Quadrotor and camera, and a detailed hardware connection diagram. For estimating the position of a base with an Embedded ArUco marker, we use the Perspective-n-Point method, known for its reliability and efficiency in real-time applications. We then propose using a PID controller to manage the UAV's horizontal position, streamlining the landing process and ensuring reliable performance in real-world scenarios. The system, implemented on a Raspberry Pi 4 Model B, was validated through practical experiments. The results demonstrate that the UAV can land accurately on the base with small error from its center, indicating potential for direct application in UAV models at the enterprise level, such as those used by HTI Group.

Keywords: Autonomous Landing System · Reconnaissance UAVs · PID · Perspective-n-Point · Embedded ArUco

1 Introduction

With the rapid advancement of technology, Unmanned Aerial Vehicles (UAVs) have become essential in both civil and defense sectors. In applications like reconnaissance and delivery, precise takeoff and landing are vital for mission success [1]. While landing is a critical task, many current solutions rely heavily on GPS, which can suffer from accuracy issues due to sensor errors. To overcome this, recent studies have integrated computer vision with control methods and trajectory planning to improve UAV landing accuracy on both fixed and moving platforms [2–9].

Practical studies [2, 3] have significantly improved the accuracy of autonomous UAV landings on target locations. Using the ArUco marker library in OpenCV, they identified and estimated the marker's position relative to the UAV and applied a virtual border technique with channel override to control its position. However, this method relies

© The Author(s), under exclusive license to Springer Nature Switzerland AG 2025
H. A. Le Thi et al. (Eds.): ICAMCS 2024, LNNS 1311, pp. 475–487, 2025.
https://doi.org/10.1007/978-3-031-90606-0_40

heavily on practical experience to choose suitable velocity values for roll (left/right) and pitch (forward/backward) movements, making it challenging in complex environments. Another study [4] utilized the Robot Operating System (ROS) to control a quadrotor's autonomous landing on an ArUco marker, employing Visual Inertial Odometry (VIO) to combine visual and inertial data for 3D pose estimation. However, this approach demands extensive ROS knowledge, adding complexity in managing messages between topics and nodes.

Optimization-based methods, such as those in [5–7], have shown promising results in UAV landing. Study [5] proposes an automatic take-off and landing system using a gradient-based local motion planning method to generate optimal trajectories, though it demands high computational resources. Similarly, optimization techniques like Model Predictive Control (MPC) [6, 7] face challenges in complex scenarios. Machine learning and reinforcement learning approaches have also been explored [8–11]. Study [8] uses LSTM networks and Kalman filters to predict platform trajectories for more accurate landings, but it lacks practical validation. Study [9] employs meta-reinforcement learning for landing on moving platforms with unknown paths. While effective, these methods require extensive training data and computational power. These drawbacks also exist in research [10, 11].

Inspired by the aforementioned works, this study aims to contribute a comprehensive process for developing an automatic landing system for Quadrotor reconnaissance UAVs. The experimental results demonstrate that the UAV can land accurately on the platform with small errors relative to the platform's center. Our main contributions are as follows: The first, simplify the landing process while ensuring high performance in practice. The proposed automatic landing system is equipped with an embedded ArUco marker [1] to enhance the accuracy and stability of platform recognition. The Perspective-n-Point method [12] is used to estimate the position of the embedded ArUco marker, providing data to the UAV for control via the PID (Proportional-Integral-Derivative) method. The PID controller directly adjusts the UAV's position along the roll (left/right) and pitch (forward/backward) axes to ensure precise landing on the platform. The automatic landing system has been validated through practical experiments conducted on a Raspberry Pi 4 Model B embedded computer. The sencond, the automatic landing system we developed has been successfully implemented on UAV models at enterprise (HTI Group). Compared to previous research, our approach offers a significant advantage in computational efficiency, with a considerably lower computational than optimization-based methods [5–7]. Additionally, the system does not require training data or the complexity associated with machine learning methods [8] and reinforcement learning [9–11], making it a simpler and more practical solution for deployment.

The paper is organized as follows: Sect. 2 provides an overview of the system, including the mathematical model of the Quadrotor UAV, the mathematical model of the camera, and the interconnection diagram. Section 3 details the design process of the autonomous landing system and the methods used. Section 4 presents and analyzes the experimental results. Finally, Sect. 5 concludes the study and suggests directions for future improvements.

2 Comprehensive Overview of the System

2.1 Mathematical Model of a Quadrotor Reconnaissance UAV

Understanding the system's mathematical model is crucial for selecting and implementing the right control algorithms for optimal performance when designing a controller. The Fig. 1 below illustrates a common configuration of a quadcopter UAV, consisting of four motors and four propellers. Technically, the vehicle can generate forces F_i for $i = 1, 2, 3, 4$ to control the motion of its rigid body through control inputs, including thrust (u_1) and torques (u_2, u_3, u_4). In this model, L represents the arm length, defined as the distance from the center of mass to each propeller, which directly affects the torques generated by the propeller forces, c_d is the force-to-torque coefficient, M is mass of quadcopter, and g is the gravitational acceleration.

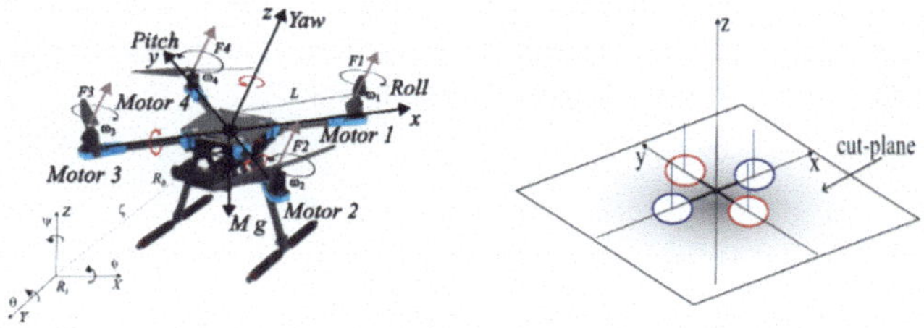

Fig. 1. Quadcopter Reconnaissance UAV model and its main axes

According to studies [13, 14], we can establish a set of equations describing the four control inputs u_i as follows:

$$\begin{cases} u_1 = F_1 + F_2 + F_3 + F_4 \\ u_2 = L(F_2 - F_4) \\ u_3 = L(F_3 - F_1) \\ u_4 = c_d(-F_1 + F_2 - F_3 + F_4) \end{cases} \tag{1}$$

The dynamic model of the quadcopter for its horizontal translation is expressed as follows [13, 14]:

$$\begin{cases} \ddot{X} = \dfrac{1}{M}(\cos\phi \sin\theta \cos\psi + \sin\phi \sin\psi)u_1 \\ \ddot{Y} = \dfrac{1}{M}(\cos\phi \sin\theta \sin\psi - \sin\phi \cos\psi)u_1 \end{cases} \tag{2}$$

and its vertical and rotary dynamics are described as [13, 14]:

$$\begin{cases} \ddot{Z} = -g + \frac{1}{M}(\cos\phi \cos\theta)u_1 \; ; \; \ddot{\phi} = \frac{J_y - J_z}{J_x}\dot{\psi}\dot{\theta} + \frac{1}{J_x}u_2 \\ \ddot{\theta} = \frac{J_z - J_x}{J_y}\dot{\psi}\dot{\phi} + \frac{1}{J_y}u_3 \; ; \; \ddot{\psi} = \frac{J_x - J_y}{J_z}\dot{\phi}\dot{\theta} + \frac{1}{J_z}u_4 \end{cases} \tag{3}$$

where $\{X, Y, Z\}$ denote the position and $\{\phi, \theta, \psi\}$ the attitude of the quadcopter in the inertial frame R_i. In addition, the moments of inertia about the X, Y, and Z axes are J_x, J_y, and J_z respectively.

The purpose of separating the dynamic model into two sets of equations, (2) and (3), is to effectively implement control algorithms. In this study, we focus on designing a PID controller to manage the dynamics in Eq. (2), thereby controlling the UAV's horizontal position $\{X, Y\}$ to achieve the desired target point. Meanwhile, the dynamics in Eq. (3) are controlled through the algorithms integrated within Ardupilot on the Pixhawk board.

2.2 Mathematical Model of a Camera

The camera is vital for detecting and estimating the landing point in automatic landing systems. Integrating it into the UAV control system requires understanding its mathematical model. The Pinhole model, shown in Fig. 2 [17], describes how 3D points are projected onto a 2D image through perspective projection. This is represented by a projection matrix that combines the camera's intrinsic parameters (focal length, pixel scaling, and principal point) with extrinsic parameters.

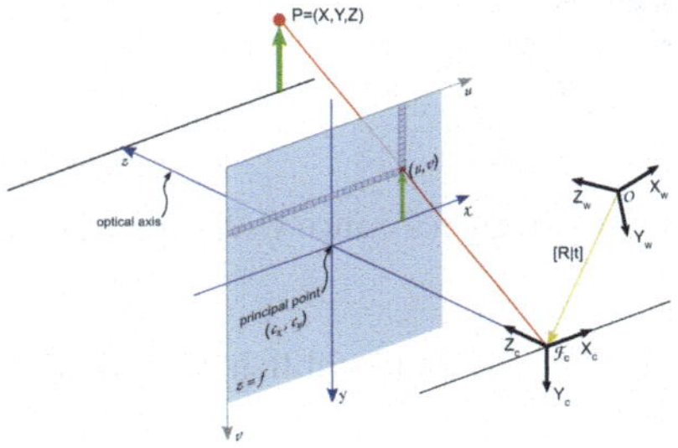

Fig. 2. Pinhole Camera Model [17]

The Pinhole model of the camera is described by the following equation [15, 16]:

$$s\begin{bmatrix} u \\ v \\ 1 \end{bmatrix} = \mathbf{K}[\mathbf{R}|\mathbf{t}]\begin{bmatrix} X \\ Y \\ Z \\ 1 \end{bmatrix} = \begin{bmatrix} f_x & 0 & c_x \\ 0 & f_y & c_y \\ 0 & 0 & 1 \end{bmatrix}\begin{bmatrix} r_{11} & r_{12} & r_{13} & t_1 \\ r_{21} & r_{22} & r_{23} & t_2 \\ r_{31} & r_{32} & r_{33} & t_3 \end{bmatrix}\begin{bmatrix} X \\ Y \\ Z \\ 1 \end{bmatrix} \tag{4}$$

where $\{X, Y, Z\}$ are the coordinates of a point in 3D space in the inertial reference frame R_i. This point is in the world coordinate system and needs to be transformed into image coordinates. (u, v) are the coordinates (in pixels) of the projection of the point $\{X, Y, Z\}$ onto the image plane. This is the result of a perspective projection from 3D space to the 2D image space. $\mathbf{K}$ is a 3×3 matrix containing the intrinsic parameters of the camera. These parameters include the focal length in the horizontal f_x and vertical f_y directions, along with the principal point coordinates (c_x, c_y) in pixels. $[\mathbf{R}|\mathbf{t}]$ is a 3×4 matrix containing the extrinsic parameters of the camera. This matrix describes the mapping from world space to camera space and includes a 3D rotation matrix and a translation vector. Additionally, s is a scale factor, commonly used to normalize the coordinates.

2.3 Detailed Connection Diagram of UAV Devices

Figure 3 illustrates the UAV reconnaissance system's connection diagram. The battery powers the system through the Power Module, which regulates voltage and monitors levels. The PDB distributes power to ESCs that control the motors. The Flight Controller (Pixhawk 6C) processes signals from the Receiver and GPS M10, sending commands to the ESCs for movement. The Receiver receives signals from the TX and relays them to Pixhawk 6C. Radio Telemetry allows two-way communication with the ground station for remote control. The PWM to PPM/SBUS Decoder is used to convert PWM signals into a PPM or SBUS format that is compatible with the Flight Controller.

The GPS M10 provides navigation data for functions like "Return to Home." Raspberry Pi 4 processes camera inputs for control algorithms, image processing, and position estimation in the autonomous landing system.

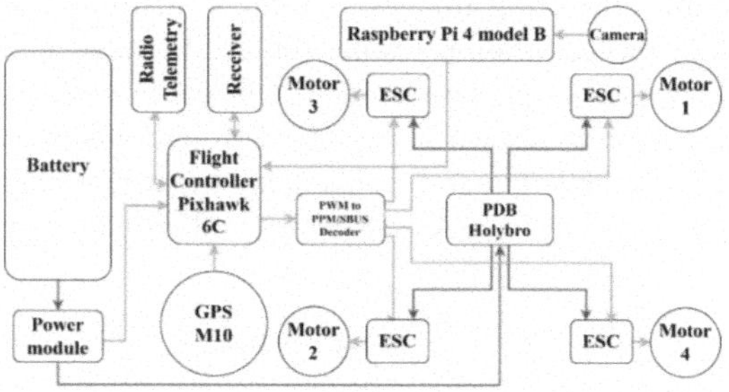

Fig. 3. Detailed Connection Diagram of UAV Devices

3 Proposed Design Process for an Autonomous Landing System for Reconnaissance UAVs

This content presents the design process of an autonomous landing system for reconnaissance UAVs, comprising three main steps: camera calibration to ensure image accuracy, detection and estimation of the Embedded ArUco marker's position in 3D space, and the application of a PID control method to effectively and stably manage the UAV's horizontal movement.

3.1 Camera Calibration

Camera calibration is the first crucial step in the design process of an autonomous landing system for UAVs. Calibration allows for determining the camera's intrinsic parameters and distortion coefficients, thereby improving the accuracy of converting point coordinates from 3D to 2D image space and vice versa. This process is typically carried out using multiple images of a checkerboard calibration pattern, as illustrated in Fig. 4, based on Zhang's method, as detailed in the study [15]. Currently, camera

Fig. 4. Checkerboard image used for camera calibration

calibration algorithms are readily integrated into the OpenCV library [12], making the calibration process easier and more convenient for users.

Our camera calibration process is carried out step-by-step as follows: capturing images of the calibration checkerboard from various angles; detecting and refining the corners of the checkerboard to ensure high accuracy for the pixel points; calculating the calibration parameters, such as the intrinsic matrix and distortion coefficients, using a calibration algorithm; correcting images to remove distortion and ensure accurate output images; storing the calibration results for use in subsequent computer vision applications; and finally, evaluating the accuracy of the calibration process by calculating the mean reprojection error, as shown in Eq. (5). Through this process, we ensure that the camera is precisely calibrated, thereby enhancing the performance and reliability of the autonomous landing system for UAVs as well as other related applications.

$$e = \frac{1}{N}\sum_{1}^{N} \|\mathbf{p}_i - \hat{\mathbf{p}}_i\|^2 \tag{5}$$

where, e is the error, $\mathbf{p}_i$ and $\hat{\mathbf{p}}_i$ are the 2D vectors representing the coordinates of the actual image points and the reprojected points from the 3D model through the corresponding camera calibration parameters, respectively, and N is the total number of image points used in the calibration proces.

3.2 Detection and Estimation of the Position of the Embedded ArUco Marker

Cameras are vital in UAV autonomous landing systems for accurately identifying landing points and providing real-time position feedback to controllers. To simplify this process, studies in aviation [18] have used ArUco markers—simple square codes easily recognized by cameras, encoding information like the marker's ID. However, traditional ArUco markers can be obscured when UAVs approach at low altitudes. To address this, some studies suggested using multiple markers of different sizes [2, 3], which takes up more space. To optimize this, study [1] introduced the Embedded ArUco marker. Building on this, we developed our design, shown in Fig. 5, featuring a small marker (ID 61) and a large marker (ID 59). This design ensures the marker stays in the camera's field of view and enhances recognition and position estimation accuracy. For more details, see the study [1].

After detecting the Embedded ArUco marker in the input image, the position and orientation of the marker are estimated using the Perspective-n-Point (PnP) algorithm [19], a popular method in computer vision. This algorithm is based on the Pinhole camera model and uses the corners of the detected Embedded ArUco marker in image

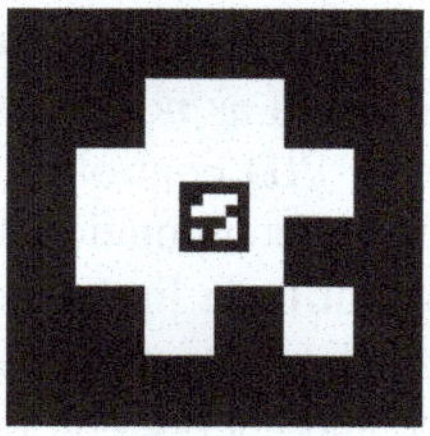

Fig. 5. Embedded ArUco Marker Design by HTI Group

space (2D) along with their actual coordinates in world space (3D) to compute the rotation vector and translation vector of the marker relative to the camera. As illustrated in Fig. 6, q1 to q4are the four corners of the marker in world space, and **p1** to **p4** are their corresponding coordinates in the image plane. This process involves several steps: detecting the Embedded ArUco marker using the OpenCV library [12] and the ArUco package, determining the corresponding corner points in world space, utilizing the Pinhole Camera model to link the 2D image points with the 3D real-world points through Eq. (4), and finally applying the PnP algorithm to calculate the rotation and translation vectors of the marker relative to the camera, using the Iterative Method provided by the OpenCV library.

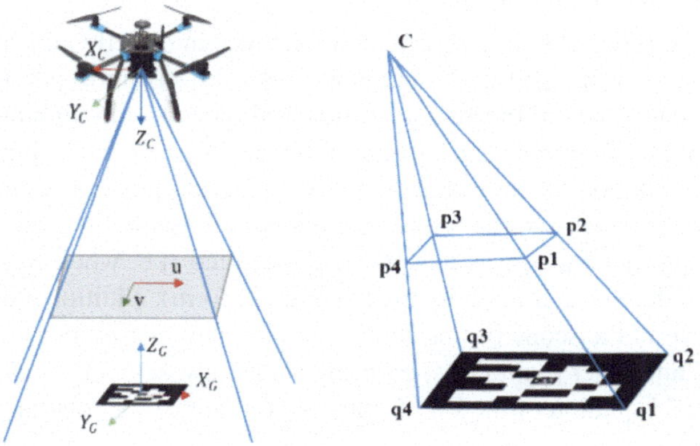

Fig. 6. Coordinate System Diagram and Embedded ArUco Marker Detection

3.3 PID Control Method for Controlling the Horizontal Position of the UAV

Our objective is to design a PID controller to control the horizontal dynamics of the UAV, as described in Eq. (2). This separation is implemented to suit practical UAV applications, where the vertical dynamics and rotational movements are typically managed by common commercial flight control boards, such as the Pixhawk 6C, using customized firmware based on ArduPilot [20], as deployed in our UAVs.

In this study, the PID (Proportional-Integral-Derivative) controller is chosen for its simplicity, proven stability, and effectiveness in controlling nonlinear dynamic systems, especially in real-world conditions with various disturbances and uncertainties. Compared to advanced control methods such as nonlinear adaptive control [21] or Model Predictive Control (MPC) [6, 7], the PID controller has the advantage of being easy to implement and tune, requiring fewer computational resources, and being seamlessly integrated with hardware systems that have limited resources. This makes it particularly suitable for UAV applications, where the computational load needs to be optimized to ensure stable flight performance and energy efficiency. The control structure of the autonomous landing system for the UAV is illustrated in Fig. 7.

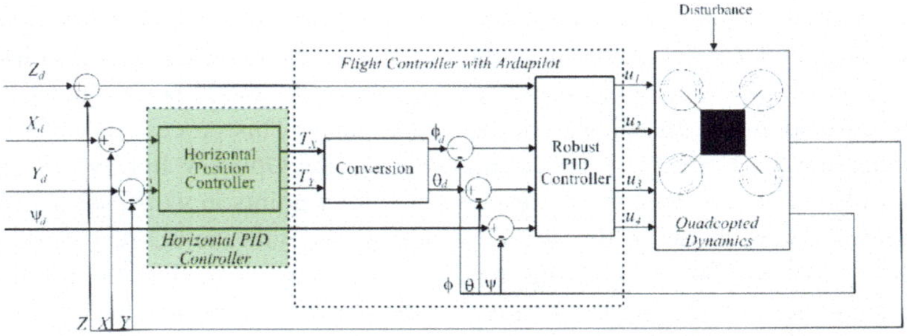

Fig. 7. Control Structure of the Autonomous Landing System for UAV

The Horizontal PID Controller plays a crucial role in directly adjusting the UAV's movement along the roll (left/right) and pitch (forward/backward) axes to ensure accurate landing on the platform, meeting position requirements with high precision during the landing process. The horizontal position values (X, Y) are fed back through the detection and estimation of the Embedded ArUco marker position using a camera, as presented in the previous section. However, it should be noted that to achieve higher accuracy in position estimation when the UAV is tilted (i.e. when the camera is not perpendicular to the plane containing the landing platform), additional techniques may be required (beyond the scope of this study). Additionally, ArduPilot manages vertical (z-axis) control and other flight tasks, ensuring stable descent and smooth landing.

Mathematical Equation of the PID Controller for Roll Axis Position Control:

$$T_X = K_P e_X(t) + K_I \int_0^t e_X(t)dt + K_D \frac{de_X(t)}{dt} \tag{6}$$

Mathematical Equation of the PID Controller for Pitch Axis Position Control:

$$T_Y = K_P e_Y(t) + K_I \int_0^t e_Y(t)dt + K_D \frac{de_Y(t)}{dt} \tag{7}$$

Notation: $e_X(t) = X_d - X$ is the error along the roll axis (x) and $e_Y(t) = Y_d - Y$ is the error along the pitch axis (y). (X_d, Y_d, Z_d) are the desired points, T_X and T_Y are the output signals of the Horizontal PID controller.

4 Experimental Results and Analysis

This section presents the experimental results of the autonomous landing system on the HTI Group's reconnaissance UAV. Figure 8 shows the UAV system setup used in the tests, including key components: the drone joystick, landing platform, laptop connected to Raspberry Pi 4 Model B via SSH, and the quadrotor UAV. The detailed connection diagram is provided in Sect. 1.3. Note that the Raspberry Pi 4 Model B connects to the Flight Controller (Pixhawk 6C) via a serial connection using the MavLink protocol. The following sections will cover the experiments evaluating the maximum altitude for marker detection and the system's effectiveness in adjusting the UAV's horizontal position for precise landing.

Fig. 8. Reconnaissance UAV System Used in the Experiment

First, the pilot controlled the UAV to ascend to the maximum altitude where it could clearly and stably detect the marker. For this study, we used a fisheye infrared (IR) camera on the Raspberry Pi with a 5-megapixel resolution and a 130-degree field of view (FoV). This wide FoV enables the system to detect the Embedded ArUco marker from various angles, improving the landing system's effectiveness. As shown in Fig. 9, the UAV can detect the Embedded ArUco marker from up to 7.5 m with the current setup. This height allows the UAV to recognize the marker from a distance and adjust its horizontal flight effectively. The detection altitude can be adjusted by altering the camera settings or image processing algorithm. However, the optimal detection height of 7.5 m is sufficient for a small reconnaissance UAV to effectively start its mission, ensuring accuracy and stability during marker detection and landing.

Next, we tested the effectiveness of the autonomous landing system. The test scenarios were designed to suit reconnaissance purposes and consisted of three main stages: Stage 1: The UAV takes off vertically from the base platform (Home point) and flies to perform reconnaissance missions in the desired area. Stage 2: After completing the mission, the UAV receives the command to return to the Home point using the RTL (Return to Launch) command. During this stage, the UAV will use the GPS system to locate and navigate back to the initial takeoff position with an error range of about 1 to 3 m. Final Stage: When the UAV approaches the Home point and the camera's field of view allows for the detection of the Embedded ArUco marker on the base platform, the autonomous landing system will be activated. The Raspberry Pi 4 Model B will calculate and send control commands to adjust the UAV's horizontal position to ensure a precise landing on the base platform, ensuring safety and effectiveness. The test results are shown in

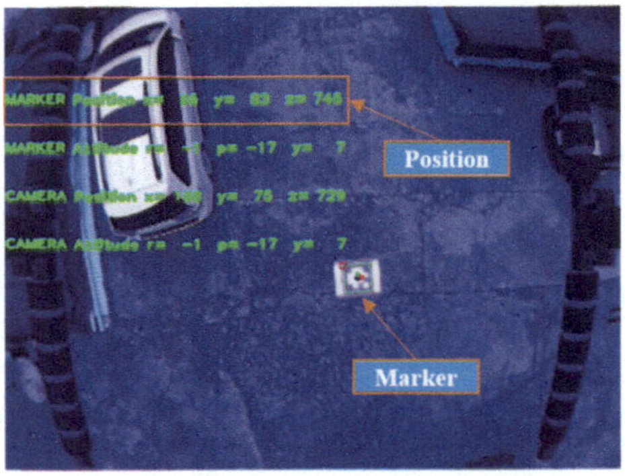

Fig. 9. Results of the maximum altitude test at which the UAV can detect the marker in real-world conditions

Fig. 10, with the controller parameters at points (6) and (7) set as follows: $K_P = 1.8$, $K_I = 0.01$, and $K_D = 0.2$.

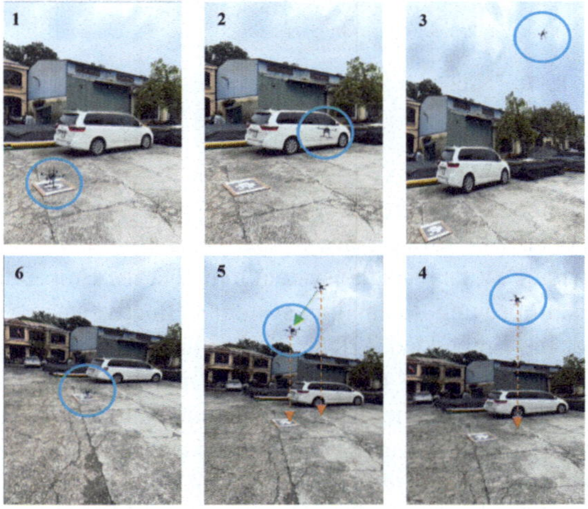

Fig. 10. Experimental Results of the Real-World Landing System

Based on the experimental results shown in Fig. 10, the UAV successfully carried out the autonomous landing on the platform with effectiveness and stability. Figures 10.1 and 10.2 illustrate Stage 1 of the test, where the UAV takes off from the platform and carries out a reconnaissance mission. Figure 10.3 depicts Stage 2, where the UAV receives a Return-to-Launch (RTL) command to return to the Home point. In Fig. 10.4, the UAV approaches the landing area but hovers 1 to 2 m away from the platform due to GPS inaccuracy, as indicated by the orange perpendicular projection, at an altitude of over 7.5 m. In the final stage, shown in Figs.10.5 and 10.6, the autonomous landing system is activated once the UAV detects the Embedded ArUco marker on the platform. The UAV then adjusts its horizontal position to align with the platform using the PID controller, following the green trajectory. Figure 10.5 shows the UAV gradually correcting its position along the X and Y axes with position feedback from the ArUco marker. Finally, the UAV smoothly lands on the platform, as demonstrated in Fig. 10.6.

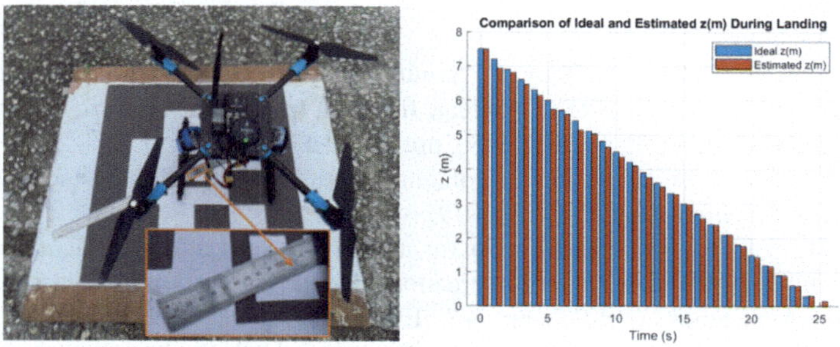

Fig. 11. Experimental error results, landing time, and estimated altitude comparison during landing

The final results show that the error between the center position of the UAV and the center of the landing platform is approximately 11 cm, as specifically illustrated in Fig. 11. This demonstrates that the system is capable of precisely and stably controlling the UAV's position, effectively meeting the requirements of real-world landing missions. Additionally, we conducted numerous similar tests, and the results consistently showed an error within 11 cm. Additionally, the landing time, and the comparison between the estimated and ideal altitudes are presented in Fig. 11. From the moment the UAV detects the marker at an altitude of approximately 7.5 m until the task is completed, it takes only 25 s, with the descent speed set to 0.3 m/s in the flight control software. Altitude data is sampled every second and represented by the orange graph (Estimated), while the ideal altitude, corresponding to a speed of 0.3 m/s, is shown by the blue bar chart. Overall, the estimated altitude shows satisfactory results, although, at the 25-s mark, there is still a residual estimate. This is because, just before touchdown, the system loses visibility but continues to record the last detected value. Based on these results, it can be concluded that our autonomous landing solution not only ensures high accuracy but also has the potential for expanded applications in reconnaissance UAVs and other autonomous control systems in the future.

5 Conclusions and Future Development Directions

This study presents an effective solution for developing an autonomous landing system for Quadrotor reconnaissance UAVs on a fixed platform with an Embedded ArUco marker. The approach includes camera calibration, marker detection, position estimation using the Pinhole camera model and PnP algorithm, and a PID controller for horizontal dynamics. Real-world tests by HTI Group show the system enables precise UAV landings with an error margin of about 11 cm from the platform center. Future work will focus on optimizing the process for even better results.

Acknowledgments. This study was funded by HTI Group, headquartered on the 15th Floor, VP2, Sun Square Building, 21 Le Duc Tho Street, My Dinh 2 Ward, Nam Tu Liem District, Hanoi.

References

1. Khazetdinov, A., Zakiev, A., Tsoy, T., Svinin, M., Magid, E.: Embedded ArUco: a novel approach for high precision UAV landing. In: 2021 International Siberian Conference on Control and Communications (SIBCON), pp. 1–6. IEEE (2021)
2. Wubben, J., et al.: Accurate landing of unmanned aerial vehicles using ground pattern recognition. Electronics **8**(12), 1532 (2019)
3. Wubben, J., et al.: A vision-based system for autonomous vertical landing of unmanned aerial vehicles. In: 2019 IEEE/ACM 23rd International Symposium on Distributed Simulation and Real Time Applications (DS-RT), pp. 1–7. IEEE (2019)
4. Daspan, A., Nimsongprasert, A., Srichai, P., Wiengchanda, P.: Implementation of robot operating system in Raspberry Pi 4 for autonomous landing quadrotor on ArUco Marker. Int. J. Mech. Eng. Robot. Res. **12**(4), 210 (2023)
5. Wang, P., Wang, C., Wang, J., Meng, M.Q.H.: Quadrotor autonomous landing on moving platform. Procedia Comput. Sci. **209**, 40–49 (2022)
6. Mohammadi, A., Feng, Y., Zhang, C., Rawashdeh, S., Baek, S.: Vision-based autonomous landing using an MPC-controlled micro UAV on a moving platform. In: 2020 International Conference on Unmanned Aircraft Systems (ICUAS), pp. 771–780. IEEE (2020)
7. Zhu, B., Zhang, B., Ge, Q.: A nonlinear model predictive control based control method to quadrotor landing on moving platform. Cogn. Comput. Syst. **5**(2), 118–131 (2023)
8. Luo, W., Ebel, H., Eberhard, P.: An LSTM-based approach to precise landing of a UAV on a moving platform. Int. J. Mech. Syst. Dyn. **2**(1), 99–107 (2022)
9. Cao, Q., Liu, Z., Yu, H., Liang, X., Fang, Y.: Autonomous landing of the quadrotor on the mobile platform via meta reinforcement learning. IEEE Trans. Autom. Sci. Eng. **22**, 2269 (2024)
10. Goldschmid, P., Ahmad, A.: Reinforcement learning-based autonomous multi-rotor landing on moving platforms. Auton. Robot. **48**(4), 1–21 (2024)
11. Ladosz, P., Mammadov, M., Shin, H., Shin, W., Oh, H.: Autonomous landing on a moving platform using vision-based deep reinforcement learning. IEEE Robot. Autom. Lett. **9**, 4575 (2024)
12. OpenCV. Open Source Computer Vision Library, OpenCV. https://opencv.org/. Accessed 09 Sept. 2024
13. Xuan-Mung, N., et al.: Quadcopter precision landing on moving targets via disturbance observer-based controller and autonomous landing planner. IEEE Access **10**, 83580–83590 (2022)
14. Zouaoui, S., Mohamed, E., Kouider, B.: Easy tracking of UAV using PID controller. Period. Polytech. Transp. Eng. **47**(3), 171–177 (2019)
15. Zhang, Z.: A flexible new technique for camera calibration. IEEE Trans. Pattern Anal. Mach. Intell. **22**(11), 1330–1334 (2000)
16. Lepetit, V., Fua, P.: Monocular model-based 3d tracking of rigid objects: A survey. Foundations and Trends® in Computer Graphics and Vision, **1**(1), 1–89 (2005)
17. Nvidia. Pinhole Camera Model. Available: https://docs.nvidia.com/vpi/appendix_pinhole_camera.html. Accessed 09 Sept. 2024
18. Nguyen, K.D., Nguyen, T.T.: Vision-based software-in-the-loop-simulation for Unmanned Aerial Vehicles using gazebo and PX4 open source. In: 2019 international conference on system science and engineering (ICSSE), pp. 429–432. IEEE (2019)

19. Marchand, E., Uchiyama, H., Spindler, F.: Pose estimation for augmented reality: a hands-on survey. IEEE Trans. Visual. Comput. Graph. **22**(12), 2633–2651 (2015)
20. ArduPilot. ArduPilot Development Documentation. Retrieved from https://ardupilot.org/dev/. Accessed 09 Sept. 2024
21. Dang, S.T., Dinh, X.M., Kim, T.D., Xuan, H.L., Ha, M.H.: Adaptive backstepping hierarchical sliding mode control for 3-wheeled mobile robots based on RBF neural networks. Electronics **12**(11), 2345 (2023)

Polynomials Without Tangent of Half Angle of Forward Kinematics of 3-Branched Planar Parallel Robots

Minh-Tuan Nguyen-Thai(✉) and Ho Dac Chung

Hanoi University of Science and Technology, Dai Co Viet Str. 1, Hanoi, Vietnam
tuan.nguyenthaiminh@hust.edu.vn

Abstract. The forward kinematics problems of 3-branched planar parallel robots have been studied extensively in the literature. The polynomial approach is one of the best ways to find all the solutions to kinematics problems. However, it is usually not easy to reduce a kinematics problem to a univariate polynomial, and tangent half-angle formulae are usually used to do this task. The disadvantage of using tangent of half angle is that the solutions where the angle equals pi cannot be found from the polynomial. In this paper, the forward kinematics problems of 3-branched planar parallel robots are classified and their polynomials are found without using tangent half-angle formulae. It is shown that there are sixteen different types of possible forward kinematics problems of 3-DOF 3-branched planar parallel robots, including 6 types that the authors have not been found in the literature. For each type of problem, a univariate polynomial is found. The variable is either the cosine of the rotational angle of the moving platform or a coordinate of the reference point of the moving platform. Illustrative examples are presented to demonstrate the effectiveness of the proposed approach. It is shown that the proposed approach is effective even when the polynomial in tangent of half angle is ineffective.

Keywords: Planar Parallel Robots · Forward Kinematics · Kinematic Polynomial

1 Introduction

Parallel robots, due to some advantages over serial robots, are now widely used in industry [1]. The parallel robot can be spatial robot such as the famous 6-DOF Stewart platform [2], spatial Delta robot [3, 4] or planar parallel mechanisms such as 3-RRR and 3-RPR planar robots [5]. The last two, especially the 3-RRR planar robot, are also the most studied 3-branched planar parallel robots [6–8], but there are still other types that see applications as well, such as 3-PRP planar robot and 2-PRP-1-PPR planar robot [9].

The forward kinematics problems of 3-branched planar parallel robots are also of great interest. Although numerical methods are always an option, analytical-based methods are preferable when they are feasible since they may efficiently produce all the solutions of the problems. Planar kinematic mapping can be used to convert the forward kinematics problem of a 3-RRR planar robot into a system of three quadratic equations

© The Author(s), under exclusive license to Springer Nature Switzerland AG 2025
H. A. Le Thi et al. (Eds.): ICAMCS 2024, LNNS 1311, pp. 488–500, 2025.
https://doi.org/10.1007/978-3-031-90606-0_41

in four homogeneous coordinates X_1, X_2, X_3, X_4 [10]. Using this method, two cases $X_4 = 0$ and $X_4 = 1$ are considered to make sure that all possible solutions are found. The most popular approach is to reduce a forward kinematics problem into a polynomial in the tangent of a half angle [11–13]. Dealing with forward kinematics of a 3-R$\underline{\text{P}}$R planar robot, Gosselin and Merlet present a method to get a polynomial of degree six in variable $t = \tan(\varphi/2)$, where φ is the rotation angle of the moving platform, then some special architects of the robot are considered [11]. Merlet also presents the polynomials in variable t for other types of 3-branched planar parallel robots but does not consider any PP passive branch [12]. Oetomo et al. use the same idea and consider some special cases of 3-RRR planar robot in which closed-form solution can be found [13]. The drawback of the use of $\tan(\varphi/2)$ as the principal variable is that $\varphi = \pi$ cannot be found from the polynomial.

In this study, we present polynomials of forward kinematics problems of all 3-branched planar parallel robots with rotational joints or prismatic joints or both, which is more general than the problems dealt by Merlet [12]. Moreover, the variable in the polynomial is the cosine of an angle to avoid the drawback of the tangent of a half angle.

The main parts of the paper are structured as follows. Section 2 introduces and classifies all the possible forward kinematics problems of all 3-branched planar parallel robots with rotational and prismatic joints. Section 3 reduces each of the forward kinematics problems into a polynomial. Section 4 illustrates the proposed polynomials with some numerical examples and compares them with the method proposed by Merlet [12].

2 Forward Kinematics Problems of All 3-Branched Planar Parallel Robots with Rotational and Prismatic Joints

2.1 General Kinematic Architecture and Classification of Branches Based on Passive Joints

Generally, each of the considered robots consists of nine joints and eight links, including a fixed platform and a moving platform as shown in Fig. 1. The nine joints are divided into three sets, each set and two links form a branch to connect the fixed platform and the moving platform. Hence, a robot has a total of three branches.

The branches can be classified into eight types based on the types of joints from the fixed platform to the moving platform: RRR, RRP, RPR, PRR, RPP, PRP, PPR and PPP. Letter “R” denotes rotational joint, and letter “P” denotes prismatic joint. Each branch has an active joint which is marked by underlining the associated letter.

Considering forward kinematics problems, only the two passive joints of a branch determine which kind of constraint that the branch sets on the moving platform. Thus, we consider only four types of branches based on passive joints: RR type, RP type, PR type and PP type.

RR Type. Six types of branches $\underline{\text{R}}\text{RR}$, $\text{R}\underline{\text{R}}\text{R}$, $\text{RR}\underline{\text{R}}$, $\text{RR}\underline{\text{P}}$, $\text{R}\underline{\text{P}}\text{R}$, $\underline{\text{P}}\text{RR}$ are all of RR type (Fig. 2a). We call the point of the first passive joint A_i and the point of the second passive joint B_i, where i is the index of the branch. Point A_i is fixed on the fixed platform and point B_i is fixed on the moving platform. The distance between A_i and B_i is a constant

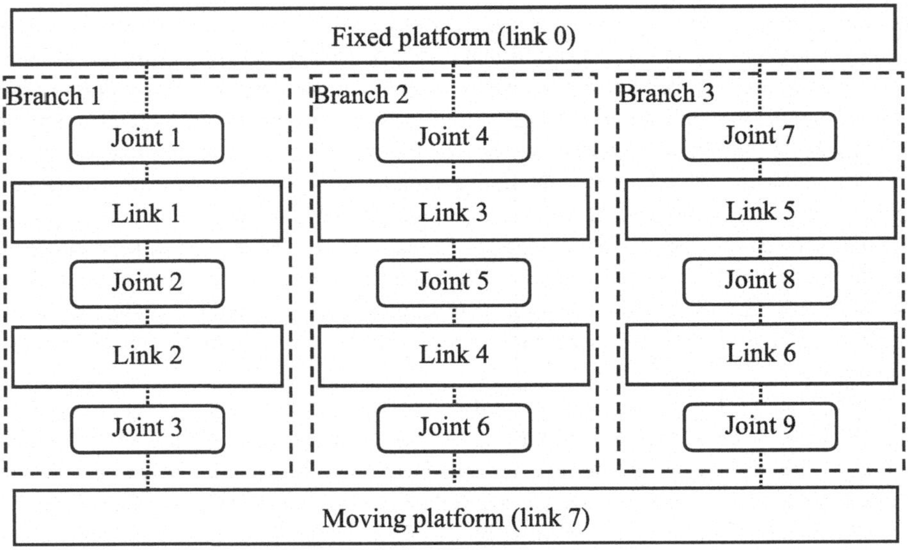

Fig. 1. General kinematic architecture of a 3-branched planar parallel robot.

l_i so the constraint is as follows

$$\left(x_{B_i} - x_{A_i}\right)^2 + \left(y_{B_i} - y_{A_i}\right)^2 = l_i^2 \tag{1}$$

where x_{A_i} and y_{A_i} are the coordinates of A_i; x_{B_i} and y_{B_i} are the coordinates of B_i. Here, x_{B_i} and y_{B_i} are unknowns.

RP Type. Six types of branches RRP, RRP, RPR, RPP, RPP, PRP are all of RP type (Fig. 2b). We call the point of the rotational joint A_i. Point B_i is a point fixed on the moving platform that lies on the axis of the prismatic joint. Point A'_i is a point on the axis of the prismatic joint so that $A_iA'_i$ is perpendicular to the prismatic joint. The coordinates of B_i $\left(x_{B_i}, y_{B_i}\right)$ and the rotational angle of the moving platform φ are constrained by

$$x_{B_i} = x_{A_i} - a_i \sin(\varphi + \alpha_i) + l_i \cos(\varphi + \alpha_i) \tag{2}$$

$$y_{B_i} = y_{A_i} + a_i \cos(\varphi + \alpha_i) + l_i \sin(\varphi + \alpha_i) \tag{3}$$

where x_{A_i} and y_{A_i} are the coordinates of A_i; a_i is the distance between A_i and A'_i; l_i is the distance between A'_i and B_i; α_i is the angle between A'_iB_i and a reference line on the moving platform. Here, x_{A_i}, y_{A_i} and α_i are constants; x_{B_i}, y_{B_i}, l_i and φ are unknowns. Multiplying Eq. (2) by $\sin(\varphi + \alpha_i)$, Eq. (3) by $\cos(\varphi + \alpha_i)$ and subtracting them, one obtains

$$\left(x_{B_i} - x_{A_i}\right)\sin(\varphi + \alpha_i) - \left(y_{B_i} - y_{A_i}\right)\cos(\varphi + \alpha_i) + a_i = 0 \tag{4}$$

PR Type. PR Type. Six types of branches RPR, PRR, PRR, PRP, PPR, PPR are all of PR type (Fig. 2c). We call the point of the rotational joint B_i. Point A_i is a point fixed on

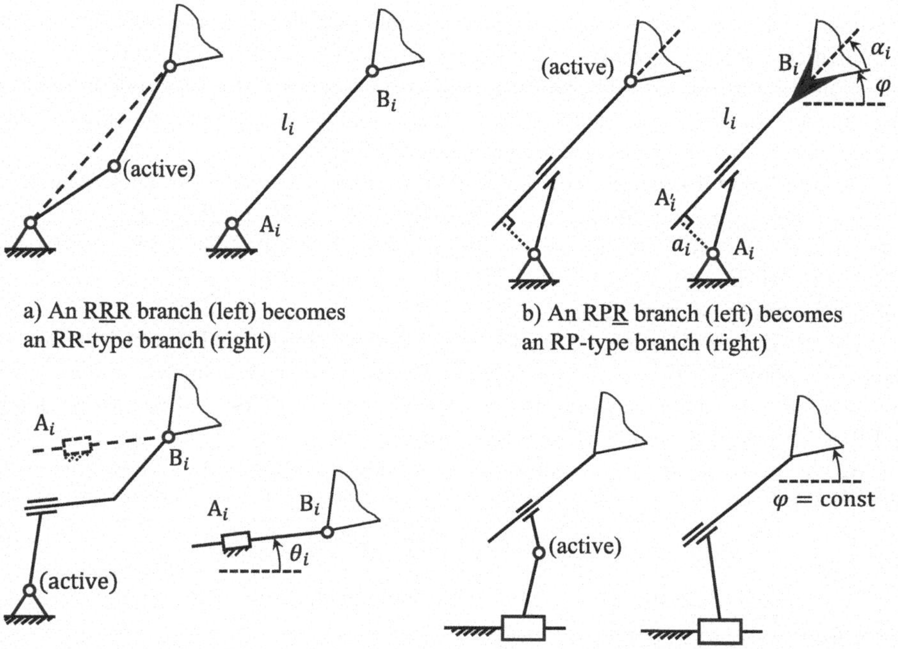

a) An RRR branch (left) becomes an RR-type branch (right)

b) An RPR branch (left) becomes an RP-type branch (right)

c) An RPR branch (left) becomes an PR-type branch (right)

d) A PRP branch (left) becomes an PP-type branch (right)

Fig. 2. Four types of branches classified based on passive joints.

the fixed platform defined as in Fig. 2c. The coordinates of B_i $\left(x_{B_i}, y_{B_i}\right)$ are constrained by

$$\left(x_{B_i} - x_{A_i}\right)\sin\theta_i - \left(y_{B_i} - y_{A_i}\right)\cos\theta_i = 0 \tag{5}$$

where x_{A_i} and y_{A_i} are the coordinates of A_i; θ_i is the rotational angle of A_iB_i. Here, x_{A_i}, y_{A_i} and θ_i are constants; x_{B_i} and y_{B_i} are unknowns.

PP Type. Six types of branches RPP, PRP, PPR, PPP, PPP, PPP are all of PP type (Fig. 2d). This type of branch is completely ignored by Merlet [12]. However, some architectures can still have this type of branch. If a 3-branched planar parallel robot has a PP type branch, the rotational angle of its moving platform is constant:

$$\varphi = \text{const} \tag{6}$$

2.2 Possible Forward Kinematics Problems

Three branches form a planar robot and, therefore, create a forward kinematics problem. To count the number of possible forward kinematics problems, it should be noted that the branches can be renumbered, and the robot can be mirrored so the order of the three branches is not important. Thus, the number of distinct forward kinematics problems is the number of 3-combinations of the types of branches with repetition. However, not all combinations of branches can form a 3-DOF planar robot. A branch of PP type constrains the orientation of the moving platform, so if two branches of PP type assembled to a moving platform, their constraints overlap. Hence, we do not consider 3-branched planar parallel robots that have two or three branches of PP type.

The number of distinct possible forward kinematics problems is 16. They are named based on the type of their branches: 3-RR, 2-RR-1-RP, 2-RR-1-PR, 2-RR-1-PP, 1-RR-2-RP, RR-RP-PR, RR-RP-PP, 1-RR-2-PR, RR-PR-PP, 3-RP, 2-RP-1-PR, 2-RP-1-PP, 1-RP-2PR, RP-PR-PP, 3-PR, 2-PR-1-PP. All six types that have a PP-type branch are not considered by Merlet [12].

Because there can be at most one PP branch and we always put PP branch at number three, four points A_1, B_1, A_2 and B_2 can always be defined in a kinematics problem. Two points A_3 and B_3 can be defined if the last branch is not of PP type. If the last branch is of PP type, A_3 and B_3 can be chosen arbitrarily.

We choose A_1 as the origin and A_1A_2 as the reference line of the fixed platform, so that (Fig. 3)

$$x_{A_1} = 0, y_{A_1} = 0, y_{A_2} = 0 \tag{7}$$

We choose B_1B_2 as the reference line of the moving platform so that φ is the rotational angle from $\overrightarrow{A_1A_2}$ to $\overrightarrow{B_1B_2}$.

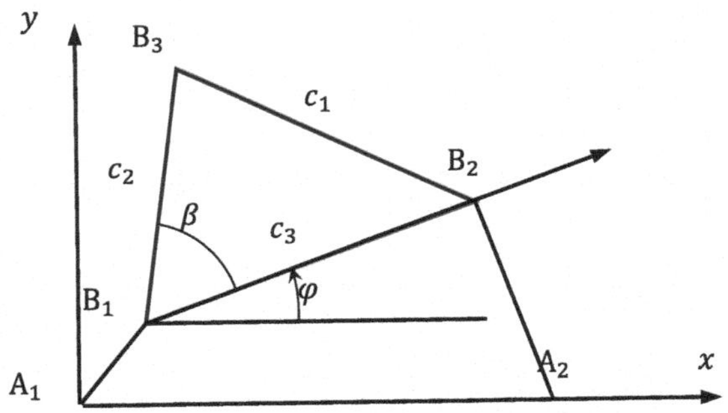

Fig. 3. Parameters of the moving platform.

The parameters of the moving platform are angle β between B_1B_3 and B_1B_2, distance c_3 between B_1 and B_2 and distance c_2 between B_1 and B_3 (Fig. 3). The constraints among three points B_1, B_2 and B_3 are

$$x_{B_2} = x_{B_1} + c_3 \cos\varphi \tag{8}$$

$$y_{B_2} = y_{B_1} + c_3 \sin\varphi \tag{9}$$

$$x_{B_3} = x_{B_1} + c_2 \cos(\varphi + \beta) \tag{10}$$

$$y_{B_3} = y_{B_1} + c_2 \sin(\varphi + \beta) \tag{11}$$

3 Polynomials of the Forward Kinematics Problems

3.1 Polynomial of 3-RR Problem

From Eqs. (1), (8)–(11), the constraints of a 3-RR problem are

$$x_{B_1}^2 + y_{B_1}^2 = l_1^2 \tag{12}$$

$$\left(x_{B_1} + c_3 \cos\varphi - x_{A_2}\right)^2 + \left(y_{B_1} + c_3 \sin\varphi\right)^2 = l_2^2 \tag{13}$$

$$\left(x_{B_1} + c_2 \cos(\varphi + \beta) - x_{A_3}\right)^2 + \left(y_{B_1} + c_2 \sin(\varphi + \beta)\right)^2 = l_3^2 \tag{14}$$

Denote

$$C = \cos\varphi, S = \sin\varphi \tag{15}$$

Using Eq. (12) to eliminate $x_{B_1}^2$ and $y_{B_1}^2$ in Eqs. (13) and (14), we obtain a system of two algebraic equations

$$\boldsymbol{A}(C, S)\begin{bmatrix} x_{B_1} \\ y_{B_1} \end{bmatrix} = \boldsymbol{b}(C, S) \tag{16}$$

where $\boldsymbol{A}(C, S)$ is a two-by-two matrix function and $\boldsymbol{b}(C, S)$ is a two-element column vector function. Solving system (16) for x_{B_1} and y_{B_1} and substituting into Eq. (12) yield a polynomial in C and S, where the highest degree of S is 4. To reduce the highest degree of S to 1, we use the following identities

$$S^2 = 1 - C^2, S^3 = \left(1 - C^2\right)S, S^4 = \left(1 - C^2\right)^2 \tag{17}$$

The obtain equation has the following form

$$P_2(C)S = P_3(C) \tag{18}$$

where $P_2(C)$ and $P_3(C)$ are polynomial of degree 2 and 3, respectively. Squaring Eq. (18) and applying (17) again completely eliminate S, and the obtain equation is a polynomial of degree 6 in C.

$$P_2^2(C)\left(1 - C^2\right) - P_3^2(C) = 0 \tag{19}$$

The forward kinematics problem is solved in the following order: solving Eq. (19) for $\cos\varphi$, solving Eq. (18) for $\sin\varphi$ and solving system (16) for x_{B_1} and y_{B_1}. Equation (18) is linear in $\sin\varphi$, and system (16) is linear in x_{B_1} and y_{B_1} so the maximum number of solutions of the forward kinematics problem equals the degree of (19).

3.2 Polynomials of 2-RR-1-RP and 2-RR-1-PR Problems

From Eqs. (1), (8), (9), the first two constraints are

$$x_{B_1}^2 + y_{B_1}^2 = l_1^2 \tag{20}$$

$$\left(x_{B_1} + c_3 \cos\varphi - x_{A_2}\right)^2 + \left(y_{B_1} + c_3 \sin\varphi\right)^2 = l_2^2 \tag{21}$$

Using Eq. (20) to eliminate $x_{B_1}^2$ and $y_{B_1}^2$ in Eqs. (21), we have

$$2\left(c_3 \cos\varphi - x_{A_2}\right)x_{B_1} + 2c_3 \sin\varphi y_{B_1} + l_1^2 - l_2^2 + c_3^2 + x_{A_2}^2 - 2c_3 \cos\varphi x_{A_2} = 0 \tag{22}$$

The third constraint, which is either (4) or (5), can be transformed using Eqs. (10) and (11). Equation (22) and the transformed third constraint are linear in variables x_{B_1} and y_{B_1} so that they are a system of two algebraic equations that has the same form as in Eq. (16). Then, applying the same process described in Eqs. (17)–(19), we also obtain polynomials of degree 6 in $\cos\varphi$ for the two considered cases.

3.3 Polynomials of 1-RR-2-RP, RR-RP-PR and 1-RR-2PR Problems

The first constraint is (20). The remaining two constraints are either (4) or (5) or both so they are linear in variables x_{B_1} and y_{B_1} after transformed by Eqs. (8)–(11), and they have the same form as in Eq. (16). Then, applying the same process described in Eqs. (17)–(19), we also obtain polynomials of in $\cos\varphi$. The degrees of the polynomials in the cases of 1-RR-2-RP and 1-RR-2PR problems are 4. The degree in the case of RR-RP-PR problem is 6.

3.4 Polynomials of 3-RP, 2-RP-1-PR, 1-RP-2PR and 3-PR Problems

In all four cases, all three constraints are linear in variables x_{B_1} and y_{B_1} so they can be rewritten in the following forms

$$\boldsymbol{g}(C, S)^{\mathrm{T}}\begin{bmatrix} x_{B_1} \\ y_{B_1} \end{bmatrix} = h(C, S) \tag{23}$$

$$\boldsymbol{A}(C, S)\begin{bmatrix} x_{B_1} \\ y_{B_1} \end{bmatrix} = \boldsymbol{b}(C, S) \tag{24}$$

where $\boldsymbol{A}(C, S)$ is a two-by-two matrix function and $\boldsymbol{b}(C, S)$ and $\boldsymbol{g}(C, S)$ are two-element column vector functions and $h(C, S)$ is a scalar function. Solving system (24) for x_{B_1} and y_{B_1} and substituting into Eq. (23) yield a polynomial in C and S. Then, applying the same process described in Eqs. (17)–(19), we also obtain polynomials in $\cos\varphi$. The degrees of the polynomials in the cases of 3-RP and 3-PR problems are 2. The degrees in the cases of 2-RP-1-PR and 1-RP-2PR problems are 4.

3.5 Polynomials of 2-RR-1-PP Problem

From Eqs. (1), (6), (8) and (9), the constraints of a 2-RR-1-PP problem are

$$x_{B_1}^2 + y_{B_1}^2 = l_1^2 \tag{25}$$

$$\left(x_{B_1} + c_3 \cos\varphi - x_{A_2}\right)^2 + \left(y_{B_1} + c_3 \sin\varphi\right)^2 = l_2^2 \tag{26}$$

Note that φ in this case is a constant. Using Eq. (25) to eliminate $x_{B_1}^2$ and $y_{B_1}^2$ in Eqs. (26) and using the denotations in (15), we obtain

$$Gx_{B_1} + H = -Ky_{B_1} \tag{27}$$

where we have in this case

$$G = 2\left(c_3 \cos\varphi - x_{A_2}\right), H = l_1^2 - l_2^2 + c_3^2 + x_{A_2}^2 - 2c_3 \cos\varphi x_{A_2}, K = 2c_3 \sin\varphi \tag{28}$$

Squaring Eq. (27) and using Eq. (25) again yield a polynomial of degree 2 in variable x_{B_1}

$$\left(G^2 + K^2\right)x_{B_1}^2 + 2GHx_{B_1} + H^2 - K^2 l_1^2 = 0 \tag{29}$$

The forward kinematics problem is solved in the following order: solving Eq. (29) for x_{B_1} and then solving system (27) for y_{B_1}. Equation (27) is linear in y_{B_1} so the maximum number of solutions equals the degree of (29), which is 2.

3.6 Polynomials of RR-RP-PP and RR-PR-PP Problems

From Eqs. (1), (4) (or (5)), (6), (8) and (9), the constraints of an RR-RP-PP (or RR-PR-PP) problem are

$$x_{B_1}^2 + y_{B_1}^2 = l_1^2 \tag{30}$$

$$Gx_{B_1} + H = -Ky_{B_1} \tag{31}$$

In the case of RR-RP-PP problem we have constant coefficients

$$G = \sin(\varphi + \alpha_2) \tag{32}$$

$$H = -x_{A_2} \sin(\varphi + \alpha_2) + y_{A_2} \cos(\varphi + \alpha_2) + c_3 \sin\alpha_2 + a_2 \tag{33}$$

$$K = -\cos(\varphi + \alpha_2) \tag{34}$$

In the case of RR-PR-PP problem we have constant coefficients

$$G = \sin\theta_2 \tag{35}$$

$$H = -x_{A_2} \sin\theta_2 + y_{A_2} \cos\theta_2 + c_3 \sin(\theta_2 - \varphi) \tag{36}$$

$$K = -\cos\theta_2 \tag{37}$$

Squaring Eq. (31) and using Eq. (30) yield a polynomial of degree 2 in variable x_{B_1} like Eq. (29).

3.7 Polynomials of 2-RP-PP, RP-PR-PP and 2-PR-PP Problems

Because of the PP-type branch, φ is a constant. Using Eqs. (8) and (9), we can transform the constraints of the first two branches into a system of two linear algebraic equations in variables x_{B_1} and y_{B_1}. It can be reduced to a polynomial of degree 1 in variable x_{B_1} (or y_{B_1}) and can be solved easily.

4 Numerical Examples and Discussion

4.1 Example 1

Consider a 3-RR problem with the following parameters (the dimensions are ignored):

$$c_1 = 40, c_2 = 42, c_3 = 50, x_{\mathrm{A}_2} = 220, x_{\mathrm{A}_3} = 80, y_{\mathrm{A}_3} = 140,$$

$$l_1 = 100, l_2 = 180, l_3 = 80.$$

Equations (19), (18) and (16) now read

$$\begin{gathered} -0.35642C^6 + 2.6732C^5 - 7.2982C^4 + 8.0493C^3 \\ -1.6688C^2 - 1.7947C - 0.0897 = 0 \end{gathered} \tag{38}$$

$$\left(-0.61121C^2 + 1.6337C - 0.82163\right)S = -1.7862C^3 + 6.9237C^2 - 6.0867C - 1,2538 \tag{39}$$

$$(140c - 440)x_{B1} + (100s)y_{B1} = 22000c - 28500 \tag{40}$$

$$\begin{gathered} (53.280C - 64.940S - 160)x_{B1} + (64.940C + 53.28S - 280)y_{B1} \\ = 13354C + 22640S - 31364 \end{gathered} \tag{41}$$

Solving polynomial (38), we have six roots

$$C_1 = -0.053331, C_2 = -0.306471$$

$$C_{3,4} = 1.926813 \pm 0.352307i, C_{5,6} = 2.003122 \pm 0.014755i$$

Substituting two real roots into (39), we receive the values of S. Then, solving Eqs. (40) and (41), we evaluate x_{B_1} and y_{B_1}. The results of the kinematics problem are

$$\varphi = \operatorname{atan2}(-0.99858, 0.053331), x_{\mathrm{B}_1} = 46.818, y_{\mathrm{B}_1} = 88.363 \tag{42}$$

and

$$\varphi = \operatorname{atan2}(0.95188, -0.30647), x_{\mathrm{B}_1} = 85.402, y_{\mathrm{B}_1} = 52.023 \tag{43}$$

4.2 Example 2

Consider a 3-RR problem with the following parameters (the dimensions are ignored):

$c_1 = 25, c_2 = 15, c_3 = 20, x_{A_2} = 20, x_{A_3} = 60, y_{A_3} = 30, l_1 = 50, l_2 = 30, l_3 = 25.$

Equations (19), (18) and (16) now read

$$\begin{aligned}&-0.25920C^6 + 0.17626C^5 - 0.41939C^4 - 1.2831C^3\\&-0.12778C^2 + 0.33982C + 0.0392947 = 0\end{aligned} \quad (44)$$

$$\left(0.28800C^2 + 0.97920C + 2.7648\right)S = 1.5840C^3 - 0.73440C^2 + 0.37440C + 2.6928 \quad (45)$$

$$(40C - 40)x_{B_1} + (40S)y_{B_1} = 800C - 2400 \quad (46)$$

$$(-30S - 120)x_{B_1} + (30C - 60)y_{B_1} = 900C - 1800S - 6600 \quad (47)$$

Solving polynomial (44), we have six roots

$$C_1 = -1, C_2 = -0.11625, C_3 = 0.49153, C_4 = -0.64394,$$

$$C_{5,6} = 0.97433 \pm 1.7807i.$$

Substituting four real roots into (45), we receive the values of S. Then, solving Eqs. (46) and (47), we evaluate x_{B_1} and y_{B_1}. The results of the kinematics problem are

$$\varphi = \text{atan2}(0, -1), x_{B_1} = 40, y_{B_1} = 30 \quad (48)$$

$$\varphi = \text{atan2}(-0.99322, -0.11625), x_{B_1} = 49.489, y_{B_1} = 7.1316 \quad (49)$$

$$\varphi = \text{atan2}(-0.87086, 0.49153), x_{B_1} = 29.399, y_{B_1} = 40.443 \quad (50)$$

$$\varphi = \text{atan2}(-0.76508, -0.64394), x_{B_1} = 23.890, y_{B_1} = 43.923 \quad (51)$$

The polynomial in $t = \tan \varphi/2$ obtained with the method used by Merlet [12] is

$$0.41472t^5 + 1.3190t^4 + 0.99072t^3 + 0.44352t^2 + 0.80640t + 0.39168 = 0 \quad (52)$$

Solving (52), we receive three real root:

$$t_1 = -1.1238, t_2 = -0.58387, t_3 = -2.1487$$

These roots agree with (49), (50), (51). However, the other solution (48) cannot be found using (52).

4.3 A Numerical Efficiency Comparison

We compare two numerical methods using Matlab. The first one is the "fsolve" method: using function "fsolve" to solve the constraint equations directly. The second one is the "roots" method: using function "roots" to find all the solutions of the kinematic polynomial and then solve the remaining linear equations.

To compare the efficiency of the two methods, two sets of kinematics problems are considered. The first set contains 3-RR problems, and the second set contains 2-RP-1-PR problems. Each set has 1000 kinematics problems that have at least one real solution. Such problems are considered to avoid prolonging the operation of "fsolve" without finding a solution.

Table 1 presents the running times of the two methods to solve all 1000 problems in each set. Note that the "fsolve" method only provides one solution for each problem while the "roots" method provides all the possible solutions. Both methods are run on the same computer with processor Intel® Core(TM) i5-8265U and provide solutions with the tolerance of 10^{-6}.

Table 1. Running times of the "fsolve" method and the "roots" method

Set of problems	Time to solve 1000 problems using the "fsolve" method	Time to solve 1000 problems using the "roots" method
3-RR	4.3 s	0.076 s
2-RP-1-PR	2.4 s	0.059 s

As shown in Table 1, the "roots" method outperforms the "fsolve" method. It shows that the proposed polynomial approach may provide a highly effective numerical tool.

5 Conclusions and Outlook

The branches of a 3-branched planar parallel robot are classified based on passive joints into four types: RR type, RP type, PR type and PP type. Based on this classification, there are 16 different possible kinematics problems, six of them are not considered by a similar study by Merlet [12].

Ten of the problems do not have a PP-type branch. They can be governed by a system of polynomials whose variables include sine and cosine of the rotational angle of the moving platform. Instead of using half-angle formulae to unify sine and cosine terms into the tangent of half an angle, we eliminate sine terms in the kinematics equations. First, the Pythagorean trigonometric identity is used to reduce the degree of sine terms to one. Then, applying the Pythagorean trigonometric identity the second time after rearranging and squaring yields a polynomial in one variable, which is the cosine of the rotational angle of the moving platform.

Six of the problems have one PP-type branch. The rotational angle of the moving platform is constrained to be constant. The variables in these problems are the coordinates of the reference point of the moving platform. The polynomial systems in these cases are easily reduced to a univariate polynomial.

Because the tangent of an angle can take any value from minus infinity to infinity, the polynomial based on tangent become ineffective in the case that the rotational angle of the moving platform is π. In contrast, the proposed polynomials do not suffer from that disadvantage.

In future studies, more complicated problems such as 4-branched robots should be considered. Spatial parallel robots should also be considered [14, 15]. The joint types of considered robots may not be limited to prismatic and rotational joints. Another development that can be made is to build software tools to solve classes of kinematics problems automatically.

Disclosure of Interests. The authors have no competing interests to declare that are relevant to the content of this article.

References

1. Gasparetto, A., Scalera, L.: From the Unimate to the Delta robot: the early decades of industrial robotics. In: Explorations in the History and Heritage of Machines and Mechanisms: Proceedings of the 2018 HMM IFToMM Symposium on History of Machines and Mechanisms, pp. 284–295. Springer International Publishing (2019)
2. Stewart, D.: A platform with six degrees of freedom. Proc. Inst. Mech. Eng. **180**(1), 371–386 (1965)
3. Vischer, P., Clavel, R.: Kinematic calibration of the parallel Delta robot. Robotica **16**(2), 207–218 (1998)
4. Rey, L., Clavel, R.: The delta parallel robot. In: Parallel Kinematic Machines: Theoretical Aspects and Industrial Requirements, pp. 401–417. Springer London, London (1999)
5. Arsenault, M., Boudreau, R.: Synthesis of planar parallel mechanisms while considering workspace, dexterity, stiffness and singularity avoidance. J. Mech. Des. **128**(1), 69–78 (2005)
6. Yu, A., Bonev, I.A., Zsombor-Murray, P.: Geometric approach to the accuracy analysis of a class of 3-DOF planar parallel robots. Mech. Mach. Theory **43**(3), 364–375 (2008)
7. Briot, S., Bonev, I.A.: Accuracy analysis of 3-DOF planar parallel robots. Mech. Mach. Theory **43**(4), 445–458 (2008)
8. Huang, M.Z., Thebert, J.L.: A study of workspace and singularity characteristics for design of 3-DOF planar parallel robots. Int. J. Adv. Manuf. Technol. **51**, 789–797 (2010)
9. Joubair, A., Slamani, M., Bonev, I.A.: A novel XY-Theta precision table and a geometric procedure for its kinematic calibration. Robot. Comput. Integrated Manuf. **28**(1), 57–65 (2012)
10. Husty, M.L.: An algorithm for solving the direct kinematics of general Stewart-Gough platforms. Mech. Mach. Theory **31**(4), 365–379 (1996)
11. Gosselin, C.M., Merlet, J.P.: The direct kinematics of planar parallel manipulators: special architectures and number of solutions. Mech. Mach. Theory **29**(8), 1083–1097 (1994)
12. Merlet, J.P.: Direct kinematics of planar parallel manipulators. In: Proceedings of IEEE International Conference on Robotics and Automation, vol. 4, pp. 3744–3749. IEEE (1996)
13. Oetomo, D., Liaw, H. C., Alici, G., Shirinzadeh, B.: Direct kinematics and analytical solution to 3RRR parallel planar mechanisms. In: 2006 9th International Conference on Control, Automation, Robotics and Vision, pp. 1–6. IEEE (2006)

14. Gosselin, C.M., Sefrioui, J., Richard, M.J.: On the direct kinematics of spherical three-degree-of-freedom parallel manipulators of general architecture. J. Mech. Des. **116**(2), 594–598 (1994)
15. Li, W., Zhang, S., Wang, B., Angeles, J., Gao, F., Guo, W.: Forward kinematics of three classes of 3-RRR spherical parallel mechanisms admitting closed-form solutions. Mech. Mach. Theory **201**, 105751 (2024)

Information Theory and Information Security

Design Index Supporting Efficient Querying on Encrypted Numerical Data Based on Stacked Bucket Loading and Order Preservation

Canh Ngoc Hoang, Thi Thu Thuy Nguyen(✉), Danh Kim Le Tran, and Huy Quang Vu

Thuongmai University, Hanoi, Vietnam
{canhhn,thuynguyenthithu,danhtlk,huyvq}@tmu.edu.vn

Abstract. Querying on encrypted data is a currently interesting area aimed at addressing security issues when deploying databases on cloud servers. The research proposes an improvement in the design method of indexes supporting the search which is based on buckets. This method allows to query efficiently the data on those indexes. The efficiency is demonstrated through criteria related to security, performance, and deploy-ability. The overlapping properties of the buckets enhance security and reduce false positive results, while the order preservation property improves search performance, facilitates query transformation, and ensures deploy-ability for all modern relational databases.

Keywords: Bucket · Overload · Order Preserving · Searchable Encryption (SE) · Proxy

1 Introduction

Recently, the use of cloud servers for deploying software services and databases has become an inevitable trend in businesses. However, cloud servers can be seen as an "honest but curious" environment which might create risks of data exposure for owners. A reliable solution is to encrypt data before deploying databases on cloud servers. While the security issue is addressed, it affects users' ability to manipulate data, as queries on clear data cannot operate on encrypted data.

There are many schemes such as SE [3–9] which have been designed on index supporting search on encrypted data. These schemes are categorized as bucket based index [4, 6, 8], hash based index [3], B^+ Tree based index [7], order preserving encryption (OPE) [9], homomorphic encryption (HE) [5], etc.

In visual point of view, querying on clear numerical data can easily be performed by comparing the user's query conditions with each record or the nodes in the database's special data structure. Therefore, most indexes built in SE schemes retain the order preservation property to facilitate comparisons and easily construct auxiliary data structures to speed up queries. For example, hash-based index solutions support equality

© The Author(s), under exclusive license to Springer Nature Switzerland AG 2025
H. A. Le Thi et al. (Eds.): ICAMCS 2024, LNNS 1311, pp. 503–514, 2025.
https://doi.org/10.1007/978-3-031-90606-0_42

comparisons well, while Order-Preserving Encryption (OPE) fully supports all operators ($>$, $<$, $=$, $\geq$, $\leq$) with a one-to-one mapping between clear data and the index. These solutions provide good query performance and simplify the process of transforming query commands but can lead to serious index information leakage. On the other hand, bucket-based indexes also fully support comparison operators but establish an n-1 relationship between clear data and the index, enhancing security compared to the previously mentioned solutions. However, bucket-based indexes may return false positive results after querying. Additionally, Homomorphic Encryption (HE) solutions focus on supporting computational operations rather than comparison operators and require significant execution costs, while B + Tree-based indexes provide a structure that enhances comparison speed on the indexes.

In this paper, we focus on leveraging the security strengths and simple design of the bucket-based index solution, combined with proposed improvements in range division, to create a more efficient bucket index based on both security criteria and minimizing data redundancy after querying.

2 Literature Review

Hacigumus et al. [6] first proposed an SE scheme using indexes based on range of clear data, subsequently using a unique value (BucketID) to represent each range. The querying process on the cloud server is performed directly on the BucketIDs, returning records belonging to the BucketID that meet the query conditions. There are several methods for designing buckets such as overlapping buckets and non-overlapping buckets. The overlapping buckets method does not preserve order, which ensure the highest safety of the index but reduce performance and the diversity of query commands due to difficulties in transforming clear data queries into encrypted data queries. Conversely, the method of non-overlapping buckets that preserve order facilitates easy transformation of query commands and achieves high query speeds on encrypted databases, but it leads to information leakage of clear data through frequency analysis attacks and exploitation of BucketID order. In addition to the aforementioned advantages and disadvantages, the presence of false positive data in the query results is another drawback that increases the decryption burden on end users.

In other research on security for buckets, Hore et al. [4] identified two metrics as Variance and Entropy, whose value changes are related to performance and security issues for buckets. The research indicates that Variance represents the data dispersion within each bucket with the larger this value, the better the resistance against dynamic information leakage attacks. Entropy, on the other hand, represents the distribution of data among buckets with the smaller this value, the more effective the resistance against static information leakage attacks. However, the solutions in Hore's study are quite distinct and lack coherence in demonstrating how to divide buckets to increase Variance while minimizing Entropy, all while ensuring the reduction of false positive results after querying.

Wang & Du [8] further pointed out existing issues with Variance. Specifically, Variance affects internal security within each bucket, where the size of the Variance is closely related to the size of the bucket (size referring to the range of values within the bucket). An increased Variance tends to disrupt the uniform distribution of data within each bucket. The research also asserts that for two buckets of equal size, the one with a more uniform distribution will be more secure. Therefore, having a large variance is insufficient; the authors introduced a new metric called "Probability Distribution Deviation" (PDV) to represent the deviation between actual values and the uniformly distributed values of the bucket – the smaller the PDV value, the more secure the bucket. The research also shown some cases of high Variance that still lead to data leakage within buckets, and then uses PDV as a supplemental measure to address the localized issue. The disadvantage of the research is that there is no solution of providing for users to design buckets that are both overlapping and order-preserving, ensuring both security and query performance.

To solve this issue, this research will propose a method for constructing buckets that are both overlapping and order-preserving to address the stated issues, serving as a foundation for creating indexes that support efficient querying on encrypted digital data.

3 Methodology

Designing indexes to support the searching of encrypted numerical data based on buckets is a technique applied in many SE schemes. Following the approach, we propose new improvements in bucket construction to enhance the security of the buckets, reduce redundancy, while ensuring query performance. The methods can be seen as follows.

- Firstly, research inherits previous publication bucket construction methods, particularly the factors affecting security and query performance on buckets. This helps to identify existing gaps and specific directions for improvement in the data ranges which are divided to construct buckets.
- Secondly, research designs buckets based on the principles of overlapping and order preservation with items in each bucket uniformly distributed. The research employs the idea of flattening and dividing clear data ranges. Flattening is based on the principle that each distinct numerical value in the clear dataset, appearing f times. This will be mapped into at least f different variants, serving as the foundation for creating uniformly distributed buckets with high security. The overlapping and order preservation of buckets is based on a method of randomly and sequentially dividing ranges from left to right in the expanded dataset.
- Then, research analyses of variance, entropy, and PVD for the created buckets to evaluate security levels compared to previous studies.

The research also runs alternative experiments with different query scenarios on the proposed bucket-based index using the TPC-H database in [10] in order to show the improvement in query efficiency.

4 Design Index to Support Efficient Querying on Encrypted Numerical Data Based on Overlapping Buckets and Order Preservation

4.1 Objectives for Design Index Based on Bucket

The design of the Effective Bucket Index (EBI) aims to achieve multiple advantages for all three issues as Security, Performance, and Usability. Specifically, the objectives to be achieved with EBI as follows:

- Protection against types of attacks: dynamic information leakage and static information leakage.
- Easy transformation of query commands from clear data to queries on EBI.
- Direct range querying support on EBI, reducing redundancy in the result set.

To realize these goals, we propose the following idea for range division to create EBI as follows.

- The clear dataset needs to be transformed into a new value set with randomly generated and uniformly distributed items, so that every value in the new set has an equal probability of occurrence.
- Using a continuous random range division approach to create buckets based on the principles of overlapping and order preservation.
- Each clear value can be represented by multiple EBIs and vice versa.

4.2 Design Effective Bucket Index (EBI)

To build an index that meets the stated objectives and design ideas, the preprocessing of the numerical dataset will be carried out based on two steps as flattening and range division. To implement data flattening, we provide some definitions and examples as follows:

Definition 1. Let $V = \{v_1, v_2, \ldots, v_z\}$ be the set of sensitive numerical clear. Then $D = \{d_1, d_2, \ldots, d_n\}(n \leq z)$ is the set of distinct items of set V if the items of D are determined as, $d_i = \begin{pmatrix} f_i \\ v_i \end{pmatrix}$, $d_{i+1} = \begin{pmatrix} f_{i+1} \\ v_{i+1} \end{pmatrix}$ where $1 \leq i \leq n-1$; v_i, v_{i+1}, $\in V$; $v_i \leq v_{i+1}$; f_i is the frequency of occurrence of v_i in V.

Example 1. Let $V = \{2, 8, 6, 17, 2, 43, 6, 17, 43, 6, 61, 17, 43, 43, 2, 6, 8, 43, 61\}$. As mentioned in definition 1, D includes 6 distinct items represented for V can be seen as follows:

$$D = \left\{\begin{pmatrix} f_i \\ v_i \end{pmatrix}\right\} = \left\{\begin{pmatrix} 3 \\ 2 \end{pmatrix}, \begin{pmatrix} 4 \\ 6 \end{pmatrix}, \begin{pmatrix} 2 \\ 8 \end{pmatrix}, \begin{pmatrix} 3 \\ 17 \end{pmatrix}, \begin{pmatrix} 5 \\ 43 \end{pmatrix}, \begin{pmatrix} 2 \\ 61 \end{pmatrix}\right\}$$

Definition 2. Let $D = \{d_1, d_2, \ldots, d_n\}$ be the distinct items of V, where $d_i, d_{i+1} \in D$. Then, $P_i = \{p_{i1}, p_{i2}, \ldots, p_{ik_i}\}(1 \leq i \leq n)$ will be called as random value set representing for $d_i = \begin{pmatrix} f_i \\ v_i \end{pmatrix}$. Similarity, P_{i+1} is the set of random value represented for $d_{i+1} = \begin{pmatrix} f_{i+1} \\ v_{i+1} \end{pmatrix}$. P_i, P_{i+1} need to be satisfied the following requirements.

- $[P_i^{left}, P_i^{right}]$ is the valued range of P_i for P_i^{left}, P_i^{right} and these values are integers satisfying $P_i^{right} - P_i^{left} \geq f_i$.
- $p_{ij} \leftarrow PRF(P_i^{left}, P_i^{right})$ where PRF is a function that generates pseudo-random number in the range of $[P_i^{left}, P_i^{right}]$, $1 \leq j \leq k_i, k_i = |P_i|, k_i \geq f_i$ và $p_{i1} \leq p_{i2} \leq \cdots \leq p_{ik_i}$.
- P_{i+1} is also defined the same as P_i, but all items in P_{i+1} is greater than items in P_i. In other word, $P_i^{righ} \leq P_{i+1}^{left}$.
- Number of items of $k_i (1 \leq i \leq n)$ in P_i will be determined as rule: $\frac{k_1}{f_1} \approx \frac{k_2}{f_2} \approx \ldots \approx \frac{k_i}{f_i} \approx \ldots \approx \frac{k_n}{f_n} \approx \varepsilon$, where ε being an integer greater than or equal to 1. ε can be seen as expansion coefficient factor.

Example 2. Consider D in **Example 1** with 6 items. In this case, the random values set of $P_i (1 \leq i \leq 6)$ with $\varepsilon = 1$ can be determined in the following steps.

Step 1. Determine the valued range for each $P_i (1 \leq i \leq 6)$:

$P_1 \in [0, 80], P_2 \in [200, 500], P_3 \in [600, 900],$
$P_4 \in [2000, 4500], P_5 \in [7000, 8300], P_6 \in [8500, 9500].$

Step 2. Create a list of random items in valued range in each P_i:

$P_1 = \{9, 35, 56\}, P_2 = \{265, 358, 396, 428\},$
$P_3 = \{719, 864\}, P_4 = \{2024, 3871, 4269\},$
$P_5 = \{7077, 7658, 7832, 7954, 8162\}, P_6 = \{8672, 9489\}.$

Definition 3. Let $V = \{v_1, v_2, \ldots, v_z\}$ be the set of sensitive numerical clear data and $D = \{d_1, d_2, \ldots, d_n\}$ is the set of distinct items of set V and $P_i (1 \leq i \leq n)$ where P_i is the set of random values representing the random values of $d_i \in D$. P can be seen as the **flat** valued set representing for V if $P = P_1 \cup P_2 \cup \cdots \cup P_i \cup \cdots \cup P_n$.

The items in P will have the same portion of probabilities of occurrence. In other words, set P has a uniform distribution.

Now we have: P = {9, 35, 56, 265, 358, 396, 428, 719, 864, 2024, 3871, 4269, 7077, 7658, 7832, 7954, 8162, 8672, 9489}.

To continue the process of grouping/range division (Bucket) on set P, the definition 4 will be proposed as follows.

Definition 4. Let $V = \{v_1, v_2, \ldots, v_z\}$ be a sensitive clear numerical dataset and P is a flat v alue representing for V. $B = \{B_1, B_2, \ldots., B_m\}$ is a set of values EBI represention for V if:

Algorithm *BuildEBI*

Input

$V = \{v_1, v_2, \ldots, v_z\}$: set of sensitive clear numerical data;

ε: expansion coefficient factor ($\varepsilon \geq 1$);

m: total of bucket ($m \leq |P|$);

Output

Set $B = \{B_1, B_2, \ldots, B_m\}$ is a set of values EBI representing for V;

Set $V^B = \{v_1^B, v_2^B, \ldots, v_z^B\}$: is a set of EBI according to each item in V (for $v_i^B = B_j$ *and* $B_j \in B, 1 \leq j \leq m$);

Steps

(1) Build $D = \{d_1, d_2, \ldots d_i, \ldots d_n\}$ is a set of distinct items of V ($1 \leq i \leq n$);

(2) Based on the coefficient factor ε, build sets $P_1, P_2, \ldots, P_i, ..$ respectively as sets of random values representing for $d_1, d_2, \ldots d_i, \ldots d_n$ ($1 \leq i \leq n$);

(3) Build a set of flat value P representing for V. Here $P = P_1 \cup P_2 \cup \ldots \cup P_i \cup \ldots \cup P_n = \{p_1, p_2, \ldots, p_{|P|}\}$;

(4) Choose the expanded boundary (EP) values for P are EP^{left} and EP^{right} where $EP^{right} \leq p_1, EP^{right} \geq p_{|P|}$;

(5) Determine value at $m-1$ intersection points of m ranges on P by as: Calculate $m-1$ random distinct values of $\{ipv_1, ipv_2, \ldots, ipv_{m-1}\}$ as follows: $ipv_t \leftarrow PRF(EP^{left}, EP^{right})$ for $1 \leq t \leq m-1$ and $ipv_t < ipv_{t+1}$;

(6) Determine a set $B = \{B_1, B_2, \ldots B_t, \ldots B_m\}$ contains values of EBI representing for V as:

$B_1 \leftarrow PRF(EP^{left}, ipv_1)$; Value B_1 represents for set EP_1 with the range of $[EP^{left}, ipv_1)$;

$B_2 \leftarrow PRF(ipv_1, ipv_2)$; Value B_2 represents for set EP_2 with the range of $[ipv_1, ipv_2)$;

...

$B_t \leftarrow PRF(ipv_{t-1}, ipv_t)$; Value B_t represents for set EP_t with the range of $[ipv_{t-1}, ipv_t)$;

...

$B_m \leftarrow PRF(ipv_{m-1}, EP^{right})$; Value B_m represents for set EP_m with the range of $[ipv_{m-1}, EP^{right})$;

(7) Determine $V^B = \{v_1^B, v_2^B, \ldots, v_z^B\}$ containing value of EBI respectively for items in V as follows.

(7.1) For each $v_j \in V$, determine item $d_i \in D$ where $d_i = v_j$ ($1 \leq j \leq z$; $1 \leq i \leq n$);

(7.2) Determine a set P_i respectively for d_i, let a random value $p \in P_i$;

(7.3) Check if $p \in EP_t$ ($1 \leq t \leq m-1$) then B_t is a value of Bucket representing for v_j, assign $v_j^B = B_t$;

(8) $V^B = \{v_1^B, v_2^B, \ldots, v_z^B\}$ for $v_j^B \in B$ ($1 \leq j \leq z$);

(9) **Return** B, V^B;

- P is divided randomly to be m continuous range ($m \le |P|$);
- The jth range is represented as a random value as B_j;
- With $j, j' \in [1, m]$ and $j < j'$ then $B_j \le B_{j'}$.

According to **Definition 4**, each $v_i \in V(1 \le i \le z)$ will be represented as a value *EBI* $B_j \in B(1 \le j \le m)$ respectively. The algorithm ***BuildEBI*** build the set of EBI for z items of V as follows.

After constructing the index set V^B representing the values in set V, the confidential information (Table 1) related to the process of generating the EBI values will be stored in a secure area to facilitate the transformation of query conditions and update the index during process of using database if there are any changes in the database.

Table 1. Confidential information generating ***EBI***

$d_i.v$	P_i	P_i^{left}	P_i^{right}	B_i	ipv_{i-1}	ipv_i
2	P_1	0	80	B_1	EP^{left}	ipv_1
2	P_1	0	80	B_2	ipv_1	ipv_2
6	P_2	200	500	B_2	ipv_1	ipv_2
6	P_2	200	500	B_3	ipv_2	ipv_3
8	P_3	600	900	B_3	ipv_2	ipv_3
8	P_3	600	900	B_3	ipv_2	ipv_3
17	P_4	2000	4500	B_4	ipv_3	ipv_4
17	P_4	2000	4500	B_4	ipv_3	ipv_4
43	P_5	7000	8300	B_5	ipv_4	ipv_5
43	P_5	7000	8300	B_5	ipv_4	ipv_5
61	P_6	8500	9500	B_6	ipv_5	EP^{right}

The *BuildEBI* algorithm should be used on set V with a complete number of distinct items to describe all possible cases of numerical data that may arise in practice. Each value in set V can be represented by multiple values in set B and vice versa. Additionally, the intervals (Buckets) on set *EP*(expanded boundary), which have been flattened and uniformly distributed, will help mitigate dynamic and static data leakage attacks on EBI.

Moreover, the *BuildEBI* algorithm performs interval division on set *EP* instead of processing directly on set V as in previous studies. This allows the created data intervals (Buckets) to potentially include only a few variants among f_i variants representing the element v_i (where $v_i \in V$), significantly reducing redundancy when querying on EBI at the cloud server.

For $v_p, v_q \in V$ and $v_p^B = B_i$, $v_q^B = B_j$ for values of *EBIs* representing v_p, v_q $(1 \le p, q \le z; 1 \le i, j \le m)$ respectively. If $v_p \le v_q$ then $v_p^B \le v_q^B$. This order preservation facilitates the transformation of range query commands from clear data to queries on EBI easier and achieving high performance.

4.3 Query on Index of Effective Bucket Index (EBI)

In practice, the values in set V are organized and stored in a field of a relational database. Consider relation R with field X_t storing the sensitive numerical values of set V. On the cloud server, R is transformed into the encrypted relation R^E with fields X_t^E and X_t^{EBI} corresponding to X_t. Here, X_t^E contains the encrypted values of X_t through traditional algorithms such as AES [1] and DES [2], while X_t^{EBI} contains the EBI indexes representing the values of X_t. The process for querying the range $[l, h]$ requested by the user through the clear command "*Select * from R where* X_t *between l and h*" is as follows.

Step 1 – Transform operands l, h: In the secure area, based on Table 1 determine values of lower bound and upper bound of *EBI*: $c_l = Min(l^B)$ and $c_h = Max(h^B)$ as follows.

(1) Determine the clear value range that satisfies the query condition of range $[l, h]$:
   - Find the largest $d_i.v$ that is less than or equal to l, denoted as v_l;
   - Find the smallest $d_i.v$ that is greater than or equal to h, denoted as v_h;

(2) Determine l^B as the list of *EBI* representing for v_l, then find $c_l = Min(l^B)$ as the smallest of *EBI* in the list of l^B;
(3) Determine h^B as the list of *EBIs* representing for v_h, then find $c_h = Max(h^B)$ as the largest *EBI* in the list of h^B;

Step 2 – Transform the clear query command to a query on EBI: the new command takes the form of "$Select *fromR^E whereX_t^{EBI} betweenMin(l^B)andMax(h^B)$". This command will be sent to cloud server.

Step 3 – Query on cloud server: Execute the query command and receive a result set containing false positive records. The results are sent back to the secure area.

Step 4 – Decrypt and Re-Query: In the secure area, decrypt the result set received from the cloud server and continue to query this dataset using the command "*Select * from ResultDecrypt where* X_t *between l and h*". The results after the query are sent back to the end user.

5 Security Analysis of the EBI Index

Based on the parameters that evaluate the security capacity of buckets, such as variance σ^2 [4], entropy (H) [4], and pvd [8], we use the data from **Example 1** to measure and compare these parameters with two methods of bucket construction: Method 1, which involves sequential and non-overlapping buckets; Method 2, which involves buckets created using the *BuildEBI* algorithm.

Method 1: Create 3 Buckets: $B1 = \left\{\binom{3}{2}, \binom{4}{6}\right\}$; $B2 = \left\{\binom{2}{8}, \binom{3}{17}\right\}$; $B3 = \left\{\binom{5}{43}, \binom{2}{61}\right\}$.

Method 2: Create 6 Buckets: $EBI1 = \left\{\begin{pmatrix}1\\-15\end{pmatrix},\begin{pmatrix}3\\3\end{pmatrix}\right\}$; $EBI2 = \left\{\begin{pmatrix}1\\3\end{pmatrix},\begin{pmatrix}2\\7\end{pmatrix}\right\}$; $EBI3 = \left\{\begin{pmatrix}1\\7\end{pmatrix},\begin{pmatrix}2\\9\end{pmatrix},\begin{pmatrix}1\\14\end{pmatrix}\right\}$; $EBI4 = \left\{\begin{pmatrix}2\\14\end{pmatrix},\begin{pmatrix}3\\25\end{pmatrix}\right\}$; $EBI5 = \left\{\begin{pmatrix}2\\25\end{pmatrix},\begin{pmatrix}1\\50\end{pmatrix}\right\}$; $EBI6 = \left\{\begin{pmatrix}1\\50\end{pmatrix},\begin{pmatrix}1\\108\end{pmatrix}\right\}$;

Calculating of σ^2, H, pvd is based on [4, 8] we have:

Table 2. Comparing of $\boldsymbol{\sigma^2, H, pvd}$ using two methods of creating Bucket

Bucket	σ^2	H	pvd
$B1$	4.5	0.986	32
$B2$	7.5	0.971	50
$B3$	148	0.869	1250
$EBI1$	81	0.811	648
$EBI2$	5.3	0.914	32
$EBI3$	8.9	1.5	389
$EBI4$	36.3	0.971	242
$EBI5$	208.3	0.915	1250
$EBI6$	1682	1	6728

From Table 2, it can be seen that most of the measured values of σ^2, *andpvd* on the buckets created by the *BuildEBI* algorithm is higher than those from traditional methods. This implies that the security capabilities of the proposed EBIs are better than those of continuously created and non-overlapping buckets. However, the entropy (H) values show little difference between the two scenarios, with some EBIs even having lower entropy than the buckets $B1$, $B2$, $B3$. The reason is that the EBIs are not divided using the equi-depth method but are entirely designed using the random overlapping technique. Nonetheless, this random overlapping allows a clear value can be existed in multiple different EBIs. It shows that if an attacker exploits clear information based on the frequency statistics of the EBIs, they would have no basis for doing so.

6 Experiment and Discussion

Data: Data in the field column L_EXTENDEDPRICE in Table LINEITEM in the database taken from TPC-H database [10] will be used as the experimental data. TPC-H provides a benchmark dataset used for performance evaluation for most solutions in the database field, it enables random data replication and large scale scaling through the DB-Gen tool. The total experimental records in LINEITEM used in the queries will be {100K, 500K, 1M, 2M, 4M, 8M}. Column L_EXTENDEDPRICE will be transformed into three indexes as $IB1$, $IB2$ và EIB which are stored in cloud server. Execute the SQL command "*Select * From LINEITEM Where L_EXTENDEDPRICE Between **l** and **h***" in three scenarios as follows.

- Scenario 1: The bucket $IB1$ that does not preserve and overlapping.
- Scenario 2: The bucket $IB2$ that preserve order and non-overlapping.
- Scenario 3: The bucket $EIBs$ that preserve order and overlapping.

Evaluation: The measurement and evaluation criteria include the filtering rate (FIR), Error rate (FAR), and total query time.

- Filtering rate: $FIR = (N - N_{bucket})/(N - N_{result})$; Error rate: $FAR = (N_{bucket} - N_{result})/N_{bucket}$.
- Total query time as: $T = T1 + T2 + T3 + T4$.

Where N is the total number of records in the LINEITEM table, N_{result} is the total number of records returned when querying on clear data, and N_{bucket} is the total number of encrypted records returned after querying on the bucket index. $T1$ is the time to transform the clear query command into a query on the bucket index, $T2$ is the time for querying on the bucket index, $T3$ is the decryption time, and $T4$ is the re-query time.

With the number of buckets created being equivalent across the data scales in each scenario, the experimental results are shown in Fig. 1.

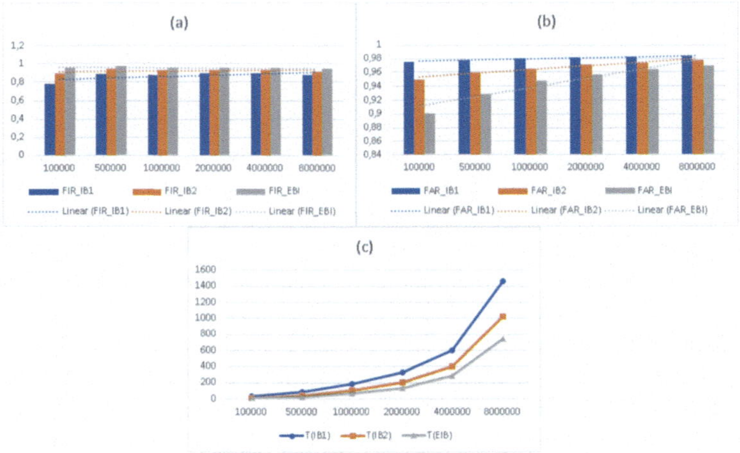

Fig. 1. Results of three scenarios of Bucket creation.

The Figure (a),(b) indicate that the filtering rate and error rate after querying for scenario 1 are the worst, while scenario 3 (which is proposed in the research) is the best. In scenario 1, despite having the highest security due to overlapping buckets and not preserving order, but it incurs a significant trade-off with a large number of false positive records returned after each query, causing FIR_IB1 to decrease and FAR_IB1 to increase.

The scenario 2 has the lowest security level. It is the easiest to design, but does not yield the best filtering effectiveness. This is because the buckets do not overlap and each numerical value belongs exclusively to one bucket. When a bucket i^{th} satisfies the query on the cloud server, the total number of returned records related to this bucket is $\sum v_i * f_i$ (where $v_i \in bucket_i$ and f_i is the frequency of v_i in the database.

In contrast, in scenario 3, the buckets overlap continuously and preserve order, ensuring that each v_i can belong to multiple buckets, so the number of records related to v_i in $bucket_i$ will be less than $\sum v_i * f_i$. This can be seen as the basis for reducing the number of false positive records returned when querying on the EBI index, thereby increasing FIR_EBI and decreasing FAR_IB1.

Figure 1(c) shows that the total query time in scenario 3 is the lowest, even though the query costs on the indexes at the cloud server for all three cases are the same. The reason for this is the lower cost of transforming the query command and the fewer false positive records compared to the other solutions, resulting in improved times for the decryption and re-query steps.

Besides the experimental method mentioned above, we continue to compare the complexity of the range query algorithm on the EBI index with several other related notable works. In detail, according to [6, 11–13], the cost of execution from range query on server is $O(N.N_b)$, $O(\log(N) + N_r)$, $O(\varepsilon.N_r.E_r)$ and $O(logN)$, where N is the total number of records, N_b is the total number of distinct buckets, N_r is the number of records satisfying the query, E_r is the encoding cost incurred, and ε is the expansion parameter. Meanwhile, in our research, by using the EBI index with its order-preserving property, we can utilize a B + Tree data structure for storage organization and this will provide higher query performance with a search cost of $O(logN_b)$.

7 Conclusion

In this paper, we propose an index to support searching on encrypted numerical data which is built based on the buckets of the clear dataset. By improving the method of overlapping range division while preserving order, the resulting buckets have higher security, return fewer false positive records after querying, as well as facilitate the transformation of query commands. Through experimental comparisons of scenarios in bucket construction, the research convincingly demonstrates improvements in filtering rates, error rates, and total query time on the proposed EBI index. Although the EBI index now cannot completely eliminate false positive records, its high query speed makes it fully integrated into the preprocessing step for filtering data in SE schemes querying numerical data.

References

1. AES: Advanced encryption standard. Nat. Inst. Sci. Technol. FIPS (2001)
2. Biryukov, A., De Cannière, C.: Data encryption standard (DES), in Encyclopedia of Cryptography and Security, H.C.A. van Tilborg (Ed.), Boston, MA: Springer US, pp. 129–135 (2005)

3. Al-Saraireh, J.: An efficient approach for query processing over encrypted database. J. Comput. Sci. **13**(10), 548–557 (2017)
4. Hore, B., Mehrotra, S., Tsudik, G.: A privacy-preserving index for range queries. In: Proceeding of International Conference on Very large data bases (VLDB'04) (2004)
5. Gentry, C.: Fully homomorphic encryption using ideal lattices. InSTOC, pp. 169–178 (2009)
6. Hacigümüs, H., Iyer, B., Li, C., Mehrotra, S.: Executing SQL over encrypted data in the database-service-provider model (2002)
7. Ho, C., Pak, K., Pak, S., Pak, M., Hwang, C.: A study on improving the performance of encrypted database retrieval using external indexing system of B+ tree structure. Procedia Comput. Sci. **154**, 706–714 (2019)
8. Wang, J., Du, X.: LOB: bucket based index for range queries. The Ninth International Conference on Web-Age Information Management pp. 86–92 (2008)
9. Agrawal, R., Kiernan, J., Srikant, R., Xu, Y.: Order preserving encryption for numeric data. In: Proceedings of the 2004 ACM SIGMOD International Conference on Management of Data, in SIGMOD '04. New York, NY, USA (2004)
10. TPC BENCHMARK H. Standard Specification Revision 3.0.0: tpc-h_v3.0.0.pdf. Accessed: (2021). https://tpc.org/TPC_Documents_Current_Versions/pdf/tpc-h_v3.0.0.pdf
11. Faber, S., Jarecki, S., Krawczyk, H., Nguyen, Q., Rosu, M., Steiner, M. : Rich queries on encrypted data: beyond exact matches. (2015). Accessed: 19 Apr. 2024. https://eprint.iacr.org/2015/927
12. Kwon, H., Hur, J., Hahn, C.: Order-hiding range query over encrypted cloud data. IEEE Access **10**, 75604–75618 (2022). https://doi.org/10.1109/ACCESS.2022.3192421
13. Peng, Y., Wang, L., Cui, J., Liu, X., Li, H., Ma, H.: LS-RQ: a lightweight and forward-secure range query on geographically encrypted data. IEEE Trans. Dependable Secure Comput. **19**(1), 388–401 (2022). https://doi.org/10.1109/TDSC.2020.2974218

Joint Resource and Delay Optimization for Training Task Offloading Scheme in MEC-Assisted Distributed Learning Using Stackelberg Game

Thanh-Dao Nguyen[1], Van-Thien Nguyen[1], Ngoc-Tan Nguyen[1(✉)], and Thi-Thu Hoang[2]

[1] VNU - University of Engineering and Technology (VNU-UET), Hanoi, Vietnam
{22028113,22028182,tan.nguyen}@vnu.edu.vn
[2] Posts and Telecommunications Institute of Technology, Hanoi, Vietnam
Thuht@ptit.edu.vn

Abstract. In the field of machine learning, distributed learning has gained significant attention as an effective approach for model training, enabling machines in different locations to collaborate in the training process without the need to transfer data to a central server. However, a key challenge is the limited computational capacity of local nodes, which restricts their ability to efficiently handle large-scale tasks. To address this issue, robust distributed learning frameworks are essential for optimizing performance. In this paper, we propose a novel approach that integrates the Multi-access Edge Computing (MEC) system with the Stackelberg game model to develop an optimal training task offloading scheme. This approach optimizes system's computing resources while jointly meeting the delay requirements of training tasks. We will conduct extensive experiments to validate the effectiveness of this method against established benchmarks, including Non-offloading and ODO methods [11].

Keywords: Distributed learning · MEC · Stackelberg game · Block coordinate descent · Linear programming

1 Introduction

The rapid growth of data in the digital age has revolutionized industries and introduced advancements in machine learning. However, traditional centralized machine learning models face a barrier of dealing with the vast amount of data distributed across various locations, along with that is the limitation in computing capacity of a single machine. To overcome these challenges, Distributed Learning (DL) has emerged as an effective solution, which distributes computational workloads across multiple machines (local nodes) and aggregating the results at a master node [1,2]. This approach offers an efficient way to process

© The Author(s), under exclusive license to Springer Nature Switzerland AG 2025
H. A. Le Thi et al. (Eds.): ICAMCS 2024, LNNS 1311, pp. 515–525, 2025.
https://doi.org/10.1007/978-3-031-90606-0_43

decentralized data. Despite its benefits, DL introduces several challenges, including resource limitations, privacy concerns, and network delays, which must be carefully managed [3].

Among these challenges, one of the most critical is ensuring efficient workload distribution across nodes, where models must be trained without causing significant delays or overloading individual machines. Recent research emphasize on enhancing communication efficiency and adaptive decision-making to improve performance and resource utilization. One notable approach is Multi-Message Communication (MMC), which allows multiple computations to be conveyed at each local node per iteration. This method, coupled with straggle avoidance techniques, addresses issues of over-computation and under-utilization by mitigating inaccuracies in predicting straggling behavior and avoiding the discarding of partial computations from straggling nodes [4]. Another innovative solution is the Distributed Deep Learning-bases Offloading (DDLO) algorithm, which leverages multiple parallel deep neural networks (DNNs) to make offloading decisions. By using a shared replay memory to store and utilize decision data, DDLO enhances collective learning and improves decision accuracy [5]. Additionally, the application of deep reinforcement learning to dynamically manage task offloading based on real-time network conditions and server loads offers a flexible approach to optimizing resource allocation [6]. While these methods present significant advancements in addressing data distribution challenges, they also come with trade-offs. MMC may still struggle with complex straggle patterns and network variability, DDLO's reliance on parallel DNNs can be computationally intensive, and dynamic reinforcement learning requires careful tuning to balance adaptability with overhead. Overall, these solutions represent a promising direction for improving distributed learning systems, though they each present unique challenges that must be carefully managed.

To overcome the challenge of workload distribution, Multi-access Edge Computing (MEC) has gained attention as an enabling technology for distributed learning. MEC shifts computational resources closer to the data source by utilizing edge servers at cellular base stations or network edges, minimizing latency and enhancing resource utilization. By offloading computations to these edge nodes, MEC reduces network congestion and improves the overall efficiency of distributed learning models. This synergy between MEC and distributed learning presents a powerful framework for real-time, scale data processing [7,8].

While MEC provides a robust platform for enhancing distributed learning performance, optimizing resource allocation remains a significant challenge. In this paper, we introduce a novel framework that employs a Stackelberg game model to analyze the interaction between Distributed Learning Operators (DLO) and high-performance servers managed by a Computing Service Provider (CSP) in a hierarchical MEC network. The Stackelberg game positions the DLO as the leader and the CSP as the follower. In this setup, the DLO can choose whether to process tasks locally or offload them to the CSP, while the CSP responds by deciding whether to accept and process the offloaded tasks. This strategic game-theory approach enables both the DLO and CSP to maximize their utility by iteratively optimizing their resource allocation strategies.

Our work introduces several key contributions:

- We propose a distributed learning framework that utilizes MEC architecture to efficiently handle the distribution of training tasks across IoT devices and access servers.
- We develop a novel offloading scheme that optimizes the profit of DLO while meeting strict delay requirements, leveraging the Stackelberg game model for optimal resource allocation.
- We provide numerical results showing the effectiveness of our offloading scheme across a variety of scenarios.

The rest of the paper is organized as follows: Sect. 2 introduces the distributed learning model assisted by a hierarchical MEC network. Section 3 formulates the Stackelberg game between the DLO and CSP, along with the proposed iterative algorithm for maximizing the leader's profit. Section 4 presents numerical results that validate the efficiency of our proposed approach. Finally, Sect. 5 concludes the paper with a summary of key findings.

2 System Model

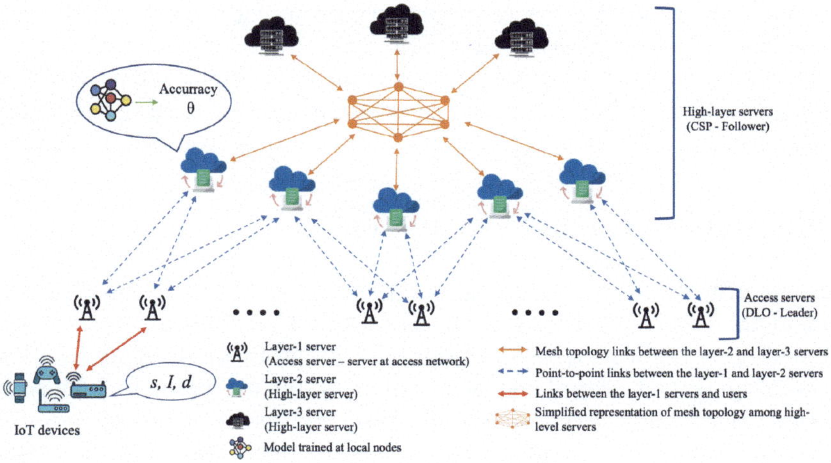

Fig. 1. The system model of the distributed learning assisted by hierarchical MEC networks.

Figure 1 illustrates our system model including IoT devices and a MEC network composed of servers organized into three layers. The 1st layer consists of access servers, managing the operation of IoT devices in their areas. Meanwhile,

2nd and 3rd layers contain high-performance servers with greater computational resources. The set of access servers are denoted as $\boldsymbol{\mathcal{A}} \triangleq \{\text{AS}_a | \forall a \in \{1, \ldots, A\}\}$. On the other hand, the set of high-layer servers in the 2nd and 3rd are denoted as $\boldsymbol{\mathcal{H}} \triangleq \{\text{HS}_h | \forall h \in \{1, \ldots, H\}\}$. Each access server is directly linked to multiple nearby second-layer servers, while the second- and third-layer servers are interconnected in a mesh topology.

The system is utilized for a distributed learning process aimed at training a machine learning model. In this setup, we assume that a Distributed Learning Operator (DLO) owns the access servers and uses data collected from IoT devices to train the model. However, the vast amount of data generated by numerous IoT devices can overwhelm the access servers, which have limited computing capacity. To handle this, the training tasks are offloaded to high-layer servers owned by Computing Service Provider (CSP) in the MEC network, which possess greater computational resources for processing.

Training tasks refer to tasks that require system nodes to train a model using data generated by IoT devices (e.g., images, text, voice recordings, etc.). When a training task is generated by an IoT device and reaches an access server, it is categorized into a task type ($\boldsymbol{T}$) based on its size (s), computational intensity (I), and delay requirement (d). The collection of all task types is represented by $\boldsymbol{\mathcal{X}} \triangleq \{\boldsymbol{T}_n = \{s_n, I_n, d_n\} | \forall n \in \{1, \ldots, N\}\}$. The access servers then decide whether to discard the tasks, offload them to high-layer servers, or process them directly. After this task distribution phase, each server in the system functions as a local node, using the data it was allocated to train a model. Each server in the system has a certain computing resource (CPU cycles/s), which is the maximum number of CPU cycles it can perform in one second. We denote R_a as computing resource of access server AS_a.

Please note that our work focuses solely on the training phase at the local nodes within the distributed learning process, excluding the aggregation step at the master node.

3 Stackelberg Game Formulation

We model a Stackelberg game to analyze the strategic interactions between the DLO and the CSP, each driven by their own profit-maximizing objectives. The DLO, acting as the leader, seeks to optimize its profit by deciding whether to process training tasks locally at access servers, offload them to high-servers, or discard them altogether. Meanwhile, the CSP, as the follower, determines whether to process the offloaded tasks or discard them. The game is evaluated over a fixed time interval, where system conditions (e.g. task arrival rates, servers' system loads) remain constant. In this study, we assume a one second time interval. The mathematical formulation of this game is detailed in the following sections.

3.1 The Leader's Strategy

The leader decides whether to process, offload, or discard training tasks by maintaining a tensor $\boldsymbol{\mathcal{P}}^{N\times A\times(H+1)} = (p^n_{a,h})$. Here, $p^n_{a,h}$ is the probability that a task of type $\boldsymbol{T}_n$ arriving at AS_a being offloaded to server HS_h. Note that, $h = H+1$ indicates the task is processed directly at AS_a. The selection of offloading server is performed by the weighted random mechanism [13].

The leader's utility is determined by the profit gained from the computation performed for training. The total computation (in CPU cycles) carried out by the leader at the access servers (ASs) during a given time period is expressed as $C^a = \sum_{a=1}^{A}\sum_{n=1}^{N} p^n_{a,H+1} S^n_a I_n$, while the total computation offloaded to HSs is represented by $C^h = \sum_{a=1}^{A}\sum_{n=1}^{N}\sum_{h=1}^{H} p^n_{a,h} S^n_a I_n$. Thus, the leader's utility is expressed as $pr^p C^a + (pr^p - pr^o)C^h$, which expands as:

$$U_L(\boldsymbol{\mathcal{P}}, pr^o) = \sum_{a=1}^{A}\sum_{n=1}^{N}\sum_{h=1}^{H+1} p^n_{a,h} S^n_a I_n pr^p - \sum_{a=1}^{A}\sum_{n=1}^{N}\sum_{h=1}^{H} p^n_{a,h} S^n_a I_n pr^o, \quad (1)$$

where S^n_a represents the total size of training tasks of type $\boldsymbol{T}_n$ arriving at AS_a. The parameters pr^p and pr^o ($\$/CPUcycle$) represent the monetary benefit the DLO receives for each CPU cycle used in processing training tasks and the cost it pays to the CSP for each offloaded CPU cycle, respectively.

3.2 The Follower's Strategy

Based on the leader's strategy, the follower, which is the CSP, decides how many offloaded training tasks it will process in order to maximize its profit. Let l_h (in CPU cycles/s) represent the maximum computational resource that HS_h can allocate to process offloaded training tasks during the given time interval. The set of computational resources is denoted as $\boldsymbol{\mathcal{L}} \triangleq \{l_h | \forall h \in \{1, \ldots, H\}\}$. The follower's utility is defined as the total profit from processing offloaded tasks:

$$U_F(\boldsymbol{\mathcal{L}}) = \sum_{h=1}^{H} [pr^o l_h - pr^e F(l_h)], \quad (2)$$

where $F(x) = \alpha_h x^2 + \beta_h x + \gamma_h$ is a quadratic function that calculates the power consumption (in Watts) of HS_h [9], and pr^e is the energy price unit that the CSP must pay for the energy consumed by the computational resource l_h. In our study, the only cost considered by the follower is the energy cost.

3.3 Solution for the Stackelberg Game

To find the solution for the proposed game, we apply the backward induction technique through the following **Theorem**.

Theorem 1. *Given a the leader's strategy (i.e., the DLO), the CSP can determine a unique optimal $\boldsymbol{\mathcal{L}}$ by calculating l_h^* values, $\forall h \in \{1, ..., H\}$ to optimize its profit at each HS_h:*

$$l_h^* = \frac{pr^o - m_h}{n_h}, \tag{3}$$

where $m_h = pr^e \beta_h$ and $n_h = 2pr^e \alpha_h$.

Proof. According to (2), the CSP's profit is the sum of the profits from servers $\text{HS}_h, \forall h \in \{1, \ldots, H\}$. This total profit is maximized when the profits of all high-layer servers $\boldsymbol{\mathcal{H}}$ are maximized. Additionally, the profit of each server HS_h is a quadratic function of l_h. which can directly derive the optimal solution in (3) by setting the derivative of this function with respect to l_h equal to zero.

According to Theorem 1, given the follower's strategy ($\boldsymbol{\mathcal{L}}$), the leader's profit optimization problem is expressed in (4). Constraints (4a) and (4b) indicate that all probabilities of training tasks offloaded to server HS_h from server AS_a must be positive, and their sum must be less than or equal to 1, respectively. The total computation processed at an access server within the time interval must not exceed its computing capacity, as specified in (4c). Similarly, the total computation offloaded to server HS_h in the time interval must not surpass its allowable computational resources, as shown in (4d). Furthermore, a task of type $\boldsymbol{T}_n$ arriving at AS_a will not be offloaded to HS_h if total delay, i.e., t_{ah}^n, exceeds the delay requirement d_n (4e). It's important to note that t_{ah}^n is a function of transmission time, computing time, and queuing time, which was defined in our previous work [11]. Lastly, for the DLO to make a profit, the offloading price must be lower than the processing price, as stated in (4f).

$$\max_{(\boldsymbol{\mathcal{P}}, pr^o)} U_L(\boldsymbol{\mathcal{P}}, pr^o), \tag{4}$$

$$\text{s.t. } 0 \leq p_{a,h}^n, \quad \forall a \in \{1, \ldots, A\}, \forall n \in \{1, \ldots, N\}, \forall h \in \{1, \ldots, H+1\}, \tag{4a}$$

$$\sum_{h=1}^{H+1} p_{a,h}^n \leq 1, \qquad \forall a \in \{1, \ldots, A\}, \forall n \in \{1, \ldots, N\}, \tag{4b}$$

$$0 \leq \sum_{n=1}^{N} p_{a,H+1}^n S_a^n I_n \leq R_a, \qquad \forall a \in \{1, \ldots, A\}, \tag{4c}$$

$$0 \leq \sum_{a=1}^{A} \sum_{n=1}^{N} p_{a,h}^n S_a^n I_n \leq \frac{pr^o - m_h}{n_h}, \qquad \forall h \in \{1, \ldots, H\}, \tag{4d}$$

$$p_{a,h}^n = 0, \quad \text{if} \quad t_{a,h}^n > d_n, \tag{4e}$$

$$0 \leq pr^o \leq pr^p. \tag{4f}$$

The objective function in (4) is a complex bilinear function involving the probabilities of training tasks offloaded to server HS_h, i.e., $\boldsymbol{\mathcal{P}}$, and the price that

Algorithm 1. Stackelberg Game-based Probabilistic Offloading

Input: The previous output $\{\boldsymbol{\mathcal{P}}^{(n-1)}, pr^{o(n-1)}\}$.
Initialize: $n = 1$, $\{\boldsymbol{\mathcal{P}}^{(0)}, pr^{o(0)}\}$, tolerance $\xi > 0$.
Compute: the leader's utility $U_L\left(\boldsymbol{\mathcal{P}}^{(0)}, pr^{o(0)}\right)$.
Repeat:
  Obtain $\boldsymbol{\mathcal{P}}^{(n)}$ for given $pr^{o(n-1)}$ by solving (4) with $U_L(\boldsymbol{\mathcal{P}}, pr^{o(n-1)})$;
  Derive the optimal value $pr^{o(n)}$ with fixed $\boldsymbol{\mathcal{P}}^{(n)}$ by solving (4) with $U_L(\boldsymbol{\mathcal{P}}^{(n)}, pr^{o})$;
  If:
    $\left|U_L(\boldsymbol{\mathcal{P}}^{(n)}, pr^{o(n)}) - U_L(\boldsymbol{\mathcal{P}}^{(n-1)}, pr^{o(n-1)})\right| < \xi$;
  Then:
    Set $\{\boldsymbol{\mathcal{P}}^{*}, pr^{o*}\} = \{\boldsymbol{\mathcal{P}}^{(n)}, pr^{o(n)}\}$ and terminate.
  Otherwise:
    Update $n \leftarrow n + 1$ and continue.
Output: The locally optimal solution $\{\boldsymbol{\mathcal{P}}^{*}, pr^{o*}\}$.

the DLO pays the CSP for processing the offloaded task, i.e., pr^o. To address this issue, we propose an iterative algorithm called Stackelberg Game-based Probabilistic Offloading (SGPO)utilizing the block coordinate descent method. First, we select an initial feasible pair of $\boldsymbol{\mathcal{P}}$ and pr^o as initialized values. Given the value of $pr^{o(n-1)}$, the optimization problem (4) becomes a linear programming problem. We then determine the optimal values of $\boldsymbol{\mathcal{P}}^{(n)}$ by using the Simplex Method to solve the problem (4). It is proven to be very effective in solving linear programming problems [12]). Subsequently, we calculate the optimal value of $pr^{o(n)}$ with the fixed values of $\boldsymbol{\mathcal{P}}^{(n)}$. This process continues until the difference in the Leader's profit between steps n and $n-1$ is less than a specified tolerance ξ. These steps are summarized in Algorithm 1.

4 Numerical Results

In this research, we measure the efficiency of the SGPO method and compare with those of two baseline methods, i.e., Optimal Delay Offloading (ODO) [11] and Non-Offloading. The metric used for measuring the efficiency of methods is: **Probability of tasks processed**, denoted as η. We have:

$$\eta = \frac{C^p}{C^r}, \tag{5}$$

where C^p (CPU cycles) is the total computation that the system performs to process training tasks in the considered time interval, including computation contributed by access servers and high-layer servers. $C^r = \sum_{a=1}^{A}\sum_{n=1}^{N} S_a^n I_n$ (CPU cycles) is the total required computation to process all the training tasks appear in the system at the considered time interval. Note that C^r is the same but C^p is different in evaluation of each method.

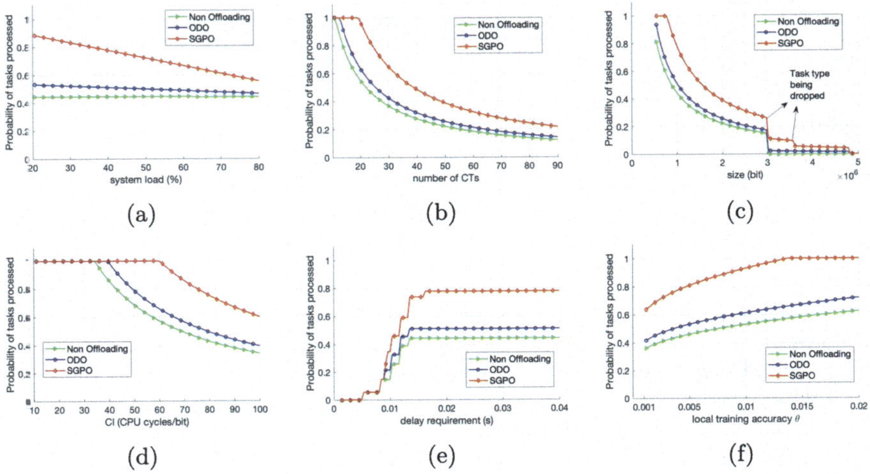

Fig. 2. Efficiencies of SGPO, ODO, and Non-offloading methods as varying (a) the high-layer servers' load, (b) the number of tasks, (c) the task size, (d) the task computational intensity, (e) the required task delay, and (f) the local training accuracy.

Default Settings: We set up our simulation environment as follows: There are 20 access servers at the 1st layer. 5 servers at layer 2 and 3 servers at layer 3. Bandwidth of links between access servers and layer-2 servers is set to 600 Mbps, and bandwidth links among high-layer servers is set to 1000 Mbps. Computing resources at 1st, 2nd, and 3rd layers are respectively 6.10^9, 12.10^9, and 30.10^9 (CPU cycles/s). The default set of task types used for evaluation consists of 7 types (s, I, d): (1000000, 30, 0.01); (1000000, 50, 0.02); (1000000, 60, 0.03); (1000000, 70, 0.03); (1000000, 80, 0.04); (1000000, 100, 0.04); (1000000, 150, 0.05). Within a time interval, each access server receives 25 tasks of each type.

We conducted six experiments designed to assess how various parameters affect the efficiency of different offloading methods as shown in Fig. 2. Each experiment modifies a single parameter while maintaining the others at default setting values.

General Assessment: In all experiments, the Non-offloading method is the least efficient, followed by the ODO method. The SGPO method shows the most efficiency among three. It is due to the fact that the Non-offloading method depends entirely on access servers, while both the ODO and SGPO methods utilize the computing power of high-layer servers. However, the ODO method offloads tasks without accounting for potential congestion at high-layer servers (e.g., multiple access servers selecting the same high-layer server for offloading). In contrast, the SGPO method prevents congestion by ensuring that the offloaded computation does not exceed the servers' available computing resources (4c), (4d).

Experiment 1: System Load Change. Figure 2a presents the efficiencies of the methods under different high-layer server loads (20%–80%). The system load of a server refers to the computing resources the server allocates to tasks other than training tasks. Since the Non-offloading method does not utilize high-layer servers, its efficiency remains constant at 0.45 across all load levels. The ODO method starts with an efficiency of approximately 0.53, staying relatively steady with a slight decline as the load increases. The SGPO method begins with an efficiency of 0.9 at lower system loads and gradually decreases to 0.54 as the load reach 80%. These efficiency trends follow linear patterns. This can be explained by the fact that a server's total computing resource usage is the sum of its system load and the resources allocated for training tasks (training resources). As the system load increases, the training resources decrease linearly. Since C^r is constant and C^p is proportional to the available training resources, the efficiency $\eta = \frac{C^p}{C^r}$ decreases linearly with increasing system load.

Experiment 2: Number of Tasks Change. Figure 2b shows the efficiency of the three methods as the number of tasks increases from 10 to 90. Throughout the experiment, the SGPO method consistently outperforms both the ODO and Non-offloading methods by approximately 50%. The curved trend lines observed in the results can be explained as follows: As the number of tasks increases, both C^p (computation used for processing tasks) and C^r (total computation required) increase. However, once C^p reaches the servers' computing resources limit, it remains constant, while C^r continues to rise. This causes the efficiency $\eta = \frac{C^p}{C^r}$ to follow a pattern similar to a reciprocal function ($1/x$).

Experiment 3: Task Size Change. Figure 2c demonstrates that the efficiency of all methods decreases as task size increases. For task sizes ranging from 0.75 to 3 Mbits, the SGPO method leads the others by approximately 30%. However, for task sizes larger than 3 Mbits, all methods show significantly lower performance, with efficiencies of 0.1 or less. The drop in efficiency at certain task sizes is due to the increase in both computation and transmission delays as the task size grows. When the total delay of a task type surpasses its requirements, access servers must discard the entire task type. Additionally, the curved line trends can be explained similarly to experiment 2 as C^p would finally reach its limit while C^r increases proportionally along with task size.

Experiment 4: Computational Intensity (CI) Change. Figure 2d shows the efficiency of the three methods as computational intensity (CI) varies from 10 to 100. When CI is low (10–35), all methods operate at full efficiency. Between CI values of 35 and 40, the Non-offloading and ODO methods start to decline, whereas the SGPO method maintains its efficiency until CI reaches 60. Similar to the previous experiments, the curved trends can be attributed to the relationship between C^p and C^r, where C^p reaches its limit while C^r continues to increase proportionally with CI.

Experiment 5: Delay Requirement Change examines the efficiencies of the three methods as the delay requirement increases from 0.001 to 0.04 s. Overall, the efficiencies of all three methods gradually rise, with the SGPO consistently outperforming the ODO and Non-offloading methods. At certain points, the efficiency rises sharply because when the delay requirements reach specific thresholds, some task types can be completed within the allowed time. As these tasks are processed, the processing of the entire task type is added up to C^p, making the overall efficiency increase significantly in a short period. Although η improves as the delay becomes less strict, it eventually reaches limits (SGPO - 0.78; ODO - 0.51; Non-offloading - 0.4). Initially, as the delay requirement loosens from 0.001 to around 0.14 s, more tasks can be processed on time. However, once C^p reaches its maximum capacity, it no longer increases despite further increases in the delay allowance.

Experiment 6: Training Accuracy Change. Optimization algorithms, like Gradient Descent, are widely used to minimize a machine learning model's loss and typically require multiple iterations to converge. In the optimization process, we introduce a critical metric: training accuracy θ at local nodes. This metric sets the stopping threshold for iterative learning tasks. The smaller θ is, the closer the solution is to optimal. According to [14], the number of iterations needed to achieve a specific accuracy θ is given by $I(\theta) = \mu \cdot \log(\frac{1}{\theta})$, where μ is a task-dependent constant.

Having established this formula for the number of iterations based on accuracy, we also identify a proportional relationship between the number of iterations and the computational intensity (CI) of the training task. Assuming that the computation (in CPU cycles) required for each iteration remains constant, the total computation needed to process a training task is directly proportional to the number of iterations. Since the total computation is the product of task size (s) and computational intensity (I), and s remains unchanged regardless of the number of iterations, it follows that the number of iterations is directly proportional to computational intensity.

With this established relationship between the number of iterations and computational intensity, we can adjust the computational intensity (CI) based on changes in θ to perform our simulation. Figure 2f illustrates the efficiencies of the three methods as the requirement for local accuracy θ varies from 0.001 to 0.2. The results show that the SGPO method significantly outperforms the others.

5 Conclusion

This paper has examined the Stackelberg game model to optimize profits for both the DLO that acts as the leader and the CSP, i.e., the follower, in a hierarchical MEC network. We have identified the linearity of the leader's optimization problem and introduced an iterative algorithm utilizing the block coordinate descent and simplex method to find the locally optimal solution for the proposed game.

Numerical results have confirmed the superior efficiency of the proposed SGPO method in comparison with other baseline methods.

Acknowledgement. This work has been supported/partly supported by VNU University of Engineering and Technology under project number CN23.15.

References

1. Cao, X., et al.: Communication-efficient distributed learning: an overview. IEEE J. Sel. Areas Commun. **41**(4), 851–873 (2023)
2. Tiffany, T., Shiqiang, W., Theodoros, S., Bong, K., Kin, L.: Demo abstract: distributed machine learning at resource-limited edge nodes. In: INFCOMW (2018)
3. Chen, M., et al.: Distributed learning in wireless networks: recent progress and future challenges. IEEE J. Sel. Areas Commun. **39**(12), 3579–3605 (2021)
4. Ozfatura, E., Ulukus, S., Gündüz, D.: Straggler-aware distributed learning: communication-computation latency trade-off. Entropy **22**(5), 544 (2020)
5. Huang, L., et al.: Distributed deep learning-based offloading for mobile edge computing networks. Mob. Netw. Appl. **27**, 1123–1130 (2022)
6. Park, J., Chung, K.: Distributed DRL-based computation offloading scheme for improving QoE in edge computing environments. Sensors **23**, 4166 (2023)
7. Multi-access Edge Computing. https://www.sciencedirect.com/topics/computer-science/multi-access-edge-computing
8. ESTI, Multi-access Edge Computing. https://www.etsi.org/technologies/multi-access-edge-computing
9. Zhang, X., Lu, J., Qin, X.: BFEPM: best fit energy prediction modeling based on CPU utilization. In: IEEE NAS, Xi'an, China, pp. 41–49 (2013)
10. Han, Z., et al.: Game Theory in Wireless and Communication Networks. https://doi.org/10.1017/CBO9780511895043
11. Nguyen, N.-T., et al.: Delay improvement in hierarchical multi-access edge computing networks. IEICE Trans. Fundam. **E107.A**(8) (2024)
12. Vanderbei, R.J.: Efficiency of the simplex method. In: Linear Programming, International Series in Operations Research & Management Science, vol. 285. Springer, Cham (2020)
13. Efraimidis, P., Spirakis, P.: Weighted random sampling. In: Encyclopedia of Algorithms. Springer, Boston (2018)
14. Luo, S., Chen, X., Wu, Q., Zhou, Z., Yu, S.: HFEL: joint edge association and resource allocation for cost-efficient hierarchical federated edge learning. IEEE Trans. Wirel. Commun. **19**(10), 6535–6548 (2020)

Multi-Document Signcryption Scheme Using RSA Cryptosystem, Merkle Hash Tree, and AES Block Cipher

Van Nghi Nguyen, Thi Hien Pham, Tu Le Minh, and Thi Hanh Tran(✉)

Academy of Cryptography Techniques, Hanoi 151090, Vietnam
{nghinv,hienpt,hanhtt}@actvn.edu.vn

Abstract. Signcryption is a cryptographic method that combines encryption and digital signature techniques to ensure both confidentiality and origin authentication for transmitted messages. Nevertheless, there is a lack of published literature on the topic of signcryption for several documents. In this paper, we propose a multi-document signcryption scheme using the RSA cryptosystem, Merkle hash trees, and AES block cipher. This research aims to present a signcryption method that offers robust security, efficient performance, and the capability to handle several documents simultaneously. The proposed scheme utilizes a multi-document signature algorithm based on Merkle hash trees. We conduct an analysis of the security and computational cost associated with the proposed scheme, and subsequently make conclusions regarding its potential for applicability.

Keywords: Multi-document · Merkle Tree · RSA · Signcryption

1 Introduction

Signcryption, a cryptographic technique established by Zheng in 1997, combines the processes of signature and encryption within a single scheme. This system offers the benefits of ensuring both the confidentiality and origin authentication of the communication, while requiring less computational resources and channel bandwidth compared to the typical sequential approach of signing and then encrypting the message. With the RSA signcryption, the average computation time is 50% smaller and the average bandwidth is 91% smaller than the traditional signing and encryption method [1].

Over the last 5 years, there has been significant focus on signcryption, as indicated by the several publications, including references [2–5]. However, these signcryption schemes are limited to the act of signing a single document or processing a single message. If the user wants to encrypt numerous documents simultaneously or multiple messages during a communication session, the only option is to execute these signcryption schemes multiple times. This makes the user have to perform many operations, which is time-consuming during the operation, and the communication bandwidth will be higher. At the same time, by attaching a signature to each document, the cryptanalyst can obtain a significant number of signatures, facilitating the determination of the secret key employed for signing.

© The Author(s), under exclusive license to Springer Nature Switzerland AG 2025
H. A. Le Thi et al. (Eds.): ICAMCS 2024, LNNS 1311, pp. 526–537, 2025.
https://doi.org/10.1007/978-3-031-90606-0_44

Merkle hash tree is a cryptographic primitive based on a cryptographic hash function that allows processing of multiple documents at the same time and can be applied to a digital signature scheme to perform multi-document digital signatures [6]. From this idea, we propose a multi-document signcryption scheme using the RSA cryptosystem, Merkle hash tree, and AES block cipher. Our proposed method effectively addresses the issue of signing numerous documents simultaneously. During the verification process, users have the option to independently verify and decrypt individual documents, or they can choose to check and decrypt numerous documents simultaneously. This novel solution enables the signing of many documents in a single operation, effectively addressing the aforementioned issue that traditional single-document signing methods have failed to resolve.

To examine the application potential of the proposed scheme, we also provide an analysis of the security and implementation cost of this method. For the security analysis of the proposed scheme, we evaluate the information security services provided by the scheme, such as confidentiality, origin authentication, integrity, non-repudiation, and forward security. For the cost analysis, we compare the proposed scheme with a single-signature scheme when they perform the same amount of text signing. The research indicates that the technique has potential applications in several real-world scenarios based on the evaluation results.

The paper is organized into five primary sections as follows: In Sect. 1, we provide an introduction. In Sect. 2, we provide a multi-document signcryption algorithm that utilizes Merkle hash trees. Section 3 presents the main content of the paper, which is to propose a multi-text signcryption scheme using the cryptographic primitives RSA, Merkle hash tree, and AES block cipher, including an analysis of the security issue and cost evaluation of the proposed scheme. To demonstrate the potential application of the scheme in practice, we have proposed an application of the scheme in electronic transactions in Sect. 4. Finally, the conclusion is presented.

2 Multi-Document Digital Signature Solutions Using Merkle Hash Trees

In this section, we present a cryptographic primitive called a Merkle hash tree and a multi-document digital signature solution using a Merkle hash tree. This is the core solution in the proposed signcryption scheme presented in Sect. 3.

2.1 Merkle Hash Tree

Merkle hash trees were introduced in 1979 by Merkle. A Merkle hash tree is an ordered binary tree constructed from a sequence of data objects $(m_1, m_2, \ldots, m_n)$ using a cryptographic hash function H. The leaf nodes of the hash tree are hash values $H(m_i)$ where $1 \leq i \leq n$. The other nodes of the hash tree will each consist of two children, a left child (*left*) and a right child (*right*). The bottom nodes of the tree are the leaves. The next nodes will contain hash values *H(left||right)*, where left is the hash value contained in the left node and right is the hash value contained in the right node and are concatenated together

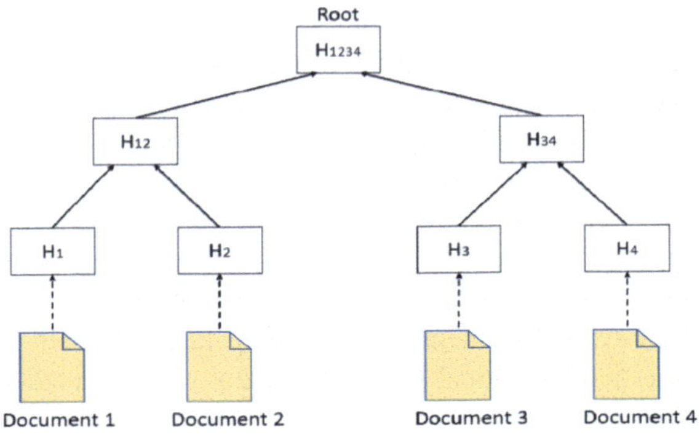

Fig. 1. Merkle hash tree.

(||). The last node of the tree is called the root node [6]. The process of constructing a Merkle hash tree is shown in Fig. 1.

The input consists of documents 1, 2, 3, and 4, which serve as the input data set. The process of constructing a Merkle hash tree involves the following steps:

1) Hash the documents 1, 2, 3, and 4 with a cryptographic hash function H to obtain the hash values H_1, H_2, H_3, and H_4, which are the hash values at the leaf node.
2) Group leaf nodes into left and right pairs; for example, if H_1 is left and H_2 is right, then the root node of this pair of leaf nodes is H_{12}, calculated by the formula $H_{12} = H(H_1||H_2)$. Similarly, for nodes H_3 and H_4 in this example, $H_{34} = H(H_3||H_4)$.
3) $Root = H_{1234} = H(H_{12}||H_{34})$.

In the general case, if the number of data points is not 2^t values where $t \in \mathbb{N}$, then the missing nodes of the Merkle tree will receive the hash value of the last data point.

2.2 Multi-Document Digital Signature Method Using a Merkle Hash Tree

From the Merkle hash tree architecture in Fig. 1, this section of the paper presents a multi-document digital signature solution with this tree architecture, including two signing and verification processes described in Fig. 2a and b [9], respectively.

Multi-Document Signing Process

1) From the documents to be signed m_i where $1 \leq i \leq n$ (n is the number of documents to be signed), build a Merkle tree.
2) Apply a digital signature scheme to sign using the root node value of the Merkle tree as input, jointly with the signer's private key, in order to obtain the signature value at the root of the Merkle tree.
3) The signature to authenticate the document will now be a pair of values consisting of: digital signature on the Merkle tree root and $PATH_i$ where $1 \leq i \leq n$. In which $PATH_i$ is called the authentication path for the document m_i, which is defined as the minimum number of nodes needed of the Merkle hash tree so that from the leaf node of document m_i it is possible to calculate the root node of the tree.

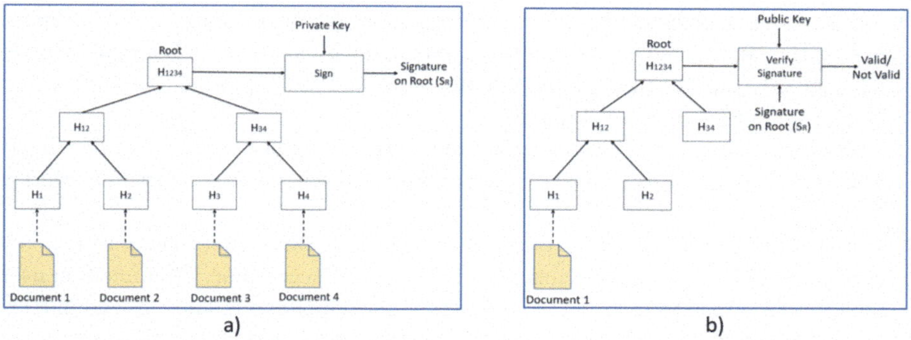

Fig. 2. Multi-document digital signature method using a Merkle hash tree.

Consider the Merkle hash tree example in Fig. 2: The authentication path for document 1 is $PATH_1 = (H_2, H_{34})$, for document 2 it is $PATH_2 = (H_1, H_{34})$, for document 3 it is $PATH_3 = (H_4, H_{12})$ and for document 4 it is $PATH_4 = (H_3, H_{12})$.

The signature for document 1 is $(S_R, (H_2, H_{34}))$, the signature for document 2 is $(S_R, (H_1, H_{34}))$, the signature for document 3 is $(S_R, (H_4, H_{12}))$, the signature for document 4 is $(S_R, (H_3, H_{12}))$.

The digital signature $(S_R, PATH_i)$ is attached to the document m_i sent to the digital signature verifier.

Multi-Document Signature Verification Process

1) Use the $PATH_i$ of the document m_i to recreate the root value of the Merkle tree.
2) Apply a particular signature verification algorithm, where the input consists of the value at the recently regenerated root node of the Merkle tree and the public key of the sender.
3) If the signature verification algorithm gives a correct value, then the signature is valid and the document is authenticated.

Figure 2b details the steps of the signature verification process with the test document being document 1 and the signature being $(S_R, (H_2, H_{34}))$.

If there is a requirement to authenticate numerous documents, we may provide them together with their respective signatures. This will allow us to reconstruct the Merkle tree. The verifier can then reconstruct the root node of the Merkle tree and verify the integrity of many documents in a single operation.

3 Multi-Document Signcryption Scheme Using RSA Cryptosystem, Merkle Hash Tree and AES Block Cipher

From the multi-document digital signature solution presented in Section II, we propose a multi-document signcryption scheme using the RSA cryptosystem, Merkle hash tree, and AES block cipher. In addition to the Merkle hash tree, the four cryptographic primitives in the proposed scheme are:

1) The RSA digital signature scheme has been standardized in Vietnamese and European national standards, and most recently we refer to the standard of the National Institute of Standards and Technology (NIST) published in 2023 [7] (some notes on secure RSA installation can be seen in [12, 13]).
2) AES block cipher is an Advanced Encryption Standard published by NIST in 2001.
3) The SHA-256 hash function is used in the proposed scheme to construct the Merkle hash tree.
4) The HKDF (HMAC-based Extract-and-Expand Key Derivation Function) is a key derivation and key expansion function from a secret seed value. This function uses the HMAC-SHA-256 function in its design architecture and is standardized in RFC 5869 [8]. This function is being used for the latest transport layer security protocol TLS1.3.

3.1 The Operation of Scheme

The proposed signcryption scheme involves two communicating parties, Alice and Bob, to transmit n documents m_i (where $1 \leq i \leq n$) from Alice to Bob. The scheme involves three processes: Key generation, multi-document signcryption, and unsigncryption.

1) RSA cryptosystem key generation process.
   1.1) Alice and Bob generate their own key pair (public key and private key).
   1.2) Alice generates the corresponding public/private key pairs (n_A, e_A) and d_A; for details on this key generation process, see [9].
   1.3) Bob generates the corresponding public/private key pairs (n_B, e_B) and d_B; for details on this key generation process.
2) Multi-document signcryption process.

   The sender Alice performs the signcryption process for n documents m_i (where $1 \leq i \leq n$).

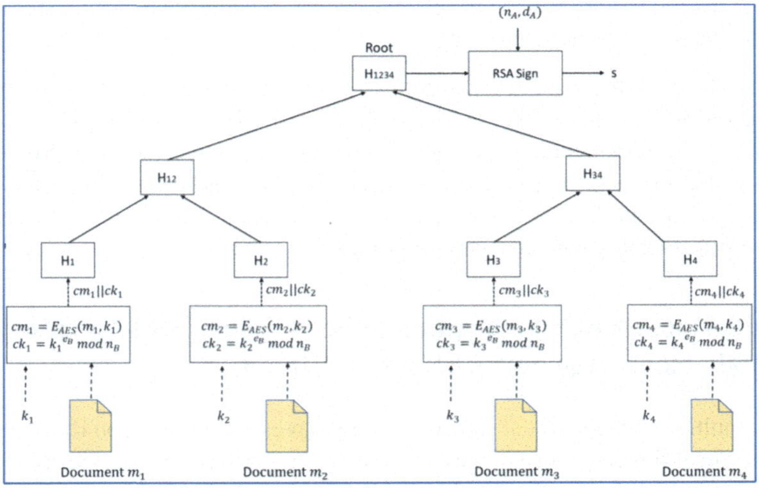

Fig. 3. The multi-document signcryption process using a Merkle hash tree with 4 documents.

The signcryption process has the following steps:

2.1) Pseudo-randomly generate secret key *seed*.
2.2) Calculate the set of n AES-256 block cipher secret keys using the HKDF key derivation and expansion function [8]:
$\{k_i\} = HKDF(seed||time, n)$ where $1 \le i \le n$.
2.3) Encrypt the document m_i to obtain the corresponding ciphertext cm_i according to the formula:
$cm_i = E_{AES}(m_i, k_i)$ with $1 \le i \le n$.
Where $E_{AES}(m_i, k_i)$ is the encryption function of the AES-256 block cipher with input document m_i and session key k_i.
2.4) Encrypt the key k_i to obtain the ciphertext ck_i according to the formula: $ck_i = k_i^{e_B} mod n_B$ where $1 \le i \le n$.
2.5) Construct the Merkle hash tree with n input data as $(cm_i||ck_i)$ where $1 \le i \le n$ to obtain the hash value at the root node, denoted as h_{root}.
2.6) Use the RSA signature algorithm to sign h_{root} with the sender's private key (n_A, d_A):
$s = (h_{root})^{d_A} mod n_A$
2.7) Return the values (s, cm_i, ck_i, $PATH_i$) where $1 \le i \le n$, which are the signed ciphertext for the message m_i and sent to Bob.

Consider the signcryption procedure from step 3 to step 6 using four input documents as described in the example provided in Sect. 2. This process will be carried out following the steps outlined in Fig. 3.

The results of the signcryption process according to Fig. 3 are the signing ciphertexts for documents 1, 2, 3, 4 respectively: (s, cm_1, ck_1, (H_2, H_{34})), (s, cm_2, ck_2, (H_1, H_{34})), (s, cm_3, ck_3, (H_4, H_{12})), (s, cm_4, ck_4, (H_3, H_{12})).

3) The unsigncryption process: The recipient Bob receives the signed message (s, cm_i, ck_i, $PATH_i$) and proceeds to decrypt it according to the following steps:
3.1) Use the parameters $(cm_i||ck_i)$ and $PATH_i$ to generate the hash value at the root node of the Merkle tree, denoted as h_{root}.
3.2) Verify the signature of h_{root} using Alice's public key (n_A, e_A): Calculate $h = s^{e_A} mod n_A$. Compare h and h_{root}. If $h = h_{root}$, then the signature is valid and go to step 4, otherwise the signature is invalid and stop.
3.3) Decrypt ck_i using the recipient's private key (n_B, d_B) to obtain the session key k_i according to the formula: $k_i = ck_i^{d_B} mod n_B$.
3.4) Decrypt the ciphertext cm_i to obtain the plaintext m_i according to the formula: $m_i = D_{AES}(cm_i, k_i)$, where $D_{AES}(cm_i, k_i)$ is the decryption function of the AES-256 block cipher with the input being the ciphertext cm_i and the session key k_i.

Consider the example in Fig. 3, suppose Bob obtains the ciphertext (s, cm_1, ck_1, (H_2, H_{34})) then the verification and decryption process with steps 1 to 3 is shown in Fig. 4.

If there is a requirement to authenticate and secure several documents, we can generate signed copies containing all the documents together with their respective digital signatures. This allows us to perform the authenticity verification in a single instance.

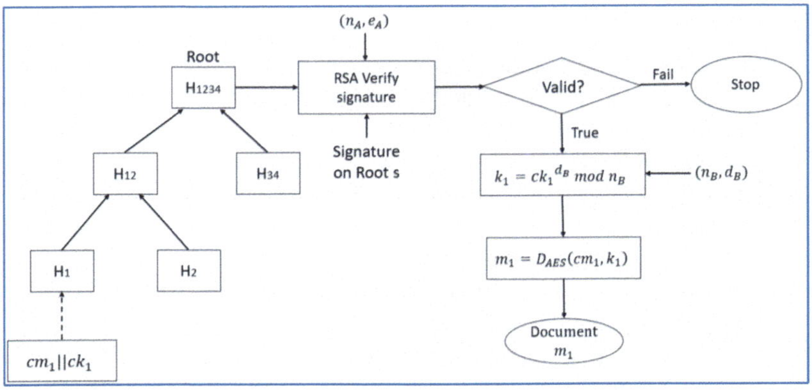

Fig. 4. The unsigncryption process for the ciphertext $(s, cm_1, ck_1, (H_2, H_{34}))$.

3.2 Security Analysis of the Proposed Scheme

In this section, we analyze the security of the proposed scheme through the common information security services of a signcryption scheme: confidentiality, authentication, integrity, non-repudiation, and forward security.

Confidentiality

The proposed signcryption scheme is constructed with two algorithms used for encryption, AES (encrypts the documents m_i), RSA is used to encrypt the keys of the AES encryption algorithm (session keys). Therefore, the confidentiality for the m_i documents is ensured by encryption with the AES symmetric encryption algorithm. For the AES block cipher, two well-known attacks are differential attack and Boomerang attack. The most effective differential attack for 7 rounds of AES-256 published has a computational complexity of $2^{250.5}$ with $2^{92.5}$ plaintext/ciphertext pairs required and 2^{153} of memory. The effective Boomerang attack for 6 rounds of AES-128 has a computational complexity of 2^{71} with 2^{71} plaintext/ciphertext pairs required and 2^{33} of memory. It is evident that attacks that can only be successful for 7 rounds are insufficient to break the whole 14-round AES-256 encryption [10].

The confidentiality of the session keys k_i is ensured by the RSA encryption algorithm. This algorithm is based on the hard problem of prime factorization of integers, which, when the parameter n is chosen large enough, cannot be solved in polynomial time.

We consider the recommendations of the National Institute of Standards (NIST) and the choice of parameters for cryptographic algorithms corresponding to the security strength and the year of use in [7]. It is found that using AES-256 with a security strength of 256 bits and the RSA cryptosystem with a length of n as 3072 bits and a security strength of 128 bits, the proposed cryptographic scheme ensures security until 2030 and beyond.

Authentication and Data Integrity

The scheme uses a Merkle hash tree and the RSA signature algorithm to ensure origin authentication and message integrity. Origin authentication is demonstrated by the RSA signature algorithm. Authentication is ensured by hashing the session key and ciphertext with the SHA-256 hash function during signing and verifying with the Merkle hash tree. Therefore, to analyze the origin authentication and message integrity assurance of the proposed scheme, we also consider the following two aspects:

Merkle hash tree spoofing attack: An attack on a Merkle hash tree aims to change or spoof with the leaf nodes or root node of the hash tree. In a Merkle hash tree built from the SHA-256 hash function, an attack on a Merkle tree is also an attack on the SHA-256 hash function. For the SHA2–256 hash function, the most common attack is a collision attack. A recently published optimal collision attack for this hash function takes 31/64 steps with a complexity of $2^{65.5}$, and in [11], the authors proposed a quantum collision attack of up to 38–39/64 steps with a computational complexity of $2^{122/\sqrt{2}}$ where $s < 212$. With attacks only effective up to 39/64 steps, the collision resistance of SHA-256 is almost absolutely secure against these attacks.

Regarding the forgery of digital signatures at the root node, it is not possible for the attacker to fabricate the signature using the RSA signature algorithm with a modulus n of 3072 bits within a reasonable amount of time.

As recommended by NIST in [7], SHA-256 has a security strength of 128 bits, and the RSA cryptosystem has a length of n of 3072 bits, which has a security strength of 128 bits. The proposed signcryption scheme ensures security until 2030 and beyond.

Forward Security

The session keys k_i are generated using the HKDF key derivation function, which takes as input the timestamp and a pseudorandom seed secret value obtained during the signcryption process. This ensures that duplication in communication sessions is highly unlikely.

Due to the non-duplication of keys in communication sessions, if the key of the previous session is exposed, the key of the next communication session will not be affected, leading to a secure forward service.

Non-Repudiation

The proposed scheme uses an RSA digital signature on the root of a Merkle hash tree with leaves $(cm_i||ck_i)$ so that the signature verification on the root is valid, and based on the security of the Merkle hash tree. The receiver can be sure that $(cm_i||ck_i)$ was sent by the sender because only the sender has the private key to sign the RSA digital signature and $(cm_i||ck_i)$ is integral. This makes it impossible for the sender to deny the information $(cm_i||ck_i)$ from the sender.

In order to assess the computational cost of our suggested signcryption scheme, we analyze the number of computations needed for the signcryption process in our proposed multi-document signcryption scheme, and compare it to that of a traditional single-document signcryption scheme (where the same signcryption scheme as described in Sect. 3 is utilized, but without the multi-document signcryption module involving a Merkle hash tree).

The number of times each document is encrypted and decrypted is the same in both cases because the encryption algorithm is called once. Therefore, using the same encryption algorithm, both cases have the same encryption cost.

The cost of signing and authentication shows that:

In the traditional solution of signing and encrypting each document, the number of documents that need to be signed is n documents, corresponding to $2^{t+1} \geq n \geq 2^t$ where $t \in \mathbb{N}$ or it can be assumed that $n \sim 2^{t+1}$ (because the difference is not much). Then the execution time is the time to sign n documents.

In the proposed approach, the execution time is the time to build the hash tree and the time to sign the root node once. The time to build the hash tree in the proposed scheme will perform $\sum_{i=0}^{t+1} 2^i$ hash operations where i is an integer and $n + 1 \geq i \geq 0$. This leads to the proposed scheme being more efficient than the traditional solution by $\left(\sum_{i=0}^{t+1} 2^i - n\right) = \left(\sum_{i=0}^{t+1} 2^i - 2^{t+1}\right) \sim \sum_{i=0}^{t} 2^i$ hash operations.

Thus, when comparing the computational costs of the two solutions, we compare the time to perform the number of hash operations, denoted as $num = \sum_{i=1}^{t} 2^i$ in the proposed solution with the time of the RSA signature algorithm for n documents in the traditional solution. It is easy to see that the value $\sum_{i=1}^{t} 2^i = 2^{t+1} - 1$ and $n \sim 2^{t+1}$ implies $num = n - 1$. In the RSA signature scheme, the process of signing a document includes a hash operation H (in this case H is SHA-256) and an exponential operation modulo n. This leads to the conclusion that the traditional solution will be slower than the proposed scheme when performing more than $n - 1$ exponential operations modulo n.

4 Application Recommendations and Experimental Performance

4.1 Application of the Proposed Signcryption Scheme in Electronic Transactions

In electronic transactions, it is common for the sender to sign and encrypt multiple text files and send them to the recipient for verification. For example, in the case of signing an economic contract, the sender must send the recipient a contract file consisting of many documents, such as a contract draft, quotation, evidence, etc. Therefore, the sender has the ability to decrypt and authenticate either the full document or a portion of it. Alternatively, the sender has the option to transmit a portion of the document collection to the recipient.

In such a case, we propose to use our multi-document signcryption scheme to facilitate the signcryption process. The proposed model can be depicted in Fig. 5.

In Fig. 5, two users, Alice and Bob, need to exchange electronic documents with confidentiality and authentication over the Internet. In this case, Alice is the sender and Bob is the receiver.

The set of electronic documents that Alice needs to sign is a set of n documents. The proposed multi-document signcryption algorithm is used with the input of n documents to be signed, the private key of the sender Alice, and the public key of the recipient Bob. To ensure security, the public key of the recipient Bob is obtained from the digital certificate of Bob provided by a trusted Certification Authority. At the end of the signing process, the sender Alice will send the signed and encrypted message corresponding to the number of documents to be sent to the recipient Bob over the Internet.

The recipient Bob receives the signed and encrypted message from the sender Alice and decrypts it using the proposed decryption algorithm, using Bob's private key and Alice's public key. Alice's public key is obtained from her digital certificate, which

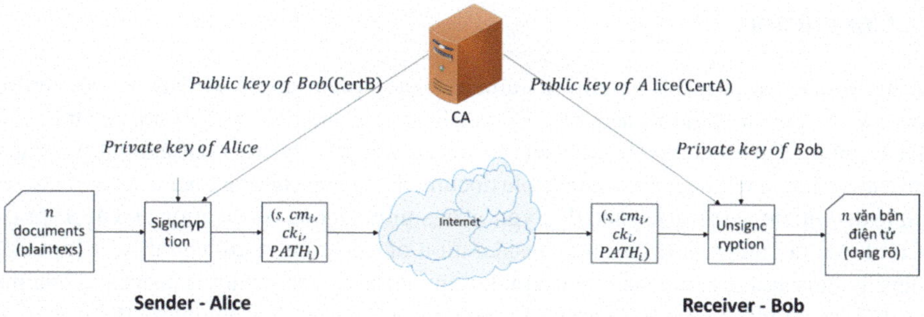

Fig. 5. Application of the proposed signcryption scheme in electronic transactions.

is supplied by a trusted Certification Authority for security purposes. As a result, the recipient Bob gets the n plaintext electronic documents that the sender Alice sent to Bob.

4.2 Experimental Results

We evaluated the experimental results of the proposed multi-document signcryption scheme and the single-document signcryption method (without using the Merkle hash tree module) with the same cryptographic primitives, RSA and AES, on the same hardware platform: a Dell Laptop with 4 GB of RAM and an Intel® Core™ i5-4300U CPU@1.90 Ghz. The experimental results are given in Table 1.

Table 1. Experimental results of the proposed scheme and the method for signing single document

No	Number of signed email files	Signcryption time of proposed scheme (second)	Single document signcryption time (second)	Unsigncryption time of proposed scheme (second)	Single document unsigncryption time (second)
1	16 files	1,0336	1,20796	1,17918	1,3655
2	32 files	2.31752	2,34107	3,4904	3,56057
3	64 files	4,60095	4.6627	6,9014	7,4265
4	128 files	9,8755	10,2381	15,5849	16,1857

Table 1 shows the results of our experiments, which encompass the signcryption procedure and the procedure of verifying signcryption. The test used four folders, each holding distinct email files, all of which were 10 MB in size. From the results in Table 1, it can be seen that the proposed multi-document signcryption scheme is faster than the single-document signcryption method. When the number of files increases, the time for signcryption and unsicryption of the proposed scheme will be much faster than that of the single-document signcryption solution.

5 Conclusion

In this paper, we propose a multi-document signcryption approach based on a multi-document signature scheme using the Merkle hash tree, the RSA algorithm, and the AES block cipher. This scheme allows users to simultaneously ensure confidentiality, origin authentication, and integrity for multiple documents in one signing operation and allows single document verification and decryption or simultaneous verification and decryption of multiple documents. From the consideration of ensuring confidentiality, origin and integrity, forward security, and non-repudiation, it can be seen that the proposed scheme is safe to use until 2030 and beyond. Regarding the cost of implementing the proposed scheme, we have demonstrated that the proposed scheme has better computational cost than the single-document signing scheme, and signing multiple documents at once will bring convenience to the signer, reduce the time required to operate, reduce the amount of information to be stored, and ease the process of signature verification. The proposed multi-document signcryption scheme has the potential for practical implementation due to its low computational cost and strong security. In addition, we have introduced a framework for implementing the suggested system in electronic transactions and demonstrated its full practicality.

Acknowledgment. This work was supported by Academy of Cryptography Techniques through the Project/Lab Grant.

References

1. Zheng, Y.: Digital signcryption or how to achieve cost (signature & encryption) $<<$ cost (signature) + cost (encryption). Adv. Cryptol.— CRYPTO (1996)
2. Sui, Z., Meer, H.: An efficient signcryption protocol for hop-by-hop data aggregations in smart grids. IEEE J. Sel. Areas Commun. **38**(1), 132–140 (2020)
3. Hoang, V., Lehtihet, E., Ghamri-Doudane, Y.: Password-Based Authenticated Key Exchange Based on Signcryption for the Internet of Things 2019 Wireless Days (WD), pp. 1–8. Manchester, UK (2019)
4. Qiu, J., Fan, K., Zhang, K., Pan, Q., Li, H., Yang, Y.: An efficient multi-message and multi-receiver signcryption scheme for heterogeneous smart mobile IoT. IEEE Access **7**, 180205–180217 (2019)
5. Zhang, P., Li, Y., Chi, H.: An elliptic curve signcryption scheme and its application. Wirel. Commun. Mob. Comput. **2022**, 7499836 (2022)
6. Liu, H., Luo, X., Liu, H., Xia, X.: Merkle tree: a fundamental component of blockchains. Int. Conf. Electron. Inf. Eng. Comput. Sci. (EIECS) (2021). https://doi.org/10.1109/EIECS53707.2021.9588047
7. Giry, D.: Recommendation for key management. Special Publication 800–57 Part 1 Rev. 5, NIST (2020). https://www.keylength.com/en/4/. Accessed 04 July 2024
8. Krawczyk, H., Eronen, P.: HMAC-based extract-and-expand key derivation function (HKDF). RFC 5869, ISSN: 2070–1721 (2010)
9. Moriarty, K., Kaliski, B., Jonsson, J., Rusch, A.: PKCS #1: RSA cryptography specifications version 2.2, RFC 8017, ISSN: 2070–1721 (2016)
10. Zodpe, H.: Shaikh: a survey on various cryptanalytic attacks on the AES algorithm. Int. J. Next Gener. Comput. **12**(2), 115–12 (2021)

11. Hosoyamada, A., Sasaki, Y.: Quantum collision attacks on reduced SHA-256 and SHA-512, Cryptology ePrint Archive (2021)
12. Phong, T.Q., Chi, D.D., Huy, T.D., Diep, N.N.: On some issues affecting the security of RSA and ECDSA digital signature schemes. J. Sci. Technol. Inf. Secur. **1**(18), 38–46 (2023). https://doi.org/10.54654/isj.v1i18.884
13. Cường, V.T., Trung, N.T., Hùng, L.Đ.: Cài đặt thuật toán sinh tham số RSA 2048 bit trên thiết bị PKI token. J. Sci. Technol. Inf. Secur. 1(11), 53–60 (2020). https://doi.org/10.54654/isj.v1i11.72

Quantum Informatics, Theory and Application

Chi Tran(✉)

Faculty of Applied Sciences, International School, Vietnam National University, Hanoi, Vietnam
tran.chi@uwm.edu.pl

Abstract. Quantum informatics builds on the counterintuitive cognitive foundations by integrating quantum physics, Hilbert space computing, and informatics. It goes beyond the realm of conventional computing and can be applied in quantum control, quantum medicine, cybersecurity, military, to build AI/ML automatic software and many other fields.

Keywords: Intuitive perception · Counterintuitive cognition · Quantum bit processing · Quantum gates · Quantum circuit · Quantum processor unit

1 Introduction

Usually, we use visual perception derived from the human sensory system, eyes, ears,..., to perceive reality. In this case, any concept is described by common language, eg. 'being or nonbeing'. Then measure an amount of information contained we use usually two alternative numbers, 1 or 0, two bits used in classical computers corresponding to classical physics and two-valued logic. In practice, we use also counterintuitive perception originated from rather the human brain, human thinking. Information contained in each counterintuitive perception is represented by quantum information in the form of quantum bit, **qbit**, 00, 01, ..., 11, corresponding to *quantum informatics* using for quantum computer and suitable to *Quantum Physics.* In nature, it describers the superposition of two intuitive perceptions and can't presented in the common language but in metalanguage form, 'being-nonbeing', which is neither being nor nonbeing but a third option that transcends both. Quantum physics is a field of progress that has had a significant impact on many areas of science and technology. In this article, we present research on a new informatics field named the Quantum Informatics, ***QI***, emphasizing to the potential correlation: *Quantum Physics, Cognitive Information Processing and Quantum computing,* which combines the power of *Quantum Computing* and Cognitive technology, allowing us to change the way disease is diagnosed, creating a powerful transformation in cybert security, in which, q*uantum physics and quantum entanglement principle that assure that the message of sender and receiver has not been intercepted, in the* quantum medicine, in the military and others.

1. **Features of quantum systems**

© The Author(s), under exclusive license to Springer Nature Switzerland AG 2025
H. A. Le Thi et al. (Eds.): ICAMCS 2024, LNNS 1311, pp. 538–553, 2025.
https://doi.org/10.1007/978-3-031-90606-0_45

Quantum feature. The quantum phenomenon governed by physical principles called *quantum feature*, which did not obey the principles of binary logic but transcending the boundary of them towards accepting the "One" philosophy. It created the so-called physical quantum world to become a bizarre world, which confirmed ancient theories and paved the way for new interpretations of reality. Many scientists with different studies related to the manipulation of individual quantum objects, photons, subatomic, etc., and their *quantum features* create the physical quantum world, basis of the quantum computer today. It has revolutionized our understanding of physics and mechanics, and opened our eyes to new possibilities.

Physical Superposition is the ability of a *quantum system* to be in *multiple states simultaneously*. For example, the flip of a coin, which consistently lands as 'heads' or 'tails'. However, when that coin is in mid-air, it is both heads and tails until it lands, heads and tails simultaneously. Thus, tossing a coin, predicting its landing state, "heads" or "tails" when the coin is in mid-air is impossible. Because, both "heads" and "tails" are equally likely to occur. It can be represented in the logic form, "heads and tails" or in meta language "heads-tails", which is neither head nor tail but a third option that transcendent both. It represents an ability of a quantum system to be in multiple states at the same time until it is measured.

Physical Entanglement means that objects created together are entangled if separated they have the ability to communicate with each other and that communication is received immediately, with no regard to traditional space and time as we know it before, regardless of how far apart they may be. Entanglement is taking objects, connecting them by permanently entangling them together. It expresses entanglement relationship between waves passing through the slots a state that cannot be decomposed into its more fundamental parts. You can see later also, for example, when adding an additional *qubit* to a quantum computer, a 100-cubit quantum machine can examine 2^{100} states simultaneously. As, the extraordinary power of quantum algorithms seems to be derived from the properties of multi particles entangled states. In addition, the entanglement of qubits (quantum bit) allows quantum computers to solve problems efficiently, finding a solution faster, with many fewer calculations.

Quantum physics is a promising field that intersects with medicine much more than originally understood. New scientific developments hold great promise for improving quality of care, clinical research, and the diagnosis and treatment of diseases. In terms of diagnosing different diseases, incorporating quantum mechanics into the study of medicine can allow for efficient diagnosis before symptoms even arise in a patient. Utilizing quantum theory in the field of medicine can help in understanding and applying treatments for a multitude of different diseases, such as Alzheimer's disease or diverse types of cancer, and even expand upon efficient and reliable diagnosis in clinical settings.

2. **Counterintuitive Perception**

The quantum perception happens also when people are in a dangerous state: let's return to the process of dealing with the boy's dangers. When eyes and ears (smart senses) receive the danger signals (fire in the real world), which are transmitted into the brain at once, and the real time decision resulted from activity of neural net and boy's experiences, depending on environmental conditions will be transferred to the foot (actuator) to run in the right direction. It is described in the image below (Fig. 1):

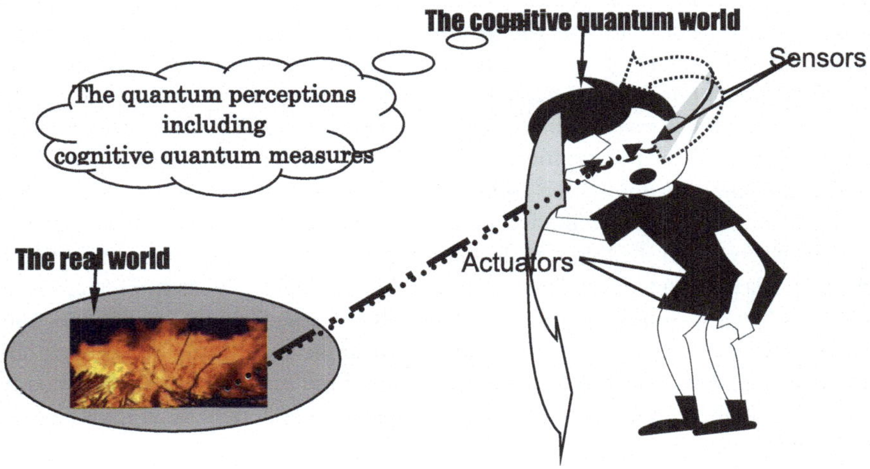

Fig. 1. Real time intelligence through Brain of Thing, BoT

In fact, danger stimulated the boy's brain to eliminate all of common thinking, towards deep contemplate quickly. Thus, the quantum perception that happens in this case also stems from a deep meditation – 'Yoga. Then, quantum perception occurs in both above cases is same originated from deep contemplation, which is a process elimination binary countervailing of common thinking towards a new way of thinking in order to reach a potential intelligence that usually not occurs in common conditions.

Physical Superposition is the ability of a *quantum system* to be in *multiple states simultaneously*. For example, the flip of a coin, which consistently lands as 'heads' or 'tails' (a very binary concept.) However, when that coin is in mid-air, it is both heads and tails until it lands, heads and tails simultaneously. It represents an ability of a quantum system to be in multiple states at the same time until it is measured.

Physical Entanglement means that objects created together are entangled if separated they have the ability to communicate with each other and that communication is received immediately, with no regard to traditional space and time as we know it before, regardless of how far apart they may be. Entanglement is taking objects, connecting them by permanently entangling them together. It expresses entanglement relationship between waves passing through the slots – a state that cannot be decomposed into its more fundamental parts.

You can see later, for example, when adding an additional *qubit* to a quantum computer, a 100-cubit quantum machine can examine 2^{100} states simultaneously. As, the extraordinary power of quantum algorithms seems to be derived from the properties of multiparticle entangled states. In addition, the entanglement of qubits allows quantum computers to solve problems efficiently, finding a solution faster, with many fewer calculations.

It's a realm where things are in potential state until they are observed, where all is connected and in this experiment, where a photon of light pas in two places at the same time. Behavior of individual photons moves at the speed of light, c, is agreement with the physical quantum principles of Max Planck and Albert Einstein, but opposed to classical physics. And, "Each photon of light passing through both slits at the same time" is impossible according to the binary philosophy – Nothing can and does not exist at the same time.

Quantum feature. The quantum phenomenon governed by physical feature called *quantum feature*, which did not obey the principles of binary logic but transcending the boundary of binary philosophy towards accepting another philosophy. Although binary philosophy is true in the syntactic aspect of the scientific world, it is not true for quantum phenomena in the semantic and pragmatic framework of the real world. Scientists such as Albert Einstein, Max Plank, Neils Bohr and Karl Heisenberg in the early part of the 20th century were investigating the *quantum feature* of particles, energy and potentiality, there is sub-atomic moved with the light speed that doesn't follow the known laws of physics. It created the so-called physical quantum world to become a bizarre world, but in all of its bizarre it confirmed ancient theories and paved the way for new interpretations of reality. Many scientists with different studies related to the manipulation of individual quantum objects, photons, subatomic, etc., and their *quantum features* create the physical quantum world, basis of the quantum revolution in the fields of quantum engineering today. Different experiments are performed in order to study particles to the extent of atomic and sub-atomic level. It has revolutionized our understanding of physics and mechanics, and opened our eyes to new possibilities. According to Niels Bohr

"Anyone who is not shocked by quantum theory has not understood it.

Today, we can see developing of a new class of materials, quantum materials, e.g., topological insulators, which constitute a new class of those materials with inverted bulk band ordering brought about by relativistic effects such as *spin-orbital* interaction. They store boundary (surfaces or edges for 3D crystals or 2D materials respectively) electronic states that are gapless (metallic with Dirac-like linear energy dispersion), helical (spin-momentum locking) and protected against disorder by certain physical symmetries. These boundary electronic states we call *quantum states* with *quantum boundaries* originated from relativistic effects of the interaction of two quantum states. These topological materials are confirmed experimentally or proposed theoretically for both canonical of topological insulators, like 3D crystals of Bi_2Se_3 or 2D HgTe/CdTe heterostructures and topological superconductors according to Y. Ando 2013, A. Bansil et all 2016.

At the frontiers of physics, remote sensing, materials science, economics and engineering, we can see another type of '*quantum boundary*' brought about by relativistic effects from the interaction of two states such as: '*physiological-psychological*' integration in the new sciences of cognition; 'hardware-software' integration in computer science; 'elastic-plastic' interactions in mechanics, 'low-medium' and 'medium-high' interactions of density of sand in geophysics; in interactions of similar signals of different diseases in medicine; in interactions of two decisive goals considered at the same time 'minimize-maximize' or interactions of two objectives: 'maximum-minimum' in optimization etc.. Nowadays, people use probability methods to solve these problems, however, the principles of probability are constructed within the framework of binary philosophy and binary logic, while quantum features are beyond their limits.

In the real world, visual/intuitive perception is based on the principles of classical physics, however, perception originated from the human brain named couterintuitive perception does not follow the known laws of classical physics too but quantum physics/quantum mechanics. Counterintuitive perception was presented fistly by Zhang Zhi, Chinese philosopher 500 BC in his the "One" phylosophy, which describers not knowledge originated from logic (logical knowledge) but rather knowledge originated from a cognitive process in the human brain. The "One" philosophy presented by *Zhang Zhi* 莊周- (369–286 B.C.), claimed that *Discrimination*, 辨, *and* the opposition of it, *Nondiscrimination*, 不 辨", are seen as oneness "辨-不辨", *Discrimination-Nondiscrimination,* referred as a third option that transcends both *Discrimination* and Non*discrimination.* It represents authors's feeling comes from a cognitive processing procces. Because author can't explain the meaning of his counterintuitive cognition by common language, Zhang Zhi had to compose a parable, "Zhang Zhi's dream". That is once Zhang Zhi dreamed he was a butterfly, a butterfly flitting and fluttering about, happy with himself and doing as he pleased. He didn't know that he was Zhang Zhi. Suddenly he woke up and there he was, solid and unmistakable Zhang Zhi. But he didn't know if he was Zhang Zhi who had dreamt he was a butterfly, or a butterfly dreaming that he was Zhang Zhi. Between Zhang Zhi and the butterfly there must be some distinction! According to author*, Discrimination* becomes *Nondiscrimination* and *Nondiscrimination* becomes *Discrimination* … we can't know where is the starting point, where is the end. We call it the meta-concept, which is graphically represented in the following Fig. 2:

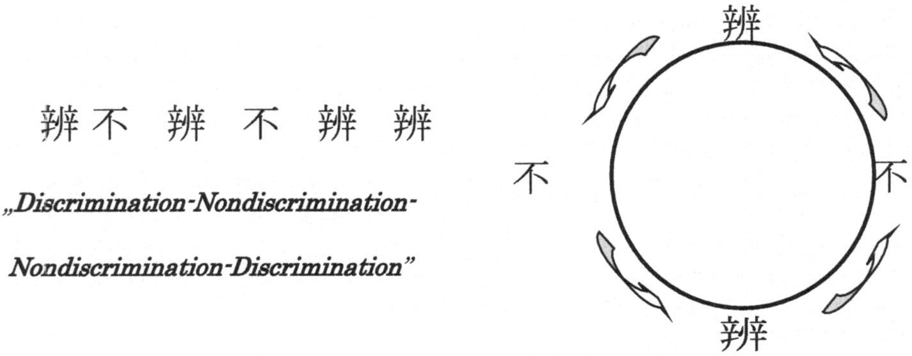

Fig. 2. 莊周-*Zhang Zhi's* idea of ***meta-concept***

This is called the Transformation of Things, ***Z. Zhi-Butterfly***. It has successfully explained phenomena such as radiation and antimatter, the workings in the world of microscopic particles, and even other phenomena in the real world. It has put forward to counterintuitive cognition and cognitive superposition and cognitive entanglement.

The appearance of two intuitive perceptions, 'Being' and 'Nonbeing' at the same time, according to Georg Wilhelm Friedrich Hegel (1770–1831), creates a new counterintuitive perception: 'Being-Nonbeing' which we call *quantum perception.* In fact, Hegel confirmed that the reality was saturated in the 'One' and we usually subdivide and rearrange it. Hegel's idea is considered a quantum philosophy (the "One" philosophy, see Tran C. (2017)). It is (Fig. 3):

"It is impossible to conceive of being without nonbeing included in the thought"

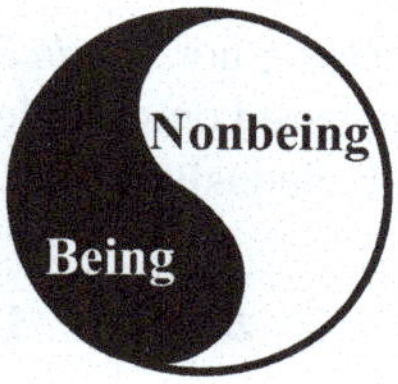

Fig. 3. The Hegel's philosophy in the sense of ontology

This philosophy is derived from the actions of the human brain's neural network to quickly adapt to changes in real environmental conditions in real time. While in the physical quantum world, subatomic being moved at the speed of light, in the human brain, information is moved with the speed of light too. It creates a special ability of the human brain to flexibly recognize changes of the real world in real time i.e., real time intelligence which we call the *Quantum Intelligence*, **QI**. Counterintuitive perception, what we can't see in the real world and can't understand under the framework of classical physics. It is the superposition of two states of consciousness, but more importantly, what the consequences of its processing giving us from the real world. They are precisely the conscious development, that determines other levels of reality, which is beyond the framework of pure matter. Graphically, through figures of Square and Hexagon in the states separated, ***Square or Hexagon***, and superposition, **'Square-Hexagon'** (Fig. 4),

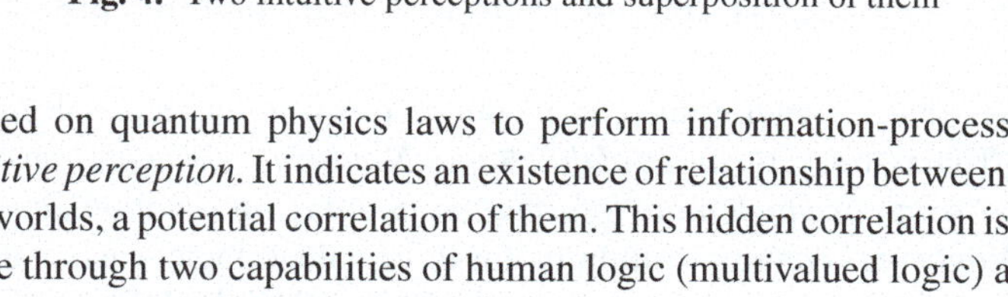

Fig. 4. Two intuitive perceptions and superposition of them

It is based on quantum physics laws to perform information-processing through *counterintuitive perception.* It indicates an existence of relationship between the physical and mental worlds, a potential correlation of them. This hidden correlation is represented in this article through two capabilities of human logic (multivalued logic) and quantum information processing using quantum computer.

2 Potential Power Exponential 2^n of Counterintuitive Perception

Let return to philosophy of Georg Wilhelm Friedrich Hegel 1770-1831, *Brain Awareness*. Here, superposition of two *intuitive perceptions*, being and nonbeing make up new concept, presented in the meta language,

'Being-Nonbeing',

which is neither ***Being*** nor ***Nonbeing*** but another option that includes both. It comes from human thinking based on so-called brain logic/multi valued logic to describe many different states simultaneously in the form: Being-Nonbeing, **B-N, '1–0':**

$$\boldsymbol{B - N = '1 - 0' = \{10,11,01,00\}} \tag{1}$$

Information traveling through the human brain at the speed of light takes on a new, counterintuitive form, measured in information units 01 (0 AND 1), rather than traditional information units, (0 OR 1), '1, 0'. Counterintuitive perception is originated from the actions of the human brain's neural network to quickly adapt to changes in in real time of real environmental conditions. It increases the power and intelligence level of the human brain.

In the framework of traditional informatics, each 'on-off' system stores bit: 'off' state = 0 and 'on' state = 1. Answer on the question 'Being OR Nobeing'?, for example, intuitive perception, presented by:

$$\text{'Being OR Nonbeing'} = \{1,0\} = 2^1 \text{ intuitive possibilities} \tag{2}$$

The two bits 1, 0 represent two-states, switche greatly miniaturized with arbitrary transformations. In contrast, the counterintuitive perception presented by the *meta concepts*, 'Being AND Nonbeing' = B*eing-Nonbeing* = '1–0' is individually being in the observer's spirit under the semantic and pragmatic aspects of them. They are precisely the conscious development – that determines other levels of reality, which is beyond the purely physical and deterministic, can be accessed. That is to say, inference based on Hegel's philosophy, *being-nonbeing*, is presented as:

$$\text{'Being AND Nonbeing'} = \{\neg BN, \neg B \neg N, BN, B \neg N\} = 2^2 \text{ quantum possibilities} \tag{3}$$

It indicates that, counterintuitive perceptions result in exponential growth, 2^2, of possibilities. That is to say, m*eaning of the '**Being-Nonbeing**' concept known as quantum consciousness contains much more information than the information contained in the '**Being OR Nonbeing**' concept.* Similarly, for three valued logic, each counterintuitive perception exists in three states simultaneously, B, N and W (both B and N), *processing* of ‘Being-Nonbeing’, is presented in the form:

$$\text{'Being AND NonBeing'} = \{\neg BN\neg W, \neg BNW, \neg B\neg N\neg W, \neg B\neg NW...\} = 2^3 \text{ quantum possibilities} \tag{4}$$

It is clear, that instead of recognizing being or non-being only, choosing between two possibilities by intuitive perception, while in observer’s mind, many new ideas could be appeared. Then, intuitive perceptions are produced by the intercourse of the body senses with the objects, originated from the observer’s physical perceptions, who can recognize quickly being or non-being by his/her eyes; the counterintuitive perceptions, ‘*being AND non-being*’ are individually being in the observer’s spirit under the semantic and pragmatic aspects of them. They are precisely the conscious development that determines other levels of reality, which is beyond the purely physical and deterministic, can be accessed.

It is clear, that the *meta-concept* ***Being-Nonbeing*** (‘*being AND non-being*’) originated from the human **C**ognitive **P**rocessing is formulated from many cases recognizing by observers through his/her internal information processing (semantic-pragmatic analysis). It represents his/her qualitative evaluations of unavoidable superposition of information, e.g., the appearance of the same symptoms of many different diseases in medicine. They oblige us to introduce new notions (2^n cases), n denotes a number of n-valued logic, into doctor’s mind rather than result derived from his/her external information processing (syntax analysis for quantitative evaluations). Thus, from the disappearance of the weak cases due to consequence of syntax analysis we must overcome its limitations through semantic-pragmatic analysis to find out more of cases from the real world into the world of assumptions. Thus, the power of quantum perception changes depending on the number of logic values. Using the Łukasiewicz’s n-valued cognitive logic, named quantum logic leading to the following results:

1. $n = 2$: *exponential growth* 2^2
2. $n = 3$: *exponential growth* 2^3
3. $n = 4$: *exponential growth* 2^4
4. $n = n$: *exponential growth* 2^n

It indicates really that quantum logic decides the potential power of counterintuitive *perception.* It is the key to increase the remarkable power exponentially, 2^n, of latent quantum intelligence.

3. Potential Power Exponential 2^n of Quantum Computing

Quantum computing is a field that combines the principles of quantum mechanics/physics with information theory to study the analysis, and transmission of information. It covers both theoretical and experimental aspects of *quantum physics.* It helps us to understand how the fundamental laws of physics can be exploited to radically improve the collection, transmission and processing of information. It turns out that the laws of quantum mechanics/physics can be exploited to perform very important and difficult information processing tasks, a promising new field of science and technology. Superposition is the ability of a system to be in multiple states simultaneously, which can be written as a superposition of eigenvalues, see Le Bellac M. 2006. It allows quantum computers to increase computational speed exponentially in which, a new concept of quantum bit, ***qubit***, used to information processing is concerned with spin of an electron under the physical nature, an ability to perform a large number of digital operations in quantum computer. Taking spin of electron as quantum bit, which is determined, according to quantum informatics by all of values in the interval [0, 1], contains a wealth of information that is incredibly closely related to the logical values of quantum logic. Superposition of 1 and 0, **'1–0'**, see Chris Bernhardt, 2019, is really represented as the quantum bit, **qubit/spin** of an electron.

- Using traditional bits to describe information, we have:

$$\text{'Being OR Nonbeing'} = \{1 \vee 0\} = 2^1 \tag{5}$$

Let x_2 represents numbers in the binary counting system and x_{10} described numbers in the decimal counting system. For example, 3, 5, in the decimal system can be write as: 3_{10} and 5_{10}, which can be changed to number belonging in the binary system, x_2, as follows:

$$3_{10} = 011_2; 5_{10} = 101_2 \tag{6}$$

And, each number in the binary system x_2, e.g., 011_2, 101_2, can be changed to number belonging in the decimal system, x_{10} as follows:

$$011_2 = 0 * 2^2 + 1 * 2^1 + 1 * 2^0 = 3_{10} \Rightarrow 011_2 = 3_{10} \tag{7}$$

$$101_2 = 1 * 2^2 + 0 * 2^1 + 1 * 2^0 = 5_{10} \Rightarrow 101_2 = 5_{10} \tag{8}$$

One of the following eight configurations for numbers from 0_{10} to 7_{10} can be stored in the classic register:

$$0_{10} = \{000_2\}, 1_{10} = \{001_2\}, 2_{10} = \{010_2\}, 3_{10} = \{011_2\} \tag{9}$$

$$4_{10} = \{100_2\}, 5_{10} = \{101_2\}, 6_{10} = \{110_2\}, 7_{10} = \{111_2\} \tag{10}$$

The triple system also allows us to store a number from 0 to 7, for example, through one of eight states:

$$0_{10} = |000\rangle, 1_{10} = |001\rangle, 2_{10} = |010\rangle, 3_{10} = |011\rangle \tag{11}$$

$$4_{10} = |100\rangle, 5_{10} = |101\rangle, 6_{10} = |110\rangle, 7_{10} = |111\rangle \tag{12}$$

- Information processing using quantum bit, ***qubit***,

Sign of the ***tensor product***, $\otimes$, for example, $|101\rangle$ is abbreviated entry for the state $|1_A\rangle$ $0_B \otimes 1_C\rangle$, where, state vectors for qubits A, B, C belong accordingly to Hilbert spaces $\mathbb{H}_A$, $\mathbb{H}_B$ and $\mathbb{H}_C$. Let |x| Let $|x\rangle$, x = 0, 1, ..., 7, means one of eight states above, for example $5 = |101\rangle$. For n, denotes a qubit number and $0 \leq x \leq 2^n - 1$, orthonormal vector $|x\rangle$ creates a computational basis in space $\mathbb{H}^{\otimes v}$ that is: *for n qubits enable us to remember 2^n numbers in the computational basis, where, n denotes qubit number.* The meaning of the *tensor product*, $\otimes$, is presented as follows. Two columns vectors $|u\rangle$ and

$|v\rangle$ of lengths m and n yield a column vector of length m.n when tensored, shown in the following equations:

$$|u\rangle \otimes |v\rangle = \begin{bmatrix} u_0 \\ u_1 \\ \cdots \\ u_m \end{bmatrix} \otimes \begin{bmatrix} v_0 \\ v_1 \\ \cdots \\ v_n \end{bmatrix} = \begin{bmatrix} u_0.v_0 \\ u_0.v_1 \\ \cdots \\ u_0.v_n \\ u_1.v_0 \\ u_1.v_1 \\ \cdots \\ u_{m-1}.v_n \\ u_m.v_0 \\ \cdots \\ \cdots \\ u_m.v_n \end{bmatrix} \tag{13}$$

We use arrow, where its direction is corresponding to the direction of spin of an electron/photon to describe the quantum bit, ***qubit*** in the vector Hilbert space:

$$|1\rangle = \begin{bmatrix} 0 \\ \hline 1 \end{bmatrix}, |0\rangle = \begin{bmatrix} 1 \\ \hline 0 \end{bmatrix} \tag{14}$$

All matter is composed of atoms, nuclei, electrons, the interactions and time evolution of atoms are governed by the laws of quantum mechanics/physics. Superconducting chips with 4–9 qubits have been demonstrated with the performance required to run quantum simulation algorithms, quantum machine learning according to D. Deutsch (see Deutsch D., 1985) and others. A practical quantum instruction set architecture is presented by Robert S. Smith, Michael J. Curtis, William J. Zeng 2016, in which authors introduce an abstract machine architecture for classical/quantum computations including compilation along with a quantum instruction language for explicitly writing these computations.

For n = 2, we have the physical state presented by,

$$|X_{00}\rangle = |0_A \otimes 0_B\rangle. \tag{15}$$

Let the first qubit, A belong to the Hilbert space $\mathbb{H}_A$, with the orthonormal basis is $\{|0_A\rangle, |1_A\rangle\}$ a second qubit B to the Hilbert space $\mathbb{H}_B$ with the orthonormal basis is $\{|0_B\rangle|1_B\rangle\}$. Let return to quantum perception processing, **'B-N'**, with two qubits, in which, superposition of all possible states are represented by basis vector $|\psi\rangle$. Using bits (numbers in binary) to describe information for traditional computer, 1_2 and 0_2, gives the corresponding result:

$$\text{qubit} = 01 = \{00_2, 01_2, 10_2, 11_2\}(4 \text{ quantum possibilities}) = 2^2 \tag{16}$$

Using qubits for quantum computer gives the corresponding result

$$01 = |0\cdot\rangle\ , |\cdot\cdot\rangle\ , |1\cdot\rangle\ , |\cdot\cdot\rangle \tag{17}$$

where, $|0.\rangle$ is abbreviated entry for state $|0_A \otimes 0_B\rangle$. Where state vector belong respectively to spaces $\mathbb{H}_A$, $\mathbb{H}_B$. We have 2^2 cognitive quantum states determined by:

$$|00\rangle = |0\rangle \otimes |0\rangle = \begin{bmatrix} 1 \\ - \\ 0 \end{bmatrix}\begin{bmatrix} 1 \\ - \\ 0 \end{bmatrix}, \quad |01\rangle = |0\rangle \otimes |1\rangle = \begin{bmatrix} 1 \\ - \\ 0 \end{bmatrix}\begin{bmatrix} 0 \\ - \\ 1 \end{bmatrix} \tag{18}$$

$$|10\rangle = |1\rangle \otimes |0\rangle = \begin{bmatrix} 0 \\ - \\ 1 \end{bmatrix}\begin{bmatrix} 1 \\ - \\ 0 \end{bmatrix}, |11\rangle = |1\rangle \otimes |1\rangle = \begin{bmatrix} 0 \\ - \\ 1 \end{bmatrix}\begin{bmatrix} 0 \\ - \\ 1 \end{bmatrix} \tag{19}$$

We have basis vector for 2-qubits system

$$|\psi_2\rangle = \tau_.|\cdot\cdot\rangle \; + \tau_.|\cdot\cdot\rangle \; + \tau_.|\cdot\cdot\rangle \; + \tau_.|\cdot\cdot\rangle \tag{20}$$

Therefore, processing quantum information for 2 qubits gives us an ***exponential power,*** 2^2.

For n = 3, we have *the Power of three qubits.* Similarly, is abbreviated entry for state $|0_A \otimes 0_B \otimes 0_C$. This state vector belong respectively to spaces $\mathbb{H}_A$, $\mathbb{H}_B$, $\mathbb{H}_C$. We have 2^3 cognitive quantum states and basis vector for 3-qubits system (n = 3) and superposition and entanglement of 3 qubits is presented by ***exponential power,*** $\mathbf{2^3}$.

$$|\psi_3\rangle = \tau_.|\cdot\cdot\cdot\rangle \; + \tau_.|\cdot\cdot\cdot\rangle \; + \tau_.|\cdot\cdot\cdot\rangle \; + \tau_.|\cdot\cdot\cdot\rangle \; + \tau_4|\cdot\cdot\cdot\rangle \; + \tau_5|\cdot\cdot\cdot\rangle \; + \tau_6|\cdot\cdot\cdot\rangle \; + \tau_7|\cdot\cdot\cdot\rangle \tag{21}$$

By similar way, we have: ***for n-qubit processing: exponential power*** $\mathbf{2^n}$. (quantum power).

It generates greatly enhanced computing power to solve various practical problems. For example, ***for n = 100, quantum machine can examine*** $\mathbf{2^{100}}$ ***states simultaneously***.

As, the extraordinary power of quantum algorithms seems to be derived from the properties of multiparticle states. Superposition/Entanglement of qubits allows quantum computers to solve problems efficiently, finding a solution faster, with many fewer calculations.

4. **Quantum Gates, Quantum Circuit.**

While in the classical computer, we have on-off system that stores one traditional bit: 'off' state = 0, and 'on' state = 1. On the quantum computer we have rather their superposition that stores quantum information, **qubit**, presented by two vectors, $|0\rangle$ and $|1\rangle$. Unlike a traditional bit, a qubit must exist in a superimposed state, where it is occupying more than one state simultaneously (quantum feature). To verify the quantum results of above logic and computation methods, we use a quantum computer, simulation of the quantum processing unit, **QPU**, (Quantum Processor) see Eric R. Johnston 2019, to process 2, 3, 4 qubits (corresponding to 2, 3, 4 logic values of brain logic). In which, we use usually *quantum* ***logic gates*** that map quantum superposition to another state that allows the evolution of the system to the desired final state.

Swap gate (in matrix term):

$$\begin{bmatrix} 1\,0\,0\,0 \\ 0\,1\,0\,0 \\ 0\,0\,1\,0 \\ 0\,0\,0\,1 \end{bmatrix}$$

Circuit representation:

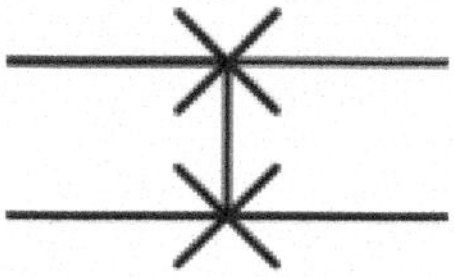

H is the Hadamard gate:

$$H = \frac{1}{\sqrt{2}}\begin{bmatrix} 1 & 1 \\ 1 & -1 \end{bmatrix} \tag{22}$$

which creates an equal superposition for quantum states: $|0\rangle$ and $|1\rangle$. The following code applies the Hadamard transform H to a qubit x and assigns the name y to the result

$$y := H(x). \tag{23}$$

Compiling this code to a circuit using H gate yields:

(24)

A possible state before applying H is $|0\rangle$x, where subscript x indicates that variable x stores 0. The following code performs a controlled application of the Hadamard transform. if x { //controlled on x, y: = H(y); //apply H to y}.

Compiling this code to a circuit using control on **x** & **H** gate yields:

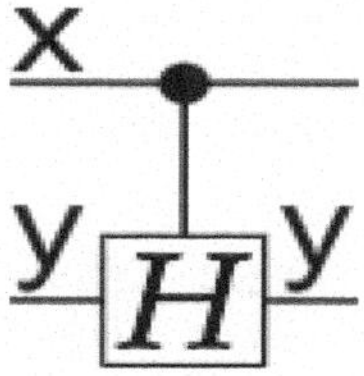

(25)

- **NOT.** (instruction QPU) is the quantum equivalent of conventional operation of same name (Figs. 5, 6 and 7)

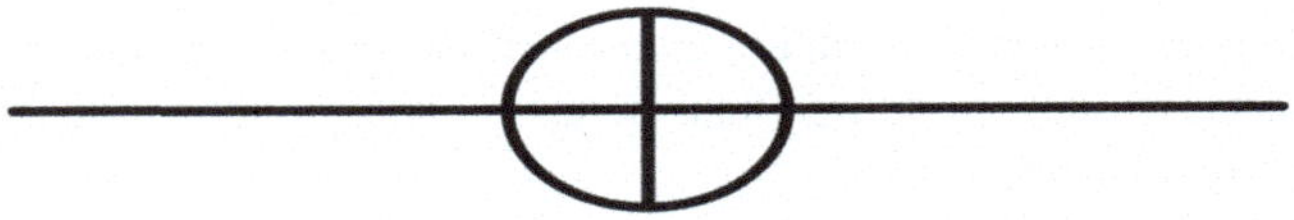

Fig. 5. Quantum circuit diagram of a NOT-gate.

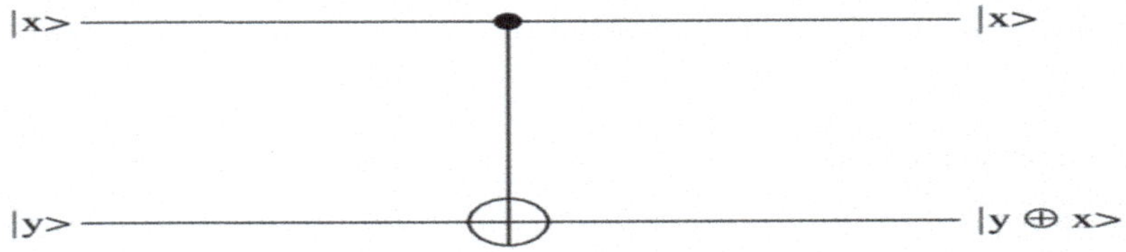

Fig. 6. Controlled-NOT quantum gate.

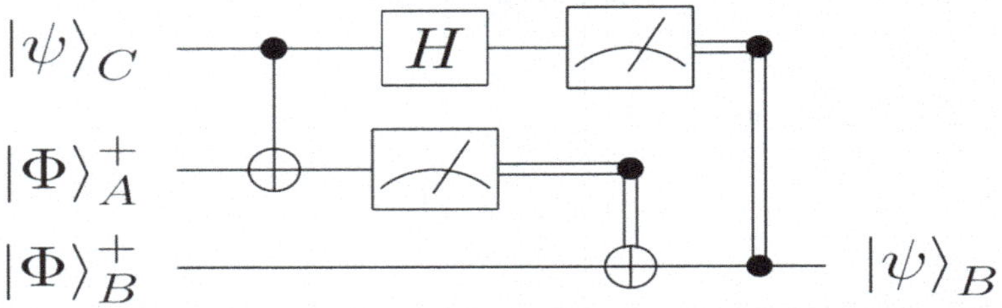

Fig. 7. Circuit that performs teleportation of a qubit.

3 Numerical Examples

3.1 Quantum Processing Using Quantum Computer (Simulation of QPU) for N Qubits.

We have the sample generates a single random byte,

qc.reset(n); // allocate some qubits qc.write(0);
// write the value zero
qc.had(); // place them all into superposition of 0 and 1 var result = qc.read();
// read the result as a digital bit qc.print('result: ' + result + '\n');qc.reset(n);
// n = 1,2,3,…, allocate one qubit. *For n = 2*, we have:

Program circuit:

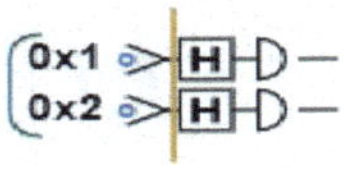

Circle notation:

Program output:

result: 0

result: 1

Thus, 2 qubit processing give us results equivalent to those obtained from quantum processing – '2-exponential' power, 2^2.
For n = 3, using above program with n = 3 we have the corresponding program circuit:

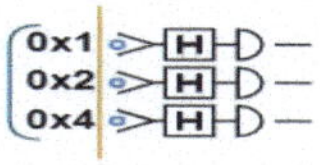

Circle notation

Program output
result: 1 result: 1
result: 2
Thus, 3 qubit processing gives us results equivalent to those obtained from quantum perception processing – '3-exponential' power, 2^3.
Similarly, for n = 4 *we have results '4 -exponential' power, 2^4, etc.* Thus, we have results equivalent to *those obtained from quantum Perception processing 'n-exponential' power, 2^n.*

In fact, the counterintuitive perception, ***being-nonbeing,*** B-N, is really represented in the informatic form, '10' of qubit, they have identical *'n-exponential' power*. This is the key to the *exponential growth* in computing power promised by qubits of quantum computer. One **qubit** alone can't provide the horsepower we really want out of quantum computers, so naturally we want to create a system that uses many, many more. The problem is once we have more than one **qubit** we're vastly increasing computing power, but we're also dealing with entanglement, wherein the qubits affect each other. Entanglement makes the measurement problem even more difficult since each qubit is affecting the superposition of the qubits around it.

3.2 Qubit Teleportation

It is a process in which ***quantum information*** can be transmitted from one location to another, with the help of the ***entanglement between the sending and receiving***. We have four Steps of *Qubit Teleportation*

Step1: ***Sender*** and ***Receiver*** share qubits of an entangled state.

- ***Receiver's*** state intimately tied to result of ***Sender's*** measurement

Step 2: ***Sender*** measures her qubit
Step 3: ***Sender*** sends her measurement bits to ***Receiver***
Step 4: ***Receiver*** applies a ***single-qubit gate*** to his qubit…

- chosen in accordance with ***Sender's*** measurement bits, ***entanglement guarantees that Receiver has recovered the message.*** A part of sample program demonstrates basic teleportation:

```
qc.reset(3); var sender = qint.new(1,
'sender'); var ep   = qint.new(1, 'ep'); var
receiver   = qint.new(1, 'receiver'); var a1
= 0; var a2 = 0;
// This will work with entangle() and sender_prep() in either order.
// Try swapping them to verify this.
entangle(); sender_prep();
sender_send(); receiver_receive();
receiver_verify();
function entangle()
.......................... ..........................
```

Program circuit, in which, **alice** = sender; **bob** = receiver.

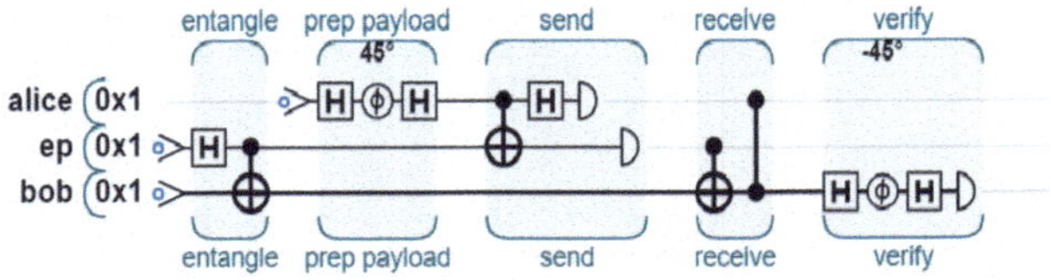

Quantum mechanics/physics and the superposition principle help Alice and Bob assure themselves that ***the message is not intercepted.***

It turns out that counterintuitive perception originated from the human brain and quantum bit originated from quantum informatics, are both based on the same physical principles. They form not only basis of 'brain-computer' connection (human brain-quantum computer) but also the key foundation of the 'human-machine' integration.

4 Conclusions

Counterintuitive perception is originated from the human brain, described in the meta-language form, ***being-nonbeing/being AND nonbeing***. It is individually being in the observer's spirit under the semantic-pragmatic aspects of the real world. Eexploiting its content is beyond the limits of comon language and classical computational methods. We need a new interdiscipline approach, quantum informatics, including quantum physics, mathematics, informatics and computer science including quantum computer. It turns out that counterintuitive perception is precisely conscious development – an ability of the so-called psychic power of consciousness to determine higher levels of reality through the powerful ability of itself, increased at the exponential level (2^n), based on the quantum computer used quantum information described by quantum bit, qubit, "1-0". The combination of the power of analytics and cognitive technology allows us to change the way disease is diagnosed, to create a powerful transformation in healthcare (quantum medicine). It could be applied to cyber and IoT security, quantum cryptography, to build the quantum cybersecurity software for digital government, to build a AI automatic software and others.

Funding. The author received no financial support for research, authorship and publication of this paper.

Declaration of Conflicting Interests. The author declared no potential conflicts of interest with respect to the research, authorship and/or publication of this article.

Ethical Approval:. This article does not contain any studies with human participants or animals performed by authors.

References

1. Furrer, F.J.: Quantum computing for everyone. Informatik Spektrum **42**(5), 372–374 (2019). https://doi.org/10.1007/s00287-019-01229-3
2. Deutsch, D.: Quantum theory, the church-turing principle and the universal quantum computer. Proc. Royal Soc. London A: Math. Phys. Eng. Sci. **400**(1818), 97–117 (1985)
3. Landauer, R.: The Physical nature of Information. Phys. Lett A 217 (1991)
4. Le Bellac, M.: Introduction a L'Information Quantique, Edition Beline – Paris (2005)
5. Smith, R.S., Curtis, M.J., Zeng, W.J.: A Practical Quantum Instruction Set Architecture, Rigetti Computing 775 Heinz Ave, p. 94710. California, Berkeley (2016)
6. Shor, P.: Quantum Computing Doc. Math. (Bielefeld) Extra Vol. ICM Berlin8, vol. I. pp. 467–486 (1998)
7. Tran, C.: (book) "Potential Quantum Intelligence", Lambert Academic Publishing in Germany 1–328 papers (2021). ISBN-10 6204729969, ISBN-13 978-620-4-79966-1
8. Tran, C.: Quantum intelligence. J. Curr. Trends Comput. Sci. Appl. (2019)
9. Tran, C.: Artificial Brain and "Human-Machine" Integration. J. Comput. Sci. Appl. Inform. Technol. (2017)

Author Index

© The Editor(s) (if applicable) and The Author(s), under exclusive license
to Springer Nature Switzerland AG 2025

H. A. Le Thi et al. (Eds.): ICAMCS 2024, LNNS 1311, pp. 555–557, 2025.
https://doi.org/10.1007/978-3-031-90606-0

The manufacturer's authorised representative in the EU is Springer Nature Customer Service Centre GmbH, Europaplatz 3, 69115 Heidelberg, Germany. If you have any concerns regarding our products, please contact ProductSafety@springernature.com

Printed and bound by CPI Group (UK) Ltd, Croydon, CR0 4YY
01/10/2026
02225851-0020